# Chemical Signals in Vertebrates 15

Benoist Schaal • Diane Rekow
Matthieu Keller • Fabrice Damon
Editors

# Chemical Signals in Vertebrates 15

Springer

*Editors*
Benoist Schaal
Development of Olfactory Communication and Cognition Laboratory
Centre des Sciences du Goût et de l'Alimentation
CNRS, Université de Bourgogne
Inrae-Institut Agro Dijon, Dijon, France

Diane Rekow
Development of Olfactory Communication and Cognition Laboratory
Centre des Sciences du Goût et de l'Alimentation
CNRS, Université de Bourgogne
Inrae-Institut Agro Dijon, Dijon, France

Matthieu Keller
Physiologie de la Reproduction et des Comportements
Inrae, CNRS, Université de Tours
Nouzilly, France

Fabrice Damon
Development of Olfactory Communication and Cognition
Centre des Sciences du Goût et de l'Alimentation
CNRS, Université de Bourgogne
Inrae-Institut Agro Dijon, Dijon, France

ISBN 978-3-031-35158-7 ISBN 978-3-031-35159-4 (eBook)
https://doi.org/10.1007/978-3-031-35159-4



This Springer imprint is published by the registered company Springer Nature Switzerland AG
The registered company address is: Gewerbestrasse 11, 6330 Cham, Switzerland

Paper in this product is recyclable.

# Preface

This 15th volume of the now famous series *Chemical Signals in Vertebrates* gathers part of the oral communications given at the 15th eponymous meeting held in Dijon, Burgundy (France) in November 2021. Since 1976,[1] an ever-rejuvenating group of scientists from the five continents has congregated every three years to share their methods, facts, theories and speculations on whether, how and why vertebrates communicate through chemical cues and signals.

This triennial pace of the CSiV meetings[1] has been disrupted by the COVID-related pandemic, however, and the expected 2020 edition had to be postponed to 2021. Furthermore, due to the then continued instability of the global health situation, the organizers did not dare inviting an in-person meeting. Instead, a substitutive online solution was found through a virtual platform which allowed attending to both live-on-screen and pre-recorded sessions in the virtual conference hall, as well as to gatherings of multiple interactants chatting around posters. It was a pleasing moment despite everybody being frustrated by lacking real live interactions around local wines. In a way, the informal CSiV network resisted the viral adversity as 102 delegates, with their dozens of co-authors, commuted – electronically – toward the three-day meeting in Dijon. The challenge was great to get participants from 20 countries spread over 18 time zones from Japan and Australia to Columbia, California and Alberta to synchronize talks and sessions. Participants were both punctual and manifested real enthusiasm as everyone – especially those of far East and of far West – woke up very early and went to bed very late. One positive side-effect is certain: we saved a lot of carbon for the Planet!

As in previous CSiV meetings, the almost complete range of vertebrate orders animated the diverse titles of the programme, although in an unbalanced way in favour of mammals, with several talks and posters devoted to the human case, a major research

[1] Previous CSiV meetings were held in Saratoga Springs, New York (1976); Syracuse, New York (1979); Sarasota, Florida (1982); Laramie, Wyoming (1985); Oxford, England (1988); Philadelphia, Pennsylvania (1991); Tubingen, Germany (1994); Ithaca, New York (1997); Krakow, Poland (2000); Corvallis, Oregon (2003); Chester, England (2006); Urbana-Champaign, Illinois (2014); Cardiff, Wales (2017).

model of the local team. Although fishes were missing, we also heard talks or read posters on amphibians, reptiles and birds along a range of themes spanning from: Methods of offline and online scent analysis; Statistical processing of complex data; Reproducibility in chemical ecology research; Chemosensory processing and olfactory memory; Glandular and microbiologic sources of chemosignals; Signals and cues in intraspecific communication; Olfactory tags of honesty; Olfactory correlates of fitness and sickness; Olfactory imprinting; Olfaction in multisensory cognition; Physiological responses to pheromones in reproductive processes; Odour-related antipredator responses and protection against disease vectors; Odour-based management of human-animal relationships for conservation and pest regulation; Odour-trained animals as biological detectors of illness; Effects of captivity on chemocommunication, etc. Five symposia were proposed on human chemical communication (reunited by Camille Ferdenzi, Lyon, France), chemoreception in neonatal organisms (Hélène Loos, Erlangen, Germany), innovative methods in semiochemistry (Marianne Gabirot, Alès, France), chemical signals in multisensory environments (Arnaud Leleu, Dijon, France) and chemical signals of health disorders (Michał Dzięcioł, Wrocław, Poland), each related with plenary talks on ‘Chemosensory communication of social connectedness in humans’ by Bettina Pause (Düsseldorf, Germany), ‘Chemical emissions and signals from human beings’ by Jonathan Williams (Mainz, Germany), ‘Using olfaction to understand intergenerational influences of parental sensory experiences’ by Brian Dias (Los Angeles, USA), ‘Evolutionary perspectives on peri-receptor events in mammalian olfaction’ by Fabrice Neiers (Dijon, France), ‘Traumatic mating in amphibians’ by Lisa Schulte (Frankfurt, Germany) and ‘Response of domestic cats to plant iridoids as defence against mosquitoes’ by Masao Miyazaki and Reiko Uenoyama (Morioka, Japan). The meeting was closed by a thoughtful dual key-note on morbidity-related odours and social behaviour by Jane Hurst and Rob Beynon (Liverpool, UK), and Guila Ganem and Flora Gouzerh (Montpellier, France).

The book begins with a selective overview and contextualization of the 15th CSiV meeting contributions by Dietland Müller-Schwarze, both a pioneer in mammalian semiochemistry and behaviour, and one of the founding fathers of the Chemical Signals in Vertebrates series, both as colloquia and reference books. Then, more than 60 junior and senior co-authors wrote chapters. We much appreciate their investment in this new volume of the Chemical Signals in Vertebrates series as an act of resistance to the ambient pressure to celebrate publication outlets only as a function of impact factor's elevation, especially when publication fees are unaffordably high. Contributions in this CSiV series follow multiple formats as data articles, exploratory attempts, syntheses or position papers, always offering stimulating summaries toward new ideas without forgetting forerunning steps and authors. All contributions were doubly reviewed by experts internal and external to our group of authors. We gratefully acknowledge the time and effort the external reviewers offered to blindly improve the quality of the contributions.

CSiV is not structured as a formal academic society, with its president-secretary-treasurer trio. It is an untold, spontaneous movement which, like children's games in the schoolyard, is transmitted from generation to generation of researchers without instructions and coaching. Some persons agree to organize the next meeting by giving it a local colour while preserving a spirit open to taxonomic diversity and comparativism, multidisciplinarity weaving chemistry, physics, biology and psychology, and generous sharing of knowledge and know-how between colleagues of all ages and expertise. CSiV has no private resources and seeks its own subsidies at each triennial recurrence. In addition to registration fees, we are therefore pleased to acknowledge sponsorship from the French National Centre for Scientific Research (CNRS), Regional Council of Bourgogne–Franche-Comté, Université de Bourgogne, CNRS Research networks MediatEC (GDR 3658) and O3 (GDR Odorant, Odeur, Olfaction), European Chemoreception Research Organization (ECRO), Société de Neuro-Endocrinologie, French Institute for Agronomic Research (Inrae, departments of Animal physiology and breeding systems and of Human alimentation), and the Center for Taste, Smell and Feeding Science (CSGA, Dijon).

Finally, we would like to express our gratitude to the members of the 'Development of Olfactory Communication and Cognition' Laboratory (Dijon) and of the 'Neuroendocrine Integration of Reproduction & Behaviour Group' (Nouzilly) who worked hard to prepare a first trial of the meeting in 2020 – sponsors, wines, dinners, music, and social program were ready! – which had to be cancelled by the imperious COVID-19. The second trial in 2021 needed a bit less people among which we address special thanks to Vincent Gigot, Benjamin Nourry and Stéphanie Busset.

The next leap of the CSiV meeting will be the 16th, prolonging the initial kick-off meeting organized by Dietland Müller-Schwarze and Robert M. Silverstein about a half-century ago! It is planned to be held in Frankfurt-am-Main, Germany, in 2024 under the supervision of Dr. Lisa Schulte and her team.

Dijon, France — Benoist Schaal
Dijon, France — Diane Rekow
Nouzilly, France — Matthieu Keller
Dijon, France — Fabrice Damon

# Contents

**The Fifteenth Volume on Chemical Signals in Vertebrates: Some Reflections** ........ 1
Dietland Müller-Schwarze

**Part I Methods from Chemistry to Memory and Honesty**

**From Swab to Spectrum: A Comprehensive Guide to Optimizing Simple Gas Chromatography-Mass Spectrometry Analysis of Vertebrate Odorants** ........ 11
Emily (Lee) Nonnamaker, Marlena Muszynska, Ian Lightcap, Christine M. Drea, and Elizabeth A. Archie

**Development of a One-Day Test of Olfactory Learning and Memory in Mice** ........ 39
Richard E. Brown, Oliver K. Schnare, Elias B. Habib, and Kyle M. Roddick

**Putting a Price on Honesty: Methods to Evaluate the Costs of Olfactory Signalling** ........ 55
Christina D. Buesching and Chris Newman

**Part II Intraspecific Communication in Non-human Vertebrates**

**Chemical Communication and Semiochemical Recognition in Frogs: From Eggs to Adults** ........ 75
Lisa M. Schulte, Konrad Lipkowski, and Diana Abondano Almeida

**Avian Chemical Signatures: An Overview** ........ 113
Tatjana Alves Soares, Barbara A. Caspers, and Helene M. Loos

**A Gull Species Recognizes MHC-II Diversity and Dissimilarity Using Odor Cues** . . . . . 139
Maxime Pineaux, Pierrick Blanchard, Léa Ribeiro, Scott A. Hatch, and Sarah Leclaire

**Exploration of Olfactory Communication in the Water Vole, *Arvicola terrestris*** . . . . . 153
Kévin Poissenot, Chantal Porte, Didier Chesneau, and Matthieu Keller

**Detection of Male Olfactory Information by Female Goats: A Calcium Imaging Approach** . . . . . 165
Maxime Meunier, Anne-Charlotte Trouillet, Matthieu Keller, and Pablo Chamero

**The Use of Excretion- and Secretion-Based Scent Communication in European Badgers (*Meles Meles*)** . . . . . 177
Tanesha M. Allen, Chris Newman, David W. Macdonald, and Christina D. Buesching

**Scent Communication Between Males and Estrous Females in Domestic Dogs (*Canis Lupus Familiaris*)** . . . . . 195
Tamako Miyazaki, Urara Onozawa, Reiko Uenoyama, and Masao Miyazaki

**Olfaction-Mediated Pathogen Avoidance in Mammals** . . . . . 207
Dagmar Schwambergová, Jitka Třebická Fialová, and Jan Havlíček

**Part III Human Chemocommunication**

**Smelling the Basis of Social Connectedness: Chemosensory Communication in Humans** . . . . . 235
Bettina M. Pause

**Handshaking and Hand-Smelling: On the Potential Role of Handshake Greeting in Human Olfactory Communication** . . . . . 257
S. Craig Roberts, Francesca P. Singleton, Paxton D. Culpepper, Piotr Sorokowski, Agnieszka Sorokowska, and Jan Havlíček

**Part IV Development**

**Early Olfactory Learning in Mammals: A Case of Imprinting?** . . . . . 273
Lucie Jelínková and Jan Havlíček

**Can Prenatal or Neonatal Exposure to Chemosensory Variety Orient the Development of Personality? Exploratory Investigations in Mice** . . . . . 305
Bruno Patris, Aurélie Dastugue, Clémence Lesimple, and Benoist Schaal

**Does the Rabbit Mammary Pheromone Attract Newborns to Maternal Faeces? A New Potential Function of the Suckling Chemosignal** . . . . . 331
Benoist Schaal, Anne-Sophie Moncomble, and Dominique Langlois

**Tips from the Nose: Odor-Driven Visual Categorization in the Developing Human Brain** . . . . . 361
Diane Rekow and Arnaud Leleu

**Part V Interspecific Communication**

**Specific Antipredator Response of Leopard Geckos (*Eublepharis macularius*) to the Smell of Snake Exuvia** . . . . . 399
Eva Landová, Petra Hnidová, Aleksandra Chomik, Jitka Jančúchová-Lásková, Petra Frýdlová, and Daniel Frynta

**Reactions of Free-Ranging African Carnivores to Single Components of Carnivore Odours: Evidence that Mammalian Semiochemicals Are Not All Irreducibly Complex** . . . . . 419
Peter Apps and J. Weldon McNutt

**Chemical Pest Defense by the Innate Response to Silver Vine and Catnip Plants in the Domestic Cat** . . . . . 435
Masao Miyazaki and Reiko Uenoyama

**Part VI Applications**

**Canine Detection of Cancer in Humans: Expectations Versus Reality** . . . . . 453
Tadeusz Jezierski and Michał Dzięcioł

**Chemical Ecology and Elephant Conservation: Examining Chili Pepper Deterrent Fences** . . . . . 489
R. Lynn Von Hagen, Sophia C. Corde, Pauline Norris, Simon Kasaine, Mwangi Githiru, Bernard Amakobe, Urbanus N. Mutwiwa, and Bruce A. Schulte

**Index** . . . . . 507

# Contributors

**Diana Abondano Almeida** Wildlife-/Zoo-Animal-Biology and Systematics, Faculty of Biological Sciences, Goethe University Frankfurt, Frankfurt am Main, Germany

**Tanesha M. Allen** Department of Biology, University of Oxford, Oxford, UK

**Tatjana Alves Soares** Department of Chemistry and Pharmacy, Friedrich-Alexander-Universität Erlangen-Nürnberg, Erlangen, Germany

**Bernard Amakobe** Wildlife Works, Voi, Kenya

**Peter Apps** BioBoundary Project, Botswana Predator Conservation, Maun, Botswana

**Elizabeth A. Archie** Department of Biological Sciences, University of Notre Dame, Notre Dame, IN, USA

**Pierrick Blanchard** Laboratoire Évolution & Diversité Biologique (UMR 5174), Université de Toulouse Midi-Pyrénées, CNRS, IRD, UPS, Toulouse, France

**Richard E. Brown** Department of Psychology and Neuroscience, Dalhousie University, Halifax, NS, Canada

**Christina D. Buesching** Department of Zoology, Irving K. Barber Faculty of Sciences, Kelowna, BC, Canada

**Barbara A. Caspers** Department of Behavioural Ecology, Bielefeld University, Bielefeld, Germany

**Pablo Chamero** UMR Physiologie de la Reproduction et des Comportements, INRAE, CNRS, IFCE, Université de Tours, Nouzilly, France

**Didier Chesneau** UMR Physiologie de la Reproduction et des Comportements, INRAE, CNRS, Université de Tours, Nouzilly, France

**Aleksandra Chomik** Ecology and Ethology Research Group, Department of Zoology, Faculty of Science, Charles University, Prague, Czech Republic

**Sophia C. Corde** Western Kentucky University, Biology Department, Bowling Green, KY, USA

**Paxton D. Culpepper** Department of Psychology, University of Stirling, Stirling, UK

**Aurélie Dastugue** Development of Olfactory Communication and Cognition Laboratory, Centre des Sciences du Goût et de l'Alimentation, CNRS, Université de Bourgogne, Inrae-Institut Agro Dijon, Dijon, France

**Christine M. Drea** Department of Evolutionary Anthropology, Duke University, Durham, NC, USA

**Michał Dzięcioł** Wrocław University of Environmental and Life Sciences, Department of Reproduction and Clinic of Farm Animals, Wrocław, Poland

**Petra Frýdlová** Ecology and Ethology Research Group, Department of Zoology, Faculty of Science, Charles University, Prague, Czech Republic

**Daniel Frynta** Ecology and Ethology Research Group, Department of Zoology, Faculty of Science, Charles University, Prague, Czech Republic

**Mwangi Githiru** Wildlife Works, Voi, Kenya

**Elias B. Habib** Department of Psychology and Neuroscience, Dalhousie University, Halifax, NS, Canada

**Scott A. Hatch** Institute for Seabird Research and Conservation, Anchorage, AK, USA

**Jan Havlíček** Department of Zoology, Faculty of Science, Charles University, Prague, Czech Republic

**Petra Hnidová** Ecology and Ethology Research Group, Department of Zoology, Faculty of Science, Charles University, Prague, Czech Republic

**Jitka Jančúchová-Lásková** Ecology and Ethology Research Group, Department of Zoology, Faculty of Science, Charles University, Prague, Czech Republic

**Lucie Jelínková** Department of Zoology, Faculty of Science, Charles University, Prague, Czech Republic

**Tadeusz Jezierski** Institute of Genetics and Animal Biotechnology of Polish Academy of Sciences, Department of Animal Behaviour and Welfare, Jastrzębiec, Poland

**Simon Kasaine** Wildlife Works, Voi, Kenya

**Matthieu Keller** Physiologie de la Reproduction et des Comportements, INRAE, CNRS, Université de Tours, Nouzilly, France

**Eva Landová** Ecology and Ethology Research Group, Department of Zoology, Faculty of Science, Charles University, Prague, Czech Republic

**Dominique Langlois** Development of Olfactory Communication and Cognition Laboratory, Centre des Sciences du Goût et de l'Alimentation, Inrae-Institut Agro Dijon, Dijon, France

**Sarah Leclaire** Laboratoire Évolution & Diversité Biologique (UMR 5174), Université de Toulouse Midi-Pyrénées, CNRS, IRD, UPS, Toulouse, France

**Arnaud Leleu** Development of Olfactory Communication and Cognition Laboratory, Centre des Sciences du Goût et de l'Alimentation, CNRS, Université Bourgogne Franche-Comté, Inrae-Institut Agro Dijon, Dijon, France

**Clémence Lesimple** Development of Olfactory Communication and Cognition Laboratory, Centre des Sciences du Goût et de l'Alimentation, CNRS, Université de Bourgogne, Inrae-Institut Agro Dijon, Dijon, France

**Ian Lightcap** Materials Characterization Facility, University of Notre Dame, Notre Dame, IN, USA

Center for Sustainable Energy at Notre Dame, University of Notre Dame, Notre Dame, IN, USA

**Konrad Lipkowski** Wildlife-/Zoo-Animal-Biology and Systematics, Faculty of Biological Sciences, Goethe University Frankfurt, Frankfurt am Main, Germany

**Helene M. Loos** Department of Chemistry and Pharmacy, Friedrich-Alexander-Universität Erlangen-Nürnberg, Erlangen, Germany

Department of Sensory Analytics and Technology, Fraunhofer Institute for Process Engineering and Packaging IVV, Freising, Germany

**David W. Macdonald** Wildlife Conservation Research Unit, Department of Zoology, University of Oxford, Recanati-Kaplan Centre, Tubney, Oxfordshire, UK

**J. Weldon McNutt** Botswana Predator Conservation, Maun, Botswana

**Maxime Meunier** UMR Physiologie de la Reproduction et des Comportements, INRAE, CNRS, IFCE, Université de Tours, Nouzilly, France

**Masao Miyazaki** Laboratory of Biomolecular Science, Department of Biological Chemistry and Food Sciences, Faculty of Agriculture, Iwate University, Iwate, Japan

**Tamako Miyazaki** Cooperative Department of Veterinary Medicine, Iwate University, Morioka, Japan

**Anne-Sophie Moncomble** Development of Olfactory Communication and Cognition Laboratory, Centre des Sciences du Goût et de l'Alimentation, Université de Bourgogne, CNRS, Inrae, Institut Agro Dijon, Dijon, France

**Dietland Müller-Schwarze** College of Environmental Science and Forestry, State University of New York, Syracuse, NY, USA

**Marlena Muszynska** Department of Biological Sciences, University of Notre Dame, Notre Dame, IN, USA

**Urbanus N. Mutwiwa** Jomo Kenyatta University of Agriculture and Technology, Nairobi, Kenya

**Chris Newman** Wildlife Conservation Research Unit, Department of Zoology, University of Oxford, Recanati-Kaplan Centre, Tubney, Oxfordshire, UK

**Emily (Lee) Nonnamaker** Department of Biological Sciences, University of Notre Dame, Notre Dame, IN, USA

**Pauline Norris** Advanced Materials Institute, Western Kentucky University, Bowling Green, KY, USA

**Urara Onozawa** Department of Biological Chemistry and Food Sciences, Faculty of Agriculture, Iwate University, Morioka, Iwate, Japan

**Bruno Patris** Development of Olfactory Communication and Cognition Laboratory, Centre des Sciences du Goût et de l'Alimentation, CNRS, Université de Bourgogne, Inrae-Institut Agro Dijon, Dijon, France

**Bettina M. Pause** Department of Experimental Psychology, Heinrich-Heine-University, Düsseldorf, Germany

**Maxime Pineaux** Laboratoire Évolution & Diversité Biologique (UMR 5174), Université de Toulouse Midi-Pyrénées, CNRS, IRD, UPS, Toulouse, France

**Kévin Poissenot** UMR Physiologie de la Reproduction et des Comportements, INRAE, CNRS, Université de Tours, Nouzilly, France

**Chantal Porte** UMR Physiologie de la Reproduction et des Comportements, INRAE, CNRS, Université de Tours, Nouzilly, France

**Diane Rekow** Development of Olfactory Communication and Cognition Laboratory, Centre des Sciences du Goût et de l'Alimentati, CNRS, Université Bourgogne, Inrae-Institut Agro Dijon, Dijon, France

**Léa Ribeiro** Laboratoire Évolution & Diversité Biologique (UMR 5174), Université de Toulouse Midi-Pyrénées, CNRS, IRD, UPS, Toulouse, France

**S. Craig Roberts** Department of Psychology, University of Stirling, Stirling, UK

Institute of Psychology, University of Wrocław, Wrocław, Poland

**Kyle M. Roddick** Department of Psychology and Neuroscience, Dalhousie University, Halifax, NS, Canada

**Benoist Schaal** Development of Olfactory Communication and Cognition Laboratory, Centre des Sciences du Goût et de l'Alimentation, CNRS, Université de Bourgogne, Inrae-Institut Agro Dijon, Dijon, France

**Oliver K. Schnare** Department of Psychology and Neuroscience, Dalhousie University, Halifax, NS, Canada

**Bruce A. Schulte** Western Kentucky University, Biology Department, Bowling Green, KY, USA

**Lisa M. Schulte** Wildlife-/Zoo-Animal-Biology and Systematics, Faculty of Biological Sciences, Goethe University Frankfurt, Frankfurt am Main, Germany

**Dagmar Schwambergová** Department of Zoology, Faculty of Science, Charles University, Prague, Czech Republic

**Francesca Singleton** Department of Psychology, University of Stirling, Stirling, UK

**Agnieszka Sorokowska** Institute of Psychology, University of Wrocław, Wrocław, Poland

**Piotr Sorokowski** Institute of Psychology, University of Wrocław, Wrocław, Poland

**Jitka Třebická Fialová** Department of Zoology, Faculty of Science, Charles University, Prague, Czech Republic

**Anne-Charlotte Trouillet** UMR Physiologie de la Reproduction et des Comportements, INRAE, CNRS, IFCE, Université de Tours, Nouzilly, France

**Reiko Uenoyama** Laboratory of Biomolecular Science, Department of Biological Chemistry and Food Sciences, Faculty of Agriculture, Iwate University, Morioka, Iwate, Japan

**R. Lynn Von Hagen** Western Kentucky University, Biology Department, Bowling Green, KY, USA

# The Fifteenth Volume on Chemical Signals in Vertebrates: Some Reflections

Dietland Müller-Schwarze[1] (✉)

[1] College of Environmental Science and Forestry, State University of New York, Syracuse, NY, USA
dmullers@windstream.net

**Abstract** The organizers of the fifteenth Chemical Signals in Vertebrates (CSiV) meeting in 2021 kindly asked me to comment on the trajectory of our field from my vantage point. As one of the founders (and the only surviving one), I have participated in, and watched the progression of our field over the years as we marched from one gathering to the next. We started the CSiV series of symposia in 1976. Initially envisioned as a one-off meeting, the gatherings have endured, alternating between the New and Old Worlds. This continuing tradition attests to the need for a forum to present and exchange data and ideas in our specific field of chemical communication.

**Keywords** Chemical ecology · Chemical signals and cues · Pheromones · Multidisciplinary and multi-site studies · Practical applications

## 1 The Trajectory

We have come a long way. The arc from the first to the fifteenth symposium on Chemical Signals in Vertebrates started with asking what animals exactly communicate with their sometimes strong smells, but also often very subtle secretions from their skin glands, as well as our first serious attempts at identifying behaviorally active compounds in vertebrates. Today the study of chemical communication operates in a "multi"-world: multicomponent, multisource, multimodal, multifunctional, multispecies, multidisciplinary, multi-author.

Picked somewhat arbitrarily, the midway point on the trajectory from then to now was the eighth Conference on Chemical Signals in Vertebrates, held at Cornell University in Ithaca, New York. The preface of its 1999 Proceedings starts with: "The field of olfactory research and chemical communication is in the early stages of revolutionary change, and many aspects of this revolution are reflected in the chapters in this book." What stage are we in now, one might ask in 2022. The preface also highlighted the interdisciplinary path we had traveled already then: "This meeting was extremely



B. Schaal et al. (eds.), *Chemical Signals in Vertebrates 15*,
https://doi.org/10.1007/978-3-031-35159-4_1

exciting because it brought together a wide range of experts, including organic chemists, biochemists, molecular biologists, zoologists, animal behaviorists, behavioral ecologists, physiologists, neuroscientists, behavioral biologists, and psychologists." This productive collaboration and interdisciplinary exchange of ideas is still the glue that binds our research community together today.

## 2 Definitions

Along the way from the first to the 15th meeting, we had to work through the definitions as they developed in this young field. Much debate has centered on the notion of pheromones. Conceptually, it is important to distinguish between signal and cue. A signal is part of an evolved communication pattern, sent by an individual and triggering a behavioral, physiological or developmental response in another. A cue can be any feature guiding the behavior of an animal, like the odor of prey, or food odor on the mouth of a group member. At the core of our research efforts, a number of vertebrate pheromones in the strict sense of the word have been identified by following a response-guided strategy: painstaking fractionation of secretions and excretions, intertwined with bioassays. Based on the findings, the pheromone definition has been updated and refined. The original formulation by Karlson and Lüscher (Karlson & Lüscher, 1959) has held up remarkably well in its essence, although based on insect research at the time: evolved signals between members of the same species trigger a behavior or developmental process, and learning is not required.

In mammals, behaviorally active secretions can be very complex. Non-volatile compounds play a large role as precursors or "keeper substances." They can also serve as pheromones themselves, as for instance darcin, a protein in the urine of male house mice that attracts females as strongly as whole urine (Roberts et al., 2010). No wonder that during some of the past meetings the notion of pheromones has triggered lively discussions, helping to clarify what is a pheromone and what it is not. Strict standards have to be applied before naming a signal a pheromone. For instance, the mammary pheromone in the rabbit, 2-methylbut-2-enal (2 MB2) met five criteria: it is a single molecular compound; elicits the complete response elicited by the original substrate (milk); is selectively active compared to other compounds; active at species level, and not in neonates from the other species assessed so far; and no prior exposure is needed for stimulus efficacy (Schaal et al., 2003). A good example for a priming pheromone is 4-ethyloctanal from the skin of the head of the male goat. It acts on the gonadotropin-releasing pulse generator in the female goat (Murata et al., 2014).

## 3 The Present

Today we have built an impressive understanding of how chemical communication in vertebrates works. We have charged forward, as shown by the spectacular findings in mice, rabbits and goats. In a landmark study, Novotny et al. (1985) not only identified

3,4-dehydro-*exo*-brevicomin and 2-*sec*-butyl-4,5-dihydrothiazole in male mouse urine as compounds stimulating male aggression and attracting females, but also that they acted synergistically. Furthermore, these compounds also were the first identified primer pheromones in mammals. They induce estrus in female mice (Jemiolo et al., 1986). The same team identified the pheromone in female urine that delays puberty in female mice as 2,5-dimethylpyrazine (Novotny et al., 1986). A major milestone was the discovery that proteins can be mammalian pheromones. Darcin, a Major Urinary Protein in male mouse urine, attracts females and stimulates a female's memory so that she later is attracted to volatile signals from only that particular male (Roberts et al., 2010).

At times we needed course corrections, as in the cases of the dog, rhesus macaque and human putative "pheromones." The last is being revisited (see Wyatt, 2015) and provides a good example of how we can get past the doldrums and gather new steam. We can focus on a different developmental stage of a mammal, or turn to techniques from other disciplines. The neonatal period is a place to look for pheromones, as olfaction is important before other sensory modalities kick in. After success in identifying the rabbit mammary pheromone, the arousal, orienting and sucking responses of human neonates to the maternal areolar gland secretion present an exciting line of research (Doucet et al., 2012). As for new techniques, established in fields like agricultural, food or atmospheric research, proton transfer reaction mass spectrometry (PTR-MS) is giving a boost to studies of human body odors (Roberts et al., 2020; also Kupferschmidt, 2019). Measuring the changing emission of volatiles in real time, this technique works like a film camera as compared to the snapshots of earlier studies like the T-shirt test for collecting axillary odor. Volatiles emitted by crowds at soccer matches (Stönner & Williams, 2016) or watching movies (Williams & Pleil, 2016; Williams et al., 2016) have been monitored in this way. Carbon dioxide and some volatile organic compound concentrations changed with "suspense" and "comedy" scenes in a film; and isoprene levels rose in a soccer stadium when spectators rose from their seats in response to a scored goal.

We now know intricate multipartite interactions, with odor an essential link. For instance, in Zika and dengue fever, body odor functions as a key link in an intriguing example of a multi-partner chain of events. Both a vertebrate and an insect, as well as microorganisms are involved. A virus manipulates the skin microbiota of mammals such as mice or humans to produce volatiles that attract mosquitoes who in turn transmit the Zika or dengue virus between host individuals. Body odor from Zika-infected mice attracted more mosquitoes than odor from healthy mice. Likewise, mosquitoes preferred odor of human dengue fever patients to that of healthy individuals. The levels of 11 volatiles were elevated. The source of the odor were skin bacteria. Normally, skin cells kill many of these microbes by means of a protein. But dengue- or Zika-infected mice produced considerably less of this protein (Zhang et al., 2022). Thus, virus, mosquito, mammal and bacteria figure as actors in this biological drama. "Electronic noses" may help in diagnosing these diseases; and treatments by suppressing the mosquito-attracting volatiles are now being explored (Leslie, 2022).

## 4 Studying More Settings

Experimental results may differ according to the setting. Field studies deal with complex settings and are harder to conduct. The controlled conditions of laboratory experiments, on the other hand, may not permit the animal to show the full potential of its behavior. New insights on these old questions were reported at CSiV 15. For instance, if an active secretion is diet-dependent, we can expect that its composition differs between laboratory and field. Indeed, comparing free-ranging and captive tamarins, Poirier et al. (Poirier et al., 2021) found that the former produced a richer mixture of more and different compounds. In some studies, field and laboratory studies confirmed each other. Free-ranging and captive owl monkeys (*Aotus spp.*) had similar secretions from the subcaudal and pectoral glands, although captive individuals differed according to their location, possibly due to diet differences (Spence-Aizenberg & Kimball, 2018). On the behavioral side, water voles in both the field and laboratory were attracted to the odors of urine and lateral gland secretions of females (Poissenot et al. 2023).

## 5 Practical Applications

Finding effective practical applications remains the Holy Grail for many researchers. When we understand a behavior well we are able to manipulate it for practical purposes. Applications include pest control, crop protection, animal conservation, animal breeding, animal welfare, wildlife management and detection of diseases in humans and animals. Both animal noses and animal secretions have important roles to play. At this meeting, several speakers reported progress using mammals as olfactory detectors.

Dogs and mice have been put to the test. Jezierski and Dzięcioł (2023) review studies of dogs "detecting" six different human cancers. Although dogs can distinguish odors of cancer patients from those of healthy persons, they cannot do so reliably in the early stages of cancer, before formal diagnosis. Furthermore, so far, no volatile organic compounds of cancer patients have been identified that dogs might cue in on. Obviously, much work lies ahead in this important area. Gadbois and Kiiroja (2021) examined whether dogs are able to detect the early onset of panic attacks in post-traumatic stress disease (PTSD) patients. Finally, searching for ways to predict epileptic seizures, Catala et al. (2021) showed that such seizures are associated with olfactory characteristics that dogs can detect as early as 4.6 h before the seizure. Maa et al. (2021a) reported dog's responses even 68 min before seizures, on average. Dogs are able to discriminate the breath odor during a seizure from other odors of the same person during calmer time periods (Catala et al., 2019). Odors during seizures have even their chemical characteristics: Maa et al. (2021b) found menthone in fear sweat and epilepsy sweat.

Mice can be trained to discriminate between the urinary odors of mice with melanoma from those that are healthy. But they do not discriminate spontaneously (Kokocińska-Kusiak et al., 2021). Similarly, female mice can discriminate the odors of male mice with pulmonary cancer from those of healthy males, even at an early stage of cancer. But the females were not attracted less to the diseased mice (Gouzerh et al., 2021).

In the field of animal welfare, animal secretions can be used to boost health. As an example, artificially raised lambs grew better when they were drinking milk from dispensers scented with inguinal wax from ewes (Alary et al., 2021, 2023).

## 6 Outlook

It would be presumptuous to try to predict the future, but some growing points offer themselves: we will understand ever better the subtleties and complexities of the chemical language of animals, embedded with other modalities, as four papers of the symposium on multisensory perception at CSiV 15 attest to. In Australian sea lions, mothers returning from the sea use visual, acoustic and olfactory cues to locate and identify their pups before nursing (Pitcher et al., 2021). Scent marking in lemurs does not only involve multiple chemical sources—genital and sternal glands, urine and tree gouging—but is also accompanied by attention-getting visual and audible behavior (Drea & Greene, 2021). In humans, odors assist the developing brain of infants: Rekow et al. (Rekow et al., 2021; see Rekow and Leleu, Chap. 18, this volume) studied brain response to images of human faces in the presence of mother's odor vs. a baseline odor. Turning to interspecific cues, among raptors, turkey vultures and southern caracaras use both vision and olfaction in foraging (Potier et al., 2021).

We will drill deeper into the biosynthesis of semiochemicals and the genetic underpinnings of behavior and chemoreception. The brain will reveal more of its workings. New methods are being developed, or adopted from neighboring disciplines. Portable, miniaturized and more sensitive equipment such as wearable electronic noses open up more uses, particularly in field studies. Currently electronic noses are put to work in fields like the food industry (coffee, wine, beer, meat), health care, air quality monitoring, agriculture, waste management, and security (Wilson & Baietto, 2009). Particularly close to our research interests, electric noses play a role in forest pathology for detection of larvae of the emerald ash borer (Wilson et al., 2019). In this case the tree volatiles are being monitored.

Practical applications have much room to grow, especially if more funding becomes available. We might be at the cusp of major new developments in the area of applied chemical ecology. Reflecting all that complexity, research teams will grow ever larger. Above all, researchers will continue to be motivated by their shared passion to explore new horizons, and they will inspire each other, aspects not expressed in sober, formal publications.

## References

Alary J, Schaal B, Patris B, Chotro G, Destrez A (2021) Painting inguinal wax on artificial nipples enhances weight gain in artificially fed lambs. Communication at the 15th meeting on Chemical Signals in Vertebrates, Dijon, France, 3–5 November 2021

Alary, J., Schaal, B., Chotro, G., Patris, B., & Destrez, A. (2023). Mother's scent for motherless neonates: responses of artificially reared lambs to ewe's inguinal wax odor. *Applied Animal Behaviour Science, 258*, 105812.

Catala, A., Grandgeorge, M., Schaff, J. L., Cousillas, H., Hausberger, M., & Cattet, J. (2019). Dogs demonstrate the existence of an epileptic seizure odour in humans. *Scientific Reports, 9*(1), 1–7.

Catala A, Cousillas H, Levasseur-Garcia C, Pagès M, Schaff JL, Till U, Vitola Pasetto L, Cattet F, Violleau M, Grandgeorge M (2021) Trained dogs and chemical analysis demonstrate the existence of an epileptic seizure odour in humans. Communication at the 15th meeting on Chemical Signals in Vertebrates, Dijon, France, 3–5 November 2021

Doucet, S., Soussignan, R., Sagot, P., & Schaal, B. (2012). An overlooked aspect of the human breast areolar glands in relation with breastfeeding pattern, neonatal weight gain, and the dynamics of lactation. *Early Human Development, 88*, 119–128. https://doi.org/10.1016/j.earlhumdev.2011.07.020

Drea CM, Greene LK (2021) The 'language' of scent within multimodal lemur signals. Communication at the 15th meeting on Chemical Signals in Vertebrates, Dijon, France, 3–5 November 2021

Gadbois S, Kiiroja L (2021) PTSD alert dogs: could dogs alert to the early onset of panic attacks and flashbacks in humans? Communication at the 15th meeting on Chemical Signals in Vertebrates, Dijon, France, 3–5 November 2021

Gouzerh F, Buatois B, Dormont L, Thomas F, Ganem G (2021) Detection and preference of mice for smells of cancerous congeners. Communication at the 15th meeting on Chemical Signals in Vertebrates, Dijon, France, 3–5 November 2021

Jemiolo, B., Harvey, S., & Novotny, M. (1986). Promotion of the Whitten effect in female mice by the synthetic analogs of male urinary constituents. *Proceedings of the National Academy of Sciences of the United States of America, 83*, 4576–4579.

Jezierski, T., & Dzięgioł, M. (2023). Canine detection of cancer in humans - expectations and reality. In B. Schaal, M. Keller, D. Rekow, & F. Damon (Eds.), *Chemical signals in vertebrates 15*. Springer.

Karlson, P., & Lüscher, M. (1959). "Pheromones": a new term for a class of biologically active substances. *Nature, 183*, 55–56.

Kokocińska-Kusiak A, Matczuk A, Jezierski T (2021) Mice training to discriminate urine odor of conspecifics with neoplasm before clinical symptoms. Communication at the 15th meeting on Chemical Signals in Vertebrates, Dijon, France, 3–5 November 2021

Kupferschmidt, K. (2019). Atmospheric scientists join pheromone quest. *Science, 364*, 112.

Leslie, M. (2022). Dengue and zika viruses turn people into mosquito bait. *Science, 377*, 137.

Maa, E., Arnold, J., Ninedorf, K., & Olsen, H. (2021a). Canine detection of volatile organic compounds unique to human epileptic seizure. *Epilepsy & Behavior, 115*, 107690.

Maa, E. H., Arnold, J., & Bush, C. K. (2021b). Epilepsy and the smell of fear. *Epilepsy & Behavior, 121*, 108078.

Murata, K., Tamogami, S., Itou, M., Ohkubo, Y., Wakabayashi, Y., Watanabe, H., Okamura, H., Takeuchi, Y., & Mori, Y. (2014). Identification of an olfactory signal molecule that activates the central regulator of reproduction in goats. *Current Biology, 24*, 681–686.

Novotny, M., Harvey, S., Jemiolo, B., & Alberts, J. (1985). Synthetic pheromones that promote inter-male aggression in mice. *Proceedings of the National Academy of Sciences of the United States of America, 82*, 2059–2061.

Novotny, M., Jemiolo, B., Harvey, S., Wiesler, D., & Marchlewska-Koj, A. (1986). Adrenal-mediated endogenous metabolites inhibit puberty in female mice. *Science, 231*, 722–725.

Pitcher BJ, Wierucka K, Harcourt R, Charrier I (2021) Multimodal recognition of the pup by Australian sea lion mothers. Communication at the 15th meeting on Chemical Signals in Vertebrates, Dijon, France, 3–5 November 2021

Poissenot, M., Porte, C., Chesneau, D., & Keller, M. (2023). Exploration of olfactory communication in the water vole *(Arvicola terrestris)*. In B. Schaal, M. Keller, D. Rekow, & F. Damon (Eds.), *Chemical signals in vertebrates 15*. Springer.

Poirier AC, Dunn JC, Watsa M, Erkenswick GA, Melin AD, Waterhouse JS, Smith AC (2021) Captivity and communication: comparative analysis of scents from wild and captive tamarins. Communication at the 15th meeting on Chemical Signals in Vertebrates, Dijon, France, 3–5 November 2021

Potier S, Duriez O, Célérier A, Bonadonna F (2021) Insight into the sensory ecology of raptors: vision and olfaction. Communication at the 15th meeting on Chemical Signals in Vertebrates, Dijon, France, 3–5 November 2021

Rekow D, Leleu A, Kiseleva A, Durand K, Schaal B, Rossion B, Baudouin JY (2021) How odors assist the developing visual system in humans. Communication at the 15th meeting on Chemical signals in vertebrates, Dijon, France, 3–5 November 2021

Roberts, S. A., Simpson, D. M., Armstrong, S. D., Davidson, A. J., Robertson, D. H., McLean, L., Beynon, R. J., & Hurst, J. L. (2010). Darcin: a male pheromone that stimulates female memory and sexual attraction to an individual male's odor. *BMC Biology, 8*, 75.

Roberts, S. C., Miszal, P. K., & Langford, B. (2020). Decoding the social volatilome by tracking rapid context-dependent odour change. *Philosophical Transactions of the Royal Society B: Biological Sciences, 375*, 20190259. https://royalsocietypublishing.org/doi/10.98/rstb.2019.0259

Schaal, B., Coureaud, G., Langlois, D., Ginies, C., Sémon, E., & Perrier, G. (2003). Chemical and behavioural characterization of the rabbit mammary pheromone. *Nature, 424*, 68–72.

Spence-Aizenberg, A., & Kimball, B. A. (2018). Chemical composition of glandular secretions from a pair-living monogamous primate: sex, age, and gland differences in captive and wild owl monkeys (Aotus spp.). *American Journal of Primatology, 80*(2), e22730.

Stönner, C., & Williams, J. (2016). Goals change crowd air chemistry. *Nature, 535*, 355. https://doi.org/10.1038/535355a

Williams, J., & Pleil, J. (2016). Crowd-based breath analysis: assessing behavior, activity, exposures, and emotional response of people in groups. *Journal of Breath Research, 10*, 032001. https://doi.org/10.1088/1752-7155/10/3/032001

Williams, J., Stönner, C., Wicker, J., Krauter, N., Derstroff, B., Bourtsoukidis, E., Klüpfel, T., & Kramer, S. (2016). Cinema audiences reproducibly vary the chemical composition of air during films by broadcasting scene specific emissions on breath. *Scientific Reports, 6*, 25464. https://doi.org/10.1038/srep25464

Wilson, A. D., & Baietto, N. (2009). Applications and advances in electronic-nose technology. *Sensors (Basel), 9*(7), 5099–5148. https://doi.org/10.3390/s90705099

Wilson, A. D., Force, L. B., Babst, B. A., & Bataineh, M. M. (2019). Detection of Emerald ash borer infestations in living green ash by noninvasive electronic-nose analysis of wood volatiles. *Biosensors (Basel), 9*(4), 123. https://doi.org/10.3390/bios9040123

Wyatt, T. D. (2015). The search for human pheromones: the lost decades and the necessity of returning to first principles. *Proceedings of the Royal Society B: Biological Sciences, 282*(1804), 20142994.

Zhang, H., Zhu, Y., Liu, Z., Peng, Y., Peng, W., Tong, L., et al. (2022). A volatile from the skin microbiota of flavivirus-infected hosts promotes mosquito attractiveness. *Cell, 185*(14), 2510–2522.

# Part I

# Methods from Chemistry to Memory and Honesty

# From Swab to Spectrum: A Comprehensive Guide to Optimizing Simple Gas Chromatography-Mass Spectrometry Analysis of Vertebrate Odorants

Emily (Lee) Nonnamaker[1] (✉), Marlena Muszynska[1], Ian Lightcap[2,3], Christine M. Drea[4], and Elizabeth A. Archie[1]

[1] Department of Biological Sciences, University of Notre Dame, Notre Dame, IN, USA
enonnama@nd.edu

[2] Materials Characterization Facility, University of Notre Dame, Notre Dame, IN, USA

[3] Center for Sustainable Energy at Notre Dame, University of Notre Dame, Notre Dame, IN, USA

[4] Department of Evolutionary Anthropology, Duke University, Durham, USA

**Abstract** Vertebrate odorants are essential to communication, but their complex chemical compositions are challenging to study. Here, we optimize a workflow for gas chromatography-mass spectrometry analysis of vertebrate odorants to generate a sensitive, broadly applicable method for detecting diverse compounds. We develop a standard mixture of 24 compounds in vertebrate odorants and use this mixture for tests of swab material, column, and solventless extraction approach. First, we test the performance of two swab materials (viscose and cotton wool) and preparation methods to minimize contamination and maximize recoverability. We find that cotton and viscose have similar contamination potentials, which are substantially reduced when "soaked-and-baked" and are equally effective at recovering compounds. We also compare two column types (polyethylene glycol (PEG) and polysiloxane) and find that the PEG column recovers higher-intensity peaks than the polysiloxane column but is susceptible to faster degradation. Finally, we compare compound recoverability for several solventless extraction approaches and adsorbent materials, including solid-phase micro-extraction (SPME) and thermal desorption (TD) on the PEG column. We find that TD outperforms SPME, providing greater flexibility to researchers lacking reliable cold chains. Together, these insights should make odorant analyses more practical, reproducible, and sensitive, helping advance research on vertebrate chemical communication.

**Keywords:** Solventless extraction · Vertebrate odorants · Vertebrate semiochemicals · Thermal desorption (TD) · Solid-phase micro-extraction (SPME)


B. Schaal et al. (eds.), *Chemical Signals in Vertebrates 15*,
https://doi.org/10.1007/978-3-031-35159-4_2

# 1 Introduction

Chemical signals are among the oldest and most varied forms of animal communication (Steiger et al., 2011). The first chemical signal, described by Karlson and Lüscher in 1959, was the pheromone bombykol, which plays a key role in reproductive communication in silk moths (Karlson & Lüscher, 1959). Pheromones are "molecules that are evolved signals…emitted by an individual and received by a second individual of the same species, in which they cause a specific reaction" (Wyatt, 2010). In 1976, Nordlund and Lewis coined the concept of "semiochemicals" as an umbrella term that includes pheromones and other chemical signals; semiochemicals are defined as "a chemical involved in the chemical interaction between organisms" (Nordlund & Lewis, 1976). Since then, much work has been done to measure the diversity and functional variability of semiochemicals across the animal kingdom (Wyatt, 2009, 2014), revealing that individuals use chemical cues to recognize many types of individuals—from members of their own species and both sexes, to individual group members and kin—with implications for reproduction, mate choice, parental care, dominance rank relations, and group cohesion (e.g., Coombes et al., 2018; Font et al., 2012; Hurst, 2009; Johnston, 2003; Keller-Costa et al., 2015; Nevitt, 2008; Scordato & Drea, 2007; Sorensen & Wisenden, 2014; Thom & Hurst, 2004; Woodley, 2015). The study of semiochemicals can yield powerful insights into vertebrate social behavior yet remains a logistically challenging endeavor.

To identify vertebrate semiochemicals, scientists must characterize the chemical profile of bodily odorants (e.g., the odorous compounds emanating from vertebrate body surfaces, excretions or secretions). Vertebrate studies often require nonlethal or noninvasive methods of repeat, in situ, odorant sampling (as opposed to one-time, whole-body extractions, performed in a laboratory), which introduces potential concerns associated with sampling methods (Zhao et al., 2022). Identifying chemical compounds in vertebrates thus presents at least five main challenges, as outlined below, some of which are relatively specific to a given system, some of which are more universal.

First, the materials used to collect vertebrate secretions will contribute natural or synthetic compounds that represent "contaminants" in the resulting chromatograms, meaning that they do not derive from the animal under study (Weiß et al., 2018). The most common choices are cotton wool and viscose, though the use of cotton has been questioned due to its potential to contain contaminants that are also found in vertebrate secretions (Birkemeyer et al., 2016). Several methods have been proposed to reduce these contaminants, including cleaning or "baking" the materials before sample collection (Birkemeyer et al., 2016; Drea et al., 2013); yet, considerable debate remains about which procedures are best. Second, the material used to collect vertebrate odorants can affect researchers' abilities to capture and retain chemical compounds; some researchers have proposed that cotton is inadequate to sample certain volatile compounds. This adequacy has yet to be tested with thermal desorption (TD) methods (Birkemeyer et al., 2016). Third, the quality of any chromatogram can be affected by the choice of gas-chromatography (GC) column or by column degradation. Each column is designed to target specific arrays of compounds, aligning target compounds from vertebrate swabs with column type is therefore crucial. For example, scientists targeting only highly volatile compounds might select a different column type than those targeting heavy compounds. Additionally, degradation issues can have profound effects on the quality of the resulting chromatograms, rendering the data from valuable samples unusable. It is therefore essential to incorporate standard methods of quality control. Fourth, because many researchers work in remote field conditions without reliable cold chain access, samples are liable to degrade between collection and analysis. Different solventless extraction methods may help stabilize these volatile compounds. Fifth, and more universally, because vertebrate odorants often contain many compounds, with diverse volatilities, polarities, acidities, and sources (Lynam, 2008), it is typically impossible to characterize a vertebrate's entire chemical profile using a single analytical method (Drea et al., 2013). As a result, researchers often make compromises to target only a subset of compounds, rather than the entire range of compounds expressed in the full odorant (Drea et al., 2013). While important patterns can be found in these subsets, recovering as many odorant components as possible is essential to clarifying our understanding of which compounds play major functional roles in vertebrate signals and cues while also augmenting comparisons across studies. More inclusive and reproducible approaches may lead to the consistent discovery of important and/or unknown compounds across taxa. Here, we test both universal analytical and more specific sampling variables that underlie these challenges, with the goal of developing a broadly applicable approach that will maximize the diversity of compounds detected in vertebrate odorants, while minimizing contamination and degradation.

To address these five challenges, we undertake four main objectives. First, we test the degree to which two common swab materials and three swab preparation methods introduce contaminants into the chromatograms of vertebrate odorants. Second, we test recoverability by comparing the performance of cotton and viscose in detecting a range of known compounds. Third, we compare the number and intensity of compounds recovered by two GC columns, as GC column type and run conditions directly affect the number of compounds detected and the degradation rate of the stationary phase in the column (Drea et al., 2013). Finally, we compare the performance of two solid-phase micro-extraction (SPME) approaches to five TD methods as well as the absorbent materials' ability to recover a variety of common vertebrate odorant compounds. Extraction methods are essential in targeting different numbers, volatilities, polarities, and intensities of compounds (Baltussen et al., 2002; Cavalli et al., 2003; Helin et al., 2020; de Koning et al., 2009; Kücklich et al., 2017; Urbanowicz et al., 2011), and TD methods have recently gained traction in analytical chemistry due to their increased sensitivity and minimal sample preparation requirements (Urbanowicz et al., 2011). To compare compound recovery and reproducibility, we developed and used a standard, biologically relevant mixture of 24 compounds, with various volatilities, polarities, and acidities. Each of the molecules in the standard mixture has been found in the odor bouquet of a wide array of vertebrates (El-Sayed, 2021). We discuss how our approach may help researchers minimize contaminants and improve the capture of biologically relevant compounds, to produce the most representative chromatogram possible for a single technique, all while being amenable to field conditions. More uniform application of such a technique could improve the quality of generalizations across studies and improve consistency within a given study system.

# 2 Methods

## 2.1 Objective 1: Testing Swab Material Choice and Preparation

We compared the presence of potential contaminant compounds for two common swab materials used to collect vertebrate odorants: cotton wool (Organic Cotton Balls; 365 Whole Foods Market, South Bend Indiana, USA) and viscose roving (Viscose Fiber, Roving, Top, Raw Color, Natural White 2, 4 oz; Etsy, Charleston, Illinois, USA). For each material, we compared raw (i.e., untreated) cotton and raw viscose to cotton and viscose prepared via three common pre-cleaning techniques.

The first technique was the "soaked" method, in which we rinsed the materials in undiluted chromatography grade methanol, followed by undiluted chromatography grade pentane, with a 12-h drying step in a fume hood between each rinse (Drea et al., 2013). The goal of this rinse was to remove compounds that would otherwise contaminate the chemical profile. Second, for the "baked" method, we evaporated potential contaminants off the swab materials by heating them in a Micro-Chamber/Thermal Extractor (Markes M-CTE250) for 210 min at 190 °C with a pressure of 30 psi of high purity nitrogen (we chose these parameters via trial and error to optimize the number of contaminants removed without burning the material; Birkemeyer et al., 2016). Third, for the "soaked-and-baked" method, we sequentially performed the "soaked" and the "baked" protocols, as described above.

For each material and treatment type, we formed five replicate swabs by rolling 0.4–0.5 g of each material into balls using tweezers rinsed in 99% ethanol ($N = 40$ swabs; i.e., 2 material types × 4 treatments, including the raw materials, × 5 replicates). We inserted swabs into conditioned direct TD tubes and ran and processed them using the procedures as outlined in direct TD (see Objective 4, below). We stored materials in glass containers that were thrice pre-rinsed with 99% ethanol and covered with tinfoil; to avoid introducing contaminants, we used nitrile gloves (Condor) and ethanol-rinsed tweezers during all handling.

## 2.2 Objective 2: Testing Recoverability

We next compared the recovery performance of cotton and viscose using our standard mixture of 24 compounds containing equal volumes of 1 mM solutions of each compound in acetone (Table 1). Our choice of compounds reflected three main goals: (i) achieving a range of volatilities, polarities, and acidities found in a wide array of vertebrates, (ii) targeting compounds that were common in the literature (selected using pherobase (El-Sayed, 2021), which provides lists of animal species that use each chemical to communicate and supporting literature), and (iii) focusing on compounds that are easily commercially available so they could be purchased and used by other researchers.

Because many semiochemicals are at least partially polar, 22 of the 24 compounds have polar functional groups attached to a nonpolar carbon chain (Table 1). We chose acetone as our solvent because it has a lower boiling point than any of the common alcohol-based solvents, such as isopropanol, and interfered the least with the

**Table 1** The list of the compounds included in our standard 24-compound mixture

| Compound | Molecular weight | Polarity | Functional group, class | Main ions | Retention time (min) |
|---|---|---|---|---|---|
| Acetone<br>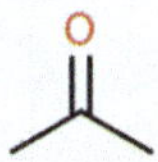<br>**Acetone** | 58.08 | Polar | Carbonyl, Ketone | 43, 58, 15 | 2.2 |
| Acetic acid<br>**Acetic Acid**<br>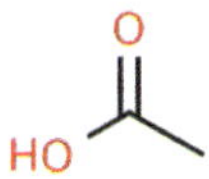 | 60.05 | Polar | Carboxyl, Carboxylic acid | 43, 45, 60 | 11.2 |
| Propanoic acid<br>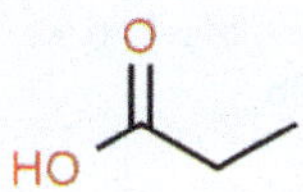<br>**Propanoic acid** | 74.08 | Polar | Carboxyl, Carboxylic acid | 74, 28, 45 | 11.6 |
| Pentan-1-ol<br><br>**Pentan-1-ol** | 88.15 | Semi-polar | Hydroxyl, alcohol | 42, 55, 41 | 8.8 |
| Isovaleric acid<br>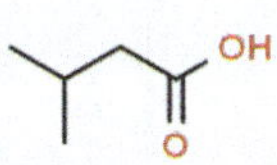<br>**Isovaleric acid** | 102.13 | Polar | Carboxyl, Carboxylic acid | 60, 43, 41 | 13.0 |
| Benzaldehyde<br>**Benzaldehyde**<br>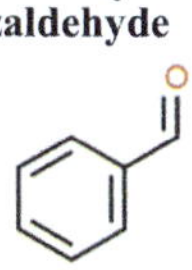 | 106.12 | Semi-polar | Formyl, Aldehyde | 77, 106, 105 | 11.9 |
| 2-Heptanone<br>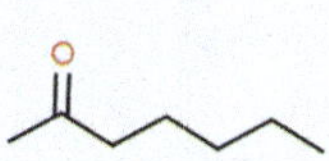<br>**2-Heptanone** | 114.19 | Polar | Carbonyl, Ketone | 43, 58, 27 | 7.4 |
| Hexanoic acid<br>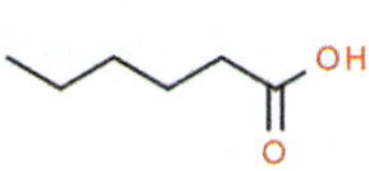<br>**Hexanoic Acid** | 116.16 | Polar | Carboxyl, Carboxylic acid | 60, 73, 41 | 14.2 |

(continued)

**Table 1** (continued)

| Compound | Molecular weight | Polarity | Functional group, class | Main ions | Retention time (min) |
|---|---|---|---|---|---|
| Methyl isovalerate 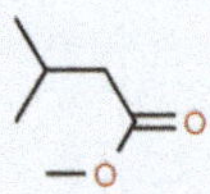 **Methyl isovalerate** | 116.16 | Polar | Ester, Fatty acid ester | 74, 43, 85 | 4.2 |
| Octanal  **Octanal** | 128.22 | Semi-polar | Formyl, Aldehyde | 43, 44, 41 | 9.4 |
| Ethyl valerate 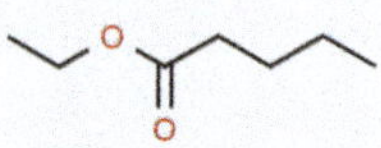 **Ethyl valerate** | 130.18 | Polar | Ester, Fatty acid ester | 29, 85, 88 | 6.4 |
| Nonanal  **Nonanal** | 142.24 | Semi-polar | Formyl, Aldehyde | 57, 41, 43 | 10.7 |
| Decanal  **Decanal** | 156.27 | Semi-polar | Formyl, Aldehyde | 43, 41, 57 | 11.7 |
| Undecane  **Undecane** | 156.31 | Nonpolar | Alkane | 57, 43, 71 | 5.8 |
| Undecanoic acid 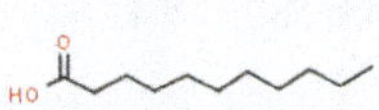 **Undecanoic Acid** | 186.29 | Polar | Carboxyl, Fatty acid | 60, 73, 43 | 17.5 |
| Dodecan-1-ol **Dodecan-1-ol** 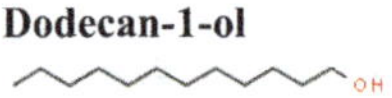 | 186.34 | Semi-polar | Hydroxyl, Fatty alcohol | 55, 43, 69 | 14.9 |

(continued)

**Table 1** (continued)

| Compound | Molecular weight | Polarity | Functional group, class | Main ions | Retention time (min) |
|---|---|---|---|---|---|
| Dodecanoic acid 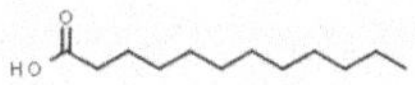 **Dodecanoic Acid** | 200.32 | Polar | Carboxyl, Fatty acid | 73, 60, 43 | 18.5 |
| Tetradecanoic acid 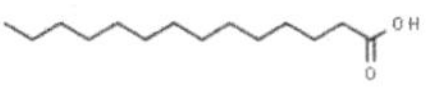 **Tetradecanoic Acid** | 228.38 | Polar | Carboxyl, Fatty acid | 73, 60, 43 | 21.2 |
| Pentadecanoic acid 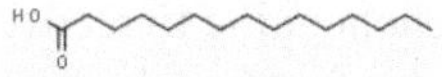 **Pentadecanoic Acid** | 242.4 | Polar | Carboxyl, Fatty acid | 73, 43, 60 | 23.5 |
| Hexadecan-1-ol **Hexadecan-1-ol**  | 242.45 | Semi-polar | Hydroxyl, Fatty alcohol | 55, 69, 83 | 17.6 |
| Hexadecanoic acid 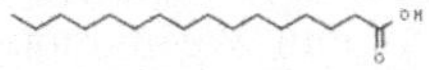 **Hexadecanoic acid** | 256.42 | Polar | Carboxyl, Fatty acid | 43, 73, 60 | 26 |
| (Z)-9-Octadecenoic acid **(Z)-9-Octadecenoic**  **Acid** | 282.47 | Polar | Carboxyl, Carboxylic acid | 41, 55, 43 | 23 |
| Eicosanoic acid **Eicosanoic acid** 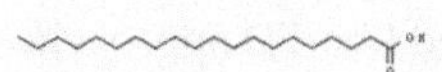 | 312.52 | Polar | Carboxyl, Fatty acid | 73, 43, 60 | NA |
| Squalene 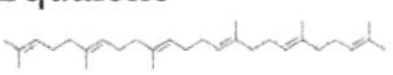 **Squalene** | 410.73 | Nonpolar | Triterpene | 69, 81, 41 | 31 |

All compounds are commonly expressed in vertebrate chemical signals and are thought to play a functional role in communication (see pherobase entries for citations supporting each compound). The first compound, acetone, served as our solvent. We selected main ions for each compound using the National Institute of Standards and Technology library (NIST: http://www.nist.gov/srd/nist1a.cfm). We calculated retention times using a direct TD technique described in Objective 4. Eicosanoic acid failed to be recovered by any method during this experiment, hence a retention time could not be calculated for this compound chromatography signal from our most volatile compounds (see row 1 of Table 1). We aliquoted 40 μL of the 24-compound mixture onto each of four cotton and four viscose swabs prepared using the "soaked-and-baked" technique described above (Objective 1; $N = 8$ swabs; i.e., 4 swabs × 2 swab materials). We then inserted the swabs into conditioned direct TD tubes and ran and processed them using the procedures as outlined below in direct TD (Objective 4).

## 2.3 Objective 3: Column Choice

We compared the performance of two commonly used columns for recovering compounds from our 24-compound mixture: a polysiloxane type DB-5MS Ultra Inert Column (Agilent Technologies; Lynam, 2008; Lynam & Smith, 2008) and a polyethylene glycol (PEG)-type DB-FATWAX Ultra Inert Column (Agilent Technologies; Dang & Vickers, 2016; Zou, 2018). The DB-FATWAX UI column was designed to handle a wide variety of compounds, including compounds that are more challenging to recover on other columns. Challenging compounds include those with polar functional groups and long-chain acids, which often interact strongly with the column, eluting as asymmetric peaks that are broad, exhibit tailing, and have low intensity. These peaks may fail to be recognized during automated integration and must be manually corrected. Broad, long-tailing, low-intensity peaks are also more likely to coelute with other peaks, which complicates integration via automated methods and reduces reproducibility (Birkemeyer et al., 2016). Thus, peak intensity is a good indicator of GC column performance, and we expected the DB-FATWAX UI column to recover compounds represented by peaks with higher intensities than those obtained with the polysiloxane column. We aliquoted 40 μL of the 24-compound mixture (Table 1) onto each of four cotton swabs prepared using the "soaked-and-baked" technique described above (Objective 1; $N = 8$; i.e., 4 cotton swabs × 2 column types). We then inserted the swabs into conditioned direct TD tubes and ran and processed them using the procedures as outlined below in direct TD (Objective 4), holding all parameters equal except the type of column used.

## 2.4 Objective 4: Comparing Solventless Extraction Approaches and Absorbent Materials

The seven solventless extraction approaches we compared included five TD approaches: (i) direct TD (Markes International), (ii) Tenax TA (Markes International), (iii) a combination of XAD and Tenax TA (produced for this experiment by Camsco), (iv) Carbograph (Markes International), and (v) a combination of Carbograph and Tenax TA (produced for this experiment by Camsco). We also tested two SPME approaches, including: (vi) a combination of Polydimethylsiloxane and Divinylbenzene (Supelco), and (vii) a combination of Polydimethylsiloxane, Divinylbenzene, and Carboxen ((Supelco) i.e., the "triple fiber approach"; Baltussen et al., 2002; Cavalli et al., 2003; de Koning et al., 2009). For all seven approaches, we aliquoted 40 μL of the 24-compound mixture (Table 1) onto each of four cotton swabs prepared using the "soaked-and-baked" technique described above (Objective 1; $N = 28$ swabs; 4 cotton swabs per method × 7 methods).

### 2.4.1 Thermal Desorption Approaches

2.4.1.1 Direct Thermal Desorption

For direct TD, we inserted the cotton swabs into direct (empty) TD tubes (Gabirot et al., 2018; Piryaei, 2019) that were cleaned via conditioning prior to use in a Markes International TC-20 conditioning station at 320 °C for 100 min with a pressure of 20 psi of high purity nitrogen. We then loaded these tubes into an ULTRA-xr TD autosampler (Markes International), connected to a UNITY-xr TD device (Markes International), equipped with a T11 universal cold trap. We chose this cold trap because it adsorbs a wide range of volatile and semi-volatile compounds ($C_4$–$C_{32}$).

To desorb these compounds off the swabs and inject them into the GC port, we followed a two-stage heating process. During the first stage, compounds were desorbed at 36 °C for 7 min with a flow of 10 mL/min, followed by 250 °C for 10 min with a flow of 20 mL/min. During the second stage, compounds were re-collected on the cold trap at 5 °C and injected by rapidly heating from 5 °C to 330 °C at the MAX heating rate into a Thermo Scientific Trace 1310 GC with a split flow of 20 mL/min. The cold trap was conditioned between each sample run. Samples were separated over a DB-FATWAX UI column ($l$ = 30.0 m; ID = 0.25 mm; film thickness = 0.25 μm; Agilent Technologies) using the following temperature control program: initial temperature of 40 °C for 0.5 min, then 5 °C/min to 75 °C for 0 min, then 20 °C/min to 230 °C for 22 min, and 20 °C/min to 250 °C for 10 min. This final temperature ramp was used to condition the column. Peaks acquired during this time were not included in peak analysis. The interface temperature to a Thermo Scientific TSQ 9000 Triple Quadropole Mass-Spectrometer (MS) was held at 300 °C with an ion source temperature of 275 °C. The data were acquired in scan mode from 20 to 420 amu at scan interval of 0.2 s and electron ionization (EI) energy of 70 eV. In addition, the GC column was conditioned every four samples using the following temperature control program: initial temperature of 120 °C for 0.5 min, then 12 °C/min to 240 °C for 8 min. Raw data were processed and integrated using GC-MS Chromeleon 7 software with a smoothing threshold of 0.40 wavelength coefficient between the retention times of 3.5 and 37 min.

2.4.1.2 *Thermal Desorption with Adsorbent Material.*

To accomplish TD with four different adsorbent materials, we desorbed compounds off the swabs and adsorbed them into TD tubes packed with either Tenax TA, a combination of 75 mg of XAD and 150 mg of Tenax (i.e. XAD/Tenax), a combination of 100 mg of Carbograph and 100 mg of Tenax (i.e., Carbograph/Tenax), or Carbograph ($N$ = 16; i.e., 4 Tenax runs, 4 XAD/Tenax runs, 40 ul/Tenax runs, and 4 Carbograph runs). The tubes were loaded with compounds by heating samples for 15 min at 230 °C with a pressure of 30 psi via a Micro-Chamber/Thermal Extractor. Once prepared, we followed the same GC-MS procedures as outlined above in direct TD.

### 2.4.2 Solid-Phase Micro-Extraction Approaches.

For the two SPME approaches, we inserted each replicate swab into a conditioned 10-mL vial (Supelco SU860100) fitted with a magnetic screw cap (Supelco SU860101). We used a TriPlusRSH autosampler to expose either a SPME fiber coated in Polydimethylsiloxane and Divinylbenzene (hereafter PMDS/DVB SPME) or a SPME fiber coated in Polydimethylsiloxane, Divinylbenzene, and Carboxen (hereafter triple fiber SPME) to the headspace of these vials for 15 min to adsorb compounds ($N = 8$; i.e., four PDMS/DVB SPME fiber runs and four triple fiber SPME runs). This autosampler has two limitations: the maximum agitator temperature for adsorbing compounds onto the SPME fibers is 200 °C (compared to 350 °C for the Micro-Chamber/Thermal Extractor), and a two-stage heating process (see section "Direct Thermal Desorption") is not possible. Via trial and error, we found that a 100 °C agitator temperature recovered the greatest number of compounds, so we used this temperature to adsorb compounds for both SPME approaches. During each run, we conditioned the SPME fibers pre- and post-desorption at 260 °C for 10 min and 30 min, respectively. We used the same GC-MS procedures as outlined above in direct TD.

## 2.5 Statistical Analyses

To compare the performance of the seven solventless extraction approaches, the cotton and viscose swab materials, and the two different column types, we generated a table containing the number of peaks and the intensity of each peak recovered from our 24-compound mixture for each sample. To do this, we manually reviewed all integrations of software-defined peaks to ensure accuracy. We scored the presence and total ion current intensity (hereafter "intensity") of each of the peaks in each sample and assigned each peak to a compound using the NIST library (http://www.nist.gov/srd/nist1a.cfm). We accepted compound identities if the search index score was ≥900. If an expected compound was absent or the SI score was below 900, we confirmed the presence or absence of these compounds by running authentic compounds using methods described above under direct TD (Piryaei, 2019).

To compare the presence of compounds that could contaminate odorant samples among raw and pre-treated cotton and viscose materials, we generated a table of the number of peaks for each sample in each method. To do this, we reviewed all integrations of software-defined peaks manually and then exported this table for data processing. To generate a table of contaminants, we tentatively identified each of the contaminant peaks in each sample and assigned each peak to a compound using the NIST library. Compound identities were accepted if the search index score was ≥800. We used this more relaxed search index parameter as contaminants are often lower-quality peaks that are more difficult to tentatively identify. To test if the cotton and viscose materials and pre-cleaning treatments differed in the number or intensities of contaminants they introduced, we used analyses of variance (ANOVAs) to compare the number and intensities of contaminants recovered from each set of samples and investigated significant differences in contaminant numbers and intensities using post-hoc Tukey's tests in R (version 4.2.0) (R Core Team, 2022).

To test for differences in compound number and intensity recovered from our 24-compound mixture by cotton and viscose, we compared the numbers and intensities of compounds recovered by each material using a two-sample $t$-test in R.

To compare the ability of each column type to recover compounds from our 24-compound mixture, we compared the numbers and intensities of compounds recovered by each column using two-sample $t$-tests in R.

To compare the performance of each solventless extraction approach and adsorbent material to recover compounds from the 24-compound mixture (Table 1), we ran an ANOVA on the number of compounds recovered from each sample. We then investigated significant differences in the number of compounds using post-hoc Tukey's tests in R.

# 3 Results

## 3.1 Objective 1: "Soaked-and-Baked" and "Baked" Methods Remove the Most Contaminants from Cotton and Viscose

The number and intensity of contaminant peaks recovered from raw cotton swabs was not significantly different from those recovered from raw viscose swabs (Figs. 1 and 2; Table 2; peak number: $\beta = 3.4, p = 0.96$; peak intensity: $\beta = -1.21 \times 10^8, p = 0.37$). Raw cotton contained an average of 13.2 compounds that could contaminate vertebrate chromatograms (mean peak intensity = $2.40 \times 10^8$), whereas the raw viscose contained an average of 16.6 contaminants (mean peak intensity = $1.20 \times 10^8$; Figs. 1 and 2). Seven contaminants were shared between cotton and viscose, but each material also harbored material-specific contaminants (Table 3). Notably, the intensity of l-(+)-Ascorbic acid 2,6-dihexadecanoate was much higher in raw cotton than in raw viscose (Fig. 3a, peak at 23 min).

In comparing "soaked," "baked," and "soaked-and-baked" cleaning methods across swab materials, the overall most effective method for reducing the intensity and number of contaminants in raw cotton and viscose was the "soaked-and-baked" approach (Figs. 1 and 2). This method removed, on average, 12.95 contaminant peaks for cotton and 11 contaminant peaks for viscose (Table 2; "soaked-and-baked" cotton vs. raw cotton: $\beta = -12.95$, $p = 0.01$; "soaked-and-baked" viscose vs. raw viscose: $\beta = -11$, $p = 0.03$). After being "soaked-and-baked," an average of 5.6 contaminant peaks remained in viscose, compared to 0.25 peaks in "soaked-and-baked" cotton. The "soaked-and-baked" method was also effective at reducing the intensity of contaminant peaks. After being "soaked-and-baked" the average intensity of contaminant peaks in cotton declined from $2.40 \times 10^8$ to $1.78 \times 10^6$ (Table 2; $\beta = -2.38 \times 10^8, p = 2.90 \times 10^{-3}$). In viscose, this method reduced contaminant peak intensity from an average of $1.20 \times 10^8$ to $2.78 \times 10^7$, however difference was not statistically significant (Table 2; $\beta = -9.06 \times 10^7, p = 0.71$).

The second-best method was the "baked" approach. This approach significantly reduced the number of contaminant peaks in cotton (raw cotton vs. "baked" cotton: $\beta = 12.4$, $p = 0.01$), with a statistical trend in the expected direction for viscose (raw viscose vs. "baked" viscose: $\beta = 9.8, p = 0.07$). The "baked" method also significantly reduced the intensity of contaminant peaks in cotton, from an average intensity of $2.40 \times 10^8$ to an average intensity of $8.05 \times 10^6$ (raw cotton vs. "baked" cotton: $\beta = 2.32 \times 10^8, p = 3.95 \times 10^{-3}$). The "baked" approach reduced the average intensity of peaks in viscose, from $2.40 \times 10^8$ to $7.80 \times 10^7$, but this difference was not significant (raw viscose vs. "baked" viscose: $\beta = 4.04 \times 10^7, p = 0.99$).

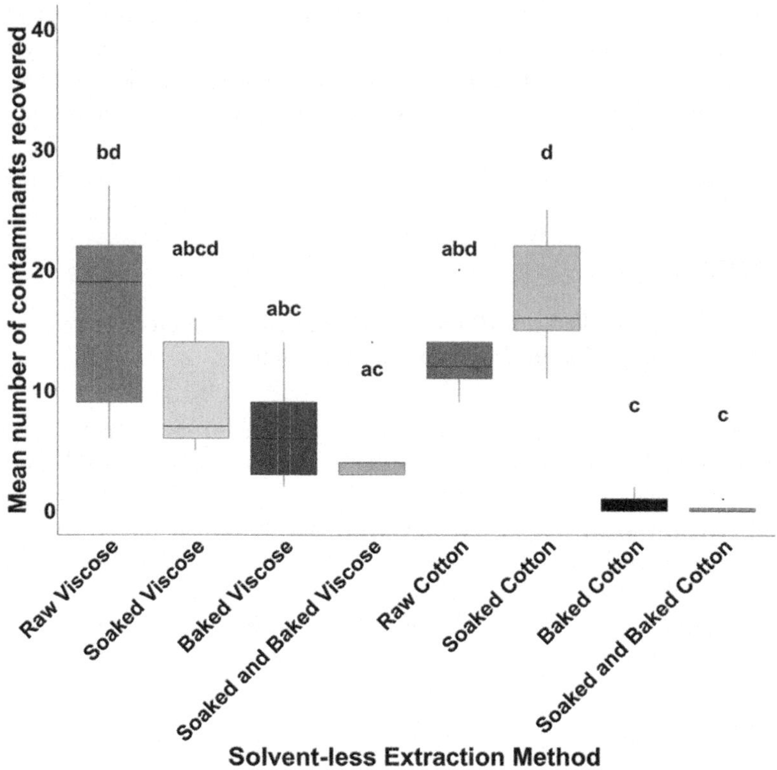


**Fig. 1** The most effective method for reducing the number of contaminants in raw cotton and raw viscose was the "soaked-and-baked" method, closely followed by the "baked" method. The *x*-axis shows all eight combinations of swab materials and treatments (in different shades of gray), each replicated five times ($N = 40$). In each boxplot, the central bar represents the median, the top and bottom of the box are the lower and upper quartiles (25% and 75%), the whiskers show the range excluding outliers, and the dot represents an outlier. Letters indicate which materials and pre-cleaning treatment combinations differed significantly, via a Tukey's test (Table 2). See Table 3 for a list of contaminants recovered

The "soaked" technique was not beneficial for removing contaminants or decreasing the intensities of contaminant peaks in either cotton or viscose, with no significant differences in the number or intensities of contaminant peaks (Table 2; "soaked" cotton vs. raw cotton peak number: $\beta = 4.6$, $p = 0.84$; "soaked" cotton vs. raw cotton peak intensity: $\beta = 6.25 \times 10^7$, $p = 0.94$; "soaked" viscose vs. raw viscose peak number: $\beta = -7$, $p = 0.39$; "soaked" viscose vs. raw viscose peak intensity: $\beta = -3.9 \times 10^7$, $p = 0.99$).

Finally, after the "soaked-and-baked" approach, cotton and viscose have similar contamination potentials, making either material an appropriate choice, given this cleaning method ("soaked-and-baked" cotton vs. "soaked-and-baked" viscose: $\beta = 5.35$, $p = 0.76$). A representative chromatogram of "soaked-and-baked" viscose shows three contaminant peaks that might interfere with vertebrate chromatograms (tentatively identified as acetic acid, formic acid, and 2-furanmethanol), while the "soaked-and-baked" cotton has none (Fig. 3b).

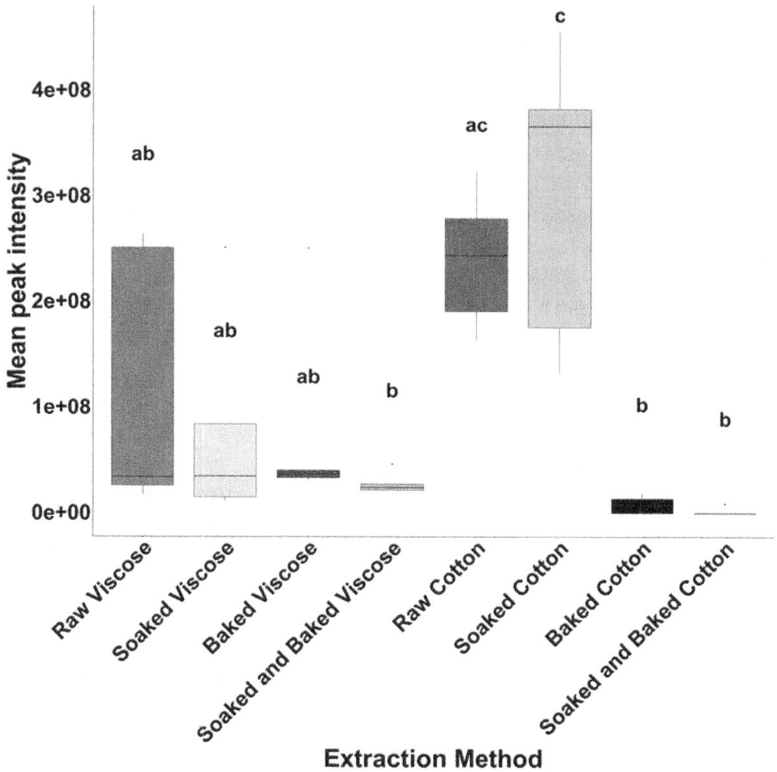


**Fig. 2** The most effective method for reducing the intensity of contaminant peaks in raw cotton and raw viscose was the "soaked-and-baked" method, closely followed by the "baked" method. The $x$-axis shows all eight combinations of swab materials and treatments (in different shades of gray), each replicated five times ($N = 40$). In each boxplot, the central bar represents the median, the top and bottom of the box are the lower and upper quartiles (25% and 75%), the whiskers show the range excluding outliers, and the dot represents an outlier. Letters indicate which materials and pre-cleaning treatment combinations differed significantly, via a Tukey's test (Table 2). See Table 3 for a list of contaminants recovered

## 3.2 Objective 2: Cotton and Viscose Recovered the Same Compounds from a Known Mixture

We exposed four replicates each of "soaked-and-baked" cotton and "soaked-and-baked" viscose to our 24-compound mixture (Table 1). We found that both materials consistently recovered the same 22 compounds. Both failed to recover propanoic acid and eicosanoic acid. Furthermore, there were no detectable differences in the intensity of compounds recovered by cotton and by viscose (two-sample $t$-test: $t(8) = 0.03$, $p = 0.98$; Fig. 4).

**Table 2** Results from a Tukey's tests on the number and intensity of contaminants recovered from eight combinations of swab materials and treatments

| | Peak number | | Peak intensity | |
|---|---|---|---|---|
| Swab material and preparation combination | $\beta$ | $P$ | $\beta$ | $P$ |
| *Comparing methods within viscose* | | | | |
| Raw viscose vs soaked viscose | −7 | 0.39 | $-3.95 \times 10^7$ | 0.99 |
| Raw viscose vs baked viscose | 9.8 | 0.07 | $4.04 \times 10^7$ | 0.99 |
| Raw viscose vs soaked and baked viscose | **−11** | **0.03** | $-9.06 \times 10^7$ | 0.71 |
| Soaked viscose vs baked viscose | 2.80 | 0.99 | $8.78 \times 10^5$ | 0.99 |
| Soaked viscose vs soaked and baked viscose | 4 | 0.91 | $5.10 \times 10^7$ | 0.98 |
| Baked viscose vs soaked and baked viscose | −1.20 | 0.99 | $-5.02 \times 10^7$ | 0.98 |
| *Comparing methods within cotton* | | | | |
| Raw cotton vs soaked cotton | 4.6 | 0.84 | $6.25 \times 10^7$ | 0.94 |
| Raw cotton vs baked cotton | **12.4** | **0.01** | $\mathbf{2.32 \times 10^8}$ | $\mathbf{3.95 \times 10^{-3}}$ |
| Raw cotton vs soaked and baked cotton | **−12.95** | **0.01** | $\mathbf{-2.38 \times 10^8}$ | $\mathbf{2.90 \times 10^{-3}}$ |
| Soaked cotton vs baked cotton | **17** | $\mathbf{2.26 \times 10^{-4}}$ | $\mathbf{2.94 \times 10^8}$ | $\mathbf{1.62 \times 10^{-4}}$ |
| Soaked cotton vs soaked and baked cotton | **17.55** | $\mathbf{3.35 \times 10^{-4}}$ | $\mathbf{3.00 \times 10^8}$ | $\mathbf{1.17 \times 10^{-4}}$ |
| Baked cotton vs oaked and baked cotton | −0.55 | 0.99 | $-6.26 \times 10^6$ | 0.99 |
| *Comparing materials (cotton and viscose) within methods* | | | | |
| Raw viscose vs raw cotton | 3.4 | 0.96 | $-1.21 \times 10^8$ | 0.37 |
| Soaked viscose vs soaked cotton | −8.2 | 0.21 | $\mathbf{-2.23 \times 10^8}$ | $\mathbf{5.95 \times 10^{-3}}$ |
| Baked viscose vs baked cotton | 6 | 0.58 | $7.00 \times 10^7$ | 0.90 |
| Soaked and baked viscose vs soaked and baked cotton | 5.35 | 0.76 | $2.60 \times 10^7$ | 0.99 |

Statistically significant results with a $p$ value <0.05 are marked in bold

### 3.3 Objective 3: The DB-FATWAX UI Column Recovered Higher-Intensity Peaks Than Did the DB-5MS UI Column

We compared the performance of two commonly used columns (DB-5MS UI and DB-FATWAX UI) for vertebrate semiochemical analysis. We found that both columns recovered similar numbers of compounds (mean of 22.8 for DB5-MS UI and 23 for DB-FATWAX UI; two-sample $t$-test: $t(6) = -1$, $p = 0.36$). The DB-5MS UI column failed to recover methyl isovalerate, whereas the DB-FAWTWAX UI column failed to recover propanoic acid; however, consistent with our expectations, the DB-FATWAX UI column recovered peaks that were significantly more intense than those recovered by the DB-5MS UI column (Fig. 5; two-sample $t$-test: $t(6) = -7.69$, $p = 2.53 \times 10^{-4}$). Importantly, whereas the DB-FATWAX UI column outperformed the DB5-MS UI column, it was prone to degradation. Specifically, while running samples for a separate experiment, we tested the efficacy of the DB-FATWAX UI column after ≥1000 runs. This analysis revealed that the efficacy of the DB-FATWAX UI column deteriorated dramatically after 1000 runs, producing substantially smaller and distorted peaks compared to those produced by a fresh column (Fig. 6).

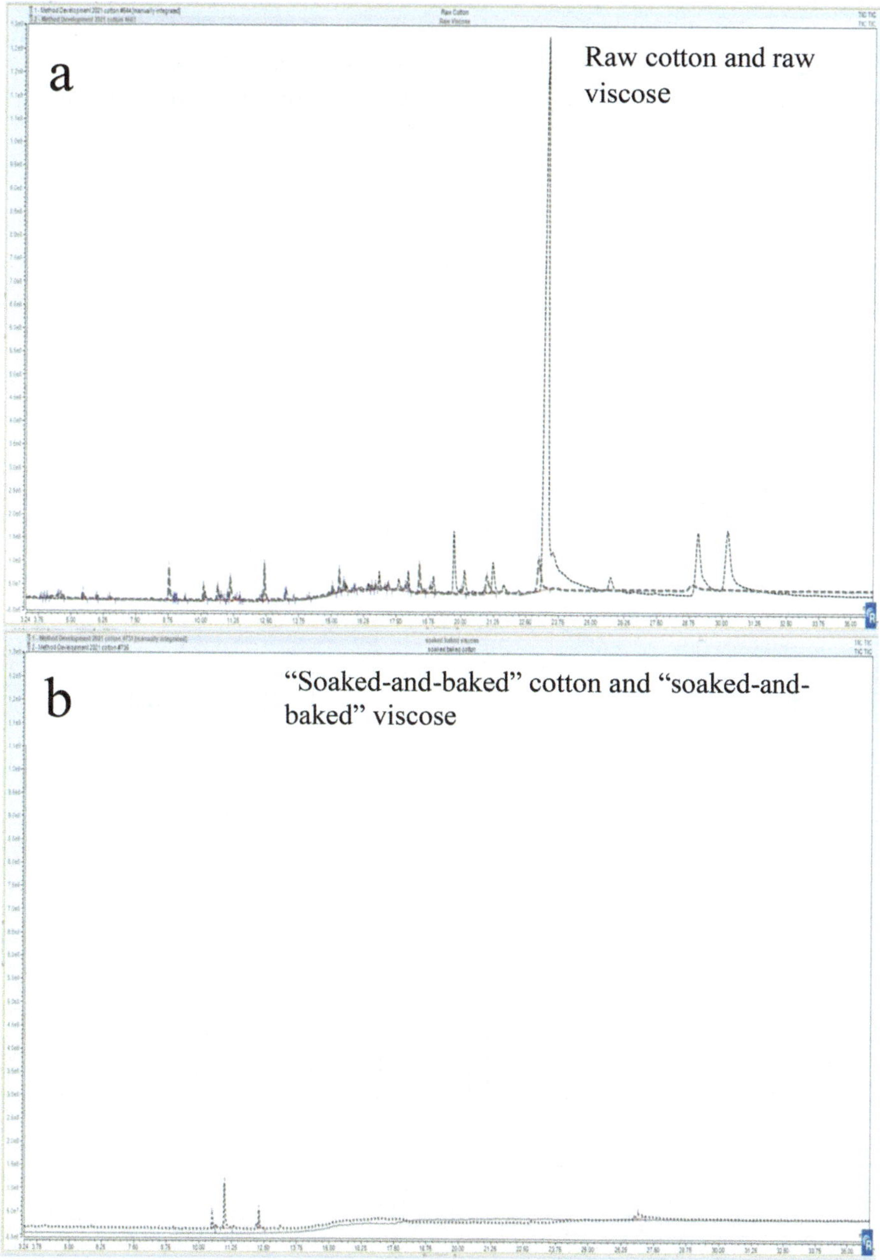


**Fig. 3** Representative total ion chromatograph comparisons. Overlaid in the following graphs are: in plot (**a**), raw viscose (gray dashed line) and raw cotton (black dashed line) are overlaid. In plot (**b**), “soaked-and-baked” viscose (gray dotted line) and “soaked-and-baked” cotton (black dotted line) are overlaid. In plot (**c**), raw viscose (gray dashed line) and the 24-compound mixture aliquoted onto “soaked-and-baked” viscose (gray solid line) are overlaid. In plot (**d**), raw cotton (black dashed line) and the 24-compound mixture aliquoted onto “soaked-and-baked” cotton

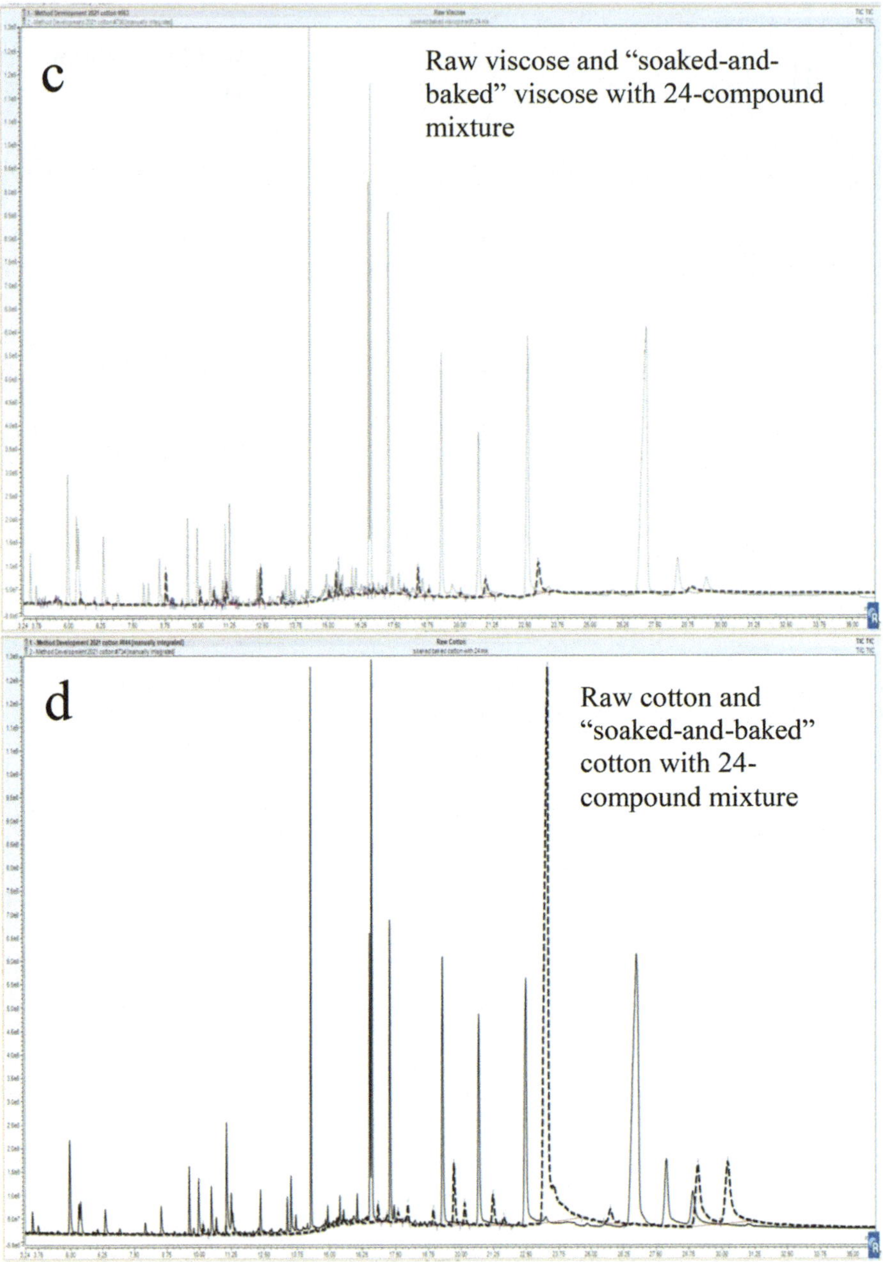


**Fig. 3** (continued) (black solid line) are overlaid. In plot (**e**), “soaked-and-baked” viscose (gray dotted line) and the 24-compound mixture aliquoted onto “soaked-and-baked” viscose (gray solid line) are overlaid. In plot (**f**), “soaked-and-baked” cotton (dotted black line) and the 24-compound mixture aliquoted onto “soaked-and-baked” cotton (solid black line) are overlaid

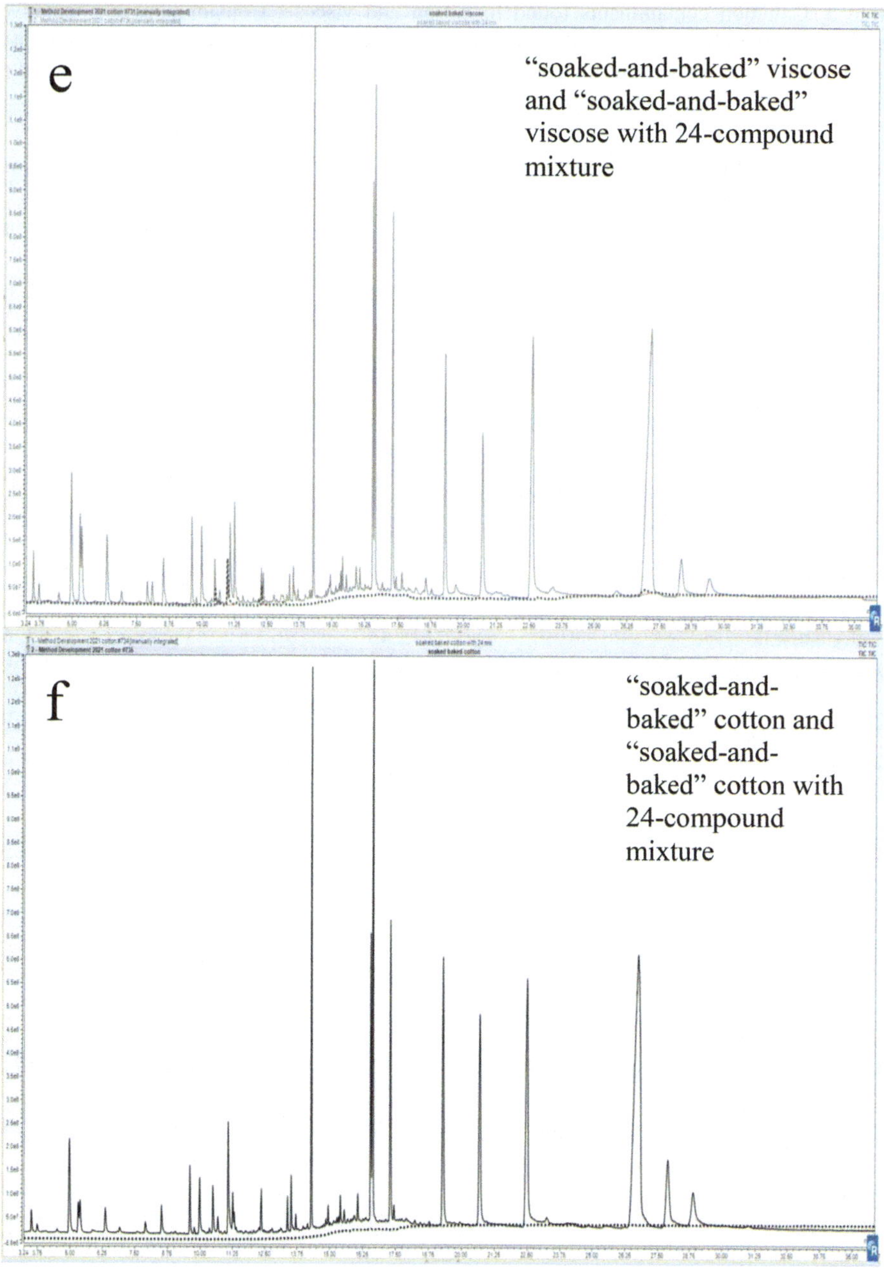


**Fig. 3** (continued)

**Table 3** Common contaminants found in cotton and viscose. Contaminants are tentatively identified via the NIST library. Only contaminants found in three or more samples and with a search index score ≥800 were included (a lower selection criterion was used to maximize the number of contaminants presented)

| Tentatively identified contaminant | Average retention time (min) | Material | Main ions |
|---|---|---|---|
| 2-Propanone, 1-hydroxy- | 9.51 | Viscose | 43, 31, 74 |
| Nonanal | 10.26 | Cotton; Viscose | 57, 41, 43 |
| Acetic acid | 11.08 | Cotton; Viscose | 43, 45, 60 |
| Formic acid | 11.98 | Viscose | 29, 46, 45 |
| 2-Furanmethanol | 12.97 | Viscose | 98, 41, 53 |
| 1,2-Cyclopentanedione | 14.08 | Viscose | 98, 55, 42 |
| Cyclopentanol | 16.22 | Viscose | 57, 44, 41 |
| Nonanoic acid | 16.40 | Cotton | 60, 73, 57 |
| 1-Hexadecanol, 2-methyl- | 16.85 | Cotton | 57, 43, 55 |
| Eicosane | 16.99 | Cotton | 57, 71, 43 |
| Heptadecanoic acid, 16-methyl-, methyl ester | 17.19 | Viscose | 74, 87, 75 |
| n-Pentadecanol | 17.96 | Cotton | 83, 55, 69 |
| Methyl stearate | 18.45 | Viscose | 74, 87, 43 |
| Octaethylene glycol monododecyl ether | 20.36 | Cotton; Viscose | 45, 89, 57 |
| Dodecanamide, N-(2-hydroxyethyl)- | 20.41 | Viscose | 85, 98, 41 |
| Tetradecanoic acid | 20.9 | Cotton | 73, 60, 43 |
| Pentacosane | 21.16 | Cotton; Viscose | 57, 71, 43 |
| Tetracosane, 11-decyl- | 21.40 | Cotton; Viscose | 57, 43, 71 |
| n-Hexadecanoic acid | 23.76 | Cotton; Viscose | 43, 73, 60 |
| Octadecanamide, N-(2-hydroxyethyl)- | 24.99 | Viscose | 85, 98, 140 |
| Pentadecanoic acid | 25.61 | Cotton | 73, 43, 60 |
| Hexanedioic acid, bis(2-ethylhexyl) ester | 26.72 | Viscose | 129, 57, 112 |
| l-(+)-Ascorbic acid 2,6-dihexadecanoate | 28.28 | Cotton; Viscose | 57, 73, 43 |
| Octadecanoic acid | 29.2 | Cotton | 43, 73, 60 |
| cis-13-Octadecenoic acid | 30.31 | Cotton | 55, 69, 41 |

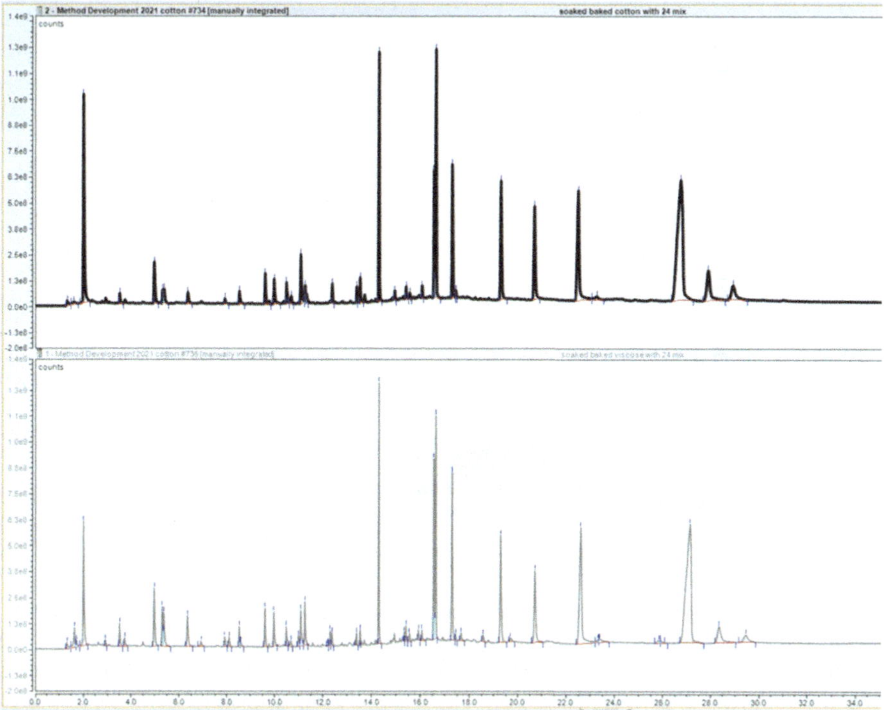

**Fig. 4** Chromatograms showing the peaks for compounds recovered from "soaked-and-baked" cotton (top trace, black solid line) and "soaked-and-baked" viscose (bottom trace, gray solid line). Both materials recovered the same 22 compounds from the 24-compound mixture (Table 1). Both materials also recovered similar intensities of compounds (two-sample *t*-test: $t(8) = 0.03$, $p = 0.98$)

## 3.4 Objective 4: Direct TD, Tenax TD, and XAD/Tenax TD All Recovered Most Compounds from a Known Mixture

Among the TD approaches, direct TD, Tenax TD, and XAD/Tenax TD all recovered the most compounds from the 24-compound mixture, recovering a mean of 21.2, 22.2, and 21.5, respectively (Tables 1 and 2.4, Fig. 7). Propanoic acid and eicosanoic acid were not detected by any of these three approaches. The Carbograph/Tenax TD and the two SPME approaches were less sensitive in the number of compounds they recovered compared to direct TD, Tenax TD, and XAD/Tenax TD (average number of compounds recovered: Carbograph/Tenax TD = 13.8, PDMS/DVB SPME = 14, triple fiber SPME = 14; Table 4). These adsorbent methods failed to recover methyl isovalerate, undecane, ethyl valerate, 2-heptanone, 1-pentanol, octanal, propanoic acid, squalene, oleic acid, and arachidic acid. The method that recovered the fewest compounds was Carbograph TD, with a mean of only 7.5 compounds (Table 4, Fig. 6). Whereas Carbograph TD recovered propanoic acid in three of the four trials, it did not consistently recover the compounds missed by the Carbograph/Tenax TD method. Carbograph

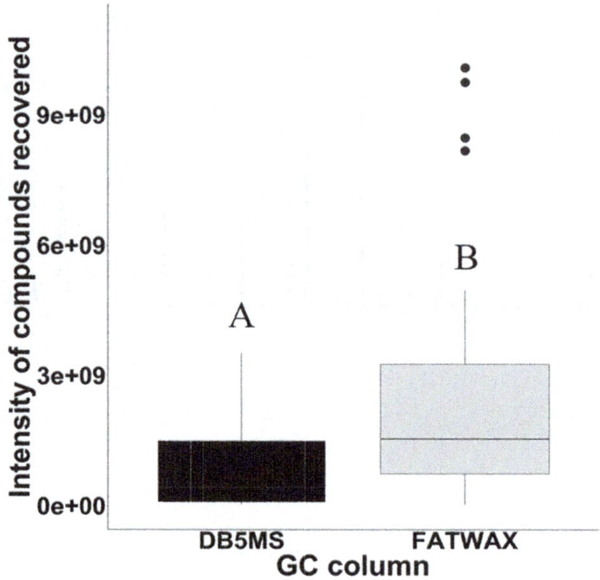


**Fig. 5** The DB-FATWAX UI column recovered higher-intensity peaks than did the DB-5MS UI column. The plot shows the distribution of peak intensities for the two column types ($N = 4$ replicates per column type for a total of eight swabs). In each boxplot, the central bar represents the median, the top and bottom of the box are the lower and upper quartiles (25% and 75%), the whiskers show the range excluding outliers, and dots represent outliers. Letters indicate significant differences in the intensities of peaks recovered via a two-sample *t*-test

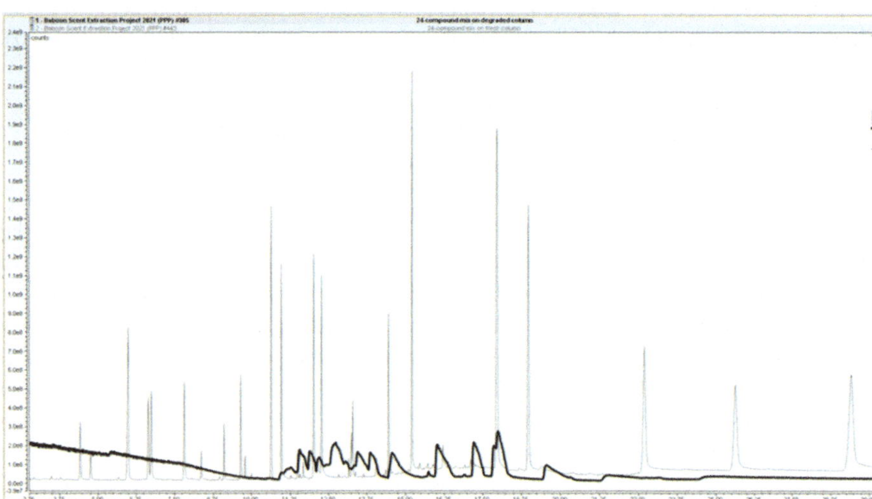

**Fig. 6** The DB-FATWAX UI column is prone to degradation following hundreds of runs. GC traces of the compounds recovered from our 24-compound mixture on "soaked-and-baked" cotton on a DB-FATWAX UI column after ~1000 runs (bold black trace) and on a fresh DB-FATWAX UI column (gray trace). Note the differences in peak numbers and intensities: peaks recovered from a fresh column were sharper and far more intense than were peaks recovered on a degraded column

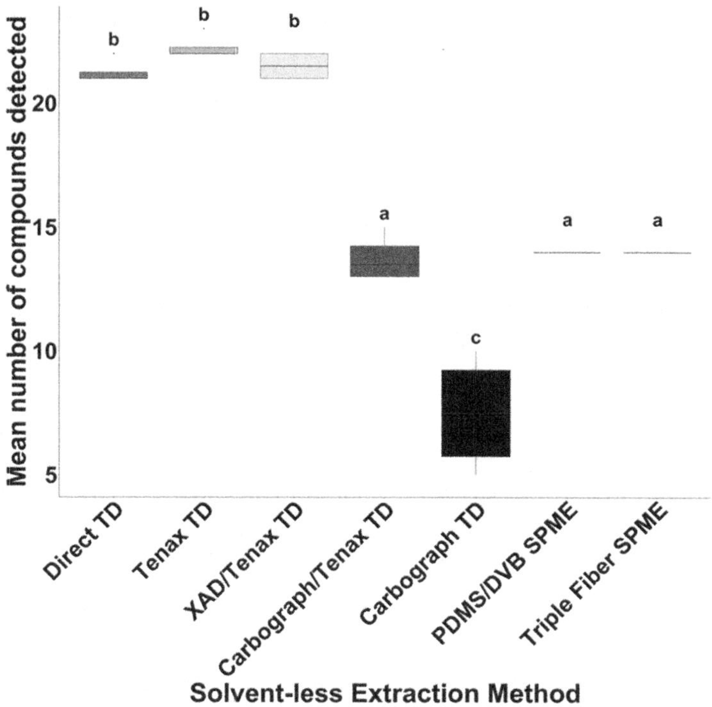


**Fig. 7** Relative to other solventless extraction methods tested, direct TD, Tenax TD, and XAD/Tenax TD recovered the most ($\approx$ 92%) compounds from our 24-compound mixture (Table 1). The $x$-axis shows all seven extraction approaches and adsorbent material combinations (shown in different shades of gray), each replicated four times ($N$ = 28). In each boxplot, the central bar represents the median, the top and bottom of the box are the lower and upper quartiles (25% and 75%), the whiskers show the range excluding outliers, and dots represent outliers. Letters indicate which approaches and material combinations differed significantly, via a Tukey's test, in the number of compounds recovered

TD additionally failed to consistently recover undecane, octanal, nonanal, decanal, benzaldehyde, hexanoic acid, and dodecan-1-ol.

All TD approaches, except Carbograph TD, recovered similar intensities of compounds (Table 4, Fig. 8). Carbograph TD recovered peaks that were significantly less intense than any of the other TD methods. The intensities of peaks recovered by this method are comparable to those recovered by the SPME approaches (Table 4, Fig. 8). Among the SPME approaches, the intensity of peaks recovered by the triple fiber SPME approach was similar to those recovered with the XAD/Tenax TD and Carbograph/Tenax TD peak intensities.

**Table 4** Results from a Tukey's test on the number and intensity of compounds recovered from seven solventless extraction approaches

| | Peak number | | Peak intensity | |
|---|---|---|---|---|
| Solventless extraction method comparison | *B* | *P* | *β* | *P* |
| Direct TD vs Tenax TD | 1.00 | 0.81 | $9.17 \times 10^{7}$ | 0.94 |
| Direct TD vs XAD/Tenax TD | 0.25 | 0.99 | $-8.41 \times 10^{7}$ | 0.96 |
| Tenax TD vs XAD/Tenax TD | −0.75 | 0.94 | $-1.76 \times 10^{8}$ | 0.44 |
| Carbograph/Tenax TD vs PDMS/DVB SPME | 0.25 | 0.99 | **$-4.18 \times 10^{8}$** | **$1.80 \times 10^{-3}$** |
| Carbograph/Tenax TD vs Triple fiber SPME | 0.25 | 0.99 | $-2.33 \times 10^{8}$ | 0.16 |
| PMDS/DVB SPME vs Triple fiber SPME | 0.00 | 1.00 | $1.85 \times 10^{8}$ | 0.39 |
| Carbograph TD vs Carbograph/Tenax TD | **6.25** | **$4.90 \times 10^{-7}$** | **$3.41 \times 10^{8}$** | **0.01** |
| Carbograph TD vs Direct TD | **13.75** | **$2.53 \times 10^{-13}$** | **$4.12 \times 10^{8}$** | **$2.10 \times 10^{-3}$** |
| Carbograph TD vs Tenax TD | **14.75** | **$9.11 \times 10^{-14}$** | **$5.04 \times 10^{8}$** | **$1.96 \times 10^{-4}$** |
| Carbograph TD vs XAD/Tenax TD | **14.00** | **$1.86 \times 10^{-13}$** | **$3.28 \times 10^{8}$** | **0.02** |
| Carbograph TD vs PDMS/DVB SPME | **6.50** | **$2.56 \times 10^{-7}$** | $-7.69 \times 10^{7}$ | 0.97 |
| Carbograph TD vs Triple fiber SPME | **6.50** | **$2.56 \times 10^{-7}$** | $1.08 \times 10^{8}$ | 0.88 |
| Direct TD vs Carbograph/Tenax TD | **7.50** | **$2.19 \times 10^{-8}$** | $7.10 \times 10^{8}$ | 0.98 |
| Tenax TD vs Carbograph/Tenax TD | **8.50** | **$2.32 \times 10^{-9}$** | $1.63 \times 10^{8}$ | 0.53 |
| PDMS/DVB SPME vs Direct TD | **−7.25** | **$3.97 \times 10^{-8}$** | **$-4.89 \times 10^{8}$** | **$2.87 \times 10^{-4}$** |
| PDMS/DVB SPME vs Tenax TD | **8.25** | **$3.99 \times 10^{-9}$** | **$5.81 \times 10^{8}$** | **$2.86 \times 10^{-5}$** |
| PMDS/DVB SPME vs XAD/Tenax TD | **7.50** | **$2.19 \times 10^{-8}$** | **$4.05 \times 10^{8}$** | **$2.54 \times 10^{-3}$** |
| XAD/Tenax TD vs Carbograph/Tenax TD | **7.75** | **$1.23 \times 10^{-8}$** | $-1.32 \times 10^{7}$ | 0.99 |
| Triple fiber SPME vs Direct TD | **−7.25** | **$3.97 \times 10^{-8}$** | **$-3.04 \times 10^{8}$** | **0.03** |
| Triple fiber SPME vs Tenax TD | **−8.25** | **$3.99 \times 10^{-9}$** | **$-3.96 \times 10^{8}$** | **$3.21 \times 10^{-3}$** |
| Triple fiber SPME vs XAD/Tenax TD | **7.50** | **$2.19 \times 10^{-8}$** | $2.20 \times 10^{8}$ | 0.21 |

Statistically significant results with a $p$ value <0.05 are marked in bold

## 4 Discussion

Through a series of experiments using a standardized mixture of common vertebrate semiochemicals, we produced two main findings. First, cotton and viscose are equally effective at recovering a range of compounds from a standard mixture. This finding builds on prior work by Birkemeyer et al. (2016), who found that cotton swabs were not appropriate for sampling volatile compounds when using liquid injections of hexane extracts, but recommended further tests using TD tubes.

Second, most TD methods significantly outperformed SPME methods in retrieving known compounds. Among the TD methods, direct TD was comparable to TD with adsorbent material. This finding allows for considerable flexibility in method design, especially for researchers interested in reducing costs or who face challenges in maintaining their samples in a cold chain. Below we discuss the implications of our findings for vertebrate semiochemical analyses.

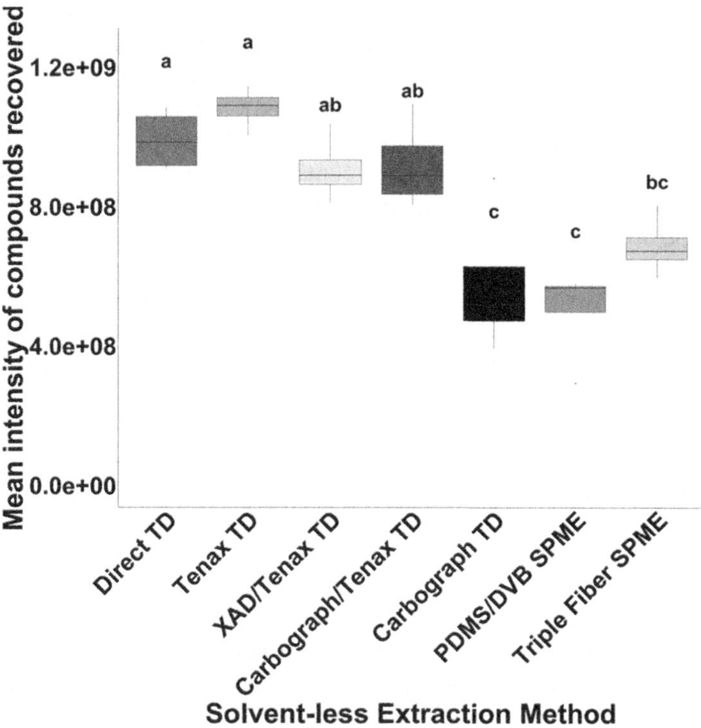


**Fig. 8** Using combinations of solventless extraction approaches and adsorbent materials, peaks recovered from our 24-compound mixture (Table 1) were found to be of higher intensity when using the direct TD and Tenax TD approaches than when using SPME approaches. The $x$-axis shows all seven extraction approaches and adsorbent material combinations (shown in different shades of gray), each replicated four times ($N = 28$). In each boxplot, the central bar represents the median, the top and bottom of the box are the lower and upper quartiles (25% and 75%), the whiskers show the range excluding outliers, and dots represent outliers. Letters indicate which approaches and material combinations differed significantly, via a Tukey's test, in the intensities of compounds recovered

## 4.1 Sampling Material and Preparation

Some researchers have proposed that cotton is a less appropriate swab material than viscose, due to cotton's potential to introduce contaminants (Weiß et al., 2018). Contrary to this idea, we found that raw cotton and raw viscose were comparable in their contamination potential, and that both materials had similar numbers of contaminants and intensities of contaminant peaks in their raw form. They also both introduced contaminants that are non-industrial in origin and might interfere with vertebrate chromatograms. This pattern contrasts somewhat with that reported in Birkemeyer et al. (2016), wherein researchers found non-industrial contaminants in cotton, but not in viscose. These differences between ours and Birkemeyer's study might originate from the raw material itself, where it was purchased, or from contamination introduced during the manufacture and packaging of these materials.

Methods for removing contaminants from swab materials have also been debated (Birkemeyer et al., 2016). We found that "baked" pre-treatment methods (i.e., evaporating potential contaminants off the swab materials by heating them) are most effective at removing contaminants: indeed, "soaked-and-baked" cotton showed virtually no contaminant trace (Fig. 3f). Furthermore, because our "soaked-and-baked" method was not significantly more effective at removing contaminants and decreasing contaminant intensities from either swab material, compared to the "baked" method alone, we suggest using the "baked" method to reduce time and material costs. Although we performed our "baked" method in a Micro-Chamber/Thermal Extractor, this process could be performed in many types of research ovens, such as a dry cabinet (Birkemeyer et al., 2016). Interestingly, our "soaked" method failed to reduce the contaminants in cotton. One explanation for this result is that the contaminants were dissolved and extracted from the cotton when "soaked," but then re-adsorbed by the cotton when the solvents evaporated. It may be possible to improve the "soaked" method by ensuring that the cotton is removed from the solvent prior to solvent evaporation, and then rinsed a second time in fresh solvent.

Finally, some researchers have suggested that cotton may be less effective than viscose at sampling a range of compounds, especially compounds with high volatility (Birkemeyer et al., 2016). Our experiments indicate that cotton is as effective as viscose material in adhering a range of volatile and semi-volatile compounds. It is worth noting that, for all our experiments, we used only one brand of cotton and one brand of viscose. Other brands may contain different types and numbers of contaminants or may adhere different odorant compounds, possibly owing to unique manufacture facility processing procedures and to the origin of the raw materials. We suggest that researchers test their own brands of materials for contaminants pre-and post-cleaning to assess the efficacy of their chosen pre-cleaning methods. No matter what combination of swab material and pre-cleaning technique a researcher chooses, researchers should collect blank swabs with no animal secretion. This step will allow researchers to control for the contaminants introduced by their swab material (Drea et al., 2013), be they many or few.

### 4.2 GC Column Choice

Researchers have a wide array of columns to choose from when sampling vertebrate odorants. Fused silica columns such as the DB5-MS UI column are widely used. Although this column has broad applicability, it underperforms in recovering long-chain acids. By contrast, the DB-FATWAX UI column is a polyethylene glycol-type column, specifically designed to separate fatty acid methyl esters, fatty acid ethyl esters, and fatty acids—compounds that routinely occur in vertebrate odorants. We were therefore not surprised to find that the DB-FATWAX UI column showed higher intensities of the recovered compounds. This finding is important because sharp, high-intensity peaks are less likely to create overlap among peaks, aiding the process of identifying compounds relevant to vertebrate communication.

Although we suggest using the DB-FATWAX UI column, we also note that this column is prone to degradation. After approximately 1000 runs, the column became unsuitable for detecting compounds. Checking column health is crucial, particularly for samples that are irreplaceable or difficult to acquire. We suggest two strategies for maintaining column health: changing to a fresh DB-FATWAX UI column after 800 runs, with the caveat that different temperature control programs might degrade the column at different rates and adopting a comparable mixture of known compounds to be used as a gauge of column health. It would be most advantageous to run such a mixture first, on a fresh column, and compare the results from this initial run to those from later runs, performed every 10 to 20 samples. Such quality control checks ensure that sample analyses are not negatively affected by column degradation.

### 4.3 Solventless Extraction Approaches

Our results highlight the importance of solventless extraction approaches and adsorbent material choices for the quality of the chromatogram produced, with implications for the efficiency of compound retention and recovery during the sampling and analysis processes, respectively. The two-stage TD heating process allowed for the recovery of a mix of volatile and semi-volatile compounds, whereas the single-stage SPME process was more limited. Within the TD approach, the choice of adsorbent material is crucial, as Tenax, XAD/Tenax, and direct TD approaches adsorbed a larger number of compounds than did the other adsorbent materials tested. Indeed, these were the only approaches sensitive enough to recover squalene, a semiochemically important compound. Direct TD and Tenax TD also produced higher-intensity peaks than any of the other solventless extraction methods, suggesting greater sensitivity.

Direct TD performed as well as the best adsorbent materials. This finding allows for considerable flexibility in method design. Direct TD tubes are far cheaper than those packed with adsorbent materials, but limit the capacity for automation. To perform direct TD, swabs with vertebrate odorants are loaded into direct TD tubes that must be run immediately to avoid sample degradation. In contrast, samples are shelf stable once they are adsorbed onto TD tubes with adsorbent materials. The latter approach allows for multiple tubes to be loaded into an automated sampler (such as the Ultra-xr TD autosampler) and run sequentially. These considerations are particularly relevant to field researchers working in remote locales for whom suboptimal sample storage and shipping increase the likelihood of degradation and, thus, present major logistical challenges to semiochemical studies. If field researchers were to adopt methods involving TD tubes with adsorbent material, swabbing and TD could be performed on the same day, without the need for a −20 °C or −80 °C freezer. Researchers could then ship their samples to a laboratory with a TD autosampler and a GC-MS without the need for dry ice. Procedures to create and validate shelf-stable samples represent a significant advancement in the sampling process.

## 5 Conclusions

In sum, the most effective methods for sampling and analyzing vertebrate odorants include using (1) "soaked-and-baked" cotton or "soaked-and-baked" viscose material, (2) a DB-FATWAX UI column, and (3) either a direct TD approach, or a TD approach with Tenax or a mix of XAD and Tenax adsorbent material(s). Additionally, we suggest the use of blank swabs to control the contaminants, and a standardized mixture of compounds to verify performance and check for column degradation. We hope that these methods will be adopted by others, particularly those working under field conditions, to improve feasibility and replicability, with the ultimate goal of furthering our understanding of vertebrate chemical communication.

**Acknowledgments** We thank Anna Matzner, Kristin Dimac, Larry Richey, and Kevin Young for aid with logistics. We thank Marylène Boulet and Anne Nonnamaker for reading earlier versions of this manuscript. We thank members of the Archie lab: Ipek Kulahci, Mauna Dasari, Chelsea Weibel, Chelsea Southworth and Gretchen Andreasen for feedback on the presentation that became this manuscript. We especially thank David Jansen for his help in statistical analyses. We thank members of the Higham lab for additional feedback: James Higham, Jessica Gunson, Amber Trujilio, Eve Cooper, Kandra Cruz, Sam Patterson, and Christina Costa. Research was supported by Sigma Xi, the Primate Society of Great Britain, The Society for the Study of Evolution, The Animal Behavior Society, and the American Society of Mammologists.

## References

Baltussen, E., Cramers, C. A. M. G., & Sandra, P. (2002). Sorptive sample preparation – A review. *Analytical and Bioanalytical Chemistry, 373*(1–2), 3–22. https://doi.org/10.1007/s00216-002-1266-2

Birkemeyer, C. S., Thomsen, R., Jänig, S., Kücklich, M., Slama, A., Weiß, B. M., & Widdig, A. (2016). Sampling the body odor of primates: Cotton swabs sample semivolatiles rather than volatiles. *Chemical Senses, 41*(6), 525–535. https://doi.org/10.1093/chemse/bjw056

Cavalli, J. F., Fernandez, X., Lizzani-Cuvelier, L., & Loiseau, A. M. (2003). Comparison of static headspace, headspace solid phase microextraction, headspace sorptive extraction, and direct thermal desorption techniques on chemical composition of French olive oils. *Journal of Agricultural and Food Chemistry, 51*(26), 7709–7716. https://doi.org/10.1021/jf034834n

Coombes, H. A., Stockley, P., & Hurst, J. L. (2018). Female chemical signalling underlying reproduction in mammals. *Journal of Chemical Ecology, 44*(9), 851–873. https://doi.org/10.1007/s10886-018-0981-x

Dang NA, Vickers AK (2016) *A new PEG GC column with improved inertness reliability and column lifetime*. Competitive comparison, Agilent Technologies

de Koning, S., Janssen, H. G., & Brinkman, U. A. T. (2009). Modern methods of sample preparation for GC analysis. *Chromatographia, 69*(S1), 33–78. https://doi.org/10.1365/s10337-008-0937-3

Drea, C. M., Boulet, M., Delbarco-Trillo, J., Greene, L. K., Sacha, C. R., Goodwin, T. E., & Dubay, G. R. (2013). The "secret" in secretions: Methodological considerations in deciphering primate olfactory communication. *American Journal of Primatology, 75*(7), 621–642. https://doi.org/10.1002/ajp.22143

El-Sayed AM (2021) The pherobase: *Database of pheromones and semiochemicals*. https://www.pherobase.com/database/compound/compounds-index.php

Font, E., Barbosa, D., Sampedro, C., & Carazo, P. (2012). Social behavior, chemical communication, and adult neurogenesis: Studies of scent mark function in Podarcis wall lizards. *General and Comparative Endocrinology, 177*(1), 9–17. https://doi.org/10.1016/j.ygcen.2012.02.015

Gabirot, M., Buatois, B., Müller, C. T., & Bonadonna, F. (2018). Odour of King Penguin feathers analysed using direct thermal desorption discriminates between individuals but not sexes. *IBIS, 160*(2), 379–389. https://doi.org/10.1111/ibi.12544

Helin, A., Hakola, H., & Hellén, H. (2020). Optimisation of a thermal desorption–gas chromatography–mass spectrometry method for the analysis of monoterpenes, sesquiterpenes and diterpenes. *Atmospheric Measurement Techniques, 13*(7), 3543–3560. https://doi.org/10.5194/amt-13-3543-2020

Hurst, J. L. (2009). Female recognition and assessment of males through scent. *Behavioural Brain Research, 200*(2), 295–303. https://doi.org/10.1016/j.bbr.2008.12.020

Johnston, R. E. (2003). Chemical communication in rodents: From pheromones to individual recognition. *Journal of Mammalogy, 84*(4), 1141–1162. https://doi.org/10.1644/BLe-010

Karlson, P., & Lüscher, M. (1959). 'Pheromones': A new term for a class of biologically active substances. *Nature, 183*(4653), 55–56. https://doi.org/10.1038/183055a0

Keller-Costa, T., Canário, A. V. M., & Hubbard, P. C. (2015). Chemical communication in cichlids: A mini-review. *General and Comparative Endocrinology, 221*, 64–74. https://doi.org/10.1016/j.ygcen.2015.01.001

Kücklich, M., Möller, M., Marcillo, A., Einspanier, A., Weiß, B. M., Birkemeyer, C., & Widdig, A. (2017). Different methods for volatile sampling in mammals. *PLoS ONE, 12*(8), e0183440. https://doi.org/10.1371/journal.pone.0183440

Lynam, K. (2008). Semivolatile analysis using an inertness-performance-tested agilent J&W DB-5ms ultra inert column. *LC GC Europe, 21*(7), 8.

Lynam, K., & Smith, D. (2008). *Polycyclic Aromatic Hydrocarbon (PAH) analysis using an agilent j&w DB-5ms ultra inert capillary GC column*. Agilent Technologies publication.

Nevitt, G. A. (2008). Sensory ecology on the high seas: The odor world of the procellariiform seabirds. *The Journal of Experimental Biology, 211*(11), 1706–1713. https://doi.org/10.1242/jeb.015412

Nordlund, D. A., & Lewis, W. J. (1976). Terminology of chemical releasing stimuli in intraspecific and interspecific interactions. *Journal of Chemical Ecology, 2*(2), 211–220. https://doi.org/10.1007/BF00987744

Piryaei, M. (2019). Direct thermal desorption technique as a very fast, easy and low-cost method for analysis of volatile components compounds by gas chromatography with mass spectrometry. *Separation Science Plus, 2*(11), 416–421. https://doi.org/10.1002/sscp.201900003

R Core Team. (2022). *R: A language and environment for statistical computing*. R Foundation for Statistical Computing, Vienna. http://www.R-project.org/

Scordato, E. S., & Drea, C. M. (2007). Scents and sensibility: Information content of olfactory signals in the ringtailed lemur, Lemur catta. *Animal Behaviour, 73*(2), 301–314. https://doi.org/10.1016/j.anbehav.2006.08.006

Sorensen, P. W., & Wisenden, B. D. (Eds.). (2014). *Fish pheromones and related cues*. Wiley.

Steiger, S., Schmitt, T., & Schaefer, H. M. (2011). The origin and dynamic evolution of chemical information transfer. *Proceedings of the Royal Society B: Biological Sciences, 278*(1708), 970–979. https://doi.org/10.1098/rspb.2010.2285

Thom, M. D., & Hurst, J. L. (2004). Individual recognition by scent. *Annales Zoologici Fennici*, 765–787. Finnish Zoological and Botanical Publishing Board.

Urbanowicz, M., Zabiegała, B., & Namieśnik, J. (2011). Solventless sample preparation techniques based on solid- and vapour-phase extraction. *Analytical and Bioanalytical Chemistry, 399*(1), 277–300. https://doi.org/10.1007/s00216-010-4296-1

Weiß, B. M., Marcillo, A., Manser, M., Holland, R., Birkemeyer, C., & Widdig, A. (2018). A non-invasive method for sampling the body odour of mammals. *Methods in Ecology and Evolution, 9*(2), 420–429. https://doi.org/10.1111/2041-210X.12888

Woodley, S. (2015). Chemosignals, hormones, and amphibian reproduction. *Hormones and Behavior, 68*, 3–13. https://doi.org/10.1016/j.yhbeh.2014.06.008

Wyatt, T. D. (2009). Fifty years of pheromones. *Nature, 457*(7227), 262–263. https://doi.org/10.1038/457262a

Wyatt, T. D. (2010). Pheromones and signature mixtures: Defining species-wide signals and variable cues for identity in both invertebrates and vertebrates. *Journal of Comparative Physiology. A, Neuroethology, Sensory, Neural, and Behavioral Physiology, 196*(10), 685–700. https://doi.org/10.1007/s00359-010-0564-y

Wyatt, T. D. (2014). *Pheromones and animal behavior: Chemical signals and signatures*. Cambridge University Press. https://doi.org/10.1017/CBO9781139030748

Zhao, Z., Zung, J. L., Hinze, A., Kriete, A. L., Iqbal, A., Younger, M. A., Matthews, B. J., Merhof, D., Thiberge, S., Ignell, R., Strauch, M., & McBride, C. S. (2022). Mosquito brains encode unique features of human odour to drive host seeking and preference. *Nature, 605*(7911), 706–712. https://doi.org/10.1038/s41586-022-04675-4

Zou, Y. (2018). *A comparison study of the analysis of volatile organic acids and fatty acids* (pp. 1–10). Agilent Technologies.

# Development of a One-Day Test of Olfactory Learning and Memory in Mice

Richard E. Brown[1] (✉), Oliver K. Schnare[1], Elias B. Habib[1], and Kyle M. Roddick[1]

[1] Department of Psychology and Neuroscience, Dalhousie University, Halifax, NS, Canada
rebrown@dal.ca

**Abstract** Heather Schellinck developed a simple and reliable test of olfactory learning and memory in mice using a Pavlovian conditioned odour preference task in which CD1 mice learned to discriminate between two odours after two daily 10-min trials with the CS+ odour (sugar reward) and two trials with the CS− odour (no sugar) over 4 days. Mice remembered the CS+ for up to 60 days after training in this study and, in other studies, male and female C57BL/6J and DBA/2J mice could remember the odour discrimination for up to 9 months after training. Mice up to 24 months of age can learn and remember the odour preferences. Likewise, the 5×FAD mouse model of Alzheimer disease showed no odour memory deficits in this task. In the present study we developed a 1-day training procedure in which male and female mice received eight CS+ and eight CS− odour presentations on 1 day and were then tested for memory 24 h, 7 days and 30 days later. Mice (C57BL/6J × SJL/J and C57BL/6J) given the 1-day training showed a conditioned odour preference on 24 h, and 7-day memory tests with no sex differences. After 30 days the odour preference was not significant. Mice (C57BL/6J) given four CS+ and four CS− trials showed no significant odour preference on 24 h and 7-day memory tests. The results suggest that different 1-day training paradigms result in different memory strength. Such a 1-day appetitive training test could be valuable for studying the neurobiological bases of short-, long-, and very long-term memory.

**Keywords** Olfaction · Learning · Memory · Mice · Mus musculus · Sex differences · Pavlovian conditioning

## 1 Introduction

Rodents live in an olfactory world, depending on their sense of smell for social and sexual behaviour, maternal-offspring interactions, spatial navigation in the environment, predator avoidance, and food selection (Adduci et al., 2021; Boone et al., 2021; Brown, 1979; Brown & Macdonald, 1985; Doty, 1986; Kinney & Antill, 1996; Lévy



B. Schaal et al. (eds.), *Chemical Signals in Vertebrates 15*,
https://doi.org/10.1007/978-3-031-35159-4_3

et al., 2004; Marin et al., 2021; Schultz & Tapp, 1973; Slotnick et al., 2005). In order to utilize olfactory information in such a wide range of behavioural responses, the olfactory system of rodents is quite extensive, integrating brain regions involved with emotion, motivation, fear, and cognitive processes. When odourant molecules bind to the olfactory receptors, the information converges onto mitral and tufted cells within the olfactory bulb. These cells project directly to the piriform cortex, amygdala and entorhinal cortex (de Castro, 2009; Emes et al., 2004; Nagayama et al., 2014; Takeuchi & Sakano, 2014). There are two principal olfactory pathways in the rodent brain: the main olfactory pathway from the primary olfactory receptors to the olfactory bulb and the accessory olfactory pathway from the vomeronasal receptors to the accessory olfactory bulb (Meredith, 1991; Scalia & Winans, 1975). These two olfactory systems communicate with each other to regulate innate and learned odour processing (Kobayakawa et al., 2007; Mucignat-Caretta et al., 2012).

Given the complexity of the rodent olfactory system, it is not surprising that rodents excel in olfactory learning and memory (Slotnick, 2001) and that many different brain areas, including the olfactory bulbs, olfactory tubercle, anterior olfactory nucleus, and olfactory (piriform) cortex have been found to be important for olfactory memory (Aqrabawi & Kim, 2020; Cansler et al., 2020; Tong et al., 2018; Wang et al., 2020; Wilson & Sullivan, 2011). The direct projections from the olfactory bulbs to the limbic system can create olfactory memories that are linked to emotional responses (Sullivan et al., 2015), while the reward value of an olfactory stimulus is encoded in the olfactory tubercle and piriform cortex (Gadziola et al., 2020). Since rodents rely so heavily on their olfactory system, the use of olfactory stimuli in both classical (Pavlovian) and instrumental conditioning procedures is a powerful and valid way to assess memory in rodents (Schellinck, 2018; Slotnick & Restrepo, 2005). With up to 5% of their total protein-coding genes dedicated to the production of olfactory receptors (Barnes et al., 2020), mice can discriminate between odours with 85% accuracy even when the odour concentration is as low as $10^{-5}$ parts per million (Roddick et al., 2016).

## 1.1 The Pavlovian Conditioned Odour Preference Task

Schellinck et al. (2001a, b) developed a simple and reliable test of olfactory learning and memory in CD1, C57Bl/6NCr/BR and DBA/2NCr/BR mice using a Pavlovian conditioned odour preference procedure in which mice received four 10-min training sessions per day for 4 days. Two daily trials were with a conditioned stimulus (CS)+ odour paired with sugar reward and two were with a CS− odour, which was not paired with reward. Thus, mice had eight CS+ and eight CS− trials over 4 days and were tested for their 24-h odour memory in a preference test on Day 5. During training, the CS+ odour was presented in an odour pot with small pieces of sugar buried in pine chip bedding (Fig. 1a). The CS− odour pot contained pine chip bedding, but no sugar reward. The training sessions occurred in a polyethylene cage, identical to their housing cages (Fig. 1b). Mice were placed on a food deprivation schedule for 4 days prior to the experiment in order to maintain 80–85% of their free-feeding weight (Schellinck et al., 2001a). Food-deprivation has a significant effect on food-searching behaviour which

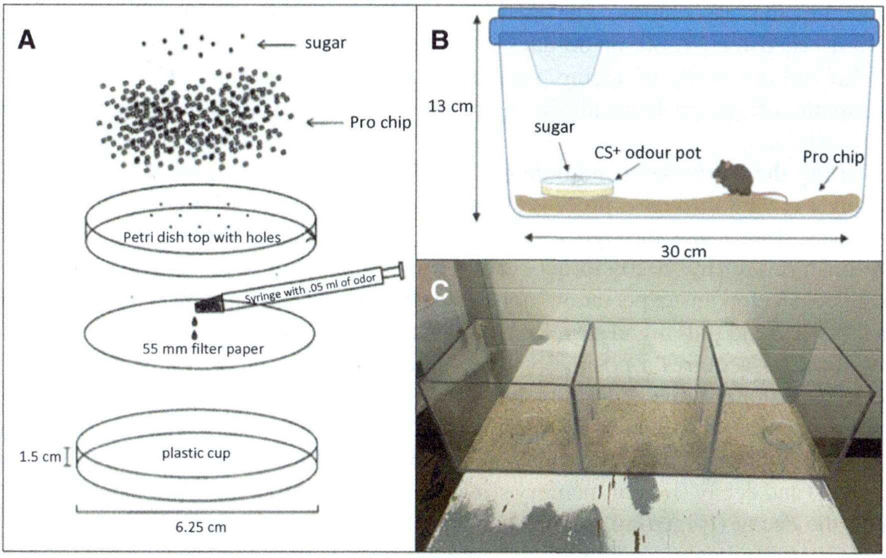


**Fig. 1** (**a**) Diagram of the odour pot (6.25 cm in diameter and 1.5 cm in height) used during both training and test trials. 0.05 mL of the odourant was placed onto the filter paper (5.5 cm in diameter). A Petri dish with holes drilled in it was placed on top of the filter paper and covered in woodchip bedding. Eight sugar pieces were placed on top of the woodchip bedding. (Image adapted from Schellinck et al. (2001a)). (**b**) Diagram of the polyethylene cage (19 cm wide, 13 cm high and 30 cm long) used for training trials. The cage included an odour pot (6.25 cm in diameter and 1.5 cm high) filled with woodchip bedding. The floor of the cage was covered in woodchip bedding. (Image created by Elias Habib). (**c**) Photograph of the Plexiglas three-chambered odour discrimination box (20 cm wide, 20 cm high and 69 cm long) used during memory tests. There was one odour pot in each of the end chambers. The floor of the box was covered in woodchip bedding. (Image taken by Oliver Schnare)

leads to strengthened acquisition of the conditioned odour preference (Forestell et al., 2001).

Following the acquisition stage, a memory test was given in a novel room in a three-chambered box (Fig. 1c), that contained a CS+ odour pot at one end and a CS− odour pot at the other end, with no sugar reward in either pot. The amount of time the mice spent digging in each odour pot was measured as the conditioned response and a preference for the CS+ was scored if the mice spent over 80% of their digging time in the CS+ odour pot. Using this procedure, Schellinck et al. (2001a) showed that CD1, C57Bl/6NCr/BR and DBA/2NCr/BR mice showed significant preferences for digging in the CS+ odour, indicating a memory for the association of that odour with the sugar reward. Mice which received just two training sessions a day for 4 days also displayed a significant preference for the CS+ odour pot on their 24-h memory test (Schellinck et al., 2001a). Brown and Wong (2007) showed that males and females of 13 different strains of mice, including those that were visually impaired, could learn the odour preference using this paradigm and O'Leary et al. (2020) showed that males and females of the 5×FAD mouse model of Alzheimer's disease up to 15 months of age could learn and remember this association. The memory for the association of the CS+ with sugar reward lasted for 14, 30, or 60 days after the last training session (Schellinck et al., 2001a) and in a study of ageing, male and female C57BL/6J and DBA/2J mice up to 24 months of age could recall this association for up to 9 months (Wong & Brown, 2007, 2013).

During the Pavlovian conditioning procedure, mice associate the CS+ odour with the sugar reward through a process of "autoshaping" (Brown & Jenkins, 1968; Cleland & Davey, 1983) and through a similar process associate the CS− with no reward (Millman & Murthy, 2020; Torquet et al., 2014). The result of this conditioning is that the mice develop an "expectation" for sugar when they encounter the CS+ odour in the memory test and thus dig in search of the sugar pellets which they expect to gain. In this way the CS+ acquires a positive valence while the CS− acquires a negative valence (Rescorla, 2003; Savage & Ramos, 2009). Memory strength is the stability of a memory or its resistance to extinction. Memory strength depends on three components: acquisition (building up memory strength), consolidation (storing and maintaining memory strength) and decay (loss of memory strength). Memory strength is measured by examining the decay (forgetting) curve over time (Koen et al., 2013; Rohrer, 1996; Vallet et al., 2016). In Pavlovian conditioning, memory strength is increased with the number of training trials and the number of rewards received and measured by the rate of extinction (Delamater, 2004; Rescorla, 1966, 1988). Synaptic strength is often considered to be the physiological basis of memory strength (Humeau & Choquet, 2019; Lisman et al., 2018; Murthy, 1998). Hebb (1949) proposed that changes in synaptic strength underlie memory strength in Pavlovian conditioning and that memories involve a network of synaptic connections between neurons which Hebb called a cell assembly (Langille & Brown, 2018). Pavlovian fear conditioning has become the most popular method for examining the neural basis of memory (Aziz et al., 2019; Kim & Jung, 2006; Maren, 2001, 2008). The Rescorla-Wagner model of Pavlovian conditioning theorizes that after conditioning occurs, the animal learns to predict the US (unconditioned stimulus) when the CS is present (Rescorla, 1988, 2003). The number of pairings of the CS and US that the subject receives is a key factor in the acquisition stage of classical conditioning (Rescorla, 1966). Increasing the number of CS-US pairings strengthens the association between them and leads to increased memory strength (Hilgard & Bower, 1975). Memory strength is measured by the duration of the memory or the speed of "forgetting" in Pavlovian fear conditioning (Palmer & Gong, 2014; Reijmers et al., 2007).

### 1.2 Rationale for Current Study

From the work of Schellinck et al. (2001a, b), Wong and Brown (2007, 2013) and O'Leary et al. (2020), it is clear that both males and females of many different strains of mice from 3 to 24 months of age are able to quickly learn olfactory discriminations using classical conditioning, and maintain olfactory memories for 60–270 days. Our aim in the present study was to determine if a 1-day training procedure would result in short- and long-term memories that could be used to study memory strength and the neurobiology of memory using an appetitive reward rather than fear learning (Galaj &

Ranaldi, 2021; Sampedro-Piquero et al., 2018). The results of three experiments are reported. These experiments examine the efficacy of a 1-day acquisition phase with eight CS+ and eight CS− trials for developing a short-term (24 h) and a long-term (7-day) memory trace (Experiment 1); whether this memory trace will last for 30 days (Experiment 2) and whether four CS+ and four CS− training trials will be sufficient to consolidate short-and long-term memories (Experiment 3). These experiments were conducted during the COVID-19 pandemic, during which we could not order new mice, so had to use aged mice available from our colony, thus the results should be considered as preliminary until they are replicated. Both male and female mice were used and the mice were all over a year old, but we have considerable evidence (see Wong & Brown, 2007, 2013; O'Leary et al., 2020) that male and female mice of various strains between 12 and 24 months of age can learn the Pavlovian odour preferences using the method of Schellinck et al. (2001a, b).

## 2 Experiment 1: The Ability of Mice to Learn a Pavlovian-Conditioned Odour Preference with One Day of Training

### 2.1 Methods

In this experiment, 18 B6SJL (C57BL/6J × SJL/J) mice (nine females and nine males) which were from 287 to 479 days old (mean age = 402.3 days) were tested. The experiments were approved by the Dalhousie University Committee on Laboratory Animals (approval number: 20-049). Mice were housed individually in polyethylene cages (30 × 19 × 13 cm) with woodchips for bedding and strips of paper for nesting. A polyethylene tube (6 cm in diameter, 8 cm long) was placed in the cages for enrichment. Water was available ad libitum and food (Prolab® RMH 3500, autoclaved) was available ad libitum, except during the 24-h food deprivation periods prior to testing. Cages had a metal-wire cover and a micro-isolator filter and the housing room was on a reversed 12:12 h light/dark cycle with lights off at 0930. All testing occurred during the dark phase, between 10:00 and 15:00.

Prior to the experiment mice were pseudo-randomly assigned (balancing the number of males and females by age in each group) to one of three groups: B+/L− ($n$ = 7), L+/B− ($n$ = 7), or control (B−/L−) ($n$ = 4). The B+/L− mice received banana odour as the CS+ (presented with sugar during training) and lemon odour as the CS− (presented without sugar during training). The L+/B− mice received lemon odour as the CS+ and banana odour as the CS−. Control mice (B−/L−) did not have sugar presented with either odour. On Day 1 the mice were placed on food deprivation for 24 h and eight pieces of sugar cube (mean total weight = 0.7 g) were placed in the home cage to introduce sugar as a reward and to overcome neophobia. On Day 2 the mice received 16 five-min training trials, eight with the CS+ odour pot and sugar, and eight with the CS− odour pot and no sugar, with the CS+ and CS− odours alternating on each trial. During the acquisition stage, training trials were conducted in polyethylene cages with woodchip bedding, identical to those used for housing (Fig. 1b). The odour stimuli were

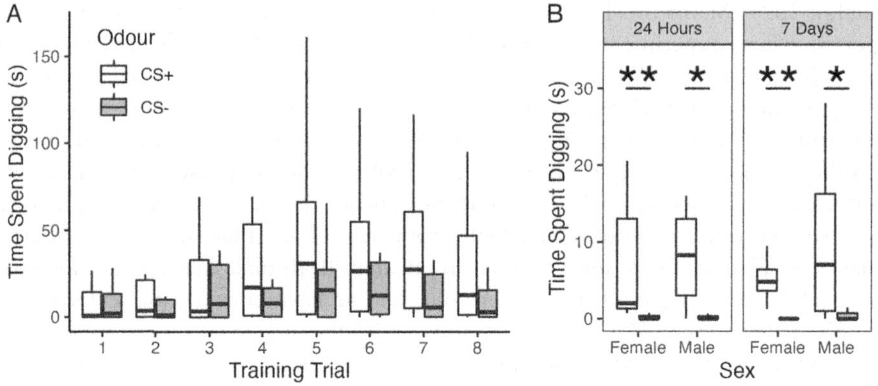


**Fig. 2** Box and whisker plots displaying the median amount of time (in seconds) spent digging by female and male mice in the CS+ and CS− odour pots in (**a**) the eight acquisition trials and (**b**) the 24-h and 7-day memory tests in Experiment 1. The boxes represent the inter-quartile range (IQR) with the bars in the middle showing the medians, the boxes showing the 25th and 75th percentiles and the whiskers extending to the furthest points within 1.5 inter-quartile ranges

banana and lemon liquids (0.05 mL, Club House brand) which were placed on filter paper (5.5 cm in diameter) which was then placed into a plastic "odour pot" (6.25 cm in diameter and 1.5 cm in height; Fig. 1a). The filter paper was covered with a petri dish with holes drilled in it to allow the odours to pass through. Two rooms were used, one for each odour, and mice were transported between rooms after each trial. After the acquisition day, the mice were kept on food deprivation, but given 2 mL of powdered food in water.

The 24-h memory test was given on Day 3 in a different room from the training trials. Mice were placed in the three-chambered, Plexiglas odour discrimination box (20 × 20 × 69 cm) with woodchip bedding (Fig. 1c) for a 2-min habituation trial with non-scented odour pots placed in the right and left chamber. The amount of time spent in each chamber was recorded with a stopwatch. Immediately after habituation, the bedding in the box and odour pots was replaced and the mouse was given a 10-min test; with the CS+ odour placed in the end chamber the mice spent the lesser time in during the habituation trial and the CS− placed in the preferred end. Neither the CS+ nor the CS− odour were presented with sugar during the memory tests. For mice in the control group, the end chambers that contained the banana and lemon odour pots were randomly selected. The amount of time spent digging in each odour pot was recorded. Digging was defined as the mouse moving the woodchip bedding in the odour pot with either its nose or front paws. After the 24-h memory test food was available ad libitum until Day 8, when the mice were returned to 24-h food deprivation. The 7-day memory test, which was identical to the 24-h test, but did not have a habituation trial, occurred on Day 9 of the experiment (Table 1).

All training and testing trials were filmed using a Canon Powershot camera. Habituation trials were not filmed, but the time spent in each chamber was measured using a stopwatch. The videos were scored for cumulative digging time for each trial using the program BORIS (Version 7.9.15; Friard & Gamba, 2016) and statistical

**Table 1** Training and testing schedule for Experiment 1

| | | |
|---|---|---|
| Acquisition stage | Day 1 | 24-h food deprivation with sugar introduction |
| | Day 2 | 8 CS+ and 8 CS− alternating training trials (5 min each) |
| Memory testing stage | Day 3 | Habituation trial (2 min) and 24-h memory test (10 min) |
| | Day 8 | 24-h food deprivation |
| | Day 9 | 7-day memory test (10 min) |

analyses conducted for data from the memory tests using the Wilcoxon Signed Ranks Test (one-tailed) for small samples (Siegel & Castellan Jr., 1988). A one-tailed test was used because we had the strong prediction from previous research that the S+ odour would be preferred to the S− odour. If a mouse failed to dig in both odour pots (CS+ = 0.0 and CS− = 0.0) it was dropped from the analysis and the sample size reduced. Data were analysed separately for both sexes in Experiment 1. Sex differences were analysed using the Wilcoxon Mann-Whitney U test for independent samples. Means are given in the text and medians in the figures. Effect sizes are reported as "*r*," based on the recommendations of Tomczak and Tomczak (2014). Statistical analyses were conducted using R version 4.2.2 (R Core Team, 2022) with the "rstatix" (Kassambara, 2021), "sjstats" (Lüdecke, 2021) and "tidyverse" (Wickham et al., 2019) packages.

## 2.2 Results of Experiment 1

**Acquisition** Across all training trials, mice in the experimental groups ($n = 14$) spent a mean of 29.59 ± 30.12 s (s) digging in the CS+ odour pot and 21.70 ± 30.22 s digging in the CS− odour pot (Fig. 2a), a difference which was not significant ($n = 13$, $T = 67$, $p = 0.073$; $r = 0.42$). The mean preference score for the CS+ odour during the acquisition stage (calculated by dividing the time spent digging in the CS+ odour pot by total time spent digging in both odour pots) for the experimental groups was 0.62 ± 0.34. There was no significant sex difference in the preference for the CS+ odour ($n = 6,7$; $W = 36$, $p = 0.063$; $r = 0.515$). The control group spent more time digging in the banana odour pot (6.13 ± 7.32 s) than the lemon odour pot (1.87 ± 1.57 s) but this difference was not significant ($n = 4$, $T = 8$, $p = 0.375$; $r = 0.548$).

**Twenty-Four-Hour Memory Test** Overall, mice in the CS+ groups spent a mean of 14.28 ± 24.4 s digging in CS+ odour pot, and 0.55 ± 1.29 s digging in the CS− odour pot, showing a significant preference for the CS+ odour ($n = 13$, $T = 87$, $p < 0.001$; $r = 0.804$), with a preference score of 0.91 ± 0.19. Female mice spent a mean of 17.56 ± 33.36 s digging in the CS+ odour pot and 0.32 ± 0.59 s digging in the CS− odour pot, a significant difference ($n = 7$, $T = 28$, $p = 0.0078$; $r = 0.894$; Fig. 2b), with a preference score of 0.92 ± 0.14 for the CS+ odour. Males spent 11.0 ± 12.54 s digging in the CS+ odour pot and 0.77 ± 1.77 s digging in the CS− odour pot, a significant difference ($n = 6$, $T = 20$, $p = 0.031$; $r = 0.813$; Fig. 2b), with a preference score for the CS+ of 0.89 ± 0.25. There was no significant sex difference in the preference scores for the CS+ ($n = 6,7$; $W = 45$, $p = 0.568$; $r = 0.158$). Mice in the control group spent a mean of 1.69 ± 1.95 s digging in the banana odour pot, and 0.38 ± 0.75 s digging in the lemon odour pot, a difference which was not significant ($n = 3$, $T = 5$, $p = 0.500$; $r = 0.316$).

**Seven-Day Memory Test** Overall, mice in the CS+ groups spent an average of 7.52 ± 8.20 s digging in the CS+ odour pot, and 0.71 ± 1.84 s digging in the CS− odour pot, a significant difference ($n = 12$, $T = 78$, $p < 0.001$; $r = 0.883$), with a mean preference score for the CS+ of 0.92 ± 0.18. Females in the CS+ group spent 5.08 ± 2.74 s digging in the CS+ odour pot, and 0.25 ± 0.66 s digging in the CS− odour pot, a significant difference ($n = 7$, $T = 28$, $p = 0.0078$; $r = 0.894$; Fig. 2b), with a preference score of 0.94 ± 0.16, which was higher than in the 24-h memory test. Males in the CS+ group spent 9.95 ± 11.14 s digging in the CS+ odour pot, and 1.18 ± 2.53 s digging in the CS− odour pot, a significant difference ($n = 5$, $T = 15$, $p = 0.0313$; $r = 0.905$; Fig. 2b), with a preference score of 0.88 ± 0.21. There was no significant sex difference in the preference scores ($n = 5,7$; $W = 41$, $p = 0.465$; $r = 0.211$). Mice in the control group spent an average of 0.25 ± 0.50 s digging in the lemon odour pot, and 0.18 ± 0.37 s digging in the banana odour pot, a non-significant difference ($n = 2$, $T = 1$, $p = 1.00$; $r = 0.316$).

## 3 Experiment 2: The Addition of a 30-Day Memory Test

This experiment included eight C57BL/6J mice (five female and three male), ranging from 330 to 348 days old (Mean = 339.1 days) that were tested in three memory tests. The procedure for the acquisition stage, 24-h and 7-day memory tests were as described in Experiment 1, but the mice in this experiment were given an additional 30-day memory test, which was identical to the 24-h and 7-day memory tests. On Day 31 the mice received 24 h of food deprivation before the 30-day memory test on Day 32 (Table 2). Since we found no sex differences in Experiment 1, the data were pooled over sex in this study.

### 3.1 Results of Experiment 2

**Acquisition Stage** During the acquisition trials, the mice spent a mean of 34.8 ± 32.6 s digging in the CS+ odour pot, and 20.09 ± 19.19 s digging in the CS− odour pot (Fig. 3a), with a mean preference score of 0.58 ± 0.44, and no difference in the time spent digging in the CS+ and CS− odour pots ($n = 8$, $T = 24$, $p = 0.231$; $r = 0.297$).

**Table 2** Training and testing schedule for Experiment 2

| | | |
|---|---|---|
| Acquisition stage | Day 1 | 24-h food deprivation with sugar introduction |
| | Day 2 | 8 CS+ and 8 CS− alternating training trials (5 min each) |
| Memory testing stage | Day 3 | Habituation trial (2 min) and 24-h memory test (10 min) |
| | Day 8 | 24-h food deprivation |
| | Day 9 | 7-day memory test (10 min) |
| Long-term memory test | Day 31 | 24-h food deprivation |
| | Day 32 | 30-day memory test (10 min) |

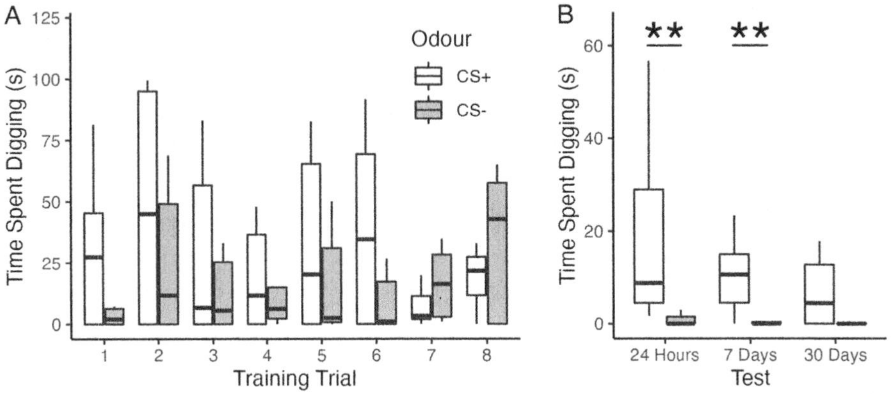


**Fig. 3** Box and whisker plots displaying the median amount of time (in seconds) spent digging by mice in the CS+ and CS− odour pots in (**a**) the eight acquisition trials and (**b**) the 24-h and 7-day and 30-day memory tests in Experiment 2. The boxes represent the inter-quartile range (IQR) with the bars in the middle showing the medians, the boxes showing the 25th and 75th percentiles and the whiskers extending to the furthest points within 1.5 inter-quartile ranges

**Twenty-Four-Hour Memory Test** In the 24-h memory test, the mice spent significantly more time digging in the CS+ odour pot (18.74 ± 20.46 s) than in the CS− odour pot (0.99 ± 1.57 s; $n = 8$, $T = 35$, $p = 0.0078$; $r = 0.842$; Fig. 3b), with a preference score of 0.89 ± 0.21.

**Seven-Day Memory Test** In the 7-day memory test the mice spent significantly more time digging in the CS+ odour pot (10.51 ± 7.8 s) than in the CS− odour pot (0.40 ± 0.74 s; $n = 7$, $T = 28$, $p = 0.0078$; $r = 0.894$; Fig. 3b), with a preference score of 0.93 ± 0.15.

**Thirty-Day Memory Test** In the 30-day memory test, the mice spent more time digging in the CS+ odour pot (6.56 ± 7.42 s) than in the CS− odour pot (0.38 ± 1.06 s; Fig. 3b), with a preference score of 0.80 ± 0.45, but this difference was not significant ($n = 5$, $T = 14$, $p = 0.0625$; $r = 0.784$).

## 4 Experiment 3: Will Mice Learn the Odour Preference with Only Four CS+ and Four CS− Acquisition Trials?

In this experiment, seven B6SJL mice (four female and three male) ranging from 326 to 406 days of age (Mean = 351.3 days) were tested using the same procedure as Experiment 1, but the number of training trials was reduced to four CS+ and four CS− trials for a total of eight training trials (Table 3). Data were pooled over sex.

### 4.1 Results of Experiment 3

**Acquisition Stage** The mean time spent digging per training trial was 16.88 ± 20.45 s in the CS+ odour pot, and 14.35 ± 12.23 s in the CS− odour pot, which did not differ ($n = 6$, $T = 9$, $p = 0.656$; $r = 0.128$), resulting in a mean preference score across the training trials of 0.49 ± 0.34 (Fig. 4a).

**Twenty-Four-Hour Memory Test** In the 24-h memory test, mice spent an average of 8.08 ± 12.5 s digging in the CS+ odour pot and 0.50 ± 1.32 s digging in the CS− odour pot, a difference which was not significant ($n = 4$, $T = 9$, $p = 0.125$; $r = 0.730$; Fig. 4b), with a preference score of 0.84 ± 0.32.

**Seven-Day Memory Test** On the 7-day memory test mice spent an average of 1.48 ± 1.98 s digging in the CS+ odour pot and 0.49 ± 1.30 s digging in the CS− odour pot which was not significant ($n = 4$, $T = 7$, $p = 0.312$; $r = 0.365$; Fig. 4b), with a mean preference score of 0.75 ± 0.50.

## 5 Summary and Conclusion

The results of Experiment 1 allow us to conclude that the Pavlovian conditioned odour preference test created by Schellinck et al. (2001a, b) was powerful enough to create significant short-term (24-h) and a long-term (7-day) memory traces with eight CS+ and eight CS− trials on one acquisition day (Fig. 2). In Experiment 1, the mice that were not presented with a CS+ odour during training displayed very little digging behaviour during the memory tests and showed no evidence on an odour preference memory. This indicates that both male and female mice in the CS+ groups were able to acquire, store, and consolidate olfactory memories since they displayed more digging behaviour than unconditioned mice. There were no significant sex differences detected in Experiment 1.

The results of Experiment 2 suggest that the 1-day acquisition with eight CS+ and eight CS− trials can result in the consolidation of a very long-term memory trace lasting 30 days, but the small sample size did not result in significant results at 30 days (Fig. 3). Finally Experiment 3 suggested that only four CS+ and four CS− trials on one acquisition day did not result in significant odour preference memories (Fig. 4). The reduced number of acquisition trials did not appear sufficient to build memory strength for the association of the odour with the sugar reward. Combined, these results suggest that four CS+ and four CS− training trials produced a weaker memory strength than

**Table 3** Training and testing schedule for Experiment 3

| | | |
|---|---|---|
| Acquisition stage | Day 1 | 24-h food deprivation with sugar introduction |
| | Day 2 | 4 CS+ and 4 CS− alternating training trials (5 min each) |
| Memory testing stage | Day 3 | Habituation trial (2 min) and 24-h memory test (10 min) |
| | Day 8 | 24-h food deprivation |
| | Day 9 | 7-day memory test (10 min) |

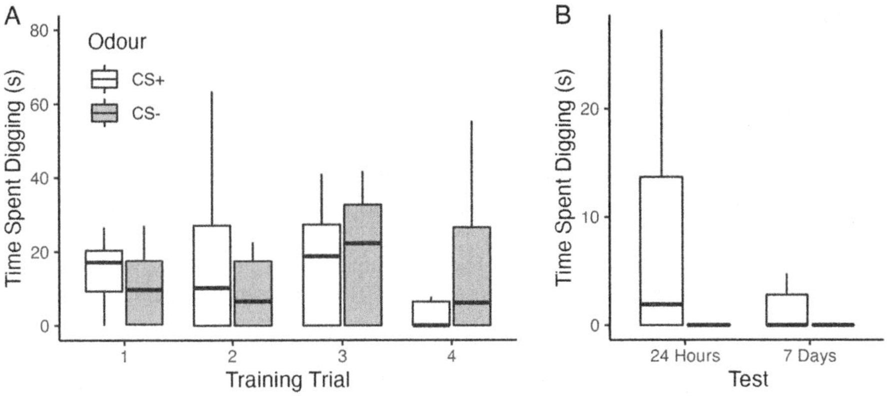


**Fig. 4** Box and whisker plots displaying the median amount of time (in seconds) spent digging by mice in the CS+ and CS− odour pots in (**a**) the four acquisition trials and (**b**) the 24-h and 7-day memory tests in Experiment 3. The boxes represent the inter-quartile range (IQR) with the bars in the middle showing the medians, the boxes showing the 25th and 75th percentiles and the whiskers extending to the furthest points within 1.5 inter-quartile ranges

eight CS+ and eight CS− training trials. This lends support to the hypothesis that increased exposure to the stimuli increases memory strength (von Hippel & Hawkins, 1994).

Prior to testing, we hypothesized that the mice would show a gradual learning curve during the acquisition stage and would dig less in the CS− and more in the CS+ with each training trial. This was not observed during the acquisition stage for any of the experiments, but the time spent digging in the CS− odour pot was greatly decreased during the memory tests compared to training trials. This suggests that the mice did learn during the acquisition stage. A possible explanation for the lack of an odour preference during acquisition trials is that the mice were only presented with one odour at a time, whereas during the memory tests the mice had the option to dig in either odour. Since there was no other stimuli present during the CS− odour training trials the mice were neither punished nor rewarded for digging in the CS− odour pot. The unexpected high amounts of digging in the CS− during the acquisition stage may have actually helped the mice on the memory tests since increased exposure to the non-reward stimulus would strengthen the discrimination learning between stimuli (von Hippel & Hawkins, 1994). A future study could present the odours simultaneously during training and should expect to see less digging in the CS− since simultaneous exposure to stimuli has been shown to make discrimination faster and stronger than successive exposure (Mundy et al., 2009; Spence, 1952). We are now repeating these experiments with larger sample sizes and mice of different ages.

## References

Adduci, L. B., León, V. A., Schlötelburg, A., Busch, M., & Fraschina, J. (2021). Avoidance behaviour in laboratory house mice (*Mus musculus*) and Norway rats (*Rattus norvegicus*) towards predator odours. *PLoS ONE, 16*(1), e0245441. https://doi.org/10.1371/journal.pone.0245441

Aqrabawi, A. J., & Kim, J. C. (2020). Olfactory memory representations are stored in the anterior olfactory nucleus. *Nature Communications, 11*(1), 1246. https://doi.org/10.1038/s41467-020-15032-2

Aziz, W., Kraev, I., Mizuno, K., Kirby, A., Fang, T., Rupawala, H., Kasbi, K., Rothe, S., Jozsa, F., Rosenblum, K., Stewart, M. G., & Giese, K. P. (2019). Multi-input synapses, but not LTP-strengthened synapses, correlate with hippocampal memory storage in aged mice. *Current Biology, 29*(21), 3600–3610.e4. https://doi.org/10.1016/j.cub.2019.08.064

Barnes, I. H. A., Ibarra-Soria, X., Fitzgerald, S., Gonzalez, J. M., Davidson, C., Hardy, M. P., Manthravadi, D., Van Gerven, L., Jorissen, M., Zeng, Z., Khan, M., Mombaerts, P., Harrow, J., Logan, D. W., & Frankish, A. (2020). Expert curation of the human and mouse olfactory receptor gene repertoires identifies conserved coding regions split across two exons. *BMC Genomics, 21*(196), 1–15. https://doi.org/10.1186/s12864-020-6583-3

Boone, M. H., Liang-Guallpa, J., & Krashes, M. J. (2021). Examining the role of olfaction in diet choice. *Cell Reports, 34*(7), 108755. https://doi.org/10.1016/j.celrep.2021.108755

Brown, R. E. (1979). Mammalian social odors: A critical review. *Advances in the Study of Behaviour, 10*, 103–162.

Brown, P. L., & Jenkins, H. M. (1968). Auto-shaping of the pigeon's key-peck. *Journal of the Experimental Analysis of Behavior, 11*, 1–8.

Brown, R. E., & Macdonald, D. W. (1985). *Social odours in mammals*. Clarendon Press.

Brown, R. E., & Wong, A. A. (2007). The influence of visual ability on learning and memory performance in 13 strains of mice. *Learning & Memory, 14*(3), 134–144.

Cansler, H. L., Wright, K. N., Stetzik, L. A., & Wesson, D. W. (2020). Neurochemical organization of the ventral striatum's olfactory tubercle. *Journal of Neurochemistry, 152*(4), 425–448. https://doi.org/10.1111/jnc.14919

Cleland, G. G., & Davey, G. C. L. (1983). Autoshaping in the rat: The effects of localizable visual and auditory signals for food. *Journal of the Experimental Analysis of Behavior, 40*, 47–56.

de Castro, F. (2009). Wiring olfaction: The cellular and molecular mechanisms that guide the development of synaptic connections from the nose to the cortex. *Frontiers in Neuroscience, 3*, 52. https://doi.org/10.3389/neuro.22.004.2009

Delamater, A. R. (2004). Experimental extinction in Pavlovian conditioning: Behavioural and neuroscience perspectives. *The Quarterly Journal of Experimental Psychology, 57B*, 97–132.

Doty, R. L. (1986). Odor-guided behavior in mammals. *Experientia, 42*, 257–271.

Emes, R. D., Beatson, S. A., Ponting, C. P., & Goodstadt, L. (2004). Evolution and comparative genomics of odorant- and pheromone associated genes in rodents. *Genome Research, 14*(4), 591–602. https://doi.org/10.1101/gr.1940604

Forestell, C. A., Schellinck, H. M., Boudreau, S. E., & LoLordo, V. M. (2001). Effect of food restriction on acquisition and expression of a conditioned odor discrimination in mice. *Physiology & Behavior, 72*(4), 559–566.

Friard, O., & Gamba, M. (2016). BORIS: A free, versatile open-source event-logging software for video/audio coding and live observations. *Methods in Ecology and Evolution, 7*(11), 1325–1330.

Gadziola, M. A., Stetzik, L. A., Wright, K. N., Milton, A. J., Arakawa, K., Del Mar Cortijo, M., & Wesson, D. W. (2020). A neural system that represents the association of odors with rewarded outcomes and promotes behavioral engagement. *Cell Reports, 32*(3), 107919. https://doi.org/10.1016/j.celrep.2020.107919

Galaj, E., & Ranaldi, R. (2021). Neurobiology of reward-related learning. *Neuroscience and Biobehavioral Reviews, 124*, 224–234.

Hebb, D. O. (1949). *The organisation of behavior*. Wiley.

Hilgard, E. R., & Bower, G. H. (1975). *Pavlov's classical conditioning. Theories of learning* (pp. 63–89). Prentice-Hall.

Humeau, Y., & Choquet, D. (2019). The next generation of approaches to investigate the link between synaptic plasticity and learning. *Nature Neuroscience, 22*(10), 1536–1543. https://doi.org/10.1038/s41593-019-0480-6

Kassambara, A. (2021). rstatix: Pipe-friendly framework for basic statistical tests. R package version 0.7.0, https://CRAN.R-project.org/package=rstatix

Kim, J. J., & Jung, M. W. (2006). Neural circuits and mechanisms involved in Pavlovian fear conditioning: A critical review. *Neuroscience and Biobehavioral Reviews, 30*(2), 188–202. https://doi.org/10.1016/j.neubiorev.2005.06.005

Kinney, N. E., & Antill, R. W. (1996). Role of olfaction in the formation of preference for high-fat foods in mice. *Physiology & Behavior, 59*(3), 475–478. https://doi.org/10.1016/0031-9384(95)02086-1

Kobayakawa, K., Kobayakawa, R., Matsumoto, H., Oka, Y., Imai, T., Ikawa, M., Okabe, M., Ikeda, T., Itohara, S., Kikusui, T., Mori, K., & Sakano, H. (2007). Innate versus learned odour processing in the mouse olfactory bulb. *Nature, 450*(7169), 503–508.

Koen, J. D., Aly, M., Wang, W. C., & Yonelinas, A. P. (2013). Examining the causes of memory strength variability: Recollection, attention failure, or encoding variability? *Journal of Experimental Psychology. Learning, Memory, and Cognition, 39*(6), 1726–1741. https://doi.org/10.1037/a0033671

Langille, J. J., & Brown, R. E. (2018). The synaptic theory of memory: A historical survey and reconciliation of recent opposition. *Frontiers in Systems Neuroscience, 12*, 52. https://doi.org/10.3389/fnsys.2018.00052

Lévy, F., Keller, M., & Poindron, P. (2004). Olfactory regulation of maternal behavior in mammals. *Hormones and Behavior, 46*(3), 284–302. https://doi.org/10.1016/j.yhbeh.2004.02.005

Lisman, J., Cooper, K., Sehgal, M., & Silva, A. J. (2018). Memory formation depends on both synapse-specific modifications of synaptic strength and cell-specific increases in excitability. *Nature Neuroscience, 21*(3), 309–314. https://doi.org/10.1038/s41593-018-0076-6

Lüdecke, D. (2021). sjstats: Atatistical functions for regression models (Version 0.18.1). https://doi.org/10.5281/zenodo.1284472. https://CRAN.R-project.org/package=sjstats

Maren, S. (2001). Neurobiology of Pavlovian fear conditioning. *Annual Review of Neuroscience, 24*, 897–931.

Maren, S. (2008). Pavlovian fear conditioning as a behavioral assay for hippocampus and amygdala function: Cautions and caveats. *The European Journal of Neuroscience, 28*(8), 1661–1666. https://doi.org/10.1111/j.1460-9568.2008.06485.x

Marin, A. C., Schaefer, A. T., & Ackels, T. (2021). Spatial information from the odour environment in mammalian olfaction. *Cell and Tissue Research, 383*(1), 473–483. https://doi.org/10.1007/s00441-020-03395-3

Meredith, M. (1991). Sensory processing in the main and accessory olfactory systems: Comparisons and contrasts. *The Journal of Steroid Biochemistry and Molecular Biology, 39*(4B), 601–614. https://doi.org/10.1016/0960-0760(91)90258-7

Millman, D. J., & Murthy, V. N. (2020). Rapid learning of odor-value association in the olfactory striatum. *The Journal of Neuroscience, 40*(22), 4335–4347. https://doi.org/10.1523/JNEUROSCI.2604-19.2020

Mucignat-Caretta, C., Redaelli, M., & Caretta, A. (2012). One nose, one brain: Contribution of the main and accessory olfactory system to chemosensation. *Frontiers in Neuroanatomy, 6*, 46. https://doi.org/10.3389/fnana.2012.00046

Mundy, M. E., Honey, R. C., & Dwyer, D. M. (2009). Superior discrimination between similar stimuli after simultaneous exposure. *The Quarterly Journal of Experimental Psychology, 62*(1), 18–25. https://doi.org/10.1080/2F17470210802240614

Murthy, V. N. (1998). Synaptic plasticity: Step-wise strengthening. *Current Biology, 8*(18), R650–R653. https://doi.org/10.1016/s0960-9822(07)00414-9

Nagayama, S., Homma, R., & Imamura, F. (2014). Neural organization of olfactory bulb circuits. *Frontiers in Neural Circuits, 8*, 98. https://doi.org/10.3389/fncir.2014.00098

O'Leary, T. P., Stover, K., Mantolino, H. M., Darvesh, S., & Brown, R. E. (2020). Olfactory memory from 3 to 15 months of age in the 5xFAD mouse model of Alzheimer disease. *Behavioural Brain Research, 393*, 112731. https://doi.org/10.1016/j.bbr.2020.112731

Palmer, J. H. C., & Gong, P. (2014). Association learning of classical conditioning as an emergent property of spatially extended spiking neural circuits with synaptic plasticity. *Frontiers in Computational Neuroscience, 8*(79), 1–10. https://doi.org/10.3389/fncom.2014.00079

R Core Team. (2022). *R: A language and environment for statistical computing*. R Foundation for Statistical Computing. https://www.R-project.org/

Reijmers, L. G., Perkins, B. L., Matsuo, N., & Mayford, M. (2007). Localization of a stable neural correlate of associative memory. *Science, 317*(5842), 1230–1233. https://doi.org/10.1126/science.1143839

Rescorla, R. A. (1966). Predictability and number of pairings in Pavlovian fear conditioning. *Psychological Science, 4*(11), 383–384.

Rescorla, R. A. (1988). Pavlovian conditioning: It's not what you think it is. *The American Psychologist, 43*(3), 151–160. https://doi.org/10.1037/0003-066X.43.3.151

Rescorla, R. A. (2003). Contemporary study of Pavlovian conditioning. *The Spanish Journal of Psychology, 6*(2), 185–195. https://doi.org/10.1017/S1138741600005333

Roddick, K. M., Roberts, A. D., Schellinck, H. M., & Brown, R. E. (2016). Sex and genotype differences in odor detection in the 3x Tg-AD and 5XFAD mouse models of Alzheimer's disease at 6 months of age. *Chemical Senses, 41*(5), 433–440.

Rohrer, D. (1996). On the relative and absolute strength of a memory trace. *Memory and Cognition, 24*(2), 188–201.

Sampedro-Piquero, P., Moreno-Fernández, R. D., Mañas-Padilla, M. C., Gil-Rodríguez, S., Gavito, A. L., Pavón, F. J., Pedraza, C., García-Fernández, M., Ladrón de Guevara-Miranda, D., Santín, L. J., & Castilla-Ortega, E. (2018). Training memory without aversion: Appetitive hole-board spatial learning increases adult hippocampal neurogenesis. *Neurobiology of Learning and Memory, 151*, 35–42. https://doi.org/10.1016/j.nlm.2018.03.023

Savage, L. M., & Ramos, R. L. (2009). Reward expectation alters learning and memory: The impact of the amygdala on appetitive-driven behaviors. *Behavioural Brain Research, 198*(1), 1–12. https://doi.org/10.1016/j.bbr.2008.10.028

Scalia, F., & Winans, S. S. (1975). The differential projections of the olfactory bulb and accessory olfactory bulb in mammals. *The Journal of Comparative Neurology, 161*(1), 31–55. https://doi.org/10.1002/cne.901610105

Schellinck, H. M. (2018). Measuring olfactory processes in *Mus musculus*. *Behavioural Processes, 155*, 19–25. https://doi.org/10.1016/j.beproc.2017.08.009

Schellinck, H. M., Forestell, C. A., & Lolordo, V. M. (2001a). A simple and reliable test of olfactory learning and memory in mice. *Chemical Senses, 26*(6), 663–672. https://doi.org/10.1093/chemse/26.6.663

Schellinck, H. M., Forestell, C., Dill, P., LoLordo, V., & Brown, R. E. (2001b). The development of a simple associative test of olfactory learning and memory. In A. Marchlewska-Koj, J. Lepri, & D. Muller-Schwarze (Eds.), *Chemical signals in vertebrates 9* (pp. 459–465). Kluwer Academic.

Schultz, E. F., & Tapp, J. T. (1973). Olfactory control of behavior in rodents. *Psychological Bulletin, 79*(1), 21–44.

Siegel, S., & Castellan, N. J., Jr. (1988). *Nonparametric statistics for the behavioral sciences*. McGraw Hill.

Slotnick, B. (2001). Animal cognition and the rat olfactory system. *Trends in Cognitive Sciences, 5*(5), 216–222. https://doi.org/10.1016/s13646613(00)01625-9

Slotnick, B., & Restrepo, D. (2005). Olfactometry with mice. *Current Protocols in Neuroscience, 8*(20), 1. https://doi.org/10.1002/0471142301.ns0820s33

Slotnick, B., Schellinck, H., & Brown, R. E. (2005). Olfaction. In I. Q. Whishaw & B. Kolb (Eds.), *The behavior of the laboratory rat: A handbook with tests* (pp. 90–104). Oxford University Press.

Spence, K. W. (1952). The nature of the response in discrimination learning. *Psychological Review, 59*(1), 89–93. https://psycnet.apa.org/doi/10.1037/h0063067

Sullivan, R. M., Wilson, D. A., Ravel, N., & Mouly, A. M. (2015). Olfactory memory networks: From emotional learning to social behaviors. *Frontiers in Behavioral Neuroscience, 9*, 36.

Takeuchi, H., & Sakano, H. (2014). Neural map formation in the mouse olfactory system. *Cellular and Molecular Life Sciences, 71*, 3049–3057.

Tomczak, M., & Tomczak, E. (2014). The need to report effect size estimates revisited. An overview of some recommended measures of effect size. *Trends in Sport Sciences, 1*(21), 19–25.

Tong, M. T., Kim, T. P., & Cleland, T. A. (2018). Kinase activity in the olfactory bulb is required for odor memory consolidation. *Learning & Memory, 25*(5), 198–205. https://doi.org/10.1101/lm.046615.117

Torquet, N., Aimé, P., Messaoudi, B., Garcia, S., Ey, E., Gervais, R., Julliard, A. K., & Ravel, N. (2014). Olfactory preference conditioning changes the reward value of reinforced and non-reinforced odors. *Frontiers in Behavioral Neuroscience, 8*, 229. https://doi.org/10.3389/fnbeh.2014.00229

Vallet, G. T., Rouleau, I., Benoit, S., Langlois, R., Barbeau, E. J., & Joubert, S. (2016). Alzheimer's disease and memory strength: Gradual decline of memory traces as a function of their strength. *Journal of Clinical and Experimental Neuropsychology, 38*(6), 648–660. https://doi.org/10.1080/13803395.2016.1147530

von Hippel, W., & Hawkins, C. (1994). Stimulus exposure time and perceptual memory. *Perception & Psychophysics, 56*(5), 525–535. https://doi.org/10.3758/BF03206949

Wang, P. Y., Boboila, C., Chin, M., Higashi-Howard, A., Shamash, P., Wu, Z., Stein, N. P., Abbott, L. F., & Axel, R. (2020). Transient and persistent representations of odor value in prefrontal cortex. *Neuron, 108*(1), 209–224.e6. https://doi.org/10.1016/j.neuron.2020.07.033

Wickham, H., Averick, M., Bryan, J., Chang, W., McGowan, L. D., François, R., Grolemund, G., Hayes, A., Henry, L., Hester, J., Kuhn, M., Pedersen, T. L., Miller, E., Bache, S. M., Müller, K., Ooms, J., Robinson, D., Seidel, D. P., Spinu, V., Takahashi, K., Vaughan, D., Wilke, C., Woo, K., & Yutani, H. (2019). Welcome to the tidyverse. *Journal of Open Source Software, 4*(43), 1686. https://doi.org/10.21105/joss.01686

Wilson, D. A., & Sullivan, R. M. (2011). Cortical processing of odor objects. *Neuron, 72*(4), 506–519. https://doi.org/10.1016/j.neuron.2011.10.027

Wong, A. A., & Brown, R. E. (2007). Age-related changes in visual acuity, learning and memory in C57BL/6J and DBA/2J mice. *Neurobiology of Aging, 28*, 1577–1593.

Wong, A. A., & Brown, R. E. (2013). Prevention of vision loss protects against age-related cognitive impairment in DBA/2J mice. *Frontiers in Aging Neuroscience, 5*, 52. https://doi.org/10.3389/fnagi.2013.00052

# Putting a Price on Honesty: Methods to Evaluate the Costs of Olfactory Signalling

Christina D. Buesching[1] (✉) and Chris Newman[2]

[1] Department of Zoology, Irving K. Barber Faculty of Sciences, The University of British Columbia, Kelowna, BC, Canada
christina.buesching@ubc.ca
[2] Wildlife Conservation Research Unit, Department of Zoology, University of Oxford, Oxford, UK
chris.newman@biology.ox.ac.uk

**Abstract** Scent-marking is broadly considered to provide an honest signal of an individual's quality. Here, we use the European badger (*Meles meles*) as a well-established model species for investigating olfactory communication, particularly with their unique subcaudal gland secretion, to discuss the various costs that may be incurred by scent-marking. Auditing energy expenditures on scent-marking is complex, and thus we evaluate a variety of methods designed to assess energetic costs connected to producing scent secretions and distributing scent marks. We infer that, for badgers, in purely energetic terms, producing subcaudal gland secretion is not a handicap advertisement. Nevertheless, the dispersion and interpretation of scent marks exert a significant draw on badger overall daily energy expenditure consistent with honest fitness advertisement.

**Keywords** European badger · *Meles meles* · Scent-marking · Secretion · Energy content · Energy expenditure · Energy budget · Subcaudal gland

## 1 Introduction

Evolutionary theory predicts that behaviours and physiological processes that increase fitness are selected for, while those that are neutral or incur costs outweighing fitness benefits will be selected against and consequently disappear over time. In the natural world, fitness is ultimately determined through the common currency of energy acquisition (Chimienti et al., 2020), and how the subsequent expenditure of that energy is budgeted. Much of this energy expenditure is needed for cryptic, homeostatic functions



B. Schaal et al. (eds.), *Chemical Signals in Vertebrates 15*,
https://doi.org/10.1007/978-3-031-35159-4_4

connected to the ongoing survival of the individual (i.e., somatic fitness: Stearns, 1992); however, surpluses permit elective investments into reproduction, perpetuating the individual's genetic lineage (i.e., reproductive fitness: Hirschfield & Tinkle, 1975; Williams, 1996). Furthermore, current success may be sacrificed to 'accrue capital' for the future (Campbell et al., 2017) in order to live longer or breed later, if this strategy is likely to enhance overall lifetime breeding success (Sugianto et al., 2021). This creates a dilemma for animals where a proportion of their spending should be conspicuous, aimed at advertising their resource holding potential (RHP *sensu* Parker, 1974; see Kilshaw et al., 2009; Tinnesand et al., 2015) or attracting mates through honest (i.e., unfakable) signals of desirability (Kodric-Brown & Brown, 1984). Even profligate spending on costly 'handicaps' has evolved through sexual selection, to demonstrate to potential mating partners that the advertising individual can survive and thrive despite otherwise wasteful commitment of energy (Zahavi, 1975). Conversely, attracting attention can draw competition, even predation (Montgomerie & Thornhill, 1989; Hughes et al., 2010), or counter-productively reveal markers of ill-health (Newman & Buesching, 2019). Ultimately, the ability to invest energy is also constrained by ecological factors, such as the range of food types the species consumes, its body size, social structure, mating system, predation pressures, home range area, etc. (González-Suárez et al., 2013; Sanders et al., 2018).

Because scent-marking, which often involves the production and distribution of specialised secretions, and elaborate scent-marking behaviours incur energetic costs that may be further exacerbated if scent-marking increases the risk of detection by predators/competitors (reviewed in Buesching & Jordan, 2022), it is widely assumed that only 'fit' individuals (i.e., healthy animals in good condition) possess sufficient surplus energy to invest in specialised scent-secretion and undertake and/or intensify marking activity while maintaining their somatic requirements (Gosling et al., 2000). Marking activity as well as the chemical composition of glandular scent-secretions should thus correlate with marker 'quality'—a concept known as 'index', allowing conspecifics to infer fitness-related information about the signaller (Számadó, 2011). Indeed, inter-individual differences in scent-marking behaviour, glandular activity and secretion composition have been shown in a wide variety of species to correlate with physiological fitness indicators (Buesching & Stankowich, 2017), corresponding with variation in conspecific responses (Petrulis, 2013).

Here, we review the costs incurred by olfactory communication and how the invested energy can be measured, highlight the limitations of these methods and discuss if scent-marking can function as an honest fitness advertisement signal using the group-living European badger, *Meles meles*, in the UK (Macdonald et al., 2015), as a model.

### 1.1 Honest Signals, Handicap Trade-Offs and Social Capital

The 'Truth in advertising model' (Kodric-Brown & Brown, 1984) proposes that sexual selection favours the display of, particularly, male traits that vary phenotypically within and among individuals in a way that maintains a positive correlation between the degree of expression of the traits and overall genetic fitness. Because the exaggerated expression of such traits requires the allocation of limited resources to reproduction, it reflects age, nutritional condition, social status and resistance to predators and pathogens. It is thus advantageous for females to choose mates with these exaggerated traits because these traits reliably advertise desirable attributes that can be passed on to both male and female offspring. Indeed, in badgers, males produce more and lighter coloured subcaudal gland secretion than females, where secretion volume and colour also correlates with male body condition (Buesching et al., 2002a).

As an elaboration of this concept, Zahavi (1975) proposed that elaborate signals that are both costly and wasteful evolved to enforce honesty: so-called signal selection. Grafen (1990) subsequently vindicated Zahavi's handicap principle with a series of mathematical models, and this became the dominant paradigm for explaining the evolution of honest signalling in the biological and social sciences (Maynard Smith, 1985). Recent re-analysis by Penn and Számadó (2020) has, however, revealed that Grafen's models do not support the handicap hypothesis, although they do corroborate that males adjust their investment into sexual signals according to their condition and ability to bear the costs as well as the risks to their survival. Thus, rather than being wasteful over-investments, honest signals evolve in this scenario because selection favours efficient and optimal investment into signal expression while minimising signalling costs—a subtle but important difference to the handicap hypothesis. Indeed, theoretical studies have shown that signalling costs paid at the equilibrium are neither sufficient nor necessary to maintain signal honesty, and that honesty can evolve through differential benefits, as well as differential costs (Számadó, 2011).

### 1.2 Assessing Costs and Benefits

But how can we test and audit the evolutionary economics of spending on scent communication? Vitally, this has three facets, because it is not only a question of (i) accounting for spending on advertising, but also (ii) needs to take into account the desired (but not guaranteed) benefits of the targeted recipient(s) (e.g., a suitable mating partner) investing time into searching out and 'reading' these scent marks as well as (iii) additional potential costs/risks of predators or competitors finding this information, where the unintended disclosure of the marker's whereabouts to non-target recipients may prove dangerous or even lethal. In terms of energetic investments, fundamentally two cost sources are involved, namely activity (i.e., scent-marking behaviour) and physiology (i.e., secretion production).

## 2 Methods to Monitor Scent-Marking Behaviour

### 2.1 Tracking

An iterative approach to quantifying movement involves actually tracking the movements of the individual and applying some average rate of energy expenditure to that activity, often transposing rates calculated for domestic analogue animals (using respirometry; McNab, 2006). Herein the problem becomes how to decipher what movement is being committed solely to scent communication and what is involved in foraging trips, seeking mates and dispersal. Important is to ascertain when and where the animal is scent-marking and whether it is actively reading scents.

This is made easier when species have highly predictable centres of activity, for example dens or latrines, which enables the use of remote monitoring technologies capable of detecting scent-marking activity at fixed places.

### 2.2 Video Surveillance

Remote infra-red video surveillance technology has advanced considerably over the past two decades and is used extensively to record natural scent-marking as well as reactions to experimental scent provisioning. For example, video surveillance of the scent-marking behaviour of individually marked badgers using VHS cassette tapes powered by a truck battery (Stewart et al., 1997a) at two communal dens (termed 'setts') between April 1996 and June 1997 recorded 442 incidences of object-marking with subcaudal gland secretion from 20 males and 22 females (Buesching & Macdonald, 2004). The frequency of object-marking varied significantly in relation to season, sex, age and reproductive status. In both sexes, relative object-marking rates were highest during the mating season, when individuals in reproductive condition marked significantly more often than non-reproductive animals. Females marked at a significantly higher rate than males during the cub rearing season, and, in both sexes, adults scent-marked significantly more frequently than younger individuals. Approximately 30% of all scent-marks received an over-mark within 24 h of their deposit, highlighting the reciprocal interest in marking. In males, over-marking behaviour was recorded only during the mating season, whereas females over-marked in all seasons at equal rates. Simultaneously, this video surveillance also revealed 3021 instances of allo-marking, highlighting the importance of scent-marking not only in the context of resource and mate defence, but also for social group-cohesion on appeasement behaviour amongst conspecifics (Buesching et al., 2003). Interestingly, 95% ($n = 2866$) of allo-markings were sequential with one badger marking the body of another, and only 5% ($n = 155$) were mutual, in which two badgers pressed their subcaudal glands together (Buesching et al., 2003). However, while sequential allo-marking was strongly seasonal and depended on social context, mutual allo-marking occurred at the same low rates between all individuals throughout the year. Because mutual allo-marking leads to an exchange of microbiota between the marking partners and the actual odour-profile of scent-marks is strongly affected by microbial activity in the secretion (Sin et al., 2012; Buesching et al., 2016a), this behaviour likely helps to create a shared group odour. It, however, also carries the risks of losing individual-specific information in scent profiles due to over-assimilation of the microbial community as well as the risk of pathogenic infection (Newman & Buesching, 2019).

Nevertheless, while remote video recordings can capture an accurate record of scent-marking activity at specific sites unencumbered by the presence of human observers, they are constricted by the field of view of the camera and behaviour outside this limited area is not being captured. Video recording does, however, also allow analysis of the time dedicated to olfactory communication (i.e., scent-marking and sniffing) at the various locations (e.g., the den, latrines, feeding sites, etc.). This approach has revealed in badgers that some individuals often make specific visits to their home sett and territorial border latrines as well as (already depleted) feeding sites several times in the same night with the sole purpose of scent-(over-)marking and sniffing previous marks, likely to gage the reaction of conspecific to their earlier marks (Buesching, unpubl. data). While scent-marking in itself does not take longer than a few milliseconds in badgers, these special journeys are likely a much larger proportion of their nightly time-budget.

Video observations also allow for targeted scent provisioning experiments to record the response behaviour of individuals to scent from conspecifics of known status. In badgers, we used high-resolution infra-red flash, rapid trigger trail cameras, to investigate how conspecifics respond to specific secretions from known donors. We presented anal gland secretions (AGS) ($n = 351$) collected from known individuals under sedation to identifiable recipients ($n = 187$), at each of 11 different social groups over 3 days, to assess response-variation according to familiarity (own-group, neighbours, strangers) and spatial context (in-context: at a shared border; out-of-context: at an unshared border/the main sett). At the group level, the durations of sniffing responses were longer when the sample was collected from less familiar individuals, being briefest to own group AGS (mean ± SD = 1.44 s ± 2.58), intermediate to neighbouring group samples (mean ± SD = 2.71 s ± 2.56), and most prolonged to stranger AGS (mean ± SD = 4.29 s ± 2.60; Noonan et al., 2019; Tinnesand et al., 2015). Similarly, the likelihood of over-marking decreased in the order: strangers, neighbours, own group. These results are congruent with the 'dear enemy phenomenon' (Műller & Manser, 2007), and again illustrate the concern and effort badgers put into their scent communication. It is, however, impossible from these recordings to ascertain how far the scent is discernible to badgers under different weather and soil conditions, and data analysis is limited to those individuals that approach the provided scent sample closely enough to be captured (and identified) on camera.

## 2.3 Active Radio Frequency Identification (aRFID)

The use of active tags that read against static detectors is well established in the security industry and a similar set-up, fitting tags to animals, can be used to read their presence at dens and latrines equipped with automated base stations. In a novel application of this approach to badger activity (Ellwood et al., 2017), involving 32 collared badgers from 7 social groups over 13 weeks, we found that each badger spends on average 12 h per week in the vicinity of latrines, far longer than perfunctory defecation and urination would require. This implies that badgers are taking time to investigate scent marks left by conspecifics. Moreover, of 1834.1 h of detections ($n = 161{,}333$ events), 43.9% were at border latrines (males: 297.6 h, females: 507.3 h), during which time badgers were detected in the simultaneous company of members of other groups for 16.1% of their co-location time (mean ± SE = 0.16 h dyad$^{-1}$ week$^{-1}$ ± 0.02). The lack of any bite wounds on subject badgers at subsequent trapping sessions revealed that these intergroup interactions were not aggressive, contrary to territorial orthodoxy. Nevertheless, without concurrent behavioural observations, aRFID does not exclude the possibility that badgers spend a disproportionate amount of time feeding in the vicinity of latrines or simply wait to meet members from the same or neighbouring groups, e.g., for mating.

### 2.4 Overall Dynamic Body Acceleration

Adopting terminology from classical flow field theory (Bassing et al., 2022), the aRFID and video surveillance approaches described above, are 'Eulerian'; that is, they involve observation of moving targets from a static point. More conventional telemetry, from Video Home System (VHS) to Global Positioning System (GPS), is 'Lagrangian', that is, moving with the trajectory of the target to trace its path (Byrne et al., 2014). But simply knowing where an animal is, outside of the context of latrine use, tells us little about what it does. Recently, however, advances in bi- and tri-axial accelerometry have enabled not only the intensity of activity to be quantified through Overall Dynamic Body Acceleration (ODBA) (Wilson et al., 2006; Shepard et al., 2008; Bright Ross, 2021), but also that particular acceleration profiles can be 'classified' through preliminary study, cross-referencing against concomitant postural changes observed on video surveillance or through instrumenting captive individuals (Ladds et al., 2017). Squatting to subcaudal mark, urinate or defecate creates a distinct 'trace' within the accelerometry record, enabling each event to be singled out and collated.

Measuring ODBA from 63 individual badgers, we found substantial variation between individuals in all seasons, with the most active badger expending 1.6 times as much as the least active in spring, 2.2-fold in summer, and 2.1-fold in autumn (Bright Ross et al., 2020). Many factors could account for these differences, but differential investment into scent-marking is likely to be amongst them as badgers with higher energy budgets also produced more subcaudal secretion. Nevertheless, the behavioural validation of ODBA patterns is labour-intensive and ODBA patterns of functionally (very) different behaviours (e.g., touching the nose to the ground to feed vs. to sniff a scent-mark) can be extremely difficult to discriminate.

## 3 Methods to Monitor Energy Used in Scent-Production

### 3.1 Respirometry

The most accurate way to measure metabolic rate is to measure oxygen consumption in a respirometry chamber (McNab, 2006). For intractable wild animals, however, this is generally not practicable; either small animals become distressed and hyperactive, or larger animals must be sedated and become subdued (McClune et al., 2015). Besides, although roughly indicative, this respirometry value only applies while the animal is constrained within the apparatus. Particularly in larger species, this precludes measurement of metabolic costs for a wide range of natural behaviours, typically also including scent-marking.

Historically, an alternative method for calculating energy budgets in the wild involved dietary analyses, attempting to quantify 'food-in' as a proxy for potential 'expenditure-out', although the accuracy of this method is susceptible to energy being banked in fat stores and improved somatic condition (Ross et al., 2021) or lost through thermoregulation or immunological activity. Recently, a new approach has been developed that could be applied to elucidate the direct physiological investments into scent production and marking behaviour: doubly labelled water (DLW).

### 3.2 Doubly Labelled Water (DLW)

While ODBA (see Sect. 2.4) correlates with total metabolism in a wide range of species (Elliott et al., 2013; Lyons et al., 2013; Miwa et al., 2015), particularly at short timescales (Dalton et al., 2014), it cannot account for energy used in homeostasis, elevated metabolic rate (respiration) and biosynthesis. A more comprehensive assessment of 'daily energy expenditure' (DEE), in kilojoules/day, can be achieved by using a metabolic isotope marker technique (Butler et al., 2004), for example 'doubly labelled water' (Speakman, 1997).

In contrast to normal water, DLW features a higher representation of heavy $H^2$ and $O^{18}$ (hence the name doubly labelled (with 650,765 ppm $O^{18}$ and 342,395 ppm $H^2$) rather than conventional $H^1$ and $O^{16}$ isotopes. The administration procedure involves injecting the animal with a harmless dose of DLW, allowing 3 h to give the isotope time to equilibrate in the blood, then drawing a first blood sample. Circa 1 week later, a second blood sample is required, and thus the recapture of that same individual in a wild scenario. The change in ratios of natural to heavy isotopes can then be calculated to estimate DEE. $H^2$ is depleted through water loss (urination, panting) and $O^{18}$ eliminated through both water loss and $CO_2$ respiration. Thus, the total $CO_2$ respired over the intervening days can reveal the amount of joules burnt. However, this method can be logistically difficult to use in the wild: if the individual is recaptured too late and all the DLW has depleted, any before and after contrast is precluded and DEE cannot be calculated. In addition, only the overall expenditure can be measured, but energy allotment to different activities (such as scent-marking) or physiological processes (such as scent production) cannot be determined. Furthermore, DLW is expensive, costing £33.18 per gram, or £110–151 to dose badgers in the 6-to-10-kg range.

Nevertheless, working with a sub-sample of the same badgers used in the ODBA assessment above, we found that DEE also varied 1.3-fold between individuals in spring (averaging = 3013 kJ $day^{-1}$ ± SD = 240 kJ $day^{-1}$), and twofold in both summer (averaging 3049 kJ $day^{-1}$ ± 630 kJ $day^{-1}$) and autumn (averaging 3483 kJ $day^{-1}$ ± 767 kJ $day^{-1}$). Heavier badgers expended up to *c.* 1500 kJ $day^{-1}$ more DEE than did lighter badgers. Pooling all seasons, DEE scaled to the 0.62 power with body mass, very close to the mammalian mean (see below). More interestingly, lighter badgers (with their fundamentally lower DEE) spent a higher proportion of their total energy expenditure, and thus stayed light; including all females that had lactated during spring, which had ODBA values on average 1.3 times higher than other badgers. Older badgers exhibited lower DEE than did younger ones (e.g., on average, 9-year-olds expended 627 kJ $day^{-1}$ less than 2-year-olds) likely due to diminished capacity to generate heat (McDonald & Horwitz, 1999) and maintain 'normothermia' (i.e., to thermoregulate autonomically; Terrien et al., 2011; Nybo et al., 2014). Differential investment into the biosynthesis of subcaudal gland secretions and the deposition of (subcaudal) scent-marks is one (among several) candidate costs that might explain this variation.

However, even if two individuals, matched by age and sex, may expend the same amount of energy, this does not imply that they are spending it on the same activities (see Noonan et al., 2014). Particularly, in the arena of scent communication the potential for non-equivalence abounds: some individuals may predominantly invest energy into scent-marking, others into seeking out and sniffing scent marks from conspecifics, where either activity may exert similar costs—particularly as both may necessitate special journeys to latrines (see above) taking time away from feeding.

## 4 Methods to Measure Energy Contained in Scent Marks

Sources of scent can be broadly classified into three categories: the use and dispersion of metabolite waste products (i.e., excreta: urine and faeces) that carry none or little production cost; the use of glandular secretions that are produced by the body for a different primary purpose not connected to scent-marking (e.g., saliva); and the use of specialised glandular secretions produced solely/primarily for the purpose of scent-marking, which require energy investment to synthesise (e.g., anal secretions; Maynard Smith, 1985; Grafen, 1990; but see Számadó, 2011).

### 4.1 Metabolic Waste Products

Implicitly, there is a fundamentally 'honest' component to excreta, where an individual obtaining insufficient food or water will suffer a compromised ability to scent-mark with faeces or urine, irrespective of the qualitative malnutrition cues revealed. Because some species may convey their resource-holding potential through the frequency/number of urinations and/or faecal deposits (e.g., mustelids: Buesching & Stankowich, 2017), for example in badgers, counting scats or recording urine-marking activity can reveal population density (e.g., Tuyttens et al., 2001) as well as temporal (e.g., Tsunoda et al., 2019) and spatial patterns of group (e.g., Buesching et al., 2016b) or individual (Kilshaw et al., 2009) marking activity.

Nevertheless, there are indications that some species may be able to purposefully alter the size of their faecal/urinary deposits by producing several mini-scats instead of a single normal-sized scat (Buesching et al., 2016b) or emptying their bladder not in a single bout but through many 'dribble urinations' (Buesching & Macdonald, 2004). Thus, the same total quantity of material may be distributed in different ways (Buesching & Jordan, 2022). Several dispersed small scent marks will invariably cost more to distribute than fewer larger ones (Buesching et al., 2016b), especially so if some aspect of territorial marking requires that individual to travel a long distance from where it is feeding. For example, in badgers, maintaining latrine-spacing patterns appears more important than the actual number of fresh faeces deposited in each latrine (Kilshaw et al., 2009) with neighbouring groups seemingly attempting to match the number of faeces deposited by the other group (Stewart et al., 2001). Consequently, where territories are small and groups large, the capacity to produce faeces may exceed the minimum need for perimeter marking, resulting in faecal redundancy and large latrines. In contrast, in larger territories, especially when occupied by smaller groups, badgers may experience faecal constraint, thus maintaining smaller latrines (Buesching et al., 2016a, b).

### 4.2 Glandular Secretions

But it is the deliberate investment of energy into producing and distributing bespoke glandular secretions that should be most suitable to convey a directed honest signal of the individual's fitness (Gosling & Roberts, 2001). Various mammals produce scent secretions from specialised facial glands (e.g., Alpine Marmots: Bel et al., 1999; Tigers: Smith et al., 1989), sternal glands (e.g., Koalas: Salamon & Davies, 1998; Mandrill: Setchell et al., 2010) and various skin glands (Mykytowycz & Goodrich, 1974), and all carnivores produce anal gland secretions deposited along with faeces (Gorman & Trowbridge, 1989). In badgers, the volume of subcaudal gland secretion in the subcaudal pouch correlates with body condition and reproductive status (Buesching et al., 2002) as well as subcaudal marking activity (Buesching & Macdonald, 2004). Indeed, individuals below a necessary threshold of somatic condition, or during immaturity and senescence, may be unable to produce scent secretions, or produce restricted quantities (Sugianto et al., 2019). Nevertheless, water content in subcaudal secretions varies considerably between individuals and sampling events (Allen et al., 2023), and thus it is likely that secretion volume does not correlate tightly with energetic investment and therefore—by extension—absolute frequency of glandular-marking activity in itself is likely not a reliable indicator of the marker's fitness, necessitating the analysis of the chemical composition of the scent.

### 4.3 Chemical Analyses

Glandular secretions typically contain a complex mixture of volatile organic compounds (VOCs), non-volatile organic compounds (non-VOCs) and proteins (Apps, 2016). Various methods (e.g., gas chromatography coupled with mass spectrometry, GC-MS) can measure the abundance of specific component classes but require preprocessing techniques for their isolation. Thus, to analyse complex chemical mixtures, a multitude of steps—often involving several different laboratory methods—are required (Drea et al., 2013; Nonnamaker et al., 2023). This necessitates large quantities of sample material, where glandular scent secretions are typically produced in small amounts and difficult to collect in vivo. Exceptions are the proximate-composition analysis (PCA) that measures the percentages of water, ash, proteins, lipids and carbohydrates, but requires five different, consecutive processes (where, based on the compound combustion equivalents, absolute energetic content can be calculated: Doyle et al., 2007), and high-performance liquid chromatography that—without any further processing—only determines presence/absence of different component classes/chemical compounds. Quantification results differ considerably between methodologies, and, importantly, all traditional methods measure relative rather than absolute compound amounts, thus requiring further statistical data normalisation and analysis (Noonan et al., 2018). Although energy content can then be calculated from the chemical structure of the individual compounds, biosynthesis costs cannot be inferred from any of these methods. In badgers, liquid extraction GC-MS and headspace GC-MS have shown that the chemical composition of (non-)VOCs of subcaudal (Buesching et al., 2002b, c) as well as anal (Tinnesand et al., 2015; Noonan et al., 2019) gland secretions are highly individual-specific, and vary according to fitness-related parameters, such as age, sex, reproductive status and body condition, where scent provisioning experiments (Allen et al., 2023; Tinnesand et al., 2015) show that badgers can decipher this information and moderate their response behaviour accordingly. But as GC-MS only measures relative, and not absolute, amounts of each compound in the secretion mixture, and is not suited to analyse proteins, it is impossible to infer the overall energetic investment into scent production in this way.

## 4.4 Bomb Calorimetry

A different approach, well established in the food industry, is bomb calorimetry. Briefly, the bomb calorimeter technique involves burning freeze-dried samples in pure oxygen, encased in a stainless steel 'bomb' placed in an insulated container holding 2424 mL of water. Heat transferred to the surrounding water is measured in °C, from which the calorific content of the secretion can be calculated, where one calorie is defined as the heat that must be absorbed by 1 gram of water to raise its temperature by 1 °C, and where one calorie = 4.184 joules. As described elsewhere in this volume (Allen et al., 2023), we adapted this approach to estimate the energy content (calorific/joule content) of badger subcaudal gland secretions, which contain predominantly long-chained fatty acids, protein and water (Buesching et al., 2002b, c). Since synthesising long-chained fatty acids and proteins requires energetic input from metabolic processes (Cui et al., 2016; Pontes et al., 2015), this composition suggests that secretion production is energetically costly. Accounting for natural levels of hydration, subcaudal gland secretion has a median value of 6525 J/mL for males or 6774 J/mL for females—around 1.5 calories (1 gram of solid undiluted fat would contain *c.* 9 calories). Typically, a healthy adult badger's subcaudal gland will contain anything from 0.5 to 2.5 mL of secretion (Buesching et al., 2002a), and this volume can be regenerated within a day. Nevertheless, although energy content of badger subcaudal gland secretion varies according to sex and reproductive status, and badgers can discriminate between samples according to relative energy richness (see Allen et al., 2023) , bomb-calorimetry neither permits to infer biosynthesis costs, nor does it permit to account for lost energy content due to microbial activity (Sin et al., 2012; Buesching et al., 2016b). It also does not provide any information on the calorific contribution of 'waste products' from other physiological processes included in the scent profiles (e.g., sex steroids).

## 5 Further Complications: Seasonality, Life-History Trade-Offs and Scale Laws

Almost without exception, the established techniques outlined above are best applied in the short term, able to audit energy expended on scent-marking through a small window of time. Yet rates of investment into olfactory communication vary according to season, reproductive investment, body condition and throughout the lifetime of the individual (Buesching et al., 2002; McLean et al., 2012; Vogt et al., 2014). Furthermore, an individual's pace of life is generally a product of those social and environmental conditions it experiences (Sugianto et al., 2019; Bright Ross et al., 2020). For example, in our badger study we found that sex ratio and sex-specific population density explained 52.8% (males) and 91.0% (females) of variance in the ratio of fertility to age at first reproduction ($F/\alpha$) among adults relative to the long-term population median $F/\alpha$. For females, low same-sex density favours early primiparity whereas up to 10% of male yearlings mated successfully at high same-sex densities. In so far as mating advertisement is achieved through scent communication in badgers (Buesching & Macdonald, 2001), these life-history factors will affect lifetime energy investment schedules profoundly.

Unpredictable events can also disrupt the supply of residual energy to invest into glandular secretions for scent-marking. For example, immune responses, particularly fever, to combat pathogenic/parasitic challenges (which vary spatially and temporally) (Lochmiller & Deerenberg, 2000), or events evoking a hypothalamic pituitary adrenal (HPA) axis response (Loucks, 2009), are extremely costly (Segerstrom, 2007), albeit with the benefit of prolonging survival for future breeding seasons, or minimising the extent to which the individual's social behaviour is compromised (Cotter & Kilner, 2010). These stressors can again be quantified, although not directly, into kilojoules without the use of respirometry. For example, in badgers a measure of chronic stress (McLaren et al., 2003; Montes, 2008) can be achieved by measuring leukocyte activation and established assays for measuring oxidative stress and damage (Bilham et al., 2018), where, interestingly, vitamin-E analogue provides a reliable gauge of non-enzymatic antioxidant capacity (Bilham et al., 2013). Given that vitamin-E is a major component of subcaudal gland secretion, likely functioning as an antioxidant preservative of its chemical profile (Sin et al., 2016), it is currently impossible to determine if it is produced and secreted specially (and thus incurring an energetic cost) to enhance the lifespan of subcaudal scent marks or if its inclusion in subcaudal secretion is purely a by-product of other physiological processes.

Beyond badgers, scale allometries must also be considered (Dial et al., 2008): a smaller animal loses heat more quickly per unit mass than a larger one (Klieber's law: Riek & Geiser, 2013, where basal metabolic rate scales as approximately $M^{0.75}$ in mammals; Duncan et al., 2007); however, a large animal requires more energy to move each unit of mass than does a smaller one (Alexander, 2013; Scaling at around M0.67 across mammals; Heusner, 1991). In turn, allometries also have implications for territory scaling, and the proportion of edge that needs marking to defend against intruder, affecting population connectivity (Hartfelder et al., 2020).

# 6 Conclusions

Is it then possible, or even feasible, to fully audit energy expended on olfactory communication to determine if this pays a rational fitness dividend? As discussed, variables are complex, units of expenditure do not always easily resolve to joules and random factors (e.g., an otherwise fit individual succumbs to a traffic fatality; inclement weather impacts offspring survival) will always test the strength of sample sizes. Nevertheless, a relative comparison of an individual's scent-marking effort and the energy content of its scent gland secretions against conspecifics is theoretically possible, along with an assessment of the general energy expenditures that individual makes to homeostasis, health and movement. In the badger model, given that the average badger expends around 3250 kJ per day (around 777 calories, a similar-sized dog would need around 400 calories per day), then the 6.6 kJ it commits to subcaudal gland secretion (not including biosynthesis costs) are trivial at around 0.2% of its DEE: that is, subcaudal gland secretion has not evolved in badgers as a means to advertise ability to withstand handicap. Likely far more significant are the 12 h per week, or 1 h 45 min per day, badgers average at latrines. With a typical nocturnal activity window of around 10 h, this suggests they spend 17.5% of their time (and thus similar energy) focused on scent communication; this is notwithstanding time spent scent-marking at setts. Moreover, those latrines badgers visit and transit between per night are often not close to their active foraging sites, possibly up to 1 km away in the Wytham Woods study population (Kilshaw et al., 2009), where marking at key latrines seems linked to coordinated group range marking and communicating with neighbours (Stewart et al., 2001), although, notably, not in the context of animosity or defence. Perhaps tellingly, almost 50% of badger cubs are fathered by males from neighbouring groups, indicating the potential importance of widespread olfactory advertising.

Ultimately, the response variable that best measures fitness is lifetime reproductive success; simply do individuals that spend more on scent-marking honestly leave more descendants than their contemporaries that invest less? However, even this is not a straightforward causal correlation, given those other manifest factors, from predation to disease, that can influence the fitness equation. For example, in badgers, mustelid (MuSGHV-1), a sexually transmitted infection, reduces fertility (Tsai et al., 2020), and so, counterproductively, (esp. mutual) scent-marking to solicit mating risks lowering the marking individual's reproductive success. It thus seems that, for scent-marking, the truth in advertising model (Kodric-Brown & Brown, 1984) is better able to resolve energetics, attempting to correlate whether individuals that spend the most on advertising actual do have the richest energy budget to draw upon, possibly even overcoming scent handicaps.

**Acknowledgements** We would like to thank Prof. David W. Macdonald for his many years managing the Wytham Badger Project and the many students and collaborators who contributed energy, enthusiasm, ideas and data to the project.

## References


Alexander, R. M. (2013). Body support, scaling, and allometry. In M. Hildebrand, D. M. Bramble, K. F. Liem, & D. B. Wake (Eds.), *Functional vertebrate morphology* (pp. 26–37). Harvard University Press.

Allen, T. M., Newman, C., Macdonald, D. W., & Buesching, C. D. (2023). The use of excretion- and secretion-based scent communication in European badgers (meles meles). In B. Schaal, D. Rekow, M. Keller, & F. Damon (Eds.), *Chemical signals in vertebrates 15*. Springer.

Apps, P. (2016). Does deconvolution help to disentangle the complexities of mammal odors? In B. A. Schulte, T. E. Goodwin, & M. H. Ferkin (Eds.), *Chemical signals in vertebrates 13* (pp. 415–433). Springer.

Bassing, S. B., Devivo, M., Ganz, T. R., Kertson, B. N., Prugh, L. R., Roussin, T., et al. (2022). Are we telling the same story? Comparing inferences made from camera trap and telemetry data for wildlife monitoring. *Ecological Applications, 33*, e2745.

Bel, M. C., Coulon, J., Sreng, L., Allainé, D., Bagnères, A. G., & Clement, J. L. (1999). Social signals involved in scent-marking behavior by cheek-rubbing in Alpine marmots (Marmota marmota). *Journal of Chemical Ecology, 25*, 2267–2283.

Bilham, K., Sin, Y. W., Newman, C., Buesching, C. D., & Macdonald, D. W. (2013). An example of life history antecedence in the European badger (Meles meles): rapid development of juvenile antioxidant capacity, from plasma vitamin E analogue. *Ethology Ecology & Evolution, 25*(4), 330–350.

Bilham, K., Newman, C., Buesching, C. D., Noonan, M. J., Boyd, A., Smith, A. L., & Macdonald, D. W. (2018). Effects of weather conditions on oxidative stress, oxidative damage, and antioxidant capacity in a wild-living mammal, the European badger (*Meles meles*). *Physiological and Biochemical Zoology, 91*(4), 987–1004.

Bright Ross, J. (2021). *The effects of energetic expenditure tactics and life-history variability on European badger (Meles meles) ecology*. Doctoral dissertation, University of Oxford

Bright Ross, J. G., Newman, C., Buesching, C. D., & Macdonald, D. W. (2020). What lies beneath? Population dynamics conceal pace-of-life and sex ratio variation, with implications for resilience to environmental change. *Global Change Biology, 26*(6), 3307–3324.

Buesching, C. D., & Jordan, N. R. (2022). The function of carnivore latrines: Review, case studies, and a research framework for hypothesis testing. In *Small carnivores: Evolution, ecology, behaviour, and conservation* (pp. 131–171). Wiley Online Library.

Buesching, C. D., & Macdonald, D. W. (2001). Scent-marking behaviour of the European badger (*Meles meles*): Resource defence or individual advertisement? In A. Marchlewska-Koj, J. J. Lepri, & D. D. Müller- Schwarze (Eds.), *Chemical signals in vertebrates 9* (pp. 321–327). Kluwer Academic/Plenum Publishers.

Buesching, C. D., & Macdonald, D. W. (2004). Variations in scent-marking behaviour of European badgers Meles meles in the vicinity of their setts. *Acta Theriologica, 49*(2), 235–246.

Buesching, C. D., & Stankowich, T. (2017). Communication amongst the musteloids: Signs, signals, and cues. In D. W. Macdonald, C. Newman, & L. A. Harrington (Eds.), *Biology and conservation of musteloids*. Oxford University Press.

Buesching, C. D., Newman, C., & Macdonald, D. W. (2002a). Variations in colour and volume of the subcaudal gland secretion of badgers (*Meles meles*) in relation to sex, season and individual-specific parameters. *Mammalian Biology, 67*, 147–156.

Buesching, C. D., Waterhouse, J. S., & Macdonald, D. W. (2002b). Gas-chromatographic analyses of the subcaudal gland secretion of the European badger (*Meles meles*) part I: Chemical differences related to individual parameters. *Journal of Chemical Ecology, 28*(1), 41–56.

Buesching, C. D., Waterhouse, J. S., & Macdonald, D. W. (2002c). Gas-chromatographic analyses of the subcaudal gland secretion of the European badger (*Meles meles*) part II: Time-related variation in the individual-specific composition. *Journal of Chemical Ecology, 28*(1), 57–69.

Buesching, C. D., Stopka, P., & Macdonald, D. W. (2003). The social function of allo-marking in the European badger (*Meles meles*). *Behaviour, 140*(8), 965–980.

Buesching, C. D., Newman, C., Service, K., Macdonald, D. W., & Riordan, P. (2016a). Latrine marking patterns of badgers (*Meles meles*) with respect to population density and range size. *Ecosphere, 7*(5), e01328.

Buesching, C. D., Tinnesand, H. V., Sin, Y., Rosell, F., Burke, T., & Macdonald, D. W. (2016b). Coding of group odor in the subcaudal gland secretion of the European badger *Meles*: Chemical composition and pouch microbiota. In B. A. Schulte, T. E. Goodwin, & M. H. Ferkin (Eds.), *Chemical signals in vertebrates 13* (pp. 45–62). Springer.

Butler, P. J., Green, J. A., Boyd, I. L., & Speakman, J. R. (2004). Measuring metabolic rate in the field: The pros and cons of the doubly labelled water and heart rate methods. *Functional Ecology, 18*(2), 168–183.

Byrne, A. W., Quinn, J. L., O'Keeffe, J. J., Green, S., Paddy Sleeman, D., Wayne Martin, S., & Davenport, J. (2014). Large-scale movements in European badgers: Has the tail of the movement kernel been underestimated? *The Journal of Animal Ecology, 83*(4), 991–1001.

Campbell, R. D., Rosell, F., Newman, C., & Macdonald, D. W. (2017). Age-related changes in somatic condition and reproduction in the Eurasian beaver: Resource history influences onset of reproductive senescence. *PLoS ONE, 12*(12), e0187484.

Chimienti, M., Desforges, J. P., Beumer, L. T., Nabe-Nielsen, J., van Beest, F. M., & Schmidt, N. M. (2020). Energetics as common currency for integrating high resolution activity patterns into dynamic energy budget-individual based models. *Ecological Modelling, 434*, 109250.

Cotter, S. C., & Kilner, R. M. (2010). Personal immunity versus social immunity. *Behavioral Ecology, 21*(4), 663–668.

Cui, W., Liang, Y., Tian, W., Ji, M., & Ma, X. (2016). Regulating effect of β-ketoacyl synthase domain of fatty acid synthase on fatty acyl chain length in de novo fatty acid synthesis. *Biochimica et Biophysica Acta - Molecular and Cell Biology of Lipids, 1861*(3), 149–155.

Dalton, A. J., Rosen, D. A., & Trites, A. W. (2014). Season and time of day affect the ability of accelerometry and the doubly labeled water methods to measure energy expenditure in northern fur seals (Callorhinus ursinus). *Journal of Experimental Marine Biology and Ecology, 452*, 125–136.

Dial, K. P., Greene, E., & Irschick, D. J. (2008). Allometry of behavior. *Trends in Ecology & Evolution, 23*(7), 394–401.

Doyle, T. K., Houghton, J. D., McDevitt, R., Davenport, J., & Hays, G. C. (2007). The energy density of jellyfish: Estimates from bomb-calorimetry and proximate-composition. *Journal of Experimental Marine Biology and Ecology, 343*(2), 239–252.

Drea, C. M., Boulet, M., Delbarco-Trillo, J. A., Greene, L. K., Sacha, C. R., Goodwin, T. E., & Dubay, G. R. (2013). The "secret" in secretions: Methodological considerations in deciphering primate olfactory communication. *American Journal of Primatology, 75*(7), 621–642.

Duncan, R. P., Forsyth, D. M., & Hone, J. (2007). Testing the metabolic theory of ecology: Allometric scaling exponents in mammals. *Ecology, 88*(2), 324–333.

Elliott, K. H., Le Vaillant, M., Kato, A., Speakman, J. R., & Ropert-Coudert, Y. (2013). Accelerometry predicts daily energy expenditure in a bird with high activity levels. *Biology Letters, 9*(1), 20120919.

Ellwood, S. A., Newman, C., Montgomery, R. A., Nicosia, V., Buesching, C. D., Markham, A., et al. (2017). An active-radio-frequency-identification system capable of identifying co-locations and social-structure: Validation with a wild free-ranging animal. *Methods in Ecology and Evolution, 8*(12), 1822–1831.

González-Suárez, M., Gómez, A., & Revilla, E. (2013). Which intrinsic traits predict vulnerability to extinction depends on the actual threatening processes. *Ecosphere, 4*(6), 1–16.

Gorman, M. L., & Trowbridge, B. J. (1989). The role of odor in the social lives of carnivores. In Carnivore behavior, ecology, and evolution (pp. 57–88). Boston, MA: Springer US.

Gosling, L. M., Roberts, S. C., Thornton, E. A., & Andrew, M. J. (2000). Life history costs of olfactory status signalling in mice. *Behavioral Ecology and Sociobiology, 48*(4), 328–332.

Gosling, L. M., & Roberts, S. C. (2001). Scent-marking by male mammals: cheat-proof signals to competitors and mates. In Advances in the Study of Behavior (Vol. 30, pp. 169–217). Academic Press.

Grafen, A. (1990). Biological signals as handicaps. *Journal of Theoretical Biology, 144*, 517–546.

Hartfelder, J., Reynolds, C., Stanton, R. A., Sibiya, M., Monadjem, A., McCleery, R. A., & Fletcher, R. J., Jr. (2020). The allometry of movement predicts the connectivity of communities. *Proceedings of the National Academy of Sciences, 117*(36), 22274–22280.

Heusner, A. A. (1991). Size and power in mammals. *The Journal of Experimental Biology, 160*(1), 25–54.

Hirschfield, M. F., & Tinkle, D. W. (1975). Natural selection and the evolution of reproductive effort. *Proceedings of the National Academy of Sciences, 72*(6), 2227–2231.

Hughes, N. K., Price, C. J., & Banks, P. B. (2010). Predators are attracted to the olfactory signals of prey. *PLoS ONE, 5*(9), e13114.

Kilshaw, K., Newman, C., Buesching, C., Bunyan, J., & Macdonald, D. (2009). Coordinated latrine use by European badgers, *Meles meles*: Potential consequences for territory defense. *Journal of Mammalogy, 90*(5), 1188–1198.

Kodric-Brown, A., & Brown, J. H. (1984). Truth in advertising: The kinds of traits favored by sexual selection. *The American Naturalist, 124*(3), 309–323.

Ladds, M. A., Thompson, A. P., Kadar, J. P., Slip, D. J., Hocking, D. P., & Harcourt, R. G. (2017). Super machine learning: Improving accuracy and reducing variance of behaviour classification from accelerometry. *Animal Biotelemetry, 5*(1), 1–9.

Lochmiller, R. L., & Deerenberg, C. (2000). Trade-offs in evolutionary immunology: just what is the cost of immunity?. *Oikos, 88*(1), 87–98.

Loucks, A. B. (2009). Is stress measured in joules? *Military Psychology, 21*(sup1), S101–S107.

Lyons, G. N., Halsey, L. G., Pope, E. C., Eddington, J. D., & Houghton, J. D. R. (2013). Energy expenditure during activity in the American lobster Homarus americanus: correlations with body acceleration. *Comparative Biochemistry and Physiology Part A: Molecular & Integrative Physiology, 166*(2), 278–284.

Macdonald, D. W., Newman, C., & Buesching, C. D. (2015). Badgers in the rural landscape - conservation paragon or farmland pariah? Lessons from the Wytham Badger Project. In D. W. Macdonald & R. E. Feber (Eds.), *Wildlife conservation on farmland, vol. 2* (pp. 65–95). Oxford University Press.

Maynard Smith, J. (1985). Sexual selection, handicaps and true fitness. *Journal of Theoretical Biology, 115*(1), 1–8.

McClune, D. W., Kostka, B., Delahay, R. J., Montgomery, W. I., Marks, N. J., & Scantlebury, D. M. (2015). Winter is coming: Seasonal variation in resting metabolic rate of the European badger (*Meles meles*). *PLoS ONE, 10*(9), e0135920.

McDonald, R. B., & Horwitz, B. A. (1999). Brown adipose tissue thermogenesis during aging and senescence. *Journal of Bioenergetics and Biomembranes, 31*, 507–516.

McLaren, G. W., Macdonald, D. W., Georgiou, C., Mathews, F., Newman, C., & Mian, R. (2003). Leukocyte coping capacity: A novel technique for measuring the stress response in vertebrates. *Experimental Physiology, 88*(4), 541–546.

McLean, S., Davies, N. W., & Wiggins, N. L. (2012). Scent chemicals of the Brushtail Possum, *Trichosurus vulpecula*. *Journal of Chemical Ecology, 38*(10), 1318–1339.

McNab, B. K. (2006). The relationship among flow rate, chamber volume and calculated rate of metabolism in vertebrate respirometry. *Comparative Biochemistry and Physiology, Part A: Molecular & Integrative Physiology, 145*(3), 287–294.

Miwa, M., Oishi, K., Nakagawa, Y., Maeno, H., Anzai, H., Kumagai, H., Okano, K., Tobioka, H., & Hirooka, H., (2015). Application of overall dynamic body acceleration as a proxy for estimating the energy expenditure of grazing farm animals: relationship with heart rate. *PloS one, 10*(6), p.e0128042.

Montes, I. (2008). *Leukocyte coping capacity and leukocyte activation as a measure of stress in wild badgers (Meles meles).* Doctoral dissertation, Coventry University

Montgomerie, R., & Thornhill, R. (1989). Fertility advertisement in birds: A means of inciting male-male competition? *Ethology, 81*(3), 209–220.

Müller, C. A., & Manser, M. B. (2007). 'Nasty neighbours' rather than 'dear enemies' in a social carnivore. *Proceedings of the Royal Society B: Biological Sciences, 274*(1612), 959–965.

Mykytowycz, R., & Goodrich, B. S. (1974). Skin glands as organs of communication in mammals. *Journal of Investigative Dermatology, 62*(3), 124–131.

Newman, C., & Buesching, C. D. (2019). Detecting the smell of disease and injury: Scoping evolutionary and ecological implications. In C. D. Buesching (Ed.), *Chemical signals in vertebrates 14* (pp. 238–250). Springer.

Nonnamaker, E., Muszynska, M., Lightcap, I., Drea, C. M., & Archie, E. A. (2023). From swab to spectrum: A comprehensive guide to optimizing simple GC-MS analysis of mammalian odorants. In B. Schaal, D. Rekow, M. Keller, & F. Damon (Eds.), *Chemical signals in vertebrates 15*. Springer.

Noonan, M. J., Markham, A., Newman, C., Trigoni, N., Buesching, C. D., Ellwood, S. A., & Macdonald, D. W. (2014). Climate and the individual: Inter-annual variation in the autumnal activity of the European badger (*Meles meles*). *PLoS ONE, 9*(1), e83156.

Noonan, M. J., Tinnesand, H. V., & Buesching, C. D. (2018). Normalizing gas-chromatography–mass spectrometry data: Method choice can alter biological inference. *BioEssays, 40*(6), 1700210.

Noonan, M. J., Tinnesand, H. V., Müller, C. T., Rosell, F., Macdonald, D. W., & Buesching, C. D. (2019). Knowing me, knowing you: Anal gland secretion of European badgers (*Meles meles*) codes for individuality, sex and social group membership. *Journal of Chemical Ecology, 45*(10), 823–837.

Nybo, L., Rasmussen, P., & Sawka, M. N. (2014). Performance in the heat--physiological factors of importance for hyperthermia-induced fatigue. *Comprehensive Physiology, 4*, 657–689.

Parker, G. A. (1974). Assessment strategy and the evolution of fighting behaviour. *Journal of Theoretical Biology, 47*(1), 223–243.

Penn, D. J., & Számadó, S. (2020). The Handicap Principle: How an erroneous hypothesis became a scientific principle. *Biological Reviews, 95*(1), 267–290.

Petrulis, A. (2013). Chemosignals, hormones and mammalian reproduction. *Hormones and Behavior, 63*(5), 723–741.

Pontes, M. H., Sevostyanova, A., & Groisman, E. A. (2015). When too much ATP is bad for protein synthesis. *Journal of Molecular Biology, 427*(16), 2586–2594.

Riek, A., & Geiser, F. (2013). Allometry of thermal variables in mammals: Consequences of body size and phylogeny. *Biological Reviews, 88*(3), 564–572.

Ross, J. G. B., Newman, C., Buesching, C. D., Connolly, E., Nakagawa, S., & Macdonald, D. W. (2021). A fat chance of survival: body condition provides life-history dependent buffering of environmental change in a wild mammal population. *Climate Change Ecology, 2*, 100022.

Salamon, M., & Davies, N. W. (1998). Identification and variation of volatile compounds in sternal gland secretions of male koalas (Phascolarctos cinereus). *Journal of Chemical Ecology, 24*, 1659–1676.

Sanders, D., Thébault, E., Kehoe, R., & van Veen, F. F. (2018). Trophic redundancy reduces vulnerability to extinction cascades. *Proceedings of the National Academy of Sciences of the United States of America, 115*(10), 2419–2424.

Segerstrom, S. C. (2007). Stress, energy, and immunity: An ecological view. *Current Directions in Psychological Science, 16*(6), 326–330.

Setchell, J. M., Vaglio, S., Moggi-Cecchi, J., Boscaro, F., Calamai, L., & Knapp, L. A. (2010). Chemical composition of scent-gland secretions in an Old World monkey (Mandrillus sphinx): influence of sex, male status, and individual identity. *Chemical Senses, 35*(3), 205–220.

Shepard, E. L. C., Wilson, R. P., Halsey, L. G., Quintana, F., Gómez Laich, A., Gleiss, A. C., Liebsch, N., Myers, A. E., & Norman, B. (2008). "Derivation of body motion via appropriate smoothing of acceleration data." *Aquatic Biology, 4*(3), 235–241.

Sin, Y. W., Buesching, C. D., Burke, T., & Macdonald, D. W. (2012). Molecular characterization of the microbial communities in the subcaudal gland secretion of the European badger (*Meles meles*). *FEMS Microbiology Ecology, 81*(3), 648–659.

Sin, Y. W., Newman, C., Dugdale, H. L., Buesching, C., Mannarelli, M. E., Annavi, G., et al. (2016). No compensatory relationship between the innate and adaptive immune system in wild-living European badgers. *PLoS ONE, 11*(10), e0163773.

Smith, J. L. D., Mcdougal, C., & Miquelle, D. (1989). Scent marking in free-ranging tigers, Panthera tigris. *Animal Behaviour, 37*, 1–10.

Speakman, J. (1997). Doubly labelled water: theory and practice. Springer Science & Business Media.

Stearns, S. C. (1992). *The evolution of life histories*. Oxford University Press, Oxford.

Stewart, P. D., Ellwood, S. A., & Macdonald, D. W. (1997a). Remote video-surveillance of wildlife–an introduction from experience with the European badger *Meles meles*. *Mammal Review, 27*(4), 185–204.

Stewart, P. D., Anderson, C., & Macdonald, D. W. (1997b). A mechanism for passive range exclusion: Evidence from the European badger (*Meles meles*). *Journal of Theoretical Biology, 184*(3), 279–289.

Stewart, P. D., Macdonald, D. W., Newman, C., & Cheeseman, C. L. (2001). Boundary faeces and matched advertisement in the European badger (*Meles meles*): A potential role in range exclusion. *Journal of Zoology, 255*(2), 191–198.

Sugianto, N. A., Newman, C., Macdonald, D. W., & Buesching, C. D. (2019). Heterochrony of puberty in the European badger (*Meles meles*) can be explained by growth rate and group-size: Evidence for two endocrinological phenotypes. *PLoS ONE, 14*(3), e0203910.

Sugianto, N. A., Heistermann, M., Newman, C., Macdonald, D. W., & Buesching, C. D. (2021). Alternative reproductive strategies provide a flexible mechanism for assuring mating success in the European badgers (*Meles meles*): An investigation from hormonal measures. *General and Comparative Endocrinology, 310*, 113823.

Számadó, S. (2011). The cost of honesty and the fallacy of the handicap principle. *Animal Behaviour, 81*(1), 3–10.

Terrien, J., Perret, M., & Aujard, F. (2011). Behavioral thermoregulation in mammals: A review. *Frontiers in Bioscience, 16*, 1428–1444.

Tinnesand, H. V., Buesching, C. D., Noonan, M. J., Newman, C., Zedrosser, A., Rosell, F., & Macdonald, D. W. (2015). Will trespassers be prosecuted or assessed according to their merits? A consilient interpretation of territoriality in a group-living carnivore, the European badger (*Meles meles*). *PLoS ONE, 10*(7), e0132432.

Tsai, M. S., Fogarty, U., Byrne, A. W., O'Keeffe, J., Newman, C., Macdonald, D. W., & Buesching, C. D. (2020). Effects of mustelid gammaherpesvirus 1 (Musghv-1) reactivation in European badger (*Meles meles*) genital tracts on reproductive fitness. *Pathogens, 9*(9), 769.

Tsunoda, M., Kaneko, Y., Sako, T., Koizumi, R., Iwasaki, K., Mitsuhashi, I., Saito, M. U., Hisano, M., Newman, C., Macdonald, D. W., & Buesching, C. D. (2019). Human disturbance affects latrine-use patterns of raccoon dogs. *Journal of Wildlife Management, 83*(3), 728–736.

Tuyttens, F. A. M., Long, B., Fawcett, T., Skinner, A., Brown, J. A., Cheeseman, C. L., Roddam, A. W., & Macdonald, D. W. (2001). Estimating group size and population density of Eurasian badgers *Melesmeles* by quantifying latrine use. *Journal of Applied Ecology, 38*(5), 1114–1121.

Vogt, K., Zimmermann, F., Kölliker, M., & Breitenmoser, U. (2014). Scent-marking behaviour and social dynamics in a wild population of Eurasian lynx *Lynx lynx*. *Behavioural Processes, 106*, 98–106.

Williams, G. C. (1996). *Adaptation and natural selection: A critique of some current evolutionary thought*. Princeton University Press.

Wilson, R. P., White, C. R., Quintana, F., Halsey, L. G., Liebsch, N., Martin, G. R., & Butler, P. J. (2006). Moving towards acceleration for estimates of activity-specific metabolic rate in free-living animals: the case of the cormorant. *Journal of Animal Ecology, 75*(5), 1081–1090.

Zahavi, A. (1975). Mate selection—a selection for a handicap. *Journal of Theoretical Biology, 53*(1), 205–214.

# Part II

# Intraspecific Communication in Non-human Vertebrates

# Chemical Communication and Semiochemical Recognition in Frogs: From Eggs to Adults

Lisa M. Schulte[1] (✉), Konrad Lipkowski[1], and Diana Abondano Almeida[1]

[1] Wildlife-/Zoo-Animal-Biology and Systematics, Faculty of Biological Sciences, Goethe University Frankfurt, Frankfurt am Main, Germany
Schulte@bio.uni-frankfurt.de

**Abstract** Semiochemicals, such as sex pheromones or predator kairomones, trigger inter- and intraspecific interactions among animals. With more than 7500 species, frogs (Anura) are a highly diverse group with complex life cycles and adaptations to aquatic and terrestrial habitats. In this review we give a comprehensive and numerical overview about the research conducted concerning semiochemical use in different anuran live stages. Analysing more than 300 studies, we found that chemical communication and kairomone recognition are best studied in tadpoles. Tadpoles of different species are well known to react for example to damage-released alarm substances or predator kairomones. This knowledge has been used in some species to elope more complex research questions about learning mechanisms or the impact of chemical pollution. But also juvenile frogs and adults have been in the focus of semiochemical research, e.g., analysing the recognition of predator kairomones, conspecific semiochemicals, substances involved in parental care, or potential courtship pheromones. However, compared to tadpoles the research efforts in these life stages are still scarce. Furthermore, the species studied regarding semiochemical use do not remotely reflect the diversity of anuran species known today. Future studies of anuran chemical communication should provide additional insight to the role of semiochemicals across different life stages and species.

**Keywords** Alarm substances · Anura · Amphibians · Breeding glands · Chemical pollution · Invasive species · Parental care · Pheromone · Predator kairomones · Tadpoles

## 1 Introduction

Chemical communication can be found in most taxonomic groups of the animal kingdom and is considered the oldest and most widespread communication form (Agosta, 1992; Wyatt, 2014). Chemical compounds involved in chemical communication are defined as semiochemicals (Law & Regnier, 1971). They include for example pheromones, i.e., chemical signals exchanged between individuals of the same species, or kairomones, i.e., chemical cues emitted by one species (e.g., a predatory species) and



B. Schaal et al. (eds.), *Chemical Signals in Vertebrates 15*,
https://doi.org/10.1007/978-3-031-35159-4_5

recognized and used to their advantage by another species (e.g., a prey species; for details regarding different semiochemicals see Nordlund, 1981; Schulte, 2016). While chemical communication and semiochemical recognition are well studied in some animal taxa (e.g., mammals and insects, Ali & Morgan, 1990; Vandenbergh, 2012), for other taxa chemical communication is (or has been) considered rather unimportant (e.g., birds; Whittaker, 2022).

With regard to amphibians, the assessments differ depending on the taxonomic group. The currently known 8527 amphibian species belong to three orders (Frost, 2022). For the order Caudata (salamanders and newts, 792 species), chemical communication is seen as the main communication mode, originally investigated due to their lack of vocal communication (in most species) and meanwhile supported by a plethora of studies (for reviews, see Bossuyt et al., 2019a; Woodley & Staub, 2021). For the smallest order, Gymnophiona (caecilians, 215 species), not much is known about their communication modes due to their hidden lifestyle. However, lacking vocal communication and living underground or in water, chemical communication has been suggested to be the most important communication form in this group too (Warbeck & Parzefall, 2001). In contrast, frogs and toads (Anura, 7520 species) produce conspicuous acoustic signals, which are generally assumed to be their primary communication mode (Gerhardt & Huber, 2002). Chemical communication in these animals has often been denied or is seen as a rather rare, secondary communication form (e.g., Madison, 1977). However, chemical communication might have been the primary mode in anurans before acoustic communication evolved (Byrne & Keogh, 2007; Van Bocxlaer et al., 2015), and may be more important in many species than currently known (Bossuyt et al., 2019b). Furthermore, these statements usually refer to adult frogs, not to larval or juvenile anurans.

### 1.1 Anuran Life Stages and Semiochemicals

The complex life cycles of anurans comprise several stages, starting with the eggs, which in most species are deposited aquatically (but see Schulte et al., 2020). After an embryonic phase, different tadpole stages follow (e.g., Gosner, 1960; Nieuwkoop et al., 2020), which usually develop in water and are free-swimming (but there are many exceptions; e.g., Westrick et al., 2022). During the late tadpole stages, limbs develop and the tail gets shorter: the tadpole metamorphoses into a small terrestrial frog (i.e., froglet), which later grows into an adult. During metamorphosis, i.e., the transition from water to land, the olfactory system goes through major transformations. While the olfactory systems of tadpoles are adapted to aquatic olfaction, adult frogs have both, a water-nose as well as an air-nose (for reviews, see Belanger & Corkum, 2009; Gascuel & Amano, 2013; Weiss et al., 2021), suggesting that olfaction and chemical communication are important both on land and water for these animals.

Adult frogs are known for a long time to use their sense of smell, e.g., in orientation or to recognize food items (Grubb, 1973; Kramer, 1933). More and more studies also suggest that adult frogs communicate chemically with each other (Schulte et al., 2022; Starnberger et al., 2013; Willaert et al., 2013) or show reactions towards potential pheromones or other semiochemicals (Schulte et al., 2015; for definitions see Table 1). Still, most research concerning adult frog communication is focused on acoustics (e.g., Escalona Sulbarán et al., 2019). This is different for the other life stages, i.e., embryos, tadpoles, and froglets, which do not communicate acoustically (with some exceptions; e.g., Natale et al., 2011; Ten Hagen et al., 2016). Tadpoles are furthermore believed to have poor vision (Eluvathingal et al., 2009). Their olfactory sense, however, becomes functional before other sensory systems (i.e., during embryonic stages; Hepper & Waldman, 1992), pointing towards its great importance. Furthermore, in many species, eggs are deposited in big clutches and tadpoles live in large sibling/conspecific groups, i.e., tadpole schools (Wassersug et al., 1981). Due to these factors, numerous studies have focused on semiochemical recognition in tadpoles, especially regarding antipredator responses mediated via prey-released substances (e.g., Pfeiffer, 1966) and predator kairomones (e.g., Nystrom & Abjornsson, 2000), as well as in the context of general conspecific/kin recognition (e.g., Waldman, 1985). But also embryos and froglets have been shown to recognize and react to different semiochemicals (e.g., Capellán & Nicieza, 2010; Flowers & Graves, 1997), which illustrates that anurans use and depend on chemical cues and signals (for definitions see Table 1) throughout their entire life cycle.

## 1.2 Literature Reviews About Semiochemicals in Anurans

Being studied since the late 1940s, chemical communication and the recognition of semiochemicals in anurans are not blank slates, and several reviews already exist in this field. Especially the reaction of tadpoles to semiochemicals has been the focus of review articles and book chapters. Most of them focus on predator recognition and damage-released alarm substances (DRAS). However, these reviews usually cover various taxonomic groups, i.e., also including fish, salamanders and/or invertebrates. Chivers and Smith (1998), for example, give an overview of different taxa (vertebrates and invertebrates) regarding the use of DRAS (and other predation-related substances), with Ferrari et al. (2010) reviewing discoveries made since then and Mitchell et al. (2017) focusing on the underlying physiological mechanisms that mediate the response. Mirza and Chivers (Mirza & Chivers, 2001) focus on DRAS in aquatic vertebrates. While the actual identity or chemical properties of these substances remain opaque, Van Buskirk et al. (2014) conducted a study and reviewed the literature regarding the rate of degradation of waterborne chemical substances that indicate predation risk. All reviews contain several examples of tadpole reactions to conspecific DRAS. Also, Rajchard (2006) gives a short review with a focus on DRAS in anurans and salamanders, including some examples of responses to cues of predators fed with conspecifics. The reviews of Chivers and Mirza (2001), Stabell (2005), and Scherer and Smee (2016) deal especially with the latter: predator diet cues and latent DRAS. Their reviews focus on fish and tadpoles, only on tadpoles, or on a variety of different taxa (including tadpoles), respectively. They all point out the importance that the diet of a predator has for the chemical recognition by the prey (e.g., if the predator fed conspecifics versus other prey).

Behavioural responses to predator odours in all animal taxa, both in water and on land, are reviewed by Kats and Dill (1998), including several studies regarding the antipredator responses of tadpoles. Both, DRAS as well as predator kairomones are

**Table 1** Technical terms used in this review, which classify chemical effects between organisms

| Terminology | Definition and functional aspects | Net benefit for sender/ receiver | Evolutionary aspects (signal/ cue) | Reference |
|---|---|---|---|---|
| Chemical signals | Chemical substances emitted by one animal, intended to provide information to another | (+ / +) | Evolved for communication | Steiger et al. (2011), Wyatt (2014), Ferrari et al. (2010) |
| Chemical cues | Chemical substances emitted by one animal, which unintentionally provide information to another | (0 / +) or (− / +) | Chemical by-products, not evolved for communication | Steiger et al. (2011), Wyatt (2014), Ferrari et al. (2010) |
| Semiochemicals | Chemical substances that evoke a behavioural or physiological response in individuals of the same or another species | | Include signals and cues | Law and Regnier (1971), Sbarbati and Osculati (2006) |
| (sex) Pheromones | Chemical substances emitted by one individual and received by another individual of the same species, which shows a specific (innate) reaction; in case of sex pheromones the two individuals are of opposite sexes | (+ / +) | Chemical signals | Karlson and Lüscher (1959), Wyatt (2014) |
| Damage-released alarm substances | Chemical substances that are either passively or actively released by injured conspecifics (or heterospecifics), that evoke antipredator responses in other prey individuals | (? / +) | Remains to be defined | Fraker et al. (2009), Ferrari et al. (2010), Hettyey et al. (2015) |
| Disturbance substances | Chemical substances that are either passively or actively released by disturbed (not injured) conspecifics (or heterospecifics), that evoke antipredator responses in other prey individuals | (? / +) | Remains to be defined | Hazlett (1985), Kiesecker et al. (1999a) |

(continued)

**Table 1** (continued)

| Terminology | Definition and functional aspects | Net benefit for sender/ receiver | Evolutionary aspects (signal/ cue) | Reference |
|---|---|---|---|---|
| Predator-borne kairomones | Chemical substances emitted by a predatory species, unintentionally providing information and thereby benefitting the prey species (evoking an antipredator response); also defined as "enemy-avoidance kairomones" | (– / +) | Chemical cues | Dicke and Sabelis (1988), Ruther et al. (2002) |
| Digestion-released substances | Chemical substances emitted by a predatory species after digesting a prey species, evoking an antipredator response in other prey individuals | (– / +) or (? / +) | Unclear (mix of the above), probably cues | Hettyey et al. (2015) |
| Foraging kairomones | Chemical substances emitted by a prey species, unintentionally providing information and thereby benefitting the predator species (food-source location) | (– / +) | Chemical cues | Ruther et al. (2002) |

+ beneficial outcome, – negative outcome, and 0 neutral outcome for sender or receiver,? net benefit remains unclear

discussed by Dodson et al. (1994). Also here tadpoles are only part of the review, which also includes other freshwater vertebrates and invertebrates. The increasing research interest in semiochemicals mediated antipredator responses in tadpoles, however, also resulted in tadpole-specific reviews. Rödin et al. (2011) systematically compiled and analysed the literature regarding antipredator responses of anuran larvae and explored behavioural patterns related to semiochemicals solely released by tadpoles (i.e., disturbance cues and DRAS). With this, the authors uncovered behavioural patterns that emerge at the family level, but also depend on the origin of released substances (i.e., from con- or heterospecifics) and the gregarious or solitary lifestyle of tadpoles. In addition, a meta-analysis regarding antipredator responses only in tadpoles (both behavioural and morphological) was conducted by Hossie et al. (2017). The authors found that several factors influenced the tadpoles' antipredator responses, including the preceding prey consumption of the predators. Even though not only studies dealing exclusively with chemical cues were included here, this review gives a great overview regarding different behavioural and morphological reactions triggered by predators from different taxonomic groups.

A few reviews also exist regarding chemical communication in adult amphibians (e.g., Rajchard, 2005). However, most of them have a strong focus on salamanders, where much more is known about (sex) pheromone use than in frogs (Houck, 2009; Kikuyama et al., 2002; Madison, 1977; Woodley, 2010). The most extensive reviews about both adult and tadpole anuran chemical communication were written by Belanger and Corkum (2009) and Weiss et al. (2021). Both give a nice overview with many examples of semiochemical studies in frogs and tadpoles and have an additional focus on the anuran olfactory system (including the anatomical differences between tadpoles and adults). Other reviews focus on the importance of semiochemicals in anuran embryos (Mirza & Kiesecker, 2005), during parental care (Buxton & Sperry, 2017; Schulte, 2016) or for kin recognition (Ogurtsov, 2004; Waldman, 1986).

The majority of the reviews about semiochemicals in anurans are already >10 years old (but see Weiss et al., 2021), and many of them only cover a fraction of the known literature, e.g., because anurans are not the main focus of the review, or because no standardized literature search parameters were used (but see Hossie et al., 2017). Furthermore, most of the reviews have a special focus on a specific category of semiochemicals (e.g., predator kairomones or sex pheromones) or a single anuran live stage (e.g., tadpoles or adults). With the present review, we want to try to cover the entire field of semiochemical recognition and chemical communication research in anurans. Our aim is to give both a comprehensive and numerical overview about the research that has been conducted in this field, compare and structure the different categories of semiochemical use for the different live stages of this highly diverse group and ultimately elaborate which aspects still need further research.

## 2 Methods

During the last 10 years we established an extensive literature collection, including several standardized searches on the "Web of Science" browser using the following terms: anura*, frog*, and tadpole* combined with chemical communication, chemical cue*, chemosignal*, kairomone*, odor*/odour*, olfaction, olfactory, pheromone*, predator recognition, scent*, semiochemical*, and smell*. The last search was conducted in August 2022. The search results were thoroughly examined: all studies were considered, that dealt with behavioural and/or physiological reactions of anurans to any type of semiochemical released by other animals (conspecifics, heterospecifics, prey animals, and predators). If it was not clear that a reaction was truly triggered by semiochemicals, but could be confounded with other cues/signals (e.g., visual cues), the study was not included here. We excluded for example more than 100 studies dealing with predator cues, where the predators were presented in a cage, i.e., visual cues were introduced alongside chemical ones. We furthermore included literature revealed from citations in these studies, thereby also including some studies presented in other languages than English.

All studies that fulfilled the requirements were classified into one or several of the following categories: DRAS/disturbance substances, predator kairomones, sex pheromones, parental care, kin and conspecific recognition, heterospecific recognition, prey recognition and disease/parasite recognition (for details regarding these categories, see Sect. 3). We further assorted them by the type of measurement (e.g., behavioural or morphological measurements, see Sect. 3.1), developmental stage, and the taxonomic rank. All studies were used for our analyses, but due to space limitations not all of them can be cited here.

# 3 Results and Discussion

## 3.1 General Results

We found a total of 308 studies dealing with semiochemicals in anurans. The earliest study we found was published in 1949 (Eibl-Eibesfeldt, 1949), followed by only five studies in the next three decades. Since then, publications regarding this matter were steadily rising (see Fig. 1). The publications we found were published in 103 journals and one book series (Chemical Signals in Vertebrates). Journals belonged to one or more of the following categories: Ecology, general Biology/Science, Zoology, Behaviour, Herpetology, Hydrology, Ecotoxicology, Chemistry, Evolution, Conservation and Physiology. Most of the publications (76%) deal with the recognition of semiochemicals in tadpoles, while much less is known about chemical communication and semiochemical recognition in anuran embryos, froglets, and adult frogs (see Fig. 2).

Most of the studies on tadpoles deal with either DRAS/disturbance substances, predator kairomones, or both (see Sects. 3.2, 3.3, and 3.4). Being the most studied topic(s) in the most studied developmental anuran stage, DRAS/disturbance substances and

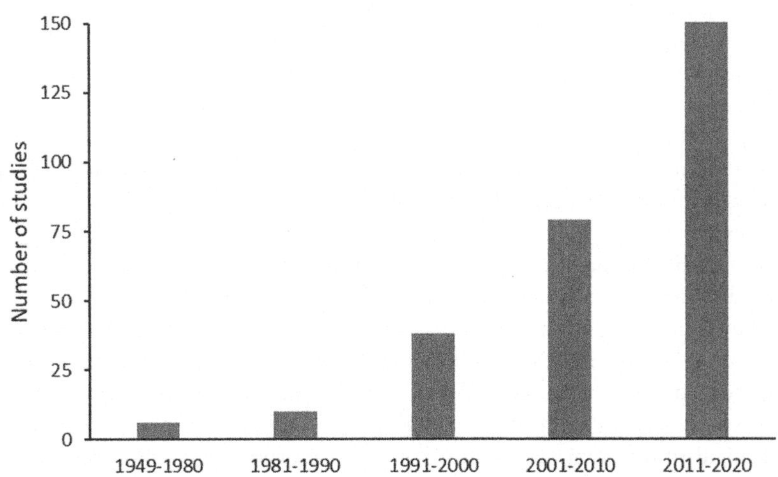


**Fig. 1** Number of studies published about chemical communication and semiochemical recognition in anurans over different decades. Notice that the first bar comprises three decades; the other ones, each one decade. Studies published in 2021 and 2022 are not considered here

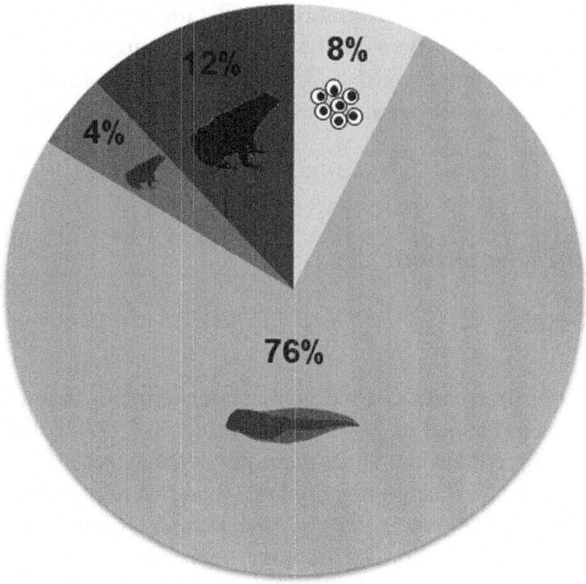


**Fig. 2** Total number and percentages of studies published on chemical communication and semiochemical recognition in different developmental life stages in anurans: embryos, tadpoles, froglets and adult frogs (from light to dark grey). A total of 308 studies have been analysed. Of these, 6.5% studied more than one developmental stage

predator kairomones (and their combination) are in total the most extensively studied of the topics defined above (see Fig. 3). The combined topics of con- and heterospecific recognition (including kin recognition) are followed by 13%, while all the other topics are much less studied. Except for the exchange of sex pheromones (being impossible during larval stages), all of the categories defined in Sect. 2 have been studied in tadpoles. Also, for adults all categories have been studied, with sex pheromones being the most studied category (with 14 studies). Embryos have mainly been studied for predator kairomones and DRAS, as well as a few studies regarding conspecifics' recognition. The same is the case for froglets, but here also the chemical recognition of prey animals has been tested.

Concerning the animals' responses to different semiochemicals, most studies focus on behavioural tests (271 studies, of which 234 exclusively measure behaviour). Here, predominantly general activity, as well as avoidance or attraction of/to semiochemicals were measured. In 52 studies, developmental measurements were taken after semiochemical exposure. All of these studies deal with semiochemical responses in embryos and tadpoles, measuring the time to hatch, the time to metamorphosis, the growth rate or the survival in different stages. In 35 studies, morphological changes triggered by semiochemical exposure were measured. These changes were all measured in tadpoles and/or embryos and were mainly associated with exposure to predator kairomones and/

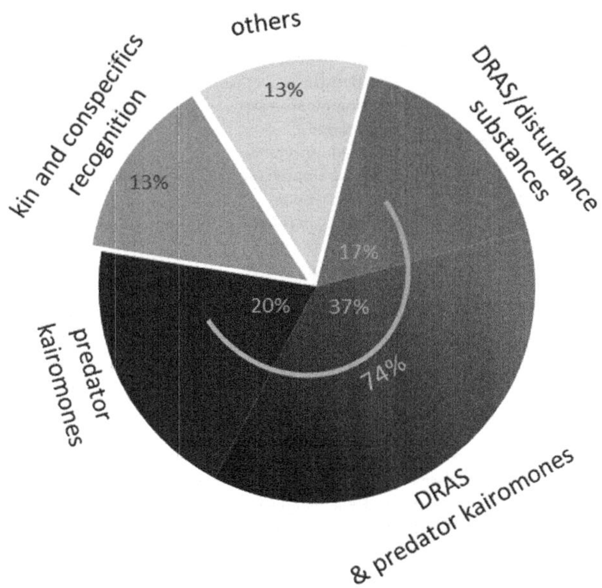


**Fig. 3** Percentages of studies conducted on different topics/semiochemicals. Of the 308 studies analysed, 7.5% covered more than one topic (excluding the ones covering both DRAS and predator kairomones, which are shown together in the figure)

or DRAS, leading to morphological adaptations associated with escape behaviours (such as deeper and stronger tails). Furthermore, seven studies were (additionally) looking at physiological changes (i.e., hormonal, metabolic and neurological changes). Of all studies listed here, 47 combined two or more measurements, for example, both behavioural and morphological changes were tested in the same study after semiochemical exposure.

With regard to the species tested in the examined studies, an unequal taxonomic distribution becomes evident. The currently known 7493 anuran species belong to 56 families (Frost, 2022; see Fig. 4). Of those families, 21 have been studied regarding semiochemical use. Of the 308 studies analysed, 57 studies examined the use of semiochemicals in more than one species and 42 include more than one anuran family. The most studied family are the True Frogs (Ranidae) with a total of 145 studies (32 of which also include species of other families), followed by Toads (Bufonidae, 96 studies, 31 including other families) and Treefogs (Hylidae, 47 studies, 19 including other families). These three families are also the only ones where each of the four life stages has been tested for semiochemical use (for details see Fig. 4). In most of the families (19), chemical signal/cue recognition has been tested in tadpoles, followed by studies testing adults (15 families). While in some families the recognition of semiochemicals has only been tested for one stage in a single species (and, in some cases, was not proven; see Fig. 4), for most families (35 of 56) not a single species has been tested regarding semiochemical use in any of the life stages.

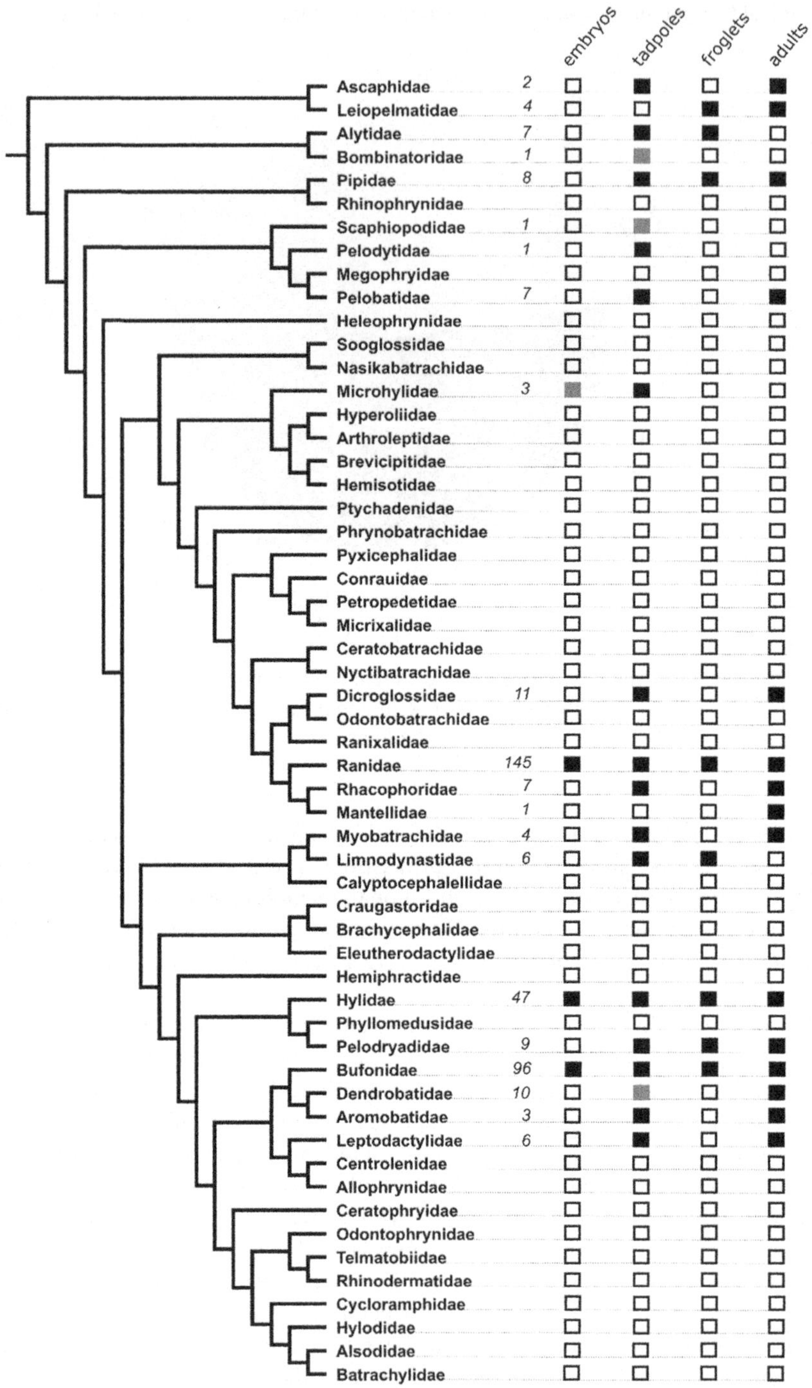


**Fig. 4** Cladogram of anuran families. The numbers behind each family stand for the number of studies on semiochemicals. The boxes show which life stage(s) has/have been investigated in each of the families. Black: reaction to semiochemicals has been shown; grey: semiochemicals have been tested but no reaction was shown; white: semiochemicals have not been investigated yet. (Modified after Pyron & Wiens, 2011, with data from Frost, 2022)

## 3.2 Damage-Released Alarm and Disturbance Substances

Among the most studied semiochemicals in anurans are damage-released alarm substances (DRAS; see Fig. 3). They are released by a prey animal when being injured by a predator and are used by conspecifics to assess the underlying predation risks and respond by adapting their behaviour (e.g., Carlson et al., 2015), morphology (e.g., Hagman et al., 2009) and life history (e.g., Chivers et al., 1999a). We are using the term "DRAS" here instead of alarm cues, signals or pheromones (for definition and evolutionary implications of terminology see Table 1; Ferrari et al., 2010, Steiger et al., 2011), because, in most species, key aspects for a proper classification remain opaque (i.e., net benefit for sender/receiver and intentional/unintentional release of substances). Only few studies show that adults and froglets avoid DRAS of conspecifics (Chivers et al., 1999b; Hagman & Shine, 2008c; Wirsing et al., 2005). The majority of experiments that measure the effects of DRAS focus on tadpoles (e.g., Hagman & Shine, 2009a, b; Manteifel, 2001; Raven et al., 2017). While not all species studied show a reaction to DRAS (Pfeiffer, 1966; Pizzatto et al., 2014), many species react to conspecific DRAS (Hagman & Shine, 2008a, b; Hews & Blaustein, 1985; Petranka, 1989; Summey & Mathis, 1998), and some even react to both, con- and heterospecific DRAS (although dependent on their taxonomic relationship; Adams & Claeson, 1998; Hrbáček, 1950; Pfeiffer, 1966). The majority of studies focus on behavioural changes triggered in the receiver tadpoles shortly after exposure, including decreased activity, avoidance behaviours or flight responses (e.g., Ferrari et al., 2007; Hagman & Shine, 2009a; Kulzer, 1954; Lucon-Xiccato, 2019). Some studies focus on morphological and developmental changes such as reduced body size or time to metamorphosis (Benard & Fordyce, 2003; Hagman et al., 2009; Jefferson et al., 2013). In these studies, DRAS are usually presented regularly over a longer period of time (e.g., daily). Such developmental studies often start during embryonic stages, leading to a delay or acceleration of time to hatch (Chivers et al., 2003; Mandrillon & Saglio, 2007a, 2009; Touchon et al., 2006).

While many advances in understanding the chemical ecology of predator-prey interactions in tadpoles have been made (Chivers & Smith, 1998; Ferrari et al., 2010), the underlying physiological mechanisms that mediate antipredatory responses remain largely opaque (Mitchell et al., 2017). This is mainly due to the lack of knowledge about the identity and chemical properties of active substances that elicit antipredator responses. Only few studies tried to identify relevant substances and determine their tissue of origin, as well as chemical properties such as degradation time and temperature stability. Early studies found that skin extracts (or different body parts with skin) elicited reactions in conspecifics, while intestines did not, suggesting that the substances of interest are released from the skin (Eibl-Eibesfeldt, 1949; Fraker et al., 2009; Hews & Blaustein, 1985; Hrbáček, 1950; Kulzer, 1954; Raices et al., 2020). In addition, it is believed that DRAS could be either proteins (Fraker et al., 2009; Raices et al., 2020) or m/z 501 anions (Austin et al., 2018; Crossland et al., 2019). Fraker et al. (2009) found two peptides in the skin that only in combination triggered a reduction in activity. They furthermore suggest that tadpoles must be fully conscious (i.e., not anesthetized) to release DRAS, thereby implying an active release mechanism of these substances. Extracts taken from individuals anesthetized with benzocaine at the time of extraction did not elicit an antipredator response. However, recent evidence suggests a negative influence of chemical anaesthetics in bioassays (Achtymichuk et al., 2022), rendering the interpretation difficult. Mirza et al. (2013) separated skin extracts of tree frog tadpoles via high-performance liquid chromatography (HPLC) into two fractions (based on retention time) and found that neither of the two fractions triggered the same behaviour as the full extract. Austin et al. (2018) described one DRAS fraction containing a single anion to change the behaviour of ranid tadpoles and suspected this anion to be a sulphated steroid (see also Ferland-Raymond et al., 2010). One of the most comprehensive attempts to characterize relevant active substances was made by Crossland et al. (2019) and resulted in the identification of over 30 molecules in bioactive fractions, of which 4 were confirmed to elicit a behavioural response in *Rhinella marina* tadpoles. Studies looking at the persistency of DRAS show that behavioural responses of tadpoles subsided 5–240 min after extraction (studied in bufonids and ranids, both during ex situ and in situ experiments; Ferrari et al., 2008a; Kulzer, 1954; Mathis et al., 2008; Petranka, 1989). While substances do not trigger behavioural reactions anymore when aged or dried, they retain their effectiveness after freezing and heating (Achtymichuk et al., 2022; Clarke et al., 2016; Fraker et al., 2009; Raices et al., 2020).

Besides DRAS, also "disturbance substances" (first proposed as "disturbance pheromones" by Hazlett, 1985) have been shown to affect anurans. Unlike for the release of DRAS, for disturbance substances, no injury is required. These substances can also be described as "pre-attack chemical information" which is released by prey after being stressed but not hurt by a predator (Giri et al,, 2012; Kiesecker et al., 1999a). Disturbance substances are generally less studied than DRAS and focus predominantly on tadpoles of the genera *Bufo* (Bufonidae), *Rana* and *Lithobates* (both Ranidae). For example, substances collected from conspecifics chased with a stick evoke a reduction in activity in wood frog tadpoles (*Lithobates sylvaticus*; e.g., Bairos-Novak et al., 2020), a reduction in activity and hiding behaviours in northern red-legged frog tadpoles (*Rana aurora*; Kiesecker et al., 1999a, b) and avoidance behaviours in common toad tadpoles (*Bufo*; Manteifel et al., 2006), while substances released by shaken conspecifics lead to panic swimming behaviours in Heckscher's frog tadpoles (*Lithobates heckscheri*; Altig & Christensen, 1981). Some of these studies found that the reactions to disturbance substances could also be triggered when ammonia was added to the water, suggesting that renal excretions may be involved and that ammonia may be a component of the disturbance substances (Kiesecker et al., 1999a; Manteifel et al., 2005, 2006; Manteifel & Kiseleva, 2011). A study that compared the reaction towards DRAS and disturbance substances in Iberian green frog tadpoles (*Pelophylax perezi*) found that even though both substances lead to decreased activity, DRAS evoked a much stronger response than disturbance substances (Gonzalo et al., 2010). Few studies have addressed disturbance substances in adults and, to the best of our knowledge, there are no studies regarding the reactions of froglets to disturbance substances. Deng et al. (2022) showed that female serrate-legged small treefrogs (*Kurixalus odontotarsus*) avoid the odour of males after they have been disturbed (i.e., squeezed). Also for the variable poison frog (*Ranitomeya variabilis*) avoidance of conspecific smells after confinement has been shown (being rather a long-term stress substance though; Schulte & Rößler, 2013).

The reaction towards DRAS and disturbance substances is very well established in some species, leading to approaches using them to study learning behaviours and conditioning (Lucon-Xiccato et al., 2016; Pueta & Perotti, 2016; Rivera-Hernández et al.,

2022), the impact of chemical pollution (Mandrillon & Saglio, 2009; Moore et al., 2015) and their potential application against invasive species (Hagman & Shine, 2008a, c, 2009b, Hagman & Shine, 2009a; for details, see Sect. 3.9). For most species and the majority of families, however, it remains unknown if DRAS and disturbance cues exist and if they are exchanged between tadpoles—especially in anurans where tadpoles do not swim in large schools and live rather solitary.

### 3.3 Predator-Borne Kairomones

Besides antipredator behaviours displayed by amphibians after experiencing DRAS, there is also a rich literature about reactions to substances presenting potential predator kairomones (i.e., predator-borne substances that benefit the receiver, but are not chemically defined yet; in the following text for simplicity called kairomones; also see Table 1). However, the study design concerning predator kairomones varies substantially. Many studies test these substances by introducing the predators in cages. Even though it is likely that chemicals are involved in the reactions, we did not take these studies into account here, unless it was certain that no additional visual or tactile cues could be perceived (see also Sect. 2). Studies using only chemical substances released by predators need to be further divided: while in some studies predators were starved or fed with non-related prey species (such as invertebrates), in others the predators were fed with conspecifics. In the latter case, predator-borne cues can be confounded with prey-borne substances (such as DRAS, see Sect. 3.2) as well as digestion-released cues (for details on the terminology see Hettyey et al., 2015 and Mitchell et al., 2017). To exclude any confounding results, we only discuss predator-borne cues here and go into detail regarding digestion-released cues and comparative studies in Sect. 3.4.

Chemical recognition of predators is especially well studied in anuran tadpoles, with kairomones originating from a variety of predatory species, including damselfly larvae (e.g., Iglesias-Carrasco et al., 2022), giant water bugs (e.g., Casillas-Barragán et al., 2016), backswimmers (e.g., Kiesecker et al., 1996), salamanders (e.g., Lefcort & Eiger, 1993), tadpoles (e.g., Sharma et al., 2008), turtles (e.g., Zhang et al., 2015) and fish (e.g., Holomuzki, 1995). The reaction of tadpoles varies highly between species and is dependent on the type of predator. Some predator-borne kairomones do not elicit any (of the measured) behavioural changes in certain species, while others trigger a reaction. For example, chemical cues collected from damselfly larvae and giant water bugs presented to bufonid tadpoles (*Rhinella ornata*) did not lead to an alternation of swimming behaviours (Costa & Nomura, 2014), and sunfish kairomones trigger avoidance behaviours in only two out of eight species (five of which encounter this predator frequently; Kats et al., 1988). A single-prey species reacts with reduced feeding activity to three out of four different native predators (Feminella & Hawkins, 1994) and some studies show that predator-borne kairomones are only recognized in populations syntopic with the predator (including invasive predators; Anderson & Lawler, 2016; Kiesecker & Blaustein, 1997). Most studies look at behavioural changes, such as avoidance behaviours (Mogali et al., 2015), refuge use (Petranka et al., 1987), background shifts (Swart & Taylor, 2004), activity changes (Lefcort, 1996) and freezing behaviours (Hickman & Watling, 2014). Others investigate the long-term effects of predator cues on morphology and development (El Balaa & Blouin-Demers, 2013; Nunes et al., 2014a). Developmental changes are also measured when looking at anuran embryos. When measuring the time to hatch, Capellán and Nicieza (2010) could not find an effect of kairomones from leeches (i.e., egg predators). However, looking at invasive egg predators (fish), the results varied between two studies (even though the same species were examined): while Dibble et al. (2009) also found no difference in hatching time, Smith and Fortune (2009) found that wood frog embryos hatched earlier when exposed to invasive mosquito fish kairomones.

Besides embryos, several studies have also focused on the abilities of tadpoles to recognize invasive predator cues, coming to very different results depending on the tested species. While some studies show that kairomones of alien predators trigger behavioural or developmental changes in tadpoles similar to those of native predators (Berec et al., 2016; Manteifel & Zhushev, 1996; Pease & Wayne, 2014; Vodrážková et al., 2020), others show that alien predator cues cannot be recognized by native species (Polo-Cavia et al., 2010; Polo-Cavia & Gomez-Mestre, 2014). However, this depends on both the prey and predator species and if or how long they coexist with each other. In studies crossing multiple species, reactions differ vastly (Nunes et al., 2013; Pearl et al., 2003). In studies examining the reactions among different populations within the same species, reactions can vary depending on the history between native prey and invasive predator—or stay the same (Gomez-Mestre & Díaz-Paniagua, 2011; Nunes et al., 2014b).

But of course not only tadpoles benefit from recognizing predator-borne kairomones. Froglets of different species are able to recognize and avoid chemicals released by predatory snakes (Flowers & Graves, 1997; Gonzalo et al., 2006; Schley & Griffiths, 1998) or predatory frogs and toads (Chivers et al., 2001; Pizzatto & Shine, 2009). The same is the case for adult frogs, avoiding snake cues (Hamer et al., 2011; Murray et al., 2004) and mammalian predators. Zamora-Camacho et al. (2019) showed that female Iberian spadefood toads (*Pelobates cultripes*), which hide in self-dug burrows, change their burrowing behaviour when confronted with boar kairomones. However, the latter might be more of a stress response. Given the variety of anuran predators (Stynoski & Porras-Brenes, 2022), research on responses to predator-borne kairomones in juvenile and adult frogs is still comparably scarce.

## 3.4 Digestion-Released Substances and DRAS-Predator-Kairomone Combinations

As described under Sects. 3.2 and 3.3, predation related substances can be either released by the prey species (and recognized by conspecifics) or the predator species (and recognized by the prey species). However, in many studies this border is not so clear-cut because either predator kairomones and DRAS were tested simultaneously (i.e., as a mix; e.g., Hettyey et al., 2019; Kurali et al., 2016; Üveges et al., 2021) or predators were fed with prey animals (e.g., McCoy et al., 2015; McIntyre & McCollum, 2000; Ramamonjisoa et al., 2018). In the latter case, DRAS from the feeding event may still be in the water and/or prey leftovers from digestion may be mixed with predator substances (so-called digestion-released substances; for further definitions, see Hettyey et al., 2015). Conspecific-fed predator substances have been used in many studies to test inducible antipredator defences against the most realistic threat (i.e., conspecifics-consuming predators; e.g., Dijk et al., 2016; Laurila, 2000; Takahara et al., 2012) or to investigate higher-order questions related to antipredator defences (Balestrieri et al., 2019; Bennett et al., 2013; Chuirazzi et al., 2021; Melo et al., 2021).

Digestion-released substances have been shown in many studies to be more effective than predator kairomones and/or DRAS alone (Chivers & Mirza, 2001; Eklöv, 2000; Mogali et al., 2011; Schoeppner & Relyea, 2005), and in some species exclusively the combination of predator and prey substances elicits responses in tadpoles (Hemnani et al., 2022; Mogali et al., 2020; Preston & Forstner, 2015; Shaffery & Relyea, 2016). In other species, digestion-released substances have the same effect as DRAS, pointing in the direction that substances of digested conspecifics are more important than the predators themselves (Kiesecker et al., 2002; Wilson & Lefcort, 1993). However, there are also cases where the reaction towards pure predator kairomones is the same as to digestion-released substances, implying that the important information is derived from the predator itself (Maag et al., 2012; Petranka & Hayes, 1998; Scribano et al,, 2020).

While most studies that include digestion-released substances in anurans involve tadpoles, there are also a few studies that test the effect of conspecific-fed predators on embryonic development (Capellán & Nicieza, 2010; Gazzola et al., 2021) and froglet behaviour (Barbasch & Benard, 2011; Belden et al., 2000).

### 3.5 Other Kairomones

Anurans are predatory animals, and it has long been believed that they only use their vision for hunting (Maturana et al., 1960). But there are some examples that for both adults and juveniles, olfaction can play a role during prey recognition, too (i.e., foraging kairomones, Ruther et al., 2002). Dole et al. (1981) showed that adult western toads (*Bufo boreas*) can recognize their prey by odour, and also subadults of aquatic African clawed frogs (*Xenopus laevis*) show changes in movements when exposed to either intact or injured prey animals (South et al., 2020). Indian bullfrog tadpoles (*Hoplobatrachus tigerinus*) are able to recognize their prey both visually and chemically (Saidapur et al., 2009), while Jerdon's bullfrog tadpoles (*Hoplobatrachus crassus*) only use visual, not chemical cues (Phuge & Phuge, 2019).

Besides predator and prey kairomone recognition, there are few studies investigating the effects of kairomones from parasites and pathogens. Marino et al. (2014) tested the reaction of tadpoles to parasitic nematode cues (in interaction with predator cues), but could not find a behavioural effect. In contrast, chemical cues of conspecifics infected with the pathogenic yeast *Candida humicola* were avoided by bullfrog tadpoles (*Lithobates catesbeianus*; Kiesecker et al., 1999b). Likewise, metabolites of the fungus *Batrachochytrium dendrobatidis*, responsible for the global amphibian chytrid pandemic, have been shown to trigger avoidance behaviour in adult southern toads (*Anyxyrus terrestris*), but only after they survived a first infection. Other frog species tested, however, were not able to recognize the metabolites of the pathogens (McMahon et al., 2021). The same pathogenic metabolites also trigger hylid tadpoles to develop faster, while they have no effects on adults (McMahon et al,, 2019). The field of chemical recognition of pathogenic metabolites is quite promising in view of disease control and has potential as a learning tool (also compare Sect. 3.9).

## 3.6 Kin, Conspecific and Heterospecific Recognition

Aside from the recognition of DRAS and disturbance substances (which can also be classified as a type of conspecific recognition mechanism), there are also several studies showing that frogs and tadpoles can recognize non-hurt and non-stressed con- (and hetero-) specifics chemically. Adult great barred-frogs (*Mixophyes fasciolatus*), for example, are attracted to their own odour, as well as to that of conspecifics and even sympatric heterospecifics (Hamer et al., 2011). Adult frogs of the genus *Leiopelma* also have a preference for their own chemical substances (compared to those of other conspecific individuals), as well as for juvenile conspecifics (Delaney & Bishop, 2007; Lee & Waldman, 2002; Waldman, 2016; Waldman & Bishop, 2004). Great Plains toad froglets (*Anaxyrus cognatus*) are attracted to semiochemicals of conspecific froglets (Graves et al., 1993), while in other species such an attraction behaviour among froglets was not observed (Belden et al., 2000; Gonzalo et al., 2006; Pizzatto et al., 2015). Among males, chemical recognition often leads to competition behaviours, such as aggression (King et al., 2005) and changes in calling behaviours (Byrne & Keogh, 2007; Deng et al., 2021). Measurements of hormonal levels also showed that male chemical stimulation leads to increased testosterone levels and, when combined with acoustic stimuli, to increased corticosterone levels in Tungara frogs (*Engystomops pustulosus*; Still et al., 2019). Poth et al. (2012) showed that male mantellids (*Mantidactylus multiplicatus*) increased their activity when confronted with two compounds isolated from conspecific males' femoral glands and were attracted to one of the two compounds (also see Sect. 3.8). While these compounds were volatile, King et al. (2005) isolated a peptide from male skin secretions that had a chemoattractive effect on conspecific males and stimulated aggressive behaviours in mountain chicken (*Leptodactylus fallax*).

Unlike adults, tadpoles of many anuran species swim in larger groups, so-called tadpole schools. Several studies have shown that they show a preference for chemicals released by their own siblings over unfamiliar conspecifics (Blaustein et al., 1993; Blaustein & O'Hara, 1982; Waldman, 1985). Depending on the study design it can (or cannot) be assured if this is due to genetic relationship or due to familiarity (for details and a review on kin recognition in tadpoles, see Waldman, 1986). Further studies showed that besides kin preferences there are also preferences of conspecifics over heterospecifics (Eluvathingal et al., 2009; Kiseleva, 1989), or only attraction to conspecifics, without differentiation between kin and non-kin (Kiseleva, 1996). Other studies, however, show that chemicals from conspecifics were avoided (Manteifel, 1995; Manteifel & Zhushev, 1996), triggered changes in activity patterns (Buttermore et al., 2011) or did not trigger any change in behaviour (Berec et al., 2016). All these tests were conducted with bufonids or ranids, families which are known to perform schooling behaviour in rather large water bodies. Tests with chemicals released by poison frog tadpoles (Dendrobatoidea), which usually live with a few conspecifics in very small water bodies, resulted neither in avoidance (Szabo et al., 2021) nor in differences in activity levels among conspecifics (Kumpulainen, 2022).

### 3.7 Chemical Cues and Signals During Parental Care

Anurans are well known for their highly diverse patterns of parental behaviours, with 28 defined parental care modes, such as egg guarding, offspring transport or tadpole feeding behaviours (Crump, 1996; Schulte et al., 2020). Even though chemical communication and semiochemical recognition during parental care are not well studied yet, there are a few examples showing that adult frogs use semiochemicals to determine where to deposit their eggs or tadpoles. Frogs that oviposit in large ponds with potential egg predators (e.g., fish) have been shown to avoid predator kairomones (which can also be defined as a type of predator recognition, see Sect. 3.3; Kloskowski, 2020; Takahashi, 2007). In Neotropical poison frogs (Dendrobatoidea), parents transport their tadpoles to smaller water bodies. In the Zimmermann's poison frog (*Ranitomeya variabilis*) small clutches are deposited in water-filled plant axils (phytotelmata) and, upon hatching, the male transports the tadpoles one by one into different phytotelmata to avoid cannibalism among them. Studies have shown that they are able to recognize chemicals released by conspecific tadpoles to avoid pools that are already occupied, i.e., contain a tadpole that could potentially cannibalize their offspring (Schulte et al., 2011; Schulte & Veith, 2014)—or, depending on the season, prefer them for tadpole deposition (Schulte & Lötters, 2013). They are further able to also recognize tadpoles of other species but react quite differently to them. Semiochemicals of tadpoles of other phytotelmata-breeding poison frogs were also avoided for tadpole deposition, but those breeding in streams were not. Semiochemicals of phytotelmata-breeding treefrog tadpoles, in contrast, were preferred for tadpole deposition and possibly seen as a food source. For egg deposition, however, the semiochemicals of all tadpoles were avoided (Schulte & Lötters, 2014). Kairomones of invertebrate predators (damselfly larvae), however, were not recognized and avoided for offspring deposition by *R. variabilis* (Schulte et al., 2013). The chemicals involved in the avoidance behaviours regarding conspecific tadpoles as well as one heterospecific tadpole were examined in detail and it turned out that both species were avoided due to a mix of two chemical compounds (conspecifics: $C_8H_7NO$ and $C_9H_{18}N_2O$; heterospecifics: $C_{13}H_{25}NO_2$ and $C_9H_{18}N_2O$), one of which was equal between the two species (Schulte et al., 2015; for an overview, see also Schulte, 2016).

In the brilliant-thighed poison frog (*Allobates femoralis*), tadpoles are also transported after hatching, but they are not cannibalistic among each other and occur in small groups in small terrestrial water bodies. Unlike in *R. variabilis,* the tadpole-transporting parents do not avoid, but are attracted to, pools containing semiochemicals of conspecific tadpoles, possibly using them as an indicator for pool quality (Pašukonis et al., 2016). Another dendrobatid species where semiochemicals were tested in the context of parental care is the strawberry poison frog (*Oophaga pumilio*). In this species, the females place their tadpoles single in small phytotelmata and feed them regularly with trophic eggs. As soon as the female enters the pool, the tadpole starts a specific begging behaviour. Stynoski and Noble (2012) showed that this behaviour is not triggered by semiochemicals of the adult frogs, but by a combination of semiochemicals and visual cues. In contrast, in another egg-feeding species of the family Rhacophoridae (Eiffinger's tree frog, *Kurixalus eiffingeri*), tadpoles did respond to adult semiochemicals in expectation of feeding with increased activity. This behaviour was even stronger when a mix of substances from adults and conspecific tadpoles was offered (Kam & Yang, 2002). In *R. variabilis* there are also first hints that tadpoles may recognize adult caregivers chemically when they approach them for transportation, but further verification tests are still needed (Schulte & Mayer, 2017). Given the high diversity of parental care behaviours in anurans, the research of semiochemicals involved in these behaviours has great potential.

### 3.8 Sex Pheromones and Breeding Glands

Even though communication via sex pheromones is one of the most studied chemical communication forms in many taxa (Symonds & Elgar, 2008; Wyatt, 2014), only relatively few studies in anurans show a reaction to chemicals from the other sex. To the best of our knowledge, the first study was conducted by Rabb and Rabb (1963) and showed agitation behaviour, increased calling and attempts to clasp in aquatic male Surinam toads (*Pipa pipa*) after being exposed to water that previously contained mating conspecifics. Here, however, it could not yet be determined what exactly the males reacted to, if their behaviour was for example triggered by female pheromones, same-sex pheromones or a combination of both. Males orienting themselves towards airborne chemicals released by gravid females have been described for American toads (*Bufo americanus*; Forester & Thompson, 1998), and serrate-legged tree frogs (*Kurixalus odontotarsus*) changed their calling behaviour when they perceived airborne substances from females (Deng et al., 2020). Byrne and Keogh (2007) tested the reaction of male terrestrial toadlets (*Pseudophryne bibronii*) to female and male-marked substrates and found that the males showed a preference for females, while they rather avoided male substrates. Females did not show a preference for one sex over the other in the same tests, but they used male marks for orientation in a y-maze. For green-and-black poison dart frogs (*Dendrobates auratus*), a preference for the other sex was shown in both males and females tested in an olfactometer (Korbeck & McRobert, 2005). In tailed frogs (*Ascaphus truei*), only males showed a spatial preference for female chemicals, but both sexes reacted to chemicals of the opposite sex with increased jumping behaviour (Asey et al., 2005). Several studies showed that females react toward male chemicals, e.g., by spending more time in an area with tissues or water holding a male before (Graham et al., 2020; Rojas et al., 2021; but see Schulte & Rößler, 2013). But until today only for one frog species a male pheromone that triggers a female reaction has been deciphered: Wabnitz et al. (1999, 2000) found a skin peptide (named splendipherin) in the males' skin secretions of the magnificent tree frog (*Ranoidea "Litoria" splendida*) which attracts females at a short range.

While splendipherin was extracted from skin glands scattered over the frogs' dorsum, in many anuran species adult males develop sexually dimorphic skin glands that hypertrophy during the breeding season and that can be located at various positions across the body (e.g., below the chin, the hands, the thighs, etc.; Brizzi et al., 2003; Thomas et al., 1993). In an aquatic pipid species (*Hymenochirus* spec.), it was shown that females orient themselves towards both water housing males and water-containing homogenized post-axillary male-breeding glands (Pearl et al., 2000). Moreover, homogenized post-axillary breeding glands of male Yunnan pond frogs (*Nidirana pleuraden*) triggered a sexual attraction in females, but not in males. However, when trypsin, a protein digesting enzyme, was added, this attraction disappeared, which hints in the direction that a protein pheromone is involved (Gong et al., 2020).

The transcriptome of sexually dimorphic skin glands of several frog species (families Hylidae, Pipidae, Nyctibatrachidae, Ranidae) was examined, looking for pheromone candidates (Bossuyt et al., 2019b; Schulte et al., 2021, 2022). Sodefrin precursor-like factor (SPF), a protein known for its pheromone function in salamanders, was found to be highly expressed in the male glandular tissue. Likewise, Amplexin, another proteinaceous potential pheromone candidate, was described in grass frogs (*Rana temporaria*) and showed similarity with plethodontid modulating factor (PMF), a courtship pheromone in terrestrial salamanders (Willaert et al., 2013). The function of these proteins remains unknown in anurans, but a possible function as sex pheromones has been suggested, given the sex-specific occurrence as well as the resemblance of glandular tissue location and forms of transmission in salamanders (Bossuyt et al., 2019b; Schulte et al., 2021, 2022; Willaert et al., 2013).

Diverse volatile compounds known as macrolides, considered to be frog pheromone candidates (Schulz et al., 2021), were identified and described in the secretions of male femoral glands of mantellid frogs (Melnik et al., 2019; Peram et al., 2017; Poth et al., 2013) and male gular glands of hyperoliid frogs (Kuhn & Schulz, 2022; Menke et al., 2016)—regions known to have specialized glands with potential to release volatile compounds (Starnberger et al., 2013; Vences et al., 2007). Furthermore, macrolides identified in frogs were found to share key constituents of macrolides with a pheromone function in butterflies, bees, or beetles (Hötling et al., 2014; Schulz et al., 2021). Noteworthy is the case of Frogolide, a macrolide widespread within Mantellinae and Hyperoliidae (Menke et al., 2018).

Macrolides have been hypothesized to act as semiochemicals, yet the potential function of the different macrolides in frog reproduction is still to be demonstrated. The main olfactory organ (MOO) of mantellid species showed neuronal reactions to macrolides, which is indicative of an olfactory perception of these volatile compounds (Nowack et al., 2017). Only the study carried out by Poth et al. (2012) tested the effect of two compounds found in the extracts of male femoral glands on the behaviour of females and males of *M. multiplicatus*. Males increased activity with both compounds and showed a directional movement towards one. Females, however, did not show any significant reaction, which suggests that the substances are important for conspecific recognition rather than working as a sexual pheromone (see Sect. 3.6). All in all, there is strong evidence in favour of sexually dimorphic glands (breeding glands) being able to produce sexual pheromones—proteinaceous or volatile—but bioassays are still necessary to determine their function.

## 3.9 Semiochemicals as a Tool to Address Learning Behaviour, Chemical Pollution Effects and Invasive Species Control

Although semiochemicals in most anuran species are still poorly studied, for some species semiochemical recognition (and the corresponding behaviour) is so well understood that it has been used as a tool for further advanced research questions. For example, the known effect of DRAS on *L. sylvaticus* tadpoles has been used to approach a multitude of questions regarding learning behaviour, such as classical conditioning and habituation (e.g., Chivers et al., 2015; Ferrari et al., 2012, 2018; Mitchell et al., 2016b). In such experiments, DRAS were used as a known risk variable, which was displayed, for example, before or simultaneously with substances from novel predators to measure if the subsequent response to the predators correlates with the previously given DRAS concentrations or frequencies (Crane & Ferrari, 2017; Ferrari et al., 2009, 2016; Mitchell et al., 2016a), or the timing of previous DRAS exposure (Crane et al., 2019; Ferrari et al., 2008b; Ferrari & Chivers, 2009c). Other experiments showed that predator recognition can be transmitted from experienced to naïve tadpoles (Ferrari et al., 2007), and that the habitat in which the conditioning takes place can have an influence on the tadpoles' behaviours (Mitchell et al., 2018). Besides tadpoles, embryos can be exposed to DRAS and/or predator cues and alter the reaction of tadpoles after hatching (e.g., Ferrari & Chivers, 2009a, b; Lucon-Xiccato et al., 2017; Mathis et al., 2008). Apart from *L. sylvaticus*, tadpoles and embryos of other anuran species have been used for learning experiments, including other ranids (Batabyal et al., 2014; Fraker, 2009; Garcia et al., 2017; Gazzola et al., 2015; Gonzalo et al., 2013; Urszán et al., 2018), bufonids (Mandrillon & Saglio, 2005; Mirza et al., 2006), hylids (Ferrari & Chivers, 2008; Garcia et al., 2019), dicroglossids (Supekar & Gramapurohit, 2017, 2022) and leptodactylids (Pueta et al., 2022). The different effects tested in the context of DRAS and predator cue learning behaviours in anuran tadpoles and embryos are plentiful, and deserve their own review article.

Another topic where the recognition of (conspecific-fed) predator cues in tadpoles is used as a research tool is the investigation of the effect of chemical pollution on anuran behaviours. Ehrsam et al. (2016), for example, showed that the behavioural effects triggered by substances from conspecific-fed predators on Cuban tree frog tadpoles (*Osteopilus septentrionalis*) disappeared when tadpoles were exposed to ecologically relevant concentrations of herbicides. Also, the tadpoles of other anuran species show impaired reactions to predator cues in the presence of chemical pollutants such as insecticides, herbicides, agricultural by-products and waste water (Curi et al., 2022; Ferreira do Amaral et al., 2018; Mandrillon & Saglio, 2007b; Polo-Cavia et al., 2016; Sievers et al., 2018; Troyer & Turner, 2015; but see Mikó et al., 2015). Most of these results show that the alteration of the chemical environment mediates the chemosensory perception in tadpoles, making them more vulnerable to predation—and thereby having potentially deadly effects even when the doses are generally considered sublethal.

Besides the threats that anurans face (such as chemical pollution), under certain circumstances anurans can also be seen as a threat themselves—i.e., when they occur in non-native habitats. One of the most successful invasive species worldwide is the cane toad (*Rhinella marina*), and a lot of measures have been taken in the attempt to control their dispersal. Among them is also the use of conspecific semiochemicals. First studies found that DRAS of cane toad tadpoles may be a promising tool to negatively influence the behaviour, development and survival rate of conspecific tadpoles (Hagman et al., 2009; Hagman & Shine, 2008c). Consecutive studies focused on the potential use of competitive larval cues that suppress development in conspecific embryos (Clarke et al., 2015; Crossland & Shine, 2012; Haramura et al., 2017). Most promising for the use in tadpole-funnel-traps, however, are waterborne cues of late-stage cane toad eggs, which are actively cannibalized by conspecific tadpoles (Crossland & Shine, 2011). Crossland et al. (2012) found that the substances released by the eggs are also produced in the paratoid toxin glands of adult cane toads, and that the adult gland content can be used to lure tadpoles into the traps (while native heterospecific species are repelled by the toxins). Further studies showed that the egg-mimicking paratoid gland content is very effective in trapping the invasive tadpoles, both in the laboratory as well as in the field (although less reliably at colder water temperatures; Crossland et al., 2021; McCann et al., 2019a, b). The results of these studies present a very promising tool in the fight against invasive cane toads, and inspired similar studies in another highly invasive species, the American bullfrog (*Lithobates catesbeianus*). While their tadpoles are attracted to substances from freshly laid eggs as well as skin secretions from conspecific adults, the effect is less strong than in cane toads, and their use in traps might therefore not have the same positive effect (Xu et al., 2017).

## 4 Conclusions

Due to their complex life stages and adaptations to life in water and on land, anurans are an intriguing group to study chemical communication and their use of semiochemicals. Our standardized search revealed that there is a rich body of literature about this general topic. However, when dividing the studies by life stages, taxonomic ranks and types of semiochemicals, some disparities became evident. While research on tadpoles is quite extensive, we still know very little about the other stages, such as adult chemical communication. Also the taxonomic distribution of studies is very uneven: although Ranidae is not the most species-rich anuran family, nearly half of the studies concerning semiochemicals include ranid species. Many other anuran families have never been considered in semiochemical investigations. However, this seems to be slowly changing, with more and more species being studied outside of Europe and North America lately.

Regarding the semiochemicals analysed, we found especially diverse approaches investigating different types of substances involved in antipredator mechanisms. Reviewing the literature dealing with this topic, however, made it evident that the use of descriptive terminology is often inconsistent (for details also see Ferrari et al., 2010; Hettyey et al., 2015; Mitchell et al., 2017). Using the correct terminology is admittedly difficult when the chemical identity and functional characteristics are mostly unknown, yet key for their classification. For this reason, we want to emphasize the importance of carefully evaluating the available knowledge and assess key aspects of chemical communication, such as: (a) net benefits for sender and receiver, (b) emission by specific sources vs general (metabolic, endocrine, immune) sources and (c) specificity in response.

Other semiochemicals are still much less investigated. Pheromones included in courtship and mating behaviours, for example, have so far mainly been studied in one-sided approaches: either the frogs' reaction to unidentified substances of the other sex is known, or substances (volatiles or proteins) have been described without confirming their properties as pheromones in behavioural or physiological assays. The growing body of literature regarding the latter is, however, promising and the pheromonal properties (for definition see Table 1) of many described substances—both in water and on land, and for different developmental stages—might be proven soon. This is especially important when considering the worldwide decline of amphibian diversity. A better understanding of the role of potential courtship pheromones involved in amphibian reproduction may help ex situ conservation efforts.

## References

Achtymichuk, G. H., Crane, A. L., Simko, O. M., Stevens, H. E. F., & Ferrari, M. C. O. (2022). The choice of euthanasia techniques can affect experimental results in aquatic behavioural studies. *Animal Behaviour, 185*, 1–8.

Adams, M. J., & Claeson, S. (1998). Field response of tadpoles to conspecific and heterospecific alarm. *Ethology, 104*(11), 955–961.

Agosta, W. C. (1992). *Chemical communication: the language of pheromones*. Henry Holt and Company.

Ali, M. F., & Morgan, E. D. (1990). Chemical communication in insect communities: A guide to insect pheromones with special emphasis on social insects. *Biological Reviews, 65*(3), 227–247.

Altig, R., & Christensen, M. T. (1981). Behavioral characteristics of the tadpoles of *Rana heckscheri*. *Journal of Herpetology, 15*(2), 151–154.

Anderson, R. B., & Lawler, S. P. (2016). Behavioral changes in tadpoles after multigenerational exposure to an invasive intraguild predator. *Behavioral Ecology, 27*(6), 1790–1796.

Asey, M. J., Harowicz, P. G., & Su, L. (2005). Chemically mediated mate recognition in the Tailed Frog (*Ascaphus truei*). In R. T. Mason, M. P. LeMaster, & D. Müller-Schwarze (Eds.), *Chemical signals in vertebrates 10* (pp. 24–31). Springer.

Austin, C. E., March, R. E., Stock, N. L., & Murray, D. L. (2018). The origin and ecological function of an ion inducing anti-predator behavior in *Lithobates* tadpoles. *Journal of Chemical Ecology, 44*(2), 178–188.

Bairos-Novak, K. R., Crane, A. L., Achtymichuk, G. H., Hsin, J., Rivera-Hernández, I. A. E., Simko, O. M., Wrynn, T. E., Chivers, D. P., & Ferrari, M. C. O. (2020). Forget the audience: tadpoles release similar disturbance cues regardless of kinship or familiarity. *Behavioral Ecology and Sociobiology, 74*(12), 1–10.

Balestrieri, A., Gazzola, A., Pellitteri-Rosa, D., & Vallortigara, G. (2019). Discrimination of group numerousness under predation risk in anuran tadpoles. *Animal Cognition, 22*(2), 223–230.

Barbasch, T., & Benard, M. F. (2011). Past and present risk: Exposure to predator chemical cues before and after metamorphosis influences juvenile wood frog behavior. *Ethology, 117*(5), 367–373.

Batabyal, A., Gosavi, S. M., & Gramapurohit, N. P. (2014). Determining sensitive stages for learning to detect predators in larval bronzed frogs: Importance of alarm cues in learning. *Journal of Biosciences, 39*(4), 701–710.

Belanger, R. M., & Corkum, L. D. (2009). Review of aquatic sex pheromones and chemical communication in anurans. *Journal of Herpetology, 43*(2), 184–191.

Belden, L. K., Wildy, E. L., Hatch, A. C., & Blaustein, A. R. (2000). Juvenile western toads, *Bufo boreas*, avoid chemical cues of snakes fed juvenile, but not larval, conspecifics. *Animal Behaviour, 59*, 871–875.

Benard, M. F., & Fordyce, J. A. (2003). Are induced defenses costly? Consequences of predator-induced defenses in western toads, Bufo boreas. *Ecology, 84*(1), 68–78.

Bennett, A. M., Pereira, D., & Murray, D. L. (2013). Investment into defensive traits by anuran prey (*Lithobates pipiens*) is mediated by the starvation-predation risk trade-off. *PLoS One, 8*(12), e82344.

Berec, M., Klapka, V., & Zemek, R. (2016). Effect of an alien turtle predator on movement activity of European brown frog tadpoles. *The Italian Journal of Zoology, 83*(1), 68–76.

Blaustein, A. R., & O'Hara, R. K. (1982). Kin recognition in *Rana cascadae* tadpoles: Maternal and paternal effects. *Animal Behaviour, 30*(4), 1151–1157.

Blaustein, A. R., Yoshikawa, T., Asoh, K., & Walls, S. C. (1993). Ontogenetic shifts in tadpole kin recognition: Loss of signal and perception. *Animal Behaviour, 46*, 525–538.

Bossuyt, F., Maex, M., Treer, D., Schulte, L. M., Van Bocxlaer, I., & Janssenswillen, S. (2019a). Chemistry between salamanders: Evolution of the SPF courtship pheromone system in Salamandridae. In C. D. Buesching (Ed.), *Chemical signals in vertebrates 14* (pp. 205–220). Springer.

Bossuyt, F., Schulte, L. M., Maex, M., Janssenswillen, S., Novikova, P. Y., Biju, S. D., Van de Peer, Y., Matthijs, S., Roelants, K., Martel, A., & Van Bocxlaer, I. (2019b). Multiple independent recruitment of sodefrin precursor-like factors in anuran sexually dimorphic glands. *Molecular Biology and Evolution, 36*(9), 1921–1930.

Brizzi, R., Delfino, G., & Jantra, S. (2003). An overview of breeding glands. In B. G. M. Jamieson (Ed.), *Reproductive biology and phylogeny of Anura* (pp. 253–317). Science Publishers, Inc.

Buttermore, K. F., Litkenhaus, P. N., Torpey, D. C., Smith, G. R., & Rettig, J. E. (2011). Effects of mosquitofish (*Gambusia affinis*) cues on wood frog (*Lithobates sylvaticus*) tadpole activity. *Acta Herpetologica, 6*(1), 81–85.

Buxton, V. L., & Sperry, J. H. (2017). Reproductive decisions in anurans: A review of how predation and competition affects the deposition of eggs and tadpoles. *BioScience, 67*(1), 26–38.

Byrne, P. G., & Keogh, J. S. (2007). Terrestrial toadlets use chemosignals to recognize conspecifics, locate mates and strategically adjust calling behaviour. *Animal Behaviour, 74*(5), 1155–1162.

Capellán, E., & Nicieza, A. G. (2010). Constrained plasticity in switching across life stages: Pre- and post-switch predators elicit early hatching. *Evolutionary Ecology, 24*(1), 49–57.

Carlson, B. E., Newman, J. C., & Langkilde, T. (2015). Food or fear: Hunger modifies responses to injured conspecifics in tadpoles. *Hydrobiologia, 743*(1), 299–308.

Casillas-Barragán, I., Costa-Pereira, R., & Peixoto, P. E. C. (2016). Perceived predation risk decreases movement and increases aggregation of Amazon milk frog (Anura, Hylidae) tadpoles throughout ontogeny. *Hydrobiologia, 765*(1), 379–386.

Chivers, D. P., & Mirza, R. S. (2001). Importance of predator diet cues in responses of larval wood frogs to fish and invertebrate predators. *Journal of Chemical Ecology, 27*(1), 45–51.

Chivers, D. P., & Smith, R. J. F. (1998). Chemical alarm signalling in aquatic predator-prey systems: A review and prospectus. *Ecoscience, 5*(3), 338–352.

Chivers, D. P., Kiesecker, J. M., Marco, A., Wildy, E. L., & Blaustein, A. R. (1999a). Shifts in life history as a response to predation in western toads (*Bufo boreas*). *Journal of Chemical Ecology, 25*(11), 2455–2463.

Chivers, D. P., Kiesecker, J. M., Wildy, E. L., Belden, L. K., Kats, L. B., & Blaustein, A. R. (1999b). Avoidance response of post-metamorphic anurans to cues of injured conspecifics and predators. *Journal of Herpetology, 33*(3), 472–476.

Chivers, D. P., Wildy, E. L., Kiesecker, J. M., & Blaustein, A. R. (2001). Avoidance response of juvenile pacific treefrogs to chemical cues of introduced predatory bullfrogs. *Journal of Chemical Ecology, 27*(8), 1667–1676.

Chivers, D. P., Kiesecker, J. M., Marco, A., DeVito, J., Anderson, M. T., & Blaustein, A. R. (2003). Predator-induced life history changes in amphibians: Egg predation induces hatching. *Oikos, 92*(1), 135–142.

Chivers, D. P., Mathiron, A., Sloychuk, J. R., & Ferrari, M. C. O. (2015). Responses of tadpoles to hybrid predator odours: Strong maternal signatures and the potential risk/response mismatch. *Proceedings of the Royal Society B: Biological Sciences, 282*, 20150365.

Chuirazzi, C., Ocampo, M., & Takahashi, M. K. (2021). Influence of prey diet quality on predator-induced traits in wood frog tadpoles (*Lithobates sylvaticus*). *Amphibia-Reptilia, 42*(3), 331–341.

Clarke, G. S., Crossland, M. R., Shilton, C., & Shine, R. (2015). Chemical suppression of embryonic cane toads *Rhinella marina* by larval conspecifics. *Journal of Applied Ecology, 52*(6), 1547–1557.

Clarke, G. S., Crossland, M. R., & Shine, R. (2016). Can we control the invasive cane toad using chemicals that have evolved under intraspecific competition? *Ecological Applications, 26*(2), 463–474.

Costa, R. N., & Nomura, F. (2014). Assessment risk and limited behavioral plasticity in tadpoles of *Rhinella ornata* (Anura, Bufonidae). *Iheringia, 104*(2), 162–167.

Crane, A. L., & Ferrari, M. C. O. (2017). Evidence for risk extrapolation in decision making by tadpoles. *Scientific Reports, 7*(1), 1–7.

Crane, A. L., Chivers, D. P., & Ferrari, M. C. O. (2019). Time-dependent latent inhibition of predator-recognition learning. *Biology Letters, 15*(5), 20190183.

Crossland, M. R., & Shine, R. (2011). Cues for cannibalism: Cane toad tadpoles use chemical signals to locate and consume conspecific eggs. *Oikos, 120*(3), 327–332.

Crossland, M. R., & Shine, R. (2012). Embryonic exposure to conspecific chemicals suppresses cane toad growth and survival. *Biology Letters, 8*(2), 226–229.

Crossland, M. R., Haramura, T., Salim, A. A., Capon, R. J., & Shine, R. (2012). Exploiting intraspecific competitive mechanisms to control invasive cane toads (*Rhinella marina*). *Proceedings of the Royal Society B: Biological Sciences, 279*, rspb20120821.

Crossland, M. R., Salim, A. A., Capon, R. J., & Shine, R. (2019). The effects of conspecific alarm cues on larval cane toads (*Rhinella marina*). *Journal of Chemical Ecology, 45*(10), 838–848.

Crossland, M. R., Salim, A. A., Capon, R. J., & Shine, R. (2021). Chemical cues that attract cannibalistic cane toad (*Rhinella marina*) larvae to vulnerable embryos. *Scientific Reports, 11*(1), 1–8.

Crump, M. L. (1996). Parental care among the amphibia. In J. S. Rosenblatt & C. T. Snowdon (Eds.), *Parental care. Evolution, mechanisms, and adaptive significance. Advances in the study of behavior* (Vol. 25p, pp. 109–144). Academic Press.

Curi, L. M., Cuzziol Boccioni, A. P., Peltzer, P. M., Attademo, A. M., Bassó, A., León, E. J., & Lajmanovich, R. C. (2022). Signals from predators, injured conspecifics, and pesticide modify the swimming behavior of the gregarious tadpole of the Dorbigny's Toad, *Rhinella dorbignyi* (Anura: Bufonidae). *Canadian Journal of Zoology, 100*(999), 19–27.

Delaney, K., & Bishop, P. J. (2007). Communication in the world's most ancestral frogs. *New Zealand Journal of Zoology, 34*(3), 262.

Deng, K., He, Q.-L., Zhou, Y., Zhu, B.-C., Wang, T.-L., Wang, J.-C., & Cui, J.-G. (2020). Male serrate-legged treefrogs adjust competition strategies according to visual or chemical cues from females. *The Journal of Experimental Biology, 223*(21), jeb229245.

Deng, K., Zhou, Y., He, Q.-L., Zhu, B.-C., Wang, T.-L., Wang, J.-C., & Cui, J.-G. (2021). Conspecific odor cues induce different vocal responses in serrate-legged small treefrogs, but only in the absence of acoustic signals. *Frontiers in Zoology, 18*(1), 1–6.

Deng, K., Zhou, Y., Zhang, H.-D., He, Q.-L., Zhu, B.-C., Wang, T.-L., Wang, J.-C., Halfwerk, W., & Cui, J.-G. (2022). Conspecific disturbance odors act as alarm cues to affect female mate choice in a treefrog. *Behavioral Ecology and Sociobiology, 76*(4), 1–8.

Dibble, C. J., Kauffman, J. E., Zuzik, E. M., Smith, G. R., & Rettig, J. E. (2009). Effects of potential predator and competitor cues and sibship on wood frog (*Rana sylvatica*) embryos. *Amphibia-Reptilia, 30*(2), 294–298.

Dicke, M., & Sabelis, M. W. (1988). Infochemical terminology: Based on cost-benefit analysis rather than origin of compounds? *Functional Ecology, 2*, 131–139.

Dijk, B., Laurila, A., Orizaola, G., & Johansson, F. (2016). Is one defence enough? Disentangling the relative importance of morphological and behavioural predator-induced defences. *Behavioral Ecology and Sociobiology, 70*(2), 237–246.

Dodson, S. I., Crowl, T. A., Peckarsky, B. L., Kats, L. B., Covich, A. P., & Culp, J. M. (1994). Non-visual communication in freshwater benthos: An overview. *Journal of the North American Benthological Society, 13*(2), 268–282.

Dole, J. W., Rose, B. B., & Tachiki, K. H. (1981). Western toads (*Bufo boreas*) learn odor of prey insects. *Herpetologica, 37*(1), 63–68.

Ehrsam, M., Knutie, S. A., & Rohr, J. R. (2016). The herbicide atrazine induces hyperactivity and compromises tadpole detection of predator chemical cues. *Environmental Toxicology and Chemistry, 35*, 2239–2244.

Eibl-Eibesfeldt, I. (1949). Über das Vorkommen von Schreckstoffen bei Erdkrötenquappen. *Experientia, 5*, 236.

Eklöv, P. (2000). Chemical cues from multiple predator-prey interactions induce changes in behavior and growth of anuran larvae. *Oecologia, 123*(2), 192–199.

El Balaa, R., & Blouin-Demers, G. (2013). Does exposure to cues of fish predators fed different diets affect morphology and performance of Northern Leopard Frog (*Lithobates pipiens*) larvae? *Canadian Journal of Zoology, 91*(4), 203–211.

Eluvathingal, L. M., Shanbhag, B. A., & Saidapur, S. K. (2009). Association preference and mechanism of kin recognition in tadpoles of the toad *Bufo melanostictus*. *Journal of Biosciences, 34*(3), 435–444.

Escalona Sulbarán, M. D., Ivo Simões, P., Gonzalez-Voyer, A., & Castroviejo-Fisher, S. (2019). Neotropical frogs and mating songs: The evolution of advertisement calls in glassfrogs. *Journal of Evolutionary Biology, 32*(2), 163–176.

Feminella, J. W., & Hawkins, C. P. (1994). Tailed frog tadpoles differentially alter their feeding behavior in response to von-visual cues from four predators. *Journal of the North American Benthological Society, 13*(2), 310–320.

Ferland-Raymond, B., March, R. E., Metcalfe, C. D., & Murray, D. L. (2010). Prey detection of aquatic predators: Assessing the identity of chemical cues eliciting prey behavioral plasticity. *Biochemical Systematics and Ecology, 38*(2), 169–177.

Ferrari, M. C. O., & Chivers, D. P. (2008). Cultural learning of predator recognition in mixed-species assemblages of frogs: The effect of tutor-to-observer ratio. *Animal Behaviour, 75*(6), 1921–1925.

Ferrari, M. C. O., & Chivers, D. P. (2009a). Latent inhibition of predator recognition by embryonic amphibians. *Biology Letters, 5*(2), 160–162.

Ferrari, M. C. O., & Chivers, D. P. (2009b). Sophisticated early life lessons: Threat-sensitive generalization of predator recognition by embryonic amphibians. *Behavioral Ecology, 20*(6), 1295–1298.

Ferrari, M. C. O., & Chivers, D. P. (2009c). Temporal variability, threat sensitivity and conflicting information about the nature of risk: Understanding the dynamics of tadpole antipredator behaviour. *Animal Behaviour, 78*(1), 11–16.

Ferrari, M. C. O., Messier, F., & Chivers, D. P. (2007). First documentation of cultural transmission of predator recognition by larval amphibians. *Ethology, 113*(6), 621–627.

Ferrari, M. C. O., Messier, F., & Chivers, D. P. (2008a). Degradation of chemical alarm cues under natural conditions: Risk assessment by larval woodfrogs. *Chemoecology, 17*(4), 263–266.

Ferrari, M. C. O., Messier, F., & Chivers, D. P. (2008b). Larval amphibians learn to match antipredator response intensity to temporal patterns of risk. *Behavioral Ecology, 19*(5), 980–983.

Ferrari, M. C. O., Brown, G. E., Messier, F., & Chivers, D. P. (2009). Threat-sensitive generalization of predator recognition by larval amphibians. *Behavioral Ecology and Sociobiology, 63*(9), 1369–1375.

Ferrari, M. C. O., Wisenden, B. D., & Chivers, D. P. (2010). Chemical ecology of predator–prey interactions in aquatic ecosystems: A review and prospectus. *Canadian Journal of Zoology, 88*(7), 698–724.

Ferrari, M. C. O., Brown, G. E., & Chivers, D. P. (2012). Temperature-mediated changes in rates of predator forgetting in woodfrog tadpoles. *PLoS One, 7*(12), e51143.

Ferrari, M. C. O., Crane, A. L., & Chivers, D. P. (2016). Certainty and the cognitive ecology of generalization of predator recognition. *Animal Behaviour, 111*, 207–211.

Ferrari, M. C. O., Brown, G. E., & Chivers, D. P. (2018). Understanding the effect of uncertainty on the development of neophobic antipredator phenotypes. *Animal Behaviour, 136*, 101–106.

Ferreira do Amaral, D., Montalvão, M. F., de Oliveira Mendes, B., de Souza, J. M., Chagas, T. Q., de Lima Rodrigues, A. S., & Malafaia, G. (2018). Insights about the toxic effects of tannery effluent on *Lithobates catesbeianus* tadpoles. *Science of the Total Environment, 621*, 791–801.

Flowers, M. A., & Graves, B. M. (1997). Juvenile toads avoid chemical cues from snake predators. *Animal Behaviour, 53*(3), 641–646.

Forester, D. C., & Thompson, K. J. (1998). Gauntlet behaviour as a male sexual tactic in the American toad (Amphibia: Bufonidae). *Behaviour, 135*(2), 99–119.

Fraker, M. E. (2009). Predation risk assessment by green frog (*Rana clamitans*) tadpoles through chemical cues produced by multiple prey. *Behavioral Ecology and Sociobiology, 63*(10), 1397–1402.

Fraker, M. E., Hu, F., Cuddapah, V., McCollum, S. A., Relyea, R. A., Hempel, J., & Denver, R. J. (2009). Characterization of an alarm pheromone secreted by amphibian tadpoles that induces behavioral inhibition and suppression of the neuroendocrine stress axis. *Hormones and Behavior, 55*(4), 520–529.

Frost, D. R. (2022). *Amphibian species of the world: An online reference*. Version 6.1. (2 Nov 2022). American Museum of Natural History, New York. https://amphibiansoftheworld.amnh.org/index.php

Garcia, T. S., Urbina, J. C., Bredeweg, E. M., & Ferrari, M. C. O. (2017). Embryonic learning and developmental carry-over effects in an invasive anuran. *Oecologia, 184*(3), 623–631.

Garcia, T. S., Bredeweg, E. M., Urbina, J., & Ferrari, M. C. O. (2019). Evaluating adaptive, carry-over, and plastic antipredator responses across a temporal gradient in Pacific chorus frogs. *Ecology, 100*(11), e02825.

Gascuel, J., & Amano, T. (2013). Exotic models may offer unique opportunities to decipher specific scientific questions: The case of *Xenopus olfactory* system. *The Anatomical Record, 296*(9), 1453–1461.

Gazzola, A., Brandalise, F., Rubolini, D., Rossi, P., & Galeotti, P. (2015). Fear is the mother of invention: Anuran embryos exposed to predator cues alter life-history traits, post-hatching behaviour and neuronal activity patterns. *The Journal of Experimental Biology, 218*(24), 3919–3930.

Gazzola, A., Balestrieri, A., Scribano, G., Fontana, A., & Pellitteri-Rosa, D. (2021). Contextual behavioural plasticity in Italian agile frog (*Rana latastei*) tadpoles exposed to native and alien predator cues. *The Journal of Experimental Biology, 224*(9), jeb240465.

Gerhardt, H. C., & Huber, F. (2002). *Acoustic communication in insects and anurans*. University of Chicago Press.

Giri, A., Yadav, S. S., Giri, S., & Sharma, G. D. (2012). Effect of predator stress and malathion on tadpoles of Indian skittering frog. *Aquatic Toxicology, 106–107*, 157–163.

Gomez-Mestre, I., & Díaz-Paniagua, C. (2011). Invasive predatory crayfish do not trigger inducible defences in tadpoles. *Proceedings of the Royal Society B: Biological Sciences, 278*(1723), 3364–3370.

Gong, Y., Zeng, Y., Zheng, P., Liao, X., & Xie, F. (2020). Structural and bio-functional assessment of the postaxillary gland in *Nidirana pleuraden* (Amphibia: Anura: Ranidae). *Zoological Letters, 6*(1), 1–16.

Gonzalo, A., Cabido, C., Galan, P., Lopez, P., & Martin, J. (2006). Predator, but not conspecific, chemical cues influence pond selection by recently metamorphosed Iberian green frogs, Rana perezi. *Canadian Journal of Zoology, 84*(9), 1295–1299.

Gonzalo, A., Lopez, P., & Martin, J. (2010). Risk level of chemical cues determines retention of recognition of new predators in Iberian green frog tadpoles. *Behavioral Ecology and Sociobiology, 64*(7), 1117–1123.

Gonzalo, A., López, P., & Martín, J. (2013). Adaptive forgetting in Iberian green frog tadpoles (*Pelophylax perezi*): Learned irrelevance and latent inhibition may avoid predator misidentification. *Journal of Comparative Psychology, 127*(1), 56.

Gosner, K. L. (1960). A simplified table for staging anuran embryos and larvae with notes on identification. *Herpetologica, 16*(3), 183–190.

Graham, B. M., O'Hearn, D. J., MacAllister, I. E., & Sperry, J. H. (2020). Behavioral responses by adult Northern leopard frogs to conspecific chemical cues. *Journal of Herpetology, 54*(2), 168–173.

Graves, B. M., Summers, C. H., & Olmstead, K. L. (1993). Sensory mediation of aggregation among postmetamorphic *Bufo cognatus*. *Journal of Herpetology, 27*(3), 315–319.

Grubb, J. C. (1973). Olfactory orientation in *Bufo woodhousei fowleri, Pseudacris clarki* and *Pseudacris streckeri*. *Animal Behaviour, 21*(4), 726–732.

Hagman, M., & Shine, R. (2008a). Australian tadpoles do not avoid chemical cues from invasive cane toads (*Bufo marinus*). *Wildlife Research, 35*(1), 59–64.

Hagman, M., & Shine, R. (2008b). Tadpoles of invasive cane toads (*Bufo marinus*) do not respond behaviourally to chemical cues from tadpoles of four species of Australian frogs. *Australian Journal of Zoology, 56*(4), 211–213.

Hagman, M., & Shine, R. (2008c). Understanding the toad code: Behavioural responses of cane toad (*Chaunus marinus*) larvae and metamorphs to chemical cues. *Austral Ecology, 33*(1), 37–44.

Hagman, M., & Shine, R. (2009a). Factors influencing responses to alarm pheromone by larvae of invasive cane toads, *Bufo marinus*. *Journal of Chemical Ecology, 35*(2), 265–271.

Hagman, M., & Shine, R. (2009b). Larval alarm pheromones as a potential control for invasive cane toads (*Bufo marinus*) in tropical Australia. *Chemoecology, 19*(4), 211–217.

Hagman, M., Hayes, R. A., Capon, R. J., & Shine, R. (2009). Alarm cues experienced by cane toad tadpoles affect post-metamorphic morphology and chemical defences. *Functional Ecology, 23*(1), 126–132.

Hamer, R., Lemckert, F. L., & Banks, P. B. (2011). Adult frogs are sensitive to the predation risks of olfactory communication. *Biology Letters, 7*(3), 361–363.

Haramura, T., Crossland, M. R., Takeuchi, H., & Shine, R. (2017). Methods for invasive species control are transferable across invaded areas. *PLoS One, 12*(11), e0187265.

Hazlett, B. A. (1985). Disturbance pheromones in the crayfish *Orconectes virilis*. *Journal of Chemical Ecology, 11*(12), 1695–1711.

Hemnani, M., Guimarães, I. S. C., Kaefer, I. L., & da Pires, T. H. S. (2022). Alarm reaction depends on multiple chemical cues in tadpoles of the cane toad (*Rhinella marina*). *Ethology Ecology and Evolution, 35*, 1–13.

Hepper, P. G., & Waldman, B. (1992). Embryonic olfactory learning in frogs. *Quarterly Journal of Experimental Psychology: Section B, 44*(3–4), 179–197.

Hettyey, A., Tóth, Z., Thonhauser, K. E., Frommen, J. G., Penn, D. J., & Van Buskirk, J. (2015). The relative importance of prey-borne and predator-borne chemical cues for inducible anti-predator responses in tadpoles. *Oecologia, 179*(3), 699–710.

Hettyey, A., Üveges, B., Móricz, Á. M., Drahos, L., Capon, R. J., Van Buskirk, J., Tóth, Z., & Bókony, V. (2019). Predator-induced changes in the chemical defence of a vertebrate. *The Journal of Animal Ecology, 88*(12), 1925–1935.

Hews, D. K., & Blaustein, A. R. (1985). An investigation of the alarm response in *Bufo boreas* and *Rana cascadae* tadpoles. *Behavioral and Neural Biology, 43*(1), 47–57.

Hickman, C. R., & Watling, J. I. (2014). Leachates from an invasive shrub causes risk-prone behavior in a larval amphibian. *Behavioral Ecology, 25*(2), 300–305.

Holomuzki, J. R. (1995). Oviposition sites and fish-deterrent mechanisms of two stream anurans. *Copeia, 1995*(3), 607–613.

Hossie, T., Landolt, K., & Murray, D. L. (2017). Determinants and co-expression of anti-predator responses in amphibian tadpoles: A meta-analysis. *Oikos, 126*(2), 173–184.

Hötling, S., Haberlag, B., Tamm, M., Collatz, J., Mack, P., Steidle, J. L. M., Vences, M., & Schulz, S. (2014). Identification and synthesis of macrolide pheromones of the grain beetle *Oryzaephilus surinamensis* and the frog *Spinomantis aglavei*. *Chemistry - A European Journal, 20*(11), 3183–3191.

Houck, L. D. (2009). Pheromone communication in amphibians and reptiles. *Annual Review of Physiology, 71*, 161–176.

Hrbáček, J. (1950). On the flight reaction of tadpoles of the common toad caused by chemical substances. *Cellular and Molecular Life Sciences, 6*(3), 100–102.

Iglesias-Carrasco, M., Cabido, C., & Ord, T. J. (2022). Natural toxins leached from *Eucalyptus globulus* plantations affect the development and life-history of anuran tadpoles. *Freshwater Biology, 67*(2), 378–388.

Jefferson, D. M., Hobson, K. A., & Chivers, D. P. (2013). Understanding the information value of repeated exposure to chemical alarm cues: What can growth patterns tell us? *Annales Zoologici Fennici, 50*, 237–246.

Kam, Y.-C., & Yang, H.-W. (2002). Female-offspring communication in a Taiwanese tree frog, *Chirixalus eiffingeri* (Anura: Rhacophoridae). *Animal Behaviour, 64*(6), 881–886.

Karlson, P., & Lüscher, M. (1959). 'Pheromones': A new term for a class of biologically active substances. *Nature, 183*(4653), 55–56.

Kats, L. B., & Dill, L. M. (1998). The scent of death: Chemosensory assessment of predation risk by prey animals. *Ecoscience, 5*(3), 361–394.

Kats, L. B., Petranka, J. W., & Sih, A. (1988). Antipredator defenses and the persistence of amphibian larvae with fishes. *Ecology, 69*(6), 1865–1870.

Kiesecker, J. M., & Blaustein, A. R. (1997). Population differences in responses of red-legged frogs (*Rana aurora*) to introduced bullfrogs. *Ecology, 78*(6), 1752–1760.

Kiesecker, J. M., Chivers, D. P., & Blaustein, A. R. (1996). The use of chemical cues in predator recognition by western toad tadpoles. *Animal Behaviour, 52*(6), 1237–1245.

Kiesecker, J. M., Chivers, D. P., Marco, A., Quilchano, C., Anderson, M. T., & Blaustein, A. R. (1999a). Identification of a disturbance signal in larval red-legged frogs, Rana aurora. *Animal Behaviour, 57*(6), 1295–1300.

Kiesecker, J. M., Skelly, D. K., Beard, K. H., & Preisser, E. (1999b). Behavioral reduction of infection risk. *Proceedings of the National Academy of Sciences, 96*(16), 9165–9168.

Kiesecker, J. M., Chivers, D. P., Anderson, M., & Blaustein, A. R. (2002). Effect of predator diet on life history shifts of red-legged frogs, *Rana aurora*. *Journal of Chemical Ecology, 28*(5), 1007–1015.

Kikuyama, S., Yamamoto, K., Iwata, T., & Toyoda, F. (2002). Peptide and protein pheromones in amphibians. *Comparative Biochemistry and Physiology Part B: Biochemistry and Molecular Biology, 132*(1), 69–74.

King, J. D., Rollins-Smith, L. A., Nielsen, P. F., John, A., & Conlon, J. M. (2005). Characterization of a peptide from skin secretions of male specimens of the frog, *Leptodactylus fallax* that stimulates aggression in male frogs. *Peptides, 26*(4), 597–601.

Kiseleva, E. I. (1989). Reactions of *Bufo bufo* L. toad tadpoles on chemical signals of individuals of the same and different species. *Zhurnal Obshchei Biologii, 50*(6), 794–798.

Kiseleva, E. I. (1996). The Chemical interaction of frog tadpoles (*Rana temporaria* L., Anura, Amphibia) with conspecific and heterospecific anuran tadpoles. *Zhurnal Obshchei Biologii, 57*(6), 740–746.

Kloskowski, J. (2020). Better desiccated than eaten by fish: distribution of anurans among habitats with different risks to offspring. *Freshwater Biology, 65*(12), 2124–2134.

Korbeck, R. G., & McRobert, S. P. (2005). Home area recognition via olfactory cues in the tropical poison frog *Dendrobates auratus*. *Russian Journal of Herpetology, 12*(3), 161–166.

Kramer, G. (1933). Untersuchungen über die Sinnesleistungen und das Orientierungsverhalten von *Xenopus laevis* Daud. *Zoologische Jahrbucher Abteilung Für Allgemeine Zoologie Und Physiologie Der Tiere, 52*, 629–676.

Kuhn, J., & Schulz, S. (2022). Cinnamomeoventrolide–double bond regioisomerism in frog semiochemicals. *Journal of Chemical Ecology, 48*(5), 531–545.

Kulzer, E. (1954). Untersuchungen über die Schreckreaktion der Erdkrötenkaulquappen (*Bufo bufo*). *Zeitschrift Für Vergleichene Physiologie, 36*, 443–463.

Kumpulainen, N. (2022). *Communication between the tadpoles of the dyeing poison frog Dendrobates tinctorius (Anura, Dendrobatidae)*. Master Thesis, University of Jyväskylä

Kurali, A., Pásztor, K., Hettyey, A., & Tóth, Z. (2016). Toxin depletion has no effect on antipredator responses in common toad (*Bufo bufo*) tadpoles. *Biological Journal of the Linnean Society, 119*(4), 1000–1010.

Laurila, A. (2000). Behavioural responses to predator chemical cues and local variation in antipredator performance in *Rana temporaria* tadpoles. *Oikos, 88*(1), 159–168.

Law, J. H., & Regnier, F. E. (1971). Pheromones. *Annual Review of Biochemistry, 40*(1), 533–548.

Lee, J. S. F., & Waldman, B. (2002). Communication by fecal chemosignals in an archaic frog, Leiopelma hamiltoni. *Copeia, 2002*(3), 679–686.

Lefcort, H. (1996). Adaptive, chemically mediated fright response in tadpoles of the southern leopard frog, Rana utricularia. *Copeia, 1996*(2), 455–459.

Lefcort, H., & Eiger, S. M. (1993). Antipredatory behaviour of feverish tadpoles: Implications for pathogen transmission. *Behaviour, 126*(1/2), 13–27.

Lucon-Xiccato, T. (2019). Chemical alarm cues allow prey to adjust their defensive behaviour to cover abundance. *Behavioural Processes, 162*, 86–89.

Lucon-Xiccato, T., Chivers, D. P., Mitchell, M. D., & Ferrari, M. C. O. (2016). Making the dead talk: alarm cue-mediated antipredator behaviour and learning are enhanced when injured conspecifics experience high predation risk. *Biology Letters, 12*(8), 20160560.

Lucon-Xiccato, T., Chivers, D. P., Mitchell, M. D., & Ferrari, M. C. O. (2017). Prenatal exposure to predation affects predator recognition learning via lateralization plasticity. *Behavioral Ecology, 28*(1), 253–259.

Maag, N., Gehrer, L., & Woodhams, D. C. (2012). Sink or swim: a test of tadpole behavioral responses to predator cues and potential alarm pheromones from skin secretions. *Journal of Comparative Physiology A, 198*(11), 841–846.

Madison, D. M. (1977). Chemical communication in amphibians and reptiles. In D. Müller-Schwarze & M. Mozell (Eds.), *Chemical signals in vertebrates* (pp. 135–168). Plenum Press.

Mandrillon, A. L., & Saglio, P. (2005). Prior exposure to conspecific chemical cues affects predator recognition in larval common toad (*Bufo bufo*). *Archiv für Hydrobiologie, 164*(1), 1–12.

Mandrillon, A. L., & Saglio, P. (2007a). Effects of embryonic exposure to conspecific chemical cues on hatching and larval traits in the common frog (*Rana temporaria*). *Chemoecology, 17*(3), 169–175.

Mandrillon, A. L., & Saglio, P. (2007b). Herbicide exposure affects the chemical recognition of a non native predator in common toad tadpoles (*Bufo bufo*). *Chemoecology, 17*(1), 31–36.

Mandrillon, A. L., & Saglio, P. (2009). Effects of single and combined embryonic exposures to herbicide and conspecific chemical alarm cues on hatching and larval traits in the common frog (*Rana temporaria*). *Archives of Environmental Contamination and Toxicology, 56*(3), 566–576.

Manteifel, Y. (1995). Chemically-mediated avoidance of predators by *Rana temporaria* tadpoles. *Journal of Herpetology, 29*(3), 461–463.

Manteifel, Y. B. (2001). Avoidance of alarm pheromone by tadpoles of the common toad, *Bufo bufo*, in nature. *Zoologicheskiĭ Zhurnal, 80*, 67–70.

Manteifel, Y. B., & Kiseleva, E. I. (2011). Ammonia as a pheromone in anuran tadpoles. *Physiological Research, 60*(1), S185–S191.

Manteifel, Y. B., & Zhushev, A. V. (1996). Avoidance of natural chemical stimuli by tadpoles of three anuran species. *Russian Journal of Zoology, 75*(6), 900–906.

Manteifel, Y. B., Kiseleva, E. I., & Margolis, S. E. (2005). An increase in ammonium concentration as a non-specific pheromone signal that is avoided by amphibian larvae. *Zoologičeskij Žurnal, 84*(10), 1289–1297.

Manteifel, Y. B., Margolis, S. E., & Kiseleva, E. I. (2006). Disturbance pheromone in the common toad (*Bufo bufo*) larvae: The possible communicative role of ammonium-containing excretions. *Doklady Biological Sciences, 406*(3), 82–83.

Marino, J. A., Holland, M. P., & Middlemis Maher, J. (2014). Predators and trematode parasites jointly affect larval anuran functional traits and corticosterone levels. *Oikos, 123*(4), 451–460.

Mathis, A., Ferrari, M. C. O., Windel, N., Messier, F., & Chivers, D. P. (2008). Learning by embryos and the ghost of predation future. *Proceedings of the Royal Society B: Biological Sciences, 275*(1651), 2603.

Maturana, H. R., Lettvin, J. Y., McCulloch, W. S., & Pitts, W. H. (1960). Anatomy and physiology of vision in the frog (*Rana pipiens*). *The Journal of General Physiology, 43*(6), 129.

McCann, S., Crossland, M., Greenlees, M., & Shine, R. (2019a). Invader control: Factors influencing the attraction of cane toad (*Rhinella marina*) larvae to adult parotoid exudate. *Biological Invasions, 21*(6), 1895–1904.

McCann, S., Crossland, M. R., & Shine, R. (2019b). Pheromones can cull an invasive amphibian without releasing survivors from intraspecific competition. *Ecosphere, 10*(12), e02969.

McCoy, M. W., Wheat, S. K., Warkentin, K. M., & Vonesh, J. R. (2015). Risk assessment based on indirect predation cues: Revisiting fine-grained variation. *Ecology and Evolution, 5*(20), 4523–4528.

McIntyre, P. B., & McCollum, S. A. (2000). Responses of bullfrog tadpoles to hypoxia and predators. *Oecologia, 125*(2), 301–308.

McMahon, T. A., Laggan, N. A., & Hill, M. N. (2019). Metabolites produced by *Batrachochytrium dendrobatidis* alter development in tadpoles, but not growth or mortality. *Diseases of Aquatic Organisms, 135*(3), 251–255.

McMahon, T. A., Hill, M. N., Lentz, G. C., Scott, E. F., Tenouri, N. F., & Rohr, J. R. (2021). Amphibian species vary in their learned avoidance response to the deadly fungal pathogen *Batrachochytrium dendrobatidis*. *Journal of Applied Ecology, 58*(8), 1613–1620.

Melnik, K., Menke, M., Rakotoarison, A., Vences, M., & Schulz, S. (2019). Identification and synthesis of luteolide, a highly branched macrolide semiochemical from the mantellid frog *Gephyromantis luteus*. *Organic Letters, 21*(8), 2851–2854.

Melo, G. R., Solé, M., & Eterovick, P. C. (2021). Invisible or fearless: Tadpole response to predator cues depends on color. *Ethology Ecology and Evolution, 33*(2), 99–107.

Menke, M., Peram, P. S., Starnberger, I., Hödl, W., Jongsma, G. F. M., Blackburn, D. C., Rödel, M.-O., Vences, M., & Schulz, S. (2016). Identification, synthesis and mass spectrometry of a macrolide from the African reed frog *Hyperolius cinnamomeoventris*. *Beilstein Journal of Organic Chemistry, 12*(1), 2731–2738.

Menke, M., Melnik, K., Peram, P. S., Starnberger, I., Hödl, W., Vences, M., & Schulz, S. (2018). Frogolide–an unprecedented sesquiterpene macrolactone from scent glands of African frogs. *European Journal of Organic Chemistry, 2018*(20–21), 2651–2656.

Mikó, Z., Ujszegi, J., Gál, Z., Imrei, Z., & Hettyey, A. (2015). Choice of experimental venue matters in ecotoxicology studies: Comparison of a laboratory-based and an outdoor mesocosm experiment. *Aquatic Toxicology, 167*, 20–30.

Mirza, R. S., & Chivers, D. P. (2001). Do chemical alarm signals enhance survival of aquatic vertebrates? An analysis of the current research paradigm. In A. Marchlewska-Koj, J. J. Lepri, & D. Müller-Schwarze (Eds.), *Chemical signals in vertebrates 9* (pp. 19–26). Springer.

Mirza, R. S., & Kiesecker, J. M. (2005). Chemically mediated life-history shifts in embryonic amphibians. In *Chemical signals in vertebrates 10* (pp. 373–380). Springer.

Mirza, R. S., Ferrari, M. C. O., Kiesecker, J. M., & Chivers, D. P. (2006). Responses of American toad tadpoles to predation cues: Behavioural response thresholds, threat-sensitivity and acquired predation recognition. *Behaviour, 143*, 877–889.

Mirza, R. S., Laraby, C. A., & Marcellus, A. M. (2013). Knowing your behaviour: the importance of behavioural assays in the characterisation of chemical alarm cues in fishes and amphibians. In M. L. East & M. Dehnhard (Eds.), *Chemical signals in vertebrates 12* (pp. 295–308). Springer.

Mitchell, M. D., Chivers, D. P., Brown, G. E., & Ferrari, M. C. O. (2016a). Living on the edge: How does environmental risk affect the behavioural and cognitive ecology of prey? *Animal Behaviour, 115*, 185–192.

Mitchell, M. D., Ferrari, M. C. O., Lucon-Xiccato, T., & Chivers, D. P. (2016b). Diet cues alter the development of predator recognition templates in tadpoles. *Behavioral Ecology and Sociobiology, 70*(10), 1707–1713.

Mitchell, M. D., Bairos-Novak, K. R., & Ferrari, M. C. O. (2017). Mechanisms underlying the control of responses to predator odours in aquatic prey. *The Journal of Experimental Biology, 220*(11), 1937–1946.

Mitchell, M. D., Crane, A. L., Bairos-Novak, K. R., Ferrari, M. C. O., & Chivers, D. P. (2018). Olfactory cues of habitats facilitate learning about landscapes of fear. *Behavioral Ecology, 29*(3), 693–700.

Mogali, S. M., Saidapur, S. K., & Shanbhag, B. A. (2011). Levels of predation modulate antipredator defense behavior and metamorphic traits in the toad *Bufo melanostictus*. *Journal of Herpetology, 45*(4), 428–431.

Mogali, S. M., Shanbhag, B. A., & Saidapur, S. K. (2015). Strong food odours mask predation risk and affect evocation of defence behaviours in the tadpoles of *Sphaerotheca breviceps*. *Journal of Ethology, 33*(1), 41–46.

Mogali, S. M., Saidapur, S. K., & Shanbhag, B. A. (2020). Behavioral responses of tadpoles of *Duttaphrynus melanostictus* (Anura: Bufonidae) to cues of starved and fed dragonfly larvae. *Phyllomedusa: Journal of Herpetology, 19*(1), 93–98.

Moore, H., Chivers, D. P., & Ferrari, M. C. O. (2015). Sub-lethal effects of Roundup on tadpole anti-predator responses. *Ecotoxicology and Environmental Safety, 111*, 281–285.

Murray, D. L., Roth, J. D., & Wirsing, A. J. (2004). Predation risk avoidance by terrestrial amphibians: The role of prey experience and vulnerability to native and exotic predators. *Ethology, 110*(8), 635–647.

Natale, G. S., Alcalde, L., Herrera, R., Cajade, R., Schaefer, E. F., Marangoni, F., & Trudeau, V. L. (2011). Underwater acoustic communication in the macrophagic carnivorous larvae of *Ceratophrys ornata* (Anura: Ceratophryidae). *Acta Zoologica, 92*(1), 46–53.

Nieuwkoop, P. D., Faber, J., Gerhart, J., & Kirschner, M. (2020). *Normal table of Xenopus laevis (Daudin): A systematical and chronological survey of the development from the fertilized egg till the end of metamorphosis*. Garland Science.

Nordlund, D. A. (1981). *Semiochemicals; their role in pest control* (Issues 04; SB951, N6.). Wiley

Nowack, C., Peram, P. S., Wenzel, S., Rakotoarison, A., Glaw, F., Poth, D., Schulz, S., & Vences, M. (2017). Volatile compound secretion coincides with modifications of the olfactory organ in mantellid frogs. *Journal of Zoology, 303*(1), 72–81.

Nunes, A. L., Richter-Boix, A., Laurila, A., & Rebelo, R. (2013). Do anuran larvae respond behaviourally to chemical cues from an invasive crayfish predator? A community-wide study. *Oecologia, 171*(1), 115–127.

Nunes, A. L., Orizaola, G., Laurila, A., & Rebelo, R. (2014a). Morphological and life-history responses of anurans to predation by an invasive crayfish: An integrative approach. *Ecology and Evolution, 4*(8), 1491–1503.

Nunes, A. L., Orizaola, G., Laurila, A., & Rebelo, R. (2014b). Rapid evolution of constitutive and inducible defenses against an invasive predator. *Ecology, 95*(6), 1520–1530.

Nystrom, P., & Abjornsson, K. (2000). Effects of fish chemical cues on the interactions between tadpoles and crayfish. *Oikos, 88*(1), 181–190.

Ogurtsov, S. V. (2004). Olfactory orientation in anuran amphibians. *Russian Journal of Herpetology, 11*, 35–40.

Pašukonis, A., Trenkwalder, K., Ringler, M., Ringler, E., Mangione, R., Steininger, J., Warrington, I., & Hödl, W. (2016). The significance of spatial memory for water finding in a tadpole-transporting frog. *Animal Behaviour, 116*, 89–98.

Pearl, C. A., Cervantes, M., Chan, M., Ho, U., Shoji, R., & Thomas, E. O. (2000). Evidence for a mate-attracting chemosignal in the dwarf African clawed frog *Hymenochirus*. *Hormones and Behavior, 38*(1), 67–74.

Pearl, C. A., Adams, M. J., Schuytema, G. S., & Nebeker, A. V. (2003). Behavioral responses of anuran larvae to chemical cues of native and introduced predators in the Pacific Northwestern United States. *Journal of Herpetology, 37*(3), 572–576.

Pease, K. M., & Wayne, R. K. (2014). Divergent responses of exposed and naive Pacific tree frog tadpoles to invasive predatory crayfish. *Oecologia, 174*(1), 241–252.

Peram, P. S., Vences, M., & Schulz, S. (2017). A synthetic dodecanolide library for the identification of putative semiochemicals emitted by mantellid frogs. *Organic & Biomolecular Chemistry, 15*(33), 6967–6977.

Petranka, J. W. (1989). Response of toad tadpoles to conflicting chemical stimuli: Predator avoidance versus optimal foraging. *Herpetologica, 45*, 283–292.

Petranka, J., & Hayes, L. (1998). Chemically mediated avoidance of a predatory odonate (*Anax junius*) by American toad (*Bufo americanus*) and wood frog (*Rana sylvatica*) tadpoles. *Behavioral Ecology and Sociobiology, 42*(4), 263–271.

Petranka, J. W., Kats, L. B., & Sih, A. (1987). Predator-prey interactions among fish and larval amphibians: Use of chemical cues to detect predatory fish. *Animal Behaviour, 35*(2), 420–425.

Pfeiffer, W. (1966). Die Verbreitung der Schreckreaktion bei Kaulquappen und die Herkunft des Schreckstoffes. *Journal of Comparative Physiology A, 52*, 79–98.

Phuge, S., & Phuge, A. (2019). Predator-prey interactions of tadpoles in different layers of the water column. *Journal of Ethology, 37*(2), 197–202.

Pizzatto, L., & Shine, R. (2009). Native Australian frogs avoid the scent of invasive cane toads. *Austral Ecology, 34*(1), 77–82.

Pizzatto, L., Stockwell, M., Clulow, S., Clulow, J., & Mahony, M. (2014). Chemical communication in green and golden bell frogs: Do tadpoles respond to chemical cues from dead conspecifics? *Chemoecology, 24*(5), 171–177.

Pizzatto, L., Stockwell, M., Clulow, S., Clulow, J., & Mahony, M. (2015). Finding a place to live: Conspecific attraction affects habitat selection in juvenile green and golden bell frogs. *Acta Ethologica, 19*(1), 1–8.

Polo-Cavia, N., & Gomez-Mestre, I. (2014). Learned recognition of introduced predators determines survival of tadpole prey. *Functional Ecology, 28*(2), 432–439.

Polo-Cavia, N., Gonzalo, A., Lopez, P., & Martin, J. (2010). Predator recognition of native but not invasive turtle predators by naive anuran tadpoles. *Animal Behaviour, 80*(3), 461–466.

Polo-Cavia, N., Burraco, P., & Gomez-Mestre, I. (2016). Low levels of chemical anthropogenic pollution may threaten amphibians by impairing predator recognition. *Aquatic Toxicology, 172*, 30–35.

Poth, D., Wollenberg, K. C., Vences, M., & Schulz, S. (2012). Volatile amphibian pheromones: Macrolides from Mantellid frogs from Madagascar. *Angewandte Chemie, International Edition, 51*(9), 2187–2190.

Poth, D., Peram, P. S., Vences, M., & Schulz, S. (2013). Macrolides and alcohols as scent gland constituents of the Madagascan frog *Mantidactylus femoralis* and their intraspecific diversity. *Journal of Natural Products, 76*(9), 1548–1558.

Preston, D. B., & Forstner, M. R. J. (2015). Houston Toad (*Bufo (Anaxyrus) houstonensis*) tadpoles decrease their activity in response to chemical cues produced from the predation of conspecifics and congeneric (*Bufo (Incilius) nebulifer*) Tadpoles. *Journal of Herpetology, 49*(2), 170–175.

Pueta, M., & Perotti, M. G. (2016). Anuran tadpoles learn to recognize injury cues from members of the same prey guild. *Animal Cognition, 19*(4), 745–751.

Pueta, M., Ardanaz, D., & Tallone, J. C. (2022). Habituation in anuran tadpoles and the role of risk uncertainty. *Animal Cognition, 25*(1), 63–72.

Pyron, A. R., & Wiens, J. J. (2011). A large-scale phylogeny of Amphibia including over 2800 species, and a revised classification of extant frogs, salamanders, and caecilians. *Molecular Phylogenetics and Evolution, 61*(2), 543–583.

Rabb, G. B., & Rabb, M. S. (1963). Additional observations on breeding behavior of the Surinam toad, Pipa pipa. *Copeia, 1963*(4), 636–642.

Raices, M., Jungblut, L. D., & Pozzi, A. G. (2020). Evidence of the peptide identity of the epidermal alarm cue in tadpoles of the toad *Rhinella arenarum*. *Herpetological Journal, 30*(4), 230–233.

Rajchard, J. (2005). Sex pheromones in amphibians: A review. *Veterinary Medicine, 50*(9), 383–389.

Rajchard, J. (2006). Antipredator pheromones in amphibians: A review. *Veterinary Medicine, 51*(8), 409–413.

Ramamonjisoa, N., Rakotonoely, H., & Natuhara, Y. (2018). Defense investments and growth responses under different predation risks and gape-limitation predation threats in a tadpole prey. *Behavioral Ecology and Sociobiology, 72*(9), 1–11.

Raven, C., Shine, R., Greenlees, M., Schaerf, T. M., & Ward, A. J. W. (2017). The role of biotic and abiotic cues in stimulating aggregation by larval cane toads (*Rhinella marina*). *Ethology, 123*(10), 724–735.

Rivera-Hernández, I. A. E., Crane, A. L., Pollock, M. S., & Ferrari, M. C. O. (2022). Disturbance cues function as a background risk cue but not as an associative learning cue in tadpoles. *Animal Cognition, 25*, 1–9.

Rödin, P., Forsman, A., & Hagman, M. (2011). Taxonomic patterns of tadpole behavioural responses to alarm cues. In J. Murray (Ed.), *Frogs: Biology, ecology and uses* (pp. 123–140). Nova Science Publishers.

Rojas, V., Labra, A., Valdés, J. L., & Velásquez, N. A. (2021). Females of the four-eyed frog, *Pleurodema thaul* (Anura, Leptodactylidae), respond behaviourally to conspecific male scent. *Herpetozoa, 34*, 115–120.

Ruther, J., Meiners, T., & Steidle, J. L. (2002). Rich in phenomena-lacking in terms. A classification of kairomones. *Chemoecology, 12*(4), 161–167.

Saidapur, S. K., Veeranagoudar, D. K., Hiragond, N. C., & Shanbhag, B. A. (2009). Mechanism of predator-prey detection and behavioral responses in some anuran tadpoles. *Chemoecology, 19*(1), 21–28.

Sbarbati, A., & Osculati, F. (2006). Allelochemical communication in vertebrates: Kairomones, allomones and synomones. *Cells, Tissues, Organs, 183*(4), 206–219.

Scherer, A. E., & Smee, D. L. (2016). A review of predator diet effects on prey defensive responses. *Chemoecology, 26*(3), 83–100.

Schley, L., & Griffiths, R. A. (1998). Midwife toads (*Alytes muletensis*) avoid chemical cues from snakes (Natrix maura). *Journal of Herpetology, 32*(4), 572–574.

Schoeppner, N. M., & Relyea, R. A. (2005). Damage, digestion, and defence: The roles of alarm cues and kairomones for inducing prey defences. *Ecology Letters, 8*(5), 505–512.

Schulte, L. M. (2016). Semiochemicals in anurans: Testing different categories with one poison frog species. In B. A. Schulte, T. E. Goodwin, & M. H. Ferkin (Eds.), *Chemical signals in vertebrates, volume 13* (pp. 339–350). Springer.

Schulte, L. M., & Lötters, S. (2013). The power of the seasons: Rainfall triggers parental care in poison frogs. *Evolutionary Ecology, 27*(4), 711–723.

Schulte, L. M., & Lötters, S. (2014). A danger foreseen is a danger avoided: how chemical cues of different tadpoles influence parental decisions of a Neotropical poison frog. *Animal Cognition, 17*(2), 267–275.

Schulte, L. M., & Mayer, M. (2017). Poison frog tadpoles seek parental transportation to escape their cannibalistic siblings. *Journal of Zoology, 303*(2), 83–89.

Schulte, L. M., & Rößler, D. C. (2013). Do poison frogs recognize chemical cues of the other sex or do they react to cues of stressed conspecifics? *Behavioural Processes, 100*, 32–35.

Schulte, L. M., & Veith, M. (2014). Is the response of a Neotropical poison frog (*Ranitomeya variabilis*) to larval chemical cues influenced by relatedness? *The Herpetological Journal, 24*(3), 189–192.

Schulte, L. M., Yeager, J., Schulte, R., Veith, M., Werner, P., Beck, L. A., & Lötters, S. (2011). The smell of success: choice of larval rearing sites by means of chemical cues in a Peruvian poison frog. *Animal Behaviour, 81*(6), 1147–1154.

Schulte, L. M., Schulte, R., & Lötters, S. (2013). Avoiding predation: The importance of chemical and visual cues in poison frog reproductive behaviour. In M. L. East & M. Dehnhard (Eds.), *Chemical signals in vertebrates 12* (Vol. 12, pp. 309–321). Springer.

Schulte, L. M., Krauss, M., Lötters, S., Schulze, T., & Brack, W. (2015). Decoding and discrimination of chemical cues and signals: Avoidance of predation and competition during parental care behavior in sympatric poison frogs. *PLoS One, 10*(7), e0129929.

Schulte, L. M., Ringler, E., Rojas, B., & Stynoski, J. L. (2020). Developments in amphibian parental care research: History, present advances, and future perspectives. *Herpetological Monographs, 34*(1), 71–97.

Schulte, L. M., Martel, A., Cruz-Elizalde, R., Ramírez-Bautista, A., & Bossuyt, F. (2021). Love bites: male frogs (*Plectrohyla*, Hylidae) use teeth scratching to deliver sodefrin precursor-like factors to females during amplexus. *Frontiers in Zoology, 18*(1), 1–14.

Schulte, L. M., Jendras, J., Twomey, E., Ramirez-Bautista, A., & Bossuyt, F. (2022). Gene expression of secretory proteins in the nuptial pads of three *Lithobates* species (Anura: Ranidae). *Amphibia-Reptilia, 1*, 1–9, online first.

Schulz, S., Poth, D., Peram, P. S., Hötling, S., Menke, M., Melnik, K., & Röpke, R. (2021). Chemical diversity of volatile macrocylic lactones from frogs. *Synlett, 32*(17), 1683–1701.

Scribano, G., Balestrieri, A., Gazzola, A., & Pellitteri-Rosa, D. (2020). Strong behavioural defensive responses of endemic *Rana latastei* tadpoles induced by a native predator's odour. *Ethology, 126*(9), 922–930.

Shaffery, H. M., & Relyea, R. A. (2016). Dissecting the smell of fear from conspecific and heterospecific prey: Investigating the processes that induce anti-predator defenses. *Oecologia, 180*(1), 55–65.

Sharma, S. S., Veeranagoudar, D. K., Shanbhag, B. A., & Saidapur, S. K. (2008). Activity of *Sphaerotheca breviceps* tadpoles in response to chemical cues of the predaceous tadpoles *Hoplobatrachus tigerinus*. *Journal of Ethology, 26*(3), 303–307.

Sievers, M., Hale, R., Swearer, S. E., & Parris, K. M. (2018). Contaminant mixtures interact to impair predator-avoidance behaviours and survival in a larval amphibian. *Ecotoxicology and Environmental Safety, 161*, 482–488.

Smith, G. R., & Fortune, D. T. (2009). Hatching plasticity of wood frog (*Rana sylvatica*) eggs in response to mosquitofish (*Gambusia affinis*) cues. *Herpetological Conservation and Biology, 4*(1), 43–47.

South, J., Botha, T. L., Wolmarans, N. J., Wepener, V., & Weyl, O. L. F. (2020). Playing with food: Detection of prey injury cues stimulates increased functional foraging traits in *Xenopus laevis*. *African Zoology, 55*(1), 25–33.

Stabell, O. B. (2005). Latent alarm signals: Are they present in vertebrates? In R. T. Mason, M. P. LeMaster, & D. Müller-Schwarze (Eds.), *Chemical signals in vertebrates 10* (pp. 381–388). Springer.

Starnberger, I., Poth, D., Schulz, S., Vences, M., Knudsen, J., Barej, M. F., Rödel, M.-O., Walzl, M., & Hödl, W. (2013). Take time to smell the frogs – Vocal sac glands of reed frogs contain chemical cues on species identity. *Biological Journal of the Linnean Society, 110*(4), 828–838.

Steiger, S., Schmitt, T., & Schaefer, H. M. (2011). The origin and dynamic evolution of chemical information transfer. *Proceedings of the Royal Society B: Biological Sciences, 278*(1708), 970–979.

Still, M. B., Lea, A. M., Hofmann, H. A., & Ryan, M. J. (2019). Multimodal stimuli regulate reproductive behavior and physiology in male túngara frogs. *Hormones and Behavior, 115*, 104546.

Stynoski, J. L., & Noble, V. R. (2012). To beg or to freeze: Multimodal sensory integration directs behavior in a tadpole. *Behavioral Ecology and Sociobiology, 66*(2), 191–199.

Stynoski, J. L., & Porras-Brenes, K. (2022). Meta-analysis of tadpole taste tests: Consumption of anuran prey across development and predator strategies. *Oecologia, 199*(4), 845–857.

Summey, M. R., & Mathis, A. (1998). Alarm responses to chemical stimuli from damaged conspecifics by larval anurans: Tests of three neotropical species. *Herpetologica, 54*(3), 402–408.

Supekar, S. C., & Gramapurohit, N. P. (2017). Can embryonic skipper frogs (*Euphlyctis cyanophlyctis*) learn to recognise kairomones in the absence of a nervous system? *Journal of Biosciences, 42*(3), 459–468.

Supekar, S. C., & Gramapurohit, N. P. (2022). Experience window influences development and retention of memory to recognize predators in the larval skipper frogs. *Current Science, 122*(8), 951–957.

Swart, C. C., & Taylor, R. C. (2004). Behavioral interactions between the giant water bug (*Belostoma lutarium*) and tadpoles of *Bufo woodhousii*. *Southeastern Naturalist, 3*(1), 13–24.

Symonds, M. R. E., & Elgar, M. A. (2008). The evolution of pheromone diversity. *Trends in Ecology & Evolution, 23*(4), 220–228.

Szabo, B., Mangione, R., Rath, M., Pašukonis, A., Reber, S. A., Oh, J., Ringler, M., & Ringler, E. (2021). Naive poison frog tadpoles use bi-modal cues to avoid insect predators but not heterospecific predatory tadpoles. *Journal of Experimental Biology, 224*(24), jeb243647.

Takahara, T., Kohmatsu, Y., Maruyama, A., Yamanaka, H., & Yamaoka, R. (2012). Inducible defense behavior of an anuran tadpole: Cue-detection range and cue types used against predator. *Behavioral Ecology, 23*(4), 863–868.

Takahashi, M. (2007). Oviposition site selection: pesticide avoidance by gray treefrogs. *Environmental Toxicology and Chemistry, 26*(7), 1476–1480.

Ten Hagen, L., Rodríguez, A., Menke, N., Göcking, C., Bisping, M., Frommolt, K.-H., Ziegler, T., Bonkowski, M., & Vences, M. (2016). Vocalizations in juvenile anurans: Common spadefoot toads (*Pelobates fuscus*) regularly emit calls before sexual maturity. *The Science of Nature, 103*(9), 1–8.

Thomas, E. O., Tsang, L., & Licht, P. (1993). Comparative histochemistry of the sexually dimorphic skin glands of anuran amphibians. *Copeia, 1993*(1), 133–143.

Touchon, J. C., Gomez-Mestre, I., & Warkentin, K. M. (2006). Hatching plasticity in two temperate anurans: Responses to a pathogen and predation cues. *Canadian Journal of Zoology, 84*(4), 556–563.

Troyer, R. R., & Turner, A. M. (2015). Chemosensory perception of predators by larval amphibians depends on water quality. *PLoS One, 10*(6), e0131516.

Urszán, T. J., Garamszegi, L. Z., Nagy, G., Hettyey, A., Török, J., & Herczeg, G. (2018). Experience during development triggers between-individual variation in behavioural plasticity. *The Journal of Animal Ecology, 87*, 1264–1273.

Üveges, B., Basson, A. C., Móricz, Á. M., Bókony, V., & Hettyey, A. (2021). Chemical defence effective against multiple enemies: Does the response to conspecifics alleviate the response to predators? *Functional Ecology, 35*(10), 2294–2304.

Van Bocxlaer, I., Treer, D., Maex, M., Vandebergh, W., Janssenswillen, S., Stegen, G., Kok, P., Willaert, B., Matthijs, S., Martens, E., Mortier, A., de Greve, H., Proost, P., & Bossuyt, F. (2015). Side-by-side secretion of Late Palaeozoic diverged courtship pheromones in an aquatic salamander. *Proceedings of the Royal Society of London B, 282*(1803) 20142960.

Van Buskirk, J., Krügel, A., Kunz, J., Miss, F., & Stamm, A. (2014). The rate of degradation of chemical cues indicating predation risk: An experiment and review. *Ethology, 120*(9), 942–949.

Vandenbergh, J. (2012). *Pheromones and reproduction in mammals*. Elsevier.

Vences, M., Wahl-Boos, G., Hoegg, S., Glaw, F., Spinelly Oliveira, E., Meyer, A., & Perry, S. (2007). Molecular systematics of mantelline frogs from Madagascar and the evolution of their femoral glands. *Biological Journal of the Linnean Society, 92*(3), 529–539.

Vodrážková, M., Šetlíková, I., & Berec, M. (2020). Chemical cues of an invasive turtle reduce development time and size at metamorphosis in the common frog. *Scientific Reports, 10*(1), 1–6.

Wabnitz, P. A., Bowie, J. H., Tyler, M. J., Wallace, J. C., & Smith, B. P. (1999). Aquatic sex pheromone from a male tree frog. *Nature, 401*(6752), 444–445.

Wabnitz, P. A., Bowie, J. H., Tyler, M. J., Wallace, J. C., & Smith, B. P. (2000). Differences in the skin peptides of the male and female Australian tree frog *Litoria splendida. European Journal of Biochemistry, 267*(1), 269–275.

Waldman, B. (1985). Olfactory basis of kin recognition in toad tadpoles. *Journal of Comparative Physiology A, 156*(5), 565–577.

Waldman, B. (1986). Chemical ecology of kin recognition in anuran amphibians. In D. Duvall, D. Müller-Schwarze, R. M. Silverstein (Eds.), *Chemical signals in vertebrates 4*, (Issue 4, pp. 225–242). Plenum Press

Waldman, B. (2016). Chemical communication in archaic New Zealand frogs. In *Chemical signals in vertebrates 13* (pp. 351–360). Springer.

Waldman, B., & Bishop, P. J. (2004). Chemical communication in an archaic anuran amphibian. *Behavioral Ecology, 15*(1), 88.

Warbeck, A., & Parzefall, J. (2001). Mate recognition via waterborne chemical cues in the viviparous caecilian *Typhlonectes natans* (Amphibia: Gymnophiona). In A. Marchlewska-Koj, J. L. Lepri, & D. Müller-Schwarze (Eds.), *Chemical signals in vertebrates 9* (pp. 263–268). Springer.

Wassersug, R. J., Lum, A. M., & Potel, M. J. (1981). An analysis of school structure for tadpoles (Anura: Amphibia). *Behavioral Ecology and Sociobiology, 9*(1), 15–22.

Weiss, L., Manzini, I., & Hassenklöver, T. (2021). Olfaction across the water–air interface in anuran amphibians. *Cell and Tissue Research, 383*(1), 301–325.

Westrick, S. E., Laslo, M., & Fischer, E. K. (2022). The natural history of model organisms: The big potential of the small frog *Eleutherodactylus coqui*. *eLife, 11*, e73401.

Whittaker, D. J. (2022). *The secret perfume of birds: uncovering the science of Avian Scent*. JHU Press.

Willaert, B., Bossuyt, F., Janssenswillen, S., Adriaens, D., Baggerman, G., Matthijs, S., Pauwels, E., Proost, P., Raepsaet, A., & Schoofs, L. (2013). Frog nuptial pads secrete mating season-specific proteins related to salamander pheromones. *The Journal of Experimental Biology, 216*(22), 4139–4143.

Wilson, D. J., & Lefcort, H. (1993). The effect of predator diet on the alarm response of red-legged frog, *Rana aurora*, tadpoles. *Animal Behaviour, 46*(5), 1017–1019.

Wirsing, A. J., Roth, J. D., & Murray, D. L. (2005). Can prey use dietary cues to distinguish predators? A test involving three terrestrial amphibians. *Herpetologica, 61*(2), 104–110.

Woodley, S. K. (2010). Pheromonal communication in amphibians. *Journal of Comparative Physiology A, 196*(10), 713–727.

Woodley, S. K., & Staub, N. L. (2021). Pheromonal communication in urodelan amphibians. *Cell and Tissue Research, 383*, 1–19.

Wyatt, T. D. (2014). *Pheromones and animal behavior: Chemical signals and signatures*. Cambridge University Press.

Xu, C., Gao, X., Crossland, M. R., Liu, Z., Wang, S., Zhu, W., Shine, R., & Li, Y. (2017). Foraging responses of the larvae of invasive bullfrogs (*Lithobates catesbeianus*): Possible implications for bullfrog control and ecological impact in China. *Asian Herpetological Research, 4*, 253–256.

Zamora-Camacho, F. J., Medina-Gálvez, L., & Zambrano-Fernández, S. (2019). The roles of sex and morphology in burrowing depth of Iberian spadefoot toads in different biotic and abiotic environments. *Journal of Zoology, 309*(3), 224–230.

Zhang, F., Zhao, J., Zhang, Y., Messenger, K., & Wang, Y. (2015). Antipredator behavioral responses of native and exotic tadpoles to novel predator. *Asian Herpetological Research, 6*, 51–58.

# Avian Chemical Signatures: An Overview

Tatjana Alves Soares[1], Barbara A. Caspers[2], and Helene M. Loos[1,3] (✉)

[1] Department of Chemistry and Pharmacy, Friedrich-Alexander-Universität Erlangen-Nürnberg, Erlangen, Germany
[2] Department of Behavioural Ecology, Bielefeld University, Bielefeld, Germany
[3] Department of Sensory Analytics and Technology, Fraunhofer Institute for Process Engineering and Packaging IVV, Freising, Germany
helene.loos@fau.de

**Abstract** Intraspecific communication in birds can involve the visual, auditory, tactile, and olfactory senses. Although the latter is one of the least studied signalling modalities in birds, research of the last decades showed that a bird's chemical profile varies according to different parameters. Here, we summarise current knowledge about chemical signatures of the volatile as well as the non-volatile fraction of preen oil, feathers and a bird's overall odour, considering the potential transfer of information regarding species, genetic distance/compatibility, individual identity, sex, breeding season, and age. We found good agreement between the results of the publications reviewed here with regard to the substance classes being involved in the related chemical signatures. However, conclusive findings of specific compounds or mixtures of compounds as key constituents of the considered chemical signatures are rare. Moreover, none of the investigated parameters was characterised by only one compound, supporting the notion that they are encoded in chemical signatures being built of multiple compounds. Based on the overview of available insights into avian chemical signatures, we finally delineate lines of future research in this emerging field of research.

**Keywords** Age signature · Genetic signature · Individual identity signature · Sex signature · Breeding season signature · Species signature · Volatilome · Olfaction · Bird odour · Infochemicals

## Glossary

- Species signature: includes compounds which contribute to the differentiation of bird species.

  Genetic signature: includes compounds being indicative of genetic similarity and/or genetic compatibility. Genetic similarity/distance is generally determined by analysis of microsatellites or single-nucleotide polymorphisms (SNPs), whereas genetic compatibility relates to the major histocompatibility complex (MHC).
- Individual identity signature: includes compounds which contribute to the differentiation between individual birds.
- Breeding season signature: includes compounds which contribute to the differentiation between breeding and non-breeding season.



B. Schaal et al. (eds.), *Chemical Signals in Vertebrates 15*,
https://doi.org/10.1007/978-3-031-35159-4_6

- Sex signature: includes compounds which contribute to the differentiation between the sexes.
- Age signature: includes compounds which contribute to the differentiation of different ages.

## 1 Introduction

Birds, as most other vertebrates, can receive information via their visual, auditory, tactile and chemical senses and communicate via these different modalities (Smith, 1977; Caro & Balthazart, 2010). Multimodal communication occurs when two or more of the sensory channels are used for the transmission of information (Higham & Hebets, 2013). Avian intraspecific communication via visual and auditory means can be easily observed by humans from afar (Caro & Balthazart, 2010). Chemical cues and signals, however, are invisible and unhearable. Thus, they may remain undiscovered by the human observer, particularly in the absence of strong odours and odour-guided behaviour such as sniffing and scent marking, which is the case in most bird species. This, together with other factors, such as a rather small olfactory bulb, led to the assumption that olfaction might not be of any function (Bang & Cobb, 1968) and that olfactory communication is not developed in birds. In the second half of the twentieth century, however, this misbelief was recognised to be wrong (for reviews, see Caro & Balthazart, 2010; Caro et al., 2015; Whittaker, 2022).

Olfactory communication in general has been defined as 'the process whereby a chemical signal is generated by a presumptive sender and transmitted (generally through the air) to a presumptive receiver who by means of adequate receptors can identify, integrate, and respond (in terms of either behaviour or physiology) to the signal' (Eisenberg & Kleiman, 1972, p. 1). Perception of chemical cues and signals in mammals is thought to be achieved mainly through the main olfactory system, but in some species chemical communication has been shown to be mediated by the vomeronasal system and further systems like the Grueneberg ganglion or the septal organ (Wackermannova et al., 2016). Therefore, and because further chemosensory systems may also be involved in chemical communication processes, the term airborne chemical communication may be more appropriate than olfactory communication when the mechanisms of detection of chemical cues and signals are not yet known. In birds, the vomeronasal organ is absent (Bang & Wenzel, 1985, in Leclaire et al., 2011). Therefore, their detection of chemical signals is thought to take place via the main olfactory system.

To denominate molecules and mixtures of molecules involved in chemical communication among individuals, several terms exist (e.g., Müller et al., 2020). The term pheromone is famous in this respect. In 1959, Karlson and Lüscher defined pheromones as 'substances which are secreted to the outside by an individual and received by a second individual of the same species, in which they release a specific reaction, for example a definite behaviour or a developmental process'. Since then, several further concepts to define pheromones have been proposed (cf. Butenandt et al., 1959; Rasmussen et al., 1996; Schaal et al., 2003; Wyatt, 2017). Amongst others, Wyatt (2010) defines them as 'evolved signals' and differentiates them from 'signature mixtures', which he defines, modified after the term 'mosaic signal' coined by Johnston (2003, 2005), as 'a variable chemical mixture learned by other conspecifics and used to recognize an animal as an individual or as a member of a particular social group such as a family, clan or colony'. Aiming to elucidate the molecular principles of olfactory communication in birds, research of the last decades showed that the chemical composition of avian odour emissions differs depending on various social and non-social parameters, such as species, genetics, breeding season and sex (Campagna et al., 2012; Caro et al., 2015).

In the following, we provide an overview of the current state of knowledge regarding the description of chemical signatures in birds. The term chemical signatures thereby designates (patterns of) chemicals which have been determined through chemical analysis as being indicative of specific traits or states. Behavioural evidence for the existence of pheromones and signature mixtures is not within the scope of this review. For recent overviews in this respect, the reader is referred to Roper (1999) and Caro et al. (2015). By providing this review, we aim also to clarify if specific traits and states are encoded through similar compounds across species or not. Whereas, on the one hand, one may hypothesise that chemical signatures are divergent between species, one may, on the other hand, also expect some chemical signatures to bear overlapping chemical information between species if the underlying biosynthetic pathways are similar. Chemical signatures were only included if at least three publications were available (see Table 1), and analysis of samples from at least ten birds per group was a prerequisite for including a publication into this review. Furthermore, different data analysis methods have been used in the reviewed publications. Whereas some authors study absolute peak areas, others normalise peak areas either by dividing the peak area of a compound by the area of an internal standard or by the area of the total chromatogram or by the area of the volatile/non-volatile fraction in a chromatogram. Here, we report results obtained from analysis of absolute and relative peak areas in a similar manner. If contradictory results emerge, the method used by the authors is specified in the text.

Table 2 provides an overview of the publications reviewed here, the investigated chemical signatures and the sample amounts analysed. Chemical signatures found in the volatile and in the non-volatile fraction of the investigated samples – preen oil, feathers and whole body odour – are summarised in two separate sections of this review (Sects. 2 and 3). For several reasons, non-volatile compounds are included into this

**Table 1** Number of available publications for different chemical signatures in the volatile and non-volatile fraction of preen oil, feathers or bird's overall body odour

| | Number of publications considering | |
|---|---|---|
| Chemical signature | Volatile fraction | Non-volatile fraction |
| Species | 6 | 4 |
| Genetics | 3 | 3 |
| Individual identity | 9 | 2 |
| Breeding season | 10 | 12 |
| Sex | 24 | 10 |
| Age | 6 | 3 |

review which focuses on olfactory communication. First, knowledge about the non-volatile fraction is essential as non-volatile compounds can be precursors of volatile compounds. Second, volatile and non-volatile compounds can interact and produce a different signal than volatile compounds alone, e.g. via matrix effects. Third, a wide continuum exists regarding volatility and less volatile compounds should not a priori be neglected when studying chemical communication. We finish the overview of available evidence on avian chemical signatures with a discussion of routes for further research.

## 2 Volatile Fraction

Research on the volatile fraction of avian samples started in the early 2000s. The volatile fraction is in gas phase at ambient temperature and includes compounds of different substance classes, notably alkanes, aldehydes, (methyl) ketones, carboxylic acids, alcohols, sterols, esters, fatty acid esters and sulphurous compounds. The characterisation of the volatile fraction has so far been mainly achieved by applying extraction techniques such as solvent extraction, stir bar sorptive extraction (SBSE) or solid-phase microextraction (SPME) in combination with gas chromatography-mass spectrometry (GC-MS) or gas chromatography coupled with flame ionisation detection (GC-FID). Whereas with SBSE and SPME it is inherent to the technique that volatile compounds are analysed, solvent extraction also gives access to compounds of lower volatility. In these cases, we decided to adopt the distinction applied by the authors (e.g., Leclaire et al., 2011) or divided the reported compounds into volatile and non-volatile substances by considering all compounds except for esters and diesters as volatile (Zhang et al., 2009; Mardon et al., 2010; Van Huynh & Rice, 2019).

### 2.1 Evidence for a Species Signature

Considering passerines, no consensus for specific compound classes being involved in species signatures has been achieved so far. While Bengalese finches, zebra finches, yellow-browed buntings and rooks (scientific names, i.e. binomial taxa of all species, reviewed here are given in Table 2) differed in the composition of linear alcohols in preen oil and feathers (Zhang et al., 2009), it was found, on the other hand, that the difference between the preen oil of white-throated sparrow and the closely related dark-eyed junco was made up of methyl ketones and carboxylic acids (Tuttle et al., 2014). Further research also reported a species-specific volatile composition of whole body odour of zebra finches and diamond firetails (Krause et al., 2014) and preen oil of Black-capped chickadees and Carolina chickadees (Van Huynh & Rice, 2019), but compounds or compound classes making up the difference were not mentioned. Existence of a species signature was also confirmed in non-passerine species. Specifically, the main difference between Antarctic prion and blue petrel preen oil was made up of fatty acid esters and alcohols between C17 and C30 (Mardon et al., 2010). Martin-Vivaldi et al. (2010) described that the preen oil of the European hoopoe and the green woodhoopoe shared several volatiles but can also be chemically differentiated by

**Table 2** Overview of the reviewed publications including the investigated signatures, the sample type and the sample amount analysed

| Reference | Bird species | Sample type | Samples' amount and composition | Signature(s) | Fraction |
|---|---|---|---|---|---|
| Amo et al. (2012) | Spotless starling (*Sturnus unicolor*) | Preen oil | 32 (9 nonbreeding season [3 m, 6 f]; 23 breeding season [4 m, 19 f]) adults; 15 nestlings | Age, season, sex | Volatile |
| Bhattacharyya and Chowdhury (1995) | Red-vented bulbul (*Pycnonotus cafer*) | Preen oil | Pre-breeding, peak of testicular activity, initiation of testicular collapse, quiescent phase of testicular cycle; each 15 m | Season | Non-volatile |
| Bonadonna et al. (2007) | Antarctic prion (*Pachyptila desolata*) | Feathers | 39 from 13 individuals (sampling 3 years in a row, no details about sex given) | Identity | Volatile |
| Caspers et al. (2021) | Blue tit (*Cyanistes caeruleus*) | Preen oil | 33 (13 m, 20 f) | Sex | Volatile |
| Diez-Fernandez et al. (2020) | Song sparrow (*Melospiza melodia*) | Preen oil | 37 (21 m, 16 f) | Sex | Volatile |
| Diez-Fernandez et al. (2021) | Common blackbird (*Turdus merula*) | Preen oil | 52 (26 m, 26 f; of them 21 juveniles, 31 adults) | Sex, age | Volatile |
| Douglas (2006) | Crested auklet (*Aethia cristatella*) | Whole body odour | 103 from 57 individuals (41 m, 16 f) | Identity, sex | Volatile |
| Fischer et al. (2017) | Herring gull (*Larus argentatus*) | Preen oil | 34 from 17 individuals (17 breeding season, 17 non-breeding season; 9 m, 8 f) | Season, sex | Non-volatile |
| Gabirot et al. (2016) | Cory's shearwater (*Calonectris borealis*) | Preen oil | 33 (16 m, 17 f) | Sex | Volatile |
| Grieves et al. (2019) | Song sparrow (*Melospiza melodia*) | Preen oil | 356 (breeding-stage: Newboro 49 m, 41 f; Cambridge 48 m, 36 f; post-breeding samples: Newboro 26 adults (24 m, 2 f), 28 juveniles (5 m, 18 f, 5 dubious: excluded from sex signature analysis); Cambridge 16 adults (11 m, 5 f), 36 juveniles (17 m, 17 f, 2 dubious: excluded from sex signature analysis)) | Age, season, sex | Non-volatile |

(continued)

**Table 2** (continued)

| Reference | Bird species | Sample type | Samples' amount and composition | Signature(s) | Fraction |
|---|---|---|---|---|---|
| Grieves et al. (2020) | Song sparrow (*Melospiza melodia*) | Preen oil | 54 (36 breeding-stage, 27 m, 9 f; 18 non-breeding-stage, 16 m, 2 f) | Season, sex | Non-volatile |
| Hagelin et al. (2003) | Crested auklet (*Aethia cristatella*) | Feathers | 16 crested auklets (8 m, 8 f), breeding season 10, non-breeding season 6; 3 parakeet auklets | Season, sex, species | Volatile |
| Haribal et al. (2005) | 11 species of passerine | Preen oil | 5 to 15 or more individuals for each species, 5 species sampled in different seasons | Season, species | Non-volatile |
| Jacob et al. (1979) | Mallard ducks (*Anas platyrhynchos*) | Preen oil | 6 from adult ducks (3 m, 3 f) at height of breeding season, 8 from ducklings (4 m, 4 f), 46 throughout a reproductive cycle for sex difference (from 3 m, 5 f) | Age, season, sex | Non-volatile |
| Jacob et al. (2014) | Great tit (*Parus major*) | Preen oil | 98 (44 m, 54 f) | Sex | Non-volatile |
| Jennings and Ebeler (2020) | Leach's storm petrel (*Oceanodroma leucorhoa*) | Feathers | 30 (15 m, 15 f) | Identity | Volatile |
| Leclaire et al. (2011) | Black-legged kittiwake (*Rissa tridactyla*) | Feathers, preen oil | 2008, 41 (20 m, 21 f); 2007, 18 (7 m, 11 f); sampling during breeding seasons | Season, sex, identity | Non-volatile |
| Leclaire et al. (2011) | Black-legged kittiwake (*Rissa tridactyla*) | Feathers, preen oil | 2008: 41 (20 m, 21 f)<br>2007: 18 (7 m, 11 f) | Identity, sex | Volatile |
| Leclaire et al. (2012) | Black-legged kittiwake (*Rissa tridactyla*) | Preen oil | 34 (19 m, 15 f), sampling in the pre-laying period | Genetic, season, sex | Non-volatile |
| Leclaire et al. (2014) | Black-legged kittiwake (*Rissa tridactyla*) | Preen oil | 39 (21 m, 18 f) | Genetic | Non-volatile |
| Leclaire et al. (2019) | Feral pigeon (*Columba livia domestica*) | Feathers | 74 (37 m, 37 f) | Sex | Volatile |
| Mardon et al. (2010) | Antarctic prion (*Pachyptila desolata*), blue petrel (*Halobaena caerulea*) | Preen oil | 54 from 36 birds (breeding BP, 38 samples from 20 birds: 2007–2008, 16 m, 4 f; 2008–2009, 12 m, 4 f; AP, 16 samples [10 m, 6 f]) | Sex, species, identity | Non-volatile |

| | | | | | |
|---|---|---|---|---|---|
| Mardon et al. (2010) | Antarctic prion (*Pachyptila desolata*), blue petrel (*Halobaena caerulea*) | Preen oil | 54 from 36 birds (breeding BP, 38 samples from 20 birds: 2007–2008, 16 m, 4 f; 2008–2009, 12 of 16 initial m, 4 f; AP, 16 samples [10 m, 6 f]) | Identity, species | Volatile |
| Mardon et al. (2011) | Blue petrel (*Halobaena caerulea*) | Feathers | 2007–2008, 20 (16 m, 4 f); 2008–2009, 16 of the 20 initial (12 m, 4 f) | Identity, sex | Volatile |
| Martin-Vivaldi et al. (2010) | European Hoopoe (*Upupa epops*) | Preen oil | 18 (11 nestlings, 7 adults including 4 m, 3 non-breeding f) | Season, species | Volatile |
| Montalti et al. (2005) | Rock dove (*Columba livia*) | Preen oil | 12 (6 m, 6 f) | Sex | Non-volatile |
| Piersma et al. (1999) | Red-knot shorebird (*Calidris canutus*) | Feathers, preen oil | 41 (9 of them followed throughout the period) | Season | Non-volatile |
| Potier et al. (2018) | Black kite (*Milvus migrans*) | Preen oil | 45 (22 m, 23 f; 26 of them juveniles, 19 adults), sampling during breeding and non-breeding seasons | Age, genetic, identity, season, sex | Volatile |
| Reneerkens et al., (2002) | 19 sandpiper species | Preen oil | 1 to 92 for each species, at different points in the annual cycle | Season | Non-volatile |
| Reneerkens et al. (2007b) | Red-knot shorebird (*Calidris canutus*), Western sandpiper (*Calidris mauri*), Temminck's stint (*Calidris temminickii*), Curlew sandpiper (*Calidris ferruginea*), Buff-breasted sandpiper (*Calidris subruficollis*), ruff (*Calidris pugnax*), red phalarope (*Phalaropus fulicarius*) | Preen oil | Between 5 and 47 samples for each sex of each species with sampling during pre-nesting and/or incubation or spring migration or wintering (10 captive ruff [5 m, 5 f] sampled on a weekly basis between April and July) | Season | Non-volatile |

(continued)

**Table 2** (continued)

| Reference | Bird species | Sample type | Samples' amount and composition | Signature(s) | Fraction |
|---|---|---|---|---|---|
| Sandilands et al. (2004) | Domestic chicken (*Gallus domesticus*) | Featers, preen oil | 12 respectively 4 per age with 3 respectively 4 age groups ranging from 10 to 30 weeks (two experiments on age effects) | Age | Non-volatile |
| Shaw et al. (2011) | Gray catbird (*Dumetella carolinensis*) | Preen oil | Age 63 (hatching year (20 m, 13 f), after hatching year (22 m, 19 f)); seasonal variation 59 (OH = summer $N = 41$; FL = winter $N = 18$) | Age, season, sex | Volatile |
| Slade et al. (2016) | Song sparrow (*Melospiza melodia*) | Preen oil | 60 (41 m, 19 f) | Genetic | Non-volatile |
| Soini et al. (2007) | Dark-eyed junco (*Junco hyemalis*) | Feathers, preen oil | 20 (10 m, 10 f), repeated sampling of individuals in both breeding and non-breeding condition | Season, sex | Volatile |
| Tuttle et al. (2014) | White-throated sparrow (*Zonotrichia albicollis*) | Preen oil | 22 non-breeding (11 white males, 7 tan males, 2 white females, 2 tan females), 18 of those breeding condition | Season, sex, species | Volatile |
| Van Huynh and Rice (2019) | Black-capped chickadee (*Poecile atricapillus*), Carolina chickadee (*Poecile carolinensis*) | Preen oil | 41 bcc and 40 cc sampled between Dec 2015 and Dec 2016 | Season, species | Non-volatile |
| Van Huynh and Rice (2019) | Black-capped chickadee (*Poecile atricapillus*), Carolina chickadee (*Poecile carolinensis*) | Preen oil | 41 bcc and 40 cc sampled between Dec 2015 and Dec 2016 | Season, species | Volatile |
| Whittaker et al. (2010) | Gray catbird (*Dumetella carolinensis*) | Preen oil | 31 (24 m, 7 f, males separated into treatment groups) | Sex | Volatile |
| Whittaker et al. (2010) | Dark-eyed junco (*Junco hyemalis*) | Preen oil | 26 (12 m, 14 f) | Identity, sex | Volatile |
| Whittaker et al. (2011) | Dark-eyed junco (*Junco hyemalis*) | Preen oil | 51 (35 m, 16 f), repeated sampling of individuals across a 4-week period in early spring | Season, sex | Volatile |

| | | | | | |
|---|---|---|---|---|---|
| Whittaker et al. (2016) | Dark-eyed junco (*Junco hyemalis*) | Preen oil | 44 (17 adults (8 m, 9 f), 27 nestlings) | Age, genetic, sex | Volatile |
| Whittaker et al. (2018) | Dark-eyed junco (*Junco hyemalis*) | Preen oil | 33 (17 m [breeding condition], 16 f [incubation stage of reproduction]) | Sex | Volatile |
| Whittaker et al. (2019a) | Lance-tailed manakin (*Chiroxiphia lanceolata*) | Preen oil | 46 (25 m, 21 f), sampling from April to June | Age, genetic, season, sex | Volatile |
| Zhang et al. (2009) | Bengalese finch (*Lonchura striata*), zebra finch (*Taeniopygia guttata*), yellow-bowed bunting (*Emberiza chrysophrys*), rook (*Corvus frugilegus*) | Feathers, preen oil | 17 (9 m, 8 f in breeding condition), sample amount of other species not mentioned | Sex, species | Non-volatile |
| Zhang et al. (2009) | Bengalese finch (*Lonchura striata*), zebra finch (*Taeniopygia guttata*), yellow-bowed bunting (*Emberiza chrysophrys*), rook (*Corvus frugilegus*) | Feathers, preen oil | 17 (9 m, 8 f in breeding condition), sample amount of other species not mentioned | Identity, sex, species | Volatile |
| Zhang et al. (2010) | Budgerigar (*Melopsittacus undulatus*) | Feathers, preen oil | 48 (24 m, 24 f) | Sex | Volatile |
| Zhang et al. (2013) | Bohemian waxwing (*Bombycilla garrulus*) | Preen oil | 32 (16 m, 16 f) | Sex | Volatile |

*m* male, *f* female

some compounds exclusively found in either woodhoopoes' or European hoopoes' preen oil. Finally, Hagelin et al. (2003) showed that compounds responsible for the season signature in crested auklet feathers are not detectable in parakeet auklet ones.

## 2.2 Evidence for a Genetic Signature

Potier et al. (2018; black kites) showed that in male-male and male-female dyads, the difference in the chemical composition of preen oil increases during breeding season with genetic distance (measured using microsatellites), i.e. individuals with a higher genetic similarity had a more similar preen oil composition. For female-female dyads no such relationship was found (Potier et al., 2018). Whittaker et al. (2016) further reported that preen oil of dark-eyed junco nestlings is chemically more similar to preen oil of their father than of their mother. In preen oil of lance-tailed manakins, two linear alcohols (C16, C17) and two methyl ketones (C16, C17) were lower in concentration in males with higher heterozygosity at microsatellite loci (Whittaker et al., 2019a).

## 2.3 Evidence for an Individual Identity Signature

Potential individual identity signatures in the volatile fraction of avian samples have been studied in preen oil (Mardon et al., 2010: Antarctic prion and blue petrel; Zhang et al., 2009: Bengalese finch; Whittaker et al., 2010: dark-eyed junco; Leclaire et al., 2011: black-legged kittiwake; Potier et al., 2018: black kite), feathers (Bonadonna et al., 2007: Antarctic prion; Mardon et al., 2011: blue petrel; Zhang et al., 2009: Bengalese finch; Leclaire et al., 2011: black-legged kittiwake; Jennings & Ebeler, 2020: Leach's storm petrel) and whole body odour (Douglas, 2006: crested auklet). Only part of these authors report specific compounds that may contribute to such individual identity signatures. While Whittaker et al. (2010) state that almost all detected volatiles varied in their concentration between individuals and therefore contribute to individual identity signatures in preen oil of the dark-eyed junco, Leclaire et al. (2011) mention 1-dodecanol as a statistically significant driver of individual identity variation in black-legged kittiwake preen oil.

## 2.4 Evidence for a Breeding Season Signature

The variation of the composition of preen oil between breeding and non-breeding season was investigated in dark-eyed juncos (Soini et al., 2007; Whittaker et al., 2011), gray catbirds (Shaw et al., 2011), spotless starlings (Amo et al., 2012), white-throated sparrows (Tuttle et al., 2014), lance-tailed manakins (Whittaker et al., 2019a) and black-capped and Carolina chickadees (Van Huynh & Rice, 2019) as passerine species and European hoopoes (Martin-Vivaldi et al., 2010) and black kites (Potier et al., 2018) as non-passerines species. The effect of the breeding/non-breeding season on the composition of feathers' volatiles has been studied in crested auklets (Hagelin et al., 2003) and dark-eyed juncos (Soini et al., 2007). With few exceptions (Potier et al., 2018; Van Huynh & Rice, 2019), carboxylic acids and/or linear alcohols were found to be affected by variation between breeding and non-breeding season. More precisely, linear alcohols from 1-decanol to 1-pentadecanol were more abundant during the breeding season in dark-eyed juncos (Soini et al., 2007; Whittaker et al., 2011) and white-throated sparrows (Tuttle et al., 2014). Abundance of 1-tridecanol increased in the course of the main breeding season in lance-tailed manakins, while 1-tetradecanol decreased (Whittaker et al., 2019a). However, this effect was only significant for males (Whittaker et al., 2019a). The amount of hexadecanoic acid was found to increase in the breeding season in dark-eyed juncos (Soini et al., 2007), yet decreased in lance-tailed manakins over the course of the main breeding season (significant only for males, Whittaker et al., 2019a). It has to be noted that Soini et al. (2007) divided the peak areas of the target compounds by the area of an internal standard, whereas Whittaker et al. (2019a) divided the area of the target compound by the area of all compounds. Additionally, depending on the bird species, aldehydes, (methyl) ketones, alkanes, sterols, esters and sulphur compounds were reported to change between breeding and non-breeding seasons. An increase of 2-tridecanone and 2-pentadecanone during breeding season was found in male (Soini et al., 2007; Whittaker et al., 2011) and female (Whittaker et al., 2011) dark-eyed juncos, male and female gray catbirds (Shaw et al., 2011) and female and male white-throated sparrows (Tuttle et al., 2014). Additionally, 2-pentadecanone was found to increase during the course of the breeding season significantly in male and not significantly in female lance-tailed manakins (Whittaker et al., 2019a). Except for the study by Shaw et al. (2011), all these studies also mentioned 2-tetradecanone to increase with breeding season. In the dark-eyed junco (Soini et al., 2007; Whittaker et al., 2011), (methyl) ketones and carboxylic acids were contributing to the breeding season signature. Aldehydes as part of the breeding season signature were until now exclusively found in non-passerine species (Hagelin et al., 2003; Martin-Vivaldi et al., 2010).

## 2.5 Evidence for a Sex Signature

For the Bohemian waxwing (Zhang et al., 2013), Cory's shearwater (Gabirot et al., 2016), crested auklet (Hagelin et al., 2003; Douglas, 2006), feral pigeon (Leclaire et al., 2019), Leach's storm petrel (Jennings & Ebeler, 2020) and the song sparrow (Diez-Fernandez et al., 2020), no sex signature was evident in preen oil, feather and whole body odour samples. Zhang et al. (2013) reported that through principal component analysis (PCA) no sex difference was identified, but in terms of relative abundance 2-hexadecanone was more abundant in females compared to males. For the Bengalese finch (Zhang et al., 2009: feathers, preen oil), black kite (Potier et al., 2018: preen oil), black-legged kittiwake (Leclaire et al., 2011: feathers, preen oil), blue petrel (Mardon et al., 2011: feathers), blue tit (Caspers et al., 2021: preen oil), budgerigar (Zhang et al., 2010: feathers, preen oil), common blackbird (Diez-Fernandez et al. 2021: preen oil), dark-eyed junco (Soini et al., 2007: feathers, preen oil; Whittaker et al., 2010, 2011: preen oil), gray catbird (Shaw et al., 2011: preen oil), lance-tailed manakin (Whittaker et al., 2019a: preen oil), spotless starling (Amo et al., 2012: preen oil) and white-throated sparrow (Tuttle et al., 2014: preen oil), a sex difference in the volatile composition of the respective samples was detected. In some cases (Mardon et al., 2011; Whittaker et al., 2011; Caspers et al., 2021), the authors did not specify which compounds constituted the respective signatures. Diez-Fernandez et al. (2021) reported males to possess more compounds compared to females and additionally lower proportions of alkanes in males compared to females. Leclaire et al. (2011) report 1-dodecanol and 1-tetradecanol to be less abundant in the volatile fraction of feathers and preen oil of male compared to female black-legged kittiwakes. For black kites (Potier et al., 2018) and white-throated sparrows (Tuttle et al., 2014), a sex signature in preen oil was detected only in the non-breeding season. Black kite males showed higher amounts of one sterol and several ketones compared to females, and the sex signature in white-throated sparrows was made up of lower amounts of linear alcohols (C12-C18) and methyl ketones (C14-C17) in males compared to females. For spotless starlings, 2-pentadecanone occurred in significantly lower proportions in male preen oil compared to female preen oil (Amo et al., 2012). On a descriptive level, the opposite pattern occurred for linear alcohols (C10-15,17,18) and 2-methyltridecanone (Amo et al., 2012). Zhang et al. (2009) found for Bengalese finches hexadecanol and a diester in higher amounts and octadecanol in lower amounts in males compared to females. In budgerigars, hexadecanoic acid and linear alcohols, especially C18-C20, were detected in higher abundances in males compared to females (Zhang et al., 2010). In lance-tailed manakins, higher amounts of 1-tridecanol, 1-hexadecanol, 1-heptadecanol, 2-pentadecanone, 2-hexadecanone and 2-heptadecanone and lower amounts of 1-tetradecanol and hexadecenoic acid were present in preen oil of males compared to females (Whittaker et al., 2019a).

For the dark-eyed junco and the gray catbird, conflicting evidence is available regarding the existence of a sex signature. Whereas Whittaker et al. (2016, 2018) did not find a sex difference in the composition of dark-eyed junco preen oil, Soini et al. (2007) and Whittaker et al. (2010, 2011) demonstrated an impact of sex on its composition, notably with respect to linear alcohols and methyl ketones (Soini et al., 2007; Whittaker et al., 2010). The underlying reason for this contrasting result is not clear. According to Whittaker and colleagues a possible explanation is that sampling in earlier studies took place at the beginning of the breeding season, whereas in the later studies preen oil was sampled when birds initiated their clutch (Whittaker et al., 2018) specifically when the offspring fledged from the nest (Whittaker et al., 2016). Slightly higher concentrations of 2-pentadecanone and significantly lower concentrations of 1-decanol were present in feathers and preen oil of breeding males compared to breeding females (Soini et al., 2007). Whittaker et al. (2010) mention methyl ketones (C11–C15) being more abundant and linear alcohols (C10, C11) and carboxylic acids (C12, C14, C16) being less abundant in male preen oil compared to female preen oil. In adult gray catbird preen oil, no sex signature was detected (Whelan et al., 2010; Shaw et al., 2011), whereas in juveniles, a significant effect of sex became evident with carboxylic acids (C4–C7) being more abundant in preen oil from females compared to preen oil from males (Shaw et al., 2011). Thus, it seems possible that sex differences in gray catbird preen oil are only present until sexual maturity. This should, however, be investigated in further studies as in several species findings about sex differences appear not so reliable between studies (see above).

In sum, mostly linear alcohols and methyl ketones appear to differ between the two sexes if a difference could be detected at all. Yet, from the available literature, no general pattern regarding the nature of the difference becomes evident.

### 2.6 Evidence for an Age Signature

Whether the preen oil composition is affected by age has been investigated by comparing preen oil from adults and juveniles (Shaw et al., 2011; Potier et al., 2018; Diez-Fernandez et al. 2021), adults and nestlings (Amo et al., 2012; Whittaker et al., 2016) or adults of different ages (Whittaker et al., 2019a). In five studies, linear alcohols were found to differ in relation to age (Shaw et al., 2011; Amo et al., 2012; Whittaker et al., 2016, 2019a; Diez-Fernandez et al. 2021). To be more specific, tetradecanol was more abundant in preen oil of adult gray catbirds compared to juveniles (Shaw et al., 2011) and also found in higher proportions in preen oil of older female (but not in male) lance-tailed manakins (Whittaker et al., 2019a). In spotless starlings, hexadecanol was detected in higher proportions in the preen oil of adults compared to nestlings (Amo et al., 2012), whereas the opposite pattern was found for dark-eyed juncos (Whittaker et al., 2016). Here, it has to be noted that Amo et al. (2012) calculated the proportion of each compound in the secretion and applied compositional analysis for correction of non-independent proportions, whereas Whittaker et al. (2016) divided the area of the target compound by the area of an internal standard. Diez-Fernandez et al. (2021) reported significantly higher relative proportions of alcohols in adults compared to juveniles. Additional compound classes found to be dependent on age were carboxylic acids (Shaw et al., 2011: gray catbirds), (methyl) ketones (Amo et al., 2012: spotless starlings; Diez-Fernandez et al., 2021: common blackbirds), steroids and ketones (both Diez-Fernandez et al. 2021: common blackbirds). For the lance-tailed manakin and for the dark-eyed junco, only linear alcohols were related to age (Whittaker et al., 2016, 2019a). Potier et al. (2018) did not detect a difference between adults' and juveniles' black kite preen oil composition.

## 3 Non-Volatile Fraction

Research related to the non-volatile fraction of bird odour started in the early 1960s. The non-volatile fraction includes different forms of esters, such as monoesters, diesters, wax esters, fatty acid esters, fatty acid methyl esters and other types of at room temperature non-volatile compounds, such as longer-chain hydrocarbons, terpenoids and steroids. For characterisation of the non-volatile fraction, the sample has generally been either dissolved in a solvent, hydrolysed or transesterified. Afterwards, the samples have been mainly analysed by thin-layer chromatography (TLC), GC-FID or GC-MS.

### 3.1 Evidence for a Species Signature

Species differences in the non-volatile fraction have been detected in different bird species from the orders Passeriformes, Anseriformes, Galliformes and Procellariiformes. Four studies named the acid part in wax esters to differ between the species, but all these studies did not mention the number of birds analysed in the method section

(Apandi & Edwards, 1964; Edkins & Hansen, 1972; Jacob & Glaser, 1975). Four further studies indicated (mono)esters to be responsible for the species signature. Haribal et al. (2005) showed for different passerines qualitative and quantitative differences in monoesters, whereas Zhang et al. (2009) identified exact compounds to be available in one species, namely, hexadecyl heptanoate to be exclusively found in Bengalese finches as well as heptadecyl butyrate, heptadecyl pentanoate, octadecyl butyrate and octadecyl pentanoate being only available in zebra finches. Mardon et al. (2010) identified fatty acid esters with a chain length of 22–30 carbon atoms to differ between blue petrels and Antarctic prions. Van Huynh and Rice (2019) confirmed a species distinction between black-capped and Carolina chickadees through significantly higher principal component (PC) 1 and PC2 values in black-capped chickadees. PC1 was mentioned to positively correlate with esters, whereas for PC2 no specifications were made.

### 3.2 Evidence for a Genetic Signature

In 2012, Leclaire and colleagues found for the black-legged kittiwake that the wax ester composition of preen oil reflected major histocompatibility complex (MHC) heterozygosity both in males and females, whereas it reflected genetic distance only in male-male dyads. In 2014, Leclaire and colleagues found that preen oil wax esters of black-legged kittiwakes correlated with MHC relatedness in male-male and male-female dyads. Slade et al. (2016) investigated preen oil of the song sparrow and identified four peaks, the combination of which led to the maximum correlation between genetic and chemical dissimilarity. The four peaks corresponded to heptadecyl pentadecanoate, tetradecyl nonadecanoate and mixtures of esters with components leading to a total carbon number of 28 and 34.

### 3.3 Evidence for a Breeding Season Signature

Differences in the non-volatile composition of preen oil and feathers between breeding and non-breeding season have been studied in mallard ducks (Jacob et al., 1979), red-vented bulbuls (Bhattacharyya & Chowdhury, 1995), red-knot shorebirds (Piersma et al., 1999; Reneerkens et al., 2002, 2007b), black-legged kittiwakes (Leclaire et al., 2011, 2012), herring gulls (Fischer et al., 2017), song sparrows (Grieves et al., 2019, 2020), black-capped and Carolina chickadees (Van Huynh & Rice, 2019) and different sandpiper (Reneerkens et al., 2002, 2007b) and passerine species (Haribal et al., 2005).

The non-volatile fraction changes between breeding and non-breeding seasons mainly in two ways. First, a relative decrease of monoesters accompanied by a more or less prominent raise of diesters has been described in preen oil and feathers with the beginning of mating season or in May for mallard ducks (Jacob et al., 1979), red-knot shorebirds (Piersma et al., 1999; Reneerkens et al., 2002, 2007b) and different sandpiper species (Reneerkens et al., 2002, 2007b). It is interesting to note that this change has partly been observed in females or males only (Jacob et al., 1979; Reneerkens et al., 2007b). This may depend on the incubation behaviour of the respective bird species (uni- or biparental incubation; see Discussion; Reneerkens et al., 2007b). Second, a qualitative and quantitative change in the monoester fraction of preen oil and feathers during breeding season has been determined for different passerines species (Haribal et al., 2005), black-legged kittiwakes (Leclaire et al., 2011, 2012), herring gulls (Fischer et al., 2017), red-vented bulbuls (Bhattacharyya & Chowdhury, 1995) and song sparrows (Grieves et al., 2019, 2020).

### 3.4 Evidence for a Sex Signature

With one exception (Montalti et al., 2005), the existence of a sex signature has been reported in the available studies for mallard ducks (Jacob et al., 1979), Bengalese and zebra finches, yellow-browed buntings and rooks (Zhang et al., 2009), blue petrels and Antarctic prions (Mardon et al., 2010), black-legged kittiwakes (Leclaire et al., 2011, 2012), great tits (Jacob et al., 2014), herring gulls (Fischer et al., 2017) and song sparrows (Grieves et al., 2019, 2020). Two out of nine studies explicitly mentioned that a sex difference was only evident during the reproductive/breeding stage but not in the non-breeding stage (Jacob et al., 1979; Grieves et al., 2019). The other studies did not mention if the sex signature was related to a timepoint in the year (Fischer et al., 2017; Grieves et al., 2020) or sampled only once within a year (Zhang et al., 2009; Mardon et al., 2010; Leclaire et al., 2011, 2012; Jacob et al., 2014). With few exceptions (Jacob et al., 1979: wax esters and diesters; Zhang et al., 2009: unidentified diester; Jacob et al., 2014, Grieves et al., 2020: no compounds mentioned) monoesters, specifically wax esters, contributed to the sex signature. Grieves et al. (2019) showed for preen oil from breeding song sparrows that dodecyl octadecanoate to dodecyl docosanoate as well as dodecyl pentacosanoate was more elevated in males and tridecyl to pentadecyl heptadecanoate less abundant in males compared to females. Additionally some further compounds were mentioned to contribute to the sex signature, but to a lower extent. In the sex signature of black-legged kittiwakes (preen oil and feathers), blue petrels and Antarctic prions (preen oil), pentadecyl-2-methyldecanoate was more elevated in females (Mardon et al., 2010; Leclaire et al., 2011, 2012).

### 3.5 Evidence for an Age Signature

Sandilands et al. (2004) found the fatty acid wax ester composition of preen oil and feathers from chicken to differ according to age. Significant age effects were found for fatty acids ranging from C10 to C14, C16 to C20 and C24, as well as three unidentified fatty acids. In a study about mallards, it was shown that the preen oil of adult males and female and male ducklings with an age of 1 month consisted mainly of branched ester waxes and a small amount of diester waxes. In adult females mainly diester waxes and to a smaller extent branched and unbranched ester waxes were detected (Jacob et al., 1979). Finally, Grieves et al. (2019) found a significant difference between preen oil composition of juvenile and adult song sparrows, but the compounds were not specified.

## 4 Discussion and Future Research Areas

The existence of avian chemical signatures reflecting species, genetics, individual identity, breeding season, sex and age has been approved both in the volatile and the non-volatile fractions in a majority of the literature summarised here. In general, however, it comes out that consistent results across species could only be rarely found. Thus, the currently available evidence suggests that the molecular basis of different chemical signatures can vary between different species.

Not surprisingly, the chemical profiles of birds' secretions and excretions relate to species, genetics and identity. These chemical signatures may be considered as putative signature mixtures sensu Wyatt (2010) and be hypothesised to be stable over time. However, considering the evidence summarised above, it becomes apparent that not enough details are known yet about the chemicals constituting these signatures in birds. With regard to the non-volatile fraction, only two studies have been conducted so far, and therefore, the criteria for including them in this review could not be fulfilled.

Besides advancing our knowledge about the chemical nature of the species, genetic and individual identity signatures, future research should also focus on further elucidating their functional value. In this respect, Whittaker et al. (2010) suggested that differences in the preen oil composition of two different dark-eyed junco populations, one from the University of California, San Diego, and one from Laguna Mountain Recreation Area (Cleveland national forest), prevents mixing of populations which can be helpful to adapt to special living conditions. For song sparrows, Grieves et al. (2019) showed that the preen oil wax ester composition of two breeding locations in Ontario, Canada, the Newboro and the Cambridge population, which are separated by a distance of 390 km, was significantly different during breeding. After breeding, no differences were found, supporting the notion that species differentiation via chemical signatures might be of relevance for reproductive isolation. Genetic influences on chemical signatures of preen oil may also reflect adaptation of a bird species to environmental conditions of its habitat and life history (e.g. in seabirds). For instance, the comparison of tropical and temperate bird species showed that the ester composition is differing with higher-molecular-weight esters being present in tropical bird species (Haribal et al., 2009).

Research regarding the breeding season signature is already well established (Grieves et al., 2022), with the first studies having been published in the 1970s, and it is generally acknowledged that the chemical signature of preen oil and feathers changes throughout the year in the vast majority of bird species. The publications summarised in this review report on effects related to breeding conditions, but additional influence of seasonal parameters (dry/wet season or winter/summer) cannot be excluded. The described switch from monoesters to less volatile diesters during the breeding season could reflect a protective mechanism against predators, following the chemical crypsis hypothesis (e.g. Reneerkens et al., 2007b). Adopting a similar line of thoughts, higher concentrations of volatile compounds at the beginning of egg laying, as found out for the dark-eyed junco by Whittaker et al. (2011), may at first glance be unexpected because this may increase the risk of a predator attack, at least if these volatiles are used as chemical cues by predators. Yet, an increase in volatile compounds may also raise attractiveness, which could be helpful for mating, i.e. following the sex semiochemical hypothesis (Grieves et al., 2022). An interesting finding in this respect is that in species where only one parent is involved in incubation, preen oil composition changes only in the incubating parent, showing a change from mono- to diesters (Reneerkens et al., 2007b). In contrast and with the exception of the Temminck's stint, it was shown that in species where both parents incubate, the change from mono- to diesters can be found in both parents (Reneerkens et al., 2007b). It has therefore also been proposed that the changing composition of avian secretions during breeding may help to prevent high parasite load during incubation (reviewed in Moreno-Rueda, 2010). Future studies in further species could contribute to shed light on the commonalities and differences in breeding season signatures and their adaptive value (see Grieves et al., 2022). Further, open questions regarding the season signature may be answered by adopting new experimental approaches trying to simulate breeding and non-breeding seasons in captive birds using different lights and temperature conditions. Soini et al. (2007) detected an increase in compounds of the substance classes linear alcohols, methyl ketones and fatty acids when dark-eyed juncos were exposed to more light (16 h instead of 8 h light exposure per 24 h), whereas on the contrary Whelan et al. (2010) found propanoic acid, 2-methylpropanoic acid and butanoic acid to decrease for gray catbirds throughout longer photoperiods. The findings are comparable with the above-mentioned studies with certain compounds or compound classes being more or less elevated in the breeding/non-breeding season depending on the bird species. Therefore, especially for less accessible bird species, a change of the composition throughout the year could be studied by applying such ecological conditions in captive birds.

The highest number of publications is available for the sex signature. In several bird species, a sex signature appears to be present only during the breeding season (reviewed in Grieves et al., 2022). This finding highlights that many of the factors for which chemical signatures were investigated are entangled with each other (Fig. 1). This needs to be considered in future research by including potential influencing factors as independent variables and ensuring that the number of birds analysed is sufficient to conduct data analysis with several independent variables. Alternatively, experimental lab studies could help, in which some parameters are kept constant to study the impact of variation in another parameter. Further, more studies are needed to establish the biochemical bases beyond sex signatures. In this respect, the role of hormones is a related, but still not well understood, area of research. Some insights have been gained regarding the influence of testosterone on the volatilome of birds. In gray catbirds, the amounts of propanoic and 2-methylpropanoic acids decreased when testosterone was administered (Whelan et al., 2010). In dark-eyed juncos, Whittaker et al. (2011) demonstrated that a testosterone treatment led to an increase in undecanol, dodecanol, tridecanol, undecanone and hexadecanoic acid. In contrast, an estradiol/testosterone treatment in ducklings did not influence the composition of unbranched, branched and diester waxes (Jacob et al., 1979). Future studies should elucidate further the influence of the hormonal status on the composition of avian chemical signatures. Furthermore, the relationship between chemical signatures and avian behaviour/other aspects in birds' life should be studied further. Whittaker et al. (2013), for instance, showed that there was a positive correlation with three methyl ketones constituting the sex signature of male dark-eyed juncos and the reproductive success in both sexes (Whittaker et al., 2013).

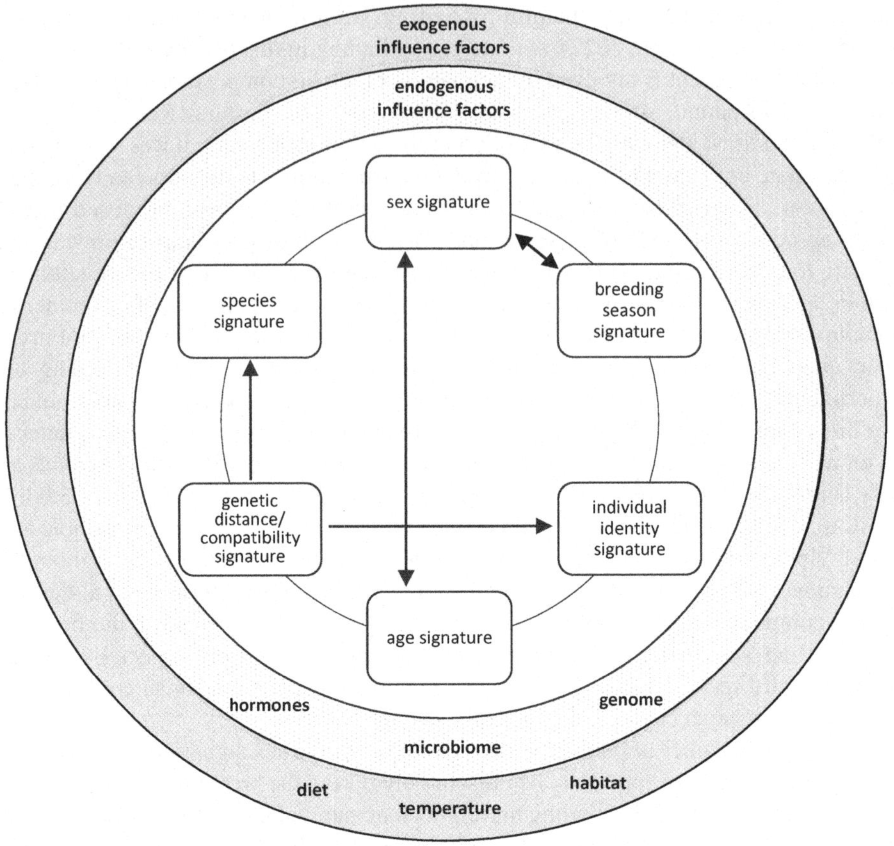


**Fig. 1** Chemical signatures – influencing factors and interrelationships

Male birds having higher and females having lower proportions of those compounds tended to have higher reproductive success (Whittaker et al., 2013).

The age signature is, besides the genetic signature, the least studied type of chemical signature. Inconsistencies in the results for the age signature may be attributed to the different age groups which have been compared so far (juveniles-adults, nestlings-adults, adults-adults). Nonetheless, several aspects can be raised in favour of the existence of an age signature. First, it is well known that ageing is accompanied by changes in the physiology of an organism, because intrinsic (e.g. hormonal status, sexual maturation, gene expression) and extrinsic (e.g. exposure to toxic substances) factors influence inflammatory and other (also secretory) processes (Rea et al., 2018; Zouboulis & Makrantonaki, 2019). Such processes can be supposed to influence also the composition of preen oil and the quality of the skin surface (cf. Nazzaro-Porro et al., 1979). Second, an age signature can be assumed to have an adaptive value, as it might be able to influence parent-offspring communication or could be helpful in finding a sexually mature mate.

Besides the parameters considered here – species, genetics, identity, season, sex, age – further influencing factors on the volatilome of birds should be studied in future research. Initial evidence has, for example, been obtained regarding an influence of captivity (Reneerkens et al., 2007a; Thomas et al., 2010), food intake (Apandi & Edwards, 1964; Thomas et al., 2010), food stress (Reneerkens et al., 2007a; Grieves et al., 2020), habitat (Haribal et al., 2009), infection (Grieves et al., 2018; Diez-Fernandez et al., 2020, 2021), microbiome (Martin-Vivaldi et al., 2010; Jacob et al., 2014; Whittaker et al., 2019b) and hormone treatment and zinc supplementation (Leclaire et al., 2019). Besides deserving further research, such factors should also be considered when studying avian chemical signatures in general, as they might be able to distort the findings when looking at other chemical signatures, especially when the study is based on a low number of individuals.

Finally, we would like to highlight three further aspects to be addressed by future research. First, it has been shown in several studies that chemical signatures and microbial signatures are intertwined (see Maraci & Caspers, 2021). Here, a fascinating research field opens up and studies using for example, a multi-omics approach or other approaches targeting specific metabolic pathways of microbiota will further unravel the potential role of the microbiome in chemical signatures and, hence, olfactory communication. Second, in a majority of studies, the volatile and non-volatile fractions of preen oil and feathers are considered separately. Of the 43 publications reviewed here, 24 considered the volatile fraction and 15 the non-volatile fraction and only 4 analysed both fractions. This is mainly due to the fact that different analytical procedures are generally employed to investigate the different fractions. Yet, it is important to jointly assess volatile and non-volatile substances to understand the coherence and mutual influences within avian chemocommunicative systems. To achieve the full picture and also to discover potential correlations between the occurrence of compounds in the different fractions, a comprehensive analytical strategy to study the different constituents of preen oil and further avian samples is therefore needed. Such approaches could be inspired by current analytical work in other fields of research, e.g. food chemistry (e.g., Zhu et al., 2010; Mehl et al., 2014).

Considering analytical strategies in general, it is essential to keep in mind that each method generates a different picture of the sample, as already discussed by Leclaire (2021). Results from the analysis of the non-volatile fraction using a transesterification during workup can, for example, not be compared to results from direct sampling without any modifications. Furthermore, results can be strongly influenced by various factors including analytical parameters such as GC column type, detector type, extraction method and sample amount analysed (for further information about this, see, for example, Rood, 2007). This should be kept in mind especially when comparing the results of different studies. Third, fast analytical methods should be developed and used to study avian chemical signatures. Comparable to the rapid evolution of molecular techniques, developments of chemical analytical methods will allow creating big data, which are needed to fully understand the high dimensionality of avian chemical signatures.

**Acknowledgements** This project was funded by the German Research Foundation (LO 2545/5-1 and CA 889/9-1).

## References

Amo, L., Aviles, J. M., Parejo, D., Pena, A., Rodriguez, J., & Tomas, G. (2012). Sex recognition by odour and variation in the uropygial gland secretion in starlings. *Journal of Animal Ecology, 81*, 605–613. https://doi.org/10.1111/j.1365-2656.2011.01940.x

Apandi, M., & Edwards, H. M. (1964). Studies on the composition of the secretions of the uropygial gland of some avian species[1,2]. *Poultry Science, 43*, 1445–1462. https://doi.org/10.3382/ps.0431445

Bang, B. G., & Cobb, S. (1968). The size of the olfactory bulb in 108 species of birds. *Auk, 85*, 55–61. https://doi.org/10.2307/4083624

Bhattacharyya, S. P., & Chowdhury, S. R. (1995). Seasonal variation in the secretory lipids of the uropygial gland of a sub-tropical wild passerine bird, *Pycnonotus cafer* (L) in relation to the testicular cycle. *Biological Rhythm Research, 26*, 79–87. https://doi.org/10.1080/09291019509360326

Bonadonna, F., Miguel, E., Grosbois, V., Jouventin, P., & Bessiere, J. M. (2007). Individual odor recognition in birds: An endogenous olfactory signature on petrels' feathers? *Journal of Chemical Ecology, 33*, 1819–1829. https://doi.org/10.1007/s10886-007-9345-7

Butenandt, A., Beckmann, R., Stamm, D., & Hecker, E. (1959). Über den Sexual-Lockstoff des Seidenspinners *Bombyx mori*. Reindarstellung und Konstitution. *Zeitschrift für Naturforschung, B, 14*, 283–284.

Campagna, S., Mardon, J., Celerier, A., & Bonadonna, F. (2012). Potential semiochemical molecules from birds: A practical and comprehensive compilation of the last 20 years studies. *Chemical Senses, 37*, 3–25. https://doi.org/10.1093/chemse/bjr067

Caro, S. P., & Balthazart, J. (2010). Pheromones in birds: Myth or reality? *Journal of Comparative Physiology A Neuroethology, Sensory, Neural & Behavioural Physiology, 196*, 751–766. https://doi.org/10.1007/s00359-010-0534-4

Caro, S. P., Balthazart, J., & Bonadonna, F. (2015). The perfume of reproduction in birds: Chemosignaling in avian social life. *Hormones and Behavior, 68*, 25–42. https://doi.org/10.1016/j.yhbeh.2014.06.001

Caspers, B. A., Marfull, R., Dannenhaus, T., Komdeur, J., & Korsten, P. (2021). Chemical analysis reveals sex differences in the preen gland secretion of breeding Blue Tits. *Journal of Ornithology, 163*, 191–198. https://doi.org/10.1007/s10336-021-01921-w

Diez-Fernandez, A., Martinez-de la Puente, J., Gangoso, L., Lopez, P., Soriguer, R., Martin, J., & Figuerola, J. (2020). Mosquitoes are attracted by the odour of *Plasmodium*-infected birds. *International Journal for Parasitology, 50*(8), 569–575. https://doi.org/10.1016/j.ijpara.2020.03.013

Diez-Fernandez, A., Martinez-de la Puente, J., Martin, J., Gangoso, L., Lopez, P., Soriguer, R., & Figuerola, J. (2021). Sex and age, but not blood parasite infection nor habitat, affect the composition of the uropygial gland secretions in European blackbirds. *Journal of Avian Biology, 52*(8), Article e02630. https://doi.org/10.1111/jav.02630

Douglas, H. D. (2006). Measurement of chemical emissions in crested auklets (*Aethia cristatella*). *Journal of Chemical Ecology, 32*, 2559–2567. https://doi.org/10.1007/s10886-006-9164-2

Edkins, E., & Hansen, I. A. (1972). Wax esters secreted by uropygial glands of some Australian waterfowl, including magpie goose. *Comparative Biochemistry and Physiology, 41*, 105–112. https://doi.org/10.1016/0305-0491(72)90012-0

Eisenberg, J. F., & Kleiman, D. G. (1972). Olfactory communication in mammals. *Annual Review of Ecology and Systematics, 3*, 1–32.

Fischer, I., Halinski, L. P., Meissner, W., Stepnowski, P., & Knitter, M. (2017). Seasonal changes in the preen wax composition of the Herring gull *Larus argentatus*. *Chemoecology, 27*(4), 127–139. https://doi.org/10.1007/s00049-017-0239-z

Gabirot, M., Raux, L., Dell'Ariccia, G., Bried, J., Ramos, R., Gonzalez-Solis, J., … Bonadonna, F. (2016). Chemical labels differ between two closely related shearwater taxa. *Journal of Avian Biology, 47*(4), 540–551. https://doi.org/10.1111/jav.00853

Grieves, L. A., Kelly, T. R., Bernards, M. A., & MacDougall-Shackleton, E. A. (2018). Malarial infection alters wax ester composition of preen oil in songbirds: Results of an experimental study. *Auk, 135*(3), 767–776. https://doi.org/10.1642/Auk-17-242.1

Grieves, L. A., Bernards, M. A., & MacDougall-Shackleton, E. A. (2019). Wax ester composition of songbird preen oil varies seasonally and differs between sexes, ages, and populations. *Journal of Chemical Ecology, 45*(1), 37–45. https://doi.org/10.1007/s10886-018-1033-2

Grieves, L. A., Bottini, C. L. J., Branfireun, B. A., Bernards, M. A., MacDougall-Shackleton, S. A., & MacDougall-Shackleton, E. A. (2020). Food stress, but not experimental exposure to mercury, affects songbird preen oil composition. *Ecotoxicology, 29*(3), 275–285. https://doi.org/10.1007/s10646-020-02171-x

Grieves, L. A., Gilles, M., Cuthill, I. C., Székely, T., MacDougall-Shackleton, E. A., & Caspers, B. A. (2022). Olfactory camouflage and communication in birds. *Biological Reviews, 97*(3), 1193–1209. https://doi.org/10.1111/brv.12837

Hagelin, J. C., Jones, I. L., & Rasmussen, L. E. L. (2003). A tangerine-scented social odour in a monogamous seabird. *Proceedings of the Royal Society B-Biological Sciences, 270*(1522), 1323–1329. https://doi.org/10.1098/rspb.2003.2379

Haribal, M., Dhondt, A. A., Rosane, D., & Rodriguez, E. (2005). Chemistry of preen gland secretions of passerines: Different pathways to same goal? why? *Chemoecology, 15*(4), 251–260. https://doi.org/10.1007/s00049-005-0318-4

Haribal, M., Dhondt, A., & Rodriguez, E. (2009). Diversity in chemical compositions of preen gland secretions of tropical birds. *Biochemical Systematics and Ecology, 37*(2), 80–90. https://doi.org/10.1016/j.bse.2008.12.005

Higham, J. P., & Hebets, E. A. (2013). An introduction to multimodal communication. *Behavioral Ecology and Sociobiology, 67*(9), 1381–1388. https://doi.org/10.1007/s00265-013-1590-x

Jacob, J., & Glaser, A. (1975). Chemotaxonomy of anseriformes. *Biochemical Systematics and Ecology, 2*(3), 215–219. https://doi.org/10.1016/0305-1978(75)90065-4

Jacob, J., Balthazart, J., & Schoffeniels, E. (1979). Sex-differences in the chemical composition of uropygial gland waxes in domestic ducks. *Biochemical Systematics and Ecology, 7*(2), 149–153. https://doi.org/10.1016/0305-1978(79)90024-3

Jacob, S., Immer, A., Leclaire, S., Parthuisot, N., Ducamp, C., Espinasse, G., & Heeb, P. (2014). Uropygial gland size and composition varies according to experimentally modified microbiome in Great tits. *BMC Evolutionary Biology, 14*. https://doi.org/10.1186/1471-2148-14-134

Jennings, S. L., & Ebeler, S. E. (2020). Individual chemical profiles in the Leach's storm petrel. *Journal of Chemical Ecology*, 46, 845–864. https://doi.org/10.1007/s10886-020-01207-0

Johnston, R. E. (2003). Chemical communication in rodents: From pheromones to individual recognition. *Journal of Mammalogy, 84*(4), 1141–1162. https://doi.org/10.1644/BLe-010

Johnston, R. E. (2005). Communication by mosaic signals: Individual recognition and underlying neural mechanisms. In R. T. Mason, M. P. LeMaster, & D. Müller-Schwarze (Eds.), *Chemical signals in vertebrates 10*. Springer. https://doi.org/10.1007/0-387-25160-X_30

Karlson, P., & Lüscher, M. (1959). 'Pheromones': A new term for a class of biologically active substances. *Nature, 183*, 55–56. https://doi.org/10.1038/183055a0

Krause, E. T., Brummel, C., Kohlwey, S., Baier, M. C., Muller, C., Bonadonna, F., & Caspers, B. A. (2014). Differences in olfactory species recognition in the females of two Australian songbird species. *Behavioral Ecology and Sociobiology, 68*(11), 1819–1827. https://doi.org/10.1007/s00265-014-1791-y

Leclaire, S. (2021). A deep dive into MHC-related odors: A comment on Schubert et al. *Behavioral Ecology, 32*(3), 375–376. https://doi.org/10.1093/beheco/arab023

Leclaire, S., Merkling, T., Raynaud, C., Giacinti, G., Bessiere, J. M., Hatch, S. A., & Danchin, E. (2011). An individual and a sex odor signature in kittiwakes? Study of the semiochemical composition of preen secretion and preen down feathers. *Naturwissenschaften, 98*(7), 615–624. https://doi.org/10.1007/s00114-011-0809-9

Leclaire, S., Merkling, T., Raynaud, C., Mulard, H., Bessiere, J. M., Lhuillier, E., et al. (2012). Semiochemical compounds of preen secretion reflect genetic make-up in a seabird species. *Proceedings of the Royal Society B-Biological Sciences, 279*(1731), 1185–1193. https://doi.org/10.1098/rspb.2011.1611

Leclaire, S., van Dongen, W. F. D., Voccia, S., Merkling, T., Ducamp, C., Hatch, S. A., et al. (2014). Preen secretions encode information on MHC similarity in certain sex-dyads in a monogamous seabird. *Scientific Reports, 4*. https://doi.org/10.1038/srep06920

Leclaire, S., Chatelain, M., Pessato, A., Buatois, B., Frantz, A., & Gasparini, J. (2019). Pigeon odor varies with experimental exposure to trace metal pollution. *Ecotoxicology, 28*(1), 76–85. https://doi.org/10.1007/s10646-018-2001-x

Maraci, O., & Caspers, B. A. (2021). Considering birds to understand the interplay of MHC, microbiota, and odor: A comment on Schubert et al. *Behavioral Ecology, 32*(3), 376–377. https://doi.org/10.1093/beheco/arab024

Mardon, J., Saunders, S. M., Anderson, M. J., Couchoux, C., & Bonadonna, F. (2010). Species, gender, and identity: Cracking petrels' sociochemical code. *Chemical Senses, 35*(4), 309–321. https://doi.org/10.1093/chemse/bjq021

Mardon, J., Saunders, S. M., & Bonadonna, F. (2011). From preen secretions to plumage: The chemical trajectory of blue petrels' *Halobaena caerulea* social scent. *Journal of Avian Biology, 42*(1), 29–38. https://doi.org/10.1111/j.1600-048X.2010.05113.x

Martin-Vivaldi, M., Pena, A., Peralta-Sanchez, J. M., Sanchez, L., Ananou, S., Ruiz-Rodriguez, M., & Soler, J. J. (2010). Antimicrobial chemicals in hoopoe preen secretions are produced by symbiotic bacteria. *Proceedings of the Royal Society B-Biological Sciences, 277*(1678), 123–130. https://doi.org/10.1098/rspb.2009.1377

Mehl, F., Marti, G., Boccard, J., Debrus, B., Merle, P., Delort, E., et al. (2014). Differentiation of lemon essential oil based on volatile and non-volatile fractions with various analytical techniques: A metabolomic approach. *Food Chemistry, 143*, 325–335. https://doi.org/10.1016/j.foodchem.2013.07.125

Montalti, D., Gutierrez, A. M., Reboredo, G., & Salibian, A. (2005). The chemical composition of the uropygial gland secretion of rock dove *Columba livia*. *Comparative Biochemistry and Physiology a-Molecular & Integrative Physiology, 140*(3), 275–279. https://doi.org/10.1016/j.cbpb.2004.10.008

Moreno-Rueda, G. (2010). Uropygial gland size correlates with feather holes, body condition and wingbar size in the house sparrow *Passer domesticus*. *Journal of Avian Biology, 41*(3), 229–236. https://doi.org/10.1111/j.1600-048X.2009.04859.x

Müller, C., Caspers, B. A., Gadau, J., & Kaiser, S. (2020). The power of infochemicals in mediating individualized niches. *Trends in Ecology & Evolution, 35*(11), 981–989. https://doi.org/10.1016/j.tree.2020.07.001

Nazzaro-Porro, M., Passi, S., Boniforti, L., & Belsito, F. (1979). Effects of aging on fatty acids in skin surface lipids. *Journal of Investigative Dermatology, 73*(1), 112–117. https://doi.org/10.1111/1523-1747.ep12532793

Piersma, T., Dekker, M., & Damste, J. S. S. (1999). An avian equivalent of make-up? *Ecology Letters, 2*(4), 201–203. https://doi.org/10.1046/j.1461-0248.1999.00078.x

Potier, S., Besnard, M. M., Schikorski, D., Buatois, B., Duriez, O., Gabirot, M., et al. (2018). Preen oil chemical composition encodes individuality, seasonal variation and kinship in black kites *Milvus migrans*. *Journal of Avian Biology, 49*(7). https://doi.org/10.1111/jav.01728

Rasmussen, L. E. L., Lee, T. D., Roelofs, W. L., Zhang, A., & Daves, G. D. (1996). Insect pheromone in elephants. *Nature, 379*(6567), 684–684. https://doi.org/10.1038/379684a0

Rea, I. M., Gibson, D. S., McGilligan, V., McNerlan, S. E., Alexander, H. D., & Ross, O. A. (2018). Age and age-related diseases: Role of inflammation triggers and cytokines. *Frontiers in Immunology, 9*. https://doi.org/10.3389/fimmu.2018.00586

Reneerkens, J., Piersma, T., & Damste, J. S. S. (2002). Sandpipers (*Scolopacidae*) switch from monoester to diester preen waxes during courtship and incubation, but why? *Proceedings of the Royal Society B-Biological Sciences, 269*(1505), 2135–2139. https://doi.org/10.1098/rspb.2002.2132

Reneerkens, J., Piersma, T., & Damste, J. S. S. (2007a). Expression of annual cycles in preen wax composition in red knots: Constraints on the changing phenotype. *Journal of Experimental Zoology Part a-Ecological Genetics and Physiology, 307a*(3), 127–139. https://doi.org/10.1002/jez.a.347

Reneerkens, J., Almeida, J. B., Lank, D. B., Jukema, J., Lanctot, R. B., Morrison, R. I. G., et al. (2007b). Parental role division predicts avian preen wax cycles. *Ibis, 149*(4), 721–729. https://doi.org/10.1111/j.1474-919X.2007.00693.x

Rood, D. (2007). *The troubleshooting and maintenance guide for gas chromatographers*. Wiley-VCH.

Roper, T. J. (1999). Olfaction in birds. *Advances in the Study of Behavior, 28*, 247–332.

Sandilands, V., Powell, K., Keeling, L., & Savory, C. J. (2004). Preen gland function in layer fowls: Factors affecting preen oil fatty acid composition. *British Poultry Science, 45*(1), 109–115. https://doi.org/10.1080/00071660410001668932

Schaal, B., Coureaud, G., Langlois, D., Giniès, C., Sémon, E., & Perrier, G. (2003). Chemical and behavioural characterization of the rabbit mammary pheromone. *Nature, 424*(6944), 68–72. https://doi.org/10.1038/nature01739

Shaw, C. L., Rutter, J. E., Austin, A. L., Garvin, M. C., & Whelan, R. J. (2011). Volatile and semivolatile compounds in gray catbird uropygial secretions vary with age and between breeding and wintering grounds. *Journal of Chemical Ecology, 37*(4), 329–339. https://doi.org/10.1007/s10886-011-9931-6

Slade, J. W. G., Watson, M. J., Kelly, T. R., Gloor, G. B., Bernards, M. A., & MacDougall-Shackleton, E. A. (2016). Chemical composition of preen wax reflects major histocompatibility complex similarity in songbirds. *Proceedings of the Royal Society B-Biological Sciences, 283*(1842). https://doi.org/10.1098/rspb.2016.1966

Smith, W. J. (1977). *The behavior of communicating. An ethological approach*. Harvard University Press.

Soini, H. A., Schrock, S. E., Bruce, K. E., Wiesler, D., Ketterson, E. D., & Novotny, M. V. (2007). Seasonal variation in volatile compound profiles of preen gland secretions of the dark-eyed junco (*Junco hyemalis*). *Journal of Chemical Ecology, 33*(1), 183–198. https://doi.org/10.1007/s10886-006-9210-0

Thomas, R. H., Price, E. R., Seewagen, C. L., Mackenzie, S. A., Bernards, M. A., & Guglielmo, C. G. (2010). Use of TLC-FID and GC-MS/FID to examine the effects of migratory state, diet and captivity on preen wax composition in White-throated Sparrows *Zonotrichia albicollis*. *Ibis, 152*(4), 782–792. https://doi.org/10.1111/j.1474-919X.2010.01050.x

Tuttle, E. M., Sebastian, P. J., Posto, A. L., Soini, H. A., Novotny, M. V., & Gonser, R. A. (2014). Variation in preen oil composition pertaining to season, sex, and genotype in the polymorphic white-throated sparrow. *Journal of Chemical Ecology, 40*(9), 1025–1038. https://doi.org/10.1007/s10886-014-0493-2

Van Huynh, A., & Rice, A. M. (2019). Conspecific olfactory preferences and interspecific divergence in odor cues in a chickadee hybrid zone. *Ecol Evol, 9*(17), 9671–9683. https://doi.org/10.1002/ece3.5497

Wackermannova, M., Pinc, L., & Jebavý, L. (2016). Olfactory sensitivity in mammalian species. *Physiol Res, 65*(3), 369–390. https://doi.org/10.33549/physiolres.932955

Whelan, R. J., Levin, T. C., Owen, J. C., & Garvin, M. C. (2010). Short-chain carboxylic acids from gray catbird (*Dumetella carolinensis*) uropygial secretions vary with testosterone levels and photoperiod. *Comparative Biochemistry and Physiology B-Biochemistry & Molecular Biology, 156*(3), 183–188. https://doi.org/10.1016/j.cbpb.2010.03.005

Whittaker, D. J. (2022). *The secret perfume of birds: Uncovering the science of avian scent.* Johns Hopkins University Press.

Whittaker, D. J., Soini, H. A., Atwell, J. W., Hollars, C., Novotny, M. V., & Ketterson, E. D. (2010). Songbird chemosignals: Volatile compounds in preen gland secretions vary among individuals, sexes, and populations. *Behavioral Ecology, 21*(3), 608–614. https://doi.org/10.1093/beheco/arq033

Whittaker, D. J., Soini, H. A., Gerlach, N. M., Posto, A. L., Novotny, M. V., & Ketterson, E. D. (2011). Role of testosterone in stimulating seasonal changes in a potential avian chemosignal. *Journal of Chemical Ecology, 37*(12), 1349–1357. https://doi.org/10.1007/s10886-011-0050-1

Whittaker, D. J., Gerlach, N. M., Soini, H. A., Novotny, M. V., & Ketterson, E. D. (2013). Bird odour predicts reproductive success. *Animal Behaviour, 86*, 697–703. https://doi.org/10.1016/j.anbehav.2013.07.025

Whittaker, D. J., Gerlach, N. M., Slowinski, S. P., Corcoran, K. P., Winters, A. D., Soini, H. A., et al. (2016). Social environment has a primary influence on the microbial and odor profiles of a chemically signaling songbird. *Frontiers in Ecology and Evolution, 4*. https://doi.org/10.3389/fevo.2016.00090

Whittaker, D. J., Rosvall, K. A., Slowinski, S. P., Soini, H. A., Novotny, M. V., & Ketterson, E. D. (2018). Songbird chemical signals reflect uropygial gland androgen sensitivity and predict aggression: Implications for the role of the periphery in chemosignaling. *Journal of Comparative Physiology a-Neuroethology Sensory Neural and Behavioral Physiology, 204*(1), 5–15. https://doi.org/10.1007/s00359-017-1221-5

Whittaker, D. J., Kuzel, M., Burrell, M. J. E., Soini, H. A., Novotny, M. V., & DuVal, E. H. (2019a). Chemical profiles reflect heterozygosity and seasonality in a tropical lekking passerine bird. *Animal Behaviour, 151*, 67–75. https://doi.org/10.1016/j.anbehav.2019.03.005

Whittaker, D. J., Slowinski, S. P., Greenberg, J. M., Alian, O., Winters, A. D., Ahmad, M. M., et al. (2019b). Experimental evidence that symbiotic bacteria produce chemical cues in a songbird. *Journal of Experimental Biology, 222*(20). https://doi.org/10.1242/jeb.202978

Wyatt, T. D. (2010). Pheromones and signature mixtures: defining species-wide signals and variable cues for identity in both invertebrates and vertebrates. *Journal of Comparative Physiology a-Neuroethology Sensory Neural and Behavioral Physiology, 196*(10), 685–700. https://doi.org/10.1007/s00359-010-0564-y

Wyatt, T. D. (2017). Pheromones. *Current Biology, 27*(15), R739–R743. https://doi.org/10.1016/j.cub.2017.06.039

Zhang, J.-X., Sun, L., & Zuo, M.-X. (2009). Uropygial gland volatiles may code for olfactory information about sex, individual, and species in Bengalese finches *Lonchura striata. Current Zoology, 55*(5), 357–365. https://doi.org/10.1093/czoolo/55.5.357

Zhang, J.-X., Wei, W., Zhang, J.-H., & Yang, W.-H. (2010). Uropygial gland-secreted alkanols contribute to olfactory sex signals in budgerigars. *Chemical Senses, 35*(5), 375–382. https://doi.org/10.1093/chemse/bjq025

Zhang, Y.-H., Du, Y.-F., & Zhang, J.-X. (2013). Uropygial gland volatiles facilitate species recognition between two sympatric sibling bird species. *Behavioral Ecology*, 24(6), 1271–1278. https://doi.org/10.1093/beheco/art068

Zhu, L., Hu, Z., Gamez, G., Law, W. S., Chen, H., Yang, S., et al. (2010). Simultaneous sampling of volatile and non-volatile analytes in beer for fast fingerprinting by extractive electrospray ionization mass spectrometry. *Analytical and Bioanalytical Chemistry, 398*(1), 405–413. https://doi.org/10.1007/s00216-010-3945-8

Zouboulis, C. C., & Makrantonaki, E. (2019). Clinical and laboratory skin biomarkers of organ-specific diseases. *Mechanisms of Ageing and Development, 177*, 144–149. https://doi.org/10.1016/j.mad.2018.08.003

# A Gull Species Recognizes MHC-II Diversity and Dissimilarity Using Odor Cues

Maxime Pineaux[1], Pierrick Blanchard[1], Léa Ribeiro[1], Scott A. Hatch[2], and Sarah Leclaire[1] (✉)

[1] Laboratoire Évolution & Diversité Biologique (EDB, UMR 5174), Université de Toulouse Midi-Pyrénées, CNRS, IRD, UPS, Toulouse, France
sarah.leclaire@univ-tlse3.fr
[2] Institute for Seabird Research and Conservation, Anchorage, AK, USA

**Abstract** The major histocompatibility complex (MHC) plays a crucial role in the resistance to parasites in vertebrates and is thus often suggested to be an important force driving social interactions, including mating preference. However, the phenotypic cues used by individuals to assess the MHC characteristics of conspecifics are generally unknown. Here, we used behavioral tests to investigate whether, in black-legged kittiwakes, females use odor cues to distinguish male MHC-II diversity and MHC-II dissimilarity. We found that females took less time to peck at an odor sample coming from a male with high or low MHC-II diversity compared to intermediate MHC-II diversity. However, this result is due to the single individual who has only one MHC-II allele. When excluding this individual, females took less time to peck at an odor sample coming from a male with high MHC-II diversity. In addition, when the odor sample came from a male with higher MHC-II dissimilarity, females took less time to peck at the sample, but once they pecked at it, they delayed the use of the sample. Altogether, our results add evidence for olfactory recognition of MHC characteristics in birds, but further studies are needed to determine whether kittiwakes use this ability to optimize fitness.

**Keywords** Birds · Olfaction · Sexual selection · Intraspecific communication · Major histocompatibility complex · Black-legged kittiwake · *Rissa tridactyla*

## 1 Introduction

Most birds do not display any apparent olfactory behavior, such as sniffing or scent-marking. Therefore, it was long thought that birds had little or no sense of smell and relied primarily on visual or acoustic cues. However, there is now abundant evidence that birds have a functional olfactory system (Bang & Cobb, 1968; Steiger et al., 2008) and use olfaction in various nonsocial contexts, including foraging, navigation, or nest



B. Schaal et al. (eds.), *Chemical Signals in Vertebrates 15*,
https://doi.org/10.1007/978-3-031-35159-4_7

sanitation (e.g., Caro & Balthazart, 2010; Abankwah et al., 2020; Potier, 2020; Bonadonna & Gagliardo, 2021). In contrast, the role of olfactory cues in avian social life is still poorly known (Caro et al., 2015), albeit being a rapidly growing research area (Whittaker & Hagelin, 2021).

As in other vertebrates, recent research suggests that the major histocompatibility complex (MHC) may be an important force involved in mate choice in birds (Zelano & Edwards, 2002; Løvlie et al., 2013). The MHC is a polymorphic group of genes that plays a critical role in immunity. In general, higher MHC diversity is considered to provide resistance to a wider range of parasites and to increase fitness (Milinski, 2006). Accordingly, species from diverse taxa have developed MHC-based mate choice, such as mate choice for MHC-dissimilar partners (i.e., MHC-compatible partners), thereby avoiding the production of offspring with low MHC diversity (Strandh et al., 2012; Kamiya et al., 2014; Hoover et al., 2018), or mate choice for MHC-diverse partners, which may provide both genetic and nongenetic benefits (Richardson et al., 2005; Griggio et al., 2011). However, high MHC diversity may also be associated with costs, such as an increased risk of autoimmune disorders or a reduction in the potential for inducing an immune reaction due to negative selection of T-cells (Nowak et al., 1992; Simmonds & Gough, 2007; Woelfing et al., 2009; Migalska et al., 2019). There is therefore increasing support for an intermediate optimum of MHC diversity (Madsen & Ujvari, 2006; Kalbe et al., 2009; Rekdal et al., 2019).

Whatever the optimum in MHC diversity, MHC characteristics need to be encoded in phenotypic cues for animals to assess the MHC characteristics of potential partners during mate choice. However, while several mammals, lizards, or fish have been shown to assess the MHC characteristics of conspecifics using odor cues (reviewed in Schubert et al., 2021), this ability in birds has been poorly explored. To the best of our knowledge, only the blue petrel (*Halobaena caerulea*), a Procellariiform seabird, and the song and house sparrows (*Melospiza melodia* and *Passer dosmesticus*, respectively), two Passeriform birds, have been tested. While blue petrels and song sparrows were shown to discriminate MHC-II characteristics using odor cues (Leclaire et al., 2017; Grieves et al., 2019), house sparrows did not discriminate MHC-I characteristics using olfaction (Amo et al., 2022). To estimate the taxonomic range of olfactory assessment of MHC in birds, more avian taxa need to be studied.

The black-legged kittiwake (*Rissa tridactyla*) is a genetically monogamous Charadriiform seabird with a functional sense of smell (Leclaire et al., 2009). In this species, preen gland secretions encode information about MHC-II dissimilarity (Leclaire et al., 2014) and offspring sex ratio varies with the MHC-II dissimilarity between parents (Pineaux et al., 2022), suggesting that kittiwakes might be able to recognize the MHC-II characteristics of conspecifics. Here, we used behavioral tests to investigate whether female kittiwakes can use odor cues to distinguish the MHC-II characteristics of males. We used a protocol inspired by Leclaire et al. (2009), which showed differential pecking response of incubating females to different odor samples (skunk, banana, fish, or control odor) put in their nest. Here, we put a single odor sample of a male to the nest of an incubating female and assessed her behavioral response. Although differential response of females cannot be interpreted as female preference, we consider that the fact that different females react on average differently along a continuum of males' MHC characteristics indicates discrimination. Because the MHC-based mating pattern of kittiwakes is unknown, we tested recognition of both MHC-II diversity and MHC-II dissimilarity and always included both the linear and the quadratic term of MHC-II diversity.

## 2 Materials and Methods

### 2.1 Study Site

The study was conducted during the 2018 incubation period (May–June) in a population of black-legged kittiwakes nesting on an abandoned US Air Force radar tower on Middleton Island, Alaska (59° 26′ N, 146° 20′ W). Nests on artificial ledges were visible from inside the building through sliding one-way glass windows (Gill & Hatch, 2002). This experiment was conducted under the approval of the USGS Alaska Science Centre Animal Care and Use Committee, in accordance with US law and under permits from the US Fish and Wildlife Service and the State of Alaska.

### 2.2 Behavioral Tests

We assessed female's response to male odor samples placed on the edge of the nest bowl. Odor samples were a mixture of cloacal samples (potentially related to fecal odor) and nest materials (potentially related to skin, plumage, and preen gland odor passively deposited on the nest when birds are sitting). We collected odor samples from 43 incubating males. We collected cloacal samples as in Leclaire et al. (2023). Briefly, after capture, we injected 1 mL of sterile phosphate-buffered saline solution into male's cloaca with a sterile needleless syringe and drew the solution out. A piece of nest material was used as the method of odor delivery to reduce perturbation of the incubating female. At the time of capture, a piece of nest material (i.e., a blend of mosses and grasses of ca. 4 cm diameter) was collected from the center of the nest bowl of the captured male, using clean nitrile gloves. The piece of nest material was immediately mixed with the cloacal sample and stored in plastic bags at −20 °C until the test. Odor samples were stored on average for 17 ± 1 days (range: 8–24 days). The time the odor sample was kept in the freezer did not affect the behavioral response of the females (Spearman's correlation test between the time the sample was kept in the freezer and the time the female took to peck at the odor sample during the test: $S = 13{,}678$, $P = 0.077$, $r = -0.28$).

Fifty-two incubating females were each tested once. They were never tested with the odor of their partner. All odor samples ($n = 43$) were used at least once. Just before the test, a small piece (ca. 2 cm diameter) of an odor sample was defrosted for

**Fig. 1** Picture of a kittiwake's nest in which an odor sample (encircled in red) was put on the edge of the nest. This picture was taken during a preliminary test on a non-incubating bird. However, in all tests considered in this study, the female was incubating her egg(s)

15 min. The remaining odor sample was kept frozen. Nine of these remaining odor samples were used in tests later on. The odor sample was placed on the edge of the nest bowl, just under the beak of an incubating female (Fig. 1). Because we had to open the sliding window of the nesting site to put the odor sample on the nest, all females, except one, took off. Behaviors were recorded by an observer as soon as the female landed back on her nest. Females returned to their nest on average ± se, 14 ± 5 s, after the odor sample was placed on the nest (range: 0–462 s). For each test, we recorded the time taken by the female to peck at the odor sample for the first time. Ninety-three percent of the females who pecked at the odor sample took the sample with their bill and moved it along the nest bowl, as they often do with nest material when maintaining the nest structure. We therefore also recorded the time taken by the female to move the odor sample for the first time. Observation lasted until the female pecked at the odor sample or for 15 min when she did not peck at the odor sample. Tests were carried out by three observers who were blind to the MHC characteristics of the males concerned.

## 2.3 MHC Analyses

Details about MHC analyses are given in Pineaux et al. (2020). Briefly, DNA was extracted from blood samples, and 258-bp fragments of exon 2 of MHC class-IIB were amplified using PCR and sequenced in four runs with an Illumina MiSeq platform. Amplicon sequences were analyzed with AmpliSAS (substitution errors, 1%; indel errors, 0.001%; minimum frequency with respect to dominant, 33%; minimum amplicon sequence frequency, 2.5%) (Sebastian et al., 2016). Considering the larger set of samples analyzed in the four Illumina MiSeq runs, we detected a maximum of six alleles per individual, suggesting the presence of a minimum of three MHC class-IIB loci. We focused on the amino acid sequences of the peptide binding region (Leclaire et al., 2014). MHC-IIB diversity and MHC-IIB dissimilarity were each estimated using two indices. First, MHC-IIB diversity was estimated using the number of MHC-IIB alleles, while MHC-IIB dissimilarity between the female and the sampled male was estimated on a pure shared/non-shared allele basis. The percent difference in MHC-IIB allele sharing was calculated as $100 \times TOT_{FM}/(N_F + N_M)$, where $TOT_{FM}$ = total number of different alleles present in the male and the female (shared allele count as 1) and $N$ = number of alleles within an individual (Strandh et al., 2012). Second, we estimated MHC-IIB diversity as the degree of functional diversity (or divergence) of alleles across all loci and MHC-IIB dissimilarity as the functional distance between two individuals. We described the chemical binding properties of each amino acid in the PBR with Sandberg's five physicochemical descriptors (z-descriptors; Sandberg et al., 1998). This Sandberg's matrix was used to construct a UPGMA dendrogram of alleles (function hclust in R), which represents clusters of functionally similar MHC-IIB sequences. The dendrogram was used as a reference to calculate the functional MHC-IIB diversity of an individual as the minimum total length of all the branches required to span its MHC-IIB alleles (function treedive in package *vegan*) (Petchey & Gaston, 2006). MHC-IIB functional dissimilarity between the male odor donor and the female was estimated using the tree distance (function treedist in package *vegan*), which is found by combining species in a common dendrogram and seeing how much of the dendrogram height is shared and how much is unique. Because five females were not genotyped at the MHC-IIB, we could not calculate the MHC-IIB dissimilarity between the female and the sampled male for five tests. The number of alleles was correlated with the functional diversity (Spearman's rank correlation test: $S = 1110$, $P < 0.0001$, $r = 0.92$, $n = 43$ males), and the percent difference in allele sharing was correlated with the functional dissimilarity (Spearman's rank correlation test: $S = 2619$, $P < 0.0001$, $r = 0.85$, $n = 47$ male-female pairs).

## 2.4 Statistics

To test for difference in females' behavioral response according to the MHC-II diversity of the male odor donor and to the MHC-II dissimilarity between the female and the male odor donor, first, we used a generalized linear mixed model (GLMM) with the probability to peck at the odor sample (0 when the female did not peck at the sample or 1 when the female pecked at the sample) as the binomial response variable. Fixed effects included MHC-II dissimilarity, male MHC-II diversity, and the quadratic term of male MHC-II diversity. Although the MHC-II diversity of the male odor donor was negatively related to the MHC-II dissimilarity to the female (Spearman rank test between the number of alleles and the percent difference in allele sharing, $S = 22{,}305$, $r = -0.29$, $P = 0.048$, and between the functional diversity and the functional dissimilarity, $S = 24{,}936$, $r = -0.44$, $P = 0.002$; $n = 47$ tests for which we have both MHC-II diversity and dissimilarity), we included the two variables together in the models as the variance inflation factors were < 1.5 (Zuur et al., 2010). Fixed effects also included the difference between the date of the test and the laying date, as female reaction to male MHC-II characteristics might potentially change across the breeding season (Manning et al., 1992; Leclaire et al., 2017). The identity of the male odor donor was included as a random effect, while the observer identity was not included as a random effect because its variance estimate was 0. Second, for females who pecked at the samples, we used a linear mixed model (LMM) with the time taken to peck at the odor sample as the response variable (log-transformed to meet normality of residuals) and the same fixed effects as above. Both the observer identity and the identity of the male odor donors were included as random effects. Third, for females who moved the odor sample, we used a binomial GLMM model to test whether the probability to delay the move of the odor sample after the first pecking (0 when the female immediately moved the sample after the first pecking or 1 when the female delayed the move of the sample) varied with MHC-II characteristics. Fixed and random effects were similar as above.

We performed all statistical analyses using the R statistical software (R Core Team, 2020). In GLMMs, we used Wald's tests to assess the significance of variables. In LMMs, significance of each predictor variable was calculated using the Satterthwaite approximation as implemented in the package *lmerTest* (Kuznetsova et al., 2017). Normality and homogeneity of variance were checked visually (Zuur et al., 2010). Data can be found at https://doi.org/10.17605/OSF.IO/H2KFN.

## 3 Results

When considering the tests for which we had both the MHC-II diversity of the male odor donor and the MHC-II dissimilarity between the male odor donor and the female ($n = 47$ tests), seven females (15%) did not peck at the odor sample. The probability of pecking at the sample did not vary with male MHC-II diversity (allele number, quadratic term, $\chi^2_1 = 0.01$, $P = 0.94$, and linear term, $\chi^2_1 = 0.17$, $P = 0.68$; functional diversity, quadratic term, $\chi^2_1 = 0.01$, $P = 0.93$, and linear term, $\chi^2_1 = 0.12$, $P = 0.73$) or MHC-II dissimilarity (percent difference in allele sharing, $\chi^2_1 = 0.06$, $P = 0.80$, and functional dissimilarity, $\chi^2_1 = 0.01$, $P = 0.94$).

Among the females who pecked at the sample ($n = 40$ tests for which we had both the male MHC-II diversity and the MHC-II dissimilarity between male and female), females took more time to peck at the odor sample when the male had an intermediate MHC-II diversity (allele number, quadratic term, $\chi^2_1 = 5.49$, $P = 0.025$, and linear term, $\chi^2_1 = 6.60$, $P = 0.015$, Fig. 2a; functional diversity, quadratic term, $\chi^2_1 = 5.97$, $P = 0.020$, and linear term, $\chi^2_1 = 3.91$, $P = 0.056$; Fig. 2b). When excluding the test where the single male odor donor had only one MHC-II allele (Fig. 2a), females took less time to peck at the odor sample when the male had higher MHC-II diversity (allele number,

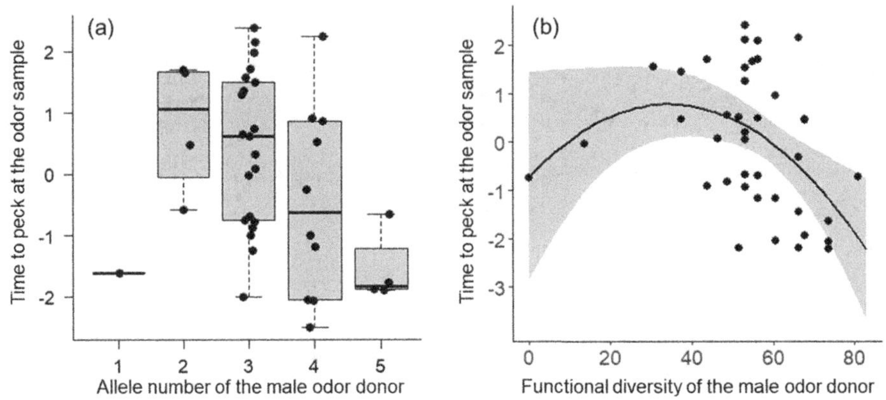


**Fig. 2** Time taken by females to peck at the odor sample according to (**a**) the number of MHC-II alleles and (**b**) the functional MHC-II diversity of the male odor donor. The time to peck at the sample was represented by the residuals of a linear mixed model in which the time to peck at the sample was explained by all variables except the MHC-II diversity of the male. The black line and the grey shadow in (**b**) represent prediction line and 95% confidence interval

linear term, $\chi^2_1 = 13.73$, $P < 0.001$, and quadratic term, $\chi^2_1 = 0.36$, $P = 0.55$; functional diversity, linear term, $\chi^2_1 = 6.98$, $P = 0.012$, and quadratic term, $\chi^2_1 = 2.43$, $P = 0.13$; Fig. 2). This result suggests that this male carrying a single MHC-II allele drives the quadratic relationship between the time taken to peck at the sample and the MHC-II diversity. Females who pecked at the sample took also less time to peck at the odor sample when they were more MHC-II dissimilar to the male odor donor (percent difference in allele sharing, $\chi^2_1 = 6.08$, $P = 0.019$, and functional dissimilarity, $\chi^2_1 = 6.89$, $P = 0.013$; Fig. 3).

Ninety-three percent of the females who pecked at the odor sample took the sample with their bill and moved it along the nest bowl ($n = 37$ tests for which we had both the MHC-II diversity and MHC-II dissimilarity). Sixty-five percent of those females ($n = 24$ females) moved the sample just after the first pecking, while the others ($n = 13$ females) delayed the move of the odor sample. The probability to delay the move of the odor sample after the first pecking increased with the dissimilarity to the male odor donor (percent difference in allele sharing, $\chi^2_1 = 5.39$, $P = 0.020$, and functional dissimilarity, $\chi^2_1 = 8.35$, $P = 0.004$; Fig. 4). The probability to delay the move of the odor sample did not vary with the MHC-II diversity of the male odor donor (all $P$ values $> 0.10$).

## 4 Discussion

Using behavioral tests, we investigated whether, in black-legged kittiwakes, females discriminated the MHC-II characteristics of males. First, we found that females took more time to peck at the odor sample when it came from a male with intermediate MHC-II diversity. However, the quadratic relationship between MHC-II diversity and

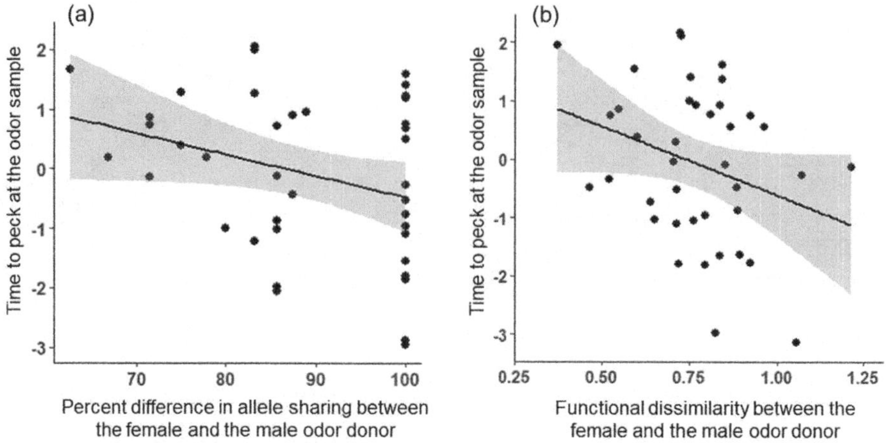


**Fig. 3** Time taken by females to peck at the odor sample according to (**a**) the percent difference in MHC-II allele sharing and (**b**) the functional MHC-II dissimilarity to the male odor donor. The time to peck at the sample was represented by the residuals of a linear mixed model in which the time to peck at the sample was explained by all variables except the MHC-II dissimilarity. The black lines and grey shadows represent the prediction line and 95% confidence interval

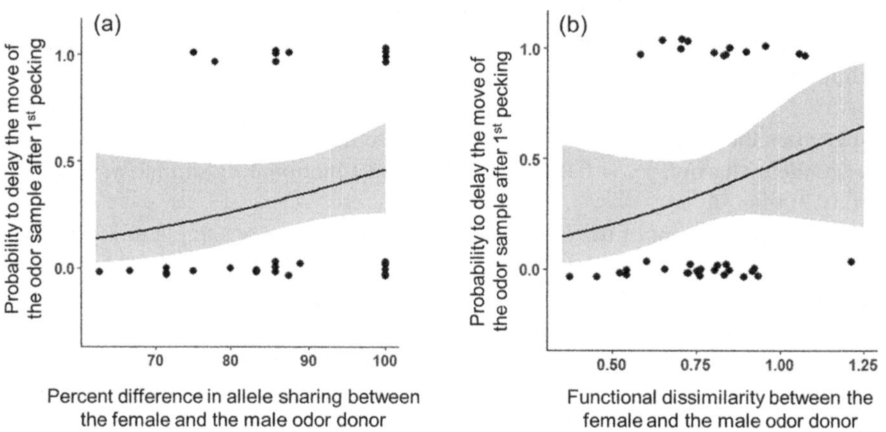


**Fig. 4** Probability for the females to delay the move of the odor sample after the first pecking (0, the female moved the sample just after the first pecking, and 1, the female delayed the move of the sample after the first pecking) according to (**a**) the percent difference in MHC-II allele sharing and (**b**) the functional MHC-II dissimilarity to the male odor donor. The black lines and grey shadows represent the prediction line and 95% confidence interval of a GLMM between the probability to delay the move of the sample after the first pecking and the MHC-II dissimilarity to the male odor donor

the time taken to peck at the sample was only due to the single male who had only one MHC-II allele (Fig. 1). When excluding this male, the relationship between male MHC-II diversity and female behavior became linear. Clearly, more tests including male odor donors carrying a single MHC-II allele would have been needed to ascertain the quadratic nature of the relationship between male MHC-II diversity and the female's reaction to the odor sample. However, in this kittiwake population, very few individuals carry a single MHC-II allele (<1% of the individuals; unpublished data), suggesting that carrying a single MHC-II allele is counter-selected. Recently, several studies in fish, snakes, and birds have tested the "golden mean hypothesis" and have shown that individuals with intermediate MHC diversity have higher reproductive success (Kalbe et al., 2009), better immunity (Rekdal et al., 2021), and lower parasite load (Wegner et al., 2004; Madsen & Ujvari, 2006). In kittiwakes, the optimum in MHC-II diversity has been tested in chicks only and was found to be at maximum diversity in female chicks (Pineaux et al., 2020). However, the optima in MHC diversity can vary according to individual traits that affect the trade-off between the benefits and costs associated with high MHC diversity (Roved et al., 2018; Pineaux et al., 2020). Therefore, the optimum in MHC-II diversity may differ between adults and chicks, because of age differences in exposure to parasites, acquired immunity, and predisposition to infection (Benskin et al., 2009). Further studies are needed to determine whether, in kittiwakes, intermediate or maximum MHC-II diversity is optimum in adult males.

We also found that females took less time to peck at an odor sample coming from a male with higher MHC-II dissimilarity. Almost all females who pecked at the odor sample moved it along the nest bowl, as they commonly do with nest material during nest maintenance. However, while some females moved the odor sample just after the first pecking, others delayed the move of the sample. We found that the probability to delay the move of the odor sample was higher when the odor sample came from a male with higher MHC-II dissimilarity. Female pecking response and sample moving response cannot be interpreted as preference or aversion for certain odors. However, our results show that female kittiwakes may use odor cues to discriminate MHC-II diversity and MHC-II dissimilarity. Further studies are needed to determine the contexts in which females may use this ability (e.g., mate choice, adjustment of parental investment, offspring sex ratio distortion). Although the role of olfaction in the recognition of MHC dissimilarity has been shown in several vertebrate species (reviewed in Schubert et al., 2021), its role in the recognition of MHC diversity has been less studied, probably because MHC diversity, by influencing health, might be reflected in any condition-dependent traits (Lie et al., 2008; Dunn et al., 2013; Slade et al., 2017). However, three-spined sticklebacks (*Gasterosteus aculeatus*), ring-tailed lemurs (*Lemur catta*), and song sparrows (*Melospiza melodia*) can discriminate MHC diversity using odor cues (Milinski et al., 2005; Grieves et al., 2019; Grogan et al., 2019).

In birds, preen gland secretions are often suggested to be the main source of body odors and have been shown to be used in social communication in a few species (Hirao et al., 2009; Zhang et al., 2010). However, in most birds, the sources of body odors are unknown (Hagelin & Jones, 2007). Here, we tested whether females could discriminate male MHC-II characteristics using a mixture of odor sources (i.e., cloacal samples mixed with nest material), thereby preventing us from identifying precisely the sources of MHC-related odors. In kittiwakes, preen secretions encode information about MHC-II dissimilarity (Leclaire et al., 2014), but studies are required to determine the role of other sources such as feces, skin, or plumage microbiota (Leclaire et al., 2019; Schubert et al., 2021). For instance, in kittiwakes, the outside edge of the nest is usually full of feces suggesting that fecal odor might play a role in communication. In addition, while our results are consistent with odor-based discrimination of MHC-II characteristics, our methods do not allow us to determine whether discrimination is based on broader genome-wide genetic variation or on other polymorphic genes whose variations covary with those in the MHC, such as the major urinary protein (MUP) genes in natural populations of mice (Sherborne et al., 2007). Finally, the development of inventive odor discrimination tests based on more relevant behavioral responses is key to better appreciate the contexts in which social odor discrimination abilities can be used in kittiwakes, but also more generally in non-burrowing large wild birds.

Altogether, our results, by adding evidence for olfactory discrimination of MHC characteristics in birds, suggest that this capacity may be widespread in this taxon. However, whether birds use it in mate choice or more generally to maximize fitness has still to be demonstrated.

**Acknowledgments** We are grateful to T. Rinaud and C. Maillotte for their help in the field. This study is part of the Laboratoire d'Excellence (LABEX) entitled TULIP (ANR-10- LABX-41). This work was supported by grants from the French Polar Institute (IPEV; Program 1162 to SL) and the Fondation Fyssen (to SL).

## References

Abankwah, V., Deeming, D. C., & Pike, T. W. (2020). Avian olfaction: A review of the recent literature. *Comparative Cognition & Behavior Reviews, 15*, 149–161

Amo, L., Amo de Paz, G., Kabbert, J., & Machordom, A. (2022). House sparrows do not exhibit a preference for the scent of potential partners with different MHC-I diversity and genetic distances. *PLoS One, 17*, e0278892. https://doi.org/10.1371/journal.pone.0278892

Bang, B. G., & Cobb, S. (1968). The size of the olfactory bulb in 108 species of birds. *Auk, 85*, 55–61.

Benskin, C. M. W. H., Wilson, K., Jones, K., & Hartley, I. R. (2009). Bacterial pathogens in wild birds: A review of the frequency and effects of infection. *Biological Reviews, 84*, 349–373. https://doi.org/10.1111/j.1469-185X.2008.00076.x

Bonadonna, F., & Gagliardo, A. (2021). Not only pigeons: Avian olfactory navigation studied by satellite telemetry. *Ethology Ecology and Evolution, 33*(3), 273–289.

Caro, S. P., & Balthazart, J. (2010). Pheromones in birds: myth or reality? *Journal of Comparative Physiology A, 196*(10), 751–766. https://doi.org/10.1007/s00359-010-0534-4

Caro, S. P., Balthazart, J., & Bonadonna, F. (2015). The perfume of reproduction in birds: Chemosignaling in avian social life. *Hormones and Behavior, 68*, 25–42. https://doi.org/10.1016/j.yhbeh.2014.06.001

Dunn, P. O., Bollmer, J. L., Freeman-Gallant, C. R., & Whittingham, L. A. (2013). MHC variation is related to a sexually selected ornament, survival, and parasite resistance in common yellowthroats. *Evolution, 67*(3), 679–687.

Gill, V. A., & Hatch, S. A. (2002). Components of productivity in black-legged kittiwakes *Rissa tridactyla*: Response to supplemental feeding. *Journal of Avian Biology, 33*(2), 113–126.

Grieves, L. A., Gloor, G. B., Bernards, M. A., & MacDougall-Shackleton, E. A. (2019). Songbirds show odour-based discrimination of similarity and diversity at the major histocompatibility complex. *Animal Behaviour, 158*, 131–138.

Griggio, M., Biard, C., Penn, D. J., & Hoi, H. (2011). Female house sparrows "count on" male genes: Experimental evidence for MHC-dependent mate preference in birds. *BMC Evolutionary Biology, 11*(44), 1–7. https://doi.org/10.1186/1471-2148-11-44

Grogan, K. E., Harris, R. L., Boulet, M., & Drea, C. M. (2019). Genetic variation at MHC class II loci influences both olfactory signals and scent discrimination in ring-tailed lemurs. *BMC Evolutionary Biology, 19*(1), 1–16.

Hagelin, J., & Jones, I. L. (2007). Bird odors and other chemical substances: A defense mechanism or overlooked mode of intraspecific communication? *Auk, 124*(3), 741–761.

Hirao, A., Aoyama, M., & Sugita, S. (2009). The role of uropygial gland on sexual behavior in domestic chicken *Gallus gallus domesticus*. *Behavioural Processes, 80*(2), 115–120.

Hoover, B., Alcaide, M., Jennings, S., Sin, S. Y. W., Edwards, S. V., & Nevitt, G. A. (2018). Ecology can inform genetics: disassortative mating contributes to MHC polymorphism in Leach's storm-petrels (*Oceanodroma leucorhoa*). *Molecular Ecology, 27*(16), 3371–3385.

Kalbe, M., Eizaguirre, C., Dankert, I., Reusch, T. B. H., Sommerfeld, R. D., Wegner, K. M., & Milinski, M. (2009). Lifetime reproductive success is maximized with optimal major histocompatibility complex diversity. *Proceedings of the Royal Society B, 276*(1658), 925–934. https://doi.org/10.1098/rspb.2008.1466

Kamiya, T., O'Dwyer, K., Westerdahl, H., Senior, A., & Nakagawa, S. (2014). A quantitative review of MHC-based mating preference: The role of diversity and dissimilarity. *Molecular Ecology, 23*(21), 5151–5163.

Kuznetsova, A., Brockhoff, P. B., & Christensen, R. H. B. (2017). lmerTest package: Tests in linear mixed effects models. *Journal of Statistical Software, 82*(1), 1–26. https://doi.org/10.18637/jss.v082.i13

Leclaire, S., Mulard, H., Wagner, R. H., Hatch, S. A., & Danchin, E. (2009). Can Kittiwakes smell? Experimental evidence in a Larid species. *Ibis, 151*(3), 584–587.

Leclaire, S., van Dongen, W. F., Voccia, S., Merkling, T., Ducamp, C., Hatch, S. A., Blanchard, P., Danchin, É., & Wagner, R. H. (2014). Preen secretions encode information on MHC similarity in certain sex-dyads in a monogamous seabird. *Scientific Reports, 4*, 6920.

Leclaire, S., Strandh, M., Mardon, J., Westerdahl, H., & Bonadonna, F. (2017). Odour-based discrimination of similarity at the major histocompatibility complex in birds. *Proceedings of the Royal Society B, 284*(1846), 20162466.

Leclaire, S., Strandh, M., Dell'Ariccia, G., Gabirot, M., Westerdahl, H., & Bonadonna, F. (2019). Plumage microbiota covaries with the major histocompatibility complex in blue petrels. *Molecular Ecology, 28*(4), 833–846.

Lie, H. C., Rhodes, G., & Simmons, L. W. (2008). Genetic diversity revealed in human faces. *Evolution, 62*(10), 2473–2486. https://doi.org/10.1111/j.1558-5646.2008.00478.x

Løvlie, H., Gillingham, M. A., Worley, K., Pizzari, T., & Richardson, D. S. (2013). Cryptic female choice favours sperm from major histocompatibility complex-dissimilar males. *Proceedings of the Royal Society London B, 280*(1769), 20131296.

Madsen, T., & Ujvari, B. (2006). MHC class I variation associates with parasite resistance and longevity in tropical pythons. *Journal of Evolutionary Biology, 19*(6), 1973–1978. https://doi.org/10.1111/j.1420-9101.2006.01158.x

Manning, C., Wakeland, E., & Potts, W. (1992). Communal nesting patterns in mice implicate MHC genes in kin recognition. *Nature, 360*, 581–583. https://doi.org/10.1038/360581a0

Migalska, M., Sebastian, A., & Radwan, J. (2019). Major histocompatibility complex class I diversity limits the repertoire of T cell receptors. *Proceedings of the National Academy of Sciences, 116*(11), 5021–5026.

Milinski, M. (2006). The major histocompatibility complex, sexual selection, and mate choice. *Annual Review of Ecology Evolution and Systematics, 37*, 159–186.

Milinski, M., Griffiths, S., Wegner, K. M., Reusch, T. B. H., Haas-Assenbaum, A., & Boehm, T. (2005). Mate choice decisions of stickleback females predictably modified by MHC peptide ligands. *Proceedings of the National Academy of Sciences USA, 102*(12), 4414–4418. https://doi.org/10.1073/pnas.0408264102

Nowak, M. A., Tarczy-Hornoch, K., & Austyn, J. M. (1992). The optimal number of major histocompatibility complex molecules in an individual. *Proceedings of the National Academy of Sciences USA, 89*(22), 10896–10899. https://doi.org/10.1073/pnas.89.22.10896

Petchey, O. L., & Gaston, K. J. (2006). Functional diversity: Back to basics and looking forward. *Ecology Letters, 9*, 741–758.

Pineaux, M., Merkling, T., Danchin, E., Hatch, S., Duneau, D., Blanchard, P., & Leclaire, S. (2020). Sex and hatching order modulate the association between MHC-II diversity and fitness in early-life stages of a wild seabird. *Molecular Ecology, 29*(17), 3316–3329.

Pineaux, M., Merkling, T., Danchin, E., Hatch, S. A., Leclaire, S., & Blanchard, P. (2022). MHC-II distance between parents predicts sex allocation decisions in a genetically monogamous bird. *Behavioral Ecology, 33*(1), 245–251.

Potier, S. (2020). Olfaction in raptors. *Zoological Journal of the Linnean Society, 189*(3), 713–721. https://doi.org/10.1093/zoolinnean/zlz121

R Core Team. (2020). R: A language and environment for statistical computing. *R Foundation for Statistical Computing*, Vienna, Austria. https://www.R-project.org/

Rekdal, S. L., Anmarkrud, J. A., Lifjeld, J. T., & Johnsen, A. (2019). Extra-pair mating in a passerine bird with highly duplicated major histocompatibility complex class II: Preference for the golden mean. *Molecular Ecology, 28*(23), 5133–5144.

Rekdal, S. L., Anmarkrud, J. A., Lifjeld, J. T., & Johnsen, A. (2021). Elevated phytohaemagglutinin-induced skin-swelling response at an intermediate number of MHC class II alleles in bluethroat nestlings. *Journal of Avian Biology, 52*(5). https://doi.org/10.1111/jav.02734

Richardson, D. S., Komdeur, J., Burke, T., & von Schantz, T. (2005). MHC-based patterns of social and extra-pair mate choice in the Seychelles warbler. *Proceedings of the Royal Society B, 272*, 759–767. https://doi.org/10.1098/rspb.2004.3028

Roved, J., Hansson, B., Tarka, M., Hasselquist, D., & Westerdahl, H. (2018). Evidence for sexual conflict over major histocompatibility complex diversity in a wild songbird. *Proceedings of the Royal Society B, 285*(1884), 20180841.

Sandberg, M., Eriksson, L., Jonsson, J., Sjöström, M., & Wold, S. (1998). New chemical descriptors relevant for the design of biologically active peptides. A multivariate characterization of 87 amino acids. *Journal of Medicinal Chemistry, 41*(14), 2481–2491.

Sarah, Leclaire Maxime, Pineaux Pierrick, Blanchard Joël, White Scott A., Hatch (2023) Microbiota composition and diversity of multiple body sites vary according to reproductive performance in a seabird Molecular Ecology 32(9) 2115–2133. https://doi.org/10.1111/mec.v32.9. https://doi.org/10.1111/mec.16398

Schubert, N., Nichols, H. J., & Winternitz, J. C. (2021). How can the MHC mediate social odor via the microbiota community? A deep dive into mechanisms. *Behavioral Ecology, 32*(3), 359–373.

Sebastian, A., Herdegen, M., Migalska, M., & Radwan, J. (2016). amplisas: A web server for multilocus genotyping using next-generation amplicon sequencing data. *Molecular Ecology Resources, 16*(2), 498–510.

Sherborne, A. L., Thom, M. D., Paterson, S., Jury, F., Ollier, W. E., Stockley, P., Beynon, R. J., & Hurst, J. L. (2007). The genetic basis of inbreeding avoidance in house mice. *Current Biology, 17*(23), 2061–2066.

Simmonds, M. J., & Gough, S. C. L. (2007). The HLA region and autoimmune disease: Associations and mechanisms of action. *Current Genomics, 8*(7), 453–465. https://doi.org/10.2174/138920207783591690

Slade, J. W. G., Watson, M. J., & MacDougall-Shackleton, E. A. (2017). Birdsong signals individual diversity at the major histocompatibility complex. *Biology Letters, 13*(11), 20170430. https://doi.org/10.1098/rsbl.2017.0430

Steiger, S. S., Fidler, A. E., Valcu, M., & Kempenaers, B. (2008). Avian olfactory receptor gene repertoires: evidence for a well-developed sense of smell in birds? *Proceedings of the Royal Society B, 275*(1649), 2309–2317.

Strandh, M., Westerdahl, H., Pontarp, M., Canback, B., Dubois, M.-P., Miquel, C., Taberlet, P., & Bonadonna, F. (2012). Major histocompatibility complex class II compatibility, but not class I, predicts mate choice in a bird with highly developed olfaction. *Proceedings of the Royal Society B, 279*(1746), 4457–4463. https://doi.org/10.1098/rspb.2012.1562

Wegner, K. M., Kalbe, M., Schaschl, H., & v& Reusch TBH. (2004). Parasites and individual major histocompatibility complex diversity—an optimal choice? *Microbes and Infection, 6*(12), 1110–1116. https://doi.org/10.1016/j.micinf.2004.05.025

Whittaker, D. J., & Hagelin, J. C. (2021). Female-based patterns and social function in avian chemical communication. *Journal of Chemical Ecology, 47*(1), 43–62. https://doi.org/10.1007/s10886-020-01230-1

Woelfing, B., Traulsen, A., Milinski, M., & Boehm, T. (2009). Does intra-individual major histocompatibility complex diversity keep a golden mean? *Philosophical Transactions of the Royal Society B, 364*(1513), 117–128. https://doi.org/10.1098/rstb.2008.0174

Zelano, B., & Edwards, S. V. (2002). An MHC component to kin recognition and mate choice in birds: Predictions, progress, and prospects. *The American Naturalist, 160*(S6), S225–S237.

Zhang, J. X., Wei, W., Zhang, J. H., & Yang, W. H. (2010). Uropygial gland-secreted alkanols contribute to olfactory sex signals in budgerigars. *Chemical Senses, 35*(5), 375–382.

Zuur, A. F., Ieno, E. N., & Elphick, C. S. (2010). A protocol for data exploration to avoid common statistical problems. *Methods in Ecology and Evolution, 1*(1), 3–14. https://doi.org/10.1111/j.2041-210X.2009.00001.x

# Exploration of Olfactory Communication in the Water Vole, *Arvicola terrestris*

Kévin Poissenot[1], Chantal Porte[1], Didier Chesneau[1], and Matthieu Keller[1] (✉)

[1] UMR Physiologie de la Reproduction et des Comportements, INRAE, CNRS, Université de Tours, Nouzilly, France
matthieu.keller@inrae.fr

**Abstract** In a large part of western Europe, water voles have a dynamic reproductive activity across the year, leading to regular multiyear outbreaks, which cause significant damage in mid-mountain grassland ecosystems. Water vole populations are currently managed with rodenticides. But their toxicity to the ecosystem imposes to find means of population control that are more environment-friendly. In this context, we began to explore here aspects of water voles' olfactory communication and tested the attractivity of several species-specific chemosignals identified in their urine or lateral scent glands' secretion. One of these odor compounds, namely, 3-ethyl-2,5-dimethyl-pyrazine, induced a strong attraction of female water voles, in terms of longer olfactory investigation. By contrast, this compound seems to be rather repulsive to males. The potential use of 2,5-dimethyl-pyrazine for the control of water vole populations in relation to outbreaks needs to be further evaluated, especially by field-trapping tests performed under different environmental conditions.

**Keywords** Sexual preference · Population control · Outbreaks · Behavior · Meadows · Field study

## 1 Introduction

The European water vole (*Arvicola terrestris*) is a rodent belonging to the Cricetidae family, a family phylogenetically close to the voles of the genus *Microtus* and *Myodes* (bank vole) and to the hamsters *Cricetulus* (Chinese hamster) and *Phodopus* (Siberian hamster). There are two known ecotypes of water voles: a semiaquatic ecotype that occupies wetlands such as river banks, with a geographic distribution covering most of Eurasia, and another burrowing ecotype, present in drier environments, with a distribution restricted mainly to mid-mountain areas. This fossorial ecotype lives in subterranean galleries and nests in open spaces such as meadows or pastures.

Morphologically, the water vole is a relatively big rodent: adults weight between 60 and 160 g for a body length of 12–18 cm from head to tail insertion, which can measure



B. Schaal et al. (eds.), *Chemical Signals in Vertebrates 15*,
https://doi.org/10.1007/978-3-031-35159-4_8

between 4 and 8 cm (Quéré, 2009) (Fig. 1). Water voles are mainly monogamous with strong male territorial defense (Airoldi, 1976; Somoano Garcia, 2017). Regarding semiochemical properties, adults of both sexes bear a bilateral aggregation of skin glands on their flanks, known as the "lateral scent glands" (LSG), which are characterized by an enlarged epithelium with abundant sebaceous structures and an absence of hair follicles (Stoddart, 1972) (Fig. 1).

In various European countries, the breeding season of the vole extends mainly from March to October with a peak between April and July (United Kingdom, Stoddart, 1971; Switzerland, Airoldi, 1978; Spain, Ventura & Gosálbez, 1990). Indeed, in adult males, there is a spectacular spring increase of the reproductive tract with a 3–4-fold increase of testis and seminal vesicle mass between January and May–June (Poissenot et al., 2021a). A similar effect is noted for the surface of seminiferous tubules and the number of spermatozoa in the epididymis, as well as for blood testosterone concentration (Poissenot et al., 2023). It is of major importance to notice that the LSG undergo a highly significant increase in weight and surface between January and the end of spring. These results are in line with Stoddart's research (Stoddart, 1972; Stoddart et al., 1975) who demonstrated that the size of the LSG is androgen sensitive, their size being reduced by castration and restituted by androgen injections.

A similar significant growth of the reproductive tract has also been found in females between January and April/June, in terms of uterus weight and uterine section surface, as well as of the number of corpora lutea on the ovaries (Poissenot et al., 2021b). This activation of reproductive function is maintained until October, with a 3-week gestation period, and females give birth to litters of four newborns on average, between three and eight times a year. The young reach sexual maturity approximately by 5 weeks (Airoldi, 1978; Poissenot et al., 2021a, b; Somoano Garcia, 2017; Ventura & Gosálbez, 1990).

These prolific reproductive abilities lead to a demography characterized by strong cyclical variations in population density, known as outbreaks. Vole density can then increase from a few individuals per hectare during the low-density phase, to more than 500 individuals per hectare during an outbreak peak (Saucy, 1994; Giraudoux et al., 1997). The average duration of these outbreak cycles is between 5 and 7 years (Truchetet et al., 2014).

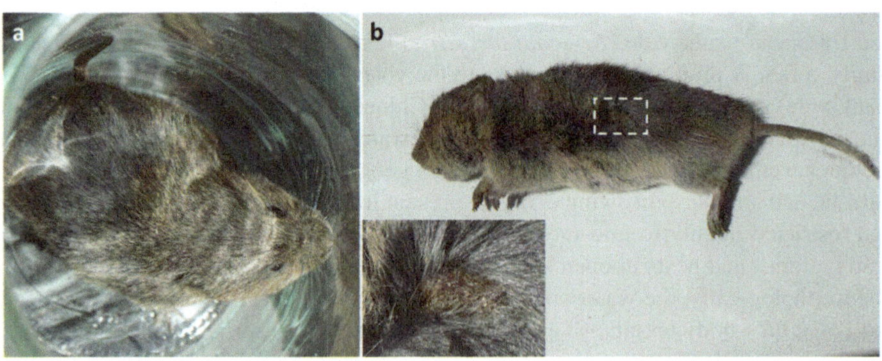


**Fig. 1** The water vole. (**a**) General aspect; (**b**) the lateral scent gland situated on each flank, mainly composed of sebaceous glands

These population outbreaks can have serious consequences for agricultural economy as they are the cause of significant damage to orchards and meadows used for growing grass in mid-mountain regions (Fig. 2). The vole is indeed a diurnal herbivore that can consume up to 80% of its weight per day, mainly in roots, bulbs, and rhizomes (Airoldi, 1979; Kopp, 1988, 1993). The consequential decrease in quality and quantity of fodder leads to potentially important economic losses for farmers (Quéré et al., 1999; Truchetet et al., 2014). These outbreaks can also represent a public health problem, the vole being a vector of several diseases transmissible to humans, including in particular alveolar echinococcosis (Houin et al., 1982). At present, the main means of controlling these pullulations are based either on the use of rodenticides, such as bromadiolone or on trapping. However, these methods have major drawbacks, viz. high environmental toxicity or a lack of specificity on nontarget species for rodenticides inducing secondary intoxication of mammalian and avian predators (Abi Khalil et al., 2021; Berny et al., 1997; Coeurdassier et al., 2014) and non-feasibility for trapping once population densities have exceeded a certain threshold. In addition, bromadiolone has recently (January 1, 2021) been banned in France. It is therefore necessary to develop new control methods or to improve existing ones, in order to limit the impact of voles on the environment.

In this context, we opted to further advance the current knowledge of the chemosensory and behavioral bases of olfactory communication in water voles with the idea that this knowledge could ultimately bring new ways to control populations using natural chemosignals. Although our understanding of olfactory communication in this species is rather limited, Stoddart (1972) has hypothesized that LSG secretions could play a notable role. In addition, olfaction is largely known to be involved in intraspecific communication in other rodents (e.g., Coombes et al., 2018; Wyatt, 2014), and voles should not be an exception (Stoddart, 1986).

**Fig. 2** Damage on a permanent meadow (Puy-de-Dôme, France) caused by fossorial water voles (Photograph by K. Poissenot)

Recently, molecular cues potentially involved in water vole chemical communication have been identified (Nagnan-Le Meillour et al., 2019). In a first step, the components of the LSG secretions were explored after solvent extraction followed by gas chromatography coupled to mass spectrometry (GC/MS). Chromatograms of both male and female LSG displayed 50 compounds among which 41 were identified in both sexes. Those compounds were long-chain fatty acid esters ranging from C7 to at least C12, cholesterol, and long-chain alcohols.

Then, the volatile organic compounds (VOCs) conveyed by vole urine were identified (Nagnan-Le Meillour et al., 2019). Twenty-nine compounds were described, mainly ketones, benzene, and pyrazine derivatives, plus a few alcohols and terpenes. These compounds composed a water vole-specific profile of urinary volatiles, close qualitatively to those found in other Cricetidae species (e.g., *Phodopus campbelli* and *Phodopus sungorus*), where pyrazines dominate when detected by solid-phase microextraction-GC/MS (Soini et al., 2005). Ultimately, biochemical analyses also revealed the occurrence in vole urine of a single urinary protein, named arvicolin, which belongs to the lipocalin family (Nagnan-Le Meillour et al., 2019). In mice, lipocalins have the ability to bind VOCs and release them gradually, a process that increases the stability over time of the odor signal (Hurst et al., 1998). The predicted three-dimensional structure of arvicolin clearly indicates the presence of a cavity in which VOCs could bind.

Based on these results, we started to explore some of the volatile molecules potentially involved in between-sex olfactory communication in water voles, especially three pyrazines, namely, 2,3,5-trimethy-pyrazine and 3-ethyl-2,5-dimethyl-pyrazine, both identified in urine, and 2-isoamyl-6-methylpyrazine which was identified in the LSG secretion. We concentrated here on these three pyrazines based on the facts that (1) previous analyses (Nagnan-Le Meillour et al., 2019) showed that those compounds were differentially expressed in males and females, suggesting that they could be attractive at least for females. For example, urinary chemical profiles significantly differ between male and female during the reproductive season (May) where the presence of 2,3,5-trimethyl-pyrazine and 2,5-dimethyl,3-ethyl-pyrazine can be noticed in a larger amount in males than females (Nagnan-Le Meillour et al., 2019). (2) These compounds were commercially available, thus avoiding lengthy chemical syntheses, and (3) these compounds were already reported as being involved in chemosensory communication in closely related species. For example, pyrazine-2,5-dimethyl delays puberty in the Cricetidae *Peromyscus californicus* (Jemiolo et al., 1994).

## 2 Material and Methods

### 2.1 Ethical Statement

This study was carried out in accordance with European directive 2010/60/UE and was approved by ethical committees for animal experimentation (C2EA-02, project 21994-109 201907510411944 and C2EA-18, project 1857-2018110717044913). Trapping of water voles was allowed in the French department of Puy-de-Dôme by prefectural authorization 19-00100.

## 2.2 Study Species and Trapping

Voles of the fossorial ecotype (*Arvicola terrestris/Arvicola amphibius*, Linnaeus, 1758) were caught with traps (Topcat®, Andermatt, France) placed into their galleries. Capture campaigns were performed during the day and the traps were inspected every ~15 min. Captures were carried out on permanent meadows on different trapping sites around the village of Valbeleix (45° 28′ 17″ N, 3° 0′ 26″ E, alt. 1130 m). Live animals were then transferred to housing facilities for laboratory tests either at the Lyon veterinary school (Rongeurs Sauvages-Risques Sanitaires et Gestion des Populations, VetagroSup Lyon, Marcy-L'Etoile) or at Université Clermont Auvergne (Génétique, Reproduction & Développement, Clermont-Ferrand). At least 2 weeks before being tested, the voles were housed individually in home-cages (length × width × height: 43 × 27 × 16 cm) disposed in an animal facility, at a controlled temperature of 20 °C, under 12 h light/dark cycle, and with food and water ad libitum.

## 2.3 Olfactory Stimuli

We tested the olfactory preference of male and female water voles towards 2,3,5-trimethy-pyrazine (Sigma) ($N = 6$ males and 7 females) and 3-ethyl-2,5-dimethyl-pyrazine (Sigma) ($N = 7$ males and 6 females) identified in urine and towards 2-isoamyl-6-methylpyrazine (CHEM-space) ($N = 9$ males and 6 females) identified in LSG. These chemicals were dissolved at a concentration of 0.1 μM in mineral oil (Sigma), which was also used as the control stimulus. Ten μl of the target compound and the control was pipetted on filter papers that were presented simultaneously on the two opposite sides on a pierced plate, to prevent direct contact with the olfactory source, placed above their own cage (Fig. 3).

## 2.4 Behavioral Tests

Male and female adult water voles were tested for olfactory investigation of the target chemosignals in two-choice tests run in their home-cage under dim light (Fig. 3). After 2 min of familiarization, the time of the vole's olfactory investigation of both stimuli was then video-recorded for 10 min for offline analyses. An animal was considered investigating a target odorant when its snout was at 1 cm (or less) of the filter papers releasing the odors. The behaviour expressed by animals was analysed using the Behavioural Observation Research Interactive Software (BORIS, University of Torino; Friard & Gamba, 2016). Results were expressed as an attraction score (PS) computed as follows: time spent investigating the odor - time spent investigating the control (mineral oil)/total time of investigation.

**Fig. 3** Device used to test olfactory attraction towards social odors in fossorial water voles. (Photograph by K. Poissenot)

# 3 Results

## 3.1 2-Isoamyl-6-Methylpyrazine

No olfactory attraction was found for this compound in both male and female water voles (Fig. 4). Indeed, in both sexes, animals exhibited a high variability in their olfactory investigation: some animals showing a strong attraction for it, others spending more time exploring the control stimulation.

## 3.2 2,3,5-Trimethy-Pyrazine

The animals of both sexes showed a weak attraction for this compound, the time spent between both stimulations being not significantly different. Here again, a quite large variability was observed between animals in both sexes.

## 3.3 3-Ethyl-2,5-Dimethyl-Pyrazine

In the attraction test, females exhibited a clear preference for this compound with an attraction score of 0.64. Indeed, five out of the six females tested investigated much longer investigation of 3-ethyl-2,5-dimethyl-pyrazine than mineral oil (sniffing duration 3-ethyl-2,5-dimethyl-pyrazine vs mineral oil; one-sample t-test, $p = 0.047$). By contrast, males exhibited rather an avoidance of the compound (mean attraction score: −0.37), with 5 out of 7 males tested sniffing longer the control stimulation than 3-ethyl-2,5-dimethyl-pyrazine.

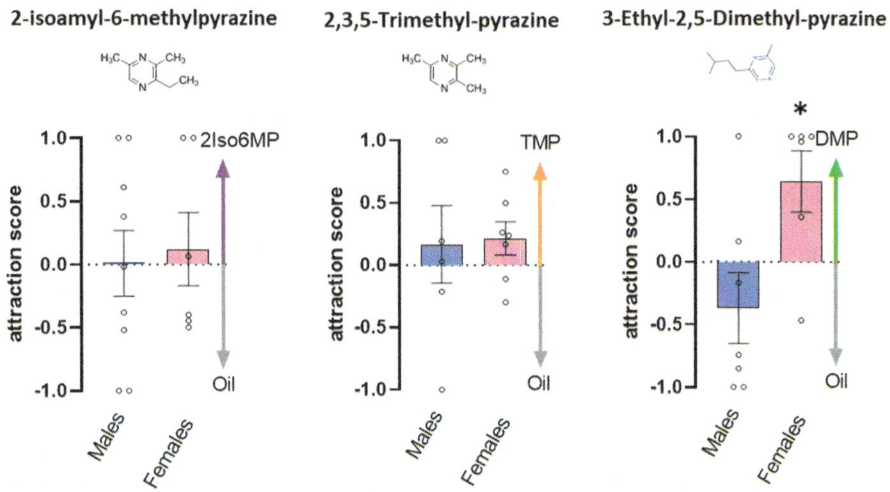


**Fig. 4** Attraction score (represented as means +/− SEM) of male and female water voles towards 2-isoamyl-6-methylpyrazine, 2,3,5-trimethy-pyrazine, and 3-ethyl-2,5-dimethyl-pyrazine when tested in their home-cage

# 4 Discussion

## 4.1 2-Isoamyl-6-Methylpyrazine

First, we found that males and females water voles did not show any differential reaction to 2-isoamyl-6-methylpyrazine, despite it was identified in the LSG secretions of males and females. This may appear surprising as LSG secretions were potential players in olfactory communication among water voles (Stoddart, 1972). However, we recently showed that no clear sexual preference was observable when assessing preference between male versus female LSG secretions (Poissenot et al. submitted). In addition, a lack of clear sex- or season-specific profiles during spring and summer in terms of the number of odor compounds identified in these secretions, including 2-isoamyl-6-methylpyrazine, may explain this result (Nagnan-Le Meillour et al., 2019). However, in other rodents bearing similar sebaceous glands, such as the hamster (*Cricetus cricetus*) or Alston's brown mouse (*Scotinomys teguina*), the animals could discriminate sex of conspecifics from these specific glands (DelBarco-Trillo et al., 2009; Johnston, 2003; Fernández-Vargas et al., 2008).

## 4.2 2,3,5-Trimethy-Pyrazine

This compound has been identified in larger concentration in male than female urine and found during the reproductive season. Our hypothesis was that 2,3,5-trimethy-pyrazine could be an attractant for females. This attraction could reflect an interest for a potential sexual partner. In many rodent species, as in hamsters, mice, prairie, and meadow voles (Ferkin & Zucker, 1991; Ferkin et al., 1994), urinary odor compounds are emitted by males to attract females. However, our results do not support our initial hypothesis as no significant attraction against the control was found.

### 4.3 3-Ethyl-2,5-Dimethyl-Pyrazine

We finally tested this compound with the hypothesis that it could act as an attractant for female voles, as it is also expressed at a higher concentration in male than female urine. Our results showed a clear and strong attraction for 3-ethyl-2,5-dimethyl-pyrazine in females. This compound might therefore serve in between-sex attraction in this species. Interestingly, water vole males show no signs of interest or attraction as attested by olfactory investigation for 3-ethyl-2,5-dimethyl-pyrazine against the control. Although nonsignificant, this apparent avoidance could reflect the strong intermale agressivity observed in this species (personal observations). Indeed, water voles are mainly monogamous with strong male territorial defense (Airoldi, 1976; Somoano Garcia, 2017).

### 4.4 Limitations of the Study

It is clear that our results are only preliminary and that many more compounds and conditions (volume and concentrations of olfactory compounds) will have to be tested before a clear view emerges about the chemosignals controlling sociosexual behavior in water voles. We only tested here attraction towards a specific compound in opposition to a neutral control (mineral oil). Other tests, including preference tests where the animal is offered the choice to investigate two different social odors, should complete this study. The need for more naturalistic studies is also mandatory. For example, animals were tested here under dim light. However, this condition could be still stressful to the voles in comparison to the darkness of their underground galleries. In this context, we are currently developing a device placed on a gallery and equipped with a video camera to monitor olfactory behavior within the natural environment of the voles.

### 4.5 Conclusions for Potential Application

Finally, field-trapping tests should be performed to evaluate the attractiveness of urinary compounds and the efficiency of the use of chemosignals for the control of water vole populations in relation to outbreaks. It should be reminded that the trapping strategy is efficient only during the early phase of an outbreak, to avoid its extension or to limit its amplitude. Once the outbreak is established, trapping does not have any significant effect given the high density of animals (>500 animals/hectare). In this context, the validation of 3-ethyl-2,5-dimethyl-pyrazine as an attractant used in conjunction with trapping seems interesting to achieve. To this end, we are planning to spray on traps some of the chemosignals used here, especially 3-ethyl-2,5-dimethyl-pyrazine, and to compare trapping efficiency with traps sprayed with a control (mineral oil).

**Acknowledgements** This work was support by the Ministère de l'Agriculture et de l'Alimentation, the DRAAF Auvergne-Rhône-Alpes, and the Région Auvergne-Rhône-Alpes. We thank Virginie Lattard (VetagroSup Lyon) and Joel Drevet (University Clermont Auvergne, Clermont Ferrand) for allowing us to perform experimentation in their animal facilities.

## References

Abi Khalil, R., Barbier, B., Fafournoux, A., Mahamat, A. B., Marquez, A., Poissenot, K., Keller, M., Desvars-Larrive, A., Fernandez-De-Simon, J., Coeurdassier, M., Benoit, E., Lefebvre, S., Pinot, A., & Lattard, V. (2021). Seasonal diet-based resistance to anticoagulant rodenticides in the fossorial water vole (*Arvicola amphibius*). *Environmental Research, 200*, 111422. https://doi.org/10.1016/j.envres.2021.111422

Airoldi, J.-P. (1976). Le terrier de la forme fouisseuse du campagnol terrestre, *Arvicola terrestris* scherman Shaw (Mammalia, Rodentia). *Revue Suisse de Zoologie, 41*, 23–42.

Airoldi, J.-P. (1978). Etude par capture et recapture d'une population de campagnols terrestre, *Arvicola terrestris* scherman shaw (Mammalia, Rodentia). *La Terre et La Vie, 32*(1), 3–45.

Airoldi, J.-P. (1979). Etude du rythme d'activité du campagnol terrestre, *Arvicola terrestris* scherman Shaw. *Mammalia, 43*(1), 25–52. https://doi.org/10.1515/mamm.1979.43.1.25

Berny, P. J., Buronfosse, T., Buronfosse, F., Lamarque, F., & Lorgue, G. (1997). Field evidence of secondary poisoning of foxes (*Vulpes vulpes*) and buzzards (*Buteo buteo*) by bromadiolone, a 4-year survey. *Chemosphere, 35*(8), 1817–1829. https://doi.org/10.1016/S0045-6535(97)00242-7

Coeurdassier, M., Riols, R., Decors, A., Mionnet, A., David, F., Quintaine, T., Truchetet, D., Scheifler, R., & Giraudoux, P. (2014). Unintentional wildlife poisoning and proposals for sustainable management of rodents. *Conservation Biology, 28*(2), 315–321. https://doi.org/10.1111/cobi.12230

Coombes, H. A., Stockley, P., & Hurst, J. L. (2018). Female chemical signalling underlying a reproduction in mammals. *Journal of Chemical Ecology, 44*, 851–873. https://doi.org/10.1007/s10886-018-0981-x

DelBarco-Trillo, J., LaVenture, A. B., & Johnston, R. E. (2009). Male hamsters discriminate estrous state from vaginal secretions and individuals from flank marks. *Behavioural Processes, 82*(1), 18–24. https://doi.org/10.1016/j.beproc.2009.03.0031

Ferkin, M. H., & Zucker, I. (1991). Seasonal control of odour preferences of meadow voles (*Microtus pennsylvanicus*) by photoperiod and ovarian hormones. *Journal of Reproduction and Fertility, 92*(2), 433–441. https://doi.org/10.1530/jrf.0.0920433

Ferkin, M. H., Ferkin, F. H., & Richmond, M. (1994). Sources of scent used by prairie voles, Microtus ochrogaster, to convey sexual identity to conspecifics. *Canadian Journal of Zoology, 72*(12), 2205–2209. https://doi.org/10.1139/z94-295

Fernández-Vargas, M., Tang-Martínez, Z., & Phelps, S. M. (2008). Olfactory responses of neotropical short-tailed singing mice, *Scotinomys teguina*, to odors of the mid-ventral sebaceous gland: Discrimination of conspecifics, gender, and female reproductive condition. *Journal of Chemical Ecology, 34*(4), 429–437. https://doi.org/10.1007/s10886-008-9428-0

Friard, O., & Gamba, M. (2016). BORIS: A free, versatile open-source event logging software for video/audio coding and live observations. *Methods in Ecology and Evolution, 7*(11), 1325–1330. https://doi.org/10.1111/2041-210X.12584

Giraudoux, P., Delattre, P., Habert, M., Quéré, J. P., Deblay, S., Defaut, R., Duhamel, R., Moissenet, M. F., Salvi, D., & Truchetet, D. (1997). Population dynamics of fossorial water vole (*Arvicola terrestris* Scherman): A land use and landscape perspective. *Agriculture, Ecosystems and Environment, 66*(1), 47–60. https://doi.org/10.1016/S0167-8809(97)80706-2

Houin, R., Deniau, M., Liance, M., & Puel, F. (1982). Arvicola terrestris an intermediate host of *Echinococcus multilocularis* in France: Epidemiological consequences. *International Journal for Parasitology, 12*(6), 593–600.

Hurst, J. L., Robertson, D. H. L., Tolladay, U., & Beynon, R. J. (1998). Proteins in urine scent marks of male house mice extend the longevity of olfactory signals. *Animal Behaviour, 55*(5), 1289–1297. https://doi.org/10.1006/anbe.1997.0650

Jemiolo, B., Gubernick, D. J., Yoder, M. C., & Novotny, M. (1994). Volatile compounds of *Peromyscus californicus*, a monogamous biparental rodent. *Journal of Chemical Ecology, 20*(10), 2489–2500.

Johnston, R. E. (2003). Chemical communication in rodents: From pheromones to individual recognition. *Journal of Mammalogy, 84*(4), 1141–1162.

Kopp, R. (1988). Les choix alimentaires de la forme fouisseuse du campagnol terrestre (Arvicola terrestris scherman): essais en terrarium. *European and Mediterranean Plant Protection Organization Bulletin, 18*(3), 393–400.

Kopp, R. (1993). *Etude de l'impact de la forme fouisseuse du Campagnol terrestre, Arvicola terrestris scherman (Shaw), sur la végétation d'une prairie*. Doctoral dissertation, University of Lausanne

Nagnan-Le Meillour, P., Descamps, A., Le Danvic, C., Grandmougin, M., Saliou, J.-M. J., Klopp, C., Milhes, M., Bompard, C., Chesneau, D., Poissenot, K., & Keller, M. (2019). Identification of potential chemosignals in the European water vole *Arvicola terrestris*. *Scientific Reports, 9*(1), 18378. https://doi.org/10.1038/s41598-019-54935-z

Poissenot, K., Chorfa, A., Moussu, C., Trouillet, A.-C., Brachet, M., Chesneau, D., Chemineau, P., Ramadier, E., Pinot, A., Benoit, E., Lattard, V., Dardente, H., Drevet, J., Saez, F., & Keller, M. (2021a). Photoperiod is involved in the regulation of seasonal breeding in male water voles (*Arvicola terrestris*). *The Journal of Experimental Biology, 224*(19), 242792. https://doi.org/10.1242/jeb.242792

Poissenot, K., Moussu, C., Chesneau, D., Ramadier, E., Abi Khalil, R., Chorfa, A., Chemineau, P., Michelin, Y., Saez, F., Drevet, J., Benoit, E., Lattard, V., Pinot, A., Dardente, H., & Keller, M. (2021b). Field study reveals morphological and neuroendocrine correlates of seasonal breeding in female water voles, Arvicola terrestris. *General and Comparative Endocrinology, 311*(July), 113853. https://doi.org/10.1016/j.ygcen.2021.113853

Poissenot, K., Moussu, C., Brachet, M., Chesneau, D., Chemineau, P., Lainé, A.-L., Migaud, M., Charbonnel, N., & Keller, M. (2023). Population density does not affect seasonal regulation of reproductive physiology in male water voles. *Biology Letters, 19*(2), 20220441. https://doi.org/10.1098/rsbl.2022.0441

Quéré, J.-P. (2009). Position systématique et éléments de la biologie du campagnol terrestre. In *Le campagnol terrestre* (pp. 27–30). Edition Quae.

Quéré, J. P., Garel, J. P., Rous, C., Pradier, B., & Delattre, P. (1999). Estimer les dégâts du Campagnol terrestre en prairie naturelle. *Fourrages (Versailles), 158*, 133–147.

Saucy, F. (1994). Density dependence in time series of the fossorial form of the water vole, Arvicola terrestris. *Oikos, 71*(3), 381–392.

Soini, H. A., Wiesler, D., Apfelbach, R., König, P., Vasilieva, N. Y., & Novotny, M. V. (2005). Comparative investigation of the volatile urinary profiles in different *Phodopus* hamster species. *Journal of Chemical Ecology, 31*(5), 1125–1143. https://doi.org/10.1007/s10886-005-4252-2

Somoano Garcia, A. (2017). *Biology and population genetics of Arvicola scherman cantabriae (Rodentia, Arvicolinae)*. Doctoral dissertation, University of Oviedo

Stoddart, D. M. (1971). Breeding and survival in a population of water voles. *The Journal of Animal Ecology, 40*(2), 487–494. https://doi.org/10.2307/3257

Stoddart, D. M. (1972). The lateral scent organs of *Arvicola terrestris* (Rodentia : Microtinae). *Journal of Zoology, 166*, 49–54.

Stoddart, D. M. (1986). The comparative responses of mice and voles to conspecific and hetero-specific odors in the field-A lesson in social behavior. In D. Duvall, D. Müller-Schwarze, & R. M. Silverstein (Eds.), *Chemical signals in vertebrates, volume 4: Ecology, evolution, and comparative biology* (pp. 533–540). Plenum Press.

Stoddart, D. M., Aplin, R. T., & Wood, M. J. (1975). Evidence for social difference in the flank organ secretion of *Arvicola terrestris* (Rodentia: Microtinae). *Journal of Zoology, 1799*, 529–540.

Truchetet, D., Couval, G., Michelin, Y., & Giraudoux, P. (2014). Genèse de la problématique du campagnol terrestre (*Arvicola terrestris*) en prairies. *Fourrages, 220*, 279–284.

Ventura, J., & Gosálbez, J. (1990). Reproductive cycle of *Arvicola terrestris* (Rodentia, Arvicoidae) in the Aran Valler, Spain. *Revue Suisse de Zoologie, 55*, 383–391.

Wyatt, T. D. (2014). *Pheromones and animal behavior*. Cambridge University Press.

# Detection of Male Olfactory Information by Female Goats: A Calcium Imaging Approach

Maxime Meunier[1], Anne-Charlotte Trouillet[1], Matthieu Keller[1] (✉), and Pablo Chamero[1] (✉)

[1] UMR Physiologie de la Reproduction et des Comportements, INRAE, CNRS, IFCE, Université de Tours, Nouzilly, France
matthieu.keller@inrae.fr; pablo.chamero-benito@inrae.fr

**Abstract** In mammals, reproductive function can be modulated by sociosexual factors. In small ruminants such as goats, phenomena such as puberty acceleration and the "male effect" have been hypothesized to depend on male olfactory information, but the source of these odorants and how they are detected is currently unclear. Calcium ($Ca^{2+}$) imaging is a widely developed approach in rodents to assess the activity of olfactory sensory neurons (OSNs) and vomeronasal neurons (VSNs), originating from the main olfactory epithelium (MOE) and the vomeronasal organ (VNO), respectively. Our study aimed to adapt this approach to primary cultures of dissociated goat OSNs and VSNs. MOE and VNO cells were stimulated with urine and fur extracts from castrated or sexually active bucks. $Ca^{2+}$ responses were observed in cells from the two olfactory systems for each stimuli. These results indicate that (1) both MOE and VNO are involved in male chemosignal detection by females during the breeding season, and (2) VSNs and OSNs from ovulatory goats are activated by male olfactory information, independently of male sexual activity. The development of this approach will be of high interest in the search for potential chemosignals involved in the regulation of sociosexual behavior in goats and associated physiological regulations.

**Keywords** Goat (*Capra hircus*) · Olfaction · Calcium imaging · Buck pheromones · Female receptivity

## 1 Introduction

Olfactory information plays a critical role in social behavior and influences reproductive function in mammals. Indeed, an individual can emit chemical compounds towards a recipient conspecific and modulate its physiology and reproductive behavior; these

Equal contributions as first and last authors, respectively.



B. Schaal et al. (eds.), *Chemical Signals in Vertebrates 15*,
https://doi.org/10.1007/978-3-031-35159-4_9

chemosignals often described as "pheromones" (Karlson & Lüscher, 1959; Wyatt, 2010) can be detected in mammals via several substructures of the olfactory system, including the main olfactory epithelium (MOE) and the the vomeronasal organ (VNO). Both the VNO and MOE contain bipolar neurons with a single dendrite and axon termed vomeronasal and olfactory sensory neurons (VSNs and OSNs, respectively). These neurons detect chemosignals through olfactory receptors (ORs) or vomeronasal receptors (VRs) which are G protein-coupled receptors (GPCRs) (Buck & Axel, 1992; Leinders-Zufall et al., 2000, 2004). In rodents, VRs have been classified into two families: the vomeronasal type 1 receptors (V1Rs) expressed by apical VSNs positive for the G protein subunit $G\alpha_{i2}$ and the vomeronasal type 2 receptors (V2Rs) expressed by basal VSNs positive for $G\alpha_o$ (Munger et al., 2009; Tirindelli et al., 2009).

Small ruminants like goats are sensitive to signals from their social environment. Although reproduction in this species is seasonal and depends on the photoperiod (Chemineau et al., 1992, 2004), the importance of sociosexual interactions has recently been reconsidered (Delgadillo et al., 2020). Two significant phenomena have been observed in this context: first, the "male effect," which consists of the induction of the sexual activity of females during the anoestrus period by sexually active males (for review, see Delgadillo et al., 2009), and second, "puberty acceleration," which is similar to the "Vandenbergh effect" in rodents (Vandenbergh, 1967, 1976), a synchronized pubertal advancement in spring- and autumn-born females induced by sexually active bucks (Chasles et al., 2019, 2018). Male olfactory information are likely involved in these two phenomena. Indeed, buck fur triggers the hypothalamic GnRH pulse generator leading to gonadotropin secretion (Hamada et al., 1996; Ichimaru et al., 1999; Murata et al., 2011, 2014; Sakamoto et al., 2013), which induces ovulation (Claus et al., 1990; Walkden-Brown et al., 1993). However, the nature of these olfactory information and the sensory neurons that detect them remain poorly characterized.

Although the VNO appears to be fully functional in small ruminants such as goats, it seems to play a minor role in pheromone detection in contrast to rodents (for review, see Keller & Lévy, 2012). First, the ratio of the volume of the goat vomeronasal neuroepithelium to the total volume of the VNO is only 8% compared to that in mice (Wakabayashi, 2002). Second, the number of the goat V1R family genes, 11 including 2 pseudogenes (Kondoh et al., 2022), is smaller than the 239 genes identified in the mouse (Francia et al., 2014). In addition, most goat V2Rs are pseudogenes and thus do not appear to be functional (Wakabayashi, 2002). Moreover, one receptor gene of the V1R family (gV1ra1) is also described to be expressed in the goat MOE (Wakabayashi, 2002; Wakabayashi et al., 2007). Several studies in ewes also emphasize the dominant role of the main olfactory system (MOS) in the integration and detection of male odor but do not exclude the involvement of the accessory olfactory system (AOS) (Gelez et al., 2004; Gelez & Fabre-Nys, 2004, 2006a, b).

Calcium imaging is a widely developed approach to monitor neuronal activity in OSNs and VSNs. This method has been extensively described for the rodent VNO using dissociated cells, tissue sections, or a whole organ mount (Chamero et al., 2007; He et al., 2008; Leinders-Zufall et al., 2000; Rivière et al., 2009; Trouillet et al., 2019; Turaga & Holy, 2012). Studies were also performed on dissociated MOE cells from rodents (Chamero et al., 2007; Na et al., 2019; Nara et al., 2011), salamanders (Leinders-Zufall et al., 1998), and domestic cats (Gomez et al., 2005). $Ca^{2+}$ imaging on dissociated sensory neurons is a relatively straightforward approach to identify and analyze OR/VSN ligands. It is based on the use of a calcium ($Ca^{2+}$)-sensitive dye that allows ratiometric quantification of changes in cytosolic $Ca^{2+}$ concentration reducing the occurrence of artefacts linked to differences in cell thickness and uneven dye loading (Chamero & Zufall, 2018). Fura-2 is often used as a $Ca^{2+}$ dye indicator since it is excitable in the UV range (340–380 nm) and can be used in combination with other markers such as fluorescent proteins of different spectral properties (Chamero & Zufall, 2018). Although Fura-2 is the most commonly used, other commercially available fluorescent dyes with different spectral properties can also be used with compatible fluorescent reporters (nonoverlapping spectra) (Chamero & Zufall, 2018). In addition, after single-cell imaging which allows the determination of activation profiles, it is possible to complete the molecular characterization of the same cells via an immunocytochemistry or single-cell RT-PCR (Chamero et al., 2007, 2011).

This manuscript describes the development of a $Ca^{2+}$ imaging approach for the search of chemosignals potentially involved in the male effect and puberty acceleration. Based on the physiological and mechanistic studies described above in small ruminants, we tested the hypothesis that fur from sexually active bucks may induce $Ca^{2+}$ responses in female goat MOE and VNO cells.

## 2 Materials and Methods

The $Ca^{2+}$ imaging protocol used here in goats (*Capra hircus*) was adapted from a standard mouse $Ca^{2+}$ imaging protocol (Chamero & Zufall, 2018). We used commercially available cell culture reagents to perform the $Ca^{2+}$ imaging protocol. All solutions were prepared under sterile conditions in a dedicated cell culture room (e.g., laminar flow hood, incubator at 37 °C, and 5% $CO_2$).

### 2.1 Dissection and Dissociation of VNO/MOE

Goats were euthanized by intravenous injection of 5 mL ketamine (Ketamidor®) and decapitated following the procedures approved by the Institutional Animal Care and Use Committee. Each skull was cut into two equal parts in the sagittal plane. Nasal bones were removed to collect the VNO and MOE (Moawad et al., 2017; Octura et al., 2018) that were kept in ice-cold phosphate-buffered saline (PBS, pH 7.4). VNOs and MOEs were dissected in separated Petri dishes filled with PBS and processed separately but in parallel. One VNO and one MOE were used for $Ca^{2+}$ imaging. The VNO was opened along the longitudinal axis then placed in another Petri dish containing 2 mL of freshly prepared dissociation buffer (1.1 mM EDTA, 5.5 mM DL-cysteine HCl, and 2.2 units/mL papain in PBS). The VNO sensory epithelium was then scratched to detach the cells. For the MOE, the sensory epithelium was scratched in a new Petri dish containing 2 mL of dissociation buffer (supplemented with 40 mM urea). For both VNO and MOE, remaining tissues were removed and then the dissociation buffer containing the cells was collected in a 15-mL tube. After incubation for 20 min at 37 °C,

2 mL of freshly prepared DNase solution (2X PCR buffer, 25 units/mL DNase in PBS) was added. Moreover, 10 mL of prewarmed Dulbecco's Modified Eagle Medium (DMEM) 1X high glucose + GlutaMAX (61965026, Gibco) supplemented with 10% fetal bovine serum (FBS) and penicillin/streptomycin was added to inactivate the DNase. After centrifugation for 5 min at 1000 rpm, the supernatant was removed with a Pasteur pipette and the pellet was resuspended in 200 μL of supplemented DMEM. Further, 25 μL of dissociated cells was plated into concanavalin-A-coated glass coverslips (0.5 μg/μL, coated overnight). Cells were incubated for 45 min to 1 h at 37 °C, 5% $CO_2$, to allow the cells to adhere to the coverslip. Then, the cells were carefully washed with freshly prepared imaging buffer consisting of Hank's Balanced Salt Solution (HBSS 10X, 14065049, Gibco) supplemented with 5 mM HEPES (2-[4-(2-hydroxyethyl) piperazin-1-yl]ethanesulfonic acid, Merck). Dissociated cells were then loaded with 250 μL of 5 μM Fura-2 for at least 30 min at room temperature.

## 2.2 $Ca^{2+}$ Imaging

Each stimulus was prepared in imaging buffer. Urine was diluted 1:100. Fur solutions were prepared by shaking 0.2 g of the sample of interest in 15 mL of imaging buffer for 2 h followed by filtration with a 0.45-μm syringe filter. The coverslip containing the cells was placed in the perfusion chamber above an Olympus IX71 microscope (10× magnification) and connected to the perfusion system to expose the cells to a constant flow of imaging buffer under a perfusion rate of 3–10 mL/min. A suction system was used to remove excessive liquid. An area with numerous Fura-2-positive cells (not too sparse and not too dense) was selected Fluorescence images were acquired using appropriate excitation wavelengths for Fura-2 (dual excitation 340–380 nm) and emitted light (500–560 nm) was recorded with a Hamamatsu C10600-10B camera.

During the recordings, perfusion for 3 min of buffer was alternated with 30 s of stimulus solution. After the last stimulus, 30 s of depolarizing high KCl solution (150 mM) was applied followed by 5 min of imaging buffer before the end of the acquisition.

## 2.3 Immunocytochemistry OMP After $Ca^{2+}$ Imaging

In order to associate this step with the $Ca^{2+}$ responses obtained previously, we performed immunostaining on the same chamber and microscopy field right after imaging. Particular care was taken to avoid any movement of the perfusion chamber during pipetting of the different reagents.

After the $Ca^{2+}$ imaging experiment, cells were fixed for 20 min with 4% paraformaldehyde (PFA). Then, cells were washed for 5 min with imaging buffer and preincubated with a solution containing Tris-buffered saline (TBS), 0.1 Triton, and 2% Normal Donkey Serum (NDS) for 30 min. Next, cells were incubated for 30 min at room temperature with a primary goat anti-OMP antibody (1:2000, Wako) diluted in preincubation solution. After a 5-min wash in imaging buffer, cells were incubated for 30 min at room temperature with a secondary donkey anti-goat Alexa Fluor 647 antibody (1:500, Invitrogen) diluted in TBS-Triton 0.1. Following another wash with imaging buffer for

5 min, Hoechst solution (1:5000) was added for 2 min and then washed again for 5 min. Fluorescence images were acquired using a ×10 magnification objective for molecular characterization of cells.

## 2.4 Data Analysis

We used ImageJ (FIJI) software for cell detection and processing. Regions of interest (ROIs) representing all Fura-2-positive cells were identified using the ROI manager plugin from FIJI. Cell responses from image stacks were measured using the Multi-Measure plugin. Data generated were exported as a .csv file for analysis with Excel. To identify ligand-induced responses, cells were considered as positive when there is a peak increase more than 1.5 times the baseline signal during the time window of stimulus application (the length of time required for the stimulant to reach the coverslip and be washed out). The baseline was calculated from the first 2 min of the acquisition during perfusion of control imaging buffer. Responses that occurred before the stimulus reached the cells and responses outside the time interest range were counted as nonspecific. The ROI list was overlaid on the fluorescence images obtained after immunostaining to determine whether a responding cell in the $Ca^{2+}$ imaging was positive for OMP.

## 2.5 Experiment: $Ca^{2+}$ Responses of MOE/VNO Cells from Adult Cyclic Goats to Male Stimuli

### 2.5.1 Collection of Stimulus Samples

Back fur, beard fur, and urine from five sexually experienced alpine bucks were sampled during the breeding period (November) and stored at −20 °C. These males were considered as "sexually active" since intact bucks display high sexual activity from late summer to late winter (Delgadillo et al., 1992; Shelton, 1978). In addition, back fur and urine from four castrated alpine bucks were collected and stored at −20 °C. These males were regarded as sexually inactive, castration having taken place 3 years before sampling, and the efficiency of castration has been previously validated by plasma testosterone assay.

### 2.5.2 Exposure of MOE/VNO Cells

We isolated MOE and VNO cells from adult cyclic goats during the luteal phase ($N = 6$). After the dissection/dissociation protocol (see Sect. 2.1), three coverslips of each olfactory organ were used for acquisition. On each coverslip, the sequence of stimuli was randomly designated and always finalized by application of KCl as a viability control. For each olfactory organ, a first coverslip was perfused with fur extracts from sexually active and castrated bucks. A second coverslip was exposed with urine from sexually active bucks as well as urine from castrated bucks. Then, a third coverslip was perfused with fur and urine from sexually active males. At the end of the acquisitions, an MOE coverslip was used to perform OMP immunocytochemistry on the same microscope field (see Sect. 2.3; $N = 2$ cyclic goats).

### 2.5.3 Data Analysis

Each coverslip-generated image stack was analyzed (see Sect. 2.4). The number of responding cells per stimulus was pooled by considering all coverslips from all animals tested. The number of OMP-positive responding cells was also collected.

## 3 Results

Dissociated cells loaded with Fura-2 elicited $Ca^{2+}$ responses following exposure to buck stimuli (Fig. 1a, b). The response percentages are shown in Fig. 1c. Approximately 3.75% of MOE cells (377/10064) and 2.20% for VNO cells (174/7921) responded to sexually active male urine. For sexually active male fur, 8.74% of MOE cells (954/10909) and 4.48% of VNO cells (333/7435) were activated. Responses were also obtained with stimuli from castrated (sexually inactive) males. Furthermore, 4.13% of MOE cells (227/5493) and 1.89% of VNO cells (79/4173) responded to castrated male urine. For castrated male fur, 8.60% of MOE cells (545/6338) and 4.67% of VNO cells (172/3687) were activated. Although more than 40% of the dissociated MOE cells (7117/16402) responded to a high concentration of KCl (150 mM), only 5% of dissociated VNO cells (579/11608) responded (Fig. 1d). Surpringsingly, less than 40% of the cells responding to buck stimuli (urine, 7/21; fur, 6/27) tested were mature OMP-positive neurons in any of the olfactory systems studied (Fig. 1e).

## 4 Discussion

This study describes the use of $Ca^{2+}$ imaging to assess the response of OSNs and VSNs to chemosignals in goats. $Ca^{2+}$ responses were observed in both goat OSNs and VSNs. In rodent models, VSNs are generally considered as the main target of pheromones that trigger physiological and neuroendocrine changes (Keller et al., 2010). Lesion and gene targeting approaches have demonstrated the critical role of VNO in various pheromonal effects, including puberty acceleration in female mice (Lomas & Keverne, 1982; Oboti et al., 2014). In small ruminants such as sheep and goats, several studies show that the MOS is deeply involved in the receptivity and processing of male olfactory information (Gelez et al., 2004; Gelez & Fabre-Nys, 2004, 2006a) without excluding the participation of the AOS (Gelez & Fabre-Nys, 2006b). The present $Ca^{2+}$ imaging results are consistent with these observations since dissociated cells of both olfactory organs from breeding female goats are able to respond to male odors.

Castration results in a decrease in sexual activity and induces a change in the olfactory profile of bucks (Hillbrick et al., 1995; Murata et al., 2014). Interestingly, the characteristic smell of bucks perceptible to humans is lost in castrated animals (Van Lancker et al., 2005). Strikingly, fur from castrated bucks fails to reactivate the reproductive axis of anovulatory goats (Claus et al., 1990; Hamada et al., 1996). Moreover, the use of a castrated buck is also ineffective in the context of puberty acceleration (Chasles et al., 2018). However, in the present study, we observed similar rates of $Ca^{2+}$ responses

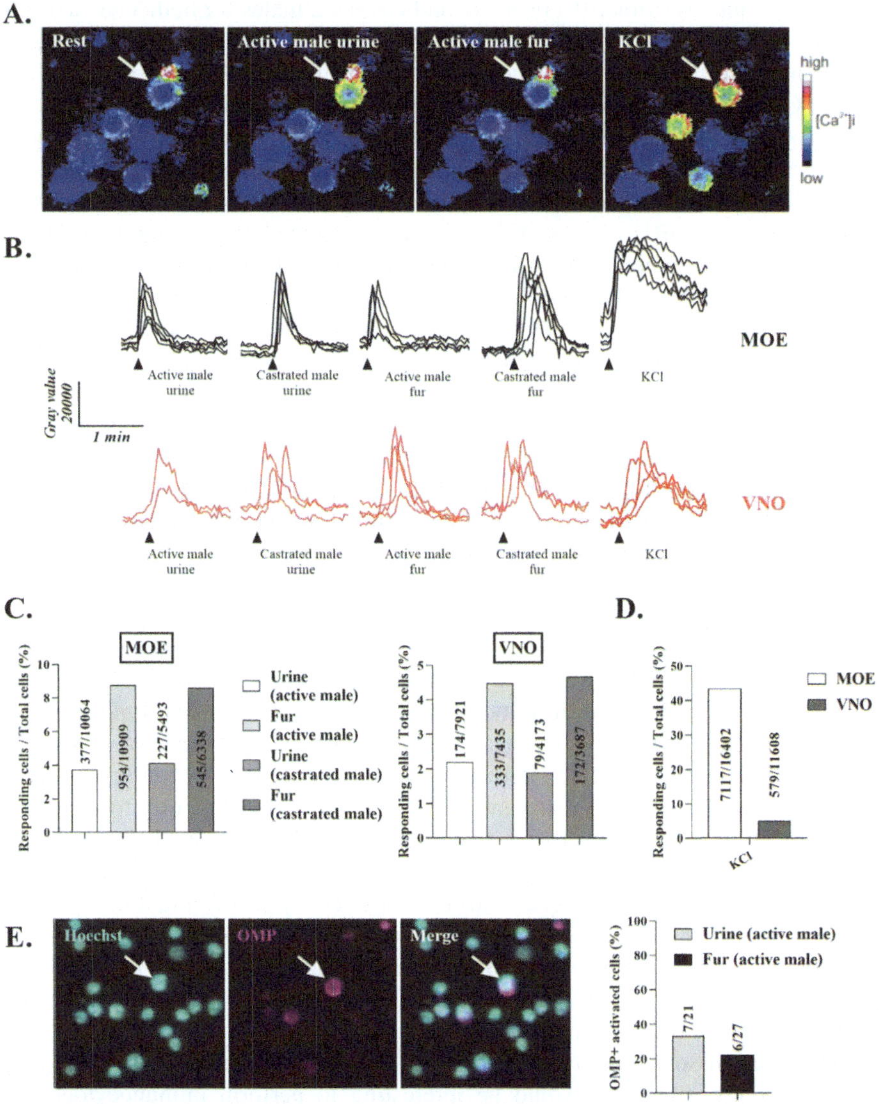


**Fig. 1** Cells in goat olfactory structures respond to olfactory stimuli from goats. (**a**) Pseudocolor images of Fura-2-loaded goat cells activated during $Ca^{2+}$ imaging recording. (**b**) Representative traces of $Ca^{2+}$ responses in dissociated MOE and VNO cells from cyclic goats (luteal phase). Cells were stimulated with urine and fur from sexually active or castrated bucks and high (150 mM) KCl solution as viability control. (**c**) Analysis of ligand-evoked $Ca^{2+}$ responses in MOE (left) and VNO (middle) dissociated cells from cyclic goats ($N = 6$). (**d**) Responses to KCl. Data of **c** and **d** are expressed as the number of responding cells to a given stimulus divided by the number of screened cells (indicated above each bar). Each stimulus was tested 1–3 times per goat ($N = 6$). (**e**) $Ca^{2+}$ imaging followed by immunohistochemistry for OMP to identify whether responsive cells are mature olfactory neurons. Data are represented as the percentage of cells showing responses to any tested stimuli that are positive for OMP ($N = 2$ cyclic goats)

between fur aqueous extracts from active and castrated males for both VNO and MOE cells. The same observation was made between urine of active and castrated males. We thus hypothesize that compounds common to both intact and castrated bucks might induce a $Ca^{2+}$ response in the dissociated cells of both olfactory organs.

As in rodents, the reproductive physiology of the goat can be profoundly influenced by sociosexual interactions, especially by exposure to a sexually active male (see review in Delgadillo et al., 2020). In the context of the buck effect, male fur is known to trigger the hypothalamic GnRH pulse generator leading to gonadotropin secretion (Hamada et al., 1996; Ichimaru et al., 1999; Murata et al., 2011, 2014; Sakamoto et al., 2013), sometimes leading to ovulation (Claus et al., 1990; Walkden-Brown et al., 1993). The ligand 4-ethyloctanal was identified as a primer pheromone, potentially responsible for the male effect in goats (Murata et al., 2014). Thus far, our approach has only focused on buck fur and urine. Thus, it would be interesting to test the ability of chemical compounds such as 4-ethyloctanal to trigger $Ca^{2+}$ responses in goat OSNs and VSNs. Moreover, $Ca^{2+}$ imaging can shed light on the olfactory responsiveness of female goats to male stimuli according to their physiological status. In female mice, the sensory perception of vomeronasal neurons has been shown to be selectively modulated by progesterone in a cycle-dependent manner (Dey et al., 2015). A similar modulatory mechanism may exist in goats, led by the social environment, or seasonal reproduction that involves drastic hormonal variations through the year.

Several ways to improve the $Ca^{2+}$ imaging approach in goats remain to be implemented. First, in the present investigation, male fur extracts were obtained by an aqueous extraction method. However, fatty fractions (fatty acids, fats, and lipids) extracted from buck fur may contain pheromonal activity in seasonally anoestrous goats (Claus et al., 1990). Thus, it would be interesting to expose MOE and VNO cells to the fat fraction of buck fur. Second, dissociated VNO cells responded poorly to the KCl (in contrast to MOE cells), even though OMP-positive cells were largely present on the coverslips (data not shown). Thus, other viability control for VSNs such as adenosine triphosphate (ATP) (Brechbühl et al., 2011; Vick & Delay, 2012) might be considered. Furthermore, the majority of MOE cells responding to our different buck stimuli were OMP-negative. In mice, immature OSNs appear to play a role in the transmission of odorants (Huang et al., 2022). Therefore, we hypothesize that these immature OSNs that express low OMP levels may be able to respond to the stimuli in the conformation obtained after dissociation. Next, to refine the molecular characterization of the responding sensory cells, it would be interesting to perform immunocytochemical labelling of the $G\alpha_{i2}$ protein present in VSNs (Takigami et al., 2000) and certain neurons of the goat MOE (Wakabayashi et al., 2007). In this investigation, the choice of stimuli was limited to buck stimuli as they were suspected to be involved in the physiological changes induced in the female as discussed above. Therefore, this choice could be more diverse by adding new control conditions such as urine and fur extracts from female goats during anestrous period.

In conclusion, this $Ca^{2+}$ imaging approach in goats offers a new direction to investigate and identify primer pheromones in small ruminants. Our method provides encouraging novel information on the responsiveness of female goats' OSNs and VSNs to buck stimuli.

**Acknowledgements** We thank Olivier Lasserre and all the keepers of the UEPAO INRAE experimental unit for providing care to the animals. We also thank S. Bourgeais and B. Marteau for carrying out the euthanasia of the animals and the recovery of the olfactory sensory organs. This work was supported by a grant from Région Centre Val de Loire (France) and INRAE (France).

## References

Brechbühl, J., Luyet, G., Moine, F., Rodriguez, I., & Broillet, M. C. (2011). Imaging pheromone sensing in a mouse vomeronasal acute tissue slice preparation. *Journal of Visualized Experiments, 58*, 1–5. https://doi.org/10.3791/3311

Buck, L. B., & Axel, R. (1992). A novel multigene family may encode odorant receptors. *Society of General Physiologists Series, 47*, 39–51.

Chamero, P., & Zufall, F. (2018). Virus-mediated overexpression of vomeronasal receptors and functional assessment by live-cell calcium imaging. *Methods in Molecular Biology, 1820*, 43–56. https://doi.org/10.1007/978-1-4939-8609-5_4

Chamero, P., Marton, T. F., Logan, D. W., Flanagan, K., Cruz, J. R., Saghatelian, A., Cravatt, B. F., & Stowers, L. (2007). Identification of protein pheromones that promote aggressive behaviour. *Nature, 450*(7171), 899–902. https://doi.org/10.1038/nature05997

Chamero, P., Katsoulidou, V., Hendrix, P., Bufe, B., Roberts, R., Matsunami, H., Abramowitz, J., Birnbaumer, L., Zufall, F., & Leinders-Zufall, T. (2011). G protein G o is essential for vomeronasal function and aggressive behavior in mice. *Proceedings of the National Academy of Sciences, 108*(31), 12898–12903. https://doi.org/10.1073/pnas.1107770108

Chasles, M., Chesneau, D., Moussu, C., Poissenot, K., Beltramo, M., Delgadillo, J. A., Chemineau, P., & Keller, M. (2018). Sexually active bucks are a critical social cue that activates the gonadotrope axis and early puberty onset in does. *Hormones and Behavior, 106*, 81–92. https://doi.org/10.1016/j.yhbeh.2018.10.004

Chasles, M., Chesneau, D., Moussu, C., Abecia, J. A., Delgadillo, J. A., Chemineau, P., & Keller, M. (2019). Highly precocious activation of reproductive function in autumn-born goats (Capra hircus) by exposure to sexually active bucks. *Domestic Animal Endocrinology, 68*, 100–105. https://doi.org/10.1016/j.domaniend.2019.01.004

Chemineau, P., Daveau, A., Maurice, F., & Delgadillo, J. A. (1992). Seasonality of estrus and ovulation is not modified by subjecting female Alpine goats to a tropical photoperiod. *Small Ruminant Research, 8*(4), 299–312. https://doi.org/10.1016/0921-4488(92)90211-L

Chemineau, P., Daveau, A., Cognié, Y., Aumont, G., & Chesneau, D. (2004). Seasonal ovulatory activity exists in tropical Creole female goats and Black Belly ewes subjected to a temperate photoperiod. *BMC Physiology, 4*, 1–11. https://doi.org/10.1186/1472-6793-4-12

Claus, R., Over, R., & Dehnhard, M. (1990). Effect of male odour on LH secretion and the induction of ovulation in seasonally anoestrous goats. *Animal Reproduction Science, 22*(1), 27–38. https://doi.org/10.1016/0378-4320(90)90035-E

Delgadillo, J. A., Leboeuf, B., & Chemineau, P. (1992). Abolition of seasonal variations in semen quality and maintenance of sperm fertilizing ability by photoperiodic cycles in goat bucks. *Small Ruminant Research, 9*(1), 47–59. https://doi.org/10.1016/0921-4488(92)90055-9

Delgadillo, J. A., Gelez, H., Ungerfeld, R., Hawken, P. A. R., & Martin, G. B. (2009). The 'male effect' in sheep and goats—revisiting the dogmas. *Behavioural Brain Research, 200*(2), 304–314.

Delgadillo, J. A., Hernández, H., Abecia, J. A., Keller, M., & Chemineau, P. (2020). Is it time to reconsider the relative weight of sociosexual relationships compared with photoperiod in the control of reproduction of small ruminant females? *Domestic Animal Endocrinology, 73*, 106468. https://doi.org/10.1016/j.domaniend.2020.106468

Dey, S., Chamero, P., Pru, J. K., Chien, M. S., Ibarra-Soria, X., Spencer, K. R., Logan, D. W., Matsunami, H., Peluso, J. J., & Stowers, L. (2015). Cyclic regulation of sensory perception by a female hormone alters behavior. *Cell, 161*(6), 1334–1344. https://doi.org/10.1016/j.cell.2015.04.052

Francia, S., Pifferi, S., Menini, A., & Tirindelli, R. (2014). Vomeronasal receptors and signal transduction in the vomeronasal organ of mammals. In *Neurobiology of chemical communication*. CRC Press/Taylor & Francis. https://www.ncbi.nlm.nih.gov/pubmed/24830038

Gelez, H., & Fabre-Nys, C. (2004). The "male effect" in sheep and goats: A review of the respective roles of the two olfactory systems. *Hormones and Behavior, 46*(3), 257–271. https://doi.org/10.1016/j.yhbeh.2004.05.002

Gelez, H., & Fabre-Nys, C. (2006a). Role of the olfactory systems and importance of learning in the ewes' response to rams or their odors. *Reproduction, Nutrition, Development, 46*(4), 401–415. https://doi.org/10.1051/rnd:2006021

Gelez, H., & Fabre-Nys, C. (2006b). Neural pathways involved in the endocrine response of anestrous ewes to the male or its odor. *Neuroscience, 140*(3), 791–800. https://doi.org/10.1016/j.neuroscience.2006.02.066

Gelez, H., Archer, E., Chesneau, D., Magallon, T., & Fabre-Nys, C. (2004). Inactivation of the olfactory amygdala prevents the endocrine response to male odour in anoestrus ewes. *The European Journal of Neuroscience, 19*(6), 1581–1590. https://doi.org/10.1111/j.1460-9568.2004.03261.x

Gomez, G., Lischka, F. W., Haskins, M. E., & Rawson, N. E. (2005). Evidence for multiple calcium response mechanisms in mammalian olfactory receptor neurons. *Chemical Senses, 30*(4), 317–326. https://doi.org/10.1093/chemse/bji026

Hamada, T., Nakajima, M., Takeuchi, Y., & Mori, Y. (1996). Pheromone-induced stimulation of hypothalamic gonadotropin-releasing hormone pulse generator in ovariectomized, estrogen-primed goats. *Neuroendocrinology, 64*, 313–319. https://doi.org/10.1016/0168-0102(96)89184-X

He, J., Ma, L., Kim, S., Nakai, J., & Yu, C. R. (2008). Encoding gender and individual information in the mouse vomeronasal organ. *Science, 320*(5875), 535–538. https://doi.org/10.1126/science.1154476

Hillbrick, G. C., Tucker, D. J., & Smith, G. C. (1995). The lipid composition of cashmere goat fleece. *Australian Journal of Agricultural Research, 46*(6), 1259–1271. https://doi.org/10.1071/AR9951259

Huang, J. S., Kunkhyen, T., Rangel, A. N., Brechbill, T. R., Gregory, J. D., Winson-Bushby, E. D., Liu, B., Avon, J. T., Muggleton, R. J., & Cheetham, C. E. J. (2022). Immature olfactory sensory neurons provide behaviourally relevant sensory input to the olfactory bulb. *Nature Communications, 13*(1), 6194. https://doi.org/10.1038/s41467-022-33967-6

Ichimaru, T., Takeuchi, Y., & Mori, Y. (1999). Stimulation of the GnRH pulse generator activity by continuous exposure to the male pheromones in the female goat. *The Journal of Reproduction and Development, 45*(4), 243–248. https://doi.org/10.1262/jrd.45.243

Karlson, P., & Lüscher, M. (1959). 'Pheromones': A new term for a class of biologically active substances. *Nature, 183*(4653), 55–56. https://doi.org/10.1038/183055a0

Keller, M., & Lévy, F. (2012). The main but not the accessory olfactory system is involved in the processing of socially relevant chemosignals in ungulates. *Frontiers in Neuroanatomy, 6*. https://doi.org/10.3389/fnana.2012.00039

Keller, M., Pillon, D., Bakker, J. (2010). Olfactory systems in mate recognition and sexual behavior. In *Vitamins and hormones*, 1st edn, vol 83, Issue C. Elsevier Inc. https://doi.org/10.1016/S0083-6729(10)83014-6

Kondoh, D., Kawai, Y. K., Watanabe, K., & Muranishi, Y. (2022). Tissue and cell artiodactyl livestock species have a uniform vomeronasal system with a vomeronasal type 1 receptor (V1R) pathway. *Tissue & Cell, 77*(June), 101863. https://doi.org/10.1016/j.tice.2022.101863

Leinders-Zufall, T., Greer, C. A., Shepherd, G. M., & Zufall, F. (1998). Imaging odor-induced calcium transients in single olfactory cilia: Specificity of activation and role in transduction. *The Journal of Neuroscience, 18*(15), 5630–5639. https://doi.org/10.1523/jneurosci.18-15-05630.1998

Leinders-Zufall, T., Lane, A. P., Puche, A. C., Ma, W., Novotny, M. V., Shipley, M. T., & Zufall, F. (2000). Ultrasensitive pheromone detection by mammalian vomeronasal neurons. *Nature, 405*(6788), 792–796. https://doi.org/10.1038/35015572

Leinders-Zufall, T., Brennan, P., Widmayer, P., Chandramani, S. P., Maul-Pavicic, A., Jäger, M., Li, X. H., Breer, H., Zufall, F., & Boehm, T. (2004). MHC class I peptides as chemosensory signals in the vomeronasal organ. *Science, 306*(5698), 1033–1037. https://doi.org/10.1126/science.1102818

Lomas, D. E., & Keverne, E. B. (1982). Role of the vomeronasal organ and prolactin in the acceleration of puberty in female mice. *Reproduction, 66*(1), 101–107. https://doi.org/10.1530/jrf.0.0660101

Moawad, U. K., Awaad, A. S., & Abedellaah, B. A. (2017). Morphological, histochemical and computed tomography on the vomeronasal organ (Jacobson's organ) of Egyptian native breeds of goats (Capra hircus). *Beni-Suef University Journal of Basic and Applied Sciences, 6*(2), 174–183. https://doi.org/10.1016/j.bjbas.2017.03.003

Munger, S. D., Leinders-Zufall, T., & Zufall, F. (2009). Subsystem organization of the mammalian sense of smell. *Annual Review of Physiology, 71*, 115–140. https://doi.org/10.1146/annurev.physiol.70.113006.100608

Murata, K., Wakabayashi, Y., Sakamoto, K., Tanaka, T., Takeuchi, Y., Mori, Y., & Okamura, H. (2011). Effects of brief exposure of male pheromone on multiple-unit activity at close proximity to kisspeptin neurons in the goat arcuate nucleus. *The Journal of Reproduction and Development, 57*(2), 197–202. https://doi.org/10.1262/jrd.10-070e

Murata, K., Tamogami, S., Itou, M., Ohkubo, Y., Wakabayashi, Y., Watanabe, H., Okamura, H., Takeuchi, Y., & Mori, Y. (2014). Identification of an olfactory signal molecule that activates the central regulator of reproduction in goats. *Current Biology, 24*(6), 681–686. https://doi.org/10.1016/j.cub.2014.01.073

Na, M., Liu, M. T., Nguyen, M. Q., & Ryan, K. (2019). Single-Neuron comparison of the olfactory receptor response to deuterated and nondeuterated odorants. *ACS Chemical Neuroscience, 10*(1), 552–562. https://doi.org/10.1021/acschemneuro.8b00416

Nara, K., Saraiva, L. R., Ye, X., & Buck, L. B. (2011). A large-scale analysis of odor coding in the olfactory epithelium. *The Journal of Neuroscience, 31*(25), 9179–9191. https://doi.org/10.1523/JNEUROSCI.1282-11.2011

Oboti, L., Pérez-Gómez, A., Keller, M., Jacobi, E., Birnbaumer, L., Leinders-Zufall, T., Zufall, F., Chamero, P. (2014). A wide range of pheromone-stimulated sexual and reproductive behaviors in female mice depend on G protein Gαo BMC Biology 12(1), 1–17. https://doi.org/10.1186/1741-7007-12-31

Octura, J. E. R., Maeda, K. I., & Wakabayashi, Y. (2018). Structure and zonal expression of olfactory receptors in the olfactory epithelium of the goat, Capra hircus. *The Journal of Veterinary Medical Science, 80*(6), 913–920. https://doi.org/10.1292/jvms.17-0692

Rivière, S., Challet, L., Fluegge, D., Spehr, M., & Rodriguez, I. (2009). Formyl peptide receptor-like proteins are a novel family of vomeronasal chemosensors. *Nature, 459*(7246), 574–577. https://doi.org/10.1038/nature08029

Sakamoto, K., Wakabayashi, Y., Yamamura, T., Tanaka, T., Takeuchi, Y., Mori, Y., & Okamura, H. (2013). A population of kisspeptin/neurokinin B neurons in the arcuate nucleus may be the central target of the male effect phenomenon in goats. *PLoS One, 8*(11), e81017. https://doi.org/10.1371/journal.pone.0081017

Shelton, M. (1978). Reproduction and breeding of goats. *Journal of Dairy Science, 61*(7), 994–1010. https://doi.org/10.3168/jds.S0022-0302(78)83680-7

Takigami, S., Mori, Y., & Ichikawa, M. (2000). Projection pattern of vomeronasal neurons to the accessory olfactory bulb in goats. *Chemical Senses, 25*(4), 387–393. https://doi.org/10.1093/chemse/25.4.387

Tirindelli, R., Dibattista, M., Pifferi, S., & Menini, A. (2009). From pheromones to behavior. *Physiological Reviews, 89*(3), 921–956. https://doi.org/10.1152/physrev.00037.2008

Trouillet, A.-C., Keller, M., Weiss, J., Leinders-Zufall, T., Birnbaumer, L., Zufall, F., & Chamero, P. (2019). Central role of G protein Gαi2 and Gαi2 + vomeronasal neurons in balancing territorial and infant-directed aggression of male mice. *Proceedings of the National Academy of Sciences, 116*(11), 5135–5143. https://doi.org/10.1073/pnas.1821492116

Turaga, D., & Holy, T. E. (2012). Organization of vomeronasal sensory coding revealed by fast volumetric calcium imaging. *The Journal of Neuroscience, 32*(5), 1612–1621. https://doi.org/10.1523/JNEUROSCI.5339-11.2012

Van Lancker, S., Van Den Broeck, W., & Simoens, P. (2005). Morphology of caprine skin glands involved in buck odour production. *Veterinary Journal, 170*(3), 351–358. https://doi.org/10.1016/j.tvjl.2004.08.017

Vandenbergh, J. G. (1967). Effect of the presence of a male on the sexual maturation of female mice. *Endocrinology, 81*(2), 345–349. https://doi.org/10.1210/endo-81-2-345

Vandenbergh, J. G. (1976). Acceleration of sexual maturation in female rats by male stimulation. *Journal of Reproduction and Fertility, 46*(2), 451–453. https://doi.org/10.1530/jrf.0.0460451

Vick, J. S., & Delay, R. J. (2012). ATP excites mouse vomeronasal sensory neurons through activation of P2X receptors. *Neuroscience, 220*, 341–350. https://doi.org/10.1016/j.neuroscience.2012.06.004

Wakabayashi, Y. (2002). A putative pheromone receptor gene is expressed in two distinct olfactory organs in goats. *Chemical Senses, 27*(3), 207–213. https://doi.org/10.1093/chemse/27.3.207

Wakabayashi, Y., Ohkura, S., Okamura, H., Mori, Y., & Ichikawa, M. (2007). Expression of a vomeronasal receptor gene (V1r) and G protein α subunits in goat, Capra hircus, olfactory receptor neurons. *The Journal of Comparative Neurology, 503*(2), 371–380. https://doi.org/10.1002/cne.21394

Walkden-Brown, S. W., Restall, B. J., & Henniawati. (1993). The male effect in the Australian cashmere goat. 2. Role of olfactory cues from the male. *Animal Reproduction Science, 32*(1–2), 55–67. https://doi.org/10.1016/0378-4320(93)90057-X

Wyatt, T. D. (2010). Pheromones and signature mixtures: Defining species-wide signals and variable cues for identity in both invertebrates and vertebrates. *Journal of Comparative Physiology. A, Neuroethology, Sensory, Neural, and Behavioral Physiology, 196*(10), 685–700. https://doi.org/10.1007/s00359-010-0564-y

# The Use of Excretion- and Secretion-Based Scent Communication in European Badgers (*Meles Meles*)

Tanesha M. Allen[1] (✉), Chris Newman[2], David W. Macdonald[2], and Christina D. Buesching[3]

[1] Department of Biology, University of Oxford, Oxford, UK
tanesha.allen@biology.ox.ac.uk
[2] Wildlife Conservation Research Unit, Department of Zoology, University of Oxford, Recanati-Kaplan Centre, Tubney House, Oxford, UK
[3] Department of Zoology, Irving K. Barber Faculty of Sciences, Kelowna, BC, Canada

**Abstract** Redundancy in animal communication enhances the likelihood that information transmission occurs between the sender and receiver. In olfactory communication, redundancy can occur, for example, when the same message is encoded in both specialised glandular secretions and in metabolic by-products. Although these two modalities often encode similar messages, few studies have compared their deployment and associated behavioural responses within the same species. Here, we used chemical analyses (i.e. gas chromatography-mass spectrometry [GC-MS], bomb calorimetry, enzyme immunoassay [EIA]) and behavioural studies (i.e. scent-presentation experiments) to indirectly compare olfactory communication utilising subcaudal gland secretions (SGSs) with signals encoded in urine from a wild population of European badgers (*Meles meles*) in Wytham Woods, Oxfordshire, UK. While both SGS and urine encoded biologically relevant information relating to season and reproductive fitness (i.e., sex, age and reproductive condition), SGS encoded additional information regarding individual identity, body condition and social group membership. Furthermore, behavioural observations indicated that age-related variations in SGS chemical composition may affect the perception of SGS energetic content (although the biological relevance of this energetic content requires further research). Overall, our findings may indicate that both SGS and urine appear to be equally important but present varying levels of fitness information.

**Keywords** Olfactory communication · Bomb calorimetry · Gas chromatography-mass spectrometry · Scent-presentation · Scent-marking · Scent secretion · Fitness advertisement · Mate selection · Reproductive behaviour



B. Schaal et al. (eds.), *Chemical Signals in Vertebrates 15*,
https://doi.org/10.1007/978-3-031-35159-4_10

# 1 Introduction: Redundancy in Olfactory Communication

In animal communication, the term 'redundancy' refers to the use of multiple signals – sometimes from different modalities – to convey a single message (Breed & Moore, 2012). Signal redundancy has likely evolved to increase the likelihood of information transmission between conspecifics rather than being a coincidental/wasteful duplication. Redundancy can occur in olfactory communication when the same information is encoded in both metabolic by-products (i.e., urine and faeces) and in various specialised glandular secretions (e.g., anal gland secretions).

Although these modalities of olfactory information transmission can encode similar information, their deployment and behavioural functions have not been fully contrasted. This comparison is important in an evolutionary context due to the different production costs of specialised glandular secretions and metabolic by-products. Unlike specialised glandular secretions, the production of metabolic by-products imposes no energetic cost since they arise necessarily as a consequence of heterotrophic diets, although species may also vary in their digestive efficiency (Sá et al., 2014). In contrast, production costs of metabolic by-products may vary. For instance, numerous species within the order Rodentia produce urine containing specially synthesised high-molecular-weight compounds (i.e. major urinary proteins or MUPs; Roberts, 2007; Phelan et al., 2014) that function as binding sites for volatile compounds (Beynon & Hurst, 2003).

# 2 Do Glandular Secretions and Metabolic By-Products Convey Similar Signals?

## 2.1 European Badgers (*Meles meles*) as a Study Species

European badgers live in pairs or in small groups throughout most of their European geographic range (Johnson et al., 2002). However, in some high-density populations, badgers live in larger social groups, especially in southern England (e.g., Wytham Woods, Oxfordshire; Macdonald et al., 2015; Bright Ross et al., 2022). Because badgers are nocturnal, and generally forage alone, even when denning communally, olfactory signals and cues predominate over visual, auditory or tactile signals/cues in their communications (Buesching & Stankowich, 2017; Buesching & Jordan, 2019).

Scent-marking allows badgers to mark contact zones between group ranges (classically interpreted as territories) arranged around more or less centrally placed communal dens or 'setts' (Stewart et al., 2002; Buesching et al., 2016a), mark recently used resources (Buesching & Stankowich, 2017) and advertise themselves in the context of mating opportunities within their polygynandrous mating system (Buesching & Macdonald, 2001, 2004; Buesching et al., 2003). Badgers use anal gland secretions, interdigital gland secretions and subcaudal gland secretions (Buesching & Stankowich, 2017), as well as urine and faeces scent marks (Stewart et al., 2002; Buesching et al., 2016a; Buesching & Stankowich, 2017). Here, in our comparison of the use of subcaudal gland secretion (SGS) and urine olfactory advertisement, we focus on the high-density badger population living in Wytham Woods, Oxfordshire, UK (51° 46′ 26″ N; 1° 19′ 19″ W; Macdonald et al., 2015), also including some information from other badger populations in southern England.

## 2.2 Spatial and Temporal Patterns of Scent Marking: Does the Pattern of SGS Marking Differ from That of Urination?

European badgers have a specialised subcaudal gland between the root of their tail and anus (Buesching & Macdonald, 2004; Buesching & Stankowich, 2017; Figs. 1 and 2). Badgers deposit SGS by briefly pressing their subcaudal pouch against the substrate (Buesching & Macdonald, 2004). Common marking sites include frequently used paths, tree trunks and latrines (Buesching et al., 2003; Buesching & Macdonald, 2004).

Conspecific marking – or 'allo-marking' – also occurs and takes two forms: sequential allo-marking, where one badger marks another individual, and mutual allo-marking, where two badgers press their subcaudal pouches against each other (Buesching et al., 2003; Buesching & Macdonald, 2004). Both forms occur among group members for social appeasement (Buesching et al., 2003), maintaining and sharing a consistent group odour (Buesching et al., 2002b) by exchanging microbiota (Sin et al., 2012; Buesching et al., 2016b) and for individual advertisement by using conspecifics as 'mobile billboards' (Buesching et al., 2003; Buesching & Stankowich, 2017).

Subcaudal gland scent-marking and secretion production (Buesching et al., 2002a; Buesching & Macdonald, 2004) increase during the post-partum mating season in late winter/early spring (January–May; Sugianto et al., 2018, 2021b). As they reach sexual

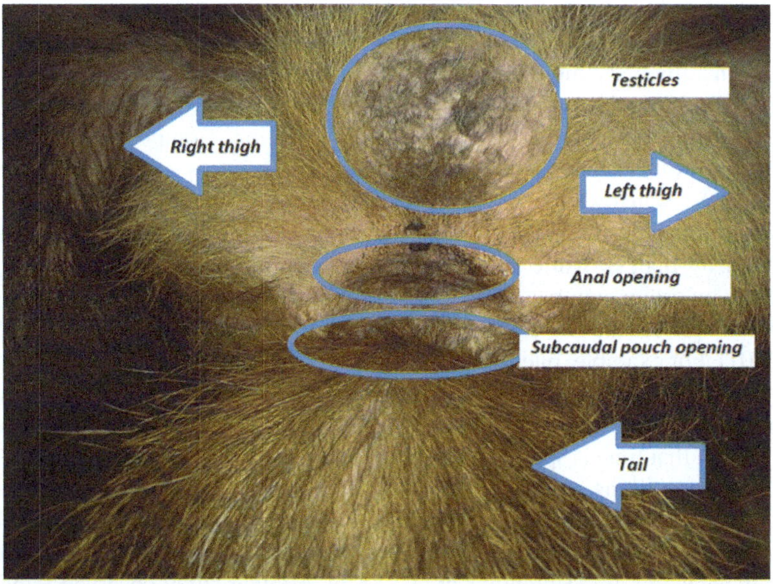


**Fig. 1** The subcaudal pouch of the European badger circled here. The badger is lying supine during a regular fieldwork session

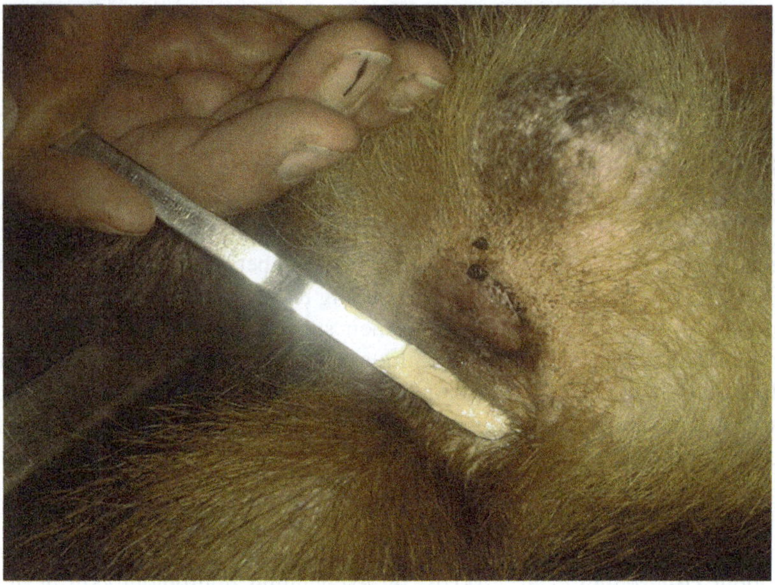

**Fig. 2** Subcaudal gland secretion that has been gently scooped out from the subcaudal pouch. This secretion has caused the yellow stain on the surrounding fur

maturity at around 2 years old (Sugianto et al., 2019), badgers increase SGS production and scent-marking frequency with peaks occurring during late winter and spring (especially among males with descended testes and oestrous females; Buesching et al., 2002a, 2003).

Latrines provide foci of olfactory information for badgers throughout the year. Badgers generally make more latrine deposits and investigate conspecific latrine deposits more often during late winter and spring (Buesching & Jordan, 2019). The categorisation of latrines depends on their content and location. Composite latrines contain SGS alongside other scent-marking deposits like urine, faeces and anal gland secretion (Buesching & Stankowich, 2017). Boundary latrines, evenly distributed along the perimeters of social group ranges, are used by badgers from neighbouring setts (and more often by males), while hinterland latrines within sett borders are used almost exclusively by group members (and more often by females and cubs; Buesching et al., 2016a). A single latrine site can be used by multiple individuals in one night (Stewart et al., 2001). For example, one study observed that 13 out of 16 latrines (81.25%) were used by several badgers in the same night (Kilshaw et al., 2009). Badgers spend, on average, about 23 ± 5 s investigating latrine deposits (predominantly faeces) per visit. About 54% of that time is spent sniffing while 16% of that time is spent crouching during faecal overmarking or subcaudal marking (Stewart et al., 2002). Reports differ about the proportion of time that badgers spend urinating at latrines. While one study proposed that 60% of urination occurs at latrines (Brown, 1993), another observed a lower percentage of 4% (Hutchings et al., 2001). Another study observed females and cubs urinating at a wider diversity of locations, more randomly positioned, compared to males (White et al., 1993).

Overall, both SGS and urine are deposited at latrines and along paths; however, because urine contains more volatile (i.e. low-molecular-weight) compounds than SGS, urine can presumably be detected from longer distances. Further research is needed to determine the distance at which SGS and urine can be detected.

### 2.3 Assessing the Chemical Composition of Substances Used for Scent-Marking: Does SGS Encode Similar Fitness Information to Urine?

Water constitutes around 41–88% of SGS composition (Gorman et al., 1984; Allen, 2020), while proteins as well as short- and long-chain fatty acids comprise the remainder (Buesching et al., 2002b; c). Gas chromatography-mass spectrometry (GC-MS) analyses have identified 110 different compounds comprising SGS, 21 of which were found in all samples (Buesching et al., 2002b, c). These GC-MS profiles indicate significant differences in SGS composition based on individual identity, sex, season, age, body condition, reproductive status and social group membership (Buesching et al., 2002b, c). Individual profiles remain consistent throughout the year (Buesching et al., 2002b, c).

Recent work (Allen, 2020) tested the use of bomb calorimetry to quantify the minimum energetic investment of SGS (i.e. not accounting for any additional biosynthesis costs that are likely high due to the high proportion of proteins and fatty acids; Bauman et al., 2006; Pontes et al., 2015). Samples ($n$ = 84; 51 males [spring = 21, summer = 8, autumn = 17, winter = 5] and 33 females [spring = 13, summer = 5, autumn = 15; no female winter samples were substantial enough for analysis]) were frozen overnight before being freeze-dried for 24 h using a ScanVac CoolSafe (LaboGene ApS, Allerød, Denmark) freeze dryer. This eliminated any water content that could interfere with calorimetric analysis. Each sample was weighed (in g) before ('wet weight') and after ('dry weight') freeze-drying. Dry matter (in %) from each sample was estimated using the following formula:

$$\frac{\text{Dried sample weight}[\text{g}]}{\text{Wet sample weight}[\text{g}]} \times 100\%$$

Moisture content (in %) was calculated by subtracting dry matter from the total percentage. Moisture content was then converted into ml by multiplying the percentage by the total secretion volume. The collected secretion volume was converted into dry matter by multiplying the percentage of dry matter by the total secretion volume.

Calorifc value (CV) analyses (in MJ/kg) were performed at Sciantec Analytical Services, Selby, UK, using a 6100 Parr bomb calorimeter (Parr Instrument Company, Illinois, USA). Freeze-dried samples were placed inside a crucible within a sealed, stainless-steel bomb containing enough oxygen to ensure complete sample combustion. The sealed, stainless-steel bomb was placed inside a larger, insulated container with 2424 ml (per manufacturer instructions; Parr Instrument Company, Illinois, USA) of water, and the combustion heat transferred to the surrounding water bath was measured in °C. Benzoic acid was used to calibrate the bombs. Sample combustion took about 10 min.

Under the assumption that (i) the moisture content from the complete sample is comparable to the moisture content within the analysed subsample, and (ii) wet mass provides a reasonable measure for secretion volume (because 1 ml of water weighs 1 g), four biological metrics were calculated:

(i) *The calorific value of freeze-dried secretion* ($CV_{dry}$; joules/gram$_{secretion}$)
(ii) *The calorific value of natural (i.e. not freeze-dried) secretion* ($CV_{natural}$; joules/ ml$_{secretion}$) – accounting for individual variation in water content: calculated as $CV_{dry}$ multiplied by moisture content (in ml)
(iii) *The calorific value of the total secretion stored in the subcaudal pouch* ($CV_{pouch}$; joules) – total CV (in J/ml) of the stored secretion in the subcaudal pouch calculated as $CV_{natural}$ × pouch secretion volume
(iv) *The calorific value of each scent mark* ($CV_{scentmark}$; joules) – total CV deposited with each scent mark (J/ml) calculated as $CV_{natural}$ × scent mark dimensions, where scent mark dimensions were calculated in mm$^3$ as pouch opening length (mm) × pouch width (3 mm) × scent mark thickness (0.1 mm; pers. obs.) before being converted to ml

$CV_{dry}$ energy values ranged from 4190 to 35,020 J/g$_{dry\ weight}$ (median = 9175 ± 6056.26 J/g$_{dry}$; 95% CI [9311.21 J/g$_{dry}$, 12825.93 J/g$_{dry}$]) in males ($n$ = 50) and 6640 to 36,660 J/g$_{dry}$ (median = 9520 ± 8468.39 J/g$_{dry}$; 95% CI [9633.47 J/g$_{dry}$, 15542.99 J/g$_{dry}$]) in females ($n$ = 34). $CV_{dry}$ content did not vary significantly based on sex, body condition (measured on a 1–3 scale with 1 = emaciated, 2 = intermediate and 3 = very good condition; Buesching et al., 2002a) or reproductive life stage (as indicated by age; Sugianto et al., 2020). Female $CV_{dry}$ content was significantly lower in spring than in autumn (linear mixed model: estimate = −0.51, $t$-value = −2.71, $p$ = 0.01; Fig. 3).

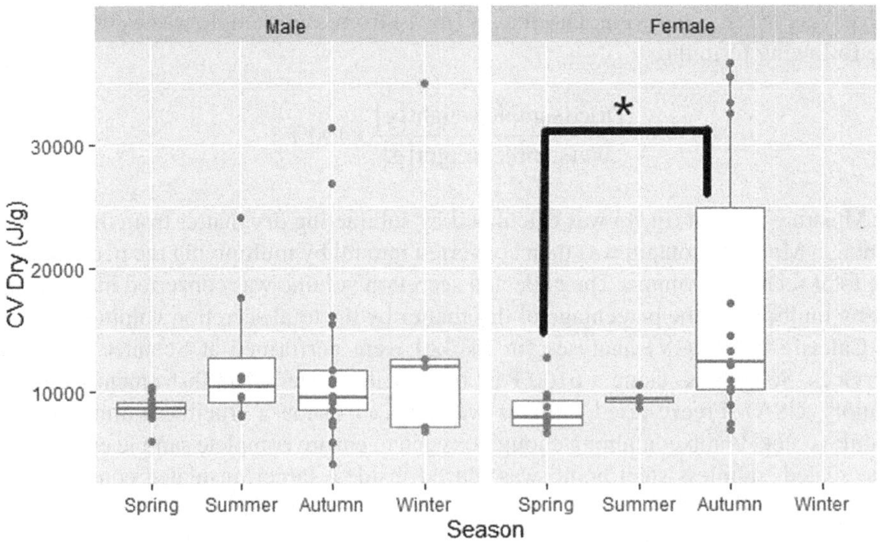


**Fig. 3** Seasonal variation in the calorific value of freeze-dried secretion ($CV_{dry}$) in males and females. Asterisks denote significance

$CV_{natural}$ energy values ranged from 3695.58 to 25914.80 J/ml$_{natural}$ (median = 6524.93 ± 2765.05 J/ml$_{natural}$; 95% CI [6396.37 J/ml$_{natural}$, 8610.44 J/ml$_{natural}$]) in males ($n$ = 50) and 5000.44 to 27366.69 J/ml$_{natural}$ (median = 6774.41 ± 5676.45 J/ml$_{natural}$; 95% CI [6690.76 J/ml$_{natural}$, 10651.97 J/ml$_{natural}$]) in females ($n$ = 34). Consistent with $CV_{dry}$, there were no significant sex-specific differences; however, season significantly impacted female $CV_{natural}$ values. Females had significantly lower $CV_{natural}$ values during spring (LMM: estimate = −0.57, $t$-value = −3.30, $p$ = 0.002) and summer than in autumn (LMM: estimate = −0.46, $t$-value = −2.06, $p$ = 0.048; Fig. 4). In females (but not males), body condition (LMM: estimate = −0.08, $t$-value = −2.92, $p$ = 0.08) and reproductive life stage (LMM: estimate = 0.09, $t$-value = 3.57, $p$ = 0.06) approached significance. $CV_{pouch}$ ranged from 2196.03 to 32116.74 J (median = 7790.36 ± 7011.78 J; 95% CI [7782.84 J, 11879.11 J]) in males ($n$ = 49) and 752.08 to 24630.02 J (median = 4095.98 ± 5830.32 J; 95% CI [4751.23 J, 8885.91 J]) in females ($n$ = 33).

Unlike $CV_{dry}$ and $CV_{natural}$, $CV_{pouch}$ showed significant sex-specific differences, with males having higher $CV_{pouch}$ content than females (LMM: estimate = 0.46, $t$-value = 2.89, $p$ = 0.005). Additionally, only male $CV_{pouch}$ content exhibited significant seasonal variations, with males having higher $CV_{pouch}$ content during spring (LMM: estimate = 0.77, $t$-value = 3.45, $p$ = 0.001; Fig. 5). Neither body condition nor reproductive life stage significantly affected $CV_{pouch}$ content in either sex.

$CV_{scentmark}$ ranged from 58.71 to 388.65 J (median = 112.11 ± 62.67 J; 95% CI [104.87 J; 141.19 J]) in males ($n$ = 50) and 64.47 to 451.55 J (median = 112.47 ± 92.97 J; 95% CI [112.16 J; 177.04 J]) in females ($n$ = 34). Season significantly affected female $CV_{scentmark}$ energy content, higher in females in autumn than in other seasons (LMM: estimate = −0.51, $t$-value = −2.31, $p$ = 0.03; Fig. 6). Sex, body condition and reproductive life stage did not affect $CV_{scentmark}$ content significantly.

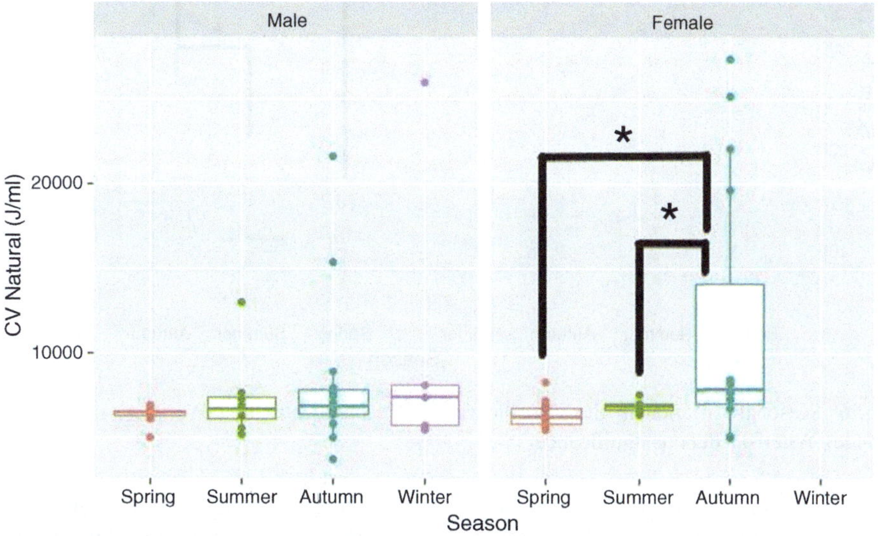


**Fig. 4** Seasonal variation in the calorific value of natural secretion ($CV_{natural}$) in males and females. Asterisks denote significance

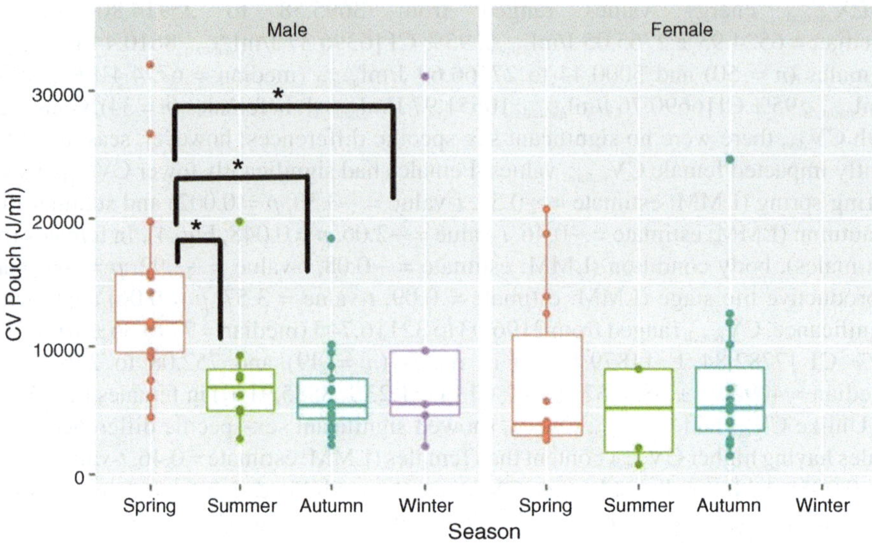


**Fig. 5** Seasonal variation in the calorific value of secretion stored in the pouch ($CV_{pouch}$) in males and females. Asterisks denote significance

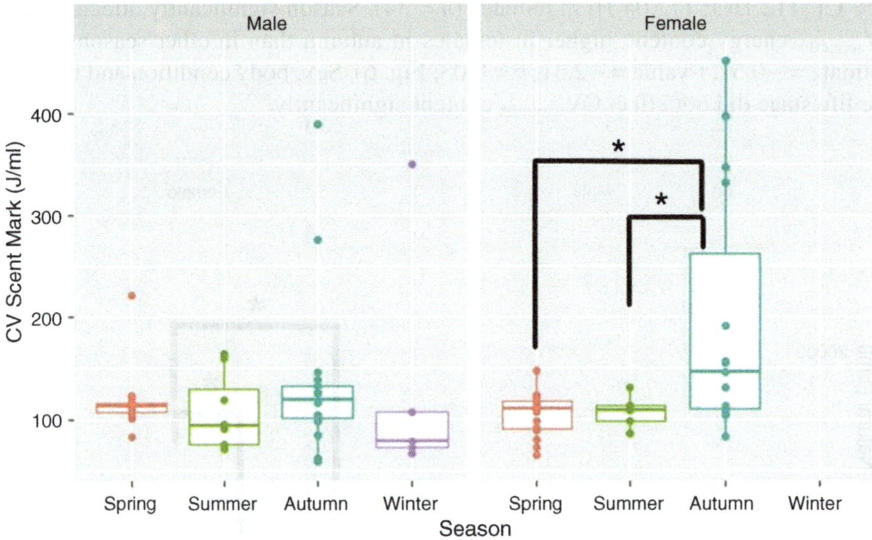


**Fig. 6** Seasonal variation in the calorific value of each scentmark ($CV_{scentmark}$) in males and females. Asterisks denote significance

Several analyses have been conducted on the chemical composition of urine. One study on a wild badger population at Woodchester Park, Gloucestershire (Service et al., 2001), collected 84 urine samples ($n = 40$ males, 44 females) from November 1994 to October 1996. About 300 chemicals were originally identified via GC-MS headspace analysis, and further analysis (i.e. calculating GC-MS peak areas and principal component analysis [PCA]) was conducted on 33 compounds that were found in at least 25 badger urine samples. Aldehydes, ketones, aromatics and pyrazines were the predominant compounds found in these samples. Furthermore, GC-MS profiles exhibited prominent seasonal trends (presumably associated with diet and reproductive seasonality) with sex-specific trends more evident during summer and autumn. Within our Wytham Woods population, seasonal variations in urine composition were corroborated with enzyme immunoassays (EIAs) measuring sex steroid metabolites (Sugianto et al., 2021a). We collected urine samples from 127 adults ($n$ = 66 males, 61 females) via gentle manual palpation of the bladder. New latex gloves were worn between individuals. Urine was collected in glass vials with Teflon-coated screw lids (Sigma-Aldrich, Dorset, UK). Plastic pipettes (one per urine sample) were used to divide samples between 5 and 10 Eppendorf tubes (2 ml), each filled with 0.5 ml of urine. Subsamples were frozen at −20 °C until further use. Blood samples (no more than 2 ml) from these same individuals were collected at the same time. Population-wide analyses found correlations between sex steroid (testosterone and oestrogens) metabolite levels within urine and circulating plasma sex steroid levels that corresponded with seasonal reproductive patterns (i.e. higher testosterone/oestrogen levels during the primary mating season in spring, lower testosterone/oestrogen levels during summer and autumn when the primary mating season has ended). Furthermore, GC-MS analyses revealed that urine from reproductively active males and oestrous females had different volatile organic compound (VOC) compositions than urine from cubs and anoestrous females and that VOC composition varied significantly between the sexes (Tinnesand et al., unpublished).

Overall, the chemical composition of both SGS and urine are correlated with season, sex and reproductive condition. The chemical composition of SGS, however, is correlated with additional information about individual identity, body condition and social group membership alongside information on SGS energetic content, but it is currently unknown if urine encodes similar supplementary information.

### 2.4 Assessing the Biological Relevance of the Chemical Composition: Do Badgers React Differently to SGS Versus Urine?

Although numerous methods have been used to assess chemical differences in scent marks, confirming the biological relevance of these variations requires a different method – scent-presentation experiments to test content-dependent reactions of conspecifics (Tinnesand et al., 2015). Presenting individuals with scent-marking products and measuring their behavioural responses (e.g. response likelihood, countermarking frequency/likelihood, sniffing, avoidance) reveals their biological significance.

In a scent-presentation experiment (Allen, 2020), we presented badgers with paired SGS samples to contrast higher against lower energy content (while controlling for sex, familiarity and sample collection date). We collected 84 SGS samples ($n$ = 51 males, 33 females) from three neighbouring social groups between 13 January 2010 and 25 November 2017. SGS secretion was gently scooped out of the subcaudal pouch with a rounded, stainless-steel spatula sterilised with ethanol (and dried) between individuals.

Secretion volume was estimated to the nearest 0.1 ml (Buesching et al., 2002a). Samples were subdivided into glass vials with Teflon-coated screw lids (Sigma-Aldrich, Dorset, UK) for bomb calorimetry analysis (1 subsample; ca. 1 ml) and scent-presentation experiments (4 subsamples; ca. 0.1 ml each) before freezing them immediately at −20 °C to prevent bacterial decay from potentially altering scent profiles and/or energetic content (Sin et al., 2012; Buesching et al., 2016b).

Scent-presentation experiments were performed at three focal setts (one sett per sampled social group) between 5th June 2018 and 15th June 2018. SGS samples collected in spring 2010, 2011, 2016 and 2017 were used because the experiment required seasonally contemporaneous samples with known energetic content. Previous studies on human biological sample storage and freeze-thaw cycles indicate that the sample composition is not negatively affected by storage length (Gika et al., 2008; Hubel et al., 2014). Because our samples were stored at −20 °C and were only thawed out shortly before the scent-presentation experiments, we are certain that the composition of our SGS samples was not altered.

Following the rules in Sugianto et al. (2019), SGS donors were categorised as 'residents' (i.e. member of the respective focal sett; $n = 26$) or 'neighbours' (i.e. residing in a social group adjacent to the focal sett; $n = 7$). To control for familiarity, samples from neighbours ($n = 3$ low-energy females, 3 high-energy females, 3 low-energy males, 3 high-energy males) were paired separately from samples from residents ($n = 2$ low-energy males, 2 high-energy males). Samples from 2010 to 2011 ($n = 2$ low-energy males, 2 high-energy males, 1 low-energy female, 1 high-energy female; all still alive at the time of the experiments) were paired together to further control for familiarity as group odour composition may have changed between 2010–2011 and 2018.

SGS samples were defrosted immediately before use. New latex gloves were worn when handling each sample to prevent cross-contamination between scents. Two SGS samples (1 low-energy, 1 high-energy as measured by $CV_{dry}$ content) were provisioned about 1–2 h before badger emergence. Samples were placed 50 cm apart (i.e. the head-body length of an adult badger) on opposite sides of frequently travelled paths near (ca. 2–8 m) each focal sett (Figs. 7 and 8). An identical sample pair was positioned near the original pair (ca. 4–8 m) to maximise encounter frequency. Vial lids were removed for scent exposure, and the vials were pushed to rim level in the soil. All vials were removed the following morning. New samples were repositioned ca. 50 cm from the previous pair with the relative positions being switched (e.g. low-energy sample on right or left of the path) to prevent familiarisation.

Low-energy female samples ($n = 3$; range: 6.64–7.22 MJ/kg) were paired with high-energy female samples ($n = 3$; range: 9.66–12.83 MJ/kg), and low-energy male samples ($n = 5$; range: 6.0–8.27 MJ/kg) were paired with high-energy male samples ($n = 5$; range: 9.42–10.45 MJ/kg). Badgers were considered as showing 'no response' when they did not approach either sample within 50 cm (i.e. 1 head body length). A 'potential response' was recorded if a badger came within a 10–50-cm radius of either sample, or an 'actual response' was recorded if an individual approached either vial within 10 cm. For all potential and actual responses, the occurrence (categorised as yes/no) and the frequency of subcaudal countermarking (i.e. scent-marking with subcaudal secretion either directly on a vial ['over-marking'] or within a 15-cm radius ['proximity

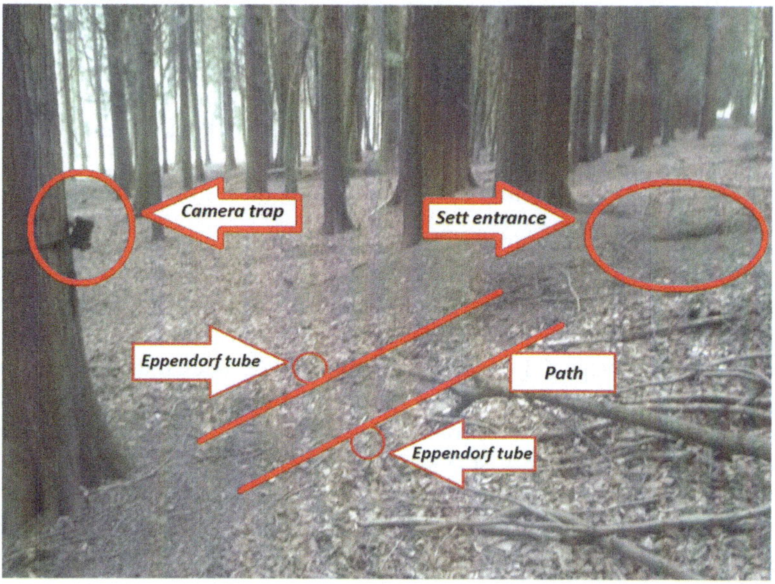


**Fig. 7** Scent-presentation set-up in Wytham Woods. A camera trap (circled) attached to a tree is pointed towards the samples

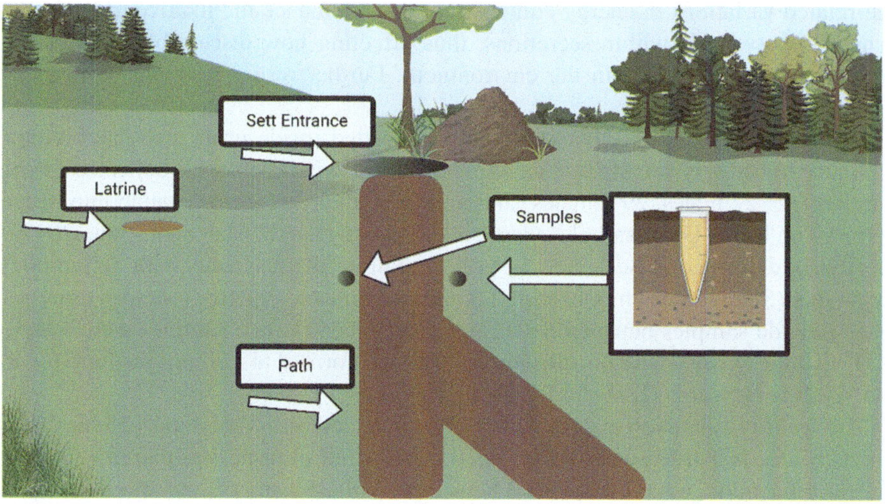


**Fig. 8** Scent-presentation set-up in Wytham Woods. Two samples in Eppendorf tubes were put into the ground. The tubes' openings were level with the soil. (Figure adapted from 'Virtual Background – Golf Course' by Biorender.com (2022). Retrieved from https://app.biorender.com/biorender-templates)

marking']) were recorded, and sniffing duration was measured in seconds. Three behaviours were analysed:

- *Response likelihood*: proportion of 'actual' responses to total responses
- *Countermarking likelihood*: proportion of proximity marking/over-marking to actual responses
- *Sniffing duration*: proportion of time on screen spent sniffing SGS samples

We recorded 246 observations with identifiable individuals (i.e. clip-marked with unique patterns exposing the white underfur; Stewart & Macdonald, 1997) responding to the SGS samples: 12 adult males, 14 adult females and 1 female yearling. While relative energy content (joules) alone did not significantly affect how badgers responded to SGS samples (LR chi-squared = 0.06, $P = 0.80$), badgers altered their behaviour based on encoded information regarding sex and age, mirroring previous observations of badger responses to anal gland secretions (Tinnesand et al., 2015). Badgers responded more frequently to lower-energy SGS samples from older (8+ years old) donors and to higher-energy SGS samples from young (2–7 years old) donors (LR chi-squared = 3.90, $P = 0.049$), while spending a longer proportion of time sniffing male SGS samples compared to female SGS. Furthermore, badgers countermarked higher-energy SGS samples from same-sex donors (15.8%; 12/76) more often than they countermarked lower-energy SGS samples from same-sex donors (7.50%; 6/80). Regardless of relative energetic content, there was no difference in countermarking rates towards SGS samples from opposite-sex donors (LR chi-squared = 4.60, $P = 0.03$). The significant interaction of donor age and SGS energetic content on response likelihood suggests that age-related variations in energy composition may influence the fixative nature of proteins and fatty acids within secretions, thus affecting how discernible low- or high-energy secretions are within the environment. Further research would be needed to confirm this.

Furthermore, we found that information encoded inside urine also has biological relevance to badgers. For similar scent-presentation experiments, we collected 51 urine samples ($n = 22$ males, 29 females) and measured their sex steroid metabolite concentration with EIAs. Urine samples were collected between 22 May and 3 June 2017. As previously described in Sect. 2.3, we collected urine in glass vials with Teflon-coated screw lids (Sigma-Aldrich, Dorset, UK) and used plastic pipettes (one per urine sample) to divide samples between 5-10 Eppendorf tubes (2 ml). Samples were stored at −20 °C prior to use. EIAs on subsamples were performed at Leibniz Institute for Zoo and Wildlife Research (IZW) in Berlin, Germany.

The scent-presentation experiments were conducted on 24 May and 12 June 2017 in Wytham Woods. An Eppendorf tube filled with 0.5 ml of urine was paired with a control (an Eppendorf tube filled with 0.5 ml of deionised water) at our three focal setts. After collating 143 videos with 26 identifiable adult responders, we found a correlation between sniffing duration and sex steroid metabolite levels. In particular, adult males spent a longer time sniffing female urine with lower oestradiol levels (Allen et al., 2019). One possible explanation for this is that the level of VOCs are indirectly – via metabolic pathways – correlated with sex steroid levels. Seasonal variation in chemical composition (Service et al., 2001) – partly due to fluctuations in reproductive receptivity – may also play a factor in sniffing. To test this, we collected 100 urine samples ($n = 47$ males, 53 females) from May 2017 to May 2018 (Allen, 2020). Contemporaneous samples were presented each season (spring = 66 samples, summer = 16 samples, autumn = 18 samples) at the three focal setts in scent-presentation experiments. We found that sniffing likelihood – calculated as the proportion of sniffing occurrences to 'actual' responses in which the badger came within 50 cm of the vial – decreased during spring. Badgers also altered their behaviour based on encoded sex- and age-specific information within urine. A pilot study indicated that adult badgers of both sexes sniffed cub urine ($n = 2$; $7.54 \pm 4.89$ s) longer than they sniffed adult urine ($n = 8$; $2.56 \pm 0.67$ s) and that adult badgers of both sexes countermarked male urine more often than female urine (Allen et al., 2019).

Overall, scent-presentation experiments suggest that fitness information encoded in both SGS and urine are biologically relevant. Seasonal differences in urine also appears to have biological relevance, but no direct comparison between this and seasonal information within SGS has, as yet, been made because season-specific responses towards SGS have not yet been analysed. Nevertheless, GC-MS analyses of SGS confirm distinct seasonality in SGS chemical composition (Buesching et al., 2002c). Moreover, the energetic content of SGS appears to have biological relevance, but further work is needed to fully ascertain what this relevance is.

### 2.5 Factors Influencing Olfactory Signal Discernment: How Does SGS Enable Conspecific Discernment Compared to Urine?

Microbial communities within subcaudal pouches metabolise primary gland products into secondary odourants (Sin et al., 2012; Ezenwa & Williams, 2014; Buesching et al., 2016b). The resulting olfactory stimulus may have attributes relating to memorability and discriminability. Subcaudal pouches mostly contain Actinobacteria and Firmicutes (mainly found in cubs) along with numerous other – but less abundant – bacterial species (Sin et al., 2012). These microbial communities vary between social groups (Buesching et al., 2016b) as well as between adults and cubs (Sin et al., 2012). Because these bacteria metabolise primary gland products, such as long-chain fatty acids, the group- and age-related variations in microbial communities directly relate to chemical composition variations within SGS (Buesching et al., 2002b, c, 2016b) thereby enabling badgers to discriminate between conspecifics based on familiarity and age. Relative energetic content, when combined with these microbiota factors, appears also to be a discriminable characteristic in SGS evidenced by the significant correlation between energetic content, donor sex and behavioural responses (Allen, 2020).

From our further experiments at Stansfeld Park, Oxfordshire, UK (Allen, 2020), badgers can discern between urine from different sources (badgers, domestic dogs, horses, vegetarian humans, non-vegetarian humans and children). We created composite, species-specific urine mixtures to control for individual variations in odour profiles. Samples were mixed proportionally to prevent sex biases within the composite scent. For each species and human diet category, 5–10 Eppendorf tubes (2 ml) were filled with 0.5 ml of the urine mixture using plastic pipettes (one pipette per urine sample) and centrifuged in an Eppendorf Centrifuge 5810 R (Eppendorf UK Limited, Stevenage, UK) for 3 min at 2000 rpm to remove any impurities. The supernatant was removed and transferred to new Eppendorf tubes before being frozen at −20 °C. To promote foraging, two 5-L buckets were filled with sand and mixed with 1 cup of peanuts. These buckets were placed at opposite sides of the sett in specially dug holes so that the bucket rims were level with the surrounding soil. Four Eppendorf tubes were positioned in a circle at 10 cm from the bucket rim. Similar to our other scent-presentation experiments, the urine samples were defrosted 1–2 h before the experiments, had their lids removed to expose the scent and were pushed into the ground so that their rim was level with the soil.

Badgers were seen within 50 cm of the samples during 85 observations. Individuals countermarked less around urine from domestic dogs ($n = 8$; Dunnett's test: estimate = −1.74, $p = 0.006$) and non-vegetarian humans ($n = 9$; Dunnett's test: estimate = −2.38, $p = <0.001$), prompting the hypothesis that this is because scents of these two provenances are aversive. Badgers also spent a shorter proportion of time sniffing the urine from these two sources. Possible VOC composition variation relating to the protein-heavy diets (Nolte et al., 1994) consumed by these urine donors could explain how badgers were able to make this discrimination.

Overall, both SGS and urine seem equally effective in terms of discriminability, particularly when considering the long-term stability of group ranges and latrine-marked intersections.

## 2.6 Behavioural Responses Towards Conspecific Scent Marks: Does Context Influence Behavioural Responses Towards SGS and Urine?

As described, badgers modify their behavioural responses based on the relative energetic content in SGS and encoded information regarding sex and age; however, responder traits (age group, sex) alone did not influence how badgers responded towards SGS samples. Instead, conspecific behaviour was influenced more by donor traits (age group, sex) and the interaction between donor and responder traits (relative sex; results detailed in previous sections).

Donor traits (familiarity, age group, sex and sexual maturity) and interactions between donor and responder traits were similar to conspecific responses to urine samples (Allen et al., 2019). However, in contrast to the findings of the scent-presentation experiments conducted with SGS, responder traits alone were also significant for urine. Adult females responded significantly more often towards urine samples than did adult males, while reproductively senescent individuals of both sexes countermarked urine samples less often than did sexually immature and reproductively active responders.

Overall, conspecific responses towards SGS and urine marks both appear context-specific in that interactions between various traits (donor characteristics and responder characteristics) influence how badgers respond. This has also been confirmed in badger reactions to anal gland secretions (Tinnesand et al., 2015; Noonan et al., 2019). However, for SGS, the energetic content of the scent-marking substance and its interactions with donor/responder traits provides another layer of intricacy.

## 3 Conclusion and Future Directions

Metabolic by-products and specialised scent secretions both have certain advantages and limitations when used for olfactory information transmission. Few studies have analysed if and how these advantages and limitations influence deployment and behavioural functions within a species, especially when the relative production costs may vary.

More research should be conducted to further understand how energetic content impacts the perception of SGS and urine scent marks. The energetic content and presumed production costs of olfactory signals may influence the extent to which redundancy is observed among different species. Species that have more opportunities to reliably send and receive messages – due to longer lifespans, higher population densities or a more stable environment – may exhibit a different modality preference than species that have fewer opportunities for reliable communication. Detection range – particularly in relation to species/nose height above the ground, olfactory membrane size and the proportion of its brain dedicated to scent processing – may also affect modality preferences. Ultimately, production costs are likely only one factor that influences modality preferences (or the lack thereof) within a species, and future studies should examine what other factors may be at play.

**Acknowledgements** We thank all members of the Wytham Badger Project who helped with fieldwork and gathering samples. We also thank Dr. Tristram Wyatt for his comments on an earlier draft.

## References

Allen, T. M. (2020). *Honest fitness advertisement in European badgers (Meles meles)*. Doctoral thesis, University of Oxford. Oxford University Research Archive. https://ora.ox.ac.uk/objects/uuid:88026af1-d8fe-4774-ace0-50cd3457f769

Allen, T. M., Sugianto, N. A., Ryder, C., Newman, C., Macdonald, D. W., & Buesching, C. D. (2019). Encoded information within urine influences behavioural responses among European badgers (Meles meles). In C. D. Buesching (Ed.), *Chemical signals in vertebrates 14* (pp. 38–59). Springer Nature.

Bauman, D. E., Mather, I. H., Wall, R. J., & Lock, A. L. (2006). Major advances associated with the biosynthesis of milk. *Journal of Dairy Science, 89*(4), 1235–1243. https://doi.org/10.3168/jds.S0022-0302(06)72192-0

Beynon, R. J., & Hurst, J. L. (2003). Multiple roles of major urinary proteins in the house mouse, Mus domesticus. *Biochemical Society Transactions, 31*, 142–146. https://www.jenner.ac.uk/lipocalin.htm

Breed, M. D., & Moore, J. (2012). *Animal behavior* (1st ed.). Academic Press.

Bright Ross, J. G., Newman, C., Buesching, C. D., & Macdonald, D. W. (2022). Preserving identity in capture–mark–recapture studies: Increasing the accuracy of minimum number alive (MNA) estimates by incorporating inter-census trapping efficiency variation. *Mammalian Biology, 102*, 1–14. https://doi.org/10.1007/s42991-021-00210-6

Brown, J. A. (1993). *Transmission of bovine tuberculosis (Mycobacterium bovis) from badgers (Meles meles) to cattle.* Doctoral thesis, University of Bristol. University of Bristol Research Portal. https://research-information.bris.ac.uk/ws/portalfiles/portal/34494708/DX178305.pdf

Buesching, C. D., & Jordan, N. (2019). The social function of latrines: A hypothesis-driven research approach. In C. D. Buesching (Ed.), *Chemical signals in vertebrates 14* (pp. 94–103). Springer Nature.

Buesching, C. D., & Macdonald, D. W. (2001). Scent-marking behaviour of the European badger (Meles meles): Resource defence or individual advertisement? In A. Marchlewska-Koj, J. J. Lepri, & D. Müller-Schwarze (Eds.), *Chemical signals in vertebrates 9* (pp. 321–327). Kluwer Academic/Plenum Publishers.

Buesching, C. D., & Macdonald, D. W. (2004). Variations in scent-marking behaviour of European badgers Meles meles in the vicinity of their setts. *Acta Theriologica, 49*(2), 235–246. https://doi.org/10.1007/BF03192523

Buesching, C. D., & Stankowich, T. (2017). Communication amongst the musteloids: Signs, signals, and cues. In D. W. Macdonald, C. Newman, & L. A. Harrington (Eds.), *Biology and conservation of musteloids*. Oxford University Press.

Buesching, C. D., Newman, C., & Macdonald, D. W. (2002a). Variations in colour and volume of the subcaudal gland secretion of badgers (Meles meles) in relation to sex, season and individual-specific parameters. *Mammalian Biology, 67*, 147–156.

Buesching, C. D., Waterhouse, J. S., & Macdonald, D. W. (2002b). Gas-chromatographic analyses of the subcaudal gland secretion of the European badger (Meles meles) part I: Chemical differences related to individual parameters. *Journal of Chemical Ecology, 28*(1), 41–56. https://doi.org/10.1023/A:1013558718057

Buesching, C. D., Waterhouse, J. S., & Macdonald, D. W. (2002c). Gas-chromatographic analyses of the subcaudal gland secretion of the European badger (Meles meles) Part II: Time-related variation in the individual-specific composition. *Journal of Chemical Ecology, 28*, 57–69.

Buesching, C. D., Stopka, P., & Macdonald, D. W. (2003). The social function of allo-marking in the European badger (Meles meles). *Behaviour, 140*, 965–980.

Buesching, C. D., Newman, C., Service, K., Macdonald, D. W., & Riordan, P. (2016a). Latrine marking patterns of badgers (Meles meles) with respect to population density and range size. *Ecosphere, 7*(5), 1–11. https://doi.org/10.1002/ecs2.1328

Buesching, C. D., Tinnesand, H. V., Sin, Y. W., Rosell, F., Burke, T., & Macdonald, D. W. (2016b). Coding of group odor in the subcaudal gland secretion of the European Badger Meles meles: Chemical composition and pouch microbiota. In B. A. Schulte, T. E. Goodwin, & M. H. Ferkin (Eds.), *Chemical signals in vertebrates 13* (pp. 45–62). Springer International Publishing. https://doi.org/10.1007/978-0-387-73945-8

Ezenwa, V. O., & Williams, A. E. (2014). Microbes and animal olfactory communication: Where do we go from here? *BioEssays, 36*(9), 847–854. https://doi.org/10.1002/bies.201400016

Gika, H. G., Theodoridis, G. A., & Wilson, I. D. (2008). Liquid chromatography and ultra-performance liquid chromatography–mass spectrometry fingerprinting of human urine: Sample stability under different handling and storage conditions for metabonomics studies. *Journal of Chromatography. A, 1189*(1–2), 314–322.

Gorman, M. L., Kruuk, H., & Leitch, A. (1984). Social functions of the sub-caudal scent gland secretion of the European badger Meles meles (Carnivora: Mustelidae). *Journal of Zoology, 203*(4), 549–559. https://doi.org/10.1111/j.1469-7998.1984.tb02349.x

Hubel, A., Spindler, R., & Skubitz, A. P. (2014). Storage of human biospecimens: Selection of the optimal storage temperature. *Biopreservation and Biobanking, 12*(3), 165–175.

Hutchings, M. R., Service, K. M., & Harris, S. (2001). Defecation and urination patterns of badgers Meles meles at low density in south west England. *Acta Theriologica, 46*(1), 87–96.

Johnson, D. D., Jetz, W., & Macdonald, D. W. (2002). Environmental correlates of badger social spacing across Europe. *Journal of Biogeography, 29*(3), 411–425.

Kilshaw, K., Newman, C., Buesching, C., Bunyan, J., & Macdonald, D. W. (2009). Coordinated latrine use by European badgers, meles meles: potential consequences for territory defense. *American Society of Mammalogists, 90*(5), 1188–1198.

Macdonald, D. W., Newman, C., & Buesching, C. D. (2015). Badgers in the rural landscape - conservation paragon or farmland pariah? Lessons from the Wytham Badger project. In D. W. Macdonald & R. Feber (Eds.), *Wildlife conservation on farmland: Conflict in the countryside* (Vol. 2, pp. 65–95). Oxford University Press.

Nolte, D. L., Mason, J. R., Epple, G., Aronov, E., & Campbell, D. L. (1994). Why are predator urines aversive to prey? *Journal of Chemical Ecology, 20*(7), 1505–1516. https://doi.org/10.1007/BF02059876

Noonan, M. J., Tinnesand, H. V., Müller, C. T., Rosell, F., Macdonald, D. W., & Buesching, C. D. (2019). Knowing me, knowing you: anal gland secretion of European badgers (Meles meles) codes for individuality, sex and social group membership. *Journal of Chemical Ecology, 45*(10), 823–837.

Phelan, M. M., McLean, L., Armstrong, S. D., Hurst, J. L., Beynon, R. J., & Lian, L. Y. (2014). The structure, stability and pheromone binding of the male mouse protein sex pheromone darcin. *PLoS One, 9*(10), 1–16. https://doi.org/10.1371/journal.pone.0108415

Pontes, M. H., Sevostyanova, A., & Groisman, E. A. (2015). When too much ATP is bad for protein synthesis. *Journal of Molecular Biology, 427*(16), 2586–2594. https://doi.org/10.1016/j.jmb.2015.06.021

Roberts, S. C. (2007). Scent marking. In J. O. Wolff & P. W. Sherman (Eds.), *Rodent societies: An ecological and evolutionary perspective* (pp. 255–267). Chicago University Press.

Sá, F. C., Silva, F. L., de Gomes, M. O. S., Brunetto, M. A., Bazolli, R. S., Giraldi, T., & Carciofi, A. C. (2014). Comparison of the digestive efficiency of extruded diets fed to ferrets (Mustela putorius furo), dogs (Canis familiaris) and cats (Felis catus). *Journal of Nutritional Science, 3*. https://doi.org/10.1017/jns.2014.30

Service, K. M., Brereton, R. G., & Harris, S. (2001). Analysis of badger urine volatiles using gas chromatography-mass spectrometry and pattern recognition techniques. *Analyst, 126*, 615–623. https://doi.org/10.1039/b100755f

Sin, Y. W., Buesching, C. D., Burke, T., & Macdonald, D. W. (2012). Molecular characterization of the microbial communities in the subcaudal gland secretion of the European badger (Meles meles). *FEMS Microbiology Ecology, 81*(3), 648–659. https://doi.org/10.1111/j.1574-6941.2012.01396.x

Stewart, P. D., & Macdonald, D. W. (1997). Age, sex, and condition as predictors of moult and the efficacy of a novel fur-clip technique for individual marking of the European badger (Meles meles). *Journal of Zoology, 241*(3), 543–550.

Stewart, P. D., Macdonald, D. W., Newman, C., & Cheeseman, C. L. (2001). Boundary faeces and matched advertisement in the European badger (Meles meles): A potential role in range exclusion. *Journal of Zoology, 255*(2), 191–198. https://doi.org/10.1017/S0952836901001261

Stewart, P. D., Macdonald, D. W., Newman, C., & Tattersall, F. H. (2002). Behavioural mechanisms of information transmission and reception by badgers, Meles meles, at latrines. *Animal Behaviour, 63*(5), 999–1007. https://doi.org/10.1006/anbe.2001.1990

Sugianto, N. A., Buesching, C. D., Heistermann, M., Newman, C., & Macdonald, D. W. (2018). Linking plasma sex steroid hormone levels to the condition of external genitalia in European badgers (Meles meles): A critical evaluation of traditional field methodology. *Mammalian Biology, 93*, 97–108. https://doi.org/10.1016/j.mambio.2018.09.007

Sugianto, N. A., Newman, C., Macdonald, D. W., & Buesching, C. D. (2019). Heterochrony of puberty in the European badger (Meles meles) can be explained by growth rate and group-size: Evidence for two endocrinological phenotypes. *PLoS One, 14*(3), 1–21.

Sugianto, N. A., Newman, C., Macdonald, D. W., & Buesching, C. D. (2020). Reproductive and somatic senescence in the European badger (Meles meles): Evidence from lifetime sex-steroid profiles. *Zoology, 125803*. https://doi.org/10.1016/j.zool.2020.125803

Sugianto, N. A., Dehnhard, M., Newman, C., Macdonald, D. W., & Buesching, C. D. (2021a). A non-invasive enzyme immunoassay (EIA) method to assess the reproductive status of the European badger (Meles meles) from urinary sex-steroid metabolites. *General and Comparative Endocrinology, 301*. https://doi.org/10.1016/j.ygcen.2020.113655

Sugianto, N. A., Heistermann, M., Newman, C., Macdonald, D. W., & Buesching, C. D. (2021b). Alternative reproductive strategies provide a flexible mechanism for assuring mating success in the European badgers (Meles meles): An investigation from hormonal measures. *General and Comparative Endocrinology, 310*. https://doi.org/10.1016/j.ygcen.2021.113823

Tinnesand, H. V., Buesching, C. D., Noonan, M. J., Newman, C., Zedrosser, A., Rosell, F., & Macdonald, D. W. (2015). Will trespassers be prosecuted or assessed according to their merits? A consilient interpretation of territoriality in a group-living carnivore, the European badger (Meles meles). *PLoS One, 10*(7), e0132432. https://doi.org/10.1371/journal.pone.0132432

White, P. C. L., Brown, J. A., & Harris, S. (1993). Badgers (Meles meles), cattle and bovine tuberculosis (Mycobacterium bovis): A hypothesis to explain the influence of habitat on the risk of disease transmission in southwest England. *Proceedings of the Royal Society London B, 253*(1338), 277–284.

# Scent Communication Between Males and Estrous Females in Domestic Dogs (*Canis Lupus Familiaris*)

Tamako Miyazaki[1] (✉), Urara Onozawa[2], Reiko Uenoyama[2], and Masao Miyazaki[2]

[1] Cooperative Department of Veterinary Medicine, Iwate University, Morioka, Japan
okatama@iwate-u.ac.jp

[2] Department of Biological Chemistry and Food Sciences, Faculty of Agriculture, Iwate University, Morioka, Japan

**Abstract** Scent signals are important for reproduction in mammals, including domestic dogs (*Canis lupus familiaris*). Male dogs use their olfactory system to detect signals emitted from the urine and vaginal secretions of female dogs and possibly to distinguish between estrous and anestrous female dogs. Previous studies have suggested that some compounds secreted by female dogs act as sex pheromones. However, the existence of these compounds in female urine, and their bioactivities in male dogs, remains controversial. In this study, we examined the estrous cycle of three females (two daughters and the mother) and sexual behavior of four sibling dogs (two sisters and two brothers) along with their mother from birth until 3 years old and compared urinary volatile compounds between estrous and anestrous females by bioassay and gas chromatography/mass spectrometry. Unlike wolves, sibling females underwent estrus, while their mother was a dominant female in the pack. Males chased, roared with excitement, and tried to mount the female siblings and their mother in estrus, but not the anestrous females. Methyl paraben (MP), a candidate canine sex pheromone, was detected from the headspace and total lipid extracts of urine from females in the estrous and anestrous stages by gas chromatography/mass spectrometry. Synthetic MP alone did not induce the characteristic responses of sniffing, licking, and lip-smacking with salivation in male dogs, which were all observed when they encountered urine from a female in estrus. Aromatic carbonyl compounds, methyl ketones, and sulfide compounds, which have been suggested to be attractants and repellents involved in the interactions between males and females in estrus, fluctuated in the headspace of urine obtained from anestrous females and those in estrus. These results indicate that female dogs in estrus probably secrete unidentified compounds as sex pheromones that strongly mediate reproductive behaviors in males, but further studies are necessary to identify the sex pheromones specific to females in estrus.

**Keywords** Domestic dog (*Canis lupus familiaris*) · Estrous urine · Urinary volatile compounds · Sex pheromone · Gas chromatography-mass spectrometry



B. Schaal et al. (eds.), *Chemical Signals in Vertebrates 15*,
https://doi.org/10.1007/978-3-031-35159-4_11

# 1 Background

Male domestic dogs (*Canis lupus familiaris*) can roam up to approximately 8 km to identify female dogs in estrus (Beaver, 1999). A previous study has suggested that airborne emissions of volatile compounds from female dogs in estrus attract male dogs without visual or auditory cue by connecting two chambers by a pipeline of 5 cm in diameter (length not stated) with unidirectional air movement using fan (Woszczyło et al., 2020). Male dogs can discriminate between the urine odors of anestrous females and those in estrus (Beach & Gilmore, 1949). Males spend more time investigating the urine and vaginal odors of females in estrus than those of females in diestrus (Doty & Dunbar, 1974) and visit female dogs in estrus for longer than female dogs in anestrous given the attractants contained in the estrous urine and vaginal secretions (Dunbar, 1977). Methyl paraben (MP) has been identified from the odor of vaginal secretions of females in estrus as a candidate canine sex pheromone that induces sexual behavior in male dogs (Goodwin et al., 1979), but its bioactivity (Kruse & Howard, 1983; Santos et al., 2013; Dzięcioł et al., 2017) and content in estrous urine (Schultz et al., 1985; Dzieciol et al., 2014) remain controversial.

Behavioral observations of male and female dogs in estrus show that scent communication between the sexes is important for females to attract males in reproduction. However, limited data are available regarding the bioactive compounds emitted from female dogs in estrus that attract male dogs. In this study, we report our recent findings regarding the sexual responses of males and females, obtained using a dog family consisting of a mother, two daughters, and two sons (Fig. 1a, b). We also examined by gas chromatography/mass spectrometry (GC/MS) the volatile compounds emitted from the urine of estrous vs. anestrous females.

# 2 Estrous Cycles of the Female Siblings and Their Mother

We first examined the estrous cycles and behavioral responses of three intact female border collie dogs from December 2019 to July 2022, i.e., a mother (female A, 5.5–8 years old; born on June 30, 2014) and her two daughters from the same litter (females B and C, 9–40 months old; born on March 12, 2019). These females lived together with two intact male dogs (males A and B) that were of the same litter. These five dogs were kept by a family with five members, including a father (M.M.), a mother (T.M.), and three children, living in a house. They had continuous access to drinking water and received dry food for dogs aged 1–6 years (Science Diet Adult dogs age 1–6; Hill's-Colgate, Tokyo, Japan) twice daily (mornings and evenings). Within the house, the five dogs had full visual, auditory, and olfactory contact with each other and used the same litter tray (with an absorbent sheet). During the estrous cycles in the three females, the dogs were occasionally kept in separate cages or rooms to prevent interactions, mating or fighting.

We followed up the timing of the estrous cycle in the three female dogs for 3 years, from December 2019 to July 2022. In female C, the first bloody discharge occurred at

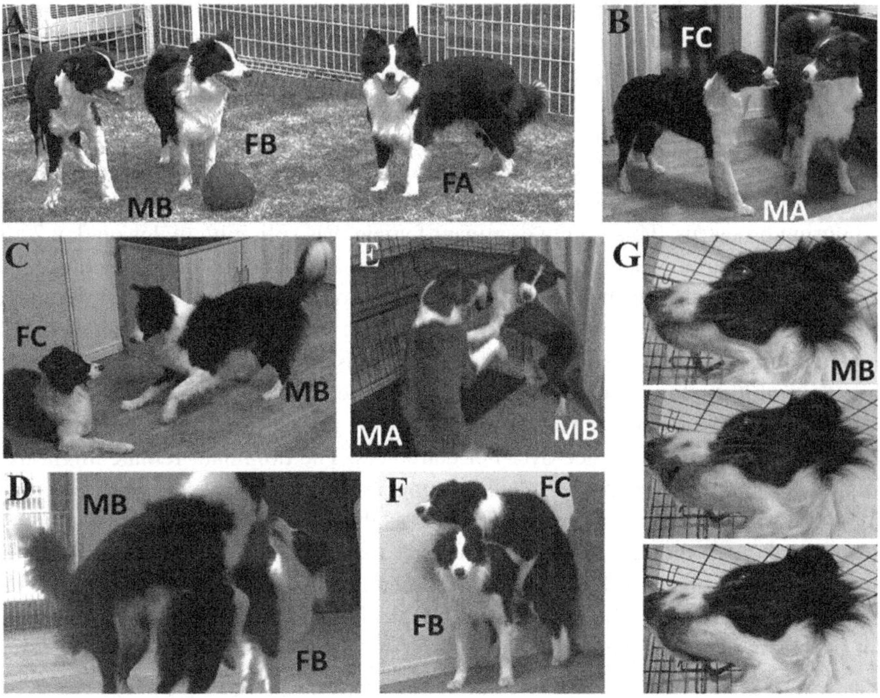


**Fig. 1** Sexual behaviors in domestic dogs. (**a, b**) A dog family used in this study: mother, FA; daughters, FB and FC; and sons MA and MB. (**b**) FC repelled MB by growling and baring the teeth. (**c**) Male B roared with excitement when he contacted estrous FB. (**d**) MB could not mount FB because of rejection by her. (**e**) Males fought each other around estrous females. (**f**) Diestrous FC mounted estrous female FB. (**g**) Characteristic lip-smacking of males toward female estrous urine. They licked the urine frequently, audibly smacked their lips as they opened, and closed their mouths rapidly

**Table 1** Estrus cycle of females A, B, and C (date format: yyyy.mm.dd)

| | Date of beginning of blood discharge | | |
|---|---|---|---|
| | Female A | Female B | Female C |
| Parturition | 2019.03.12 | | |
| First estrus | 2020.02.12 | 2019.12.28 | 2019.12.25 |
| Second estrus | 2020.09.01 | 2020.06.26 | 2020.06.24 |
| Third estrus | 2021.03.09 | 2020.12.31 | 2020.12.28 |
| Fourth estrus | 2021.09.21 | 2021.07.21 | 2021.07.11 |
| Fifth estrus | 2022.03.27 | 2022.01.18 | 2022.01.11 |
| Sixth estrus | | 2022.08.06 | 2022.07.25 |

9 months old (December 25, 2019), and a discharge from female B was detected soon thereafter (December 28, 2019). Subsequently, their mother entered estrous on February 12, 2020 (11 months after parturition). All three females continued to cycle twice per year (Table 1). The duration of proestrus and estrus was approximately 2 weeks based

on continuous bloody discharge in urine following the initial discharge, mucus that dripped on the stairs, and mounting behavior between the female siblings (Fig. 1f) simultaneous with aggressive displays toward the male siblings (Fig. 1b).

In wolves (*Canis lupus*), which are the progenitors of dogs, reproductive activity is usually controlled by only one female in estrus in the pack (Packard & Mech, 1980). One study reported that a 1-year-old wolf in proestrus was dominant over her sibling females, as evidenced by bloody discharge immediately after her 7-year-old mother died during proestrus. This young dominant female was soliciting males on day 7 after the bloody discharge, was in estrus from days 18 to 27, and delivered two pups on day 84. The other females in the pack remained anestrous without elevated progesterone levels (Seal et al., 1979). Considering this report on wolves, we hypothesized that the two daughter females (B and C) examined in the present study might remain anestrous, while their mother dog (female A) entered estrus, because as in wolves, female A was dominant and often made her pups to submit her by growling, baring teeth, and biting, which indicated her dominance over her daughters. The dominant female might use a primer pheromone to keep the related younger females in anestrus. However, this hypothesis was not supported by our findings as the two B and C daughters expressed estrous at about 9 months old, indicating that the mechanisms of control of female estrous cycles in a pack differ markedly between dogs and wolves. Based on this case study, we therefore suggest that dominant female dogs do not emit scent signals to prevent estrous in other female dogs.

## 3 Male-Specific Behavior Toward Estrous Females and Their Urine

From the first time they experienced female estrus at 9 months of age, the males were very interested in the mother and her female siblings when they were estrous (Fig. 1c). They chased the females, roared with excitement, and tried to mount the females in estrus even though they were siblings (Fig. 1d). In contrast, the female dogs, including the mother, always repelled the male dogs by growling and baring their teeth (Fig. 1b). Our observations indicated that the males became reactive to conspecific urine at least a day before detecting the bloody discharge in the females' urine. They carefully explored the urine spot left by an estrous female. Their interest appeared to increase as estrus proceeded. Both males became aggressive to each other, occasionally fought each other (Fig. 1e), and lost their appetite during the females' fertile breeding period 10–13 days after the detection of their bloody discharge. These male behaviors were not recorded when the females exhibited no such characteristic excretory signs during diestrus and anestrus. Although we did not examine such responses toward estrous females in males younger than 9 months old, our two males did not exhibit such responses before encountering estrous female urine, suggesting that the responses are related to sexual maturity. These data strongly suggest that estrous female urine contains chemosignals that induced characteristic responses in males that are at least older than 9 months. Interestingly, both males showed motivation to mate with their estrous female siblings regardless of their genetic relatedness (Fig. 1d), in contrast to the females rejecting the males (Fig. 1b).

When exposed to urine samples from proestrous and estrous females, the male dogs exhibited a series of characteristic responses. They sniffed the estrous urine sample with considerable interest for 15.15 s (median) compared to 2.25 s for anestrous urine (examining two male dogs' behaviors toward two females' urine of estrus and anestrus once for each pair; Wilcoxon matched paired test; $p = 0.0304$). They licked the urine frequently, audibly smacked their lips as they opened, and closed their mouths rapidly (Fig. 1g), often accompanied by salivation, and showed over-marking by urination. These behaviors were always observed for approximately 20–40 s when the male dogs encountered estrous urine in their litter. Such lip-smacking has also been referred to as *teeth chattering* (Dunbar, 1977), and as *tonguing*, which has been characterized as a flehmen-like response in dogs through which they introduce the licked urine into the vomeronasal organ (Dzięcioł et al., 2016). Considering the lip-smacking as a typical behavior induced by estrous female urine and vaginal secretions, it might be used as a criterion to track the occurrence of sex pheromones in the secretions. Since there is no pheromone identified in dogs to date, we assume that the identification of compound(s) inducing lip-smacking will allow further steps toward that aim.

## 4 Methyl Paraben Contents and the Male Response to Synthetic Methyl Paraben

Methyl *p*-hydroxybenzoate or methyl paraben (MP, Fig. 2a) was first identified by analyzing the odor of a cotton applicator containing vaginal smears from female beagles using GC/MS in 1979. MP emerged as a candidate sex pheromone in dogs because synthetic MP applied to the vagina of female beagles, German shepherd, and dalmatian in anestrus induced sexual behavior in male beagles, beagle-spits crosses, and German shepherd (Goodwin et al., 1979). Several studies have attempted to reproduce this phenomenon and to detect MP in estrous urine. The interactions between mature intact male and anestrous female beagles were compared between three treatments of females: untreated control, insertion of saline-soaked cotton applicator in the vagina, and insertion of a saline-soaked cotton applicator that was rolled over MP in the vagina (Kruse & Howard, 1983). Mounting attempts occurred with roughly equal frequency between the saline- and MP-treated females, but no mounting attempts occurred in response to MP. Santos et al. (2013) exposed males to spayed females of mongrel dogs (whose vaginas had been impregnated with MP) before electroejaculation and reported a 100% antegrade ejaculation rate, with high sperm motility; however, they did not observe excitement in the males during their interactions with the females. Moreover, in another study, no sign of excitation or changes in penile perfusion were observed in males after they had interacted with females in anestrous to which a commercially available synthetic pheromone containing MP (Eau D'Estrus; Synbiotics Corp., Kansas City, MO, USA) had been applied in Bedlington Terrier, Yorkshire Terrier, Labrador, and mixed breed (Dzięcioł et al., 2017).

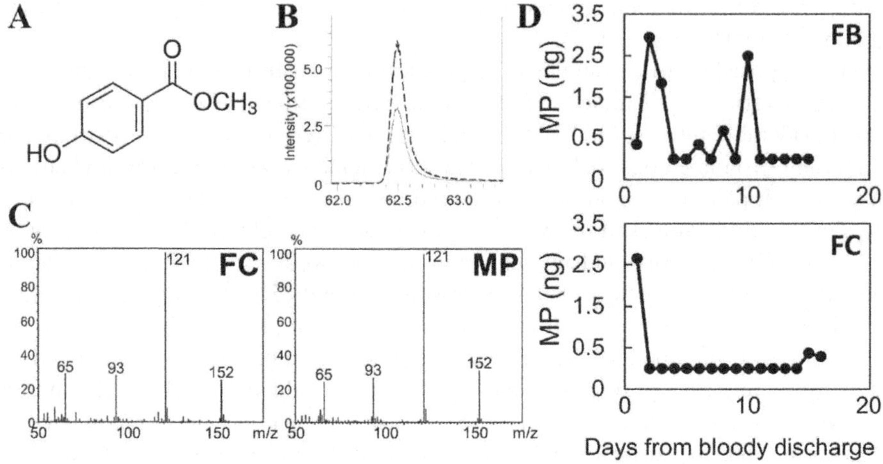


**Fig. 2** Methyl paraben content in the headspace of estrous and anestrous urine. (**a**) Chemical structure of methyl paraben (MP). (**b**) Comparison of GC/MS total ion chromatograms (*m/z* = X) between synthetic methyl paraben (dashed line) and a urine sample (gray line) obtained from an estrous female urine sample. X-axis and Y-axis indicate retention time (min) and intensity, respectively. (**c**) Comparison of the mass spectra of synthesized MP and the compound detected at 62.5 min in the headspace of urine. (**d**) Temporal changes in methyl paraben (MP) contents in the headspace of urine samples from females B (FB) and C (FC) in estrous. X-axis and Y-axis indicate days from bloody discharge and MP contents (ng), respectively

Some chemical studies of urinary volatile compounds have been run on female dogs. In thermal desorption (TD)-GC/MS analysis of the urinary headspace, which was concentrated in an adsorbent (Tenax), three major compounds, i.e., acetone, methyl propyl sulfide, and methyl butyl sulfide, and nine minor compounds (but not MP) were detected in female beagle urine during estrus (Schultz et al., 1985). MP was not detected in any samples by a solid phase microextraction GC/MS analysis of the urinary headspace of female dogs in various stages of the ovarian cycle (Dzieciol et al., 2014). These studies suggest that MP is not present in the urine of estrous females.

To further evaluate whether MP is excreted as a sex pheromone in the urine of females in estrus, we analyzed estrous and anestrous urine samples collected from female A, B, and C using TD-GC/MS (TD-20, GCMS Ultra; Shimadzu Co., Tokyo, Japan) equipped with a DB-WAX column (60 m × 0.25 mm × 0.25 μm; Agilent Technologies, Santa Clara, CA, USA). Once in estrus and anestrus (at least 2 months from estrus), the headspace of the urine samples (5 ml aliquots) was concentrated in Tenax-TA adsorbent (GL science Inc., Tokyo, Japan) by purging nitrogen gas (60–65 ml/min) for 2 h at 37 °C. Moreover, the headspace of the urine samples (5 ml aliquots), which was collected every day for 2 weeks from proestrus females when bloody discharge was observed, were concentrated in Tenax-TA adsorbent in the same manner to measure the MP content. MP was detected in urine from female A during estrus and anestrus, female B only during anestrus, and female C during estrus and anestrus (Fig. 2b, c). MP contents fluctuated between 0 and 3 ng in female B during estrus, whereas MP was not detectable (except on days 0, 14, and 15 at concentrations of 2.7, 0.3, and 0.2 ng, respectively) in female C (Fig. 2d).

We also quantified the urinary contents of MP in total lipid extracts of estrous and anestrous urine samples (40 ml aliquots), which were prepared using the Bligh and Dyer method (Bligh & Dyer, 1959). MP was detectable in samples equivalent to 200 μl of urine obtained from three females at some point in estrus and once in anestrus. The urinary MP concentrations were approximately 138.1 and 120.4 μg/mL in female B during estrus and anestrus, respectively.

Finally, we examined the behavioral response of male dogs toward 100 ng of synthetic MP and anestrous urine containing 2 mg/mL MP. Our analysis showed that the males sniffed both samples but did not lick or smack their lips (which was observed when they were presented with estrous urine as a positive control). After sniffing the urine sample with MP, males also did not exhibit the characteristic responses observed to estrous females, such as chasing them, roaring to them, and trying to mount them.

Our analyses indicated that MP is a minor component in the urine of female dogs and is not an artifact of highly popular industrial plasticizers, as suggested in a previous report (Woszczyło et al., 2020). However, MP was not specific to the urine of females in estrous, and its content varied between individuals and days during estrus and anestrus. MP did not induce the characteristic behavior responses observed in males when they encounter estrous females or their secretions. Furthermore, MP applied to the vulvas of anestrous or spayed female beagles, German Shepherd, and dalmatian induced sexual arousal in male beagles, beagle-spits crosses, and German shepherd (Goodwin et al., 1979), but this phenomenon was not verified in later studies using male beagles (Kruse & Howard, 1983), mongrel dogs (Santos et al., 2013), Bedlington Terrier, Yorkshire Terrier, Labrador, and mixed breed (Dzięcioł et al., 2017). Thus, MP alone had no bioactivity which induced the characteristic mating behaviors in male dogs of any breed. Based on these results, we concluded that MP is present in the female urine of dogs but does not act as a releaser pheromone as reported previously.

## 5 Other Putative Bioactive Compounds in Estrous and Anestrous Urine

A recent study reported that the urinary content in aromatic carbonyl compounds and methyl ketones increased, while sulfide compounds decreased in female beagles during estrous, suggesting the concurrent occurrence of attractants and repellents influencing the interactions between male and female dogs in estrous (Dzięcioł et al., 2018). Based on that study, we examined the temporal changes in the amounts of aromatic carbonyl compounds, methyl ketones, and sulfide compounds in the headspace above 5 mL of urine using the Tenax-TA adsorbent. Calibration curves for synthetic acetophenone, 2-pentanone, and 1-methylthiopropane were generated using 0.39, 1.56, 6.25, 25.00, 100.00, 200.00, and 400.00 ng/μL Tenax-TA. Calibration curves for synthetic benzaldehyde, 2-hexanone, 2-octanone, 1-methylthiobutane, 1-methylthiopentane, and dimethyl trisulfide were generated using 0.39, 1.56, 6.25, 25.00, and 100.00 ng/μL Tenax-TA. We obtained fresh urine samples from two sibling female dogs and confirmed the bioactivity of the estrous urine in male dogs every day. In our analysis, the amount of acetophenone, benzaldehyde, 2-pentanone, 2-hexanone, 2-octanone, 1-methylthiopropane, 1-methylthiobutane, 1-methylthiopentane, and dimethyl trisulfide varied among individuals, as well as daily during estrus (Fig. 3), which contradicts a previous report (Dzięcioł et al., 2018). Thus, further investigations with more urine samples of various breeds are necessary to fully understand whether and how these compounds vary in concentration over time.

## 6 Conclusion

Dogs have a strong olfactory sense, which is applied for scent communication between conspecific individuals, and for inspecting metabolic waste products from the body such as urine and feces. Most mammals smell urine spots containing a variety of compounds, some of which function as odor cues or signals providing information regarding the "emitter" (Hurst & Beynon, 2004; Apfelbach et al., 2005). However, these innate behaviors in dogs are often controlled by humans, because dogs are domesticated, and many aspects of their lives, including reproduction, are supervised by humans. Therefore, studies on the behavior of domestic dogs in packs, particularly their intraspecific communication, are rare (Cafazzo et al., 2012; Siniscalchi et al., 2018).

Our study on a single "pack" of dogs showed that young female dogs entered estrus even when the dominant female was in the pack. Male dogs tried to mount their female siblings and mother to mate, while all female dogs showed repelling behavior toward them. A characteristic sequence of responses, such as sniffing, licking, and lip-smacking, was seen when the male dogs encountered urine from a female in estrus. These typical behavioral sequences, especially lip-smacking, might thus be used as indicators in olfactory bioassay designed to assess the reactogenic potency of canine substrates, their extracts, or isolated components.

We demonstrated that the candidate canine pheromone MP was contained in estrous and anestrous urine but did not act as a pheromone in isolation for domestic dogs (based on reevaluation of its contents in estrous and anestrous urine and its bioactivity toward males). Furthermore, temporal changes in other putative scent signals of estrous females, such as aromatic carbonyl compounds, methyl ketones, and sulfide compounds, were not specific to females in estrus. These results indicate that scent signals and bioactive compounds, such as pheromones, must be identified from secretions and excretions by females in estrous to improve our understanding of scent communication between male and female dogs during reproduction. Thus, we are currently trying to identify scent signals and sex pheromones from urine and vaginal secretions from estrous females, utilizing lip-smacking of males as an indicator of special behavioral salience.

**Fig. 3** (continued) 1-methylthiopropane, 1-methylthiobutane, 1-methylthiopentane, and dimethyl trisulfide in the headspace of urine samples from estrous and anestrous females B (FB, black circles) and C (FC, gray squares). X-axis and Y-axis indicate days from bloody discharge (estrus, day1–16 for female B and C; anestrus, day 82–84 for female B; and day 92–94 for female C) and compound contents (ng), respectively

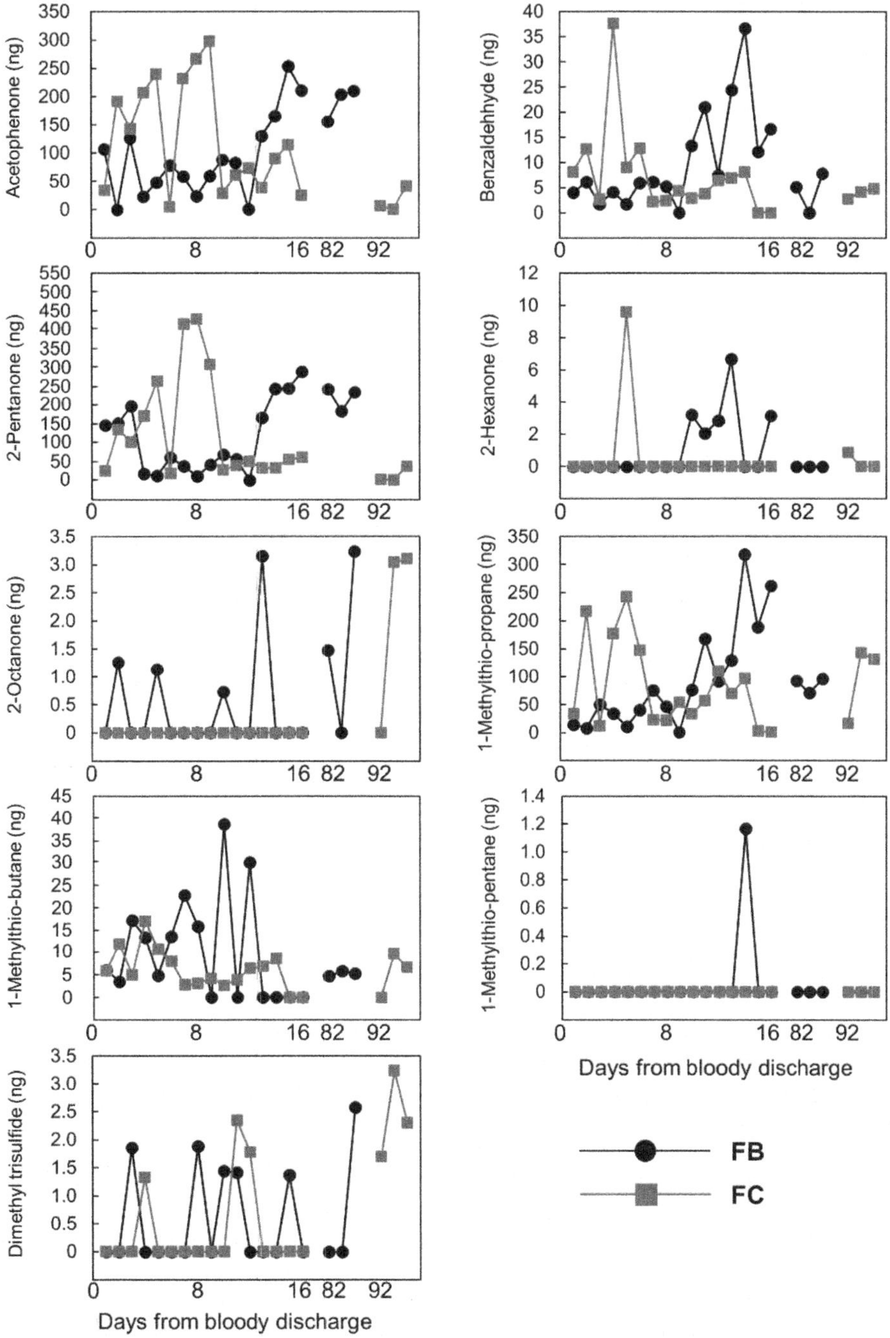


**Fig. 3** Temporal changes in the contents of aromatic carbonyl compounds, methyl ketones, and sulfide compounds in the headspace of estrous and anestrous female urine. Temporal changes in the contents of acetophenone, benzaldehyde, 2-pentanone, 2-hexanone, 2-octanone,

**Conflict of Interest** The authors declare that they have no conflict of interest.

**Ethical Approval** All applicable international, national, and/or institutional guidelines for the care and use of animals were followed. All procedures performed in studies involving animals were approved by the animal research committee of Iwate University (Approval numbers A202243).

## References

Apfelbach, R., Blanchard, C. D., Blanchard, R. J., Hayes, R. A., & McGregor, I. S. (2005). The effects of predator odors in mammalian prey species: a review of field and laboratory studies. *Neuroscience and Biobehavioral Reviews, 29*(8), 1123–1144.

Beach, F. A., & Gilmore, R. W. (1949). Response of male dogs to urine from females in heat. *Journal of Mammalogy, 30*(4), 391–392.

Beaver, B. V. (1999). *Canine behavior: a guide for veterinarians*. WB Saunders.

Bligh, E. G., & Dyer, W. J. (1959). A rapid method of total lipid extraction and purification. *Canadian Journal of Biochemistry and Physiology, 37*(8), 911–917.

Cafazzo, S., Natoli, E., & Valsecchi, P. (2012). Scent-marking behaviour in a pack of free ranging domestic dogs. *Ethology, 118*, 955–966.

Doty, R. L., & Dunbar, I. (1974). Attraction of beagles to conspecific urine, vaginal and anal sac secretion odors. *Physiology & Behavior, 12*(5), 825–833.

Dunbar, I. F. (1977). Olfactory preferences in dogs: the response of male and female beagles to conspecific odors. *Behavioral Biology, 20*(4), 471–481.

Dzieciol, M., Politowicz, J., Szumny, A., & Nizanski, W. (2014). Methyl paraben as a sex pheromone in canine urine–is the question still open? *Polish Journal of Veterinary Sciences, 17*(4), 601–605.

Dzięcioł, M., Szumny, A., Niżański, W., Wrzosek, M., Stańczyk, E., & Polikowska, J. (2016). Semiochemical communication in dogs in the context of reproductive behavior. In *Chemical signals in vertebrates 13* (pp. 381–389). Springer.

Dzięcioł, M., Niżański, W., Jezierski, T., Szumny, A., Godzińska, E. J., Ochota, M., Stańczyk, E., Najder-Kozdrowska, L., Woszczyło, M., & Pieczewska, B. (2017). The efficiency of synthetic sex pheromones in sexual arousal stimulation in domestic dogs. *Polish Journal of Veterinary Sciences, No 3*, 429–437.

Dzięcioł, M., Woszczylo, M., Szumny, A., Jezierski, T., Kupczyński, R., Godzińska, E. J., Pieczewska, B., & Niżański, W. (2018). Identification of putative volatile sex pheromones in female domestic dogs (*Canis familiaris*). *Animal Reproduction Science, 197*, 87–92.

Goodwin, M., Gooding, K., & Regnier, F. (1979). Sex pheromone in the dog. *Science, 203*(4380), 559–561.

Hurst, J. L., & Beynon, R. J. (2004). Scent wars: the chemobiology of competitive signalling in mice. *BioEssays, 26*(12), 1288–1298.

Kruse, S. M., & Howard, W. E. (1983). Canid sex attractant studies. *Journal of Chemical Ecology, 9*(11), 1503–1510.

Packard, J. M., & Mech, L. D. (1980). Population regulation in wolves. In M. N. Cohen, R. S. Malpass, & H. G. Klein (Eds.), *Biosocial mechanism of population regulation* (pp. 135–150). Yale University Press.

Santos, I. P., Ramos, C., Ramos, J. L. G., Oliveira, R. F., & Cunha, I. C. N. (2013). Efficient Association between PGF2α and Methyl 4-hydroxybenzoate sex pheromone prior to electroejaculation in Dogs. *Reproduction in Domestic Animals, 48*(1), 160–164.

Schultz, T. H., Kruse, S. M., & Flath, R. A. (1985). Some volatile constituents of female dog urine. *Journal of Chemical Ecology, 11*(2), 169–175.

Seal, U., Plotka, E., Packard, J., & Mech, L. (1979). Endocrine correlates of reproduction in the wolf. I. Serum progesterone, estradiol and LH during the estrous cycle. *Biology of Reproduction, 21*(5), 1057–1066.

Siniscalchi, M., d'Ingeo, S., Minunno, M., & Quaranta, A. (2018). Communication in Dogs. *Animals, 8*, 131.

Woszczyło, M., Jezierski, T., Szumny, A., Niżański, W., & Dzięcioł, M. (2020). The role of urine in semiochemical communication between females and males of domestic dog (Canis familiaris) during estrus. *Animals, 10*(11), 2112.

# Olfaction-Mediated Pathogen Avoidance in Mammals

Dagmar Schwambergová[1] (✉), Jitka Třebická Fialová[1], and Jan Havlíček[1]

[1] Department of Zoology, Faculty of Science, Charles University, Prague, Czech Republic
schwambd@natur.cuni.cz

**Abstract** Avoidance of pathogens is crucial for one's survival. The mechanisms responsible for pathogen avoidance are known as the 'behavioural immune system', which consists of various perceptual, affective, cognitive, and executive processes. In this chapter, we review current knowledge on pathogen avoidance in mammals mediated by odour cues. We start with an overview of changes in the body odour caused by pathogens. Then, we summarize current evidence regarding the avoidance of odours of infected individuals with emphasis on rodents and humans, and discuss the effects of multimodal integration on avoidance behaviour. We also review possible physiological mechanisms of pathogen avoidance, such as a decrease in pain sensitivity and hormonal involvement, including oxytocin, vasopressin, cortisol, and the sex hormones. Finally, we discuss ontogenetic development of pathogen avoidance. In conclusion, we outline some directions for future studies and point to the need to study a broader range of species.

**Keywords** Behavioural immune system · Disgust · Immunity · Infection · Mammals · Odour

## 1 Introduction

One of the key components to organisms' survival is pathogen defence. It is especially important in social species with high population densities, because such conditions sharply increase the risk of pathogen transmission (Altizer et al., 2003). The physiological immune system (PIS) and structural barriers, such as the skin and mucosae, are the body's main defence mechanism against pathogens. PIS plays a fundamental role in the recognition of structures of own versus foreign origin and in the activation of an adequate response. In vertebrates, PIS consists of innate and adaptive subsystems, which differ in their specificity as well as speed. Innate PIS is a nonspecific defence mechanism; as such, it is capable of providing an immediate response to pathogens. Adaptive PIS, on the other hand, is highly specific, but it may take days or even weeks to develop a response. Importantly, adaptive immunity develops 'memory', which can considerably accelerate the response after repeated exposure to a specific pathogen. It should be noted that in most infections, the innate and adaptive PIS operate in concert, often



B. Schaal et al. (eds.), *Chemical Signals in Vertebrates 15*,
https://doi.org/10.1007/978-3-031-35159-4_12

forming a complex interconnected network (Parkin & Cohen, 2001). But the system is energetically costly because recognition of a pathogen can trigger a cascade of metabolically demanding immune reactions (Sheldon & Verhulst, 1996). The energy used to generate an immune response might be instead used for other purposes, such as reproduction or securing of food sources. Moreover, a physiological immune response often has a negative impact on the organism's functioning. Many symptoms of infection, including fever or fatigue, are not caused directly by pathogens: they are part of the immune system's response to pathogens. PIS should therefore be activated only in the presence of a relevant and actual threat. Moreover, given that PIS is activated only after the organism is actually infected, what should primarily be favoured is avoidance of potential sources of pathogens. It has been proposed that this is the role of the behavioural immune system (BIS), which thus serves as the first line of defence (Schaller & Duncan, 2016).

The BIS contains various psychological mechanisms which help organisms avoid contagion from conspecifics, other animals, and inanimate objects (Schaller & Park, 2011). More specifically, the BIS consists of complex processes, including perception and various affective, cognitive, and executive components. The detection of potentially threatening stimuli is based on visual, olfactory, gustatory, acoustic, and tactile cues, which may activate a negative affective response (in humans described as 'disgust'). An interplay of these components may then lead to avoidant behaviour.

Operation of the BIS can be conceived of in terms of the error management theory, which predicts that a system should favour less energy-demanding errors (Johnson et al., 2013). In other words, the theory claims that there is a higher susceptibility to false-positive errors in order to reduce the probability of false negatives. The BIS thus operates analogously to a smoke detector, which is designed to be highly sensitive: it is triggered at the slightest hint of smoke, which can lead to frequent false positive alarms (Nesse, 2005; Schaller & Park, 2011). On the other hand, this sensitivity is balanced by a drastic reduction of false-negative errors that could lead to activation of the energy-demanding reactions of PIS or have other negative, even fatal, consequences. As a result, the BIS often evaluates some cues as potentially dangerous and triggers an avoidance response even where there is no real risk. Moreover, unlike other threats, pathogens usually cannot be directly detected by ones senses due to their size. Their activity can be discerned based on indirect cues, such as the smell of rotten food. This is another factor which contributes to the high alertness of the BIS.

The presence of a pathogen threat can alter the behaviour of individuals before they become infected. The BIS has been extensively studied in humans (Schaller & Park, 2011), but various studies conducted on nonhuman mammals, such as rodents and primates, indicate that various mechanisms belonging to this system are shared (Curtis et al., 2004; Kavaliers & Choleris, 2018). This may also apply to disgust, which used to be thought to be unique to humans (Rozin et al., 2015), but recent studies conducted on great apes suggest that disgust is a part of pathogen avoidance in other species as well. For instance, a large survey among fieldwork researchers who work with great apes investigated behaviours resembling human disgust and found that nonhuman apes avoid bad-tasting food and pungent odours, they wash or wipe preferred food in case it was previously contaminated, and they try to avoid the body excretions of conspecifics. However, the study did not find sufficient evidence of disgust-like behaviour towards sick conspecifics (Case et al., 2020). Moreover, it has been found that while food contaminated with faeces or dirt tends to be avoided by chimpanzees, some individuals will still eat it (Sarabian et al., 2018). Yet, it should be noted that in humans, culture and the cognitive component of the BIS play an important role, and these aspects are rarely studied in nonhuman mammals, although they cannot be excluded as non-existent.

Recent studies on pathogen avoidance are usually formulated within the conceptual framework of the BIS (Kavaliers & Choleris, 2018; Sarabian et al., 2017; Poirotte et al., 2019). This theoretical framework was originally formulated in human research – and applied in the context of mate choice – but the underlying mechanisms seem to have a broader applicability to other mammals and contexts. In the following, we thus do not limit ourselves to the context of mate choice: we assume that pathogen avoidance is an intrinsic response to an organism's ecology.

The main aim of this chapter is to review the role of chemical cues in pathogen avoidance in mammals. In the introduction, we focused on the proposed mechanism of such pathogen avoidance, namely, the BIS. Now we will turn to the possible causes of changes in body odour during infection and to the perception of chemical cues in both nonhuman and human mammals. Having this in place, we shall review the physiological mechanisms, hormonal influences, and ontogeny of avoidance behaviour.

## 2 Disease-Related Changes in the Body Odour

Many infectious diseases are accompanied by characteristic changes in body odour. There are several possible ways (and their combinations) in which this can happen. First of all, an ongoing infection can change the metabolic circuits as a direct effect of the pathogen itself, for example, a skin infection. These metabolic changes can be subsequently perceptible in the body excretions, such as urine, faeces, breath, or sweat from different areas of the body (Havlíček et al., 2017; Shirasu & Touhara, 2011). Secondly, an ongoing infection can change the composition of the microbiome, which can in turn alter the odour quality (Schellinck et al., 1991; Penn & Potts, 1998). Skin microbiome in particular changes throughout life and might be able to reflect rapid changes in metabolites (Li et al., 2019).

Thirdly, and perhaps most importantly, activation of an immune response can lead to changes in the body odour. Although the details of particular mechanisms remain largely unknown, we assume that immune response activation (leading, for instance, to an inflammation) affects the metabolic circuits. Volatile products of such metabolic changes can be perceptible in body excretions. Lipopolysaccharide (LPS), widely used in studies on disease-related changes in odour, may serve as an illustrative example of this phenomenon. LPS is a major component of Gram-negative bacteria's surface membrane, and it highly effectively triggers the innate immune response. It is the main component encountered by the immune system, and it plays a significant role in pathogenicity, i.e. an organism's ability to cause a disease (Bertani & Ruiz, 2018). Injection with an antigen, such as the LPS, can activate a cascade of metabolic immune reactions, and these metabolic changes may in turn have an impact on the substances which bodies excrete in urine, faeces, or sweat. In rodents, it has been shown that the production of darcin (a mice male sex pheromone) and other urinary proteins drop significantly after activation of the immune system (e.g. Litvinova et al., 2005; Lopes & König, 2016). Nevertheless, the particulars of the mechanisms which produce these immunity-related metabolites in excretions are still mostly unknown.

Finally, an immune response to infection can also lead to changes in the levels of steroid hormones, which can likewise affect body odour. Androgens, mainly testosterone, have an immunosuppressive effect (Folstad & Karter, 1992), which is why it might be adaptive to decrease their levels during infection (Wingfield et al., 2001; Trigunaite et al., 2015). The decrease can be adaptive because the energy thus spared can preferentially be used for immune responses, which increases the likelihood of success in fighting the pathogen (Wedekind & Folstad, 1994). Still, the details of association between testosterone and immune system remain unclear (for a review, see Roberts et al., 2004). It has been observed that in male mice whose immune response was activated by an injection of sheep red blood cells (SRBC), concentrations of major urinary proteins (MUPs) had significantly dropped (Litvinova et al., 2005), whereby it is relevant to note that MUPs bind to and stabilise a variety of volatile scent substances. In this manner, they are responsible for the release of pheromones in air from urine scent marks (Brennan & Kendrick, 2006). The drop in their concentration led to a decrease in scent attractiveness in both castrated and control males treated with SRBC. In contrast, in males treated with high levels of testosterone, no decline in scent attractiveness was observed (Litvinova et al., 2005). Similarly, LPS-treated male mice had significantly lower testosterone levels than controls and were less preferred by females. Moreover, after immunoactivation, male mice were less active, and their vocalisation patterns had changed (Lopes & König, 2016). LPS injections led to an increase in corticosterone and decrease in testosterone in male rats, but the avoidance response to the odour of LPS-injected males was not associated with changes in the above-mentioned hormones (Arakawa et al., 2010a).

Levels of stress hormones can also change during an ongoing infection. For example, the influenza virus in infected mice increases corticosterone levels in the plasma (Dunn et al., 1989). In mice, even viruses with a different host, such as the Newcastle disease virus, led to an increase in corticosterone levels (Smith et al., 1982). In humans, cortisol levels likewise increase during an infectious disease, as documented, for instance, by Reichenberg et al. (2001) in their experiments which used bacterial endotoxin.

Nevertheless, the results of studies which investigated the association between steroid hormones and body odour qualities in healthy humans are inconsistent. For instance, one study showed a positive association between cortisol levels and perceived odour attractiveness (Rantala et al., 2006), while another study reported a negative relationship (Butovskaya et al., 2013), and yet other research showed that cortisol does not predict body odour attractiveness in men at all (Thornhill et al., 2013). Similar discrepancies have been reported for the association between the quality of body odour and testosterone levels (Rantala et al., 2006; Thornhill et al., 2013). Clearly, further research is needed to disentangle these inconsistencies and enable some robust conclusions about infection-related changes in body odour mediated by steroid hormones.

Interestingly, not just the levels of steroid hormones but also scent-marking as such can provide (in mice) sufficient cues to current health status. Males infected with *Salmonella enterica* have reduced the area of marking, although the total number of scent marks did not differ from noninfected individuals (Zala et al., 2004). The infection can affect behaviour in a broader sense, by activating so-called sickness behaviour, which is triggered by pro-inflammatory cytokines produced by activated neutrophils and macrophages after contact with pathogens (Konsman et al., 2002). This set of adaptive behavioural changes consists of decreased food intake, social withdrawal, lethargy, and overall decrease in motor activity (Szentirmai & Krueger, 2014).

Odour changes have been considered merely a side effect of pathogen activity and host's response. But it might also be a direct adaptive activity of the pathogen (Poulin, 1995). According to the manipulation hypothesis, pathogens purposefully and specifically alter the behaviour or attractiveness of their hosts so as to increase the likelihood of their transmission to an uninfected host. For instance, it might be advantageous to attract an insect vector using an attractive chemical stimulus. In the case of malaria, the parasite (*Plasmodium falciparum*, which is transmitted by the *Anopheles* mosquito) increases attractiveness of the hosts' odour for the vector, apparently to facilitate parasite transmission to other hosts (Lacroix et al., 2005). To the best of our knowledge, molecular mechanisms are unknown, but *P. falciparum* produces a repertoire of plant-like volatile compounds (parasite-produced terpenes), which may be responsible for the higher attraction of vector mosquitoes to the hosts (Kelly et al., 2015). A change in odour can also be a warning that the host is already infected, and another parasite should search for another host, because otherwise they would compete for a single source (Prugnolle et al., 2009).

## 3 Olfactory Detection of Pathogen Threat

Detection of a potential source of threat is a crucial component of the BIS, without which a subsequent avoiding reaction cannot be executed. Assessment of pathogen threat from conspecifics or a potential source of contagion in the environment can rely on various visual, acoustic, olfactory, gustatory, and tactile cues, or their combinations.

Although for humans, the visual modality is undeniably important in assessing the current health status of other people, for various nonhuman mammals, it is olfaction that plays the key role. The benefits of chemical perception include its ability to detect at relatively long distances, which decreases the likelihood of pathogen transmission. Moreover, the relative stability of sources: some substances, depending on their chemical properties such as volatility and degradability, can remain perceptible even for several days. Olfaction is especially useful in dark and noisy environments, where visual and auditory perceptions are limited, while chemical substances can persist and remain detectable even in sensorily overloaded environments (Wyatt, 2003).

### 3.1 Olfactory Detection and Discrimination of Pathogen Threat in Nonhuman Mammals

Various nonhuman animals can detect infected conspecifics and tend to prefer the healthy over the infected ones. So far, most studies on this subject have been conducted on rodents. It should be further noted that a majority of studies on the discrimination between healthy and infected conspecifics were performed in the context of mate choice. Still, avoidance of sick conspecific is relevant in other social contexts as well, and it extends to same-sex individuals and juveniles. Another consequence of the above-mentioned focus on mate choice is that female rodents are disproportionally more often used as assessors because of their high level of mate selectivity.

The detection of and reactions to odour cues of infected rodents are usually tested by a preference test (e.g. total time spent next to the odour sample) or by avoidance behaviour. Preference and investigation tests are conducted mostly in Y-mazes or in special chambers where individual animals are released and their behaviour recorded. Animals can sniff odour samples through a wire mesh or through special inner walls between the prepared chambers, whereby samples take the form of soiled bedding or collected urine and associated secretions presented on a piece of filter paper. The target individuals are experimentally infected with a pathogen, while control individuals are treated with a saline or a water solution.

In a study addressing this issue, female mice preferred as their first choice, spent more time investigating, and made more contact with samples of urine and associated secretions of noninfected male mice ($n = 15$) compared to samples from males infected with *Eimeria vermiformis* ($n = 15$) (Kavaliers & Colwell, 1995a). *E. vermiformis* is an intracellular protozoan which infects the small intestine epithelium in murine species; their oocysts are shed in faeces and can infect other individuals during feeding and grooming (Figueiredo-Campos et al., 2018). In most animals, the infection is asymptomatic but, depending on the infection dose, can cause diarrhoea (Perrot-Minnot & Cézilly, 2010). Samples were collected 5 days after infection, while oocyst shedding starts on day 8, with maximum output around day 10 after infection (Kavaliers & Colwell, 1995a). The same effect was observed in another study which compared sham-infected males and males infected with *E. vermiformis*. Preferences of oestrous females ($n = 5$ per group) for a blank odour sample compared to an infected male sample were dependent on the stage of infection. Pre-infective males on day 4 of infection were preferred to blank odour samples; around the onset of infectibility (on day 7), there was no significant preference for either odour sample, and on day 10, when individuals were highly infective, preference for the infected males had significantly dropped (Kavaliers et al., 1997). A similar pattern was found even when the infection agent was a nonreplicating strain of *Salmonella enterica*, a gram-negative, intracellular bacteria which causes gastroenteritis, fever, and headaches, but can also lead to a serious infection of the central nervous system (Wickham et al., 2007). Females ($n = 28$) preferred the urine odour of sham-infected males ($n = 16$) to the urine odour of experimentally infected males ($n = 19$), although the initial choice or number of visits did not significantly differ (Zala et al., 2004). In another study, female rats spent more time with urine odour samples from sham-infected males ($n = 30$) compared to males infected with *Hymenolepis diminuta* ($n = 39$), but the first-visited arm of the maze was not different than expected by chance (Willis & Poulin, 2000). *H. diminuta* is a cestode living in the small intestine of rats; it reduces their digestive efficiency, and its intermediate host are coprophilic arthropods. From these results, one can conclude that naïve individuals (i.e. who did not come into earlier direct contact with an infected conspecific) display a relatively straightforward preference for healthy conspecifics.

Preference for healthy individuals can be modulated by various factors, including previous experience and the mating system (Kavaliers et al., 2003a; Klein et al., 1999). For example, preferences can be altered by a brief pre-exposure of females ($n = 45$; 15 per condition) to the urine odour of male mice infected with *Heligmosomoides polygyrus* (Kavaliers et al., 2003a), a naturally occurring intestinal helminth parasite of rodents with a direct life cycle and high infectability (Johnston et al., 2015). In mice, it leads to a chronic infection where multiple immunomodulatory mechanisms suppress the host's immune response (Maizels et al., 2012). The initial choice and overall preference were in favour of the urine odour of sham-infected males, but when females were pre-exposed to the odour of an infected male, they subsequently preferred that odour (Kavaliers et al., 2003a). Interestingly, when female mice are pre-exposed to cadaverine ($n = 20$), a substance that can serve as a cue to the presence of a decaying corpse, they subsequently decrease the number of contacts with males injected with LPS ($n = 17$), which activates the immune response of an organism (Renault et al., 2008). It seems therefore that environmental odour cues can significantly modulate social contacts and possible mate choice.

Furthermore, the urine odour of male mice infected with murine louse (*Polyplax serrata*) ($n = 5$) was less preferred than the urine odour of clean males ($n = 5$) on 2 consecutive days, regardless of familiarity with the odour of louse-infected males (Kavaliers, Fudge, Colwell, & Choleris, 2003c). *P. serrata* is a louse of mice and rats, and the relatively low level of infestation (Murray, 1990), which has been controlled for in the study, did not cause in the mice any symptoms, such as weight loss or poor grooming. The females ($n = 5$ per condition) preferred the odour of urine collected from new and clean males on the first day, and their preference did not change on the second day of the experiment, although the louse-infected males' odour was familiar. In short, when presented with two new clean males on the second day, they preferred the odour of the new individuals (Kavaliers, Fudge, Colwell, & Choleris, 2003c). It seems therefore that familiarity with urine odour may play a role in the preference, but this depends on the type of parasite and its infectability. Moreover, a longer pre-exposure (3 days prior to testing) to the urine odour of an unfamiliar or infected male with *H. polygyrus* enhanced in the females ($n = 10$ per condition) their subsequent discrimination against the odour of infected males. Female mice were pre-exposed for 15 minutes to the odour of urine collected from an infected familiar male, infected unfamiliar male, uninfected familiar male, and uninfected unfamiliar male, or to the odour of clean bedding. Females displayed a significant preference for the urine odour of uninfected males regardless of familiarity with the pre-exposed male, whereby the odours of an unfamiliar or infected male enhanced the above-mentioned discrimination against the odour of infected males (Kavaliers et al., 2014). In mice, familiarity via pre-exposure to an odour seems to be an important modulating factor of preference. When females are briefly pre-exposed to a male urine odour, they later tend to prefer the familiar odour regardless of the infection status of the male. On the other hand, a longer pre-exposure can enhance the discrimination against the odour of infected males.

Interestingly, the females of prairie voles (*Microtus ochrogaster*) ($n$ = 20) and meadow voles (*Microtus pennsylvanicus*) ($n$ = 20) differ in their odour preference when choosing between the urine odour of males either not infected ($n$ = 20 per species) or infected with *Trichinella spiralis* ($n$ = 20 per species) (Klein et al., 1999). *T. spiralis* is a nematode parasite which causes trichinosis in a wide range of vertebrate species, whereby the infection is often asymptomatic (Golab et al., 2009). Female meadow voles spent more time with the bedding from noninfected males, while female prairie voles spent a comparable amount of time with the bedding of both infected and noninfected males (Klein et al., 1999). It should be noted that meadow voles are polygynous, while prairie voles are a monogamous species. This suggests that species with polygynous and promiscuous mating systems (such as mice or rats), which are more susceptible to pathogen transmission, are more sensitive to various markers of the health status of potential partners. In another experiment, male mice ($n$ = 20) were less likely to mount and copulate with female mice infected with *T. spiralis* ($n$ = 20) than with uninfected females ($n$ = 10). On the other hand, when males experienced and unexperienced in terms of previous copulation were observed separately, the experienced males showed significantly more investigatory activity towards infected females, while responses of the unexperienced males were not affected by infection in the females (Edwards & Barnard, 1987). The preference for uninfected individuals is evident, but factors such as the previous experience are important as well. Experienced individuals seem to be more cautious about approaching an infected conspecific, but it is not known whether this caution is reflected in reproductive success.

It has been shown that the influenza virus decreases the attractiveness of male urine odour to female mice: females ($n$ = 80) spent more time investigating post-infection and pre-infection odours than odours collected from currently infected males ($n$ = 15) (Penn et al., 1998). On the other hand, male mice ($n$ = 30) infected with a subclinical dose of tick-borne encephalitis virus (viral RNA was detected after RT-PCR control) had higher levels of blood testosterone, and oestrus females ($n$ = 90) spent more time with their bedding than with the bedding of males from a control group ($n$ = 30) who received a saline vehicle (Moshkin et al., 2002).

In rodents, odour can be affected not only by parasites or viruses. Experiments that used a non-replicating bacterial strain (Zala et al., 2004) or foreign antigens (sheep red blood cells) (Moshkin et al., 2002) provide evidence that immune activation on its own can reduce the attractiveness of male urine odour. Similarly, activation of the immune system by an LPS treatment also alters the quality of odour. Odour samples from male rats treated with LPS ($n$ = 24) were less sniffed, and their bedding was more frequently buried and avoided by both female ($n$ = 9) and male rats ($n$ = 18) (Arakawa et al., 2009). In another study, male rats ($n$ = 10) also exhibited an aversive response to LPS-treated males ($n$ = 10); this was expressed by reduced sniffing, increased overall time spent on the other side of the cage, and by burying the bedding (Arakawa et al., 2010a). This pattern was not, however, observed in prepubertal males in whom the level of sniffing or avoidance did not differ between LPS-treated conspecifics ($n$ = 4) and the control group ($n$ = 4) (Arakawa et al., 2009). Furthermore, the aversive properties of urine odour collected from LPS-treated individuals ($n$ = 12) were blocked by treatment with interleukin-10 (IL-10), an anti-inflammatory cytokine central to the inflammatory processes involved in the production of illness-related urinary odour cues (Arakawa et al., 2010a). A similar effect has been demonstrated in mice treated with keyhole limpet haemocyanin (KLH), a T-cell-dependent antigen commonly used in mammals and birds. KLH-treated males ($n = 19$) were less preferred, which manifested itself in fewer sniffs and shorter sniff duration. Interestingly, males with a stronger immune response (antibody response) were preferred to males with a weaker reaction to KLH (Gerlinskaya et al., 2012).

As shown above, most studies on discrimination between infected and healthy individuals were conducted in laboratory conditions. Data based on wild vertebrate populations are rare but show similar patterns of preference in favour of less parasitised conspecifics. Temporarily isolated mandrills ($n = 25$) spent less time in the proximity (less than 1 metre) of a highly parasitised faecal sample than in the proximity of a less parasitised faecal sample (Poirotte et al., 2017). It seems therefore that mandrills can discriminate, based on faecal odours, between more and less parasitised social partners. Wild mandrills also avoided grooming conspecifics who had been infected with orofecally transmitted parasites, mainly *Balantidium coli*, which causes dysentery. Another evidence shows that the feeding behaviour of chimpanzees (*Pan troglodytes troglodytes*) ($n = 20$) is influenced by possible contaminants, mainly conspecific faeces. Animals were less likely to feed and tended to leave the area, when food was associated with the odour of faeces. On the other hand, their feeding behaviour did not differ when they were exposed to the odours of blood or semen (Sarabian et al., 2017), which need not be linked to pathogen avoidance but rather to antipredator behaviour or reaction to conspecific aggression. This possibility was explored in a study where female Holtzman rats ($n = 15$) escaped faster from the experimental area where they were exposed to the blood of rats or mice compared to a slower reaction to human blood or distilled water (Hornbuckle & Beall, 1974).

But not all studies show avoidance of infected individuals. In the highly social species of banded mongooses (*Mungos mungo*) ($n = 6$ troops, 5 to 65 individuals per troop), a study found no avoidance of individuals with clinical signs of disease (caused by *Mycobacterium mungi*) during focal observations, although the sick banded mongooses showed typical sickness behaviour, such as reduced activity, which may serve as a sufficient cue (Fairbanks et al., 2015). Moreover, mongooses have a very good sense of smell, which helps them identify individual conspecifics. Nevertheless, it should be noted that *M. mungi* is a novel pathogen and it is characterised by a unique transmission and exposure routes. It is usually transmitted via infected urine and anal gland secretions, which are used in olfactory communication of banded mongooses (Alexander et al., 2018).

It has been suggested that odour preferences are a reliable proxy to mate choice (Egid & Brown, 1989), but other evidence indicates that mate preference and actual reproductive success may differ. In mice, although the initial choice of females ($n = 24$) favoured sham-infected males, the numbers of offspring did not significantly differ between the sham-infected males ($n = 9$) and males infected with *S. enterica* ($n = 9$) (Zala et al., 2015). A similar research showed that female mice ($n = 5$ per arena) did not differentiate in their mate choice between sham-infected males ($n = 2$ per arena) and males infected with *H. polygyrus* ($n = 1$ per arena) in four arena settings (Ehman & Scott, 2004).

From these studies, we can conclude that nonhuman mammals, mainly rodents, are capable of distinguishing between odour of healthy and infected conspecifics in laboratory conditions. These effects are influenced by familiarity of the conspecifics and by novelty of the pathogen in the studied species. It also seems that actual partner choice does not always follow prior odour preferences, because the number of offspring did not differ between healthy and parasitised males. Further studies should test differences between reproductive success, avoidant reaction to the odour of infected individuals, and the effect of pathogenicity. The decision to copulate with a parasitised individual may be flexible and depend on the availability of alternative partners as well as hazardousness and infectability of a particular pathogen. For example, if the risk of contracting the infection is lower than the benefits of possible reproduction, it may be advantageous to copulate with such an individual regardless of its health status.

## 3.2 Olfactory Detection and Discrimination of Sick Conspecifics in Humans

Experimental studies which investigated human ability to discriminate between body odour samples from healthy and sick individuals show similar patterns as the previously presented studies conducted on nonhuman mammals. Moreover, in humans, we can investigate affective states, such as disgust. Similar studies in nonhuman animals are rare, although a display of stereotypical facial expressions associated in mice with toxin disgust has been recently documented (Dolensek et al., 2020).

Various diseases are accompanied by changes of body odour: for example, bacterial vaginosis is frequently associated with a fishy odour caused by the production of trimethylamine (Wolrath et al., 2005), while the diarrhoea caused by *Vibrio cholerae* has a specific sweet odour (Garner et al., 2009). Overall, it has been demonstrated that various infectious diseases do affect the odour emitted from skin, breath, urine, or faeces (see reviews by Shirasu & Touhara, 2011; Havlíček et al., 2017).

Behavioural tests assessing the attraction to/avoidance of various odours, commonly used in rodent research, are rarely used in human studies. In humans, researchers usually rely on hedonic ratings of odour samples and on self-reporting of affective states. Research has shown that axillary body odours collected from individuals infected with gonococcus (*Neisseria gonorrhoeae*) ($n = 13$) were rated by female raters ($n = 14$) as less pleasant and more putrid than samples collected from healthy ($n = 16$) or recovered donors ($n = 5$) (Moshkin et al., 2012). Interestingly, samples of recovered donors were more likely to be described as floral. Another study collected samples from donors ($n = 23$) who were currently sick with a naturally occurring acute respiratory infection (suffering from cough, sore throat, shortness of breath, coryza, fever, headache, malaise, or myalgia). The odours were rated by both sexes ($n = 46$) as nominally (but not formally significantly) less pleasant, less healthy, and more intense and disgusting than samples collected from the same donors when they were in full health (Sarolidou et al., 2020b). People are also capable of detecting the early olfactory cues of sickness. Within a few hours after activation of immune system by an LPS injection, body odour samples collected from donors ($n = 8$) were perceived by raters ($n = 40$) as less pleasant, less healthy, and more intense than samples from the same donors after a placebo injection (Olsson et al., 2014). Moreover, after LPS administration, the natural decrease of averseness of healthy urine was disrupted, and both pyrrole and acetophenone (distinctive urine volatiles used as biomarkers for various diseases) were more abundant in urine collected from LPS-treated individuals ($n$ = 20) than in the urine collected from placebo-treated controls ($n$ = 13). Non-disrupted natural rhythm, meanwhile, is characterised by a decrease in urine odour aversiveness through the day (Gordon et al., 2018).

In a recent study, we activated the immune system by vaccination against hepatitis A/B and *meningococcus* (combination of immunisation against a viral and bacterial disease) in men ($n$ = 21). Interestingly, we found that body odour samples collected 14 days after vaccination were rated by females ($n$ = 88) as more attractive, healthier, and less intense than samples collected before vaccination (Schwambergová et al., submitted). In contrast to previous studies based on the use of LPS, we did not perform the rating shortly after vaccination but 2 weeks after it. It is therefore possible that a drop in odour attractiveness is followed by its increase – and our study captured only the latter. The mechanisms of these alterations remain unknown, and future studies should investigate their dynamics in more detail.

Besides the direct effect of odour collected from sick individuals on the perceiver, chemical cues may also have modulatory effects. When participants were presented with a facial photograph simultaneously with body odour collected after LPS treatment, their ratings of liking the presented person had significantly decreased (Regenbogen et al., 2017).

Although there is only a handful of studies on the discrimination of health status based on odour cues in humans, we can tentatively conclude that they show similar patterns as in nonhumans, namely, preference for healthy individuals. This suggests that pathogen disgust could be the central motivational component of avoidance behaviour, which is what is proposed by the theoretical framework of the BIS (reviewed in Oaten et al., 2009; Kavaliers et al., 2020). Responses of the BIS include not only the affective experience of disgust but also the activation of aversive cognitions into working memory. Subsequently, the motivational system guides the behavioural response to minimise the risk of infection (e.g. leaving a crowded room with sneezing individuals) (Schaller & Park, 2011). Future studies should employ experimentally induced avoidance behaviour, for example, in a room with unpleasant smell, and measure disgust to uncover the possible mechanisms of the behavioural response.

### 3.3 Multisensory Integration

Most studies on pathogen avoidance focus on a single modality, but in everyday life, individuals are often exposed to stimuli that can be perceived by multiple senses. An interplay of cues from various modalities significantly increases the likelihood of detection and adequate evaluation of potential danger. Moreover, a response can be amplified (*multisensory enhancement*), meaning that a response to stimuli from multiple modalities exceeds the response to either of the stimuli presented alone (Stein & Stanford, 2008). This is attested, for instance, by responses to jointly presented odour of faeces and a depiction of faeces, or a decrease in reported liking of a person when the facial photograph and 'sick' body odour are presented together (Regenbogen et al., 2017). Moreover, the multisensory percept can be qualitatively different from the mere sum of the individual components and convey unique information. On the other hand, a decrease in response (*multisensory depression*) implies that response to cross-modal stimuli is weaker than the response to the most effective of its component stimuli (Stein & Stanford, 2008). For example, because visual cues create expectations of flavour, when the colour of salsa is experimentally adjusted to be less red, thus in our eyes less spicy, the spiciness rating after tasting the product is also depressed (Shermer & Levitan, 2014).

To the best of our knowledge, studies conducted on nonhuman animals have not yet investigated the multimodal perception of infection cues. What we do know, however, is that it is not only the urine odour of male mice that changes its quality after LPS treatment. Behavioural changes in injected individuals also play a significant role. For example, males ($n = 21$) were also less active, less motivated to engage with females, and they changed their vocalisation patterns after LPS injection (Lopes & König, 2016). This indicates that there was a variety of cues available, although the olfactory one might be most important. A previous human study asked participants ($n = 30$) to indicate their liking of a person whose picture was displayed while their body odour was simultaneously presented during an fMRI session (Regenbogen et al., 2017). Facial photographs after LPS treatment were rated as less likeable, less healthy, less attractive, and less desirable for social interaction, and the ratings were even lower if that person's odour after LPS treatment was presented at the same time. Sickness status also increased the neural activation of odour perception networks, specifically the superior temporal sulcus, which is a core area for multisensory integration, and the orbitofrontal cortex, which is part of the olfactory network. In another study, participants ($n = 77$) were exposed to facial pictures and body odours from the same individuals ($n = 22$) either before or after the LPS treatment. Both facial images and odour collected after LPS treatment were rated as less likable, but facial electromyography showed no activity of muscles that would indicate a disgust response (Sarolidou et al., 2020a). Future studies should, however, put more emphasis on the multisensory perception of cues related to the health status to test their interactions and relative importance.

## 4 Physiological Mechanisms of Odour-mediated Avoidance Behaviour

In the previous sections, we have reviewed preferences for and discrimination between odours from infected and healthy individuals. In the following, we focus on the physiological and/or affective reactions of odour perceivers. In animal research, studies of affective states are mostly limited to stress or fear reactions. Humans, on the other hand, can report their feelings (e.g. how attractive or disgusting they find a particular odour) and thus provide data that cannot be collected from nonhuman animals, where researchers must rely on indirect indicators. The most common method for measuring physiological reactions in rodents is the level of analgesia (reduced pain sensitivity). Pain sensitivity is usually assessed on a hot plate (50–60 °C), where vigorous foot flutter or licking is recorded. Subsequently, researchers observe thermal latencies. Analgesia is a defensive response to threatening stimuli: it helps the organism to prepare for a fight or flight reaction. It has been found that exposure to the odour of urine of parasitised males induced analgesia in females and their reaction differed depending on the length of exposure (Kavaliers et al., 2000). Prolonged exposure (around 15 minutes) induced an endogenous opioid-mediated analgesia, which is associated with stress responses, while a short exposure (under 1 minute) elicited a shorter non-opioid-mediated analgesia associated with anxiety-related anticipatory defence reactions.

Previous studies showed that female mice ($n = 6$ per case) exposed to the urine odour of males infected with *E. vermiformis* ($n = 15$) displayed higher thermal response latencies, which were measured as delayed vigorous foot flutter or lick of foot when put on an experimental hot plate. The latency lasted at least 15 minutes after a 30-minute exposure to the odour of parasitised males. Maximum analgesic responses were induced by the odour of pre-infective (5 days after infection) and infective males (10 days after infection) (Kavaliers & Colwell, 1993). The results were similar when female mice were exposed to the urine odour of males infected with *H. polygyrus* ($n = 30$) compared to sham-infected males. Once again, females ($n = 5$ per case) showed higher latencies, and a longer exposure to odour of infected males led to a greater analgesia (Kavaliers & Colwell, 1995b). On the other hand, when female mice ($n = 5$ per case) were infected with the same parasite as the males, their analgesic response to the odour of urine of infected males was reduced compared to exposure to physically stressed males, which normally leads to a lower analgesia than the urine odour of infected males (Kavaliers et al., 1998a). The lower analgesic response to physically stressed males suggests that an infected conspecific may represent a higher threat. Analgesic responses are also influenced by the level of sexual experience. Specifically, sexually experienced male mice ($n = 5$ per case) displayed a significantly greater level of analgesia compared to sexually naïve males ($n = 5$ per case) when exposed to the urine odour of females infected with *H. polygyrus* (Kavaliers et al., 1998b). Similarly, prior familiarity lowers analgesia in female mice ($n = 15$) in reaction to familiar louse-infected males but not to novel louse-infected males (Kavaliers et al., 2003c). These results indicate that rodents find the odour of parasitised conspecifics threatening and/or stressful, whereby lower pain sensitivity may facilitate a subsequent escape. Mice are also capable of maintaining their behavioural avoidance based on prior experience (familiarity) with infected individuals. This may reduce the cost associated with prolonged stress, which is immunosuppressive, as opposed to acute stress, which may temporarily boost immune response (Elenkov & Chrousos, 1999). Future studies should test physiological response during experiments in arena settings, where mice may be in direct contact. This approach may help to disentangle the importance of trade-offs between approaching (risking the contagion) and fleeing from a potentially harmful situation. Moreover, it would be beneficial to collect blood samples before and during the experiment to monitor the levels of cytokines and hormones.

### 4.1 Hormonal Influences on Pathogen Avoidance

The processing of social information responsible for subsequent pathogen avoidance depends on various neurobiological and endocrinological regulatory mechanisms (Choleris et al., 2009). Social recognition and discrimination are crucial for distinguishing individuals who represent a potential pathogen threat. For example, unfamiliar individuals may carry novel pathogens and thus represent a higher health risk. The most prominent hormones which play a role in social recognition and pathogen avoidance are oxytocin, oestrogens, arginine-vasopressin, and testosterone. Interplay between these hormones allows individuals to respond rapidly and appropriately to social information derived from potential pathogen threats (Kavaliers & Choleris, 2018).

Oxytocin is a nonapeptide hormone facilitating prosocial and sexual behaviours in mammals (Veening et al., 2015). Oxytocin and its receptors play a role in modifying cognitive and motivational processes in reaction to social environment. In this way, oxytocin mediates approach and avoidance of social information (Shamay-Tsoory & Abu-Akel, 2016). Mice in whom the gene responsible for oxytocin production was knocked out failed to recognise familiar conspecifics after a repeated exposure and that despite normal olfactory and learning abilities (Ferguson et al., 2000, 2001). It has also been demonstrated that mice which received antisense DNA targeted against oxytocin receptors were as impaired in social recognition as were mice with knocked-out gene for oxytocin (Choleris et al., 2003).

The importance of oxytocin in pathogen avoidance has been demonstrated in studies on mice with a knocked-out oxytocin gene. Female mice with this condition did not show preference for the urine odour of either healthy or louse-infected males. In contrast, wild-type females and females heterozygous in the oxytocin gene ($n$ = 20 per genotype) exhibited a clear initial preference for healthy males. Females whose oxytocin gene has been knocked out were also unable to distinguish between novel and familiar males regardless of their health status, and they displayed lower analgesic responses (Kavaliers et al., 2003b). It should be noted that females with a knocked-out oxytocin gene showed normal associative olfactory conditioning to predator odours and were capable of distinguishing between the urine odours of intact and castrated males or stressed vs. non-stressed males (Kavaliers et al., 2003b). These results suggest that oxytocin plays a specific role in pathogen avoidance, but not in general olfactory associative learning.

A similar pattern was found when females ($n$ = 20 per genotype) were choosing between the urine odour of healthy males and males infected with *H. polygyrus* (Kavaliers et al., 2003b). It can thus be concluded that at least one copy of the gene which encodes oxytocin is sufficient for discrimination between healthy and infected individuals and for a modulation of subsequent analgesic responses and avoidance behaviour. Deletion of the oxytocin gene leads only to slight analgesic responses. This was demonstrated by the fact that female mice with knocked-out oxytocin gene were unable to modulate their analgesic response: it remained on the same level, and these mice did not perform increased investigation of urine odours of novel mice. On the other hand, wild-type female and heterozygous female mice ($n$ = 20 per genotype) are able to modulate their analgesic responses based on prior familiarity: they display a reduced analgesic response to the odours of familiar males and an enhanced response to novel males (for review, see Kavaliers & Choleris, 2018).

In humans, researchers often use oxytocin nasal spray, which has an impact on social cognition and early detection of emotionally charged social cues (see Graustella & MacLeod, 2012). Oxytocin also enhances the aversive response to threat presented in the form of negative facial expressions, whereby the response is stronger in women (Luo et al., 2017). To the best of our knowledge, there is as yet no study that tested the effect of oxytocin on the ability to discriminate between odours based on the health status in humans.

The processing of social information is also dependent on oestrogens, partly due to their modulating effect on oxytocin. It has been shown that knocking out genes for ER-α and ER-β receptors impairs social cognition and behaviour. Social recognition in ER-α knocked-out mice was impaired to a comparable degree to mice with knocked-out genes for oxytocin. Oestrogens acting through the ER-β can directly regulate oxytocin gene expression, but social impairment in this condition is only partial (for a review, see Ervin et al., 2015). Choleris et al. (2003) proposed that oestrogens regulate oxytocin secretion in the paraventricular nucleus of the hypothalamus through the ER-β. Through the neurons of the paraventricular nucleus, oxytocin reaches the amygdala where oestrogens regulate the expression of oxytocin receptors through the ER-α. In fact, it has been shown that male mice knocked out in ER-β and ER-α displayed impaired recognition and aversive responses to the urine odour of males infected with *H. polygyrus* (Kavaliers et al., 2004).

Another important component of olfaction-mediated social recognition is nonapeptide arginine-vasopressin. Male rats ($n$ = 8 per group; 3 groups) exposed to the urine odour of a conspecific with immune system stimulated by LPS administration (Arakawa et al., 2010b) activate their arginine-vasopressin receptors, mainly in the medial amygdala. Arginine-vasopressin appears to be more essential for social recognition in males than in females, because its expression is usually higher in male brains. Oxytocin and arginine-vasopressin can interact with the same three types of receptors (OT, V1A, and V1B) (Choleris et al., 2009). It should be noted, though, that the main effect of abovementioned hormones is not in its separate contribution to social recognition but rather in the interplay between arginine-vasopressin, oxytocin, and the sex hormones (such as oestrogens or testosterone). Oestrogens facilitate social recognition through their receptors by regulating oxytocin, while arginine-vasopressin is regulated by testosterone, which is why the effects of androgens on social recognition could be mediated by arginine-vasopressin, but the exact mechanism is yet to be discovered (see Gabor et al., 2012).

The compensatory prophylaxis hypothesis (CPH) suggests that progesterone enhances pathogen avoidance, in particular disgust, in women (Fessler et al., 2005). Disgust sensitivity is expected to be high when progesterone levels are elevated, for instance, during pregnancy when the physiological immune system is downregulated to prevent it from rejecting the embryo. Moreover, during the mid-luteal phase of the menstrual cycle, when progesterone levels are high, disgust sensitivity is higher than during the menstrual phase ($n$ = 30) (Żelaźniewicz et al., 2016). The same pattern was observed for pathogen disgust in women in their luteal phase who lately had an infection ($n$ = 75) but not in healthy women ($n$ = 477) (Milkowska et al., 2019). In early pregnancy, the decreased levels of pro-inflammatory cytokines are compensated by an increase in contamination disgust (Kaňková et al., 2022). Female mice treated either with allopregnanolone, a metabolite of progesterone, or with a low or high dose of progesterone ($n$ = 10 per group), spent significantly less time near the odour of males ($n$ = 30) infected with *H. polygyrus* (Kavaliers et al., 2021; Bressan & Kramer, 2022).

We can conclude that avoidance behaviour is affected by a coordinated action of oxytocin, vasopressin, oestrogen, and testosterone. To fully capture the proximate mechanisms of pathogen avoidance, future studies should also consider progesterone and test its interactions with the above-mentioned hormones. The interaction between arginine-vasopressin and androgens likewise deserves further investigation, especially in association with avoidance of potential threat.

## 4.2 Association Between the BIS and the PIS Mechanisms

The BIS and the PIS do not operate independently: they tend to modulate each other. After exposure to a possible source of contamination, the activation of the BIS may lead to a preparation of the PIS (Schaller et al., 2010). Interactions between these two systems involve opioid compounds, which are responsible for greater analgesia after a longer exposure to odours of infected males but have also immunomodulatory properties. They may thus play a role in an adaptive preparation of the PIS in case of subsequent pathogen transmission (Stefano et al., 1996). Another study has shown that after injecting rats with LPS, their social partners ($n$ = 32) displayed modulated cytokines responses. In particular, they displayed increased levels of interleukin-1β (IL-1β), interleukin-6 (IL-6), and tumour necrosis factor-α (TNF-α) (Hamasato et al., 2017). Cytokine IL-1β is a key mediator of the inflammatory response produced by cells of the innate immune system, and a similar role is played by IL-6, which is likewise pro-inflammatory and thus active during inflammation, while contributing to the maturation of B cells. TNF is a pro-inflammatory cytokine with an important role in cell survival, proliferation, differentiation, and death (Gulati et al., 2016). All these cytokines display activity during the preparation phase of the PIS in organisms that come into proximity of a potential source of contamination, such as a sick conspecific or body secretions thereof.

Disgust, a widely studied aspect of human psychology, has an effect analogous to opioid analgesia in rodents. Persons ($n$ = 32) who were exposed to disgusting pictures depicting food or non-food stimuli had a higher core body temperature and manifested an oral inflammatory response, increased levels of pro-inflammatory cytokines (salivary TNF-a and albumin), and a downregulation of immunoglobulin A (IgA) (Stevenson et al., 2012). This suggests that a potential pathogen threat may lead to a specific immune response responsible for sufficient preparation of the immune system. On the other hand, the upregulated production of salivary IgA may be lost as saliva is flushed from the gaping mouth. It has been hypothesised that disgust evolved from distaste in terrestrial vertebrates and that it is responsible for spitting out potentially harmful and toxic substances (Rozin et al., 2008). Moreover, when participants ($n$ = 28) were exposed to photographs depicting symptoms of an infectious disease, they produced significantly more pro-inflammatory IL-6 than a group exposed to photographs depicting guns as a control threat (Schaller et al., 2010). Recently, Juran et al. (2023) showed that disgusting odours as compared to neutral odours elevate TNF-α levels.

These results indicate that the BIS may also prepare the organism for subsequent infection after exposure to potential contaminants. The above-mentioned opioid peptides have immunomodulatory properties and contribute to communication with the PIS; they thus play a role similar to IL-1β, IL-6, and TNF-α, which show elevation after contact of healthy rats with LPS-treated conspecifics or after exposure to images depicting potential harmful and disgusting stimuli in humans. Future studies should address this issue and test whether olfactory cues to diseases have an effect similar to that of visual stimuli depicting symptoms of infectious disease. Moreover, one could monitor the strength of reaction when participants are exposed to multimodal stimuli compared to solely visual or olfactory stimuli.

## 5 The Ontogeny of Odour-mediated Pathogen Avoidance

To the best of our knowledge, there is currently no direct evidence regarding the development of olfaction-related pathogen avoidance. It seems that pathogen avoidance depends on learning via interaction with conspecifics and the environment. In naïve mice, for instance, exposure to biting flies led to heightened analgesia and self-burying to avoid being bitten. One day later, the biting parts of flies were removed, but mice behaved just as they did the day before, although they could not be actually bitten. Moreover, naïve mice which witnessed other mice being attacked by biting flies exhibited analgesia and self-burying as well (Kavaliers et al., 2001). Early learning might work on the same principle. Still, the learning of pathogen avoidance might take place rather early and could be to some extent independent of any previous experience. For instance, it has been shown that even prepubertal, 21-day-old, rats naïve to the 'infection' odour were able to distinguish between LPS-treated and healthy adult rats. They showed a preference in favour of healthy individuals: the LPS-treated ones were sniffed for a shorter time (Arakawa et al., 2009). This initial preference for healthy individuals might be further modulated by familiarity or sexual experience as discussed above. Furthermore, a study on great apes and their willingness to eat contaminated food had shown that juvenile bonobos were less contamination-risk averse than the adults (Sparks et al., 2018).

Young children tend to tolerate the smell of decomposition or faeces, which can be a potential source of pathogen contagion. In the first 2 years of life, it is common for children to put in their mouths objects most adults consider disgusting, quite regardless of the nature of the object and the possibility of infection. Nevertheless, the caretakers regulate children's activities by guiding them in what they can put in their mouth and what they should avoid (Rozin et al., 1985). From 2 years on, the perception of pathogen disgust increases: in one study, for instance, images of maggot-like animals elicited disgust and avoidance in 2.5-year-old children (Stevenson et al., 2010). On the other hand, even very young infants show an aversive reaction to certain tastes and smells generally found disgusting by adults, such as butyric acid (Soussignan et al., 1997). These results support the idea that disgust evolved from distaste and that it helps in the identification of potentially harmful and toxic substances that can be subsequently spat out (Rozin et al., 2008).

The ontogeny of disgust and aversive behaviour deserves further intensive investigation. It would be important to obtain longitudinal data to capture the development of the above-mentioned responses. Future studies should test the differences in the olfactory avoidance behaviour of adults and prepubertal individuals. Human research should similarly focus on the reactions of infants and preschool children to odours which are considered disgusting by adults.

## 6 Conclusions

There is convincing evidence that various mammals are able to discriminate between healthy and infected conspecifics. On the other hand, the reproductive success of infected and healthy males is – at least in rodents – comparable, which suggests that preferences are also modulated by other factors, such as familiarity or dominance. Instead of simple avoidance, the BIS may rather prepare the individual for potential infection. The odour of an infected conspecific leads to heightened analgesia and increased levels of opioids and immune markers in the perceiver. Future studies should address the mechanisms of odour changes in the emitters, which are still not well understood. Chemical and microbial analyses would be particularly informative. Research should also investigate the interplay between hormones which affect social recognition and pathogen avoidance, as well as functional links between the PIS and the BIS. Similarly, multimodal perception and integration also deserve further investigation. Furthermore, without a better insight into the ontogeny of pathogen avoidance, our understanding of the underlying processes remains greatly limited. Finally, studies should include a variety of mammals to enable wider generalisations regarding pathogen avoidance. Preferentially, such studies should be conducted on wild populations because, as we have seen, odour preference in the laboratory may differ from actual mate choice. The ecology of the animals concerned, in particular their social structure, might also play an important role in pathogen avoidance. In the past two decades, we have clearly witnessed an enormous progress in our understanding of olfaction-mediated pathogen avoidance, but the list of unknowns remains long, and we impatiently await further discoveries in this fascinating field.

**Acknowledgements** We are grateful to Anna Pilátová, Ph.D., for her highly valuable comments and proofreading. JTF & JH are supported by Charles University Research Centre (UNCE) programme UNCE/ HUM/ 025(204056). All authors are supported by a grant of the Czech Science Foundation (GAČR 21-29772S).

## References


Alexander, K. A., Laver, P. N., Williams, M. C., Sanderson, C. E., Kanipe, C., & Palmer, M. V. (2018). Pathology of the emerging *Mycobacterium tuberculosis* complex pathogen, *Mycobacterium mungi*, in the banded mongoose (*Mungos mungo*). *Veterinary Pathology, 55*(2), 303–309. https://doi.org/10.1177/0300985817741730

Altizer, S., Nunn, C. L., Thrall, P. H., Gittleman, J. L., Antonovics, J., Cunningham, A. A., et al. (2003). Social organization and parasite risk in mammals: Integrating theory and empirical studies. *Annual Review of Ecology, Evolution, and Systematics, 34*(1), 517–547. https://doi.org/10.1146/annurev.ecolsys.34.030102.151725

Arakawa, H., Arakawa, K., & Deak, T. (2009). Acute illness induces the release of aversive odor cues from adult, but not prepubertal, male rats and suppresses social investigation by conspecifics. *Behavioral Neuroscience, 123*(5), 964–978. https://doi.org/10.1037/a0017114

Arakawa, H., Arakawa, K., & Deak, T. (2010a). Sickness-related odor communication signals as determinants of social behavior in rat: A role for inflammatory processes. *Hormones and Behavior, 57*(3), 330–341. https://doi.org/10.1016/j.yhbeh.2010.01.002

Arakawa, H., Arakawa, K., & Deak, T. (2010b). Oxytocin and vasopressin in the medial amygdala differentially modulate approach and avoidance behavior toward illness-related social odor. *Neuroscience, 171*(4), 1141–1151. https://doi.org/10.1016/j.neuroscience.2010.10.013

Bertani, B., & Ruiz, N. (2018). Function and biogenesis of lipopolysaccharides. *EcoSal Plus, 8*(1). https://doi.org/10.1128/ecosalplus.ESP-0001-2018

Brennan, P. A., & Kendrick, K. M. (2006). Mammalian social odours: Attraction and individual recognition. *Philosophical Transactions of the Royal Society, B: Biological Sciences, 361*(1476), 2061–2078. https://doi.org/10.1098/rstb.2006.1931

Bressan, P., & Kramer, P. (2022). Progesterone does raise disgust. *Hormones and Behavior, 137*, 104937. https://doi.org/10.1016/j.yhbeh.2021.104937

Butovskaya, M. L., Veselovskaya, E. V., Rostovtseva, V. V., Selverova, N. B., & Ermakova, I. V. (2013). Mechanisms of human reproductive behavior: Olfactory markers of male attractiveness. *Biology Bulletin Reviews, 3*, 196–208. https://doi.org/10.1134/S2079086413030031

Case, T. I., Stevenson, R. J., Byrne, R. W., & Hobaiter, C. (2020). The animal origins of disgust: Reports of basic disgust in nonhuman great apes. *Evolutionary Behavioral Sciences, 14*(3), 231–260. https://doi.org/10.1037/ebs0000175

Choleris, E., Clipperton-Allen, A. E., Phan, A., & Kavaliers, M. (2009). Neuroendocrinology of social information processing in rats and mice. *Frontiers in Neuroendocrinology, 30*(4), 442–459. https://doi.org/10.1016/j.yfrne.2009.05.003

Choleris, E., Gustafsson, J. Å., Korach, K. S., Muglia, L. J., Pfaff, D. W., & Ogawa, S. (2003). An estrogen-dependent four-gene micronet regulating social recognition: A study with oxytocin and estrogen receptor-α and-β knockout mice. *Proceedings of the National Academy of Sciences of the USA, 100*(10), 6192–6197. https://doi.org/10.1073/pnas.0631699100

Curtis, V., Aunger, R., & Rabie, T. (2004). Evidence that disgust evolved to protect from risk of disease. *Proceedings of the Royal Society of London. Series B: Biological Sciences, 271*, S131–S133. https://doi.org/10.1098/rsbl.2003.0144

Dolensek, N., Gehrlach, D. A., Klein, A. S., & Gogolla, N. (2020). Facial expressions of emotion states and their neuronal correlates in mice. *Science, 368*(6486), 89–94. https://doi.org/10.1126/science.aaz9468

Dunn, A. J., Powell, M. L., Meitin, C., & Small, P. A., Jr. (1989). Virus infection as a stressor: Influenza virus elevates plasma concentrations of corticosterone, and brain concentrations of MHPG and tryptophan. *Physiology & Behavior, 45*(3), 591–594. https://doi.org/10.1016/0031-9384(89)90078-4

Edwards, J. C., & Barnard, C. J. (1987). The effects of *Trichinella* infection on intersexual interactions between mice. *Animal Behaviour, 35*(2), 533–540. https://doi.org/10.1016/S0003-3472(87)80278-6

Egid, K., & Brown, J. L. (1989). The major histocompatibility complex and female mating preferences in mice. *Animal Behaviour, 38*(3), 548–550. https://doi.org/10.1016/S0003-3472(89)80051-X

Ehman, K. D., & Scott, M. E. (2004). Microsatellite analysis reveals that female mice are indiscriminate when choosing infected or dominant males in an arena setting. *Parasitology, 129*(6), 723–731. https://doi.org/10.1017/S0031182004006195

Elenkov, I. J., & Chrousos, G. P. (1999). Stress hormones, Th1/Th2 patterns, pro/anti-inflammatory cytokines and susceptibility to disease. *Trends in Endocrinology & Metabolism, 10*(9), 359–368. https://doi.org/10.1016/S1043-2760(99)00188-5

Ervin, K. S., Lymer, J. M., Matta, R., Clipperton-Allen, A. E., Kavaliers, M., & Choleris, E. (2015). Estrogen involvement in social behavior in rodents: Rapid and long-term actions. *Hormones and Behavior, 74*, 53–76. https://doi.org/10.1016/j.yhbeh.2015.05.023

Fairbanks, B. M., Hawley, D. M., & Alexander, K. A. (2015). No evidence for avoidance of visibly diseased conspecifics in the highly social banded mongoose (*Mungos mungo*). *Behavioral Ecology and Sociobiology, 69*, 371–381. https://doi.org/10.1007/s00265-014-1849-x

Figueiredo-Campos, P., Ferreira, C., Blankenhaus, B., & Veldhoen, M. (2018). *Eimeria vermiformis* infection model of murine small intestine. *Bio-Protocol, 8*(24), e3122–e3122. https://doi.org/10.21769/BioProtoc.3122

Ferguson, J. N., Aldag, J. M., Insel, T. R., & Young, L. J. (2001). Oxytocin in the medial amygdala is essential for social recognition in the mouse. *The Journal of Neuroscience, 21*(20), 8278–8285. https://doi.org/10.1523/JNEUROSCI.21-20-08278.2001

Ferguson, J. N., Young, L. J., Hearn, E. F., Matzuk, M. M., Insel, T. R., & Winslow, J. T. (2000). Social amnesia in mice lacking the oxytocin gene. *Nature Genetics, 25*, 284–288. https://doi.org/10.1038/77040

Fessler, D. M., Eng, S. J., & Navarrete, C. D. (2005). Elevated disgust sensitivity in the first trimester of pregnancy: Evidence supporting the compensatory prophylaxis hypothesis. *Evolution and Human Behavior, 26*(4), 344–351. https://doi.org/10.1016/j.evolhumbehav.2004.12.001

Folstad, I., & Karter, A. J. (1992). Parasites, bright males, and the immunocompetence handicap. *The American Naturalist, 139*(3), 603–622. https://doi.org/10.1086/285346

Gabor, C. S., Phan, A., Clipperton-Allen, A. E., Kavaliers, M., & Choleris, E. (2012). Interplay of oxytocin, vasopressin, and sex hormones in the regulation of social recognition. *Behavioral Neuroscience, 126*(1), 97–109. https://doi.org/10.1037/a0026464

Garner, C. E., Smith, S., Bardhan, P. K., Ratcliffe, N. M., & Probert, C. S. J. (2009). A pilot study of faecal volatile organic compounds in faeces from cholera patients in Bangladesh to determine their utility in disease diagnosis. *Transactions of the Royal Society of Tropical Medicine and Hygiene, 103*(11), 1171–1173. https://doi.org/10.1016/j.trstmh.2009.02.004

Gerlinskaya, L. A., Shnayder, E. P., Dotsenko, A. S., Maslennikova, S. O., Zavjalov, E. L., & Moshkin, M. P. (2012). Antigen-induced changes in odor attractiveness and reproductive output in male mice. *Brain, Behavior, and Immunity, 26*(3), 451–458. https://doi.org/10.1016/j.bbi.2011.11.010

Golab, E., Rozej, W., Wnukowska, N., Rabczenko, D., & Masny, A. (2009). Detection of *Trichinella spiralis* DNA in mouse faeces during the early stage of infection. *Journal of Microbiological Methods, 78*(2), 213–215. https://doi.org/10.1016/j.mimet.2009.05.019

Gordon, A. R., Kimball, B. A., Sorjonen, K., Karshikoff, B., Axelsson, J., Lekander, M., et al. (2018). Detection of inflammation via volatile cues in human urine. *Chemical Senses, 43*(9), 711–719. https://doi.org/10.1093/chemse/bjy059

Graustella, A. J., & MacLeod, C. (2012). A critical review of the influence of oxytocin nasal spray on social cognition in humans: Evidence and future directions. *Hormones and Behavior, 61*(3), 410–418. https://doi.org/10.1016/j.yhbeh.2012.01.002

Gulati, K., Guhathakurta, S., Joshi, J., Rai, N., & Ray, A. (2016). Cytokines and their role in health and disease: A brief overview. *MOJ Immunology, 4*(2), 00121. https://doi.org/10.15406/moji.2016.04.00121

Hamasato, E. K., Lovelock, D., Palermo-Neto, J., & Deak, T. (2017). Assessment of social behavior directed toward sick partners and its relation to central cytokine expression in rats. *Physiology & Behavior, 182*, 128–136. https://doi.org/10.1016/j.physbeh.2017.10.011

Havlíček, J., Fialová, J., & Roberts, S. C. (2017). Individual variation in body odor. In A. Buettner (Ed.), *Springer handbook of odor* (pp. 125–126). Springer. https://doi.org/10.1007/978-3-319-26932-0_50

Hornbuckle, P. A., & Beall, T. (1974). Escape reactions to the blood of selected mammals by rats. *Behavioral Biology, 12*(4), 573–576. https://doi.org/10.1016/S0091-6773(74)92531-0

Johnson, D. D., Blumstein, D. T., Fowler, J. H., & Haselton, M. G. (2013). The evolution of error: Error management, cognitive constraints, and adaptive decision-making biases. *Trends in Ecology & Evolution, 28*(8), 474–481. https://doi.org/10.1016/j.tree.2013.05.014

Johnston, C. J., Robertson, E., Harcus, Y., Grainger, J. R., Coakley, G., Smyth, D. J., et al. (2015). Cultivation of *Heligmosomoides polygyrus*: An immunomodulatory nematode parasite and its secreted products. *Journal of Visualized Experiments, 98*, e52412. https://doi.org/10.3791/52412

Juran, S. A., Tognetti, A., Lundström, J., Kumar, L., Stevenson, R. J., Lekander, M., & Olsson, M. J. (2023). Disgusting odors trigger the oral immune system. *Evolution, Medicine, and Public Health, 11*(1), 8–17. https://doi.org/10.1093/emph/eoac042

Kaňková, Š., Takács, L., Krulová, M., Hlaváčová, J., Nouzová, K., Hill, M., et al. (2022). Disgust sensitivity is negatively associated with immune system activity in early pregnancy: Direct support for the compensatory prophylaxis hypothesis. *Evolution and Human Behavior, 43*(3), 234–241. https://doi.org/10.1016/j.evolhumbehav.2022.02.001

Kavaliers, M., & Colwell, D. D. (1993). Aversive responses of female mice to the odors of parasitized males: Neuromodulatory mechanisms and implications for mate choice. *Ethology, 95*(3), 202–212. https://doi.org/10.1111/j.1439-0310.1993.tb00471.x

Kavaliers, M., & Colwell, D. D. (1995a). Discrimination by female mice between the odours of parasitized and non-parasitized males. *Proceedings of the Royal Society of London. Series B: Biological Sciences, 261*(1360), 31–35. https://doi.org/10.1098/rspb.1995.0113

Kavaliers, M., & Colwell, D. D. (1995b). Odours of parasitized males induce aversive responses in female mice. *Animal Behaviour, 50*(5), 1161–1169. https://doi.org/10.1016/0003-3472(95)80032-8

Kavaliers, M., Colwell, D. D., Ossenkopp, K. P., & Perrot-Sinal, T. S. (1997). Altered responses to female odors in parasitized male mice: Neuromodulatory mechanisms and relations to female choice. *Behavioral Ecology and Sociobiology, 40*, 373–384. https://doi.org/10.1007/s002650050353

Kavaliers, M., Colwell, D. D., & Choleris, E. (1998a). Parasitized female mice display reduced aversive responses to the odours of infected males. *Proceedings of the Royal Society of London. Series B: Biological Sciences, 265*(1401), 1111–1118. https://doi.org/10.1098/rspb.1998.0406

Kavaliers, M., Colwell, D. D., & Choleris, E. (1998b). Analgesic responses of male mice exposed to the odors of parasitized females: Effects of male sexual experience and infection status. *Behavioral Neuroscience, 112*(4), 1001–1011. https://doi.org/10.1037/0735-7044.112.4.1001

Kavaliers, M., Colwell, D. D., & Choleris, E. (2000). Parasites and behaviour: An ethnopharmacological perspective. *Parasitology Today, 16*(11), 464–468. https://doi.org/10.1016/S0169-4758(00)01786-5

Kavaliers, M., Choleris, E., & Colwell, D. D. (2001). Learning from others to cope with biting flies: Social learning of fear-induced conditioned analgesia and active avoidance. *Behavioral Neuroscience, 115*(3), 661–674. https://doi.org/10.1037/0735-7044.115.3.661

Kavaliers, M., Colwell, D. D., Braun, W. J., & Choleris, E. (2003a). Brief exposure to the odour of a parasitized male alters the subsequent mate odour responses of female mice. *Animal Behaviour, 65*(1), 59–68. https://doi.org/10.1006/anbe.2002.2043

Kavaliers, M., Colwell, D. D., Choleris, E., Ågmo, A., Muglia, L. J., Ogawa, S., & Pfaff, D. W. (2003b). Impaired discrimination of and aversion to parasitized male odors by female oxytocin knockout mice. *Genes, Brain, and Behavior, 2*(4), 220–230. https://doi.org/10.1034/j.1601-183X.2003.00021.x

Kavaliers, M., Fudge, M. A., Colwell, D. D., & Choleris, E. (2003c). Aversive and avoidance responses of female mice to the odors of males infected with an ectoparasite and the effects of prior familiarity. *Behavioral Ecology and Sociobiology, 54*, 423–430. https://doi.org/10.1007/s00265-003-0631-2

Kavaliers, M., Choleris, E., Ågmo, A., & Pfaff, D. W. (2004). Olfactory-mediated parasite recognition and avoidance: Linking genes to behavior. *Hormones and Behavior, 46*(3), 272–283. https://doi.org/10.1016/j.yhbeh.2004.03.005

Kavaliers, M., Colwell, D. D., Cloutier, C. J., Ossenkopp, K. P., & Choleris, E. (2014). Pathogen threat and unfamiliar males rapidly bias the social responses of female mice. *Animal Behaviour, 97*, 105–111. https://doi.org/10.1016/j.anbehav.2014.09.006

Kavaliers, M., & Choleris, E. (2018). The role of social cognition in parasite and pathogen avoidance. *Philosophical Transactions of the Royal Society, B: Biological Sciences, 373*(1751), 20170206. https://doi.org/10.1098/rstb.2017.0206

Kavaliers, M., Ossenkopp, K. P., & Choleris, E. (2020). Pathogens, odors, and disgust in rodents. *Neuroscience and Biobehavioral Reviews, 119*, 281–293. https://doi.org/10.1016/j.neubiorev.2020.09.037

Kavaliers, M., Bishnoi, I. R., Ossenkopp, K. P., & Choleris, E. (2021). Differential effects of progesterone on social recognition and the avoidance of pathogen threat by female mice. *Hormones and Behavior, 127*, 104873. https://doi.org/10.1016/j.yhbeh.2020.104873

Kelly, M., Su, C. Y., Schaber, C., Crowley, J. R., Hsu, F. F., Carlson, J. R., & Odom, A. R. (2015). Malaria parasites produce volatile mosquito attractants. *MBio, 6*(2), e00235–e00215. https://doi.org/10.1128/mBio.00235-15

Klein, S. L., Gamble, H., & Nelson, R. J. (1999). *Trichinella spiralis* infection in voles alters female odor preference but not partner preference. *Behavioral Ecology and Sociobiology, 45*, 323–329. https://doi.org/10.1007/s002650050567

Konsman, J. P., Parnet, P., & Dantzer, R. (2002). Cytokine-induced sickness behaviour: Mechanisms and implications. *Trends in Neurosciences, 25*(3), 154–159. https://doi.org/10.1016/S0166-2236(00)02088-9

Lacroix, R., Mukabana, W. R., Gouagna, L. C., & Koella, J. C. (2005). Malaria infection increases attractiveness of humans to mosquitoes. *PLoS Biology, 3*(9), e298. https://doi.org/10.1371/journal.pbio.0030298

Li, M., Budding, A. E., van der Lugt-Degen, M., Du-Thumm, L., Vandeven, M., & Fan, A. (2019). The influence of age, gender and race/ethnicity on the composition of the human axillary microbiome. *International Journal of Cosmetic Science, 41*(4), 371–377. https://doi.org/10.1111/ics.12549

Litvinova, E. A., Kudaeva, O. T., Mershieva, L. V., & Moshkin, M. P. (2005). High level of circulating testosterone abolishes decline in scent attractiveness in antigen-treated male mice. *Animal Behaviour, 69*(3), 511–517. https://doi.org/10.1016/j.anbehav.2004.05.014

Lopes, P. C., & König, B. (2016). Choosing a healthy mate: Sexually attractive traits as reliable indicators of current disease status in house mice. *Animal Behaviour, 111*, 119–126. https://doi.org/10.1016/j.anbehav.2015.10.011

Luo, L., Becker, B., Geng, Y., Zhao, Z., Gao, S., Zhao, W., et al. (2017). Sex-dependent neural effect of oxytocin during subliminal processing of negative emotion faces. *NeuroImage, 162*, 127–137. https://doi.org/10.1016/j.neuroimage.2017.08.079

Maizels, R. M., Hewitson, J. P., Murray, J., Harcus, Y. M., Dayer, B., Filbey, K. J., et al. (2012). Immune modulation and modulators in *Heligmosomoides polygyrus* infection. *Experimental Parasitology, 132*(1), 76–89. https://doi.org/10.1016/j.exppara.2011.08.011

Milkowska, K., Galbarczyk, A., & Jasienska, G. (2019). Disgust sensitivity in relation to menstrual cycle phase in women with and without an infection. *American Journal of Human Biology, 31*(3), e23233. https://doi.org/10.1002/ajhb.23233

Moshkin, M., Gerlinskaya, L., Morozova, O., Bakhvalova, V., & Evsikov, V. (2002). Behaviour, chemosignals and endocrine functions in male mice infected with tick-borne encephalitis virus. *Psychoneuroendocrinology, 27*(5), 603–608. https://doi.org/10.1016/S0306-4530(01)00096-8

Moshkin, M., Litvinova, N., Litvinova, E. A., Bedareva, A., Lutsyuk, A., & Gerlinskaya, L. (2012). Scent recognition of infected status in humans. *The Journal of Sexual Medicine, 9*(12), 3211–3218. https://doi.org/10.1111/j.1743-6109.2011.02562.x

Murray, M. D. (1990). Influence of host behaviour on some ectoparasites of birds and mammals. In C. F. Barnard (Ed.), *Parasitism and host behaviour* (pp. 290–315). Taylor & Francis Group.

Nesse, R. M. (2005). Natural selection and the regulation of defenses: A signal detection analysis of the smoke detector principle. *Evolution and Human Behavior, 26*(1), 88–105. https://doi.org/10.1016/j.evolhumbehav.2004.08.002

Oaten, M., Stevenson, R. J., & Case, T. I. (2009). Disgust as a disease-avoidance mechanism. *Psychological Bulletin, 135*(2), 303–321. https://doi.org/10.1037/a0014823

Olsson, M. J., Lundström, J. N., Kimball, B. A., Gordon, A. R., Karshikoff, B., Hosseini, N., et al. (2014). The scent of disease: Human body odor contains an early chemosensory cue of sickness. *Psychological Science, 25*(3), 817–823. https://doi.org/10.1177/0956797613515681

Parkin, J., & Cohen, B. (2001). An overview of the immune system. *Lancet, 357*(9270), 1777–1789. https://doi.org/10.1056/NEJM200007063430107

Penn, D., & Potts, W. K. (1998). Chemical signals and parasite-mediated sexual selection. *Trends in Ecology & Evolution, 13*(10), 391–396. https://doi.org/10.1016/S0169-5347(98)01473-6

Penn, D., Schneider, G., White, K., Slev, P., & Potts, W. (1998). Influenza infection neutralizes the attractiveness of male odour to female mice (*Mus musculus*). *Ethology, 104*(8), 685–694. https://doi.org/10.1111/j.1439-0310.1998.tb00102.x

Perrot-Minnot, M. J., & Cézilly, F. (2010). Parasite infection and host's behavior. In G. Koob, M. Le Moal, & R. Thompson (Eds.), *Encyclopedia of behavioral neuroscience* (pp. 7–13). Elsevier.

Poirotte, C., Massol, F., Herbert, A., Willaume, E., Bomo, P. M., Kappeler, P. M., & Charpentier, M. J. (2017). Mandrills use olfaction to socially avoid parasitized conspecifics. *Science Advances, 3*(4), e1601721. https://doi.org/10.1126/sciadv.1601721

Poirotte, C., Sarabian, C., Ngoubangoye, B., MacIntosh, A. J., & Charpentier, M. (2019). Faecal avoidance differs between the sexes but not with nematode infection risk in mandrills. *Animal Behaviour, 149*, 97–106. https://doi.org/10.1016/j.anbehav.2019.01.013

Poulin, R. (1995). "Adaptive" changes in the behaviour of parasitized animals: A critical review. *International Journal for Parasitology, 25*(12), 1371–1383. https://doi.org/10.1016/0020-7519(95)00100-X

Prugnolle, F., Lefèvre, T., Renaud, F., Møller, A. P., Missé, D., & Thomas, F. (2009). Infection and body odours: Evolutionary and medical perspectives. *Infection, Genetics and Evolution, 9*(5), 1006–1009. https://doi.org/10.1016/j.meegid.2009.04.018

Rantala, M. J., Eriksson, C. P., Vainikka, A., & Kortet, R. (2006). Male steroid hormones and female preference for male body odor. *Evolution and Human Behavior, 27*(4), 259–269. https://doi.org/10.1016/j.evolhumbehav.2005.11.002

Regenbogen, C., Axelsson, J., Lasselin, J., Porada, D. K., Sundelin, T., Peter, M. G., et al. (2017). Behavioral and neural correlates to multisensory detection of sick humans. *Proceedings of the National Academy of Sciences of the USA, 114*(24), 6400–6405. https://doi.org/10.1073/pnas.1617357114

Reichenberg, A., Yirmiya, R., Schuld, A., Kraus, T., Haack, M., Morag, A., et al. (2001). Cytokine-associated emotional and cognitive disturbances in humans. *Archives of General Psychiatry, 58*(5), 445–452. https://doi.org/10.1001/archpsyc.58.5.445

Renault, J., Gheusi, G., & Aubert, A. (2008). Changes in social exploration of a lipopolysaccharides-treated conspecific in mice: Role of environmental cues. *Brain, Behavior, and Immunity, 22*(8), 1201–1207. https://doi.org/10.1016/j.bbi.2008.05.008

Roberts, M. L., Buchanan, K. L., & Evans, M. R. (2004). Testing the immunocompetence handicap hypothesis: A review of the evidence. *Animal Behaviour, 68*(2), 227–239. https://doi.org/10.1016/j.anbehav.2004.05.001

Rozin, P., Fallon, A., & Augustoni-Ziskind, M. (1985). The child's conception of food: The development of contamination sensitivity to "disgusting" substances. *Developmental Psychology, 21*(6), 1075–1079. https://doi.org/10.1037/0012-1649.21.6.1075

Rozin, P., Haddad, B., Nemeroff, C., & Slovic, P. (2015). Psychological aspects of the rejection of recycled water: Contamination, purification and disgust. *Judgment and Decision Making, 10*(1), 50–63.

Rozin, P., Haidt, J., & McCauley, C. R. (2008). Disgust. In M. Lewis, J. M. Haviland-Jones, & L. F. Barrett (Eds.), *Handbook of emotions* (pp. 757–776). The Guilford Press.

Sarabian, C., Belais, R., & MacIntosh, A. J. (2018). Feeding decisions under contamination risk in bonobos. *Philosophical Transactions of the Royal Society, B: Biological Sciences, 373*, 20170195. https://doi.org/10.1098/rstb.2017.0195

Sarabian, C., Ngoubangoye, B., & MacIntosh, A. J. (2017). Avoidance of biological contaminants through sight, smell and touch in chimpanzees. *Royal Society Open Science, 4*(11), 170968. https://doi.org/10.1098/rsos.170968

Sarolidou, G., Axelsson, J., Kimball, B. A., Sundelin, T., Regenbogen, C., Lundström, J. N., et al. (2020a). People expressing olfactory and visual cues of disease are less liked. *Philosophical Transactions of the Royal Society B, 375*, 20190272. https://doi.org/10.1098/rstb.2019.0272

Sarolidou, G., Tognetti, A., Lasselin, J., Regenbogen, C., Lundström, J. N., Kimball, B. A., et al. (2020b). Olfactory communication of sickness cues in respiratory infection. *Frontiers in Psychology, 11*, 1004. https://doi.org/10.3389/fpsyg.2020.01004

Schaller, M., & Duncan, L. A. (2016). The behavioral immune system. In D. M. Buss (Ed.), *The handbook of evolutionary psychology* (pp. 206–224). Wiley.

Schaller, M., Miller, G. E., Gervais, W. M., Yager, S., & Chen, E. (2010). Mere visual perception of other people's disease symptoms facilitates a more aggressive immune response. *Psychological Science, 21*(5), 649–652. https://doi.org/10.1177/0956797610368064

Schaller, M., & Park, J. H. (2011). The behavioral immune system (and why it matters). *Current Directions in Psychological Science, 20*(2), 99–103. https://doi.org/10.1177/0963721411402596

Schellinck, H. M., Brown, R. E., & Slotnick, B. M. (1991). Training rats to discriminate between the odors of individual conspecifics. *Animal Learning & Behavior, 19*(3), 223–233. https://doi.org/10.3758/BF03197880

Schwambergová, D., Pátková, Ž., Třebická Fialová, J., Třebický, V., Stella, D., & Havlíček, J. (submitted). Immunoactivation affects perceived body odor and facial but not vocal attractiveness. *Evolutionary Psychology.*

Shamay-Tsoory, S. G., & Abu-Akel, A. (2016). The social salience hypothesis of oxytocin. *Biological Psychiatry, 79*(3), 194–202. https://doi.org/10.1016/j.biopsych.2015.07.020

Sheldon, B. C., & Verhulst, S. (1996). Ecological immunology: Costly parasite defences and trade-offs in evolutionary ecology. *Trends in Ecology & Evolution, 11*(8), 317–321. https://doi.org/10.1016/0169-5347(96)10039-2

Shermer, D. Z., & Levitan, C. A. (2014). Red hot: The crossmodal effect of color intensity on perceived piquancy. *Multisensory Research, 27*(3–4), 207–223. https://doi.org/10.1163/22134808-00002457

Shirasu, M., & Touhara, K. (2011). The scent of disease: Volatile organic compounds of the human body related to disease and disorder. *Journal of Biochemistry, 150*(3), 257–266. https://doi.org/10.1093/jb/mvr090

Smith, E. M., Meyer, W. J., & Blalock, J. E. (1982). Virus-induced corticosterone in hypophysectomized mice: A possible lymphoid adrenal axis. *Science, 218*(4579), 1311–1312. https://doi.org/10.1126/science.618374

Sparks, A. M., Fessler, D. M., Chan, K. Q., Ashokkumar, A., & Holbrook, C. (2018). Disgust as a mechanism for decision making under risk: Illuminating sex differences and individual risk-taking correlates of disgust propensity. *Emotion, 18*(7), 942–958. https://doi.org/10.1037/emo0000389

Soussignan, R., Schaal, B., Marlier, L., & Jiang, T. (1997). Facial and autonomic responses to biological and artificial olfactory stimuli in human neonates: Re-examining early hedonic discrimination of odors. *Physiology & Behavior, 62*(4), 745–758. https://doi.org/10.1016/S0031-9384(97)00187-X

Stefano, G. B., Scharrer, B., Smith, E. M., Hughes, T. K., Jr., Magazine, H. I., Bilfinger, T. V., et al. (1996). Opioid and opiate immunoregulatory processes. *Critical Reviews in Immunology, 16*(2), 109–144. https://doi.org/10.1615/CritRevImmunol.v16.i2.10

Stein, B. E., & Stanford, T. R. (2008). Multisensory integration: Current issues from the perspective of the single neuron. *Nature Reviews. Neuroscience, 9*, 255–266. https://doi.org/10.1038/nrn2331

Stevenson, R. J., Hodgson, D., Oaten, M. J., Moussavi, M., Langberg, R., Case, T. I., et al. (2012). Disgust elevates core body temperature and up-regulates certain oral immune markers. *Brain, Behavior, and Immunity, 26*(7), 1160–1168. https://doi.org/10.1016/j.bbi.2012.07.010

Stevenson, R. J., Oaten, M. J., Case, T. I., Repacholi, B. M., & Wagland, P. (2010). Children's response to adult disgust elicitors: Development and acquisition. *Developmental Psychology, 46*(1), 165–177. https://doi.org/10.1037/a0016692

Szentirmai, É., & Krueger, J. M. (2014). Sickness behaviour after lipopolysaccharide treatment in ghrelin deficient mice. *Brain, Behavior, and Immunity, 36*, 200–206. https://doi.org/10.1016/j.bbi.2013.11.017

Thornhill, R., Chapman, J. F., & Gangestad, S. W. (2013). Women's preferences for men's scents associated with testosterone and cortisol levels: Patterns across the ovulatory cycle. *Evolution and Human Behavior, 34*(3), 216–221. https://doi.org/10.1016/j.evolhumbehav.2013.01.003

Trigunaite, A., Dimo, J., & Jørgensen, T. N. (2015). Suppressive effects of androgens on the immune system. *Cellular Immunology, 294*(2), 87–94. https://doi.org/10.1016/j.cellimm.2015.02.004

Veening, J. G., De Jong, T. R., Waldinger, M. D., Korte, S. M., & Olivier, B. (2015). The role of oxytocin in male and female reproductive behavior. *European Journal of Pharmacology, 753*, 209–228. https://doi.org/10.1016/j.ejphar.2014.07.045

Wedekind, C., & Folstad, I. (1994). Adaptive or nonadaptive immunosuppression by sex hormones? *The American Naturalist, 143*(5), 936–938. https://doi.org/10.1086/285641

Wickham, M. E., Brown, N. F., Provias, J., Finlay, B. B., & Coombes, B. K. (2007). Oral infection of mice with *Salmonella enterica* serovar Typhimurium causes meningitis and infection of the brain. *BMC Infectious Diseases, 7*, 65. https://doi.org/10.1186/1471-2334-7-65

Willis, C., & Poulin, R. (2000). Preference of female rats for the odours of non-parasitised males: The smell of good genes? *Folia Parasitologica, 47*(1), 6–10. https://doi.org/10.14411/fp.2000.002

Wingfield, J. C., Lynn, S. E., & Soma, K. K. (2001). Avoiding the 'costs' of testosterone: Ecological bases of hormone-behavior interactions. *Brain, Behavior and Evolution, 57*(5), 239–251. https://doi.org/10.1159/000047243

Wolrath, H., Ståhlbom, B., Hallén, A., & Forsum, U. (2005). Trimethylamine and trimethylamine oxide levels in normal women and women with bacterial vaginosis reflect a local metabolism in vaginal secretion as compared to urine. *APMIS (Acta Pathologica, Microbiologica et Immunologica Scandinavica), 113*(7–8), 513–516. https://doi.org/10.1111/j.1600-0463.2005.apm_175.x

Wyatt, T. D. (2003). *Pheromones and animal behaviour*. Cambridge University Press. https://doi.org/10.1017/CBO9780511615061

Zala, S. M., Bilak, A., Perkins, M., Potts, W. K., & Penn, D. J. (2015). Female house mice initially shun infected males, but do not avoid mating with them. *Behavioral Ecology and Sociobiology, 69*, 715–722. https://doi.org/10.1007/s00265-015-1884-2

Zala, S. M., Potts, W. K., & Penn, D. J. (2004). Scent-marking displays provide honest signals of health and infection. *Behavioral Ecology, 15*(2), 338–344. https://doi.org/10.1093/beheco/arh022

Żelaźniewicz, A., Borkowska, B., Nowak, J., & Pawłowski, B. (2016). The progesterone level, leukocyte count and disgust sensitivity across the menstrual cycle. *Physiology & Behavior, 161*, 60–65. https://doi.org/10.1016/j.physbeh.2016.04.002

# Part III

# Human Chemocommunication

# Smelling the Basis of Social Connectedness: Chemosensory Communication in Humans

Bettina M. Pause[1] (✉)

[1] Department of Experimental Psychology, Heinrich-Heine-University, Düsseldorf, Germany
Bettina.Pause@uni-duesseldorf.de

**Abstract** Humans have highly developed social and olfactory competencies. It is the aim of this review to summarize the empirical evidence and theoretical considerations, indicating that both of these performances are intimately linked and possibly coevolved during mammalian evolution. In detail, the necessity of fine-tuned social stimulus processing and flexible social behavior for humans' self-definition, well-being, health, and survival rate is described. Further, the highly developed chemosensory competencies in humans are summarized. Specifically, I will focus on the feasibility of harm avoidance through chemosensory communication of stress, fear, anxiety, and aggression. In addition, possible specific peripheral and central processing systems are discussed, and the findings of specific emotional stimulus responses summarized. Further, individual differences in the perception of and response to social chemosensory signals are introduced. Regarding psychological disorders, recent studies point to important clinical implications due to deviant chemical communication. Finally, I outline that olfactory performances are linked to social network size and social skills. I conclude that the individual adjustment to social volatiles is a core feature of social intelligence.

**Keywords** Aggression · Anxiety · Chemosensory communication · Fear · Friendship · Olfaction · Social communication · Social skills · Stress

## 1 Introduction

Probably all humans, from birth to old age, and across genders and cultures, are seeking happiness and well-being on the one hand, and safety on the other hand. Both of which are directly linked to social connectedness, forming an inherent motive of humans. In order to establish and maintain social connections, individuals need to communicate their needs, emotions, and intentions. In addition to the aspiration to receive mental and physical support, and to avoid harm by being supported by others, maybe the most important need in human social encounters is receiving information about the communicator's degree of honesty. A valid acquisition of trustworthiness is



B. Schaal et al. (eds.), *Chemical Signals in Vertebrates 15*,
https://doi.org/10.1007/978-3-031-35159-4_13

the basis for differentiating cheaters from friends and thus the foundation for the maintenance of a successful social network. The need to differentiate dishonest from honest others is intimately related to the ability to differentiate fairness from unfairness. Recently, a large number of studies investigated the desire to reject unequal reward in several species, i.e., the rejection of a reward in case a conspecific receives a larger reward for the same action (Brosnan & De Waal, 2003). So far, inequity avoidance has been described in humans, monkeys, apes, caniforms, and rodents (Oberliessen & Kalenscher, 2019).

Just recently, it has become evident that among the human senses, chemosensory perception may play a fundamental role for the successful screening of the social environment. Social cues and signals based on volatile molecules (molecule mixtures) have several advantages. Like auditory or tactile stimuli, they can be used in darkness; however, their spreading across long distances (high volatile molecules) as well as their longevity (low volatile molecules) can be remarkably higher than in other modalities (Wyatt, 2003). Intriguingly, chemosensory communication may serve an optimum social communication by delivering honesty signals, directly affecting physiological responses in the receiver, bypassing cognitive control (Pause, 2012, 2017). Whereas humans can easily fake their intentions by adjusting their facial expression, body posture, gesture, intonation, and verbal information content, no human being seems to be able to adjust the expression of molecules deriving from body fluids and serving as social information. Even by adding perfume, the activity of body fluids producing glands cannot be changed. In accordance with empirical reports (Richard Ortegón et al., 2022), it is highly unlikely that adding artificial molecules changes the informational content of social molecules which developed over thousands or millions of years of evolutionary shaping. Social volatiles inform the receiver about numerous biological processes (Lübke & Pause, 2015) related to parent-child bonding (e.g., through genetic relatedness), mate selection (e.g., through immunogenetic relatedness, status of sexual hormones, nutritional and health status), potential environmental or social harm (e.g., through chemosensory signals of stress, fear, and anxiety), food quality (e.g., through chemosignaling of disgust), and probably also about social safety (e.g., through chemosignaling happiness and friendship matching). The adjustments to these chemical signals and cues in the receiver cover processes of the perceptional, muscular, and neural systems, affecting cognitive processes and overt behavior. In addition, according to numerous studies, social volatiles exert their effects on the perceiver, even though they are present at or below the perceptual threshold. It has therefore been concluded that chemosensory signal processing does not require attentional resources (for an overview, see Pause, 2017). Accordingly, social volatiles are often not perceived as odors, and receivers are hardly able to cognitively control or regulate their responses towards them. Social chemosensory information, whether or not being associated with an odor, reaches the amygdala directly via neurons of the olfactory bulb, without necessarily passing thalamic or neocortical brain structures (Patin & Pause, 2015). The amygdala in turn affects the activity of different systems (LeDoux, 2007), e.g., adjusting motor behavior (via the ventral striatum), hormone production (via the hypothalamus), and the activity of the autonomic nervous system (via the hypothalamus and the nervus vagus).

It is here proposed that successful social actions strongly benefit from relying on nasal chemoreception. Accordingly, it has been speculated that social group formation could be one reason for the keenness of the olfactory system in humans (Bastir et al., 2011). Humans outperform many other species in their capacity to detect and differentiate odors (Laska, 2017). Accordingly, signal sharpening within the olfactory bulb seems to be more effective in humans than in mice (Maresh et al., 2008; McGann, 2017). It has been suggested that the functional significance of social chemical communication is related to reproductive success and an increased survival rate due to successful harm avoidance (Lübke & Pause, 2015). The significance of chemosensory communication in reproductive success, e.g., through mate choice and partnership formation, has been highlighted elsewhere (Ferdenzi et al., 2020; Havlíček et al., 2020); therefore, in the following, I will focus on the role of chemosensory communication in harm avoidance. Afterward, theoretical considerations and empirical evidence will be presented, linking social and chemosensory performances. It will be discussed whether neocortical brain development in humans is due to the manifestation of broadened chemosensory capabilities.

## 2 Harm Avoidance Through Human Chemosensory Communication

For socially living mammals, dangerous situations arise from the attack of predator animals or out-group members, e.g., in case of limited resources. Further, in-group members can also pose a physical risk by trying for a higher social rank associated with more social power. Other health affecting threats are independent of the degree of social behavior in a certain species, e.g., the danger of environmental disasters like fires, earthquakes, or flooding.

Most of these threads can be avoided through successful social communication. Aggression by predators or out-group members can best be reduced through increased bonding effort among the in-group members. Hereby, attention to a conspecific's need increases, accompanied by increased prosocial behavior. In humans, chemosensory communication between mother and child supports bonding and health in both the offspring and the mother (Lübke & Pause, 2015). Olfactory kin recognition may support kin selection in altruistic helping behavior (Griffin & West, 2003). As the effects of olfactory kin recognition and parent infant communication have recently been reviewed (Schaal et al., 2020), I will here focus on the effects of chemosensory communication of emotions on avoiding threat of in-group members and environmental danger.

## 2.1 Stress, Fear, and Anxiety

Stress refers to an imbalance of physiological messenger systems (e.g., neurotransmitters, neuropeptides, hormones, cytokines, etc.). If chronically persistent, stress can result in the collapse of organ functions and, in the long run, can cause the organism's death (McEwen, 2007). Therefore, stress protective mechanisms are important prerequisites for ontogenetic survival. Stress experienced by a certain individual can be caused by environmental hazards, attacks of predator species, or threat posed by in-group or out-group members. In most of these cases, it is most adaptive for surrounding conspecifics to be immediately informed about the potential danger. Hereby, they can promptly adjust their body systems to the danger, preparing an optimum response behavior. Emotional contagion between members of a species is a successful way to protect surrounding others. The stress experienced by one individual will spread to conspecific individuals, and the entire group is able to initialize coping behavior. However, coping behavior could be associated with fight or flight responses, but in case of high familiarity with the signal sender also to an increase of alliance formation, strengthening of bonding and emotional concern with the signal sender (Bartal et al., 2011; Brandl et al., 2022).

### 2.1.1 Chemical Communication of Stress

Stress contagion is commonly based on chemical cues and signals. The chemosensory communication of stress seems to be ubiquitous among the animal kingdom, reported in invertebrates (Suh et al., 2004) and vertebrates, e.g., birds (Parejo et al., 2012), fish (Keller-Costa et al., 2015; Von Frisch, 1941), and mammals (Zalaquett & Thiessen, 1991), including primates (Klailova & Lee, 2014), and humans (Gomes & Semin, 2021). Furthermore, chemosensory stress communication is also existent in plants (Dudareva et al., 2006). Recent studies indicate that chemosensory communication of stress or fear is not limited to intraspecies communication but also occurs from humans to different animal species. In domestic animals like dogs (D'Aniello et al., 2018) and horses (Sabiniewicz et al., 2021; Semin et al., 2019), both having a long coevolutionary history with humans, the chemical transmission of emotionally significant signals may ease bonding between humans and the nonhuman animal species. However, human chemosensory stress signals might indicate potential harm for farm animals and rodents, and their ability to respond to these signals may have evolved in order to protect them from human-caused dangerous situations (Destrez et al., 2021).

Stress experience is probably related to a negative emotional state in most, maybe all animals, whether invertebrates or vertebrates (De Waal & Andrews, 2022). In mammals, negative emotional states related to stress or an imbalance of autonomic physiological systems can be defined as specific emotional states, each based on specific neuronal circuits (Panksepp, 2011). Only in case of the activation of several (negative) emotional systems due to strong or unknown stress situations, the term "stress" seems still most appropriate. However, following most stressors, depending on the context and coping strategies, mammals can respond with specific emotions, either aggression, fear, or anxiety. Aggression is a common emotion, often elicited by social frustration, but varying with the dominance status of the frustrated individual. Fear and anxiety on the other hand are evoked in response to individuals or environmental situations, whenever harm is expected and coping strategies are insufficient. The experience of fear and anxiety relies on different neuronal circuits, with the amygdala as the core regulating system for fear and the bed nucleus of the stria terminalis having a central role in regulating the physiological processes associated with anxiety (LeDoux & Pine, 2016). Fear and anxiety can further be separated by the specific type of emotion evoking stimuli (Walker & Davis, 1997). Fear is considered to be an unconditioned state, evoked by an unconditioned stimulus, which appears immediate or imminent. In contrast, anxiety is considered to be a conditioned state, evoked by a conditioned stimulus, or an unconditioned stimulus whose appearance is expected, but uncertain in space and time (Perusini & Fanselow, 2015).

In various animal species, chemosensory communication of negative emotional states has been linked almost exclusively to the experience of stress. Vertebrates perceiving chemical stress (alarm) signals of conspecifics show increased alertness and behavioral withdrawal from the odor source (Mackay-Sim & Laing, 1981; Zalaquett & Thiessen, 1991). Accordingly, in the context of chemosensory stress signals, autonomic motor systems related to behavioral withdrawal are activated, resulting in an increased acoustic startle response (Inagaki et al., 2008, 2014). Furthermore, stress signal perceivers' physiology resembles that of the stressed conspecific: Receivers show hyperthermia, an increased release of hormones of the hypothalamus-pituitary-adrenal (HPA) axis (especially corticotropin-releasing hormone), analgesia, and a changed immune response (Barcellos et al., 2011; Fanselow, 1985; Kikusui et al., 2001; Kiyokawa et al., 2005; Moynihan et al., 2000; Sanches et al., 2015; Sterley et al., 2018).

### 2.1.2 Peripheral Processing of Social Volatiles

The receptors transducing social volatiles, including volatiles associated with stress, fear, or anxiety, most probably belong to several different sensory systems (Pause, 2017). When social volatiles are perceived above the olfactory sensory threshold, olfactory sensory neurons will be involved in signal processing, and signal information will be transferred to the primary olfactory cortex (olfactory bulb, anterior olfactory nucleus, olfactory tubercle, piriform cortex, cortical nucleus of the amygdala). In the event that social volatiles include several acids (fatty acids, lactic acids, or uric acids), further stimulus processing will also occur through the trigeminal system. In addition to these two systems, social volatiles are probably recognized by receptors responding specifically to volatile molecules of social origin, such as trace amine-associated receptors (TAARs) and neurons within the Grueneberg ganglion (GG). TAARs evolved in early vertebrates and have originally been considered to be involved in trace amine transmitter processing within the mammalian limbic brain and to a lower extent in amine detection within several peripheral tissues (Borowsky et al., 2001). However, shortly after their discovery, it became evident that most TAAR alleles are expressed within the olfactory epithelium, representing a distinct class of chemosensory receptors, specified for detecting social chemical cues (Liberles & Buck, 2006). Meanwhile, chemosensory TAARs have been detected in several vertebrates including humans (Carnicelli et al., 2010). Neurons expressing TAARs project to discrete glomeruli within the dorsomedial olfactory bulb, and from there probably to the cortical amygdala, their activation being associated with prompting behavioral withdrawal or behavioral approach (Han et al., 2020; Liberles, 2015).

Another chemosensory subsystem, the GG, might be specialized for behaviorally relevant volatiles derived from predator species or from conspecific informing about potential harm (Fleischer, 2021). The GG has been detected in several mammalian orders (e.g., Rodentia, Insectivora, Edentata, and Carnivora; see Fleischer, 2021) and has been considered also to be existent in humans (Grüneberg, 1973). The GG represents neural cell clusters in the anterior nasal region without direct contact to the nasal epithelium. Receptors on GG neurons entail G protein-coupled receptors, including some TAARs, as well as transmembrane guanylyl cyclase (subtype G) receptors. The sensory neurons project to specific cell clusters within the olfactory bulb, called Grueneberg or necklace glomeruli (Koos & Fraser, 2005). Unlike olfactory glomeruli, they respond in a nonspecific way to GG-activating molecules (even though acting species-specific, most of them represent heteroaromatic or heterocyclic compounds, such as pyrazines, pyridines, thiazoles, and thiazolines). It is likely that the activation of almost all Grueneberg ganglions by almost all GG-activating molecules refers to the fact that the GG system is responsible for prompt behavioral adjustments in response to chemosensory alarm signals rather than for stimulus discrimination, which is a prominent role of the olfactory glomeruli (Brechbühl et al., 2008; Fleischer, 2021).

### 2.1.3 Brain Activations and Contagious Processes in Chemocommunication

Brain activations in humans through chemosensory volatiles signaling harm depend on the specific emotion of the signal sender. The experience of extreme stress, e.g., through skydiving or high rope climbing, covers high arousal emotional states ranging from extreme positive (happiness, pleasure) to extreme negative emotions (anxiety, fear; Hetland & Vittersø, 2012). Accordingly, chemosensory stress signals derived from individuals performing extreme sports have been shown to solely active the amygdala, a relatively ancient brain area responsible for behavioral adjustments in highly arousing situations (Mujica-Parodi et al., 2009). The experience of anxiety, without significant activation of other accompanying emotions, can be observed in students awaiting their final university examination. Anxiety sweat activates a brain cluster of empathy-related brain areas (insular cortex, precuneus, cingulate gyrus) as well as the fusiform gyrus, known to process social signals of anxiety (Prehn-Kristensen et al., 2009). In comparison to extreme sports, stress induced by the Trier-Social-Stress Test (TSST, consisting of a mock job interview and an arithmetic task) is associated with negative emotions only, especially anger and anxiety (Moons et al., 2010). However, in comparison to the university examination condition, two distinct emotions are evoked simultaneously, instead of one. Sweat volatiles obtained from individuals undergoing the TSST are processed within the amygdala and the anterior cingulate cortex (Maier et al., 2019). It can be concluded that nonspecific chemosignals of stress seem to activate the amygdala, while social brain areas seem to be involved in stimulus processing whenever specific emotions are perceived. In addition, the activation of social brain areas related to empathy (Prehn-Kristensen et al., 2009; Maier et al., 2019) indicates that, in analogy to other animal species, emotions are contagiously transferred via chemosignals.

This conclusion is supported by several other studies: Participants receiving social volatiles obtained from sweat donors watching fear-evoking movies respond themselves with a fearful facial expression (de Groot et al., 2012, 2014). Additionally, chemosensory anxiety signals obtained from participants awaiting their university examination amplify the startle reflex to sudden loud tones (Pause et al., 2009; Prehn et al., 2006), confirming analog effects in rodents (Inagaki et al., 2008; Inagaki et al., 2014). Additionally, contagious processes of emotional chemosignals can be observed on a behavioral level, however, varying with the emotion in question. The perception of sweat from highly stressed individuals performing a high-rope course increases risk-taking behavior, with individuals making most risky choices in a computerized decision-making task (Haegler et al., 2010). In contrast, the perception of anxiety-related sweat decreases risky decisions in a bargaining task (Meister & Pause, 2021). These divergent results are in accordance with the findings that stressed individuals show an increased risk behavior in the lab (Bendahan et al., 2017) as well as in everyday life (Algren et al., 2018). On the contrary, anxiety is associated with pessimistic risk appraisals, as evidenced by the avoidance of risky decisions (Maner & Schmidt, 2006). In sum, brain and behavioral studies refer to the conclusion that chemosignals of stress, fear, or anxiety activate contagious processes between sender and receiver.

#### 2.1.4 Individual Differences

Emotional contagion is considered to be an evolutionarily old empathic response to conspecifics and is a core feature of in groups living mammals (De Waal & Preston, 2017), e.g., enabling bonding between individuals. Understanding the mechanisms of chemical communication in humans could aid in explaining individual differences in social skills in healthy individuals as well as in patients with distinctive deviations in social behavior. As already pointed out, a strong advantage in investigating chemical communication instead of communication through other sensory channels is the near to zero involvement of cognitive control processes in the perception of social volatiles. Several studies report clear effects in the receivers, even though the social chemosignals related to stress, fear, and anxiety were presented at or below the perceptual threshold (de Groot et al., 2014, 2018; Lübke et al., 2017a; Mujica-Parodi et al., 2009; Pause et al., 2004, 2010; Prehn-Kristensen et al., 2009). Thus, it is possible to investigate emotional communication with negligible effects of cultural norms, which are usually a profound confound in studying individual differences in social behavior. Empirical data confirm that chemical fear transmission has comparable effects in western Caucasian and in eastern Asian individuals (de Groot et al., 2018).

Strong gender differences have been reported regarding the response to chemical anxiety and fear signals; however, responses to chemical stress signals seem to be similar in men and women. For example, functional brain imaging revealed no gender differences regarding the brain responses to sweat obtained from skydiving individuals (Mujica-Parodi et al., 2009). In contrast, as revealed by event-related potential data, brain responses to chemosignals of anxious sweat donors were detectable in women but not in men (Pause et al., 2010). Similarly, only women respond to chemosensory fear signals with facial mimicry (de Groot et al., 2014). These gender differences are also found with regard to behavioral responses. Male participants show reduced trust behavior toward a robot doll while being exposed to stress-related sweat from first-time skydivers (Endevelt-Shapira et al., 2018). However, only women respond to chemosensory anxiety signals, obtained during an anxiety-provoking task, with changes in risk or trust behavior in a bargaining task (Meister & Pause, 2021). In accordance with the notion that stress signals inform about potential danger, both sexes need to prepare for escape in order to survive. However, the social function of fear or anxiety signals sometimes is rather related to promoting social support and helping behavior, and women seem to respond to social signals of fear and anxiety with a stronger tendency toward affiliative behavior than men (Taylor et al. 2000).

The hormonal status may affect the processing of chemosensory anxiety signals in humans, depending on whether anxiety transmission could be deleterious instead of protective. Stress experience in pregnant women can adversely affect fetal development, including lasting effects on infant's health status, immune system, and neurocognitive and brain development (Ruiz & Avant, 2005). Accordingly, it has been discovered that pregnant women show an attenuated physiological response to psychological and physiological stressors (de Weerth & Buitelaar, 2005). The gradual decrease of stress sensitivity with advancing pregnancy is considered to be protective for the fetus' health. In accordance with these considerations, brain responses of pregnant women evoked by chemosensory anxiety signals, obtained from students awaiting the final university exam, strongly decline during the course of pregnancy. In fact, during the third trimester, such brain responses are scarcely detectable (Lübke et al., 2017a). Here, again, it has been demonstrated that investigating pre-attentive processing of social volatiles provides a very transparent insight into the biological foundations of social communication. However, it seems that hormonal changes during puberty may have less or no effects on the processing of anxiety-related chemosignals: In prepubertal girls, the activation of defensive motor reflexes in the context of anxiety-related chemosignals is comparable to that of adults (Lübke et al., 2017b). Accordingly, it is concluded that the protective effects of chemical emotion transmission are as important for children as for adults.

### 2.1.5 Clinical Implications

A large variety of mental and physical diseases are related to or based on dysfunctional social networks and social communication strategies (Eisenberger & Cole, 2012; Holt-Lunstad, 2021). Several studies show that individuals with mental disorders connected to deviant social behavior show profound changes in the chemosensory communication of negative emotional states, such as stress and anxiety. In individuals who score high in social interaction anxiety and who are prone to develop social phobia, early pre-attentive brain processing in response to chemosensory anxiety signals is faster and recruits more neuronal resources than in low-anxiety individuals (Pause et al., 2010). In addition, in the context of chemosensory anxiety signals, socially anxious individuals respond with a stronger increase of the acoustic startle reflex amplitude than non-anxious individuals (Pause et al., 2009). Finally, socially anxious participants show a more pronounced attention allocation (late brain potentials), as well as larger startle responses, toward fearful faces in the context of chemosensory anxiety signals than participants scoring low in social anxiety (Adolph et al., 2013). These findings are in line with studies showing a processing bias toward threat-related social information in social anxiety or social phobia (Hirsch & Clark, 2004). Furthermore, these data show that the effects of social anxiety on social perception can be detected even though the social volatiles are presented below the perceptual threshold (Pause et al., 2009, 2010).

Both, individuals with childhood maltreatment and patients with panic disorder have been described to show an attentional bias toward (social) threat-related information.

Brain imaging analyses revealed that this phenomenon also occurs in response to social volatiles obtained from stressed individuals: Patients with panic disorder show an increased activation in the gyrus frontalis inferior while smelling their own stress sweat (Wintermann et al., 2013). Individuals with childhood maltreatment show amygdala hyperreactivity, and increased functional connectivity between the amygdala and the hippocampus, medial orbitofrontal cortex, and the anterior cingulate cortex, while perceiving chemosensory stress signals (Maier et al., 2020). The results of both of these studies are discussed to confirm the perceptual deviations in social threat perception in panic disorder and childhood maltreatment.

Another psychological disorder strongly affecting social behavior and social attachment is autism. In individuals with autism, the subliminal presentation of stress-derived social volatiles does not reduce the degree of trust toward a robot doll, as shown in healthy controls (Endevelt-Shapira et al., 2018).

In sum, individuals with mental disorders associated with reduced social satisfaction seem also to show a deviant perception of, and response to, social volatiles signaling harm. As chemosensation is among the phylogenetically oldest senses and can occur without attentional mediation, future studies may even consider whether mental disorders associated with reduced social skills are functionally related to reduced chemosensory skills.

## 2.2 Aggression

Intraspecies signals of aggression may have evolved in the context of defending or obtaining resources and are considered to increase fitness by inducing a defense response in the receiver in order to avoid harmful fighting (Laidre & Johnstone, 2013; Lopez & Martin, 2002; Nelson & Trainor, 2007). In most animal species, scent marks inform conspecifics about the competitive ability or dominance of the signal sender (Rich & Hurst, 1998; Stowers et al., 2013). Whether the signal receiver reacts with aggression or anxiety depends on its own social status and experience and the context of odor exposure (Hurst, 1993; Mucignat-Caretta et al., 2004).

So far, few studies investigated chemical communication of aggression in humans. In one study, sweat was sampled during and after a badminton match from male winners of an important tournament, with the match outcome being crucial for the club's table position (Adolph et al., 2010). The donors' trait aggression and anxiety were inconspicuous, and their depression scores low. During the match, they felt happy, accompanied by high testosterone levels. In male and female receivers, skin conductance response increased during the application of the winners' sweat. This effect was more pronounced in receivers with higher trait social anxiety. It was hypothesized that contest winning, being associated with high testosterone levels, represents feelings of aggression and dominance. Accordingly, it was concluded that chemosensory signals of aggression/dominance, similar to angry facial expressions (Öhman et al., 2001), activate the sympathetic nervous system in order to induce arousal and orientation toward biologically relevant stimuli. Being more prominent in socially anxious participants, the effect might represent a generalized hyperreactivity of socially anxious individuals toward social threat.

Two other studies obtained aggression-related sweat from males hitting a punching bag after being frustrated through negative bogus feedback. The sweat donors reported an increase of anger and an increased readiness to aggress but also reported to feel outraged, irritable, vexed, disappointed, and annoyed. Thus, in contrast to the badminton study, here, aggression was paired with frustration instead of dominance. The perception of the aggression-related sweat was associated with increased reaction times to anxiety-related words in an emotional Stroop task, discussed as an anxiety-related stress response (Mutic et al., 2016). Brain activations in response to aggression-related sweat obtained from individuals, who were being frustrated, are found in thalamic, hypothalamic, and other limbic brain structures, indicating preferential attentional processing of aggression-related volatiles (Mutic et al., 2017).

Another study investigated brain responses to aggression-related sweat by means of chemosensory event-related potentials (CSERPs, Pause et al., 2020). Here, sweat was obtained from male and female donors being exposed to the Point Subtraction Aggression Paradigm (PSAP). During the PSAP, the participants' task is to collect as many points as possible via button presses, while a fictitious opponent simultaneously steals these points. Sweat donors here could choose between three behavioral strategies, one of which related to overt aggressive behavior against their opponent. In fact, almost all donors showed overt aggressive behavior, accompanied by an increase of anger and of salivary testosterone levels. CSERPs in response to male aggression sweat were larger than to female aggression sweat in male and female participants. Furthermore, the strongest brain responses were detected in women responding to male aggression sweat. In conclusion, first, male aggression sweat is preferentially processed by both sexes in comparison to female aggression sweat. Compared to female aggression, male aggression is most often expressed as physical aggression (Archer, 2004) and thus can threaten physical health or even life. Therefore, for both sexes, successful survival depends on a preferential processing of such signals in order to initiate prompt reactions. Second, apart from this main effect, women's responses to male chemosensory signals of aggression were stronger than men's, with differentially activated (medio-frontocentral) and deactivated (fronto-lateral) clusters of neuronal networks (as indicated by current source density maps). The proposedly strongest activations were evident within the dorsomedial prefrontal cortex (DMPFC), which is considered to be involved in the translation of negative social experiences into threat-related physiological responding (Eisenberger & Cole, 2012). It is suggested that male aggression could potentially be more harmful for women than for men, as women commonly are physically inferior to men. Accordingly, worldwide, about one-third of ever partnered women have experienced physical and/or sexual intimate partner violence (Devries et al., 2013).

In sum, these studies point to a communicative function of chemosensory aggression signals, affecting the activity of the central and autonomic nervous system, as well as decisions and behavior. The studies reveal that aggression chemosignals are arousing, prompting the organism for fast but cautious responses. Subtle, peri-liminal signals of male aggression seem to be important to process for both sexes, but they appear most important for women.

## 3 The Smell of Connectedness

The previous part was devoted to our now profound knowledge about the chemical transmission of stress, fear, and anxiety and the informative nature of chemosensory aggression signals. In addition, recent studies have been conducted, indicating that disgust (Zheng et al., 2018), sadness (Gelstein et al., 2011), and happiness (de Groot et al., 2015) might also be transmitted through chemosensory communication. Thus, the chemical senses inform us about several emotional states, important for harm detection (stress, fear, anxiety, aggression, and disgust) and for strengthening social bonds (sadness, happiness). Cues of happiness inform group members about social safety and cues of sadness about the need for social support. In fact, central functions of human friendship formation are access to social support and social safety (Dunbar, 2018). The following section will consider whether there is a direct connection between friendship formation and chemoreception.

### 3.1 The Function of Friendship

Through social ties, humans develop and maintain their self-esteem, their self-concept, and their personality. The reflection of oneself in others' opinions helps to define the core and the limits of personal interests, hobbies, worldview, political and religious convictions, in sum, the way how humans define their self-concept. Positive feedback from others, received by a smile, a hug, or verbally, is the fundament of our self-esteem. Not surprisingly, social connectedness is a critical factor explaining the degree of happiness in humans (Helliwell & Aknin, 2018a, b; Steptoe, 2019). Further, social connectedness seems to be a very prominent predictor of health and survival in humans (Eisenberger & Cole, 2012) and nonhuman mammals, like hyraxes, ungulates, whales, dolphins, and primates (Snyder-Mackler et al., 2020). As such, it seems that the survival advantage of social affiliation in mammals is evolutionarily framed since about tens of millions of years. In humans, social isolation relates to up to a 50% increase in mortality risk (Holt-Lunstad et al., 2015) due to stress-related diseases such as cardiovascular diseases (Leigh-Hunt et al., 2017), and increased inflammation rates as well as immune dysfunctions (Hawkley & Cacioppo, 2010). These consequences are probably mediated by chronically elevated cortisol and epinephrine levels, and an inappropriate gene expression in leucocytes (Cole et al., 2011, 2015). Friendships, defined as engagement in bidirectional affiliative interactions with a high frequency and consistency, can be observed in various taxa (birds, ungulates, cetaceans, and primates; Brent et al., 2014). However, besides the desire to affiliate, especially in humans, another second component is necessary for friendship formation. This second component comprises cognitive processes, like mentalizing or mind reading, and the assessment of trustworthiness, the rapid and intuitive computation of friendship reliability (Dunbar, 2018). Accordingly, in humans, friendships can be long-lasting like in other animals, but their social networks are by far more dynamic, and friendships can decay due to anticipated or real deceptive behavior or selfish lies. Throughout human life, the loss of friendships is usually compensated by new friendship formations (Dunbar, 2020).

According to the social brain hypothesis, predation risk (through outgroup conspecifics or predator animals) was the ultimate cause for group living in primates (Dunbar & Shultz, 2007, 2017). Social complexity, in turn (e.g., group size, grooming clique size, formation of coalitions, and the prevalence of social play), might have been the main promoter of cognitive capacities related to mentalizing and of the neocortex size of anthropoid primates. In anthropoid primates, friendships involve a special form of bonding that is only found elsewhere in reproductive pair bonds. As a result, the formation of long-term but flexible friendships appears unique to humans and other anthropoid primates. It is suggested that, in these species, close friendships act as coalitions, one function of which is to buffer the individual against the stress that arises from living in groups in close spatial proximity (Dunbar, 2018).

## 3.2 Formation of Friendship Through Chemosensory Cues

It is well known that the perception of body odors of related or beloved individuals is associated with calming effects on mood and physiological stress reduction. Such effects have been demonstrated repeatedly, e.g., for mother-infant relations (Schaal et al., 2020) as well as for romantic partners (Granqvist et al., 2019; Hofer et al., 2018). Accordingly, it is to be expected that olfaction also plays an important role in friendships.

So far, two independent studies indicate a direct link between friendship and chemosensation: In the first study, genotypes at the single nucleotide polymorphism level in 17,413 genes were assessed in 1932 subjects who were in one or more of 1367 friendship pairs as well as in 1,196,429 stranger pairs (Christakis & Fowler, 2014). Across the entire genome, friends' genotypes were positively correlated (homophilic). Pairs of friends are, on average, as genetically similar to one another as fourth cousins. By far, the most homophilic system was an olfactory gene set, including genes coding for olfactory receptors. The authors concluded from these data that friends may be a kind of "functional" kin, who perceive or cope with the environment in a similar way. The prominent contribution of the olfactory system in explaining the genetic homophily suggests that this system may play a significant role in the formation or maintenance of friendship ties. Usually, any two human individuals have major functional differences at many of their odorant receptor alleles, explaining the well-known interindividual differences in odor perception (Mainland et al., 2014).

The second study investigated the body odor profile of friends who initiated their friendship immediately during their first meeting, so-called click friendships (Ravreby et al., 2022). It was found that their body odor quality was rated to be more similar than that of two randomly assigned individuals. This result was confirmed by odor analyses through an electronic nose. Furthermore, complete strangers who smelled more similar had more positive nonverbal dyadic interactions than those whose body odor was less similar. The authors speculate that humans, smelling continuously their self-body odor, unconsciously form a template of their body odor; when meeting others, they compare their self-odor template to that of strangers and select friends whenever the templates match.

Both studies thus underline that friends are more alike to each other than expected by chance. This finding relates to odor perception as well as to the release of odorous volatiles. Just recently, it has been shown that neural networks of the social brain are also more similar in friends than in non-friends (D'Onofrio et al., 2022). In sum, successful friendship formation seems to be related to chemosensory skills: Successful friendships might be selected on the basis of smell, establishing friendship pairs who perceive the chemical (social) world alike. The question arises whether socially skilled and integrated individuals show pronounced chemosensory skills in social contexts.

### 3.3 Social Skills Related to Chemosensory Perception

As expected, independent research groups found that olfactory performance of individuals correlates with their social network size and social integration. The first study reported that individuals with a larger social network size also show a higher olfactory sensitivity toward a rose odor (2-Phenylethanol; Zou et al., 2016). Here, the social network size was assessed by summarizing the total number of people that the participants had regular contact with. This study was confirmed by a second study, based on a relatively large sample with about 3000 participants (Boesveldt et al., 2017). Here, it was observed that individuals with a larger number of friends and family members they feel close to, and a high frequency of socializing, show higher olfactory identification scores (tested for the smell of rose, leather, orange, fish, and peppermint). However, this effect seems to be more pronounced in women than in men. A third study investigated loneliness in older adults (Desiato et al., 2021). Again, it was found that olfactory skills relate to social integration: The higher the participants' loneliness score, the worse their olfactory functioning, especially their olfactory discrimination performance. Here, the discrimination performance was assessed for various monomolecular laboratory substances, representing different smells like fruity, flowery, spicy, fresh, etc.

These three studies show that interindividual differences in social success are related to olfactory performances. First hints on the nature of social skills responsible for social success as well as for its relationship to olfaction have been observed. Socially open individuals, enjoying getting in contact with others, respond to human body odor with brain activations within the inferior frontal gyrus and the caudate nucleus (Lübke et al., 2014). While the inferior frontal gyrus is discussed to be a part of the human mirror neuron system, the caudate nucleus is involved in social reward processing. Thus, it is suggested that body odor processing is inherently rewarding for socially open individuals, and further that body odor processing might be related to empathetic skills. The latter conclusion is supported by the findings that olfactory identification performance correlates with general self-reported empathy (Spinella, 2002) and that chemosensory anxiety signals are processed in brain areas related to empathy (Prehn-Kristensen et al., 2009).

Empathy, however, is not a one-dimensional construct. It consists of several interwoven layers. De Waal and Preston (2017) differentiate empathic skills according to their evolutionary level. Whereas emotional contagion and motor mimicry can be seen in a variety of vertebrates, and in certain invertebrate species, the ability for experiencing emotional concern, accompanied by the motivation to care about a distressed conspecific, is observed mainly in mammals and in some birds (rooks). Targeted helping behavior requires the ability to take the perspective of others as well as the ability for emotional self-regulation. Targeted helping has been described only in humans, apes, and monkeys, as well as in dolphins and elephants.

To my knowledge, so far only two studies investigated specific empathic skills, assessing empathic performances and related them to olfaction. Zhou and Chen (2009) assessed women's ability to olfactorily recognize known individuals (roommates) and their ability for perspective taking regarding other individuals' emotions, as well as their accuracy in identifying facial expressions of emotions. They found that high performances in body odor identification were related to high performances in perspective taking and to high performances in the identification of social emotions. This study was confirmed by Lübke et al. (2022), who reported that olfactory discrimination performance correlates with performances in perspective taking. Here, the odors to be discriminated against were standard olfactory test substances, representing different smells, e.g., rose, vanilla, and clove. Perspective taking was assessed as the ability to understand others' intentions by viewing only the eye region of the face. However, the relationship between olfactory discrimination performance and perspective taking was demonstrated in women only and could not be observed in men.

Thus, both studies indicate that chemosensory competencies are linked to evolutionarily new and sophisticated empathic skills, such as perspective taking. Whether these findings relate to women only, or for both sexes, requires further investigations.

## 4 Conclusion

It is here proposed that during human evolution, increased competencies in chemosensory communication might have been one major prerequisite for improved social skills. In humans, social contacts are required for well-being, mental, and physical health. Highly developed social skills in humans seem to be necessary for being integrated in supportive social networks and may have forced neocortical brain development (Dunbar & Shultz, 2007). Similarly, human olfactory capacities seem to be extremely well developed (McGann, 2017). Human chemosensory cues and signals convey information about the status of the autonomic nervous system (e.g., about arousal level), the central nervous system (e.g., about emotional state), the endocrine system (e.g., about reproductive and stress state), the genetic system (e.g., about sex of the sender), the immune system (e.g., about infectious and health status; Olsson et al., 2014), and the metabolic system (e.g., about dietary status; Pause et al., 2021). Here, I focused on the expanding knowledge on human chemosensory communication strategies related to emotional states, especially stress, fear, anxiety, and aggression. Regarding friendship formation, social network strength, and mentalizing skills, further empirical evidence was presented concerning a proposedly direct link between social and olfactory competencies.

During vertebrate evolution, encephalization seems to have expanded alongside improvement in olfactory capacities. This has been proposed for early vertebrate as well as for early mammalian development (Pause, 2012). Bastir et al. (2011) even found that unique features of the brain of *Homo sapiens* in comparison to the brain of Neanderthals relate to more sophisticated olfactory functions in modern humans as compared to Neanderthals. According to these authors, highly developed olfactory functions might have helped modern humans to be socially skilled at an optimum level.

Summarizing, it is here speculated that the proximate cause of human chemosensory communication is related to important advantages in comparison to the communication through other sensory channels (e.g., fully functional in darkness, high longevity, and honesty quality of the signal, see Introduction). The ultimate cause for human chemosensory communication on the other side might be the prevention of predation. This can be targeted by chemocommunication by affecting the bonding strength in human relationships, e.g., by aiding friendship formation, by promoting successful reproduction, and harm avoidance in groups.

**Acknowledgment** I thank Katrin Lübke for fruitful discussions on the manuscript.

## References

Adolph, D., Schlösser, S., Hawighorst, M., & Pause, B. M. (2010). Chemosensory signals of competition increase the skin conductance response in humans. *Physiology & Behavior, 101*(5), 666–671.

Adolph, D., Meister, L., & Pause, B. M. (2013). Context counts! Social anxiety modulates the processing of fearful faces in the context of chemosensory anxiety signals. *Frontiers in Human Neuroscience, 7*, 283.

Algren, M. H., Ekholm, O., Nielsen, L., Kjær Ersbøll, A., Kronborg Bak, C., & Tanggaard Andersen, P. (2018). Associations between perceived stress, socioeconomic status, and health-risk behaviour in deprived neighbourhoods in Denmark: A cross-sectional study. *BMC Public Health, 18*, 250.

Archer, J. (2004). Sex differences in aggression in real-world settings: A meta-analytic review. *Review of General Psychology, 8*(4), 291–322.

Barcellos, L. J. G., Volpato, G. L., Barreto, R. E., Coldebella, I., & Ferreira, D. (2011). Chemical communication of handling stress in fish. *Physiology & Behavior, 103*(3–4), 372–375.

Bartal, I. B. A., Decety, J., & Mason, P. (2011). Empathy and pro-social behavior in rats. *Science, 334*(6061), 1427–1430.

Bastir, M., Rosas, A., Gunz, P., Peña-Melian, A., Manzi, G., Harvati, K., et al. (2011). Evolution of the base of the brain in highly encephalized human species. *Nature Communications, 2*, 588. https://doi.org/10.1038/ncomms1593

Bendahan, S., Goette, L., Thoresen, J., Loued-Khenissi, L., Hollis, F., & Sandi, C. (2017). Acute stress alters individual risk taking in a time-dependent manner and leads to anti-social risk. *The European Journal of Neuroscience, 45*(7), 877–885.

Boesveldt, S., Yee, J. R., McClintock, M. K., & Lundström, J. N. (2017). Olfactory function and the social lives of older adults: A matter of sex. *Scientific Reports, 7*, 45118.

Borowsky, B., Adham, N., Jones, K. A., Raddatz, R., Artymyshyn, R., Ogozalek, K. L., et al. (2001). Trace amines: Identification of a family of mammalian G protein-coupled receptors. *Proceedings of the National Academy of Sciences, 98*(16), 8966–8971.

Brandl, H. B., Pruessner, J. C., & Farine, D. R. (2022). The social transmission of stress in animal collectives. *Proceedings of the Royal Society B, 289*(1974), 20212158.

Brechbühl, J., Klaey, M., & Broillet, M. C. (2008). Grueneberg ganglion cells mediate alarm pheromone detection in mice. *Science, 321*(5892), 1092–1095.

Brent, L. J., Chang, S. W., Gariépy, J. F., & Platt, M. L. (2014). The neuroethology of friendship. *Annals of the New York Academy of Sciences, 1316*(1), 1–17.

Brosnan, S. F., & De Waal, F. (2003). Monkeys reject unequal pay. *Nature, 425*(6955), 297–299.

Carnicelli, V., Santoro, A., Sellari-Franceschini, S., Berrettini, S., & Zucchi, R. (2010). Expression of trace amine-associated receptors in human nasal mucosa. *Chemosensory Perception, 3*, 99–107.

Christakis, N. A., & Fowler, J. H. (2014). Friendship and natural selection. *Proceedings of the National Academy of Sciences, 111*(supplement 3), 10796–10801.

Cole, S. W., Hawkley, L. C., Arevalo, J. M., & Cacioppo, J. T. (2011). Transcript origin analysis identifies antigen-presenting cells as primary targets of socially regulated gene expression in leukocytes. *Proceedings of the National Academy of Sciences, 108*(7), 3080–3085.

Cole, S. W., Capitanio, J. P., Chun, K., Arevalo, J. M., Ma, J., & Cacioppo, J. T. (2015). Myeloid differentiation architecture of leukocyte transcriptome dynamics in perceived social isolation. *Proceedings of the National Academy of Sciences, 112*(49), 15142–15147.

D'Aniello, B., Semin, G. R., Alterisio, A., Aria, M., & Scandurra, A. (2018). Interspecies transmission of emotional information via chemosignals: From humans to dogs (Canis lupus familiaris). *Animal Cognition, 21*(1), 67–78.

D'Onofrio, P., Norman, L. J., Sudre, G., White, T., & Shaw, P. (2022). The anatomy of friendship: Neuroanatomic homophily of the social brain among classroom friends. *Cerebral Cortex, 32*, 3031–3041.

de Groot, J. H., Smeets, M. A., Kaldewaij, A., Duijndam, M. J., & Semin, G. R. (2012). Chemosignals communicate human emotions. *Psychological Science, 23*(11), 1417–1424.

de Groot, J. H., Semin, G. R., & Smeets, M. A. (2014). I can see, hear, and smell your fear: Comparing olfactory and audiovisual media in fear communication. *Journal of Experimental Psychology. General, 143*(2), 825–834.

de Groot, J. H., Smeets, M. A., Rowson, M. J., Bulsing, P. J., Blonk, C. G., Wilkinson, J. E., & Semin, G. R. (2015). A sniff of happiness. *Psychological Science, 26*(6), 684–700.

de Groot, J. H., van Houtum, L. A., Gortemaker, I., Ye, Y., Chen, W., Zhou, W., & Smeets, M. A. (2018). Beyond the west: Chemosignaling of emotions transcends ethno-cultural boundaries. *Psychoneuroendocrinology, 98*, 177–185.

De Waal, F. B., & Andrews, K. (2022). The question of animal emotions. *Science, 375*(6587), 1351–1352.

De Waal, F., & Preston, S. D. (2017). Mammalian empathy: Behavioural manifestations and neural basis. *Nature Reviews. Neuroscience, 18*(8), 498–509.

de Weerth, C., & Buitelaar, J. K. (2005). Physiological stress reactivity in human pregnancy - a review. *Neuroscience and Biobehavioral Reviews, 29*, 295–312.

Desiato, V. M., Soler, Z. M., Nguyen, S. A., Salvador, C., Hill, J. B., Lamira, J., et al. (2021). Evaluating the relationship between olfactory function and loneliness in community-dwelling individuals: A cross-sectional study. *American Journal of Rhinology & Allergy, 35*(3), 334–340.

Destrez, A., Costes-Thiré, M., Viart, A. S., Prost, F., Patris, B., & Schaal, B. (2021). Male mice and cows perceive human emotional chemosignals: A preliminary study. *Animal Cognition, 24*(6), 1205–1214.

Devries, K. M., Mak, J. Y., Garcia-Moreno, C., Petzold, M., Child, J. C., Falder, G., et al. (2013). The global prevalence of intimate partner violence against women. *Science, 340*(6140), 1527–1528.

Dudareva, N., Negre, F., Nagegowda, D. A., & Orlova, I. (2006). Plant volatiles: Recent advances and future perspectives. *Critical Reviews in Plant Sciences, 25*(5), 417–440.

Dunbar, R. I. (2018). The anatomy of friendship. *Trends in Cognitive Sciences, 22*(1), 32–51.

Dunbar, R. I. M. (2020). Structure and function in human and primate social networks: Implications for diffusion, network stability and health. *Proceedings of the Royal Society A, 476*, 20200446.

Dunbar, R. I. M., & Shultz, S. (2007). Evolution in the social brain. *Science, 317*(5843), 1344–1347.

Dunbar, R. I. M., & Shultz, S. (2017). Why are there so many explanations for primate brain evolution? *Philosophical Transactions of the Royal Society, B: Biological Sciences, 372*(1727), 20160244.

Eisenberger, N. I., & Cole, S. W. (2012). Social neuroscience and health: Neurophysiological mechanisms linking social ties with physical health. *Nature Neuroscience, 15*(5), 669–674.

Endevelt-Shapira, Y., Perl, O., Ravia, A., Amir, D., Eisen, A., Bezalel, V., et al. (2018). Altered responses to social chemosignals in autism spectrum disorder. *Nature Neuroscience, 21*, 111–119.

Fanselow, M. S. (1985). Odors released by stressed rats produce opioid analgesia in unstressed rats. *Behavioral Neuroscience, 99*(3), 589–592.

Ferdenzi, C., Richard Ortegón, S., Delplanque, S., Baldovini, N., & Bensafi, M. (2020). Interdisciplinary challenges for elucidating human olfactory attractiveness. *Philosophical Transactions of the Royal Society B, 375*(1800), 20190268.

Fleischer, J. (2021). The Grueneberg ganglion: Signal transduction and coding in an olfactory and thermosensory organ involved in the detection of alarm pheromones and predator-secreted kairomones. *Cell and Tissue Research, 383*(1), 535–548.

Gelstein, S., Yeshurun, Y., Rozenkrantz, L., Shushan, S., Frumin, I., Roth, Y., & Sobel, N. (2011). Human tears contain a chemosignal. *Science, 331*, 226–230.

Gomes, N., & Semin, G. R. (2021). The function of fear chemosignals: Preparing for danger. *Chemical Senses, 46*. https://doi.org/10.1093/chemse/bjab005

Granqvist, P., Vestbrant, K., Döllinger, L., Liuzza, M. T., Olsson, M. J., Blomkvist, A., & Lundström, J. N. (2019). The scent of security: Odor of romantic partner alters subjective discomfort and autonomic stress responses in an adult attachment-dependent manner. *Physiology & Behavior, 198*, 144–150.

Griffin, A. S., & West, S. A. (2003). Kin discrimination and the benefit of helping in cooperatively breeding vertebrates. *Science, 302*(5645), 634–636.

Grüneberg, H. (1973). A ganglion probably belonging to the N. terminalis system in the nasal mucosa of the mouse. *Zeitschrift für Anatomie und Entwicklungsgeschichte, 140*(1), 39–52.

Haegler, K., Zernecke, R., Kleemann, A. M., Albrecht, J., Pollatos, O., Brückmann, H., & Wiesmann, M. (2010). No fear no risk! Human risk behavior is affected by chemosensory anxiety signals. *Neuropsychologia, 48*(13), 3901–3908.

Han, P., Weber, C., & Hummel, T. (2020). Brain response to intranasal trimethylamine stimulation: A preliminary human fMRI study. *Neuroscience Letters, 735*, 135166.

Havlíček, J., Winternitz, J., & Roberts, S. C. (2020). Major histocompatibility complex-associated odour preferences and human mate choice: Near and far horizons. *Philosophical Transactions of the Royal Society B, 375*, 20190260.

Hawkley, L. C., & Cacioppo, J. T. (2010). Loneliness matters: A theoretical and empirical review of consequences and mechanisms. *Annals of Behavioral Medicine, 40*(2), 218–227.

Helliwell, J. F., & Aknin, L. B. (2018a). Expanding the social science of happiness. *Nature Human Behaviour, 2*(4), 248–252.

Helliwell, J. F., & Aknin, L. B. (2018b). Publisher correction: Expanding the social science of happiness. *Nature Human Behaviour, 2*(4), 313–313.

Hetland, A., & Vittersø, J. (2012). The feelings of extreme risk: Exploring emotional quality and variability in skydiving and BASE jumping. *Journal of Sport Behavior, 35*(2), 154–180.

Hirsch, C. R., & Clark, D. M. (2004). Information-processing bias in social phobia. *Clinical Psychology Review, 24*, 799–825.

Hofer, M. K., Collins, H. K., Whillans, A. V., & Chen, F. S. (2018). Olfactory cues from romantic partners and strangers influence women's responses to stress. *Journal of Personality and Social Psychology, 114*(1), 1–9.

Holt-Lunstad, J. (2021). Loneliness and social isolation as risk factors: The power of social connection in prevention. *American Journal of Lifestyle Medicine, 15*(5), 567–573.

Holt-Lunstad, J., Smith, T. B., Baker, M., Harris, T., & Stephenson, D. (2015). Loneliness and social isolation as risk factors for mortality: A meta-analytic review. *Perspectives on Psychological Science, 10*(2), 227–237.

Hurst, J. L. (1993). The priming effects of urine substrate marks on interactions between male house mice, *Mus musculus domesticus* Schwarz & Schwarz. *Animal Behaviour, 45*, 55–81.

Inagaki, H., Kiyokawa, Y., Kikusui, T., Takeuchi, Y., & Mori, Y. (2008). Enhancement of the acoustic startle reflex by an alarm pheromone in male rats. *Physiology & Behavior, 93*(3), 606–611.

Inagaki, H., Kiyokawa, Y., Tamogami, S., Watanabe, H., Takeuchi, Y., & Mori, Y. (2014). Identification of a pheromone that increases anxiety in rats. *Proceedings of the National Academy of Sciences, 111*(52), 18751–18756.

Keller-Costa, T., Canário, A. V., & Hubbard, P. C. (2015). Chemical communication in cichlids: A mini-review. *General and Comparative Endocrinology, 221*, 64–74.

Kikusui, T., Takigami, S., Takeuchi, Y., & Mori, Y. (2001). Alarm pheromone enhances stress-induced hyperthermia in rats. *Physiology & Behavior, 72*(1–2), 45–50.

Kiyokawa, Y., Kikusui, T., Takeuchi, Y., & Mori, Y. (2005). Mapping the neural circuit activated by alarm pheromone perception by c-Fos immunohistochemistry. *Brain Research, 1043*(1–2), 145–154.

Klailova, M., & Lee, P. C. (2014). Wild western lowland gorillas signal selectively using odor. *PLoS One, 9*(7), e99554.

Koos, D. S., & Fraser, S. E. (2005). The Grueneberg ganglion projects to the olfactory bulb. *Neuroreport, 16*(17), 1929–1932.

Laidre, M. E., & Johnstone, R. A. (2013). Animal signals. *Current Biology, 23*, R829–R833.

Laska, M. (2017). Human and animal olfactory capabilities compared. In A. Buettner (Ed.), *Springer handbook of odor* (pp. 667–681). Springer International Publishing.

LeDoux, J. (2007). The amygdala. *Current Biology, 17*(20), R868–R874.

LeDoux, J. E., & Pine, D. S. (2016). Using neuroscience to help understand fear and anxiety: A two-system framework. *American Journal of Psychiatry, 173*(11), 1083–1093.

Leigh-Hunt, N., Bagguley, D., Bash, K., Turner, V., Turnbull, S., Valtorta, N., & Caan, W. (2017). An overview of systematic reviews on the public health consequences of social isolation and loneliness. *Public Health, 152*, 157–171.

Liberles, S. D. (2015). Trace amine-associated receptors: Ligands, neural circuits, and behaviors. *Current Opinion in Neurobiology, 34*, 1–7.

Liberles, S. D., & Buck, L. B. (2006). A second class of chemosensory receptors in the olfactory epithelium. *Nature, 442*(7103), 645–650.

Lopez, P., & Martin, J. (2002). Chemical rival recognition decreases aggression levels in male Iberian wall lizards, Podarcis hispanica. *Behavioral Ecology and Sociobiology, 51*, 461–465.

Lübke, K. T., & Pause, B. M. (2015). Always follow your nose: The functional significance of social chemosignals in human reproduction and survival. *Hormones and Behavior, 68*, 134–144.

Lübke, K. T., Croy, I., Hoenen, M., Gerber, J., Pause, B. M., & Hummel, T. (2014). Does human body odor represent a significant and rewarding social signal to individuals high in social openness? *PLoS One, 9*(4), e94314.

Lübke, K. T., Busch, A., Hoenen, M., Schaal, B., & Pause, B. M. (2017a). Pregnancy reduces the perception of anxiety. *Scientific Reports, 7*(9213), 1–11.
Lübke, K. T., Busch, A., Hoenen, M., Schaal, B., & Pause, B. M. (2017b). Chemosensory anxiety signals prime defensive behavior in prepubertal girls. *Physiology & Behavior, 173*, 30–33.
Lübke, K. T., Blum, T. C., & Pause, B. M. (2022). Reading the mind through the nose: Mentalizing skills predict olfactory performance. *Brain Sciences, 12*, 644.
Mackay-Sim, A., & Laing, D. G. (1981). Rat's responses to blood and body odors of stressed and non-stressed conspecifics. *Physiology & Behavior, 27*(3), 503–510.
Maier, A., Scheele, D., Spengler, F. B., Menba, T., Mohr, F., Güntürkün, O., et al. (2019). Oxytocin reduces a chemosensory-induced stress bias in social perception. *Neuropsychopharmacology, 44*(2), 281–288.
Maier, A., Heinen-Ludwig, L., Güntürkün, O., Hurlemann, R., & Scheele, D. (2020). Childhood maltreatment alters the neural processing of chemosensory stress signals. *Frontiers in Psychiatry, 11*, 783.
Mainland, J. D., Keller, A., Li, Y. R., Zhou, T., Trimmer, C., Snyder, L. L., et al. (2014). The missense of smell: Functional variability in the human odorant receptor repertoire. *Nature Neuroscience, 17*(1), 114–120.
Maner, J. K., & Schmidt, N. B. (2006). The role of risk avoidance in anxiety. *Behavior Therapy, 37*(2), 181–189.
Maresh, A., Rodriguez Gil, D., Whitman, M. C., & Greer, C. A. (2008). Principles of glomerular organization in the human olfactory bulb–implications for odor processing. *PLoS One, 3*(7), e2640.
McEwen, B. S. (2007). Physiology and neurobiology of stress and adaptation: Central role of the brain. *Physiological Reviews, 87*(3), 873–904.
McGann, J. P. (2017). Poor human olfaction is a 19th-century myth. *Science, 356*, eaam7263.
Meister, L., & Pause, B. M. (2021). It's trust or risk? Chemosensory anxiety signals affect bargaining in women. *Biological Psychology, 162*, 108114.
Moons, W. G., Eisenberger, N. I., & Taylor, S. E. (2010). Anger and fear responses to stress have different biological profiles. *Brain, Behavior, and Immunity, 24*(2), 215–219.
Moynihan, J. A., Karp, J. D., Cohen, N., & Ader, R. (2000). Immune deviation following stress odor exposure: Role of endogenous opioids. *Journal of Neuroimmunology, 102*(2), 145–153.
Mucignat-Caretta, C., Cavaggioni, A., & Caretta, A. (2004). Male urinary chemosignals differentially affect aggressive behavior in male mice. *Journal of Chemical Ecology, 30*, 777–791.
Mujica-Parodi, L. R., Strey, H. H., Frederick, B., Savoy, R., Cox, D., Botanov, Y., et al. (2009). Chemosensory cues to conspecific emotional stress activate amygdala in humans. *PLoS One, 4*(7), e6415.
Mutic, S., Parma, V., Brünner, Y. F., & Freiherr, J. (2016). You smell dangerous: Communicating fight responses through human chemosignals of aggression. *Chemical Senses, 41*(1), 35–43.
Mutic, S., Brünner, Y. F., Rodriguez-Raecke, R., Wiesmann, M., & Freiherr, J. (2017). Chemosensory danger detection in the human brain: Body odor communicating aggression modulates limbic system activation. *Neuropsychologia, 99*, 187–198.
Nelson, R. J., & Trainor, B. C. (2007). Neural mechanisms of aggression. *Nature Reviews. Neuroscience, 8*, 536–546.
Oberliessen, L., & Kalenscher, T. (2019). Social and non-social mechanisms of inequity aversion in non-human animals. *Frontiers in Behavioral Neuroscience, 13*, 133.
Öhman, A., Lundqvist, D., & Esteves, F. (2001). The face in the crowd revisited: A threat advantage with schematic stimuli. *Journal of Personality and Social Psychology, 80*(3), 381–396.
Olsson, M. J., Lundström, J. N., Kimball, B. A., Gordon, A. R., Karshikoff, B., Hosseini, N., et al. (2014). The scent of disease: Human body odor contains an early chemosensory cue of sickness. *Psychological Science, 25*(3), 817–823.

Panksepp, J. (2011). The basic emotional circuits of mammalian brains: Do animals have affective lives? *Neuroscience and Biobehavioral Reviews, 35*(9), 1791–1804.

Parejo, D., Amo, L., Rodríguez, J., & Avilés, J. M. (2012). Rollers smell the fear of nestlings. *Biology Letters, 8*(4), 502–504.

Patin, A., & Pause, B. M. (2015). Human amygdala activations during nasal chemoreception. *Neuropsychologia, 78*, 171–194.

Pause, B. M. (2012). Processing of body odor signals by the human brain. *Chemosensory Perception, 5*(1), 55–63.

Pause, B. M. (2017). Human chemosensory communication. In A. Buettner (Ed.), *Springer handbook of odor* (pp. 987–1002). Springer International Publishing.

Pause, B. M., Ohrt, A., Prehn, A., & Ferst, L. R. (2004). Positive emotional priming of facial affect perception in females is diminished by chemosensory anxiety signals. *Chemical Senses, 29*, 797–805.

Pause, B. M., Adolph, D., Prehn-Kristensen, A., & Ferstl, R. (2009). Startle response potentiation to chemosensory anxiety signals in socially anxious individuals. *International Journal of Psychophysiology, 74*(2), 88–92.

Pause, B. M., Lübke, K., Laudien, J. H., & Ferstl, R. (2010). Intensified neuronal investment in the processing of chemosensory anxiety signals in non-socially anxious and socially anxious individuals. *PLoS One, 5*, e10342.

Pause, B. M., Storch, D., & Lübke, K. T. (2020). Chemosensory communication of aggression: Women's fine-tuned neural processing of male aggression signals. *Philosophical Transactions of the Royal Society B, 375*, 20190270.

Pause, B. M., Schäfer, A. S., Hoenen, M., Lübke, K. T., & Stockhorst, U. (2021). You see what you smell: Preferential processing of chemosensory satiety cues and its impact on body shape perception. *Brain Sciences, 11*, 1152.

Perusini, J. N., & Fanselow, M. S. (2015). Neurobehavioral perspectives on the distinction between fear and anxiety. *Learning & Memory, 22*(9), 417–425.

Prehn, A., Ohrt, A., Sojka, B., Ferstl, R., & Pause, B. M. (2006). Chemosensory anxiety signals augment the startle reflex in humans. *Neuroscience Letters, 394*, 127–130.

Prehn-Kristensen, A., Wiesner, C., Bergmann, T. O., Wolff, S., Jansen, O., Mehdorn, H. M., et al. (2009). Induction of empathy by the smell of anxiety. *PLoS One, 4*(6), e5987.

Ravreby, I., Snitz, K., & Sobel, N. (2022). There is chemistry in social chemistry. *Science Advances, 8*(25), eabn0154.

Rich, T. J., & Hurst, J. L. (1998). Scent marks as reliable signals of the competitive ability of mates. *Animal Behaviour, 56*, 727–735.

Richard Ortegón, S., Fournel, A., Carlos, O., Kawabata Duncan, K., Hirabayashi, K., Tagai, K., et al. (2022). And I'm feeling good: Effect of emotional sweat and perfume on others' physiology, verbal responses, and creativity. *Chemical Senses, 47*, 1–10.

Ruiz, R. J., & Avant, K. C. (2005). Effects of maternal prenatal stress on infant outcomes: A synthesis of the literature. *Advances in Nursing Science, 28*(4), 345–355.

Sabiniewicz, A., Białek, M., Tarnowska, K., Świątek, R., Dobrowolska, M., & Sorokowski, P. (2021). A preliminary investigation of interspecific chemosensory communication of emotions: Can humans (Homo sapiens) recognise fear-and non-fear body odour from horses (Equus ferus caballus). *Animals, 11*(12), 3499. https://doi.org/10.3390/ani11123499

Sanches, F. H. C., Miyai, C. A., Pinho-Neto, C. F., & Barreto, R. E. (2015). Stress responses to chemical alarm cues in Nile tilapia. *Physiology & Behavior, 149*, 8–13.

Schaal, B., Saxton, T. K., Loos, H., Soussignan, R., & Durand, K. (2020). Olfaction scaffolds the developing human from neonate to adolescent and beyond. *Philosophical Transactions of the Royal Society B, 375*, 20190261. https://doi.org/10.1098/rstb.2019.0261

Semin, G. R., Scandurra, A., Baragli, P., Lanatà, A., & D'Aniello, B. (2019). Inter-and intra-species communication of emotion: Chemosignals as the neglected medium. *Animals, 9*(11), 887. https://doi.org/10.3390/ani9110887

Snyder-Mackler, N., Burger, J. R., Gaydosh, L., Belsky, D. W., Noppert, G. A., Campos, F. A., et al. (2020). Social determinants of health and survival in humans and other animals. *Science, 368*(6493), eaax9553.

Spinella, M. (2002). A relationship between smell identification and empathy. *The International Journal of Neuroscience, 112*(6), 605–612.

Steptoe, A. (2019). Happiness and health. *Annual Review of Public Health, 40*(1), 339–359.

Sterley, T. L., Baimoukhametova, D., Füzesi, T., Zurek, A. A., Daviu, N., Rasiah, N. P., et al. (2018). Social transmission and buffering of synaptic changes after stress. *Nature Neuroscience, 21*(3), 393–403.

Stowers, L., Cameron, P., & Keller, J. A. (2013). Ominous odors: Olfactory control of instinctive fear and aggression in mice. *Current Opinion in Neurobiology, 23*(3), 339–345.

Suh, G. S., Wong, A. M., Hergarden, A. C., Wang, J. W., Simon, A. F., Benzer, S., et al. (2004). A single population of olfactory sensory neurons mediates an innate avoidance behaviour in Drosophila. *Nature, 431*(7010), 854–859.

Von Frisch, K. (1941). Über einen Schreckstoff der Fischhaut und seine biologische Bedeutung. *Zeitschrift für Vergleichende Physiologie, 29*, 46–145.

Walker, D. L., & Davis, M. (1997). Double dissociation between the involvement of the bed nucleus of the stria terminalis and the central nucleus of the amygdala in startle increases produced by conditioned versus unconditioned fear. *The Journal of Neuroscience, 17*(23), 9375–9383.

Wintermann, G. B., Donix, M., Joraschky, P., Gerber, J., & Petrowski, K. (2013). Altered olfactory processing of stress-related body odors and artificial odors in patients with panic disorder. *PLoS One, 8*(9), e74655.

Wyatt, T. D. (2003). *Pheromones and animal behaviour*. Cambridge University Press.

Zalaquett, C., & Thiessen, D. (1991). The effects of odors from stressed mice on conspecific behavior. *Physiology & Behavior, 50*(1), 221–227.

Zheng, Y., You, Y., Farias, A. R., Simon, J., Semin, G. R., Smeets, M. A., & Li, W. (2018). Human chemosignals of disgust facilitate food judgment. *Scientific Reports, 8*(17006), 1–10.

Zhou, W., & Chen, D. (2009). Sociochemosensory and emotional functions: Behavioral evidence for shared mechanisms. *Psychological Science, 20*(9), 1118–1124.

Zou, L. Q., Yang, Z. Y., Wang, Y., Lui, S. S., Chen, A. T., Cheung, E. F., & Chan, R. C. (2016). What does the nose know? Olfactory function predicts social network size in human. *Scientific Reports, 6*, 25026.

# Handshaking and Hand-Smelling: On the Potential Role of Handshake Greeting in Human Olfactory Communication

S. Craig Roberts[1,2] (✉), Francesca P. Singleton[1], Paxton D. Culpepper[1], Piotr Sorokowski[3], Agnieszka Sorokowska[3], and Jan Havlíček[4]

[1] Department of Psychology, University of Stirling, Stirling, UK[1]
craig.roberts@stir.ac.uk
[2] Being Human Lab, University of Wrocław, Wrocław, Poland
[3] Institute of Psychology, University of Wrocław, Wrocław, Poland
[4] Department of Zoology, Faculty of Science, Charles University, Prague, Czech Republic

**Abstract** In many species, meetings between individuals are characterised by olfactory investigation. This enables individual discrimination and shapes subsequent social decisions. A recent study claimed that human handshake greetings have a similar role, suggesting that people often smell their hands after a handshake. Here we describe two studies that aimed to further test this idea. We observed differences in face-touching frequency following a handshake, apparently influenced by social context. In a public situation, during a graduation ceremony, rates of face-touching were low compared to a more private setting following a social interaction. In both contexts, however, face-touching was more frequent with the non-shaking (left) hand than the shaking (right) hand. In the private setting, nose touching was more common with the left hand. These results do not lend strong support to the idea that hand-smelling is a common form of olfactory assessment following a handshake greeting. In addition, perceptual tests suggest that individual discrimination of hand odour may be less effective than for axillary odour. However, we found that individual differences in hand odours do overlap with those in axillary odours, suggesting the potential for hand odour to contribute to individual assessment on those occasions when hands are smelled.

**Keywords** Odour/Odor · Smell · Sniffing · Social perception · Human behaviour

## 1 Introduction

Meetings between individuals are an opportunity for social exchange, whether this involves some form of competition, cooperation, or mating. As such, it is important to communicate and receive any relevant information that will facilitate subsequent social interaction. Especially in mammals, social interactions often begin with a period of mutual inspection, usually involving olfactory assessment (Brown & Macdonald, 1985;



B. Schaal et al. (eds.), *Chemical Signals in Vertebrates 15*,
https://doi.org/10.1007/978-3-031-35159-4_14

Eisenberg & Kleiman, 1972; Gosling, 1990). Mutual sniffing frequently focuses on the ano-genital area, since this is a rich source of information emanating from specialised secretory glands, urine and faeces. Such information may include individual identity, sex, dominance or reproductive status, health, genetic information and other factors involved in assessment of competitors and mates (Gosling & Roberts, 2001; Hurst & Beynon, 2004).

Some human greetings also appear to involve olfactory assessment. The Inuit *kunik* involves sniffing the forehead and cheek of other individuals (Stoddart, 1990). In some of what Schaal and Porter (1991) describe as 'odor-seeking cultures', such as the Amazonian Yanomamö or the Kanum-Irebe tribe from New Guinea, it is common for one individual to touch another's body and then to sniff that person's smell from their fingers. In the former group, males prepare olfactory charms which are then carried on their arms and which are intended to attract the attention of potential partners (Schaal & Porter, 1991); similar examples are seen in various other societies, such as, in Elizabethan England, the slices of apple impregnated with a lady's armpit odour and offered as a keepsake to potential suitors (Stoddart, 1990). Some traditional forms of dancing appear to involve the opportunity for olfactory display to potential partners (Schaal & Porter, 1991; Stoddart, 1990). Kissing or hugging would also offer similar potential (Wlodarski & Dunbar, 2013), further evidenced by close colexification of the words smell and kiss found in many languages, especially in Oceania (François, 2021; Schapper, 2019).

However, one widespread form of initial greeting – handshaking – appears to be different. By definition, handshake greetings involve touching of extremities, well away from the nose. Despite this, Frumin et al. (2015) suggested that even handshakes might have an olfactory component. They covertly filmed human participants as they arrived for a laboratory study, both before and after a greeting from one of the experimenters. Their data suggested that the rate of hand-to-face contact (specifically the area around the nose) increased after a verbal greeting with an accompanying handshake, compared to participants in the control condition who received only the verbal greeting. Frumin et al. therefore argued that people subconsciously sniff their hand following a handshake, potentially obtaining from this action some form of socially useful information.

Intrigued by this report and its potential importance in human greetings, we set out to obtain further measures of the incidence and importance of hand-smelling. (It is important to note that these studies were conducted before the recent COVID-19 pandemic. As a result of the pandemic, people now tend to be more aware of the risks of handshakes as a means of transmitting infectious disease than they were when this work was conducted.) First, we directly observed the frequency of hand-smelling following a handshake in a university graduation ceremony, a setting in which many people shake hands with the same person. Second, we conducted a smaller but conceptually similar study to that of Frumin et al., in which we covertly filmed people after engaging in a handshake. Finally, we explored the extent to which hand odour could conceivably be an effective means of transmitting certain important social information, through two perceptual tasks. First, we measured how effectively naïve smellers could discriminate odour individuality using hand odours and compared this with the same task using axillary odours, which have already been shown to be an important source of personal information. Second, we assessed whether an individual's hand odour is perceptually similar to the same person's axillary odour.

## 2 Methods

### 2.1 Handshaking at a Graduation Ceremony (Study 1)

We observed students during two graduation ceremonies at the University of Stirling in November 2017. During the ceremony, students approached the university's chancellor (a man, aged 66) in single file with an interval between students of approximately 15 seconds. Each of them shook hands with the chancellor and then continued past him, down a ramp from the stage to a table where they received their graduation certificate, and then around the perimeter of the seating area to return to their seat, a walk that lasted approximately 35 seconds. Three observers watched the ceremony from an elevated viewpoint to the side of the stage, which enabled us to keep each student in view. Observers took it in turns to watch a student (i.e., the first observer watched students 1, 4, 7..., while the second observer watched students 2, 5 and 8; and the third observer watched students 3, 6 and 9; and so on). Occasionally, a student was missed because they shook hands with the chancellor, while an observer was still watching the previous person; when this happened, observers quietly communicated to each other so that they did not watch the same person. In total, we observed 443 students (201 men, 242 women, most aged 22–25) from the moment they shook hands until the moment they sat down. We recorded the students' apparent gender and whether or not they "sniffed" their hand with either their left or right hand. Following Frumin et al., we scored a hand "sniff" (they describe it as a "sample") if the hand was brought in contact with the face somewhere between the eyebrows and the chin. We did not count, for example, touching of the underside of the chin or the forehead, nor sweeping hair from the face.

### 2.2 Covertly Filmed Handshakes (Study 2)

We conducted these observations in April 2018 at the Glasgow Science Centre, a public science communication and exhibition space. Data were collected during one evening at a closed, adult-only event, where visitors were able to spend time browsing the exhibits or taking part in one or more interactive displays hosted by research teams. On one floor of the exhibition space, we set up a temporary testing area, screened from view. The area consisted of two chairs placed on either side of a small table, on which were placed three numbered 500 ml conical flasks, each containing a plain white cotton wool pad and topped with an aluminium foil cap. About 2 m away from the table, on a smaller and lower table, we placed a laptop computer, whose camera was used for video-recording the participant when they were seated at the table. Next to the laptop computer, a tripod was placed without any camera.

Potential participants were approached from the exhibition space and asked whether they would like to take part in 'a smell task'. They were informed that they would do this out of view because we wanted to film them and they shouldn't be distracted. On this basis, they had to sign informed consent to participate and to be filmed. They were asked to wait for a minute, while we cleared the space; during this time, one experimenter turned on the camera of the laptop computer and then presented a plain black PowerPoint slide in full-screen mode, so that the computer appeared to not be in use. The small light that indicated the camera was switched on was always obscured using a small square of black tape, so that the camera appeared to be off. Once this was done, the participant was brought into the testing area (Fig. 1, upper panel) where they were greeted by a confederate who was seated next to the tripod, with a video camera in hand. The confederate first asked the participant to sit at the table, so that the participant believed they would be smelling the flasks on the table. The confederate then quickly introduced himself or herself and shook hands with the participant (an example is shown in Fig. 1, lower panel). Immediately after, the confederate explained that the camera battery needed to be changed and that they would leave briefly to fetch a new one. We did this in order for the participant to believe that they were not being filmed, even though they had been told that they would be during the study. After 40 seconds, the confederate returned and explained that the study was over; participants were debriefed, so that the nature of the study, including the fact that they had been filmed, was explained; they then were asked to sign a further informed consent form so that the data could be used (if they had declined, the video of their interaction would have been immediately deleted). All the participants gave this additional consent, perhaps because of the nature of the event: all attendees were open to learning about science. Most expressed surprise and amusement at the design.

In total, we filmed 34 interactions between a participant and an experimenter, of which 16 involved the same male confederate and 18 involved the same female confederate. We scored interactions depending on whether the participant touched their face (with the left and/or right hand) at least once during the interaction. At the same time, we marked the precise location of any face touch by mouse-clicking (on a separate computer) at the same location on a standard base face of the same sex; these were then visualised using the Heatmap feature using Qualtrics.

## 2.3 Perceptual Tests (Study 3)

We first collected axillary odours for subsequent assessment by odour raters using methods that are commonly used for odour rating studies (Ferdenzi et al., 2009; Havlicek et al., 2011; Lenochova et al., 2009). All odour donors provided informed consent. Samples were collected from 13 men (age range 18–33) and 12 women (age range 21–34). Donors were non-smokers. Each was asked to shower on the night before donation, using a non-scented soap (Simple™) that was provided by the researchers. Each donor provided a sample from the non-dominant axilla (i.e., from the left side in right-handed individuals), both for the convenience of the donors and because this standardised odour intensity which can vary between left and right axillae (Ferdenzi et al., 2009). Furthermore, we also collected hand odour from the palm of the same non-dominant hand from each individual. Three individuals donated twice. Collection took place over 24 h. The donor first met with the experimenter, who fixed both cotton pads

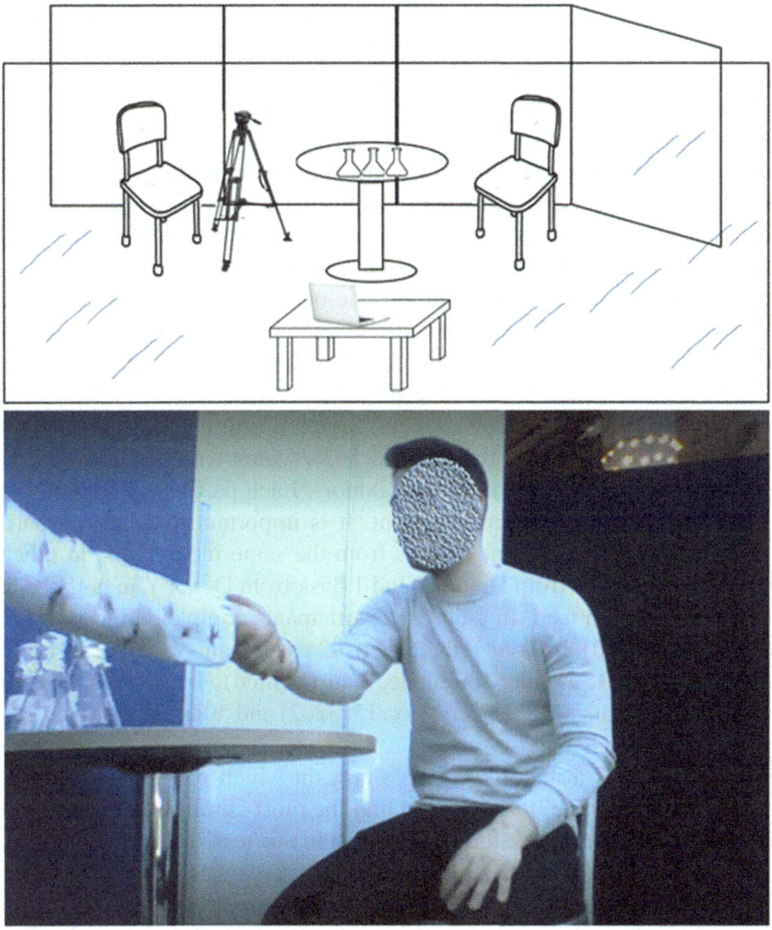

**Fig. 1** The experimental setting used in Study 2. The *upper panel* depicts a schematic of the experimental area. At the beginning of the interaction, the experimenter is positioned on the left of the picture, next to the tripod. The participant enters from behind the screens on the right-hand side and takes the chair on the right. After the handshake greeting, the experimenter removes the video camera from the tripod and exits to the right, leaving the participant seated. On the round table are three conical flasks containing cotton pads – the decoy smell task. On a low table next to a wall is the laptop from which participants were filmed. The *lower panel* shows a screenshot from one of the greetings (the participant's face is obscured to preserve anonymity)

in position. These consisted of ½ of a beauty pad (area = 22 $cm^2$) secured in place using Tegaderm™ film. Once the pads were secured, the donor was asked not to shower for the next 24 hours, avoid alcohol consumption, smoking and sharing a bed. After 24 h, the donor returned to the lab and had the pads removed. The experimenter then placed the pads in labelled plastic bags and froze them (at −20 °C) until testing (freezing does not affect the ability to detect social information from odours: Lenochova et al., 2009; Roberts et al., 2008). We then conducted two perceptual tests of the donated body odours at the Glasgow Science Centre, Scotland. Participants were visitors to the Centre during April 2018. They provided informed consent and did one of the two tests described below, but not both.

The first test compared discrimination of hand and axillary odours, in a series of triangle tests (see Allen et al., 2016). Participants were presented with four sets of triplets of amber 100-ml flasks, in each of which was an odour sample. Each triplet contained two samples from one individual and one sample from a different individual; the task was to decide which one was the odd one out. The four triplet sets consisted of odours from: male axillae only, male hands only, female axillae only and female hands only. However, we recognised the need to allow for the possibility of results being influenced by one individual having an especially distinctive odour, which could result in arbitrary or biased results (see Wedekind, 2002). For this reason, we had three versions of these samples (we call these Trials 1–3), which were the same as described above but contained odours from different donors. Each participant did only one trial, so that the trials were entirely independent. It is important to note that participants always assessed axillary and hand odours from the same individuals: in other words, there were always 2 flasks from Donor X and 1 flask from Donor Y in both the hand and axillary versions in each trial. In total, 141 participants completed the tasks (42 male, 96 female, 3 unspecified). In Trial 1, there were 42 participants (age, mean = 31.9 years; range, 23–57 years; sex, 9 male, 33 female, 0 unspecified). There were 49 participants in Trial 2 (age, 29.1 years; 20–51 years; sex, 15, 32, 2) and 50 in Trial 3 (age, 28.7 years; 20–55 years; sex, 18, 31, 1).

The second test was a match-to-sample task, in which participants were asked to sniff one flask ("the sample") and then to find the most similar odour from those contained in a set of four alternative flasks. Note that in this test, the sample was always the hand smell from one individual, and the four alternatives were always axillary odours, one of which came from the same individual as the sample. Success in this trial therefore indicates whether odour individuality carries across both hands and axillae. Each participant did this task for a set of male odours and a set of female odours. In total, 81 individuals participated (40 male, 41 female). In Trial 1, there were 30 participants (age, mean = 33.9 years; range, 19–80 years; sex, 13 male, 17 female). There were 30 participants in Trial 2 (age, 31.3 years; 20–62 years; sex, 14, 16) and 21 in Trial 3 (age, 31.7 years; 21–49 years; sex, 13, 8).

### 2.4 Ethical Permissions

The work described in this paper was approved by the General University Ethics Panel of the University of Stirling. In the case of Study 2, which involved covert filming of participants, permission required that we collect informed consent twice. Thus, we informed participants, beforehand, that they might be filmed during the study; afterwards, they were told that they were in fact filmed and were asked to provide further consent for the video-recording to be used.

# 3 Results

## 3.1 Graduation Ceremony (Study 1)

We observed 443 individuals immediately after shaking hands as they walked from the stage to their seats. In total, 32 (7.2%) individuals touched their face, while they were being observed, 13 of which were men (6.5%) and 19 women (7.9%). There was no significant difference between men and women (chi-square = 0.276, $p = 0.60$), but it is worth noting that most of these instances involved the left hand (10, 77% in men and 13, 68% in women), not the right hand with which the shake took place.

## 3.2 Covertly Filmed Handshakes (Study 2)

Table 1 shows the frequencies of interactions in which participants touched their face at least once during the 40s following a handshake. Overall, participants touched their face with either or both hands in 20 (58.8%) of the 34 covertly filmed interactions. Of these, they touched their face with their right hand in 9 interactions (26.5%), somewhat fewer than with their left (14 interactions, 41.1%; note that the sum of these interactions is more than 20 because participants touched their face with both hands in 3 cases).

The location of facial touches by male and female participants can be seen in Fig. 2. There is clear focus on the nose in both men and women, but men were more likely to touch elsewhere in their face than women were (often around the chin, especially in bearded men, suggesting non-olfactory scratching). In both men and women, touches with the right hand were more likely to be elsewhere on the face than touches with the left hand, which were more strongly focused on the nose. When the data are visualised separately for interactions involving a male or female confederate (Fig. 3), touches with the left hand again seem more focused on the nose than those with the right hand, at least after a handshake with a female confederate (bottom row).

**Table 1** Frequency of facial touches, using either hand, by participants after shaking hands with a confederate

| | | | Number of interactions: facial touches with... | | | % interactions with touch using... | |
|---|---|---|---|---|---|---|---|
| Participant | Confederate | Number of interactions | Left hand only | Right hand only | Both hands | Left hand | Right hand |
| Male | Male | 7 | 4 | 2 | 0 | 57.1 | 28.6 |
| Male | Female | 9 | 1 | 1 | 1 | 22.2 | 22.2 |
| Female | Male | 10 | 4 | 2 | 2 | 60.0 | 40.0 |
| Female | Female | 8 | 2 | 1 | 0 | 25.0 | 12.5 |
| Total | | 34 | 11 | 6 | 3 | 41.1 | 26.5 |

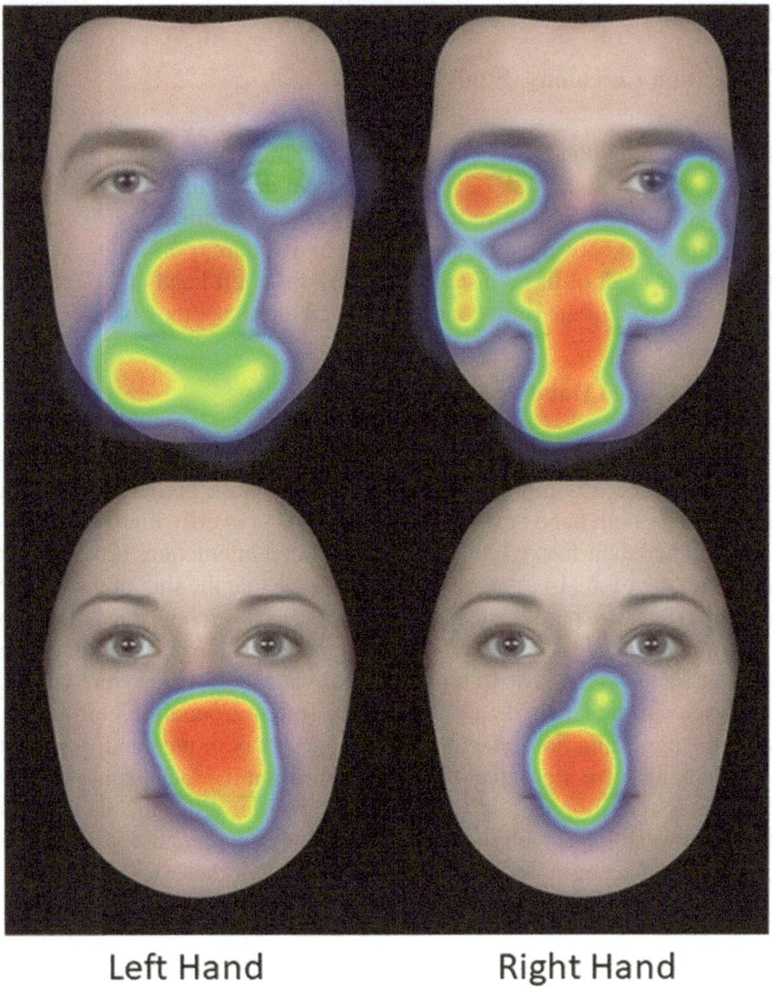


**Fig. 2** Heatmap of facial touching in male (*upper panel*) and female (*lower panel*) participants, using their left or right hands, following a handshake

### 3.3 Perceptual Tests (Study 3)

In the first task, participants attempted to discriminate the odd sample from a triplet of odours, where two came from one individual and the other was from a different individual. Across the three trials for each odour source, participants were able to correctly identify both male and female axillary odours at rates better than chance (61 and 56%, respectively), but they were not able to do this for hand odour in either sex (38 and 36%, Table 2).

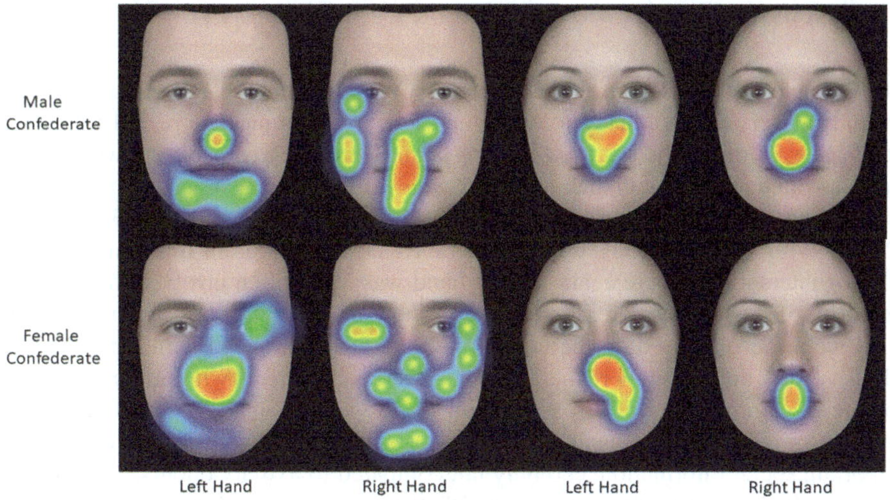


**Fig. 3** Heatmap of facial touches by participants (males on *left*, females on *right*) following a handshake, dependent on hand (left, right) used to touch the face and the sex of the handshaking confederate

**Table 2** Number and overall percentage of correct answers in the perceptual tests

| Odour source | Trial 1 | Trial 2 | Trial 3 | % correct | *P* |
|---|---|---|---|---|---|
| *Triangle tests* | | | | | |
| *n* | 42 | 49 | 50 | | |
| Male axilla | 27 | 37 | 22 | 61.0 | <0.001 |
| Male hand | 18 | 13 | 23 | 38.3 | 0.076 |
| Female axilla | 15 | 26 | 38 | 56.0 | <0.001 |
| Female hand | 5 | 23 | 38 | 36.2 | 0.201 |
| *Match-to-sample* | | | | | |
| *n* | 30 | 30 | 21 | | |
| Male donors | 13 | 9 | 7 | 35.8 | 0.020 |
| Female donors | 4 | 15 | 11 | 37.0 | 0.011 |

Statistically significant results ($P < 0.05$; binomial tests) in the triangle tests indicate that participants can discriminate differences in individual odour at rates better than chance (33%). In the matching-to-sample task, the sample was hand odour of one individual and the alternatives included the axillary odour of the same individual and from three others. Statistically significant results (binomial tests) indicate that participants can match hand and axillary odours at rates better than chance (25%)

In the second task, participants attempted to match one individual's hand odour with their axillary odour (Table 2). For male odours, mean success rate was 35.8% across the three trials, which was significantly better than chance (25%, binomial test: $p = 0.017$). Success rate was marginally higher in the assessment of female odours, at 37% ($p = 0.011$). These results suggest that there is some consistency in odour quality across an individual's hand and axilla that is perceivable by human smellers, at least in the tested conditions.

## 4 Discussion

The first two studies assessed people's proclivity to smell their hand after a handshake. In study 1, very few graduates touched their face in the immediate period following a handshake; of these, the left (non-shaking) hand was involved between two and three times as often as the right (shaking) hand. Study 2 involved a somewhat more relaxed and private setting, with a design more closely based on the study by Frumin et al. (2015). While face touching was considerably more common in this context (nearly 60% of interactions), here again the left hand was more often involved than the right hand (by 55%); furthermore, the heatmaps show that face touches involving the left hand were more focused on the area around the nose compared with the right hand. These results do not support the idea that face-touching following a handshake is a routine or common way of assessing another individual's smell. Furthermore, the triangle tests indicated that hand odours do not enable smellers to easily discriminate important social information, certainly not compared to axillary odours, for which the same smellers were much more successful. However, the match-to-sample tests indicated that an individual's hand odour could be matched with the same individual's axillary odour at rates higher than chance, suggesting that hand and axillary odours are at least to some extent perceptually similar.

Although inspired by the study of Frumin et al., our first two studies differ from theirs in numerous methodological ways which could explain the differences in findings. Clearly, our observation of a graduation ceremony is extremely different and can be criticised on several grounds. First, the graduating students tended to be young and shook hands with a much older man, with whom they presumably have no expectation of meeting again; there might be very little motivation to seek social information about him in this context, and it is hard to generalise from this single individual. Second, even if they had such motivation, the situation was not conducive to receive such information, as they were emotionally aroused, aware they were being observed by the audience and had two specific tasks to achieve (to collect their certificate and walk back to their seat in a prompt and formal manner). Third, because everyone received a certificate and carried this in one or both hands, their hands were already somewhat occupied. On the other hand, the study provides valuable real-world data. It could also be argued that if handshake-smelling is an important behaviour, then the context provides a unique opportunity to sample a person who is very high in the social hierarchy, and participants should exhibit the behaviour spontaneously even though they are being observed and have something in their hand.

The design of study 2 avoided such issues. Aspects in its favour include that it was conducted in a more relaxed and personable setting and that participants were left alone following the handshake, thus being unaware that they were being observed. However, the nature of the event where they were recruited meant that our sample size of interactions were both much fewer and necessarily shorter (approximately 40 seconds compared with approximately 3 minutes) than in Frumin et al.'s study and involved only one confederate of either sex. Furthermore, some participants spent some of the time with objects in their hands (five had a drink in one hand, five used a mobile phone and one did both). While Frumin et al. experienced the same problem with phone use (they excluded 12.5% of their sample for this reason), our sample size was insufficient to allow us to do the same. A further difference between our study and theirs is that we conducted a handshake in every interaction and only observed participants afterwards, while they had roughly half of their interactions without a handshake (control condition) and they observed participants both before and after the handshake, allowing a more powerful analysis.

We accept that these features of our two designs mean that they provide a rather different test of the proposition compared with Frumin et al. (2015). However, we think that it is noteworthy and important that the proportion of face touches involving the left hand was considerably higher than those with the right, in both studies, despite their difference in absolute abundance of face touching. It is also worth noting that this is unlikely to be due to the observation that the participants in our studies had objects in their hands. For example, if we exclude the 11 participants mentioned above who had a drink or phone in their hand, then the number of remaining interactions that included a face touch is 10 with the left hand and 6 with the right hand. This means that the left hand was then even more likely to be used: 66% more likely, compared with 55% when we include all participants.

Perhaps there may be important differences in handshaking with same and opposite-sex people. Frumin et al. reported a pattern whereby right-hand exploration increased after a greeting with individuals of the same sex, while left-hand exploration increased after shaking hands with opposite-sex individuals. Inspection of data in Table 1 shows that our data do not appear to be consistent with this pattern; although we cannot measure increase or decrease in exploration, the frequency of right-hand touches was around half the frequency of left-hand touches in both male-male and female-female interactions. Furthermore, while the rate of right-hand touches in men did not substantially change in same versus opposite-sex interactions, left-hand touches were notably less likely to occur in opposite-sex encounters (by less than half) – again inconsistent with their results. The only finding which was consistent with their results was that left-hand touches were more common in female participants during opposite-sex than same-sex interactions, but we remain unconvinced that left-hand touching has a great deal to do with inter-individual communication. A tendency to touch the face more often with the left hand, which was always the non-shaking hand, is precisely the opposite to what would be expected if handshaking was an important means of olfactory assessment. Indeed, this higher rate of left-hand face touching appears more likely to be due to some other cause, such as a form of non-olfactory displacement behaviour, as participants may have been stressed by the unusual situation in which they found themselves. Alternatively, it could have been a form of comfort-seeking by smelling their own odour because familiar odours can reduce social stress (Granqvist et al., 2019; Hofer et al., 2018).

Nonetheless, our subsequent perceptual tasks aimed to explore whether hand odours that might be transferred during a handshake have the capacity to convey important socially relevant information. Compared with axillary odour, which is well-known to carry a rich variety of social cues (Havlíček et al., 2017), including about individuality (Allen et al., 2015; Hold & Schleidt, 1977; Schleidt, 1980), we found that hand odours of different individuals were not discriminable in our triangle tests. We do not know the reason for this. One possibility is that they were not sufficiently intense compared to axillary samples. However, the match-to-sample tasks did indicate that hand, and axillary odours are at least somewhat similar perceptually, suggesting that hand odours were at least to some extent discriminable and that hand-smelling could provide some useful social cues. The difference between these results may simply reflect the relative difficulty of the different tasks. It is worth noting that in an earlier study, Wallace (1977) found that participants could discriminate the palm odours of individuals (contrary to our results). However, initial performance in her study was poor and consistently improved only over the course of 30 trials, suggesting a learning pattern. Thus handshake-smelling, if it occurred, could potentially convey useful information. Of course, we recognise that individuality is only one of many forms of social information that might be transferred, as there is a wealth of other social cues in human odour that we did not test (e.g., sex, age, health, social status and so on: Havlicek et al., 2017) and that could be explored in future studies. Either way, such communication might be more usual or effective in populations where people use their sense of smell to a greater extent than in those that have been described as WEIRD (Henrich et al., 2010) or ODD (Roberts et al., 2020).

In summary, we have explored in three studies the proposal that humans routinely smell their hands following a handshake, as a means of gathering social information about other individuals. Although our methodology has weaknesses and the sample size in the second study is relatively small, we found in both studies 1 and 2 that the left hand was much more often used to touch the face than the right hand, following a handshake. In study 2, touches with the left hand also appear to be more oriented towards the nose area than the right hand. These results are at odds with the initial study by Frumin et al. (2015). Further study is thus needed, but our current results call into question the extent to which handshaking is an important mechanism for olfactory communication during human greetings, at least in Western social settings.

**Acknowledgements** We are grateful to Vikki Martin, Konstantina Soulou and Rachel Vooght for their assistance in data collection, to the Glasgow Science Centre for hosting our research study, and to all our participants. JH is supported by Czech Science Foundation (GAČR 21-29772S) and by the Charles University Research Centres UNCE/HUM/032 (UNCE 204056). SCR, AS, and PS are supported by a University of Wrocław Incubator grant (Being Human).

## References

Allen, C., Havlíček, J., & Roberts, S. C. (2015). Effect of fragrance use on discrimination of individual body odor. *Frontiers in Psychology, 6*, 1115. https://doi.org/10.3389/fpsyg.2015.01115

Allen, C., Cobey, K. D., Havlíček, J., & Roberts, S. C. (2016). The impact of artificial fragrances on the assessment of mate quality cues in body odor. *Evolution and Human Behavior, 37*(6), 481–489. https://doi.org/10.1016/j.evolhumbehav.2016.05.001

Brown, R. E., & Macdonald, D. W. (Eds.). (1985). *Social odours in mammals*. Clarendon Press.

Eisenberg, J. F., & Kleiman, D. G. (1972). Olfactory communication in mammals. *Annual Review of Ecology and Systematics, 3*, 1–32.

Ferdenzi, C., Schaal, B., & Roberts, S. C. (2009). Human axillary odor: Are there side-trelated perceptual differences? *Chemical Senses, 34*(7), 565–571. https://doi.org/10.1093/chemse/bjp037

François, A. (2021). Olfactory words in northern Vanuatu: Langue vs. parole. In L. Jędrzejowski & P. Staniewski (Eds.), *The linguistics of olfaction: Typological and diachronic approaches to synchronic diversity* (pp. 277–304). John Benjamins Publishing Company.

Frumin, I., Perl, O., Endevelt-Shapira, Y., Eisen, A., Eshel, N., Heller, I., et al. (2015). A social chemosignaling function for human handshaking. *eLife, 4*, e05154. https://doi.org/10.7554/eLife.05154

Gosling, L. M. (1990). Scent marking by resource holders: Alternative mechanisms for advertising the costs of competition. In D. W. Macdonald, S. Natynczuk, & D. Müller-Schwarze (Eds.), *Chemical signals in vertebrates 5* (pp. 315–328). Oxford University Press.

Gosling, L. M., & Roberts, S. C. (2001). Scent-marking by male mammals: Cheat-proof signals to competitors and mates. *Advances in the Study of Behaviour, 30*, 169–217.

Granqvist, P., Vestbrant, K., Döllinger, L., Liuzza, M. T., Olsson, M. J., Blomkvist, A., & Lundström, J. N. (2019). The scent of security: Odor of romantic partner alters subjective discomfort and autonomic stress responses in an adult attachment-dependent manner. *Physiology & Behavior, 198*, 144–150. https://doi.org/10.1016/j.physbeh.2018.08.024

Havlíček, J., Lenochová, P., Oberzaucher, E., Grammer, K., & Roberts, S. C. (2011). Does length of sampling affect quality of body odor samples? *Chemosensory Perception, 4*(4), 186–194. https://doi.org/10.1007/s12078-011-9104-6

Havlíček, J., Fialová, J., & Roberts, S. C. (2017). Individual variation in body odor. In A. Buettner (Ed.), *The Springer Handbook of Odor* (pp. 949–961). Springer.

Henrich, J., Heine, S. J., & Norenzayan, A. (2010). The weirdest people in the world? *The Behavioral and Brain Sciences, 33*(2–3), 61–83. https://doi.org/10.1017/S0140525X0999152X

Hofer, M. K., Collins, H. K., Whillans, A. V., & Chen, F. S. (2018). Olfactory cues from romantic partners and strangers influence women's responses to stress. *Journal of Personality and Social Psychology, 114*, 1–9. https://doi.org/10.1037/pspa0000110

Hold, B., & Schleidt, M. (1977). The importance of human odour in non-verbal communication. *Zeitschrift für Tierpsychology, 43*, 225–238.

Hurst, J. L., & Beynon, R. J. (2004). Scent wars: The chemobiology of competitive signalling in mice. *Bioessays, 26*(12), 1288–1298.

Lenochova, P., Roberts, S. C., & Havlicek, J. (2009). Methods of human body odor sampling: The effect of freezing. *Chemical Senses, 34*(2), 127–138. https://doi.org/10.1093/chemse/bjn067

Roberts, S. C., Gosling, L. M., Carter, V., & Petrie, M. (2008). MHC-correlated odour preferences in humans and the use of oral contraceptives. *Proceedings of the Royal Society B: Biological Sciences, 275*(1652), 2715–2722. https://doi.org/10.1098/rspb.2008.0825

Roberts, S. C., Havlíček, J., & Schaal, B. (2020). Human olfactory communication: Current challenges and future prospects. *Philosophical Translations of the Royal Society, London, Biological Sciences, 375*(1800), 20190258. https://doi.org/10.1098/rstb.2019.0258

Schaal, B., & Porter, R. H. (1991). Microsmatic humans revisited - the generation and perception of chemical signals. *Advances in the Study of Behaviour, 20*, 135–199.

Schapper, A. (2019). The ethno-linguistic relationship between smelling and kissing. A Southeast Asian case study. *Oceanic Linguistics, 58*, 92–109. https://doi.org/10.1075/tsl.131.10fra

Schleidt, M. (1980). Personal odor and nonverbal communication. *Ethology and Sociobiology, 1*(3), 225–231.

Stoddart, M. (1990). *The scented ape*. Cambridge University Press.

Wallace, P. (1977). Individual discrimination of humans by odor. *Physiology & Behavior, 19*, 577–579.

Wedekind, C. (2002). The MHC and body odors: Arbitrary effects caused by shifts of mean pleasantness. *Nature Genetics, 31*, 237.

Wlodarski, R., & Dunbar, R. I. (2013). Examining the possible functions of kissing in romantic relationships. *Archives of Sexual Behavior, 42*(8), 1415–1423.

# Part IV

# Development

# Early Olfactory Learning in Mammals: A Case of Imprinting?

Lucie Jelínková[1] (✉) and Jan Havlíček[1]

[1] Department of Zoology, Faculty of Science, Charles University, Prague, Czech Republic
lucie.kapicova@natur.cuni.cz

**Abstract** In many mammals, various social interactions are guided by olfactory cues, whereby olfactory preferences often form early in ontogeny. Early olfactory learning is usually investigated within the theoretical framework of olfactory imprinting, which assumes that early olfactory preferences are acquired upon a short exposure and with no need for further reinforcement. It is also expected that such preferences are resistant to change. Olfactory imprinting has been investigated using a cross-fostering experimental design or by early exposure to artificial odours and subsequent assessment of preference for these odours. Existing findings indicate that mammals develop olfactory preferences during a limited period of time in early postnatal development, and once acquired, these preferences remain stable at least until weaning. On the other hand, there is currently no evidence to the effect that juvenile odour preferences are acquired at the same time as adult odour preferences. Stability of adult odour preferences also depends on the stimuli used in experiments, because some stimuli (e.g. natural odours) seem to elicit stronger preferences than others (e.g., artificial odours). Moreover, some reinforcement is needed for early olfactory learning, which is why it is difficult to disentangle the particular learning mechanisms involved. Despite the wide range of studies on the subject, it thus remains unclear what mechanisms underpin early odour learning.

**Keywords** Cross-fostering · Olfaction · Olfactory imprinting · Olfactory preferences · Mammals · Multimodality · Sensitive period

## 1 Introduction

In most mammals, social interactions include an assessment of olfactory cues coming from conspecifics. Adult olfactory social preferences seem to be shaped by early experience. Research on this issue peaked in the 1970s, whereby most studies were conducted within the theoretical framework of a learning mechanism known as *imprinting*. The main aim of this chapter is to review the findings of studies on early olfactory learning in mammals and to critically assess whether they indeed meet the definition of imprinting. We start with a short description of imprinting in birds and its characteristic features. Then, we review various cross-fostering studies and research based on



B. Schaal et al. (eds.), *Chemical Signals in Vertebrates 15*,
https://doi.org/10.1007/978-3-031-35159-4_15

experimental manipulation of the olfactory environment. These sections are followed by a critical examination of this material, with a focus on whether it indeed provides support for the imprinting hypothesis, mainly in terms of substantiating the assumption of a sensitive period and irreversibility of early olfactory learning. We end the chapter by suggesting some avenues for future research.

## 2 Beginning with Birds: The Imprinting Paradigm

In 1935, Konrad Lorenz performed several cross-fostering experiments with juvenile greylag goose (*Anser anser*) and observed that goslings tend to follow the first moving object they see after hatching (in his case, a human). Goslings preferred following this object over following members of their own species, although they were otherwise reared by their conspecifics. Based on these experiments, Lorenz (1937) defined a new learning mechanism that came to be known as imprinting. It is characterised by a brief acquisition period and apparently no need for further reinforcement. The learned stimulus subsequently triggers a specific behaviour. Importantly, though, Lorenz did not use the term 'learning': he spoke about a 'process' to emphasise the difference between imprinting and other forms of learning, especially associative learning. In the following, we therefore use the term 'acquisition'. When referring to such acquisition of a potential preference for certain features of an object ('potential', because the animal needs not encounter such object in later life), we speak in a kind of shorthand about an 'acquisition of preference'.

Imprinting is characterised by several features: (i) a sensitive period for acquisition, (ii) preference that does not need to be reinforced to be effective, and (iii) a low malleability of the acquired preference. The key and indeed the defining feature of imprinting are that the acquisition phase is restricted to a specific developmental period. Lorenz (1937) called this period 'critical', but later scholars opted for the term 'sensitive' (Bateson, 1979; Bischof, 1979; Hess, 1959). The main difference between the two terms is the following: when the relevant stimulus is presented outside the specific developmental window, acquisition of preference by imprinting is either impossible (critical period) or merely hampered (sensitive period) (Bateson, 1979). Preference acquisition may vary depending on the species and the type of association. The length of the sensitive period also varies across species, but it tends to coincide with early postnatal development, which is a period of neural plasticity of the parts of the brain relevant to the particular type of association (Knudsen, 2004). Usually, it occurs during the first days after hatching in birds or after several days after birth in mammals. For example, in chicks (*Gallus gallus domesticus*), it has been determined that the sensitive period for filial imprinting coincides with the first 3 days after hatching (Yamaguchi et al., 2012).

In contrast to associative learning, the acquisition phase for imprinting occurs after a single exposure to the stimulus with no need for further reinforcement (Bateson, 1966; Bolhuis et al., 1990). Some authors (Bateson, 1966; Immelmann, 1972; Bolhuis et al., 1990) distinguish between conventional reinforcement (retraining) and external reinforcement (e.g. food, warmth), but once again, there is no consensus on how these terms ought to be used. However, under natural conditions, reinforcement can be atypical or unexpected and hence difficult to plan experimentally; for example, the presence of siblings or even just being in the soft and warm nest can function as a reinforcing situation. It is therefore often difficult to distinguish imprinting from other types of learning. To test acquisition by imprinting without subsequent reinforcement, one must conduct laboratory experiments, that is, expose subjects to as much as possible controlled circumstances.

Finally, imprinting is characterised by low malleability (Lorenz, 1937). Early researchers even suggested that preferences acquired by imprinting are irreversible, but nowadays, it seems that one should rather say that preferences acquired during a sensitive period are stable, entrenched, and difficult to change once this period has passed. They can, however, be modified when exposure to relevant stimuli occurs within the window of a sensitive period.

Depending on the type of behaviour associated with the acquired preferences, Lorenz (1937) distinguished between filial and sexual imprinting. Filial imprinting is used for recognition of parents by the offspring, while sexual imprinting is used for the formation of mate preferences, where parental characteristics may serve as a template (Bateson, 1966). There are several other important differences between the two types of imprinting: in filial imprinting, the acquired preferences are used shortly after preference is formed. In sexual imprinting, there is a time gap between the acquisition of preference, which usually occurs early in ontogeny, and the expression of the specific behaviour, which typically occurs after puberty.

Imprinting is further characterised by consolidation of the acquired preferences, a process underpinned by underlying neural pathways specific to the imprinting processes, which enable the storage of preferences in long-term memory with no need for reinforcement (Bischof, 1979; Oetting et al., 1995). In filial imprinting, this consolidation stage either coincides with the sensitive period or takes place shortly after it (Suge & McCabe, 2004). For example, in domestic chicks, the exposure to a relevant stimulus during the sensitive period leads to an increased activation of the intermediate and medial mesopallium (IMM). After about 6 hours, the acquired preferences are stored in a yet unknown brain region (but other than the IMM). In contrast to other types of learning, such as associative learning, visual imprinting in birds does not involve the hippocampus as a temporary storage location before transfer to the long-term memory (Nakamori et al., 2010; Rattenborg et al., 2011).

In sexual imprinting, the consolidation period is generally longer, and the process takes place just before or shortly after the individual reaches maturity (Bischof, 1979). The sensitive period in sexual imprinting is often called the ‘acquisition phase’ to distinguish it from the sensitive period in filial imprinting. For example, in zebra finch (*Taeniopygia guttata*), the acquisition phase spans between 10 and 60 days after hatching, with acquisition being most likely to occur at around 20 days post-hatching. Consolidation partly coincides with the acquisition phase and occurs at 40–100 days after hatching with a peak at around 75 days post-hatching (Bischof, 2003). The delay of consolidation of sexual imprinting until early adulthood enables a modification of preferences acquired in early development by later social interactions experienced during adolescence. During consolidation, preferences for a specific individual are generalised to other individuals with similar characteristics (Oetting et al., 1995). If there is an accessible partner resembling the imprinted template, the acquired preferences are consolidated. If such partner is not available, preferences can be modified in the light of later experiences. The timing and length of consolidation vary across species but may also depend on the object, because some objects are acquired faster than others (Lemaire et al., 2021; see below).

Aside from geese, filial imprinting has also been observed in several other nidifugous birds, such as the mallard duck (*Anas platyrhynchos*, Ramsay & Hess, 1954) or domestic junglefowl (Bolhuis, 1999). It seems, however, that filial imprinting is not present in all nidifugous birds: it was, for instance, not found in the common curlew (*Numenius arquata*), who strictly prefer to follow own species and do not filially imprint on other species or artificial objects (Lorenz, 1937).

Filial imprinting is not limited to the nidifugous species of birds: it has also been found in some nidicolous birds, for example, in the ringdove (*Streptopelia risoria*, Klinghammer & Hess, 1964) or song sparrow (*Melospiza melodia*, Nice, 1950). Nidicolous birds do not leave the nest until they can fly, which is why it is difficult to test the following behaviour or the sensitive period for filial imprinting. In these species, imprinting has been described in terms of forming a social relationship with a human caretaker shortly after leaving the nest. Klinghammer and Hess (1964) suggested that the timing of preference acquisition in birds depends on their level of visual and locomotor activity (developmental stage) after hatching. In nidicolous birds, the sensitive period and acquisition of preferences therefore occur later than in nidifugous birds (e.g. in the ringdove at 7–9 days after hatching (Klinghammer & Hess, 1964). Based on these results, one may expect analogous differences in the imprinting process between altricial and precocial mammals.

Sexual imprinting was also first observed in birds, for example, in Muscovy ducks (*Cairina moschata*) cross-fostered by a greylag goose. Adult male Muscovy ducks then preferred to mate with greylag females rather than with their own species (Lorenz, 1937). Both filial and sexual imprinting are also well documented in mammals, for example, in sheep (*Ovis aries*) and goat (*Capra aegagrus hircus*, Kendrick et al., 1998), Steller sea lion (*Eumetopias jubatus*, Lynn et al., 2010), California sea lion (*Zalophus californianus*, Schusterman et al., 1992), but also in guinea pigs (*Cavia porcellus*, Carter & Marr, 1970) and mice (*Mus musculus*, Hayashi & Kimura, 1978).

## 2.1 Learning Predisposition

Although imprinting is characterised by the acquisition of preference for certain features of an object following exposure, animals tend to prefer certain stimuli even in the absence of any previous direct experience with them: this phenomenon is called 'predisposition' (Bolhuis et al., 1990; Horn & McCabe, 1984). It indicates that there are two processes underlying the acquisition of the preference, namely, predisposition and learning. In particular, it seems that there exists a predisposition to preferring objects that resemble conspecifics (Bolhuis et al., 1990). For example, although chicks prefer following the first moving object they see after hatching, they also have a tendency to follow a dummy resembling their own species, that is, an object with a differentiated head and neck region, and they prefer it to other tested stimuli (e.g. a rotating box, Johnson et al., 1985). This preference increases after each round of testing and does not extend to stimuli not resembling own species (Johnson et al., 1985). It has been shown that chicks reared individually in darkness showed no increased preference for either dummies resembling conspecifics or for artificial stimuli (Johnson et al., 1985). One could thus assume that although some predisposition is in place even without any previous exposure to the (potentially preferred) stimuli, other nonspecific stimulation is required for its activation (Hampton et al., 1995). Moreover, when domestic chicks are trained to react to a stimulus similar to one they are predisposed for, the acquired preference is more stable. This indicates that imprinting might be irreversible especially when both perceptual mechanisms, i.e., both the predisposition and acquisition of preference are involved (Bolhuis & Trooster, 1988). Neurophysiological studies support the notion that predisposition is a separate process from imprinting (in chicks), because it does not involve the IMM (Horn & McCabe, 1984). Predisposition thus does not seem to direct the acquisition, but it does play a supportive role in the imprinting process (Bolhuis & Honey, 1998).

Interestingly, even preference for certain artificial objects can be acquired more easily than preference for other artificial objects. The key question is whether some particular characteristics of the object influence how easily a preference for it is acquired. Candidate characteristics are the size, shape, colour, and especially mobility of the object, because moving objects are often preferred to immobile stimuli (Bateson, 1966). For example, domestic chicks hatched in the dark (to avoid exposure to colours) are more likely to follow red, blue, orange, and green stimuli rather than yellow stimuli (e.g. in an experiment using coloured balls, Schaefer & Hess, 1959). Similarly, Salzen et al. (1971) found that chicks prefer to follow a red disc, while a yellow disc was the least preferred object to follow. Preference for the red colour could be influenced by a higher contrast with the background or permeability of red light used during hatching. Gaston et al. (1969) trained guinea pig pups for different objects every 2 hours during their first day of life. Pups' preferences were then tested at different timepoints (2–17 days after birth), and the results showed that pups were more likely to follow stimuli to which they had been previously exposed rather than novel stimuli. Interestingly, the sound, colour, shape, or material of the stimulus did not seem to make any difference: the pups reacted to all types of stimuli in the same way (Gaston et al., 1969).

Importantly, acquired preferences are not restricted to the visual modality. For example, in a case study with Steller sea lion (*Eumetopias jubatus*), a pup bottle-fed by humans followed not only its caretaker but also other presented auditory stimuli (Lynn et al., 2010). Three caretakers, including the bottle-feeding one, emulated sea lion's maternal calls, while one used human language. The young sea lion reacted to all human calls, but its reaction to calls from the bottle-feeding caretaker was stronger. In several bird species, it has been shown that the acquired preferences for visual features depend on auditory cues. Ducklings of the common sheldrake (*Tadorna tadorna*) exposed only to auditory stimuli (one of the two calls roughly resembling the calls made by females) during the sensitive period followed these sounds only rarely. Their success rate slightly increased when the sounds they were exposed to resembled the voices of their conspecifics (Klopfer, 1959). But when the ducklings were exposed during the sensitive period to both visual (experimenter) and auditory stimuli, they showed an increased preference for sounds accompanied by the visual stimuli (Klopfer, 1959). Smith and Bird (1964) exposed 1-day-old chicks of the domestic fowl separately to artificial visual (a rotating red box) and auditory stimuli (repetitive pure tones) on 2 consecutive days. On the third day, they were exposed to visual and auditory stimuli simultaneously. Chicks reacted to the visual stimuli faster when they were presented together with auditory stimuli. In another experiment, chicks exposed only to auditory stimuli displayed only a weak reaction (Van Kampen & Bolhuis, 1991). Some modalities thus seem to be more essential for the preference acquisition than others, which can nevertheless play a complementary role. This suggests that imprinting may rely on multimodal stimulation rather than on a single modality.

In the social life of mammals, olfaction seems to be of a pivotal importance. It is well possible that in mammals, early olfactory experiences play in the imprinting process a role analogous to birds' preference for certain visual stimuli.

## 3 Early Olfactory Learning in Mammals

Early olfactory learning has been studied mainly in rodents, insectivores, and ungulates. These studies tend to rely on two methodological approaches: (i) rearing juveniles by a foster mother of a different species (cross-fostering studies) and (ii) rearing juveniles by their own mother or a dummy mother perfumed with a novel odour. Alternatively, a novel odour can be spread on the bedding or in the surrounding air. Subsequently, the olfactory preferences of target animals are tested at different ages and compared with controls reared under unaltered conditions. Preferences are usually operationalised as the time spent in proximity of the test stimulus.

In general, olfactory preferences can develop throughout life, but olfactory preferences acquired by imprinting are established during a limited period of exposition early in life, and this pertains to social odours. In contrast, general olfactory preferences are mostly linked to environmental stimuli, including predators and food odours. Most studies work within the imprinting paradigm, and their aim is to determine the duration of the imprinting period in question. While some studies tested the effect of early odour experiences on odour preferences expressed during the juvenile period (filial imprinting), other studies focused on the influence of early odour experiences on adult preferences (sexual imprinting).

In the following paragraphs, we review studies on early olfactory learning which were conducted in precocial and altricial mammals using the cross-fostering design. Next, we introduce studies which investigated the development of olfactory preferences using artificial odours. This section is further divided depending on the main question the studies aimed to answer. We describe all studies in some detail to facilitate subsequent critical assessment of their findings.

## 3.1 Cross-Fostering Studies

Similar to studies on visual imprinting, the effect of early experience with social odours on later olfactory preferences was tested using the cross-fostering design in different species of both precocial and altricial mammals. The social or sexual olfactory preferences of animals reared by a foster mother or foster parents were compared to preferences of a control group, mostly individuals reared by their own (i.e. biological) mother. Cross-fostering studies do not enable determination of the exact timing of the sensitive period. Instead, they tend to examine preferences for odours experienced before weaning and possible sex differences in these preferences. Several studies have also tested whether early odour preferences can be modified by later odour experiences.

### 3.1.1 Cross-Fostering in Precocial Mammals

Beauchamp and Wellington (1981) compared the effect of early experiences on olfactory preferences in wild guinea pig pups (*Cavia aperea*) reared by the domestic guinea pig and control group of wild guinea pig pups reared by their own mother. In their study, male wild guinea pigs (females were not tested) were separated from their mothers several hours after birth and reared by nulliparous females of either the domestic guinea pig (treatment group) or the wild guinea pig (control group) for 3 or 16 weeks. Subsequently, the males were caged individually in a room with other animals of both sexes and, after reaching maturity at 4 months, tested for their olfactory preference. Each individual was tested in a series of four two-choice tests of preference for the urine odours of: (i) same-sex wild vs. domestic guinea pigs, (ii) opposite-sex wild vs. domestic guinea pigs, (iii) wild male vs. wild female guinea pigs, and (iv) domestic male vs. domestic female guinea pigs. Males reared by domestic guinea pig females preferred the urine odour of domestic over wild guinea pig females, and, similarly, males reared by wild females preferred the urine odour of wild over domestic female guinea pigs. The same pattern was found for same-sex urine odour preferences in both the treatment and the control groups. Still, it turned out that early odour experiences had a stronger impact on opposite-sex preferences than on same-sex preferences. Males reared by females of domestic guinea pigs preferred the urine odour of domestic guinea pig females to that of domestic guinea pig males. Males reared by wild females preferred the urine odour of domestic guinea pig females over that of domestic guinea pig males but showed some preference for the odour of wild guinea pig female urine as well. For males reared by wild females and their preferences for wild male vs. female urine odour samples, the results were similar: these males preferred the urine odour of wild guinea pig females to that of wild guinea pig males. Males reared by a domestic female showed only a weak preference for wild female as opposed to wild guinea pig male urine odour. These results indicate that male preferences in each of the alternative choice tests with female odour of both species were influenced by the males' early odour experiences, which were shaped by their rearing condition. Still, the males tended to prefer female to male odour regardless of their rearing condition, which indicates that while the rearing conditions do affect the preferences, preferences can be modified based on accessibility of the acquired preferred characteristics. The length of odour experience had no effect on the resulting preferences (Beauchamp & Wellington, 1981).

### 3.1.2 Cross-Fostering in Altricial Mammals

Albonetti and D'udine (1986) carried out a cross-fostering experiment with the SEC and C57 strains of inbred female laboratory mice *Mus musculus* (males were not tested). The control group consisted of SEC pups reared by their own parents. The treatment group was comprised of SEC female pups reared by C57 parents from birth until 21 days of age. After this period, the females were kept in same-sex groups until they reached sexual maturity (at 120 days), when they were tested for their opposite-sex odour preferences for different bedding, each soiled by the odour of C57 or SEC males. Both the SEC pups reared by SEC females and the SEC pups reared by C57 females spent more time in the compartment with bedding odourised by C57 males. Subsequently, the females were kept individually for 7 days in a cage-like device connected with four compartments with either SEC or C57 males. These cages permitted olfactory and acoustic interactions between the female and the males, meaning they were able to experience the opposite-sex odour. Subsequently, the females' preferences were tested once again. The results showed that both the control and treatment SEC females exposed to SEC males in later life preferred to spend more time in a compartment with SEC male bedding, while both the treatment and the control group females exposed to C57 males preferred the bedding from C57 males. This indicates that early odour preferences can be modified by later odour experiences, because cross-fostered SEC females preferred the odour of SEC males after exposure to SEC male odour before testing. It seems that for opposite-sex odour preferences in adulthood, the later odour experience can override the early experience. Unfortunately, this study tested only the development of female preferences and one cannot extrapolate from it to either male preferences or to possible sex differences.

Both male and female preferences in mice were tested by Quadagno and Banks (1970), who performed reciprocal cross-fostering between pygmy mouse (*Baiomys taylori ater*) and inbred laboratory mice. At 2 days after birth, half of the pups of each species were transferred to foster mothers (treatment group). Each female reared two of her own pups and one fostered pup. In the control group, litters were reduced to three young who were reared by their own mother. Both groups were reared by the respective females until 21 days of age and then placed individually in cages standing next to each other in the same room. It is therefore possible that some olfactory and auditory cues were available to the animals even after the separation. The mice were then tested for their olfactory, social, and sexual preferences. Olfactory preferences were tested after reaching maturity at 75–85 days after birth in a Y-shaped maze with vaginal smears of dioestrus pygmy mice and laboratory mice as stimuli. While the control group preferred the odour of their own species, cross-fostered laboratory mice of both sexes and cross-fostered females (but not males) of pygmy mouse showed preference for the odour of the species that fostered them. Subsequently, social preferences were tested at 100–115 days after birth using small cages where males were paired with males and oestrus females of their own or foster species. The results showed that the cross-fostered animals reacted more positively (a composite score of behaviours such as higher frequency of approaches and physical contacts, the number of encounters, and duration of allogrooming) and less negatively (a score of behaviours such as the frequency of avoidance behaviour, fights and threats, tail rattling, leg bracing, and the number of escape leaps) to individuals of their foster species regardless of their sex. This positive effect was stronger in cross-fostered females than in cross-fostered males, especially in laboratory mice. In contrast, animals in the control group reacted more positively to individuals of their own species. But when sexual preferences were tested at 110–120 days after birth in both the control and the treatment group of laboratory mice who were paired with their opposite-sex conspecifics, researchers found no differences in the females' positive reaction (exhibiting lordosis vs. frequency of mounting or attempted mounting in males) to the males. These results indicate that the responsiveness of mice to experienced odours varies between species. In contrast to laboratory mouse males, pygmy mouse males were less influenced by the early odour experience with the odour of their foster mother, indicating that pygmy mouse might be more prone to the modification of their early life olfactory experience than the laboratory mouse is. The development of preferences for olfactory cues experienced in early life may be influenced by sex differences, but it is also possible that these differences are the result of same-sex vs. opposite-sex preferences (the stimulus was female, i.e. vaginal smears).

Sex differences in the reactions of young animals to the odour of a foster species were demonstrated also by Sokolov and Kotenkova (1987), who performed a cross-fostering study with laboratory mice reared by a female rat (*Rattus norvegicus*). Adult mouse females, but not males, cross-fostered at 1–3 days after birth preferred the female urine odour of their foster species, which indicates that females might have been influenced by their early odour experience to a higher extent than the males. On the other hand, the authors used only female odour as the testing stimulus and did not provide a male alternative, i.e. they did not test the opposite-sex preferences in females. The observed pattern might thus be due to differences in reacting to same-sex vs. opposite-sex odours as in the previous study (Quadagno & Banks, 1970).

Further, no sex difference were observed in reciprocally cross-fostered southern grasshopper mouse (*Onychomys torridus*) and white-footed mouse (*Peromyscus leucopus*, McCarty & Southwick, 1977). Pups of both species were reared by a foster female from 2 to 24–26 days after birth. The control group included animals of both species reared by their own mothers. The animals were tested for their preference for bedding soiled by opposite-sex southern grasshopper and white-footed mouse. The results showed that control grasshopper mice of both sexes preferred bedding with the odour of their opposite-sex conspecifics over bedding odourised by white-footed mice. In the treatment group, though, both sexes of the cross-fostered grasshopper mouse showed a weak tendency to prefer the odour of their foster species. White-footed mice in both the control and the treatment group preferred bedding odourised by their foster species. These results were consistent across eight tests performed at 30–100 days of age.

Kirchhof-Glazier (1979) cross-fostered female (males were not tested) albino house mice 0–12 hours after birth with both parents of prairie deer mouse (*Peromyscus maniculatus bairdii*). The control group for every test group consisted of female albino house mice cross-fostered to non-related parents of their own species. After 18 days, the females were kept separately in a male-free room. The animals were tested for their olfactory preferences as either juveniles (30 days) or adults (44 days) or repeatedly at the age of 30 and 44 days using a T-maze. The test stimuli were either sexually experienced males of both species or their soiled bedding. The control group of females, which was tested at 30 days after birth, displayed a significantly higher preference for deer mouse bedding odour than the cross-fostered group did. Similar results were found for cross-fostered females tested at 44 days after birth, where the stimulus was a random sample of bedding. It is possible, however, that these samples were insufficiently saturated with the male odour, and the results might also be explained as the effect of novelty. In the third group, when retested at 44 days, there was no difference between the control and testing group. Experiments with live males used as olfactory stimuli did not show any significant differences between the cross-fostered animals and the control groups.

LeRoy et al. (1982) tested preferences for huddling with rat or mouse stimuli (live animals) in three groups of young animals aged 36 days. One group consisted of young mice (*Mus domesticus*) reared by their own mother in a mouse colony. The second group consisted of young mice reared by their own mother but in rat colony, and the third group consisted of young mice reared by rat foster mothers in a rat colony. Pups were reared by their mothers until weaning (21 days) and then kept separately. All groups preferred to huddle with their own species, except for young mice reared by mouse mothers in a rat colony, who tended to huddle with both rats and mice. Sex differences were not tested.

Zippelius (1972) carried out experiments to test specifically filial olfactory imprinting in the bicolored shrew (*Crocidura leucodon*), whose pups are born without fully developed vision and engage in so-called 'caravanning' behaviour. This means that the first pup joins the mother by gripping, with its mouth, near the base of her tail, the next pup joins the caravan by gripping the sibling in front in the same way, and so on, until all pups follow their mother. They engage in this behaviour if the nest is disturbed after the age of 8 days. The researcher separated the shrew pups from their mother at the age of 8 days until 15 days after birth. Subsequently, the mother's odour was presented on a cloth to the pups without any sound. In this case, the pups displayed caravanning behaviour and followed their mother's odour. In contrast, they did not follow a dummy perfumed with a mixture of maternal and artificial odour. In a follow-up study, a litter of young shrews was cross-fostered at 8–14 days after birth by a bank vole (*Myodes glareolus*). At the age of 15 days (i.e. 7 days after last exposure to their mother's odour), the shrew pups were returned to their biological mother, but although their biological mother started to feed them, they still followed their foster mother instead of their own mother (Zippelius, 1972).

Kendrick et al. (1998) performed reciprocal cross-fostering between sheep and goats to test the stability of their social and sexual preferences. Lambs were cross-fostered immediately after birth, and their social behaviour towards both their conspecifics and the species that fostered them was subsequently followed. As juveniles, both lambs and kids preferred being in the presence of their foster species, and their playing and grooming behaviour likewise resembled the behaviour of their foster species. In adulthood, cross-fostered males of both species preferred to mate and socialise with females of the species that fostered them, and this persisted even after 3 years of being only in the company of their own species. In contrast, the females of both species preferred males of the cross-fostering species only until the age of 1 or 2 years. In the third year of life, all cross-fostered females preferred to mate with individuals of their own species. Interestingly, the same pattern has been demonstrated for visual stimuli: both sexes preferred images of the species that fostered them, but this preference was stronger in males than in females. Unfortunately, the authors did not provide details on either the duration of cross-fostering or the age at which the lambs and kids were placed back among animals of their own species.

Finally, Müller-Schwarze and Müller-Schwarze (1971) tested filial olfactory imprinting in black-tailed deer (*Odocoileus hemionus columbianus*). Male fawns were separated from their mothers 1 hour after birth and reared by a moving dummy with a bottle holder, which also emitted a low bleating sound. The dummy was perfumed with the ischiadic scent of a male pronghorn (*Antilocapra americana*). The fawns' odour preferences were subsequently tested at three different ages (15, 33, and 75 days). Three types of odour stimuli presented side by side were used for perfuming the dummy: (i) pronghorn odour (i.e. the rearing condition), (ii) female black-tail deer tarsal scent, and (iii) female mule deer (*Odocoileus hemionus*) tarsal scent. At 75 days, the test was repeated with the scent of female pronghorn, black-tail deer, and mule deer as the olfactory stimuli. A series of experiments showed that of these three stimuli, *Odocoileus* fawns preferred the ischiadic male scent of the pronghorn, to which they were exposed early in their development.

### 3.2 Olfactory Preferences for Artificial Odours

An alternative approach to studying the development of olfactory preferences consists in exposing animals to a novel odour which they cannot encounter under natural conditions and which interacts with their olfactory rearing environment in a social context. Various studies used as odour stimulus various artificial odours spread over the bedding, in the air, or rubbed on the mother. In most cases, the resulting odour was a mixture of natural and artificial odour.

#### 3.2.1 The Sensitive Period for Learning Olfactory Preferences

Carter and Marr (1970) conducted a series of experiments which tested the development of olfactory preferences until early adulthood. They perfumed the mothers and young domestic guinea pigs on their backs and sides with either ethyl benzoate or acetophenone. Pups in the control group were reared by non-perfumed mothers. Subsequently, the animals' preferences were tested by a three-choice test. In experiment 1, the pups were reared by their mother from 24 hours after birth for 15 or 22 days and then tested after 24 hours of isolation. Pups from both the odour-treated and control group preferred the rearing odour (same-sex animal rubbed with the odour). In experiment 2, the rearing conditions were identical, but the animals were retested on day 16 and 22. Between the testing days, the pups were kept with their biological mother but with no ethyl benzoate or acetophenone perfume (the mother was rubbed with ethanol to remove the rearing odour). Results of the day 16 test were the same as in experiment 1 in both the control pups and the pups reared by ethyl benzoate-perfumed mother, while pups reared by an acetophenone-perfumed mother showed a preference for either acetophenone-perfumed or unperfumed individuals. In the test on day 22, all pups preferred individuals in line with their initial rearing conditions. These results suggest that early odour preferences acquired by exposure to an odour during the first 15 days of life remain stable even after a week of exposure to the natural guinea pig odour. In experiment 3, different groups of pups were exposed to their mother rubbed with experimental odours on days 1–6, 7–12, 10–15, and 16–22 after birth. Subsequently, their preferences were assessed repeatedly 16, 23, and 60 days after birth. During the period between odour exposure and testing, the pups were housed with their unperfumed mother. Pups from both the treatment groups and the control groups preferred unperfumed individuals, especially at 60 days of age, which suggests that initial odour preferences can be modified by later odour experiences. But these results also indicate that, at least in this species, 1–6 days after birth may constitute a sensitive period for developing olfactory preferences, because odour exposure during this period had the greatest impact on the pups' tendency to prefer individuals with a given odour.

While in the studies above, the animals were exposed to the odour stimuli for several days before testing, Porter and Etscorn (1974) tested olfactory preferences after just a short odour exposure. At 1 to 12 hours after birth, spiny mouse (*Acomys cahirinus*) pups and their parents were placed for 24 hours in a cage with bedding scented with either ground cinnamon or cumin. Immediately after this odour exposure, the pups were tested for their cinnamon or cumin preference in a two-choice test. The results showed that the pups preferred the odour they were previously exposed to. In a subsequent experiment, the pups were exposed to the experimental odour for just 1 hour 2–12 hours after birth, and preference tests were performed 24 hours after exposure to the odour had stopped. During the exposure period, the pups were removed from their parents and placed together in an experimental chamber. This time, the cinnamon or cumin odour was not presented on bedding but diffused in the air and delivered from an olfactory chamber through a filter to the experimental chamber. Irrespective of the length or manner of odour exposure, the results were the same as in the first experiment, which indicates that young spiny mice develop olfactory preferences shortly after birth and even a brief exposure to an odour is sufficient for establishing an odour preference.

In a follow-up experiment, spiny mouse pups were exposed to cinnamon odour diffused in the air 2–5 days after birth and tested on the day following the odour exposure. Their preferences were compared with a control group of naïve animals. The results showed that preference for a previously encountered odour was established during the first 3 days after birth, because pups exposed to the experimental odour 4–5 days after birth showed no preference for it and behaved much like pups in the control group (Porter & Etscorn, 1976). Nevertheless, a later study which used a multiple choice test showed that odour preferences of spiny mouse pups can be established even later, up to 16 days after birth (Janus, 1989). In this study, an entire litter of spiny mouse juveniles was exposed to cinnamon or cumin odour for 1.5 hours in a cage without their parents. Control animals were treated in the same way but in a cage with clean bedding. They were exposed to the stimuli at 2, 4, 6, 12, 16, and 18 days after birth, and testing for odour preference was conducted 48 hours after the end of odour exposure. Preference for the treatment odour was observed in animals exposed to it 2–16 days after birth. Spiny mice exposed to the test odour 18 days after birth did not show any preference for it. The control group, as expected, showed no preference for the test odours at any tested age.

Altricial rodents, on the other hand, may need a longer period of odour exposure to develop an olfactory preference. Brunjes and Alberts (1979) tested rat pups reared either by an artificially perfumed biological mother (treated with a Wild Musk spray) or by a control mother treated only with water. Animals were exposed to the odour from the day of birth until the testing day. Using a two-choice test at 5, 10, 15, or 20 days of age, these pups were tested for their huddling preferences with anesthetised prepubertal pups of both sexes sprayed with either the perfume or water. The results showed that preference for perfumed stimuli (the above-mentioned conspecifics) rapidly increased at day 15 and only slightly decreased from day 20. A follow-up experiment tested rat juveniles' preference for an odour they had been previously exposed to as compared to an unfamiliar odour at 15 and 20 days of age. The results showed that all pups spent more time with individuals scented with a previously experienced odour. Brunjes and Alberts (1979) suggest that the first 15 days after birth may coincide with the period when rat pups also start reacting to the pheromones produced by their mother (Leon & Moltz, 1972).

Voznessenskaya et al. (1999) predicted that the duration of early exposure to an odour also increases pups' sensitivity to the odour in question. They exposed young rats and mice to the odour 0.5 ml of conspecific urine spread on familiar cage bedding twice a day during three periods of development, i.e. 1–14, 15–28, and 29–42 days after birth. These periods were selected in view of the generally accepted stages of brain development (brain growth, synaptogenesis, and myelinogenesis, respectively; Dmitrieva & Kassil, 1982). Five weeks after odour exposure, the pups were tested for their sensitivity (reactivity to a sample by searching and sniffing to it) to the urine stimulus presented repeatedly in decreasing concentrations. Both juvenile and adult rats exposed to the odour after opening of their eyes, i.e. 15–28 days after birth, were sensitive to low urine concentrations during testing, while pups exposed to urine odour either earlier or later (after day 28) detected only higher concentrations. The authors suggest that exposure to a urine sample at 15–28 days after birth increases sensitivity to the experimental odour later in life as a result of changes in the periphery or the central nervous system (e.g. new or sensitised pathways within or beyond the olfactory bulb) during the sensitive period.

A similar length of the sensitive period was reported by Marr and Lilliston (1969), who reared rat littermates by their mother but rubbed their backs, bellies, and sides with ethyl benzoate or acetophenone at 3–10, 11–18, or 22–29 days of age. Between exposures to the artificial odours, the pups were rubbed with a cloth smelling of natural rat litter odour. After 30 days, the pups were housed individually in a rat colony. Their odour preferences were tested between 45 and 60 days in a three-arm maze. Each arm contained either a non-perfumed individual or an individual scented with ethyl benzoate or acetophenone. Results showed that the treatment pups preferred individuals rubbed with the odour they had experienced during rearing. The most effective period for developing olfactory preferences was at 3–10, followed by 11–18 days after birth. These results indicate that in rodents, the sensitive period depends on the developmental stage of pups at birth, which varies widely across rodent species.

Differences in the timing and duration of the sensitive period might be due to the differing trajectories of sensory and motor development in the various species and, in particular, due to the level of altriciality at birth (Cornwell-Jones, 1979). Precocial rodents (such as spiny mouse, which are born with well-developed fur, locomotory abilities, and a functioning visual system) are sensitive to exposure several days after birth, while altricial rodents (such as rats, which are born naked, with little locomotion, and non-functional vision) are sensitive to exposure 2 weeks later, at a time when they are usually weaned (Cornwell-Jones, 1979; see also Sects. 3.1.1 and 3.1.2). Moreover, the sensitive period in rodents might be linked to neural plasticity, mainly in the olfactory neural cortex (Leon et al., 1984). This is in line with neuroanatomical studies, which showed an increased development of olfactory bulbs in both precocial (spiny mouse) and altricial (laboratory mouse) rodents during the first 10 days after birth. In precocial rodents, however, the development of olfactory bulbs (maturation of the glomerular layer, external plexiform layer, and the combined internal plexiform, internal granule, and ventricular zone) was almost twice as fast as in altricial species (Brunjes, 1983). Moreover, the end of the tentative sensitive period for the formation of olfactory preferences coincides with reaching the adult developmental stage of olfactory bulbs (Leon et al., 1984). This has been demonstrated in rats, mice, hamsters (Leon et al., 1984), sheep (Salazar et al., 2003), and pigs (Salazar et al., 2004).

### 3.2.2 Formation of Olfactory Preferences by Early Learning

Márquez et al. (2015) tested the development of odour preferences in the common degu (*Octodon degus*), a caviomorph rodent whose pups are born precocial. The pups were reared by their biological mothers and with their littermates, who were all rubbed ventrally and laterally with eucalyptol during the first month after birth. Subsequently, the pups were reared with their littermates until testing. Pups in the control group were reared by non-perfumed biological mothers. At the age of 5–7 months, the adult degus were tested for same-sex odour preferences using either perfumed or non-perfumed animals presented simultaneously in a two-choice Y-maze. Individuals reared by eucalyptol-perfumed mothers showed no preference for either eucalyptol or non-perfumed individuals. Animals in the control group showed a systematic preference for non-perfumed individuals. The authors do not report odour preferences separately for each sex, but the results might suggest that exposure to eucalyptol does not lead to the formation of preferences which would last into adulthood.

The effect of early odour experiences on later preferences was also tested in rats. Moore et al. (1996) perfumed rat dams either with citral (treatment group) or saline (non-perfumed control group) from day 1 until 17 days after birth. From day 25 onwards, the male pups (females were not tested) were reared in small, same-sex groups. On day 95 after birth, they were tested in a two-choice test for their preference of citral-perfumed or non-perfumed oestrous and non-oestrous females. It turned out that males from both the treatment and the control group preferred females perfumed with citral, and this held regardless of the female's oestrous state. Copulatory tests with perfumed and non-perfumed females showed no effect of the earlier citral treatment: both groups of males copulated with perfumed and non-perfumed females similarly often. In a follow-up experiment, the early rearing condition was prolonged until weaning, i.e. to 25 days after birth. A two-choice test showed that males reared by a mother perfumed with citral showed no preference for either citral-perfumed or non-perfumed females, but males reared by a non-perfumed mother preferred non-perfumed females. The hormonal state of the females had no effect on the preferences. Interestingly, while previous odour exposure had no effect on the number of copulations, males in both groups tried to copulate first with a female perfumed with the odour they had been previously exposed to (i.e. citral or the saline control). This indicates that while the males used the opportunity to copulate with various females irrespective of their odour, their preferences were to a certain extent influenced by the early odour exposure (and this showed in their first choice).

Nyby et al. (1978) tested early odour preferences and their modification in later life. They separated male juvenile mice (females were not tested) reared by their biological mothers in four groups depending on their rearing and later experiences: (i) juveniles reared by a mother scented with perfume (Wild Musk) and later exposed to a female scented with the same odour, (ii) reared by a mother scented with perfume but later exposed to a female perfumed with ethanol, (iii) reared by a mother scented with ethanol and later exposed to a female scented with perfume, and (iv) reared by a mother scented with ethanol and later exposed to a female scented with the same odour. The males had only a limited experience of natural female odours. Specifically, they had been exposed only to their scented mother until 21 days after birth and to the odour of an unrelated scented female for several minutes a day for 8 days, approximately at the age of 90 days. For the rest of the time, they were housed either in a same-sex group or individually. Two days after the latter exposure (at the age of 92 days), the males were tested for their preferences for ethanol, perfume-treated females, and female and male urine odour. Males from all four groups displayed preference for female urine odour and not for male urine or ethanol odour. Males exposed to a female scented with the perfume preferred, regardless of their mother's odour, the perfume over the odour of ethanol. Nevertheless, the preference was the strongest in males who had been reared by a perfumed mother and subsequently exposed to a female scented with the same perfume. In contrast, males reared by a perfumed mother and later exposed to an ethanol-scented female showed no preference for either the perfume or ethanol. Males reared by a mother scented with ethanol and later exposed to an ethanol-scented female preferred the odour of female urine and showed no preference for the odour of ethanol, perfume, or male urine. This suggests that adult mouse odour preferences are not determined solely by the conditions of early learning: they are also influenced by later experiences.

### 3.2.3 The Effect of Odour Placement and Reinforcement on the Development of Olfactory Preferences

Fillion and Blass (1986) investigated a possible effect of odour placement on the female rat's body on the acquisition of preference. They had three groups of dams: (i) those perfumed with citral around the nipples and the vagina, (ii) those treated with isotonic saline in the same places, and (iii) those perfumed with citral on their backs. Mate choice of male rats (females were not tested) reared under varying odour conditions was tested 90–120 days after birth. After weaning at 28 days, the males were kept in small same-sex groups without exposure to the artificial odour. Males reared by mothers with citral- or saline-perfumed nipples and vaginal region preferred mating with females perfumed in the perivaginal area with the odour they had been previously exposed to. In contrast, the males reared by mothers perfumed with citral on their back did not show any preference for citral- or saline-treated females. The effect of odour placement on the female's body on pup's preferences was further investigated by Sevelinges et al. (2009). They applied banana oil either on rat mothers' nipples and bellies or within the nest cage twice a day from 2 days before birth until weaning at

25 days, when the animals were tested for their odour preferences using a two-choice test with banana and sunflower oil (control). Preference for the banana oil was found only in the pups whose mothers' nipples and bellies had been perfumed, but not in those who had been exposed to the odour in the nesting cage. This and previous studies (Fillion & Blass, 1986) indicate that methodological discrepancies between studies might, at least in part, be responsible for variations in the results.

Galef (1982) formed two treatment groups of rat pups, which were either reared by a foster mother whose back was perfumed with peppermint or a foster mother who was unperfumed but kept with the pups in a peppermint-scented cage. The control group consisted of pups reared by their biological mother without further odour exposure. Pups from the two treatment groups were reared by foster mothers between 2 and 19 days after birth for different lengths of time (0, 2, 3, 4, 5, 7, 12, 16, 18 days). When not with the foster mother, they were reared by their own parents. The pups were tested for their preference for peppermint odour diffused in the air of the test chamber as compared to a no-odour condition presented sequentially. The results showed that both treatment groups preferred the peppermint odour and their preference increased with the length of previous exposure to it. In contrast, the control group preferred the no-odour condition. In a second experiment, the rat pups were exposed to peppermint odour at 2–19 days after birth and then kept in a rat colony with no exposure to peppermint until the testing day. Animals were tested 19, 20, 26, 34, 50, or 65 days after birth. Animals exposed to the odour on their mother's back showed a greater preference for the peppermint odour than the group that had been exposed to a scented cage. In the group exposed to the scented cage, peppermint odour preference rapidly decreased after 19 days of age. In contrast, the preference of pups who had been exposed to the peppermint odour on the mother's back remained constant until 50 days of age and slightly decreased only 65 days after birth. As in the previous experiment, the control pups preferred the non-odour condition. In a third experiment, pups were either exposed to the peppermint odour on their foster mother or placed in a foster family cage scented with peppermint extract between 10–19 days and 21–30 days of age and tested immediately after odour exposure. Pups which had been exposed to the peppermint odour between 10 and 19 days of life showed a greater preference for the odour than those exposed to it on days 21–30, and this was the case regardless of the odour placement. In the group exposed to the peppermint odour at 21–30 days of age, odour placement influenced the odour preference: animals exposed to the peppermint odour on their mother's back showed a greater preference for it than animals exposed to the scented cage (Galef, 1982). Contrary to previous studies (Fillion & Blass, 1986; Sevelinges et al., 2009), this series of experiments indicates that odour presented on the mother's back and belly (but not in the home cage) can lead to a preference for the experimental odour.

Blais et al. (2006) tested the effect of reinforcement on odour preferences in rats. Rat pups were reared by a mother perfumed with anise odour on her belly at 1–20 days after birth. The control group was reared by a mother sprayed with water. From day 21 until testing, the pups were kept in small same-sex groups. Preference tests were performed at 21 days and repeated at 40 days of age using a two-choice Y-maze. The results showed a preference for the previously experienced odour in both the anise-treated group and the control group, but only in the earlier test at 21 days of age. At 40 days, animals from both groups showed no systematic preference for either the anise odour or water. The results were the same for both sexes. In the following experiment, carried out during 4 days following the second preference test, the anise-treated and control rats were presented with a Y-maze compartment with anise odour and reinforced with food rewards to enter. Each day, their preference for anise was tested. Both the anise odour-treated and the control group preferred the anise odour. Preference for the anise odour increased faster in the anise-reared group than in the control group. The final experiment aimed at testing the alleged irreversibility of odour preferences after reinforcement at 2 months after birth. To do so, the animals were divided in five groups and retrained in a Y-maze at either 5, 6, 7, 8, or 9 months after the original Y-maze training. The results indicated that the animals which had experienced anise odour before weaning preferred that odour to a higher extent than those in the control group, although both groups had been trained to prefer the anise odour at the age of 2 months. This suggests that olfactory learning that takes place during the sensitive period in combination with reinforcement is more effective for acquisition of long-term odour preferences than reinforcement alone: reinforcement thus seems to support or boost the stability of preferences acquired during early development.

### 3.2.4 Suitability of Artificial Odours

Marr and Gardner (1965) investigated rat pups reared either by mothers perfumed with methyl salicylate or with a commercial perfume (Yardley's Red Roses) or by non-perfumed mothers during the first 21 days after birth. Subsequently, the juvenile rats were tested for their odour preferences at the age of 30–45 days using a two-choice test (with live animals behind a screen as stimuli). Rats reared by a mother perfumed with methyl salicylate or by a non-perfumed mother preferred non-perfumed individuals to individuals perfumed with either methyl salicylate or Yardley's Red Roses. Animals reared by mothers scented with Yardley's Red Roses preferred perfumed individuals to naturally or methyl salicylate-smelling animals. The sex of the stimulus animals was not reported, and sex differences in the perceiver rats were not tested. Similar results were found in a mounting test at 75 days of age. Females reared by non-perfumed mothers copulated significantly more often with non-perfumed males. In contrast, females reared by mothers perfumed either with methyl salicylate or Yardley's Red Roses showed significantly less sexual response to non-perfumed males (Marr & Gardner, 1965). This indicates that some artificial odours (here Yardley's Red Roses) are more likely than certain other artificial odours (methyl salicylate) to lead to long-lasting preferences.

Caza and Spear (1984) exposed rat pups aged 15 days for 30 or 60 minutes to a cotton pad scented either with lemon or with methyl salicylate. The control group was exposed to an unscented cotton pad. Immediately after exposure, the pups were tested for their odour preferences in a two-choice preference test. The pups exposed to a perfumed pad preferred the experienced odour. Interestingly, pups from the control group preferred methyl salicylate to the lemon-scented or unscented pad. This might be due to the novelty effect or some inherently alluring properties of methyl salicylate. The results were similar after a 3-, 9-, and 27-minute-long exposure. In a follow-up experiment, the animals were exposed to lemon odour for 3, 27, or 81 minutes at the age of 10, 15, or 60–75 days after birth. A test of preference for the odour of lemon or methyl salicylate was performed immediately after the end of odour exposure. As in the previous experiment, the group exposed to lemon preferred the lemon odour, but the control group preferred methyl salicylate. Duration of the exposure did not increase the preference for lemon odour in any of the age groups. The experiment was performed once again, at the age of 10 or 15 days, with exposure to lemon lasting either 3 or 60 minutes. The preference test was done at three different points: immediately after exposure, 1 hour, and 24 hours after exposure. The results showed that preference for the lemon odour persisted for 24 hours after exposure, and it was most pronounced after a 60-minute exposure in 10-day-old animals.

Galef and Kaner (1980) separated rat pups from their mother and siblings 36–48 hours after birth, anesthetised them, and provided them with a stomach cannula. The pups were subsequently reared individually with no odour exposure and fed through the cannula until 19 days after birth. At the age of 25–27 days, the pups were tested for their preference for conspecific (maternal anal excreta) or artificial (peppermint) odours, neither of which they had previously experienced. The odours were presented sequentially, and preference was operationalised as the time spend sniffing the presented odour. The results of the artificially reared group were compared to those of a control group where pups were reared with their mother and siblings of both sexes. All groups preferred the odour of their conspecifics to the artificial odour. It seems therefore that preference for at least some conspecific odour (maternal excreta) might be due to a predisposition (for discussion details, see Sect. 4.5 *Predisposition to olfactory preferences*).

# 4 A Critical Assessment

## 4.1 The Sensitive Period

It has been predicted that animals learn their olfactory preferences during a specific stage of life known as the 'sensitive period' (Lorenz, 1937; Fabricius, 1964; Bateson, 1979). Based on visual imprinting studies in birds, it has been hypothesised that the sensitive period for developing olfactory preferences coincides with the early postnatal stage of development.

To examine the sensitive period, one needs to expose animals of different ages to the target olfactory stimulus and then test for preference of that stimulus later in life. If there is a sensitive period, exposure at a certain age (but not at any other age) leads to the establishment of a preference. Several studies showed that exposure several days after birth, but not later, led to the establishment of preferences in spiny mice and guinea pigs (Carter & Marr, 1970; Porter & Etscorn, 1976; Janus, 1989). In rats and mice, on the other hand, exposure to odours several days after birth did not lead to the formation of a preference. Such preference was established only if the exposure occurred 10–21 days after birth (Marr & Lilliston, 1969; Voznessenskaya et al., 1994).

Most studies on the development of olfactory preferences agree that odour learning tends to be restricted to a short period after birth (e.g. Carter & Marr, 1970; Porter & Etscorn, 1976), but none of these studies differentiate between filial and sexual imprinting. Previous studies in birds indicate that filial and sexual imprinting involve activation of different areas of the brain (for a review, see Bischof, 1983), which suggests that they might be two distinct processes. For example, the sensitive period for filial imprinting in some birds occurs just few hours after hatching (in domestic chicks, Suge & McCabe, 2004), while the sensitive period for sexual imprinting seems to be much longer (in zebra finches, Bischof, 2003).

To the best of our knowledge, no study so far tested the sensitive period for olfactory sexual imprinting in mammals. Studies on the development of odour preferences in adulthood investigated the sensitive period for juvenile olfactory preferences, i.e. the sensitive period for filial imprinting, instead of focusing on the sensitive period for developing adult odour preferences (e.g. Albonetti & D'udine, 1986; Quadagno & Banks, 1970). We cannot, therefore, distinguish whether adult olfactory preferences develop at the same time as the early olfactory preferences. In other words, it is not known whether adult olfactory preferences are the result of early odour experiences that took place during the sensitive period in early development, or whether there is also a different sensitive period for adult preferences. Moreover, it is important to note that even if there is a sensitive period when individuals are more receptive to learning new odours, this does not necessarily mean that they cannot learn new olfactory preferences later in life. If the latter is the case, though, such acquisition of olfactory preferences later in life has all the hallmarks of other learning processes: except for modification of preferences in sexual imprinting (see Sect. 4.4 *Consolidation*).

### 4.2 No Reinforcement

The key difference between associative learning and imprinting is that in the latter, preference for a particular object or quality is established after a short exposure with no need for further reinforcement (Bolhuis et al., 1990). The 'mere exposure' effect is well-documented in both birds and mammals (Alberts, 1981). For instance, spiny mouse pups exposed to cumin or cinnamon odour for an hour within the first 12 hours after birth showed subsequent preference for the experienced odour 24 hours after exposure to it (Porter & Etscorn, 1974). In a similar way, in other studies, animals exposed to an odour for a limited time with no reinforcement preferred the experienced odour at least until weaning (Caza & Spear, 1984; Janus, 1989). But it has also been shown that the formation of olfactory preferences does require some type of external trigger, such as a simulation of stimulation provided by the mother (e.g. stimulation of the genital area with a brush, which emulates licking; Pedersen et al., 1982). Moreover, some authors assume that odours placed on the mother, especially on her belly and nipples, lead to a stronger preference than odours spread in the cage or on the mother's back (Galef, 1982; Fillion & Blass, 1986; Sevelinges et al., 2009). The pups can associate the odour with conventional reinforcers linked with feeding or maternal care, which can be seen as a form of external reinforcement. For example, rat pups reared by a mother whose nipples and vaginal area were scented with citral continued to prefer citral odour in adulthood, although they were after weaning reared together with their non-perfumed conspecifics. In contrast, pups reared by a mother scented with citral on her back did not display any systematic preference pattern (Fillion & Blass, 1986). One could thus argue that under natural conditions, early odour learning in mammals involves some reinforcement in the form of nursing or physical interaction with littermates or nest materials. Similarly, under laboratory conditions, it is possible that experimenters neglect some conditions which can reinforce the animals. Preferences formed after a short exposure with no further apparent reinforcement have been reported, but it is unclear how long they persist. For instance, Blais et al. (2006) showed in rat pups that reinforcement strengthens the preference for anise odour especially in animals who experienced the odour in early development. But without reinforcement, this odour preference did not last 20 days after exposure to the odour, as indicated by the results of testing the pups 40 days after birth. This indicates that the formation of stable preferences that last from early life into adulthood requires some reinforcement.

### 4.3 Irreversibility

Imprinting studies in birds indicate that preference for sensory characteristics acquired early in life is difficult or even impossible to modify (see Sect. 2 *Beginning with birds: The imprinting paradigm*). This feature is commonly known as irreversibility (Lorenz, 1937), whereby it is assumed that after consolidation, preferences should be irreversible. A modification of preferences during sexual imprinting can be, however, compatible with irreversibility of the acquired preferences.

Interestingly, there are only a handful of studies which tested the irreversibility of preferences during sexual imprinting. Kendrick et al. (1998) tested the preference of cross-fostered goats and sheep for their own and foster species during four mating seasons. While females preferred to mate with the foster species during the first mating season and then preferred their own species, male preferences for females of the foster species persisted for three mating seasons. During the fourth season, the males showed no preference for either their own or the foster species. This suggests that irreversibility is not as absolute as originally assumed and, moreover, that the stability of preferences may vary between the sexes. On the other hand, the stability of preferences in males could also be influenced by rejection by females of the foster species, which is why it is unfortunate that the authors did not provide details regarding the females' behaviour towards the males. One can thus only speculate whether the males reoriented their preferences as the result of weakening of preferences or due to some other factor. To the best of our knowledge, no other study so far monitored olfactory sexual preferences over several mating seasons. Studies on rodents tested their odour preferences until 120 days after birth, but only a handful of studies tested the animals repeatedly (Marr & Gardner, 1965; Quadagno & Banks, 1970; Galef, 1982; Albonetti & D'udine, 1986). Further, a study by Quadagno and Banks (1970) indicated that irreversibility of preferences may vary between species. Social preferences persisted in laboratory mice, but not in pygmy mice, between 75 and 115 days after birth, but a test for sexual preferences at 120 days of age showed no preference in either species. Irreversibility might thus depend both on the species and the context. More research is needed to determine to what extent preferences are indeed irreversible.

On the other hand, considerably more studies tested early odour preferences repeatedly before the animals reached maturity (e.g. Carter & Marr, 1970; Voznessenskaya et al., 1994). Brunjes and Alberts (1979) tested odour preferences of rat pups reared by perfumed or non-perfumed mothers at 5, 10, 15, or 20 days of age before weaning. Their results indicate the highest preference for the experienced odour at 15 days of age, while at 20 days of age, the preference had slightly decreased. It is possible that filial imprinting is more significant until weaning, that is, during a period when the young are fully dependent on maternal care. After weaning, preference for the maternal odour is less important and may slowly fade. In other words, it is likely that the irreversibility of preference for certain characteristics acquired during filial imprinting applies only to a period until weaning. Future studies should directly test this speculation.

### 4.4 Consolidation

Imprinting is further characterised by the consolidation process. During consolidation, the acquired characteristics are transferred to the long-term memory. In filial imprinting, this takes place several hours after acquisition (Suge & McCabe, 2004), but in sexual imprinting, acquisition is followed by a large time gap, and consolidation takes place later (Bischof, 1979; Oetting et al., 1995). This later consolidation in sexual imprinting enables experience-based preference modification – mainly after reaching maturity. In general, this leads to a consolidation of optimised preferences for available partners, because some animals may not encounter individuals resembling the initially acquired template later in life (Oetting et al., 1995). If there is a sufficient number of partners corresponding to the template, the previously acquired preferences are consolidated in long-term memory, but in case such partners are not present, the preference can be modified based on the available potential mates.

The testing of consolidation of sexual imprinting requires long-term studies, where animals are tested for their preferences twice: once early after exposure to the relevant odour and later after reaching maturity, that is, after the animal experiences various types of odours. Moreover, to estimate the duration of the consolidation phase, one must control for modification of the early preferences at different ages (in separate groups). In studies where animals were kept individually after the initial odour exposure and showed the same early and later preferences, one can assume consolidation of the previously acquired preferences without modification. On the other hand, this could be also the effect of additional exposure, which could lead to artificially induced preferences. If animals are kept together with their conspecifics after their first exposure to an odour, one can expect that their preferences can be modified. In short, the animals might prefer the more recently experienced odour.

For example, animals reared by perfumed or cross-foster mothers until weaning and then kept with conspecifics showed a preference for the odour of both the perfumed/cross-fostered species and their conspecifics (e.g. Carter & Marr, 1970; Galef, 1982; Márquez et al., 2015). This holds for both social (same-sex) and sexual (opposite-sex) odour preferences. On the other hand, control animals who experienced only the natural odour appropriate for their species preferred the natural smell of their conspecifics (Carter & Marr, 1970; Galef, 1982; Márquez et al., 2015). Albonetti and D'udine (1986) cross-fostered two strains of female mice. In early adulthood, the females were familiarised with males of their own strain and showed preference for their own rather than the foster strain, which indicates a modification of initial preferences based on a later exposure. Similar results were reported by Nyby et al. (1978), who reared mice males by perfumed mothers, later exposed them again to perfumed females, and the males had later shown a clear preference for the previously experienced perfume. A repeated experience with perfumed females probably consolidated the initially acquired preference. On the other hand, animals reared by an ethanol-scented mother and later exposed to a perfumed female also showed a preference for the perfume. Still, these studies did not test olfactory preference repeatedly, i.e. shortly after initial exposure to the odour and then after the later experience. One thus cannot determine whether the early preference was maintained until adulthood or modified by the later experience.

Some studies also suggest that odour intensity could be crucial for modification of early preferences, because animals housed individually but exposed to odours from surrounding cages later preferred the odour experienced early in life, while animals exposed to a new odour directly preferred the odour they experienced later (Carter & Marr, 1970; Nyby et al., 1978; Beauchamp & Wellington, 1981).

## 4.5 Predisposition to Olfactory Preferences

There are some stimuli which animals tend to prefer even with no previous experience: this indicates a predisposition for reacting to a particular stimulus (see Sect. 2.1. *Learning predisposition*). In birds, researchers found a predisposition for acquiring a preference for visual stimuli in the form of a bird-shaped dummy: chicks preferred following such dummies even with no previous exposure to them, but they were conditioned by other types of nonspecific stimulation (Bolhuis, 1999). Later, they formed preferences for other objects as well, but their initial preference for the dummy was more dominant and stable (Johnson et al., 1985).

To assess the evidence regarding predispositions and to test their possible role in the development of early olfactory preferences, animals should be reared under different conditions, exposed to either artificial odours or the natural odour of a related species, and then tested repeatedly for preference of the target or conspecifics odours. If there is a predisposition, animals should prefer the experienced stimuli but with each round of testing, preference for the conspecific odour should increase more that preference for the artificially introduced stimuli, which will thus lag behind (see Johnson et al., 1985). This is because under natural conditions, a predisposition should direct the development of odour preference towards the odour of conspecifics. Additional exposure can lead to an increase of preference, but the rate of increase should differ between the tested stimuli, and a greater increase should be observed in preference for the odour related to the predisposition. These results should be compared with a control group reared under natural conditions, and this group should show a greater preference for the experienced odour (i.e. the odour of conspecifics) than the treatment groups. Another possible design for testing predispositions is by comparing preferences between several groups exposed to different conditions at the same age. A group treated with conspecific odour should display a greater preference for this odour than a group treated with an artificial odour. Moreover, such design would dispense with the need for repeated testing.

None of the studies described above tested the evidence for a predisposition directly. Differences in the preference for novel and natural odours were tested only cross-sectionally (Carter & Marr, 1970; Kirchhof-Glazier, 1979), not repeatedly across several sessions. On the other hand, Müller-Schwarze and Müller-Schwarze (1971) tested odour preferences repeatedly. They reared black-tailed deer with a dummy perfumed with pronghorn male scent and then tested the animals twice for their preference for black-tailed and pronghorn male scent. At both testing points, the animals preferred the dummy perfumed with pronghorn odour to the black-tailed deer scent, and there was no increase in their preference for conspecific odour between the testing sessions. These results did not indicate a predisposition for olfactory learning in black-tailed deer. In some cases, though, a tendency towards a predisposition has been indicated. For instance, Galef and Kaner (1980), who investigated in rat juveniles the development of preferences with no previous experience with any odour at 19 days after birth, had shown that the pups naturally preferred rat odour (maternal anal excreta) to an artificial (peppermint) odour. Because the animals were not indifferent in their preferences for two novel odours (conspecific vs. peppermint), this might be seen as evidence of some predisposition to avoid trigeminal stimuli. On the other hand, mammals can learn odour characteristics also prenatally (e.g. Semke et al., 1995; Schaal et al., 2004; Wells & Hepper, 2006), and they are familiar with the odour of their own excreta. For instance, pups which were separated from their mother 36–48 hours after birth thus had the opportunity to experience maternal odour during that period. It is therefore difficult to say whether the tested animals were completely naïve to the odour of their conspecifics. Moreover, the Galef and Kaner (1980) study used as a stimulus a slightly aversive odour (peppermint scent), which may have lowered its attractiveness. It might be more suitable to use some natural odour, such as anise, which is believed to be perceived as pleasant by rodents (Cornwell, 1975), or some less irritating odour.

## 4.6 Sex Differences

Many studies indicate that odour preferences acquired in early development are more robust and last longer in females than in males. Sex differences have been demonstrated in laboratory mice cross-fostered by rats (Sokolov & Kotenkova, 1987). Mice were tested for their preferences for the odour of female rat and mouse urine. The results showed a preference for female urine odour of the foster species – but only in the cross-fostered females. Cross-fostered males showed the same level of preference for female rat and female mice urine. Similar results were found in pygmy mouse females cross-fostered to mice, who preferred the vaginal smears of their foster species, while males preferred the vaginal smears of their own species (Quadagno & Banks, 1970). Both of these studies, however, used only female-derived stimuli, and female reaction to a female odour is same-sex (i.e., social) stimulation, while opposite-sex stimulation in males can be viewed as linked to sexual interactions. Several studies which reported sex differences in olfactory preferences for opposite-sex individuals did not test both sexes simultaneously (Marr & Gardner, 1965; Moore et al., 1996). One therefore cannot directly compare these results. Nevertheless, while these studies found no systematic preferences in males, they did describe stable early preferences in females (Marr & Gardner, 1965; Moore et al., 1996).

Opposite-sex preferences were tested in both sexes of cross-fostered white-footed mice. Both sexes preferred bedding odourised by their foster species (McCarty & Southwick, 1977). In contrast, Kendrick et al. (1998) reported different results in cross-fostered sheep and goats, where the cross-fostered males of both species – but not cross-fostered females – preferred to mate with the foster species. In contrast to rodents, the mating preferences were stronger and lasted longer in the males than in the females. This could be due to the parental system of ungulates, where male contribution is limited or absent. The mother therefore need not serve as a sufficient model for forming female attraction towards male odours. These studies indicate that sex differences in olfactory preferences are context dependent.

Further, it is important to note that sex differences were found mainly in studies which tested preferences in adulthood. Most studies on juvenile preferences do not test for sex differences even when both sexes are included in the experimental sample (e.g. Marr & Gardner, 1965; Caza & Spear, 1984). Further research on sex differences in the development of olfactory preferences at different ages is therefore needed.

## 4.7 The Suitability of Olfactory Stimuli

Further, it has been found that some stimuli lead to the formation of stronger preferences than others. This has been shown, for example, in birds' preference for red objects or auditory stimuli resembling birds' voices (Klopfer, 1959; Schaefer & Hess, 1959).

The acquisition of preference for artificial and natural odours varying in chemical characteristics should be investigated to test the suitability of particular olfactory stimuli for research into early olfactory learning. Preference for characteristics of the more suitable stimuli should be acquired more readily than preference of characteristics of the less suitable stimuli and thus also lead to the formation of stronger and longer-lasting preferences. On top of that, it ought to be borne in mind that suitability of the odour stimuli also depends on the medium used for its presentation.

The testing of olfactory preferences has been conducted using a wide range of natural odours of animal origin, such as faeces, urine, vaginal smears, and fur odour. Researchers also use various odours which animals do not usually encounter, either odours of natural origin, such as spices, or artificial odours, such as various perfumes. The ease of acquisition may, however, vary across the stimuli. For example, Carter and Marr (1970) found that guinea pigs form a preference for ethyl benzoate but not for acetophenone. Similar results have been reported for eucalyptol preference in degus. Several months after the initial exposure to this odour, animals showed no systematic preference for either the conspecific or eucalyptol odour (Márquez et al., 2015). In a study by Nyby et al. (1978), animals were exposed early in life to ethanol but later in life preferred female urine or a perfume over ethanol, which indicates that certain artificial odours may form less stable preferences and it is easier to modify them. Some stimuli, such as eucalyptol, methyl salicylate, or peppermint, lead to a substantial activation of the trigeminal nerve, eliciting an irritation (Marr & Gardner, 1965; Galef & Kaner, 1980). Such stimuli are less suitable for olfactory learning because animals tend to avoid irritants. On the other hand, there are also artificial odours – mostly perfumes based on natural components (e.g. the Wild Musk perfume) – which animals seem to find even more attractive than the natural odours of their conspecifics (Nyby et al., 1978).

Furthermore, the medium of odour presentation (excreta, bedding, air stream, a live animal) may also affect the suitability of the stimulus (Johnston, 1981). Galef (1982) showed that rats exposed to peppermint odour on their mother's back showed a greater preference for peppermint than the group exposed to a scented cage. In contrast, Kirchhof-Glazier (1979) who reared female house mice with deer mouse parents reported no difference between preference for a live animal or bedding. Similarly, there was no difference in spiny mouse preferences for cumin or cinnamon sprinkled on bedding or sprayed in the air (Porter & Etscorn, 1974).

## 4.8 Multimodality

So far, we focused on olfactory preferences. In real life, however, most objects can be perceived by multiple senses. It is therefore of great importance to investigate how perception by different senses contributes to the overall preference. Several previous studies reported that objects presented in a form enabling multimodal stimulation led to the formation of stronger preferences. For example, birds are more likely to follow an object if both visual and auditory cues are involved during acquisition of the object's characteristics (Klopfer, 1959). Further, it has been observed in birds that acoustic stimuli on their own lead to the formation of weaker preferences than visual stimuli do (Van Kampen & Bolhuis, 1991). In contrast, in sea lions, visual and acoustic stimuli lead to the formation of similarly strong preferences (Lynn et al., 2010). This difference between birds and sea lions could be due to differences in their way of life. Female sea lions communicate with their pups over long distances while hunting in the ocean (Schusterman et al., 1992). It seems, therefore, that the relative significance of particular modalities may depend on the ecology and mode of communication of the species in question.

To specifically test early olfactory learning, animals should be exposed solely to olfactory stimuli with no contribution from other modalities. Similarly, olfactory preferences should be tested without the involvement of other modalities.

Because newborn mammals are fully dependent on nursing from their mother, it is most unlikely that under natural conditions, their experience is limited to olfaction. The only way of disentangling this issue is to experimentally isolate the newborn animals immediately after birth. In this context, we should note that is has been shown that some stimulation imitating maternal behaviour is needed to evoke a preference in later life (Pedersen et al., 1982). In a majority of studies outlined above (for notable exceptions, see Galef & Kaner, 1980; Caza & Spear, 1984; Janus, 1989), the acquisition of olfactory characteristics was based on multimodal stimulation. For example, Kendrick et al. (1998) cross-fostered goats and sheep and in adulthood tested their social and mating preferences in an open field. In this setup, however, it remains unclear whether the preferences displayed were in fact based on olfactory preferences.

No involvement of other modalities is guaranteed in studies that used solely olfactory stimuli, such the odour of excreta on bedding (e.g. Beauchamp & Wellington, 1981; Albonetti & D'udine, 1986; Sokolov & Kotenkova, 1987; Zippelius, 1972). In contrast, exposure to live animals – even from a distance – probably involves other modalities (e.g. Marr & Lilliston, 1969; Carter & Marr, 1970; LeRoy et al., 1982), because the animals can produce ultra/sounds, which can in turn influence the odour preferences.

## 5 Conclusions

Despite extensive research on the development of olfactory preferences, there is no clear conclusion as to whether these preferences in mammals are established based on olfactory imprinting or some other mechanism. If imprinting (as postulated by Lorenz) is involved, olfactory preferences should be acquired within a sensitive period with no need for reinforcement, and, once consolidated, they should be irreversible or at least stable until adulthood.

Most studies reviewed here indicate that mammals (mostly rodents) acquire olfactory preferences during a limited period in their early development. This suggests the existence of a sensitive period. But it remains unclear whether the sensitive periods for juvenile and adult odour preferences (filial vs. sexual olfactory imprinting) coincide, and, so far, no studies tested specifically the existence or timing of a sensitive period in olfactory sexual imprinting.

Olfactory preferences can be acquired with no apparent reinforcement, but some form of reinforcement might be needed to form stable preferences. The situation is further complicated by the fact that natural conditions do not meet the no-reinforcement requirement, because maternal care or feeding can mediate associative learning. The no-reinforcement condition may also not be fully controlled in laboratory experiments. Moreover, similar to the acquisition of visual preferences in birds, olfactory learning in mammals may require some nonspecific stimulation (e.g. maternal care), and non-conventional reinforcing can be highly effective in facilitating early learning of odour stimuli. The development of olfactory preferences and behaviour observed under natural conditions could thus be influenced by several processes which are difficult to disentangle.

Studies on the early development of odour preferences often show a stability of preference until weaning or shortly afterwards. But these studies should also test odour preferences repeatedly, after maturity, to examine their irreversibility and their relation to adult odour preferences. The results of studies on the stability of odour preferences in adulthood are highly dependent on the methods used, in particular on the duration of odour exposure, treatment of the animals after exposure, and the type of exposure to the stimuli. Moreover, stability of adult odour preferences may also vary across species and among the sexes within species. For instance, it seems that in females, odour preferences are more stable than in males.

Only a handful of studies tested the consolidation and modification of adult odour preferences. The stability of preferences also depends on the suitability of the stimuli used: both conspecific and more intense odour stimuli (e.g. urine samples) tend to have a greater effect on the formation of preferences than artificial or less intense odours (e.g. randomly collected bedding) do. Species-specific suitability of odour stimuli together with stability of odour preferences could also be related to a predisposition to learn, but studies on predispositions in olfactory learning deliver inconsistent results.

Moreover, it seems that early olfactory learning in mammals might be part of multimodal learning rather than a phenomenon based solely on olfaction (e.g. Marr & Lilliston, 1969; Carter & Marr, 1970; LeRoy et al., 1982). Under natural conditions, other modalities are usually involved as well. In fact, experimental setups where animals are tested for their preferences with live animals as stimuli could be easily influenced by other modalities.

In experiments on olfactory imprinting, it is important to appreciate the complexity of this phenomenon. Imprinting is usually divided in two separate phenomena: filial and sexual imprinting. Unfortunately, only a handful of studies have tested both. For example, studies that directly test filial olfactory imprinting tend to focus on the sensitive period or stability of preference until weaning, but not on the other imprinting conditions (e.g. the no-reinforcement condition). Similarly, studies investigating sexual imprinting tend to test only adult preferences but not the other features of imprinting (e.g. the sensitive period or irreversibility). Moreover, the variety of methods used does not permit a direct comparison, and this applies especially to sexual imprinting.

Finally, it seems that filial olfactory imprinting could be involved in the development of early odour preferences, because rodents prefer odour characteristics acquired in early development. Still, evidence of any involvement of sexual olfactory imprinting in the acquisition of later preferences is as yet missing. Further systematic studies in mammals (especially taxonomic groups other than rodents) are thus needed to differentiate between the different learning mechanisms, including imprinting, involved in the development of olfactory preferences.

**Acknowledgements** We are grateful to Anna Pilátová, Ph.D., for her highly valuable comments and proofreading. Both authors are supported by the Czech Science Foundation (GACR 18-15168S).

## References

Alberts, J. R. (1981). Ontogeny of olfaction: Reciprocal roles of sensation and behavior in the development of perception. In R. N. Aslin, J. R. Alberts, & M. R. Petersen (Eds.), *Development of perception: Psychobiological perspectives, Volume 1* (pp. 321–357). Academic Press. https://doi.org/10.1016/b978-0-12-065301-0.50022-4

Albonetti, M. E., & D'udine, B. (1986). Social experience occurring during adult life: Its effects on socio-sexual olfactory preferences in inbred mice, *Mus musculus*. *Animal Behaviour, 34*(6), 1844–1847. https://doi.org/10.1016/S0003-3472(86)80270-6

Bateson, P. (1966). The characteristics and context of imprinting. *Biological Reviews, 41*(2), 177–217. https://doi.org/10.1111/j.1469-185X.1966.tb01489.x

Bateson, P. (1979). How do sensitive periods arise and what are they for? *Animal Behaviour, 27*(2), 470–486. https://doi.org/10.1016/0003-3472(79)90184-2

Beauchamp, G. K., & Wellington, J. L. (1981). Cross-species rearing influences urine preferences in wild guinea pigs. *Physiology and Behavior, 26*(6), 1121–1124. https://doi.org/10.1016/0031-9384(81)90218-3

Bischof, H. J. (1979). A model of imprinting evolved from neurophysiological concepts. *Zeitschrift für Tierpsychologie, 51*(2), 126–139. https://doi.org/10.1111/j.1439-0310.1979.tb00677.x

Bischof, H. J. (1983). Imprinting and cortical plasticity: A comparative review. *Neuroscience and Biobehavioral Reviews, 7*(2), 213–225. https://doi.org/10.1016/0149-7634(83)90016-7

Bischof, H. J. (2003). Neural mechanisms of sexual imprinting. *Animal Biology, 53*(2), 89–112. https://doi.org/10.1163/157075603769700313

Blais, I., Terkel, J., & Goldblatt, A. (2006). Long-term impact of early olfactory experience on later olfactory conditioning. *Developmental Psychobiology, 48*(7), 501–507. https://doi.org/10.1002/dev.20176

Bolhuis, J. J. (1999). Early learning and the development of filial preferences in the chick. *Behavioural Brain Research, 98*(2), 245–252. https://doi.org/10.1016/S0166-4328(98)00090-4

Bolhuis, J. J., & Honey, R. C. (1998). Imprinting, learning and development: From behaviour to brain and back. *Trends in Neurosciences, 21*(7), 306–311. https://doi.org/10.1016/S0166-2236(98)01258-2

Bolhuis, J. J., & Trooster, W. J. (1988). Reversibility revisited: Stimulus-dependent stability of filial preference in the chick. *Animal Behaviour, 36*(3), 668–674. https://doi.org/10.1016/S0003-3472(88)80149-0

Bolhuis, J. J., De Vos, G. J., & Kruijt, J. P. (1990). Filial Imprinting and associative learning. *The Quarterly Journal of Experimental Psychology, 42B*(3), 313–329. https://doi.org/10.1080/14640749008401886

Brunjes, P. C. (1983). Olfactory bulb maturation in *Acomys cahirinus*: Is neural growth similar in precocial and altricial murids? *Developmental Brain Research, 8*(2-3), 335–341. https://doi.org/10.1016/0165-3806(83)90016-0

Brunjes, P. C., & Alberts, J. R. (1979). Olfactory stimulation induces filial preferences for huddling in rat pups. *Journal of Comparative and Physiological Psychology, 93*(3), 548–555. https://doi.org/10.1037/h0077571

Carter, C. S., & Marr, J. N. (1970). Olfactory imprinting and age variables in the guinea-pig, *Cavia porcellus*. *Animal Behaviour, 18*(2), 238–244. https://doi.org/10.1016/S0003-3472(70)80033-1

Caza, P. A., & Spear, N. E. (1984). Short-term exposure to an odor increases its subsequent preference in preweanling rats: A descriptive profile of the phenomenon. *Developmental Psychobiology, 17*(4), 407–422. https://doi.org/10.1002/dev.420170407

Cornwell, C. A. (1975). Golden hamster pups adapt to complex rearing odors. *Behavioral Biology, 14*(2), 175–188. https://doi.org/10.1016/S0091-6773(75)90174-1

Cornwell-Jones, C. A. (1979). Olfactory sensitive periods in albino rats and golden hamsters. *Journal of Comparative and Physiological Psychology, 93*(4), 668–676. https://doi.org/10.1037/h0077596

Dmitrieva, N. I., & Kassil, V. G. (1982). Effect of various environmental factors on brain development. *Arkhiv Anatomii, Gistologii i Embriologii, 83*(9), 84–90.

Fabricius, E. (1964). Crucial periods in the development of the following response in young nidifugous birds. *Zeitschrift für Tierpsychologie, 21*(3), 326–337. https://doi.org/10.1111/j.1439-0310.1964.tb01198.x

Fillion, T. J., & Blass, E. M. (1986). Infantile experience with suckling odors determines adult sexual behavior in male rats. *Science, 231*(4739), 729–731. https://doi.org/10.1126/science.3945807

Galef, B. G. (1982). Acquisition and waning of exposure-induced attraction to a nonnatural odor in rat pups. *Developmental Psychobiology, 15*(5), 479–490. https://doi.org/10.1002/dev.420150510

Galef, B. G., & Kaner, H. C. (1980). Establishment and maintenance of preference for natural and artificial olfactory stimuli in juvenile rats. *Journal of Comparative and Physiological Psychology, 94*(4), 588–595. https://doi.org/10.1037/h0077693

Gaston, M. G., Stout, R., & Tom, R. (1969). Imprinting in guinea pigs. *Psychonomic Science, 16*(1), 53–54. https://doi.org/10.3758/BF03331912

Hampton, N. G., Bolhuis, J. J., Horn, G., & Honey, R. (1995). Induction and development of a filial predisposition in the chick. *Behaviour, 132*(5–6), 451–477. https://doi.org/10.1163/156853995X00667

Hayashi, S., & Kimura, T. (1978). Effects of exposure to males on sexual preference in female mice. *Animal Behaviour, 26*(1), 290–295. https://doi.org/10.1016/0003-3472(78)90030-1

Hess, E. H. (1959). Imprinting. An effect of early experience, imprinting determines later social behavior in animals. *Science, 130*(3368), 133–141. https://doi.org/10.1126/science.130.3368.133

Horn, G., & McCabe, B. J. (1984). Predispositions and preferences. Effects on imprinting of lesions to the chick brain. *Animal Behaviour, 32*(1), 288–292. https://doi.org/10.1016/S0003-3472(84)80349-8

Immelmann, K. (1972). Sexual and other long-term aspects of imprinting in birds and other species. In D. S. Lehrman, R. A. Hinde, & E. Shaw (Eds.), *Advances in the study of behavior* (Vol. 4, pp. 147–174). Academic Press. https://doi.org/10.1016/S0065-3454(08)60009-1

Janus, C. (1989). The development of olfactory preferences for artificial odors briefly experienced by the precocial spiny mouse young. *Behavioral and Neural Biology, 52*(3), 430–436. https://doi.org/10.1016/S0163-1047(89)90577-3

Johnston, R. E. (1981). Attraction to odors in hamsters: An evaluation of methods. *Journal of Comparative and Physiological Psychology, 95*(6), 951–960. https://doi.org/10.1037/h0077840

Johnson, M. H., Bolhuis, J. J., & Horn, G. (1985). Interaction between acquired preferences and developing predispositions during imprinting. *Animal Behaviour, 33*(3), 1000–1006. https://doi.org/10.1016/S0003-3472(85)80034-8

Kendrick, K. M., Hinton, M. R., Atkins, K., Haupt, M. A., & Skinner, J. D. (1998). Mothers determine sexual preferences. *Nature, 395*(6699), 229–230. https://doi.org/10.1038/26129

Kirchhof-Glazier, D. A. (1979). Absence of sexual imprinting in house mice cross-fostered to deermice. *Physiology and Behavior, 23*(6), 1073–1080. https://doi.org/10.1016/0031-9384(79)90299-3

Klinghammer, E., & Hess, E. H. (1964). Imprinting in an altricial bird: The blond ring dove (*Streptopelia risoria*). *Science, 146*(3641), 265–266. https://doi.org/10.1126/science.146.3641.265

Klopfer, P. H. (1959). The development of sound-signal preferences in ducks. *The Wilson Bulletin, 71*(3), 262–266.

Knudsen, E. I. (2004). Sensitive periods in the development of the brain and behavior. *Journal of Cognitive Neuroscience, 16*(8), 1412–1425. https://doi.org/10.1162/0898929042304796

Lemaire, B. S., Rucco, D., Josserand, M., Vallortigara, G., & Versace, E. (2021). Stability and individual variability of social attachment in imprinting. *Scientific Reports, 11*(1), 7914. https://doi.org/10.1038/s41598-021-86989-3

Leon, M., & Moltz, H. (1972). The development of the pheromonal bond in the albino rat. *Physiology and Behavior, 8*(4), 683–686. https://doi.org/10.1016/0031-9384(72)90095-9

Leon, M., Coopersmith, R., Ulibarri, C., Porter, R. H., & Powers, J. B. (1984). Development of olfactory bulb organization in precocial and altricial rodents. *Developmental Brain Research, 12*(1), 45–53. https://doi.org/10.1016/0165-3806(84)90175-5

LeRoy, J. R., Roy, M. A., & Briley, S. M. (1982). Social preferences in mice influenced by ambient stimuli in the general rearing environment. *Animal Behaviour, 30*(3), 939–941.

Lorenz, K. Z. (1937). The companion in the bird's world. *The Auk, 54*(3), 245–273. https://doi.org/10.2307/4078077

Lynn, B. L., Reichmuth, C., Schusterman, R. J., & Gulland, F. M. D. (2010). Filial imprinting in a Steller sea lion (*Eumetopias jubatus*). *Aquatic Mammals, 36*(1), 79–83. https://doi.org/10.1578/AM.36.1.2010.79

Márquez, N., Martínez-Harms, J., Vásquez, R. A., & Mpodozis, J. (2015). Early olfactory environment influences social behaviour in adult octodon degus. *PLoS ONE, 10*(2), e0118018. https://doi.org/10.1371/journal.pone.0118018

Marr, J. N., & Gardner, L. E. (1965). Early olfactory experience and later social behavior in the rat: Preference, sexual responsiveness, and care of young. *Journal of Genetic Psychology, 107*(1), 167–174. https://doi.org/10.1080/00221325.1965.10532774

Marr, J. N., & Lilliston, L. G. (1969). Social attachment in rats by odor and age. *Behaviour, 33*(3–4), 277–282. https://doi.org/10.1163/156853969X00116

McCarty, R., & Southwick, C. H. (1977). Cross-species fostering: Effects on the olfactory preference of *Onychomys torridus* and *Peromyscus leucopus*. *Behavioral Biology, 19*(2), 255–260. https://doi.org/10.1016/S0091-6773(77)91542-5

Moore, C. L., Jordan, L., & Wong, L. (1996). Early olfactory experience, novelty, and choice of sexual partner by male rats. *Physiology and Behavior, 60*(5), 1361–1367. https://doi.org/10.1016/S0031-9384(96)00249-1

Müller-Schwarze, D., & Müller-Schwarze, C. (1971). Olfactory Imprinting in a precocial mammal. *Nature, 229*(5279), 55–56. https://doi.org/10.1038/229055a0

Nakamori, T., Sato, K., Atoji, Y., Kanamatsu, T., Tanaka, K., & Ohki-Hamazaki, H. (2010). Demonstration of a neural circuit critical for imprinting behavior in chicks. *Journal of Neuroscience, 30*(12), 4467–4480. https://doi.org/10.1523/JNEUROSCI.3532-09.2010

Nice, M. M. (1950). Development of a Redwing (*Agelaius phoeniceus*). *The Wilson Bulletin, 62*(2), 87–93.

Nyby, J., Whitney, G., Schmitz, S., & Dizinno, G. (1978). Postpubertal experience establishes signal value of mammalian sex odor. *Behavioral Biology, 22*(4), 545–552. https://doi.org/10.1016/S0091-6773(78)92745-1

Oetting, S., Prove, E., & Bischof, H. J. (1995). Sexual imprinting as a 2-stage process – Mechanisms of information-storage and stabilization. *Animal Behaviour, 50*(2), 393–403. https://doi.org/10.1006/anbe.1995.0254

Pedersen, P. E., Williams, C. L., & Blass, E. M. (1982). Activation and odor conditioning of suckling behavior in 3-day-old albino rats. *Journal of Experimental Psychology: Animal Behavior Processes, 8*(4), 329–341. https://doi.org/10.1037/0097-7403.8.4.329

Porter, R. H., & Etscorn, F. (1974). Olfactory imprinting resulting from brief exposure in *Acomys cahirinus*. *Nature, 250*(5469), 732–733. https://doi.org/10.1038/250732a0

Porter, R. H., & Etscorn, F. (1976). A sensitive period for the development of olfactory preference in *Acomys cahirinus*. *Physiology and Behavior, 17*(1), 127–130. https://doi.org/10.1016/0031-9384(76)90278-X

Quadagno, D. M., & Banks, E. M. (1970). The effect of reciprocal cross fostering on the behaviour of two species of rodents, *Mus musculus* and *Baiomys taylori ater*. *Animal Behaviour, 18*(2), 379–390. https://doi.org/10.1016/S0003-3472(70)80051-3

Ramsay, A. O., & Hess, E. H. (1954). A laboratory approach to the study of imprinting. *The Wilson Bulletin, 66*(3), 196–206.

Rattenborg, N. C., Martinez-Gonzalez, D., Roth, T. C., & Pravosudov, V. V. (2011). Hippocampal memory consolidation during sleep: A comparison of mammals and birds. *Biological Reviews, 86*(3), 658–691. https://doi.org/10.1111/j.1469-185X.2010.00165.x

Salazar, I., Lombardero, M., Alemañ, N., & Sánchez Quinteiro, P. (2003). Development of the vomeronasal receptor epithelium and the accessory olfactory bulb in sheep. *Microscopy Research and Technique, 61*(5), 438–447. https://doi.org/10.1002/jemt.10362

Salazar, I., Sánchez Quinteiro, P., Lombardero, M., Aleman, N., & Fernández de Trocóniz, P. (2004). The prenatal maturity of the accessory olfactory bulb in pigs. *Chemical Senses, 29*(1), 3–11. https://doi.org/10.1093/chemse/bjh001

Salzen, E. A., Lily, R. E., & McKeown, J. R. (1971). Colour preference and imprinting in domestic chicks. *Animal Behaviour, 19*(3), 542–547. https://doi.org/10.1016/S0003-3472(71)80109-4

Schaal, B., Hummel, T., & Soussignan, R. (2004). Olfaction in the fetal and premature infant: Functional status and clinical implications. *Clinics in Perinatology, 31*(2), 261–285. https://doi.org/10.1016/j.clp.2004.04.003

Schaefer, H. H., & Hess, E. H. (1959). Color preferences in imprinting objects. *Zeitschrift Für Tierpsychologie, 16*(2), 161–172. https://doi.org/10.1111/j.1439-0310.1959.tb02051.x

Schusterman, R. J., Gisiner, R., & Hanggi, E. B. (1992). Imprinting and other aspects of pinniped-human interactions. In H. Davis & D. Balfour (Eds.), *The inevitable bond* (pp. 334–356). Cambridge University Press.

Semke, E., Distel, H., & Hudson, R. (1995). Specific enhancement of olfactory receptor sensitivity associated with foetal learning of food odors in the rabbit. *Naturwissenschaften, 82*(3), 148–149. https://doi.org/10.1007/BF01177279

Sevelinges, Y., Lévy, F., Mouly, A. M., & Ferreira, G. (2009). Rearing with artificially scented mothers attenuates conditioned odor aversion in adulthood but not its amygdala dependency. *Behavioural Brain Research, 198*(2), 313–320. https://doi.org/10.1016/j.bbr.2008.11.003

Smith, F. V., & Bird, M. W. (1964). The correlation of responsiveness to visual and auditory stimuli in the domestic chick. *Animal Behaviour, 12*(2–3), 259–263. https://doi.org/10.1016/0003-3472(64)90010-7

Sokolov, V. E., & Kotenkova, E. V. (1987). Reaction of house mice to the smell of individuals of the same species: The effect of early olfactory experience. *Izvestiia Akademii Nauk SSSR. Seriia Biologicheskaia, 2*, 165–172.

Suge, R., & McCabe, B. J. (2004). Early stages of memory formation in filial imprinting: Fos-like immunoreactivity and behavior in the domestic chick. *Neuroscience, 123*(4), 847–856. https://doi.org/10.1016/j.neuroscience.2003.11.002

Van Kampen, H. S., & Bolhuis, J. J. (1991). Auditory learning and filial imprinting in the chick. *Behaviour, 117*(3–4), 303–319. https://doi.org/10.1163/156853991X00607

Voznessenskaya, V. V., Parfyonova, V. M., & Wysocki, C. J. (1994). Induced olfactory sensitivity in rodents: A general phenomenon. In R. Apfelbach (Ed.), *Advances in biosciences* (Vol. 93, pp. 399–406). Pergamon Press Ltd.

Voznessenskaya, V. V., Wysocki, C. J., Chukhrai, E. S., Poltorack, O. M., & Atyaksheva, L. F. (1999). Long-lasting effects of chemical exposures in mice. In R. E. Johnston, D. Müller-Schwarze, & P. W. Sorensen (Eds.), *Advances in chemical signals in vertebrates* (pp. 563–571). Springer. https://doi.org/10.1007/978-1-4615-4733-4_50

Wells, D. L., & Hepper, P. G. (2006). Prenatal olfactory learning in the domestic dog. *Animal Behaviour, 72*(3), 681–686. https://doi.org/10.1016/j.anbehav.2005.12.008

Yamaguchi, S., Aoki, N., Kitajima, T., Iikubo, E., Katagiri, S., Matsushima, T., & Homma, K. J. (2012). Thyroid hormone determines the start of the sensitive period of imprinting and primes later learning. *Nature Communications, 3*(1), 1081. https://doi.org/10.1038/ncomms2088

Zippelius, H.-M. (1972). Die Karawanenbildung bei Feld- und Hausspitzmaus. *Zeitschrift für Tierpsychologie, 296*(1971), 279–296.

# Can Prenatal or Neonatal Exposure to Chemosensory Variety Orient the Development of Personality? Exploratory Investigations in Mice

Bruno Patris[1] (✉), Aurélie Dastugue[1], Clémence Lesimple[1], and Benoist Schaal[1]

[1] Development of Olfactory Communication and Cognition Laboratory, Centre des Sciences du Goût et de l'Alimentation, CNRS, Université de Bourgogne, Inrae-Institut Agro Dijon, Dijon, France
bruno.patris@u-bourgogne.fr

**Abstract** Odorant compounds that mammalian females ingest during pregnancy/lactation permeate to their perinatal offspring, impinging on progeny's developing chemoreception and shaping subsequent preferences and behaviour. Long-term effects of these earliest experience have been repeatedly found with single flavour qualities, which induce postnatal selectivity in increasing preferences or reducing aversions. The two studies presented here attempted to go a step further with the laboratory mouse as model. Study 1 aimed examining whether prenatal only, postnatal only, or pre- and postnatal (transnatal) exposures to a single flavour through gestating/lactating dams' drinking water orients differentially selective responsiveness for that flavour. While perinatal exposure to the flavour appeared to affect positively its exploration by 6-day-old pups, only post and transnatal exposure modified its consumption in drinking flavoured water after weaning. Study 2 aimed to go beyond exposure to a single flavour in assessing whether experiencing prenatally or postnatally a variety of flavours can influence later non-selective responsiveness to chemosensory and visual novelty. Specifically, weanling mice born to pregnant/lactating dams fed a regimen of flavour variety will be compared with weanlings exposed to a control regimen of flavour monotony. 'Chemo-variety' exposure consisted in odourising gestating/lactating dams' drinking water with six flavourants changed every other day over a period of 6 days before or after birth. Control gestating/lactating dams were offered non-flavoured water. After weaning, offspring of both groups were tested for differential reactivity to novel olfactory or visual environments. While perinatal exposure to chemo-variety appeared to somehow affect 'anxious' reactivity and neophobic behaviour in weanlings, ingestion neophobia appeared relatively insensitive to early experience. Some sex differences emerged in these conditions, with females appearing more sensitive to the impact of perinatal chemosensory enrichment.

**Keywords** Flavour · Odour · Perinatal learning · Foetus · Neonate · Chemosensory variety · Laboratory mouse (*Mus musculus musculus*)



B. Schaal et al. (eds.), *Chemical Signals in Vertebrates 15*,
https://doi.org/10.1007/978-3-031-35159-4_16

# 1 Introduction

The ways animals deal with challenges of all sorts become organised along ontogeny into more or less stable behavioural tendencies, termed character, personality, temperament or behavioural syndrome (e.g. Cloninger, 1994; Carere & Maestripieri, 2013). Among determinants of personality, genetic transmission is a major player, but gene expression strongly interacts with the environment parents afford to their offspring from (before) conception to autonomy. In mammals, parents 'translate' their environment to their progeny, mothers literally embodying the environment of their progeny. The multiple facets by which parents expose their offspring to the environment (i.e. creating secure niches; providing sensory stimulation and nutrients; transferring immunity, hormones and microbiota; favouring communicative abilities and species-typical sociality; disciplining performance; serving as coping models and demonstrators for learning, etc.) can shape various aspects of their offspring's behavioural phenotype. These personality-modulating effects of parenting are particularly influential in early development, when the brain is most plastic and when offspring entirely depend on parental care (e.g. Wiedenmayer, 2010; Worthman et al., 2010). Thus, early life is critical for the assimilation of parent-mediated environmental information into the offspring's behavioural phenotype (Stamps & Groothuis, 2010; Cabrera et al., 2021).

We will focus here on the chemosensory stimuli that are consubstantial with mother-to-offspring transfers of materials, contexts and commodities. By incorporating the local chemistry through their diet, atmosphere and skin, mammalian mothers do indeed engender a chemical ecology that constitutes her offspring's developmental niches (e.g. Smotherman & Robinson, 1985; Alberts, 1986; Coppola & Millar, 1997; Chotro & Arias, 2003; Schaal, 2005). This perinatal chemosensory ecology not only modulates the progeny's immediate responsiveness to such cues but also often determines more stable behavioural proclivities that become facets of their personality.

Chemosensation, the array of olfactory, taste and trigeminal sensors that react to odours and flavours, is shaped by chemical inputs from gestation onwards (e.g., Cadiou et al., 2014; Todrank et al., 2011). From then on, foetuses develop sensing abilities of odours and flavours (Coppola & Millar, 1997), and related brain networks sustain odour learning and long-term memory (Nolte & Mason, 1995; Schaal et al., 2000; Coureaud et al., 2002; Gruest et al., 2004; Johnson et al., 2013). Thus, the formative effect of the chemical ecology created by the mother sets on as early as embryonic life, and it appears vital in the transition towards the postnatal environment where the same chemostimuli are met in milk and skin secretions. Suckling-associated odours/flavours are then also learned as positive and attractive (Blass, 1990; Nolte & Provenza, 1991; Delaunay et al., 2006). This early imprint has repercussions on later responsiveness, particularly in terms of stable preferences for familiar stimuli and aversion to novel ones. Prenatal, postnatal and transnatal (i.e. pre- + postnatal) exposures to given odours/flavours engender preferences (or reduced aversion) for stimuli that can last, depending on the span of research, over days (Schaal et al., 2000), weeks (Wells & Hepper, 2006), months (Mennella et al., 2001) or years (Delaunay et al., 2010; Hepper et al., 2013). These early-acquired selective responses will then rule in the ingestive domain (e.g. Portella et al., 2012; Dalle Molle et al., 2016), but also in wider domains of social life, such as affiliative preferences, mate choice and sexual performance (e.g., Mainardi et al., 1965; Fillion & Blass, 1986).

The evidence of these early-acquired chemosensory preferences opens up a myriad of issues awaiting further investigation about mechanisms (e.g. proximal underpinnings from molecular to neural and perceptual levels; conditions for foetal/neonatal odour acquisition), function (e.g. adaptive value of foetal learning in solving postnatal challenges) and development (e.g. how odours/flavours learned pre-/postnatally contribute to developing cognition and hedonic appraisals). Among these themes, two were considered here for further exploratory approaches: first, the relative strength of prenatal vs. postnatal odour exposure to influence later offspring's behaviour when facing a familiar vs. novel odour dilemma and, second, the effects of prenatal exposure to a multi-flavoured, as opposed to mono-flavoured, environment on offspring's behavioural proclivities assessed in spatial and ingestive tests.

The first topic of interest, the role of timing of perinatal chemosensory exposure, was already addressed in different model species, with the aim to explore early sensitive periods (Mennella et al., 2001; Echandia et al., 1982). Some studies found prenatal or neonatal exposure sufficient to canalise stable selective behaviour in subsequent life (e.g., Hai et al., 2014), prenatal induction being sometimes stronger than postnatal induction (e.g., Oostindjer et al., 2010; Wells & Hepper, 2006) or postnatal exposure being more effectual on subsequent preference than prenatal exposure (e.g. Galef & Henderson, 1972; Mennella et al., 2001). Other studies found that a continued prenatal *and* neonatal exposure was more optimal to affect postnatal behaviour (e.g., Hepper et al., 2012), leading to the hypothesis that, to be effective in the long run, prenatal learning needs some kind of postnatal confirmation. This issue was reconsidered here in Study 1 in exposing three groups of laboratory mouse to odorant A during gestation only, lactation only or gestation *and* lactation and assessing later consequences on offspring's neophobia.

The second topic of interest concerned the effect of prenatal and/or postnatal enrichment in environmental variety on later wariness towards novelty, a basic trait of personality (Kagan et al., 1987). For example, spatial enrichment (tunnels, bridges, toys) in the cages of pregnant mice (relative to standard conditions) led to less anxious adult female offspring (Maruoka et al., 2009). Converging results were reached when such enrichment was chemosensory, all other aspects being kept equal. Thus, exposing 25-day-old rats to variable aromas in their drinking water induced them to drink more of a novel flavour as compared to non-variety exposed weanlings (Capretta et al., 1975). Likewise, rat pups suckled by dams fed a variably flavoured diet (relative to pups of dams fed a mono-flavoured diet) were more prone to eat newly flavoured pellets (Myers & Pickering, 2007). Similar results were obtained in other species (e.g., Gerrish & Mennella, 2001; Maier et al., 2008; Trickett et al., 2009; Villalba et al., 2011; Catanese et al., 2012; Maier-Nöth et al., 2016; Middelkoop et al., 2018), suggesting the generality of the effects of experiencing changing vs. monotonous 'flavourscapes' in early *postnatal* development of brain and behaviour in mammals. Backing on the fact that mammalian foetuses are already sensing odours/flavours in the amniotic environment, Gerald Turkewitz (1994) suggested that the exposure to chemosensory variety in utero could also affect postnatal reactivity to novel odours/flavours. He additionally expressed the idea that the early experience of chemo-variety could affect reactiveness to novelty in other sensory modalities. In line with these earlier investigations and ideas, Study 2 aimed to explore the impact of early chemosensory variety on the reactivity of post-weaning mice towards chemosensory and visual novelty.

## 2 Study One: Effects of Prenatal, Postnatal or Transnatal Chemosensory Exposure on Selective Odour/Flavour Responsiveness in Murine Neonates and Weanlings

In this experiment, mouse foetuses and neonates were exposed to a single flavour ingested by three groups of dams differing in the timing of this exposure (prenatal only, postnatal only and pre- and postnatal = transnatal). The magnitude of the neophobic response of the offspring was then assessed in a series of tests comparing their responses to the three flavour-exposed conditions and a control non-exposed condition.

### 2.1 Methods

#### 2.1.1 Animals and Breeding Conditions

Balb/C female mice ($n = 32$) were paired with 32 Balb/C males (Charles River, France), couples being raised in standard polycarbonate cages (28 × 17 × 13 cm) lined with sawdust. They had ad libitum access to water and pelleted food (Scientific Animal Food & Engineering, Ionisos Aube, Champagne) and were kept under a constant 12:12 h light-dark cycle (lights on at 12:00 hours) and ambient air temperature (20 °C ± 2 °C) and hygrometry. In the days following partum, the litters were culled to six pups (surnumerary pups being fostered to another female). The couples were raised with their offspring in these cages until weaning (postnatal day (PND) 21). At this age, male and female siblings were housed in separate cages. In this study, 120 pups of multiparous females (2–3 parturitions) were tested (experimental groups are detailed below).

#### 2.1.2 Experimental Groups

Pups were randomly assigned to one of four exposure groups: (1) *prenatal exposure*, pups exposed to cinnamaldehyde (CIN) in utero only, CIN being administered from conception to birth through the females' drinking water; (2) *postnatal exposure*, pups exposed to CIN after birth until PND 17 through milk (CIN being conveyed to the lactating females via drinking water; pups start consuming water on their own on PND 18–19 in early weaning,); (3) *transnatal exposure*, pups exposed to CIN during whole gestation and after birth until PND 17, CIN being given through the female's drinking water during gestation and lactation up to PND 17; and (4) *control*, pups never exposed to CIN in utero and *in lacto*.

As exposure to CIN was obtained through pregnant/lactating female's ingestion, all tested pups underwent adoption at birth to control for the exclusivity of pre- vs. postnatal exposure. At the expected term, the pregnant females, with or without CIN in their drinking water, were checked hourly from the morning to pick their pups as early after birth as possible and to foster them (when the last pups was born) to an unfamiliar female who (1) gave birth on the same day and (2) drank CIN-aromatised or plain water (depending on the exposure group). Thirty pups per group (totalising 120 tested pups), from six to eight cross-fostered litters per group, were all evaluated in two behavioural tests run on PND 5 and 22–23.

#### 2.1.3 Chemosensory Stimuli

Odorants were purchased at Sigma-Aldrich, Saint-Quentin-Fallavier, France. Cinnamaldehyde (CIN), cinnamon odour to humans, was used as exposure odour, and benzaldehyde (BEN), bearing bitter almond odour qualities to humans, was used as control. The concentration of CIN in the mother's drinking water was set at 1 g/L, which bears a clear smell to the human nose and remains below toxic concentration for daily exposure in mice (according to the purchaser's information). The placental and mammary transfer of CIN was not instrumentally quantified, but we assumed that the resulting concentration in amniotic fluid and milk were attenuated relative to the concentration ingested by the gestating and lactating dams (e.g. Désage et al., 1996; Debong et al., 2021). During the postnatal test sessions, CIN was presented against BEN, both odorants being set at the concentration of 0.5 g/L. At this dilution level, both compounds bear equivalent subjective intensity and trigeminal potency to humans (judgement test run on ten non-smoking human adults).

The choice of CIN, more than BEN, as exposure odour was justified by its use in previous studies (Blavi et al., 2016) and by a pilot study conducted on 30-day old naïve mice ($n = 10$) simultaneously exposed to CIN- and BEN-aromatised bottles for 24 hours. As a result, a 0.5-g/L solution of BEN was significantly more ingested than the 0.5-g/L CIN solution (mean ± SE of CIN vs. BEN: 1.50 ± 0.48 vs. 5.22 ± 0.47 mL; paired t-test: $t_9 = 20.07$; $p < 0.001$). Thus, we hypothesised that an exposure to CIN would induce an increase of its hedonic value relative to BEN, either equalising or inverting the preference pattern between CIN and BEN.

#### 2.1.4 Behavioural Variables

Behavioural measurements were derived from two different tests run on PND 5 and 22–23 for all pups. PND-5 mouse pups underwent an *olfactory two-choice test* run in a test arena (Fig. 1) consisting in a polypropylene cylinder (diameter, 10 cm; height, 7 cm) with a wire mesh bottom (mesh size: 1 mm$^2$) apposed on a Petri dish. Below the wire mesh (1 cm below the bottom), five drops of BEN and CIN solutions were deposited in a standardised way in each half of the Petri dish, divided by a polypropylene separation to avoid the blending of odorants below the mesh (Fig. 1). The testing device was set on a warming platform (Gestigkeit, Germany) maintained at 37 °C.

Testing of litters was run 2 to 4 hours after the start of lights on in the breeding room, following an isolation period of 20 min before the test. To evaluate whether the experimental conditions affected bodily growth, each pup was first weighed. Then, each pup was introduced into the arena with its midline aligned on the middle of the wire mesh. After releasing the pup, a 2-min observation session was started. All tests were video-recorded to be subsequently analysed using the Observer software (Noldus, Wageningen, NL). The time spent by the pup in each compartment of the arena was measured during the 2-min test. A pup was considered as oriented on one side of the arena if its entire

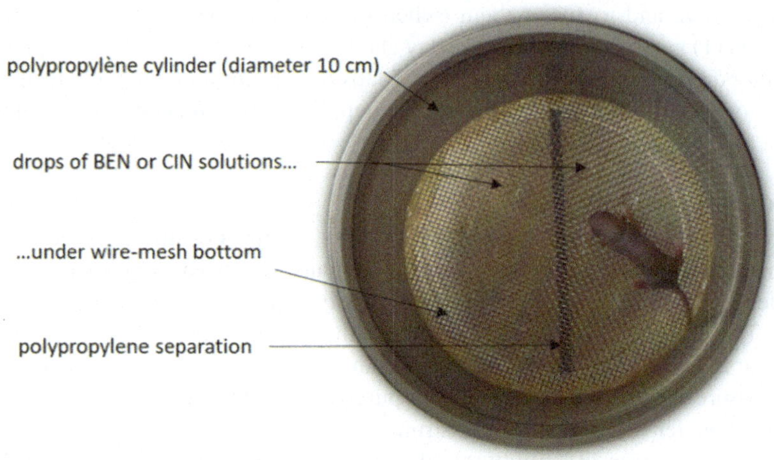


**Fig. 1** The two-odour choice arena used to measure the chemosensory preferences of 5-day-old mouse pups

head (i.e. the line between insertion of both ears) was on that side. To control for a potential lateral bias, the odour stimuli were first randomly assigned to the left or right side and then alternated after each trial in rotating the arena by 180°. At the end of the test, the pup was returned in its home cage. When the last pup in a litter was tested, the parents were returned to their home cage. When a pup took more than 30 sec before crossing the median line, urinated or stayed inactive for more than 1 min during the 2-min test, it was considered not to be in the right arousal state and accordingly excluded from the analyses. This concerned 6 pups in the prenatal exposure group, 8 in the postnatal exposure group, 7 in the transnatal exposure group, and 7 in the control group, resulting in groups composed of 24, 22, 23 and 23 pups, respectively.

Later, on PND 22–23, the *flavour preference test* was run. First, all the pups were weighed and sexed on PND 21 to assess whether the pups' sex affected growth and water intake at weaning. Then, from PND 21 to 22–23, the pups were isolated for 3 days in standard cages with ad lib access to food pellets and two bottles. On PND 21, both bottles contained plain water to measure baseline volume of drinking. On PND 22–23, the pups were given paired bottles of water flavoured with either CIN or BEN (concentration of 0.5 g/L in both cases) to measure differential intake. In each group of pups, the position of the two odourised bottles was randomised at first and shifted after 24 h. After each of these three 24-h periods, both bottles were weighed to compare consumption of plain water vs. both flavoured solutions. At the end, all related pups were reunited into a standard cage. Due to some mortality between PND 5 and 21 affecting all groups, 26/30 (females/males: 14/12), 27/30 (14/13), 24/30 (11/13) and 28/30 pups (16/12) were tested in the groups exposed to CIN prenatally, postnatally or transnatally, and in the non-exposed controls, respectively. All tested pups were included in the analyses.

#### 2.1.5 Statistical Analyses

To assess potential main or interaction effects of Litter (on PNDs 5 and 21) and Sex (on PND 21) on growth and behaviour, several one-way ANOVAs were run on Weight on PNDs 5 and 21, Exploration time on PND 5 and Amounts of water consumed on PND 21. As neither main, nor interaction, effects reached significance on any of these dependent variables, we excluded the Sex and Litter factors from further analyses. For the two-odour choice tests run on PND 5, a one-way ANOVA on pups' weight assessed the impact of the odour exposure group on growth. Pup differential responsiveness to the exposure (CIN) and reference (BEN) odorants in the two-choice arena was examined using Student's one-sample t-tests, comparing the time spent in one compartment with the random level of orientation over both compartments (i.e. 60 s). In the two-bottle choice-test run at PND 22–23, one-way ANOVAs were run on the PND-21 weight and the baseline amount of water consumed to evaluate the impact of exposure treatments on growth and odourised water intake. Then, for each exposure group, the relative preference for the CIN- vs. BEN-odourised solutions after 24 and 48 h of exposure was compared using Student's paired t-tests. All statistical analyses were two-tailed and realised with the Statistica® software (Statsoft, Maison-Alfort, France).

### 2.2 Results

#### 2.2.1 Effect of Perinatal Flavour Exposure on Progeny's Weight Gain

No main effect of the CIN exposure group reached significance on pups' body weight on PND 5 ($F_{3,88} = 1.21$, $p = 0.309$; prenatal exposure, 3.61 ± 0.09 g; postnatal exposure, 3.73 ± 0.08 g; transnatal exposure, 3.86 ± 0.13 g; non-exposed control, 3.77 ± 0.09 g). However, on PND 21, the pups' weight differed significantly among the four exposure groups ($F_{3,101} = 9.06$, $p < 0.001$). Post hoc LSD comparisons revealed that the average weight of the postnatallyexposed group (9.64 ± 0.15 g) was significantly lower than those of transnatally exposed (10.52 ± 0.24 g), prenatally exposed (10.67 ± 0.2 g) or control groups (11.18 ± 0.26 g). The transnatally exposed pups were on average lighter than the control pups.

#### 2.2.2 Odour Preference on PND 5

The control pups did not behave differentially over both odourised compartments of the test arena ($t_{23} = 0.43$, $p = 0.660$; Fig. 2). In contrast, those of the three groups born to CIN-consuming females and/or fostered to CIN-consuming lactating females roamed significantly longer over the CIN compartment in the test arena (prenatally exposed pups, $t_{24} = 2.50$, $p = 0.019$; postnatally exposed pups, $t_{22} = 3.35$, $p = 0.003$; transnatally exposed pups, $t_{23} = 2.45$, $p = 0.022$, all relative to the control pups). Thus, for at least 5 days after birth, the odour of CIN has acquired an attractive value in the CIN-exposed pups.

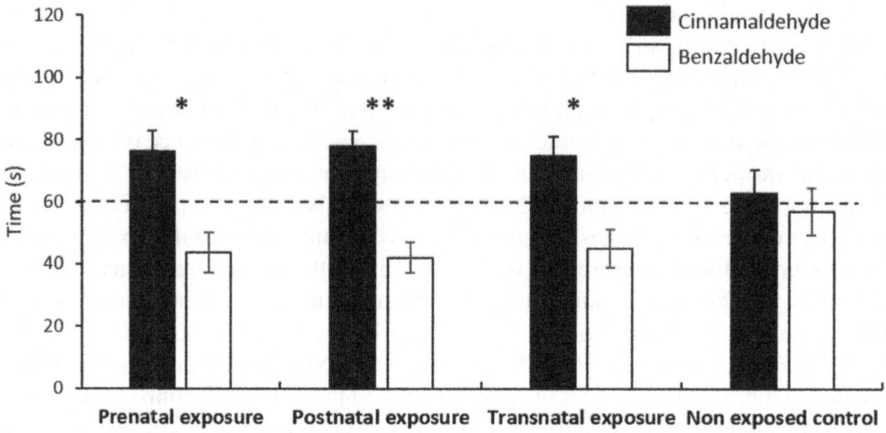


**Fig. 2** Mean time (+ SE) spent in each half of the two-odour choice arena in PND-5 mouse pups exposed to cinnamaldehyde prenatally only ($n = 24$), postnatally only ($n = 22$), transnatally ($n = 23$) or non-exposed ($n = 23$). One sample t-tests: $*p < 0.05$; $**p < 0.01$ (dashed line corresponds to random orientation)

### 2.2.3 Flavour Preference on PND 22–23

Regarding baseline water consumption on PND 21, both bottles were equally solicited within groups. But unexpected between-group differences were noted in terms of total consumption ($F_{3,101} = 20.80$, $p < 0.001$). Post hoc LSD comparisons yielded significant differences between groups, with increasing values from the transnatally exposed group (Mean ± SE = 3.75 ± 0.12 ml), control group (4.62 ± 0.24 ml), postnatally exposed group (5.28 ± 0.18 ml) to prenatally exposed group (5.84 ± 0.19 ml).

As regards differential intake from either flavoured beverage over two consecutive 24-h periods, the following pattern of results came out (Fig. 3a, b): (1) control pups consumed significantly more BEN- than CIN-flavoured water after 24 h ($t_{28} = 5.39$, $p < 0.001$) and 48 h ($t_{28} = 8.53$, $p < 0.001$), confirming a spontaneous preference for BEN over CIN flavours seen in the pilot test reported in the Methods section (cf. 2.1.3). (2) Prenatally exposed pups were similar to the controls, taking in more BEN than CIN water after 24 h ($t_{26} = 22.55$, $p < 0.001$) and 48 h ($t_{26} = 14.49$, $p < 0.001$). This suggests that prenatal exposure to CIN was unable to reverse the spontaneous preference for BEN over CIN. (3) However, postnatal ($t_{27} = 1.63$, $p = 0.341$) and transnatal ($t_{24} = 0.43$, $p = 0.673$) exposure to CIN abolished the spontaneous pattern of preference between CIN and BEN after a 24-h choice test on PND 22, and the pups from these two groups drinking the same amount of CIN- and BEN-flavoured water. However, this modification in preferential intake appeared short-lived, as by PND 23 BEN-flavoured water was again significantly more consumed by the postnatally and transnatally exposed groups ($t_{27} = 5.59$, $p < 0.01$ and $t_{24} = 13.27$, $p < 0.001$, respectively).

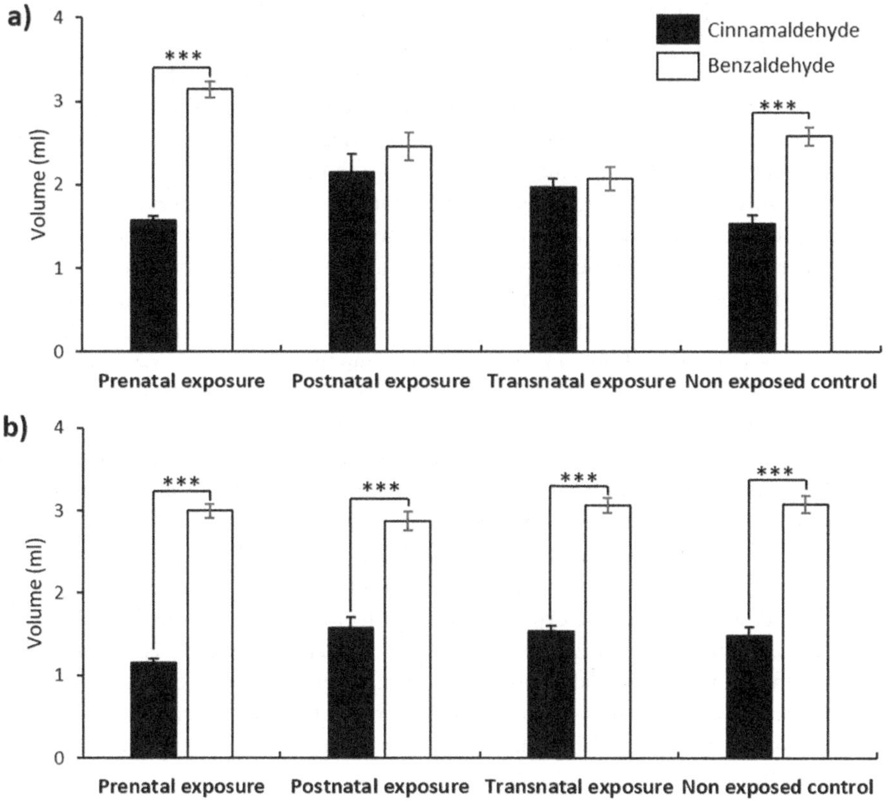


**Fig. 3** Mean volume (+SE) of cinnamaldehyde- and benzaldehyde-flavoured water consumed during two 24-h two-bottle choice test: (**a**) on PND 22 and (**b**) on PND 23 as a function of the timing of the cinnamaldehyde exposure groups: prenatal only ($n = 26$), postnatal ($n = 27$), transnatal ($n = 24$) and control ($n = 28$). (***$p < 0.001$, student paired t-test)

## 2.3 Discussion

The preferential orientation test run at PND 5 revealed that the odours of CIN vs. BEN were equally attractive to the pups of the non-exposed control group. Both odorants had thus similar reinforcing value to naïve neonate pups. However, when the mothers ingested CIN during pregnancy and/or during lactation, the offspring stayed clearly longer over the CIN relative to the BEN compartment of the two-choice device. They were thus attracted to the CIN odour to which they were exposed in utero and/or in lacto. Alternatively, they may have been more repulsed from the unfamiliar BEN odour, resulting in more orientation towards the CIN side. In any case, being exposed to CIN prenatally only, postnatally only or transnatally induced a differential behaviour in 5-day-old pups resulting in a greater orientation towards the familiar stimulus.

However, the preference for CIN at PND 5 engendered by prenatal, postnatal or transnatal exposures was short-lived. When re-examined for their relative preference towards the flavours of CIN and BEN around weaning (PND 22–23), but in a different two-choice ingestion test, the results were not aligned with those of the preferential orientation test at PND 5. Several explanations may account for this outcome. *First*, contrary to the exposure-naïve neonates who were equally attracted to CIN vs. BEN, the naïve weanlings were more positive towards the BEN- than the CIN-flavoured water (in line with the pilot test on adult mice reported in Sect. 2.1.3). Obviously, both preference assessments should be compared cautiously as they mobilised different sensory inputs (i.e. only nasal chemoreception in the neonatal test vs. oral plus nasal chemoreception in the weaning test) and different behavioural systems ('exploration' vs. ingestion). Otherwise, some unanticipated changes may have happened between PND 5 and 22–23 that rendered CIN unpalatable relative to BEN. It cannot be excluded that chemosensory processing may undergo developmental functional changes induced by maturation or exposure effects or that CIN has non-specific sensory or post-ingestive effects. The odorant CIN is indeed described as pungent to the human nose (Artcander, 1969) and may therefore be a trigeminal stimulant. In mice, trigeminal stimuli induce more or less avoidance (e.g. Galliot et al., 2012). CIN has also biological effects beyond chemosensation: its inhalation or ingestion has respiratory, metabolic, anti-inflammatory, neurotrophic and analgesic consequences in mice (Churihar et al., 2016; Clapp et al., 2017; Jiang et al., 2017; Mahapatra et al., 2020; Zuo et al., 2017). CIN has also antimicrobial properties which may have interfered with the establishment of the gut microbiota (Xiao et al., 2022) at the critical transition from milk to non-milk food (Hooper et al., 2002). Such an effect might explain the suboptimal body weight of weanling pups in the postnatal and transnatal exposure groups. Overall, exposure to CIN may thus produce a complex pattern of unforeseen sensory and physiological outcomes, some nevertheless considered as fitness inducing in neonatal mice (e.g. anti-inflammatory, protective effects; e.g. Yang et al., 2019).

*Secondly*, the spontaneous avoidance towards CIN noted in the naïve control pups was attenuated in the pups exposed to CIN postnatally only and transnatally at the 24-h monitoring of differential drinking. Postnatal exposure to CIN explains this avoidance-reducing effect in weanling pups, as those pups exposed to CIN prenatally remained repelled to it against BEN. But, *thirdly*, the effects of postnatal only or of pre- + postnatal exposure to CIN on the rate of CIN-flavour ingestion at weaning was transitory, as it was not confirmed at the 48-h measurement of differential ingestion. After 48 h of CIN vs. BEN flavour choice, the three CIN-exposed groups ingested indeed twice more BEN than CIN flavour. Thus, the prenatal vs. postnatal timing of CIN exposure did not significantly affect the long-term palatability of CIN-flavoured water. *Fourthly*, the timing of exposure (pre-, post-, transnatal) did not interact with its duration (21, 17 and 38 days in the prenatal, postnatal and transnatal groups, respectively). Specifically, the clearly longer exposure to CIN in the transnatally exposed group did not produce any advantage in terms of short- or long-term attractiveness to CIN as an odour (in the orientation test) or as a flavour (in the ingestive preference test).

To sum up, being exposed to CIN prenatally (relative to the non-exposed control condition) strengthens the approach tendencies of mouse pups to CIN odour at PND 5. But this positive effect is wiped out around weaning, as shown in the PND 22–23 retest when pups reduce their ingestion of CIN. The postnatal behavioural consequences of prenatal only exposure vs. postnatal only/transnatal exposures to CIN are thus not aligned in the present study, in contradiction with previous studies on the long-term effects of prenatal odour acquisition in the mouse (e.g. Dewaele et al., 2020; Nolte & Mason, 1995; Todrank et al., 2011). Thus, if perinatal CIN exposure altered offspring behaviour, it was in the sense of short-term attraction in newborns and long-term aversion in weanlings, but not in terms of a stable phenotypic alteration. We do not understand yet the cause of this discontinuity of perinatal exposure effects, hypothesising that it may be due to some secondary post-ingestive effect of postnatal exposure to CIN on neonatal physiology. This conclusion prevented us from using CIN in the following study and in reducing the concentrations of the alternative flavourants used to 'imprint' progeny's chemoreception by maternal ingestion during pregnancy and lactation.

# 3 Study Two: Prenatal and Postnatal Exposure to Chemosensory Variety and Weanling Mice Reactivity to Novelty

In this study, foetuses and newborns were exposed to chemosensory variety, and then, as weanling mice, their emotional responsiveness was compared with that of control animals exposed only to the standard chemosensory monotony of the lab. Three tests were used to assess neophobic reactivity, defined as an adaptive reluctance to approach or incorporate an unfamiliar, potentially dangerous, situation or stimulus (Griebel et al., 1993). At PND 26, 5 days after weaning, the mice were subjected to a T-maze test opposing a familiar vs. nonfamiliar odour. At PND 29, two open-field tests were run to evaluate the animals' level of 'boldness' when placed in an unfamiliar arena, as a function of their chemosensory variety antecedents. Finally, at PND 33, the mice could choose to drink either their habitual beverage (plain water) or strangely flavoured water. Our general expectation over this series of assessments was that earlier exposure to chemosensory variety should be followed by reduced wariness, or increased 'boldness', towards approaching or incorporating novelty. However, due to the exploratory nature of the study, no clear predictions were made regarding the effects of pre- vs. postnatal exposure to chemosensory variety or of the animals' sex.

## 3.1 Methods

### 3.1.1 Animals and Breeding Conditions

Twenty-three pairs of adult male and female C57BL/6 J mice were installed into polycarbonate cages (28 × 17 × 13 cm), with ad libitum access to water and food. The rearing room was maintained under a 12:12 h circadian cycle (lights on from 0:00 to 12:00 h), at a constant temperature (22 ± 0.5 °C) and hygrometry, and with continued air renewal. To standardise maternal care between litters and to optimise chemosensory exposures, only litters (culled to six pups) of multiparous females were analysed in the present study. The males were removed from the cages 2 days before partum to avoid postpartum fertilisation. At weaning age (PND 21), the pups were marked with scentless ink so that they could be followed up individually; they were then placed in sibling cages before behavioural testing began 4 to 5 days later. In this study, 115 weanling mice of multiparous females (two or three parturitions) were tested (experimental groups are detailed below).

#### 3.1.2 Experimental Groups

Foetuses (prenatal exposure) or neonates (postnatal exposure) were exposed to experimental chemostimuli through the flavouring of their gestating/lactating mothers' drinking water. From the dams' drinking water, the flavours were assumed to permeate into amniotic fluid and milk. The murine olfactory system being functional from embryonic day (ED) 15 (Coppola & Millar, 1997), prenatal exposure occurred from ED 15 to 21 and postnatal exposure from PND 0 (day of birth) to 6. Chemosensory exposure consisted in a variability regimen (VAR, chemosensory variety) where the dams were exposed to six flavours (in random order) over 6 days, and a control regimen (CON, chemosensory monotony) made of unadulterated drinking water over 6 days. Manipulations in both exposure regimens consisted in changing bottles and nipples daily, and in renewing entirely the cages on the day of birth in VAR and CON prenatal exposure groups and 6 days after birth in VAR and CON postnatal exposure groups. From the end of the exposure periods to behavioural testing right after weaning age, all animals accessed the same standard diet and beverage (plain water). The 115 pups participating to this study were randomly assigned to one of four exposure regimens: (1) prenatal VAR exposure ($N$ = 31), (2) prenatal CON exposure ($N$ = 25), (3) postnatal VAR exposure ($N$ = 32) and postnatal CON exposure ($N$ = 27).

#### 3.1.3 Chemosensory Stimuli

The six flavourants used in the VAR regimen consisted in butanoic acid (cheese odour to humans), allyl disulphide (garlic), 2-Methoxy-4-methylphenol (bacon), vanillin (vanilla), limonene (citrus) and trans-anethole (anise) (all purchased from Sigma-Aldrich, St Quentin Fallavier, France). To avoid acute toxicity, the concentration of these odorants was set below 10% to the murine lethal dose 50 (LD50) of each compound. Further, to render them nearly iso-intense (as pretested with three human participants), these odorants were mixed in water at 0.08, 0.004, 0.02, 0.01, 0.13 and 0,05% of the pure compound, respectively.

Regardless of the flavour of their drinking water, pregnant and lactating dams of the VAR group consumed daily as much water as the CON females. The different flavourants were thus not rejected by pregnant or lactating dams, and their offspring can be assumed as having been exposed accordingly to the various odorants in utero or through milk. During the prenatal exposure period, we noted a decrease, by an average of 9 to 6 ml between ED 15 and ED 20, in the females' drinking, regardless of the flavour treatment of the water. In late pregnancy, dams do indeed reduce their movements and drink less. In contrast, after partum, drinking increases by an average of 6 to 12 ml per day between P0 and P6 in relation with ongoing lactation and offspring solicitations.

Several additional chemosensory stimuli are introduced in the behavioural tests and are described below.

### 3.1.4 Behavioural Testing and Measured Variables

When reaching weaning age, the 115 foetuses and neonate pups exposed to the VAR or CON regimens were tested behaviourally. Testing was run under red light during the animals' nocturnal phase. The behavioural monitoring of the animals was carried out automatically using the EthoVision software (version XT®, Noldus, Wageningen, NL). This software tracks the movements of the pups using the x-y coordinates of three points of their body (tip of muzzle, gravity centre and insertion of tail); it also allows simultaneous video capture. Four behavioural tests were applied in the same order for all mice around the age of weaning: an odourised T-maze test run on PND 26 (± 1 day), two open-field tests on PND 29 (± 1 day) and a novel flavour palatability test on PND 33 (± 1 day).

#### 3.1.4.1 *Odourised T-Maze Test*

This first test consisted in putting weaning mice into a T-maze where a familiar odorant in one arm was opposed to a novel odorant in the other arm. Pilot-testing this setting with 30-day-old mice indicated longer exploration of the arm containing the familiar soiled bedding relative to the arm containing a novel odour (acetophenone). This test thus allowed measuring a mouse relative neophobia between a familiar and unfamiliar odour. The T-maze was a three-armed PVC device (Fig. 4): the starting area is a chamber at the end of the first arm, which is closed by a hatch that is lifted at test beginning. The two other arms end with walls pierced with a 1-cm diameter hole in which two bottles are affixed, one diffusing the familiar odour of soiled cage bedding of the mouse

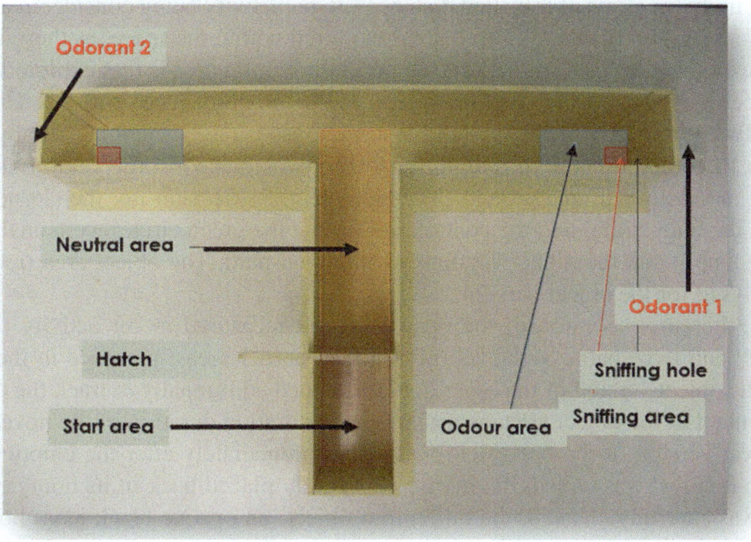


**Fig. 4** The T-maze device with the different exploration zones considered in the analyses. The familiar (soiled bedding) and unfamiliar (acetophenone) odours were positioned as Odorant 1 and 2 and counterbalanced at every trial

to be tested and the other diffusing an unfamiliar acetophenone solution (on 4 × 4-cm blotting paper soaked with 2 mL of 0.1% dilution in water). The lateral position of these two bottles was counterbalanced across trials. Procedurally, a tested mouse was placed in the starting area, the video acquisition software was launched, and then the guillotine hatch was removed. The recording of the animal's movements began when the mouse passed through the neutral area for the first time and lasted 10 minutes.

The variables derived from the T-maze test were (i) the total distance travelled in the whole device, as an index of an animal's activity, assumed to be a proxy for stress (hyperactivity); (ii) the time spent in either odourised arm (by tracking of the centre of gravity); and (iii) the time spent in the sniffing area close to either odour source (by tracking of the muzzle at least 1 cm from the holes). Computations of these proportions of time were used to assess each animal's relative attractiveness towards novelty (or neophobia) depending on their chemosensory exposure.

#### 3.1.4.2 *Open-Field Tests: Empty Versus Object Versions*

Open-field tests are commonly used to assess anxiety in mice (e.g. Walsh & Cummins, 1976; Choleris et al., 2001). The open-field device used in the present study was made up of four opaque PVC panels (length × width × height: 40 × 40 × 30 cm), placed on a PVC plate of the same colour. The mouse behaviour in the open-field was analysed using the EthoVision software. The video acquisition began when the mouse was placed in a corner of the open-field using a beaker. The acquisition of the animal's movements began 12 s after the launch of the video (the time to put the animal down and get out of its visual field) and ended after 5 min.

The open-field test made it possible to derive multiple variables: (i) an animal's motor activity (in terms of travelled distance), hyperactivity being considered a reliable index of stress, and (ii) the different areas explored within the space, as these bear different anxiogenic levels (e.g. the time spent in the central area is less explored by anxious mice, while the corners are more explored by anxious mice; Walsh & Cummins, 1976). Two versions of the open-field test were used: one was empty of any object, while in the other version, an unfamiliar object was placed in the central area, thus defining an additional area around the object (Fig. 5). To sum up, the Corners' area (outside the blue circle) and the centre area (inside the green circle) were analysed for both the empty and the object versions of the open-field. The object area (red circle) was analysed for the object version.

Thus, the empty version of the open-field test measured motor activity (distance travelled) and two spatial variables (presence in corner areas; presence in the central area). The object version of the open-field test aimed additionally to track the presence of the animal in the object area, assessing the exploration duration of the novel object. The object version of the open field was tested immediately after the empty version: once the first test was ended, the mouse was gently placed back in its home cage, the device was carefully cleaned, the unfamiliar object (an opaque black glass bottle 7.5-cm high by 4-cm wide) was placed in its centre, and the mouse was placed again in the open field for a 5-min test. All mice were subjected to these two open-field tests on PND 29 (± 1 day).

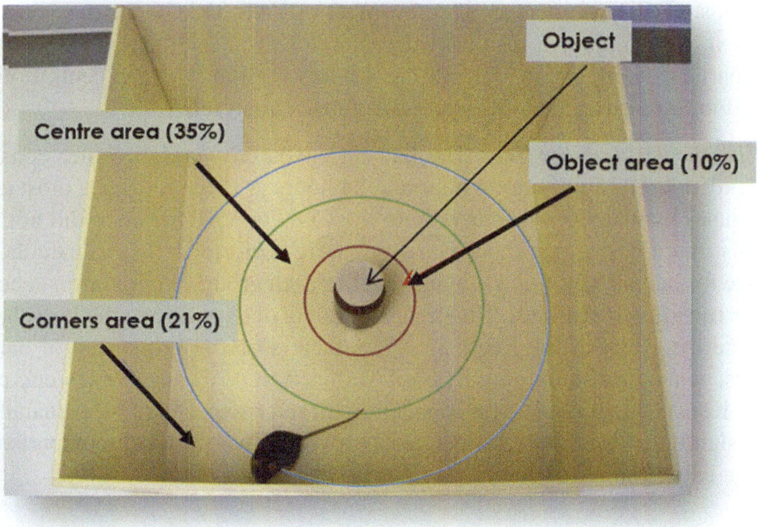


**Fig. 5** The open-field device with the defined areas (corners, central, object) analysed using the movement tracking system (percentages represent the proportions of time in the different areas if the animals behaved at random in the arena)

3.1.4.3 *Novel Flavour Palatability Test*

This test aimed to assess the degree of ingestive wariness to novelty in mice facing paired bottles in their home cage, one delivering banana-flavoured water (novel stimulus for VAR and CON) and the other delivering plain water (familiar). The banana flavour was a 0.11% dilution of amyl acetate (purchased at Sigma-Aldrich; corresponding to approximately 10% of the LD50 for mice). Procedurally, the mice were first isolated on PND 33 (± 1 day) in a cage with two bottles of plain water for 24 h (habituation). Then, after a 5-h restriction of drinking water, a bottle of plain water and a bottle of banana-flavoured water were made accessible for drinking. Differential intake was measured by weighing both bottles at three time points: 15 min, 1 h and 24 h.

### 3.1.5 Statistical Analyses

***Odourised T-Maze Test*** Data obtained for these three variables (total distance travelled and time spent in arms and sniffing areas) followed a normal distribution (Shapiro-Wilk normality test). Two-factor ANOVAs (mode of exposure × sex) were performed on these variables, independently for mice exposed prenatally and postnatally to VAR or CON regimens. In case of a significant interaction effects, two-by-two post hoc comparisons were carried out.

***Open-Field Tests*** All data obtained for the three variables (total distance travelled, time in corners and time in central areas) followed a normal distribution. First, student t-tests were performed to assess the time spent by each group of mice in each area. The observed duration of an animal's presence in one area was assessed against the theoretical duration computed as if mice behaved randomly. Then two-way ANOVAs (mode of exposure × sex) were run to compare the effect of VAR and CON exposure regimens. In the event of a significant result for the main effect or interaction, two-by-two post hoc comparisons were carried out. In the T-maze version with the central object, the time spent in the area around the object was further analysed.

***Novel Flavour Palatability Test*** The banana flavoured vs. plain water consumption were recorded after 15 minutes, 1 hour and 24 hours at PND-23. Since most groups of males/females exposed pre-/postnatally to the CON or VAR regimen did not follow a normal distribution, non-parametric analyses were performed. To evaluate the relative preference for banana solution vs. plain water in each group, Wilcoxon-matched paired tests were run after 15 minutes, 1 hour and 24 hours of exposure to the two-choice test. To evaluate the effect of the Regimen × Sex factors on the proportion of banana flavoured vs. water consumption, Kruskal-Wallis ANOVAS analyses were run after 15 min, 1 h and 24 h of choice, independently for mice exposed pre-/postnatally. In the event of significant main/interaction effects, two-by-two post hoc comparisons were carried out using Mann & Whitney U-tests.

## 3.2 Results

### 3.2.1 Mice Behaviour in the Odourised T-Maze as a Function of Early Exposure Regimen

In the T-maze tests run at PND-26, prenatal and postnatal VAR and CON mice spent about twice more time in the arm odourised with the familiar soiled bedding than in the one with unfamiliar acetophenone odour: it is therefore a good behavioural test to observe a possible effect of VAR exposure on novelty reaction.

In the *prenatal exposure*, the distance travelled in the entire T-maze was not modulated by the main effects of Exposure regimen (VAR vs. CON) or Sex factors. However, the prenatal Exposure by Sex interaction reached significance ($F_{3,52} = 14.35$; $p < 0.001$), in that the VAR female mice moved significantly more than the CON female mice ($t = -4.24$; $p < 0.001$) and the VAR male mice ($t = 4.27$; $p < 0.001$; Fig. 6a). The proportion of time spent in the acetophenone odour area (relative to total time spent in both odour areas) yielded neither main nor interaction effects of the prenatal Exposure and Sex factors. However, considering the mice' presence in the sniffing areas, closest to the odour sources, a significant Exposure by Sex interaction emerged ($F_{3,52} = 5.63$; $p = 0.021$): CON female mice spent less time close to the novel acetophenone odour than VAR female mice ($t = -2.15$; $p = 0.039$) and CON males ($t = -2.40$; $p = 0.025$; Fig. 6b).

Regarding the *postnatal* regimens of VAR vs. CON exposure, the ANOVA carried out on motor activity (distance covered in the entire maze) did not detect significant main effects of the Exposure regimen or Sex factors. However, a significant interaction effect of these factors emerged ($F_{3,54} = 7.28$; $p < 0.01$), indicating that CON female mice covered significantly more distance in the T-maze than VAR female mice ($t = 3.45$; $p < 0.01$) and CON male mice ($t = 2.15$; $p = 0.04$; Fig. 7). The proportion of time spent in the acetophenone arm of the maze did not differ significantly for animals' presence in either the odour or sniffing areas.

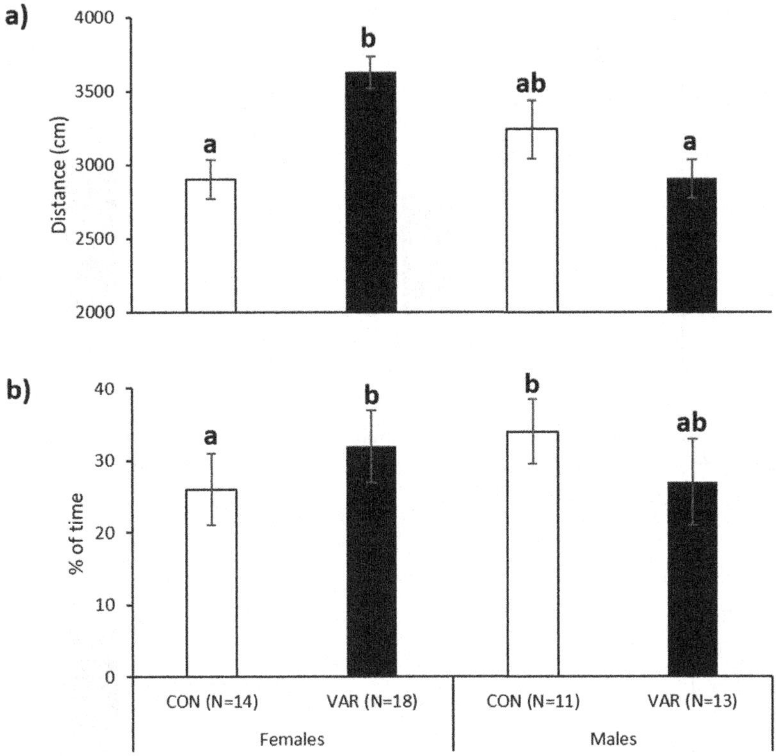


**Fig. 6** (**a**) Mean (±SE) distance travelled in the T-maze and (**b**) percentage of time (±SE) spent in the acetophenone odour area compared to total time spent in the two odour areas, by PND-26 mice born to females exposed to the control (CON) or variety (VAR) regimen during gestation (letters indicate post hoc significant differences between groups)

### 3.2.2 Mice Behaviour in the Open-Field Tests as a Function of Early Exposure Regimen

In the empty version of the open-field test, regardless of their prenatal or postnatal exposure regimens, PND-29 mice spent expectedly more time in the corner areas and less in the central area than they would spend if they behaved at random (Fig. 8).

In the groups of mice born to females exposed to the VAR vs. CON regimen during gestation, a two-way ANOVA on the distance covered in the entire device, with Exposure regimen (VAR vs. CON) and Sex as factors, did not reveal any difference between the VAR and CON exposure groups. However, a significant main effect of the Exposure factor emerged on the time spent in the central zone of the open field ($F_{3,52} = 4.47$, $p = 0.039$). Thus, the mice prenatally exposed to chemosensory variety behaved more centrally in the open field than those of the CON group (mean ± SE = 40.83 ± 2.01 vs. 31.88 ± 3.21 s). However, the time spent in the open field's corners did not significantly differ as a function of the prenatal exposure regimen (VAR or CON) or of sex.

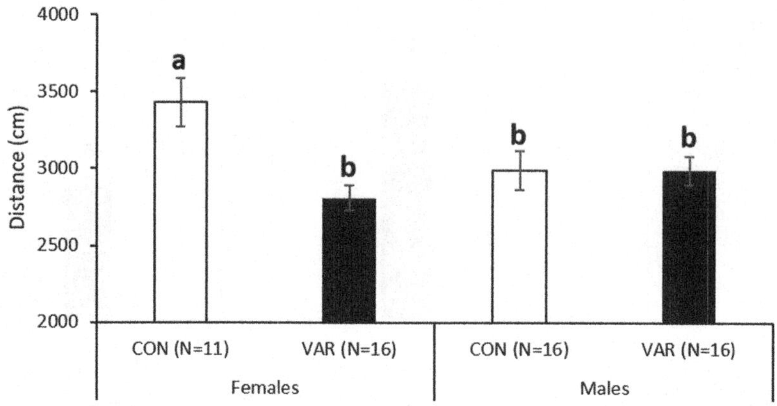


**Fig. 7** Mean (±SE) distance travelled in the T-maze by PND-26 mice born to females exposed to the control (CON) or variety (VAR) regimen during lactation (letters indicate post hoc significant differences between groups)

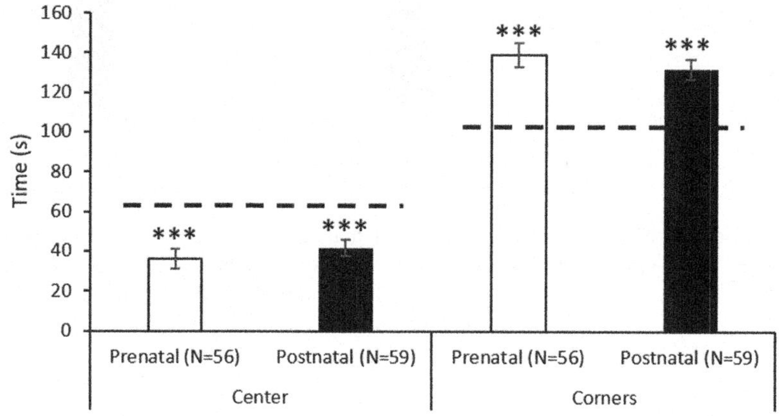


**Fig. 8** Mean time (± SE; seconds) spent by PND-29 mice in the center and corner areas of the open field without object, after prenatal or postnatal exposure to variety. *** indicates significant differences ($p < 0.001$; student-t test) with the theoretical duration of presence in the considered areas if mice behaved randomly (dashed lines)

In this same device, the postnatal exposure to VAR vs. CON conditions did not appear to induce any differential effects in terms of either total distance travelled in the entire open field or time spent in its centre or corners.

In the object version of the open-field test, all groups of PND-29 mice regardless of their exposure regimen spent significantly more time in the corners than if they followed a random pattern of spatial behaviour. In those PND-29 mice exposed *prenatally* to either VAR vs. CON regimens, distance covered over the entire surface did not differ in both male and female mice in the VAR vs. CON groups. In the same way, no main or interaction effects reached significance as a function of prenatal exposure to the VAR vs. CON regimen or of Sex on the time the mice roamed in the corners, central zone or object zone of the open field.

Concerning PND-29 mice exposed *postnatally* to the VAR or CON regimens, no main or interaction effects of Exposure or of Sex was detected on the motor activity over the entire surface of the open field. Two-way ANOVAs performed on the time spent in the open field's corners or central zone did not reveal any main or interaction effect of the effect of VAR vs. CON regimen and/or Sex. However, a main effect of the Exposure factor emerged on the time spent in the zone defined around the object ($F_{3,54} = 5.64$, $p = 0.021$). VAR mice stayed significantly longer in the area around the object than CON mice (23.18 ± 3.11 vs.12.94 ± 2.07 sec).

### 3.2.3 Mice Ingestion of a Novel Flavour as a Function of Early Exposure Regimen

The two-bottle test run at PND 33 evaluated the differential drinking rate of novel banana-flavoured water vs. the habitual plain water after 15 min, 1 h and 24 h of choice in groups of mice exposed prenatally or postnatally to VAR or CON regimens (Fig. 9).

In the groups exposed *prenatally* to VAR or CON regimens, the proportion of banana-flavoured vs. plain water consumed after 15 min ($H_{3,56} = 3.42$, $p = 0.333$), 1 h ($H_{3,56} = 4.99$, $p = 0.172$) and 24 h $H_{3,56} = 0.882$ $p = 0.829$) did not differ as a function of the variety regimen or sex. Thus, most subgroups of mice ingested both beverages equivalently at most time points. Some erratic sex differences emerged in favour of plain water in VAR females after 15 min and 1 h of the test and in CON males after 1 h (Fig. 9).

In pups exposed *postnatally* to the VAR or CON regimens, the proportion of banana-flavoured water consumed after 15 min ($H_{3,59} = 3.384$, $p = 0.338$) and 1 h ($H_{3,59} = 4.748$, $p = 0.191$) did not differ as a function of the variety regimen or sex. Thus, most of the

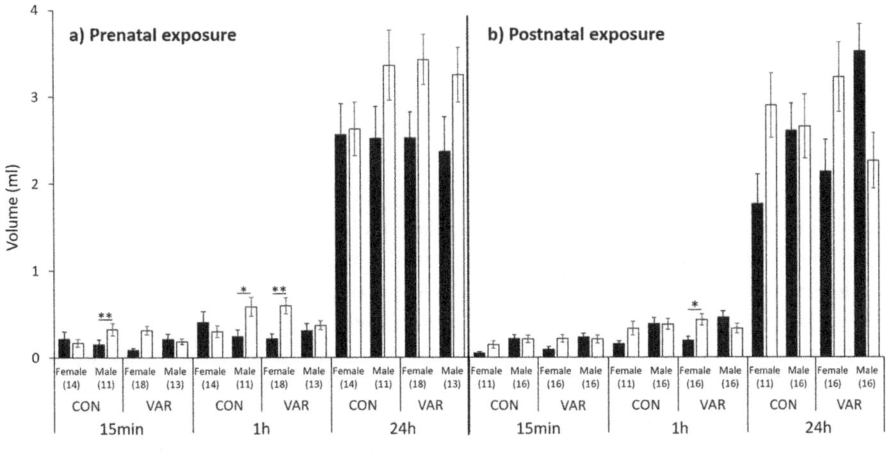


**Fig. 9** Mean volume consumed (± SE, mL) by PND-33 mice in a two-choice bottle test opposing unfamiliar banana-flavoured water (black bars) with habitual plain water (empty bars) after a choice period of 15 min, 1 h and 24 h. These mice were exposed a) prenatally or b) postnatally to a regimen of chemosensory variability (VAR) or to a standard control (CON) regimen. * ($p < 0.05$) ** ($p < 0.01$) indicates significant differences (Wilcoxon T-test)

subgroups of mice consumed as much of the plain water as of the unknown banana-flavoured water, except the VAR female pups after 1 h of test (Fig. 9). However, the proportion of banana-flavoured water consumed after 24 h tended to differ as a function of the variety regimen or sex ($H_{3,59} = 7.568$, $p = 0.058$), but only because CON females tended to ingest a smaller proportion of the banana-flavoured water than the VAR males.

## 3.3 Discussion

The effect of prenatal or postnatal exposure to chemosensory variety (VAR group) as compared to prenatal or postnatal standard conditions (CON group) was assessed using three behavioural tests. ***First***, the **T-maze test** measured the tendency of mice to approach or withdraw from an unfamiliar odorant presented against the highly familiar odour of soiled bedding from own home cage. Among those mice *exposed prenatally* to the VAR vs. CON regimen, the VAR females were most affected: (1) in showing higher locomotor activity in the maze than (i) the CON females and (ii) their VAR male counterparts and (2) in staying longer near the nonfamiliar odour than (i) the CON females and (ii) their VAR male counterparts. Thus, not only did the *prenatal* VAR exposure appear more efficient than the *prenatal* CON exposure in reducing the typical avoidance towards a novel odour at weaning age, but it did so only in female pups. Accordingly, a daily change in flavours assumed to be passed from mother to foetuses during the last 6 days of pregnancy sufficed to reduce of the typical wariness of mice towards olfactory novelty, and this effect was stronger in female offspring. It does not seem that such a sex effect of *prenatal* exposure in favour of females was previously noted with *food flavours* (e.g. Spahn et al., 2019), although similar sex-biased prenatal effects were reported in response to compounds of nutritional (e.g. in rats: Bayol et al., 2007, 2008) or pharmacologic importance (e.g. in humans: Kandel et al., 1994), or to general physical properties of early environment (e.g. Maruoka et al., 2009). However, *postnatal exposure* to chemosensory variety in the first 6 days following birth did not lead to the same outcome: CON females were more active than CON males and VAR females, but without a clear polarity in relation with the non-familiar odour. In sum, prenatal and postnatal flavour enrichment did not bear the same effects on the typical neophobia of young mice towards olfactory novelty.

*Second*, the *open-field test*, run in two versions (with and without a central object), intended to assess anxiety of young mice in a novel arena as a function of their earlier experience with chemosensory enrichment in the womb or in milk. An open field is classically assumed to incite increasing anxiety from the periphery to the centre, and more so when a nonfamiliar object is put in the centre (Prut & Belzung, 2003). It came out that, regardless of their sex, mice exposed *prenatally* to chemosensory variety, but not those exposed *postnatally*, roamed longer over the most anxiogenic central zones of the open field (empty version) than those of the CON group. When this central zone was rendered even more anxiogenic by adding a novel (visual) object, those mice exposed *postnatally* to the VAR regimen (but not those exposed *prenatally*) approached more the novel object. Exposure to the VAR regimen during prenatal or postnatal development resulted in aligned outcomes: depending on the test situation, prenatal and postnatal exposures to the VAR regimen led mice to display bolder behaviour in an unfamiliar environment than exposure to the CON regimen; prenatally VAR-exposed mice entered indeed more the most anxiogenic zone, and postnatal VAR-exposed mice increased even more their entries into the even more anxiogenic unfamiliar object placed in the central zone. Thus, postnatal exposure to chemosensory variety may have a stronger effect of anxiety attenuation, or produced stronger 'self-reassurance', than prenatal exposure in overcoming the wariness inherent to novelty. It is noteworthy that chemovariety exposure during the previsual period of development induced such an effect when mouse behaviour is driven by both visual and chemosensory cues.

*Finally*, the *paired-bottle test* assessed weanling mice's differential drinking of an unfamiliar flavour after previous experience of chemosensory variety during the 6 last foetal days or during the 6 first days post-birth. Among those exposed *prenatally*, most of the groups of male or female consumed more water than the unfamiliar banana solution, but the proportion of banana-flavoured water consumed did not differ as a function of the VAR or CON regimen. Among the mice exposed *postnatally*, only a sex difference emerged in the VAR group: females consumed fewer of the newly flavoured water after 1 and 24 h of choice, when presented against plain water. Thus, these results run again our expectations of reducing ingestive neophobia after experiencing flavour variety in utero or in lacto. Not only were the CON-exposed pups more prone to consume the newly flavoured beverage, but the VAR-exposed (especially females) avoided drinking of the newly flavoured water. At least for the postnatal VAR exposure group, these data are not in line with those of earlier studies who found that postnatal exposure to variety favours later intake of novel flavours (in the rat: Capretta et al., 1975; Myers & Pickering, 2007). This may be explained by the nature and duration of the chemosensory variety exposure (from birth to weaning vs. only 6 days in the present study) or by the delay between the end of the variety exposure and the test (immediate testing vs. testing differed of 27 days).

Taken together, the three tests above suggest that an enrichment in chemosensory variety of the foetal or neonatal environment has some consequences on behaviour around weaning. Both 'exploration' tests indicate that the tendency of young mice to keep away from (olfactory) novelty is affected by their prenatal or postnatal repeated exposition to chemosensory variety. However, in the present conditions of the ingestive test, prenatal or postnatal flavour variety did not consistently affect the typical refusal of mice to ingest flavour novelty.

## 4 Conclusions

Despite a few limitations (small sample sizes, unanticipated sensory thresholds and secondary effects of odorants used as stimuli), these two studies raise some insights on how perinatal chemosensory exposure might orient the development of one facet of personality, the tendency to pursue familiarity and the wariness to approach novelty. They constitute preliminary investigations to build upon.

In Study 1, all mother-mediated ways of perinatal (prenatal only, postnatal only and transnatal) exposure to a dominant odour/flavour induced a positive effect on exploratory behaviours towards the familiar odour in the days following birth. The familiar chemostimulus was preferred over the unfamiliar, which is in line with existing literature. However, this short-term neonatal effect of pre-, post-, and pre + post-exposures on the approach tendency towards the familiar chemostimulus did not extend in a long-term effect on the ingestion of the familiar flavour around weaning. Weanling mice of the postnatal (postnatal only and transnatal) exposure regimens ingested indeed familiar and nonfamiliar flavours alike in a first 24-h observation period, suggesting that a week of postnatal exposure to a given odorant may attenuate the repulsive effects of novelty. But this latter effect was also short-lived as they drank more of the newly flavoured water 24 h later. Thus, the reinforcing value engaged by prenatal or early postnatal experience with given chemostimuli can be rapidly superseded by more recent experiences with other chemostimuli and their associations with reinforcing events that do (perhaps) not occur in utero (e.g. sucking, gastric filling, satiation). Finally, another hypothetical driver of selective behaviour around weaning could be novelty seeking (e.g. Denenberg & Grota, 1964; Trickett et al., 2009), or at least seeking for some variability in ingested flavours to escape the boring effects of chemosensory monotony.

Study 2 found several behavioural effects of perinatal exposure to chemosensory variety created by six odorants over 6 days before or after birth, as opposed to the relative chemosensory monotony of standard lab conditions. Specifically, perinatal chemo-variety exposure was followed around weaning by an increased attendance of mice in the potentially most anxiogenic sector of the open-field arena, which novelty was even enhanced by a novel object. But no increased ingestion of an unfamiliarly flavoured beverage was noted. These results are thus only partly in line with Turkewitz's (1994) study suggesting (it was unfortunately only published as an abstract) that rat pups exposed to three flavours rotating every day during late gestation and through the suckling period will approach more a novel object and will ingest more of a novel flavour. The present results strengthen indeed the effect of perinatal exposure to chemo-variety on the murine apprehension of a novel space, prenatal and postnatal exposures having discriminable consequences in terms of reactivity to different grades of novelty intensity (empty vs. object open field). They further corroborate Turkewitz's idea that a perinatal exposure to chemosensory variety could affect vision-driven behaviour towards a novel object, suggesting that chemo-variety-based reduction of anxiety may generalise to other, later-developing sensory modalities. However, in our conditions, prenatal or postnatal chemo-variety exposure had no clear effect on ingestive behaviour of the novel flavour. This second study also points to a shallow sex difference in sensitivity to perinatal exposure to chemo-variety, as in given tests females tended to be more reactive.

In sum, the evidence derived from the present study is suggestive of behavioural consequence of being exposed to chemosensory variety during a period of maximal cerebral plasticity. Chemosensory variability of early developmental niches created by the mother may thus be one among many other factors that 'program' individual phenotypes (Weaver et al., 2004; Dalle Molle et al., 2016). To evaluate the effects of the early environment on stable phenotypic traits, similar research needs, however, to be replicated with larger samples of individuals, and the effects and mechanisms of early experience need to be ascertained by longitudinal studies reaching into adulthood.

**Acknowledgements** We thank Michel Tavan for his skills in building experimental devices and Anne Lefranc for her help in taking care of the animals. The present work was funded by the Centre National de la Recherche Scientifique and the Conseil Régional de Bourgogne-Franche-Comté. The authors dedicate this work to the memory of Professor Gerry Turkewitz, who inspired it in great part.

## References

Alberts, J. R. (1986). Postnatal development of olfactory-guided behavior in rodents. In W. Breipohl (Ed.), *Ontogeny of olfaction* (pp. 227–235). Springer.

Artcander, S. (1969). *Perfume and flavor chemicals: Aroma chemicals* (Vol. 1). Arctander Editor.

Bayol, S. A., Farrington, S. J., & Stickland, N. C. (2007). A maternal 'junk food' diet in pregnancy and lactation promotes an exacerbated taste for 'junk food' and a greater propensity for obesity in rat offspring. *British Journal of Nutrition, 98*(4), 843–851.

Bayol, S. A., Simbi, B. H., Bertrand, J. A., & Stickland, N. C. (2008). Offspring from mothers fed a 'junk food' diet in pregnancy and lactation exhibit exacerbated adiposity that is more pronounced in females. *The Journal of Physiology, 586*(13), 3219–3230.

Blass, E. M. (1990). Suckling: Determinants, changes, mechanisms, and lasting impressions. *Developmental Psychology, 26*(4), 520–533.

Blavi, L., Solà-Oriol, D., Mallo, J. J., & Pérez, J. F. (2016). Anethol, cinnamaldehyde, and eugenol inclusion in feed affects postweaning performance and feeding behavior of piglets. *Journal of Animal Science, 94*(12), 5262–5271.

Cabrera, D., Nilsson, J. R., & Griffen, B. D. (2021). The development of animal personality across ontogeny: A cross-species review. *Animal Behaviour, 173*, 137–144.

Cadiou, H., Aoudé, I., Tazir, B., Molinas, A., Fenech, C., Meunier, N., & Grosmaître, X. (2014). Postnatal odorant exposure induces peripheral olfactory plasticity at the cellular level. *Journal of Neuroscience, 34*(14), 4857–4870.

Capretta, P. J., Petersik, J. T., & Stewart, D. J. (1975). Acceptance of novel flavours is increased after experience of diverse tastes. *Nature, 254*, 689–691.

Carere, C., & Maestripieri, D. (Eds.). (2013). *Animal personalities: Behavior, physiology, and evolution*. University of Chicago Press.

Catanese, F., Distel, R. A., Provenza, F. D., & Villalba, J. J. (2012). Early experience with diverse foods increases intake of nonfamiliar flavors and feeds in sheep. *Journal of Animal Science, 90*(8), 2763–2773.

Choleris, E., Thomas, A. W., Kavaliers, M., & Prato, F. S. (2001). A detailed ethological analysis of the mouse open field test: Effects of diazepam, chlordiazepoxide and an extremely low frequency pulsed magnetic field. *Neuroscience & Biobehavioral Reviews, 25*(3), 235–260.

Chotro, M. G., & Arias, C. (2003). Prenatal exposure to ethanol increases ethanol consumption: A conditioned response? *Alcohol, 30*(1), 19–28.

Churihar, R., Solanki, P., Vyas, S., Hemant Tanwani, H., & Shubham Atal, S. J. I. J. P. (2016). Analgesic activity of cinnamaldehyde per se and it's interaction with diclofenac sodium and pentazocine in swiss albino mice. *International Journal of Phamacognosy, 3*, 97–102.

Clapp, P. W., Pawlak, E. A., Lackey, J. T., Keating, J. E., Reeber, S. L., Glish, G. L., & Jaspers, I. (2017). Flavored e-cigarette liquids and cinnamaldehyde impair respiratory innate immune cell function. *American Journal of Physiology-Lung Cellular and Molecular Physiology, 313*(2), L278–L292.

Cloninger, C. R. (1994). Temperament and personality. *Current Opinion in Neurobiology, 4*, 266–273.

Coppola, D. M., & Millar, L. C. (1997). Olfaction in utero: Behavioral studies of the mouse fetus. *Behavioural Processes, 39*(1), 53–68.

Coureaud, G., Schaal, B., Hudson, R., Orgeur, P., & Coudert, P. (2002). Transnatal olfactory continuity in the rabbit: Behavioral evidence and short-term consequence of its disruption. *Developmental Psychobiology, 40*, 372–390.

Dalle Molle, R., Bischoff, A. R., Portella, A. K., & Silveira, P. P. (2016). The fetal programming of food preferences: Current clinical and experimental evidence. *Journal of Developmental Origins of Health and Disease, 7*(3), 222–230.

Debong, M. W., N'Diaye, K., Owsienko, D., Schöberl, D., Ammar, T., Lang, R., et al. (2021). Dietary linalool is transferred into the milk of nursing mothers. *Molecular Nutrition & Food Research, 65*(23), 2100507.

Delaunay, M. E., Marlier, L., & Schaal, B. (2006). Learning at the breast: Preference formation for an artificial scent and its attraction against the odor of maternal milk. *Infant Behavior and Development, 29*(3), 308–321.

Delaunay, M.E, Soussignan, R., Patris, B., Marlier, L., & Schaal, B. (2010). Long-lasting memory for an odor acquired at the mother's breast. Developmental Science, 13(6), 849-863.

Denenberg, V. H., & Grota, L. J. (1964). Social-seeking and novelty-seeking behavior as a function of differential rearing histories. *Journal of Abnormal and Social Psychology, 69*(4), 453.

Désage, M., Schaal, B., Soubeyrand, J., Orgeur, P., & Brazier, J. L. (1996). Gas chromatographic-mass spectrometric method to characterise the transfer of dietary odorous compounds into plasma and milk. *Journal of Chromatography B: Biomedical Sciences and Applications, 678*(2), 205–210.

Dewaele, A., Badonnel, K., Persuy, M. A., Durieux, D., Bombail, V., Favreau-Peigné, A., & Baly, C. (2020). Effect of environmental exposure to a maternally-learned odorant on anxiety-like behaviors at weaning in mice. *Animal Cognition, 23*, 881–891.

Echandia, E. R., Foscolo, M., & Broitman, S. T. (1982). Preferential nesting in lemon-scented environment in rats reared on lemon-scented bedding from birth to weaning. *Physiology & Behavior, 29*(1), 47–49.

Fillion, T. J., & Blass, E. M. (1986). Infantile experience with suckling odors determines adult sexual behavior in male rats. *Science, 231*, 729–731.

Galef, B. G., & Henderson, P. W. (1972). Mother's milk: A determinant of the feeding preferences of weaning rat pups. *Journal of Comparative and Physiological Psychology, 78*(2), 213–219.

Galliot, E., Laurent, L., Hacquemand, R., Pourié, G., & Millot, J. L. (2012). Fear-like behavioral responses in mice in different odorant environments: Trigeminal versus olfactory mediation under low doses. *Behavioural Processes, 90*(2), 161–166.

Gerrish, C. J., & Mennella, J. A. (2001). Flavor variety enhances food acceptance in formula-fed infants. *The American Journal of Clinical Nutrition, 73*(6), 1080–1085.

Griebel, G., Belzung, C., Misslin, R., & Vogel, E. (1993). The free-exploratory paradigm: An effective method for measuring neophobic behaviour in mice. *Behavioural Pharmacology, 4*, 637–644.

Gruest, N., Richer, P., & Hars, B. (2004). Emergence of long-term memory for conditioned aversion in the rat fetus. *Developmental Psychobiology, 44*(3), 189–198.

Hai, P. V., Schonewille, J. T., Dam-Van, T., Everts, H., & Hendriks, W. H. (2014). Exposure to a novel feedstuff by goat dams during pregnancy and lactation versus pregnancy alone does not further improve post-weaning a cceptance of this feedstuff by their kids. *Journal of the Science of Food and Agriculture, 96*(6), 2215–2219.

Hepper, P. G., Wells, D. L., Millsopp, S., Kraehenbuehl, K., Lyn, S. A., & Mauroux, O. (2012). Prenatal and early sucking influences on dietary preference in newborn, weaning, and young adult cats. *Chemical Senses, 37*(8), 755–766.

Hepper, P. G., Wells, D. L., Dornan, J. C., & Lynch, C. (2013). Long-term flavor recognition in humans with prenatal garlic experience. *Developmental Psychobiology, 55*(5), 568–574.

Hooper, L. V., Midtvedt, T., & Gordon, J. I. (2002). How host-microbial interactions shape the nutrient environment of the mammalian intestine. *Annual Review of Nutrition, 22*, 283–307.

Jiang, J., Emont, M. P., Jun, H., Qiao, X., Liao, J., Kim, D. I., & Wu, J. (2017). Cinnamaldehyde induces fat cell-autonomous thermogenesis and metabolic reprogramming. *Metabolism, 77*, 58–64.

Johnson, M. C., Biju, K. C., Hoffman, J., & Fadool, D. A. (2013). Odor enrichment sculpts the abundance of olfactory bulb mitral cells. *Neuroscience Letters, 541*, 173–178.

Kagan, J., Reznick, J. S., & Snidman, N. (1987). Temperamental variation in response to the unfamiliar. In N. A. Krasnegor, E. M. Blass, M. A. Hofer, & W. P. Smotherman (Eds.), *Perinatal development, A psychobiological perspective* (pp. 421–440). Academic Press.

Kandel, D. B., Wu, P., & Davies, M. (1994). Maternal smoking during pregnancy and smoking by adolescent daughters. *American Journal of Public Health, 84*(9), 1407–1413.

Mahapatra, S., Roy, S., Chakraborty, P., Chakraborty, N., Paul, U., Chatterjee, P., & Bhattacharya, M. (2020). Cinnamaldehyde, the major component of cinnamomum zeylanicum, affects inflammatory pathways. *International Journal of Pharmaceutical and Sciences Research, 11*(11), 5788–5791.

Maier, A. S., Chabanet, C., Schaal, B., Leathwood, P. D., & Issanchou, S. (2008). Breastfeeding and experience with variety early in weaning increase infants' acceptance of new foods for up to two months. *Clinical Nutrition, 27*(6), 849–857.

Maier-Nöth, A., Schaal, B., Leathwood, P., & Issanchou, S. (2016). The lasting influences of early food-related variety experience: A longitudinal study of vegetable acceptance from 5 months to 6 years in two populations. *Plos One, 11*(3), e0151356.

Mainardi, D., Marsan, M., & Pasquali, A. (1965). Causation of sexual preferences of the house mouse. The behaviour of mice reared by parents whose odour was artificially altered. *Atti della Societa Italiana di Scienze Nationali e del Museo Civico di Storia Naturale di Milano, 104*, 325–338.

Maruoka, T., Kodomari, I., Yamauchi, R., Wada, E., & Wada, K. (2009). Maternal enrichment affects prenatal hippocampal proliferation and open-field behaviors in female offspring mice. *Neuroscience Letters, 454*(1), 28–32.

Mennella, J. A., Jagnow, C. P., & Beauchamp, G. K. (2001). Prenatal and postnatal flavor learning by human infants. *Pediatrics, 107*(6), E88.

Middelkoop, A., Choudhury, R., Gerrits, W. J., Kemp, B., Kleerebezem, M., & Bolhuis, J. E. (2018). Dietary diversity affects feeding behaviour of suckling piglets. *Applied Animal Behaviour Science, 205*, 151–158.

Myers, K. P., & Pickering, C. (2007). Sensory variety during nursing increases young rats' acceptance of novel flavors after weaning. *Appetite, 49*, 272–341.

Nolte, D. L., & Mason, J. R. (1995). Maternal ingestion of ortho-aminoacetophenone during gestation affects intake by offspring. *Physiology & Behavior, 58*, 925–928.

Nolte, D. L., & Provenza, F. D. (1991). Food preferences in lambs after exposure to flavors in milk. *Applied Animal Behaviour Sciences, 32*, 381–389.

Oostindjer, M., Bolhuis, J. E., Van den Brand, H., Roura, E., & Kemp, B. (2010). Prenatal flavor exposure affects growth, health and behavior of newly weaned piglets. *Physiology & Behavior, 99*, 579–586.

Portella, A. K., Kajantie, E., Hovi, P., Desai, M., Ross, M. G., Goldani, M. Z., et al. (2012). Effects of in utero conditions on adult feeding preferences. *Journal of Developmental Origins of Health and Disease, 3*(3), 140–152.

Prut, L., & Belzung, C. (2003). The open field as a paradigm to measure the effects of drugs on anxiety-like behaviors: A review. *European Journal of Pharmacology, 463*(1–3), 3–33.

Schaal, B. (2005). From Amnion to Colostrum to Milk: Odor Bridging in Early Developmental Transitions. In B. Hopkins & S. P. Johnson (Eds.), *Prenatal development of postnatal functions* (pp. 51–102). Praeger Publishers/Greenwood Publishing Group.

Schaal, B., Marlier, L., & Soussignan, R. (2000). Human foetuses learn odours from their pregnant mother's diet. *Chemical Senses, 25*, 729–737.

Smotherman, W. P., & Robinson, S. R. (1985). The rat fetus in its environment: Behavioral adjustments to novel, familiar, aversive, and conditioned stimuli presented in utero. *Behavioral Neuroscience, 99*(3), 521.

Spahn, J.M., Callahan, E.H., Spill, M.K., Wong, Y.P., Benjamin-Neelon, S.E., Birch, L., ... & Casavale, K.O. (2019). Influence of maternal diet on flavor transfer to amniotic fluid and breast milk and children's responses: a systematic review. *The American Journal of Clinical Nutrition, 109*(Supplement 1), 1003S–1026S.

Stamps, J., & Groothuis, T. G. (2010). The development of animal personality: Relevance, concepts and perspectives. *Biological Reviews, 85*(2), 301–325.

Todrank, J., Heth, G., & Restrepo, D. (2011). Effects of in utero odorant exposure on neuroanatomical development of the olfactory bulb and odour preferences. *Proceedings of the Royal Society B: Biological Sciences, 278*, 1949–1955.

Trickett, S. L., Guy, J. H., & Edwards, S. A. (2009). The role of novelty in environmental enrichment for the weaned pig. *Applied Animal Behaviour Science, 116*(1), 45–51.

Turkewitz, G. (1994). Developmental precedence and intersensory influence. *Infant Behavior and Development*, 17 (special ICIS issue), 59.

Villalba, J. J., Bach, A., & Ipharraguerre, I. R. (2011). Feeding behavior and performance of lambs are influenced by flavor diversity. *Journal of Animal Science, 89*(8), 2571–2581.

Walsh, R. N., & Cummins, R. A. (1976). The open-field test: a critical review. *Psychological Bulletin, 83*(3), 482.

Weaver, I. C., Cervoni, N., Champagne, F. A., D'Alessio, A. C., Sharma, S., Seckl, J. R., Dymov, S., Szyf, M., & Meaney, M. J. (2004). Epigenetic programming by maternal behavior. *Nature Neuroscience, 7*, 847–854.

Wells, D. L., & Hepper, P. G. (2006). Prenatal olfactory learning in the domestic dog. *Animal Behaviour, 72*(3), 681–686.

Wiedenmayer, C. (2010). Sensitive periods in the behavioral development of mammals. In C. M. Worthman, P. M. Plotsky, D. S. Schechter, & C. A. Cummings (Eds.), *Formative experiences: The interaction of caregiving, culture, and developmental psychobiology* (pp. 82–105). Cambridge University Press.

Worthman, C. M., Plotsky, P. M., Schechter, D. S., & Cummings, C. A. (Eds.). (2010). *Formative experiences: The interaction of caregiving, culture, and developmental psychobiology*. Cambridge University Press.

Xiao, Y., Zhang, F., Xu, H., Yang, C., Song, X., Zhou, Y., & Miao, J. (2022). Cinnamaldehyde microcapsules enhance bioavailability and regulate intestinal flora in mice. *Food Chemistry: X, 15*, 100441.

Yang, G., Jin, T., Yin, S., Guo, D., Zhang, C., Xia, X., & Shi, C. (2019). Trans-Cinnamaldehyde mitigated intestinal inflammation induced by *Cronobacter sakazakii* in newborn mice. *Food and Function, 10*(5), 2986–2996.

Zuo, J., Zhao, D., Yu, N., Fang, X., Mu, Q., Ma, Y., et al. (2017). Cinnamaldehyde ameliorates diet-induced obesity in mice by inducing browning of white adipose tissue. *Cellular Physiology and Biochemistry, 42*(4), 1514–1525.

# Does the Rabbit Mammary Pheromone Attract Newborns to Maternal Faeces? A New Potential Function of the Suckling Chemosignal

Benoist Schaal[1] (✉), Anne-Sophie Moncomble[1], and Dominique Langlois[1]

[1] Development of Olfactory Communication and Cognition Laboratory, Centre des Sciences du Goût et de l'Alimentation, CNRS, Université de Bourgogne, Inrae, Dijon, France
benoist.schaal@u-bourgogne.fr

**Abstract** The 'mammary pheromone' (2-methyl-but-2-enal, 2MB2-al) regulates the interactions between rabbit (*Oryctolagus cuniculus*) females and their newborn pups. Released into milk, it first elicits pups' sucking of nipples. A second function is its ability to engage learning of contingent odorants, which become then as potent as the mammary pheromone itself to elicit sucking. The present study evaluates the possibility of a third function of this chemosignal. Lactating rabbit does release some faecal pellets (FP) into their nest after suckling. These FP disappear progressively, raising the hypothesis that the pups can trace them in the dark nest. Systematic follow-up of FP dropped into the nest, and the pups' behaviour towards them suggests that they ingest them from postnatal days 2–3. Preference tests exposing pups to FP from various kinds of conspecifics show that they are attractive irrespective of the sex, identity and physiological state of the donor. But when exposed to FP sampled before and after their anal excretion, pups show more attraction to excreted FP, suggesting that the secretion of anal glands (AG) is involved. Dynamic headspace extraction and GC-MS analyses of non-excreted vs. excreted FP revealed different volatile profiles. In particular, 2MB2-al and its alcohol form, 2-methyl-but-2-enol (2MB2-ol), were detected in the effluvium of excreted FP but not of FP before excretion. Further, the headspace of AG from females contained both 2MB2-al and 2MB2-ol. Thus, in the process of anal transit, hard FP are charged with 2MB2-al and 2MB2-ol emitted from AG. Subsequent testing of 2MB2-ol showed its behavioural inactivity in pups, signifying that attraction to FP is explainable by 2MB2-al's presence on their surface. The rabbit mammary pheromone is thus released from two distinct body sites in radically different substrates to mediate specific approach and ingestive behaviour satisfying very different adaptive functions in offspring: milk intake and coprophagy.

**Keywords** European rabbit (*Oryctolagus cuniculus*) · Olfaction · Mammary pheromone · 2-methylbut-2-enal · Neonate · Milk · Faeces · Coprophagy · Caecotrophy · Microbiota · Vertical transmission · Learning



B. Schaal et al. (eds.), *Chemical Signals in Vertebrates 15*,
https://doi.org/10.1007/978-3-031-35159-4_17

# 1 Introduction

During the first weeks of their nidicolous life, newborns of the European rabbit (*Oryctolagus cuniculus*) make rapidly sense of their environment. It is for them a matter of survival to detect the right effluvia at the right time – the lactating mother being accessible for less than 5-min once per day – in a profusely fragrant nest among allied-competing littermates. On the one hand, pups pass 99% of their time to huddle with their littermates who copiously urinate into the nest (Hudson & Distel, 1982) made of aromatic hay (or wood shavings) and scented hair from the mother's abdomen (Hudson et al., 1996, 1999; Bilko et al., 2022). In the days following parturition, remnants of blood, uneaten placentae and dead littermates may also add their particular note. Both nest and offspring produce thus their own evolving effluvium, which newborn pups learn to discern and then update as it decays with their own excretions (Hudson et al., 2003; Serra & Nowak, 2008). On the other hand, during her brief suckling visits, the lactating female exposes her offspring to her various odour glands, including primarily the milk-ejecting nipples they vitally need to latch on once a day. But she bears many other scent sources, often structured as glands, located in the oral (saliva, breath), facial (lachrymal, Harderian), submandibular (chin), inguinal, urogenital and rectal-anal regions (e.g. Mykytowycz, 1970; Schalken, 1976; Merritt et al., 1982; Bell, 1985; Melo & Gonzales-Mariscal, 2010). By her intensive self-grooming, the rabbit doe also anoints herself with saliva, urine or glandular secretions, which may odourize her fur or paws in a way that is discernible to pups (Val-Laillet & Nowak, 2008). During pregnancy and the suckling period, the female finally drops scented faecal pellets (FP) into the nest. Chemicals from all these sources are known to support communication between adults (Mykytowycz & Dudzinski, 1972), such as territorial marking and hierarchy (Mykytowycz, 1968, 1972; Mykytowycz, et al., 1976), signalling of sexual status or reproductive state (Wales & Ebling, 1971; Melo & Gonzalez-Mariscal, 2010) or of individual or group identity (Goodrich & Mykytowycz, 1972). These secretions are also involved in regulating mother-infant interactions, although their roles are then less well understood (e.g. Mykytowycz & Ward, 1971; Hudson & Distel, 1994; Hudson et al., 1999; Schaal et al., 2008), especially concerning FP and related odour cues.

To navigate this complex and dynamic odour mosaic, rabbit pups are predisposed to prioritize certain chemosignals over others. In the first postnatal weeks, their top urgency is to get milk to survive, and the most salient chemosignals certainly are then those emanating from the nipples (Schley, 1977; Hudson & Distel, 1983). Among them, the mammary pheromone (2-methyl-but-2-enal, 2 MB2-al) is especially effective in releasing pups' sucking during the period of exclusive lactivory (Schaal et al., 2003, 2008). In addition to this sucking-releasing function, the pheromone boosts learning of any associated odour cue conveyed in milk or on maternal fur, which then becomes functionally similar to the mammary pheromone itself (Coureaud et al., 2004; Patris et al., 2008; Jouhanneau et al., 2016). By ~postnatal day 15, offspring begin eating increasingly significant amounts of non-milk substrates (hay) in the nest and then start to exit the nest to get weaned at ~25–31 days of age (Schley, 1985; Varga, 2013; Gonzales-Mariscal et al., 2022). But they are appealed by non-milk elements well before day 15, specifically by nest materials or FP left by mothers after suckling. Mykytowytcz and Ward (1971) first noticed that FP dropping into the nest was a component of maternal behaviour in rabbits. The phenomenon was then documented more in depth by several research groups (Hudson et al., 1996; Moncomble et al., 2004; Moncomble, 2006; Kovacs et al., 2006; Gidenne et al., 2013; Combes et al., 2014). Thus, observations of pups approaching and gnawing these FP suggest a second priority of newborn rabbits in the nest: to contact and incorporate FP dropped by their mothers right after suckling.

How do neonatal rabbits sense these FP in the dark and olfactorily noisy nest? As they can manifestly only use their sense of smell (and perhaps of touch) to trace them, on which chemical cues or signals do they rely? FP have the potential to convey multiple olfactory indicators, such as common volatiles derived from the mother's diet or metabolic wastes (Bilko et al., 1994), or specialized volatiles released by secretions from the gut or anal glands that coat FP (Hesterman & Mykytowycz, 1968). Here, we further attempt to understand which semiochemicals are involved in the pups' guidance to FP. We will first summarize the current knowledge on the dynamics of FP emission into the nest by lactating females. Second, pups' behavioural responsiveness to FP and its modulation by offspring-related variables (feeding cycle and postnatal development) or female-related variables (identity, physiological state) will be assessed. Third, the involvement of gut or anal glands' secretions will be evaluated in the pups' attraction to FP. Finally, some chemical correlates of offspring's attraction to FP will be characterized.

## 2 Faecal Pellets in the Nest

### 2.1 Dynamics of FP Deposition by Lactating Females

We first recapitulate our earlier results on the everyday amount of FP dropped into the nest by rabbit females and on the amount of FP accessible to pups during their first 2 weeks of life (Moncomble et al., 2004; Moncomble, 2006). These results were obtained in the rabbit breeding colony of ENESAD (Etablissement National d'Enseignement Supérieur Agricole de Dijon). The rabbits (New Zealand x Californian cross breed) were raised in cages (Length × width × height: 62 × 46 × 27 cm) organized in flat deck in a ventilated facility. Nest boxes (42 × 25 × 25 cm) were appended to the females' cages 2 days before parturition. The breeding facility was under a fixed lighting schedule (day/night: 14/10 h, lights on at 7 a.m.), with ad lib fodder and water (with a coccidiostatic agent added). Nursing females were subjected to a regimen of controlled access to the nest box, mimicking the species typical once per day nursing visit (Zarrow et al., 1965). Accordingly, nest boxes were opened once a day at fixed time for 15 min. After weaning, the rabbits were destined to commercial outcomes. The present series of studies were agreed by Dr. Guy Perrier, professor at ENESAD in authority for the rabbit breeding unit of this institution. All housing and handling conditions of the animals complied with the French and European directives for animal research.

In a first observation round, ten lactating does and their litter were surveyed from parturition (day 0) up to postpartum day 13. Nest inspections in search for FP were carried out daily after closure of nest boxes (i.e. after suckling) between 8 and 11 a.m.

After parturition, the nest boxes were cleared of any FP, uneaten placentas, dead pups or blood-soiled matters. The inspection of FP number was then run daily after each feed. Inspection was not made beyond day 13, as pups then suck increasingly outside nest boxes (i.e. in the breeding cage, making collection of FP difficult). For the next 24 h (viz., until next suckling), the rabbits were not handled, and the nest boxes remained closed. FP dropped daily into the nest by each female immediately after suckling were counted. The average daily amount of FP accessible to the pups in the nest from days 1 to 13 was obtained by adding the number of FP of the day with the number of FP accumulated from previous days.

The average emission curve of FP by females, regardless of dry/hard vs. soft types,[1] is presented in Fig. 1. The number of dropped FP varied significantly over the period. Females released an average of ~2 FP into the nest from days 1 to 8. This production then decreased to an average of 1 FP from days 9 to 11, and became negligible beyond day 12. On the ten females surveyed, seven dropped only hard FP, two emitted both hard and soft FP, while one did not deposit any FP during the 13-day follow-up. The cumulative amount of FP accessible to the pups also varied significantly until day 13 postpartum (Fig. 1). From an average of 2 FP available to the pups on day 2, this number increased to peak at 12.5 FP on day 8, then decreasing to 3.5 FP by day 13.

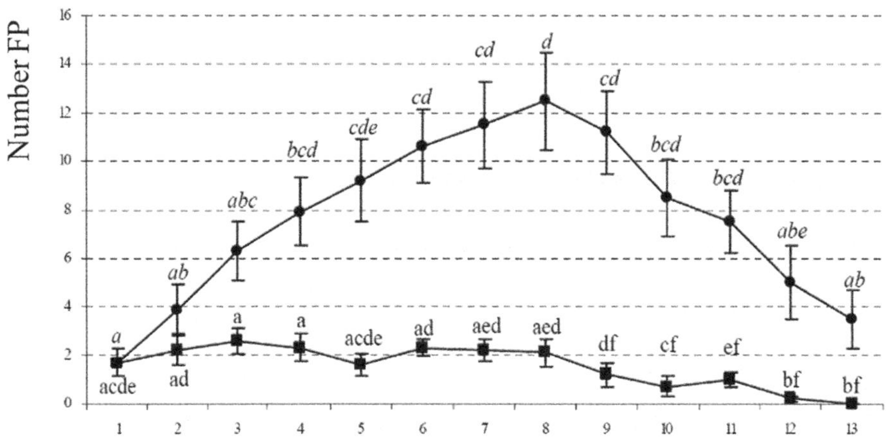


**Fig. 1** Day-to-day kinetics of the count of faecal pellets (FP) emitted by lactating rabbits into the nest (■), and cumulated number of FP accessible daily to neonate rabbit pups (●). Follow-up carried out on 10 females and their offspring (8–10 rabbits/litter) between postpartum days 1–13 (Means ± SE denoted by different letters indicate significant within-variable differences between days; Fischer's post hoc test)

[1]Lagomorphs excrete two types of FP (Björnhag, 1994; Björnhag & Snipes, 1999; Pickard & Stevens, 1972; Takagi & Tagawa, 1961): (1) the 'soft' pellets or caecotrophs, which originate from vegetal matters fermented in the caecum, are rich in microbial proteins and vitamins; re-ingested without chewing, these FP are essential to the healthy development of rabbits (de Blas & Wiseman, 1990); (2) the 'hard' or 'dry' pellets, made up of large particles partially digested in the colon, are sometimes also re-ingested (coprophagy), but in much lesser proportion than caecotrophs (Hirakawa, 2001).

To sum up, females emitted ~2 FP in the nest after each suckling during the first postpartum week. During this first week as well, the cumulative number of FP accessible to the pups increased steadily, to then decrease from day 8 to become negligible by the end of the second week. Finally, FP emission varied between females in both amount and type and as a function of timing relative to the females' digestive processes (caecotrophy).

The above observations confirmed the phenomenon of FP emission into the nest by lactating does, reported by earlier and later studies (Mykytowytcz & Ward, 1971; Hudson et al., 1996; Kovacs et al., 2006; Gidenne et al., 2013; Combes et al., 2014). These studies unanimously found that at the end of each suckling episode within the first weeks, rabbit does deposit up to 2-4 FP into the nest. The present study noted a daily deposition of ~2 FP by females until postpartum day 11. Beyond that point, post-sucking FP dropping appeared to disappear. The apparent cessation of FP emission in the nest beyond days 12-13 corroborates other studies (Combes et al., 2014). It may be due to lactation-related physiological changes in females or to the fact that pups are increasingly avid of FP, disintegrating or ingesting them more rapidly than the does produce them. It may also be due to breeding conditions, growing pups initiating suckling increasingly outside of nest boxes where suckling-related FP emission by does is difficult to reliably quantify. Finally, the amount and frequency of does' FP deposition may vary between studies, but this maternal behaviour is consistent up to postpartum day 10 across several studies using diverse breeds and breeding conditions (Hudson et al., 1996; Kovacs et al., 2006; Combes et al., 2014; present study). Also, as reported in previous studies, the does dropped prevalently FP of the hard type into the nest, although soft FP were sometimes also emitted.

## 2.2 Turnover of Maternal FP in the Nest

This second observation evaluated the fate of FP in the nest each day over the first 2 weeks postpartum and estimated the dynamics of FP disappearance according to their age of emission and age of pups. Ten multiparous, lactating non-pregnant females and their litters (8–10 pups/litter) were followed up between postpartum days 1–13. On day 1, as soon as the doe leaved the nest after suckling, FP were collected to be marked with scentless threads of different colours on each consecutive day in order to trace their fate within the nest. Every day, from day 1 to 13 and for each litter, 10 min before controlled suckling, nest material was searched to monitor the remaining colour-marked FP. The nest was then reconstituted with all FP from previous days, and pups replaced in its centre. After a delay of 1–2 min, nest boxes were opened to the does who immediately suckled their offspring without disturbance. After this suckling bout, the newly deposited FP were colour-marked as well. The counts of the colour-marked FP allowed estimating their persistence/disappearance over postnatal days 1–13.

The daily variation of the number of FP disappearing from the nests is depicted in Fig. 2. The rate of missing FP was varying during the first 13 postpartum days, ranging from 0 FP/nest on day 1 to 2–3.5 FP/nest on days 8 to 11. The rate of FP disappearance declined thereafter. The average time for colour-marked FP to vanish varied according to both their age of emission and age of pups. Disappearance of the FP emitted on days

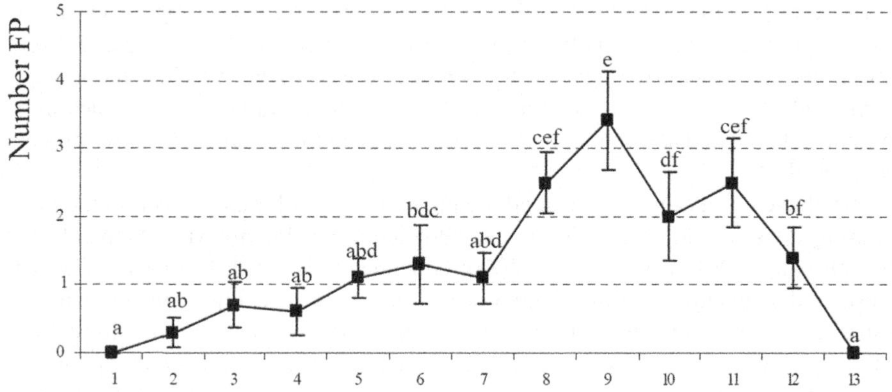


**Fig. 2** Number of FP that disappeared daily from the nest between days 1 and 13 (observations made on ten females and their litters, eight to ten rabbits/litters) (means ± SE denoted by different letters indicate significant differences between days at $p < 0.05$; Fischer's post hoc test)

5–11 was faster (38% in 24 h; 70% in 48 h) than those emitted on days 3–4 (20% in 24 h; 30% in 48 h), and even more so than those dropped on days 1–2 (20% in 7 days). In sum, the FP that does dropped into their nests disappeared faster with advancing age of the pups. The most rapid disappearance of FP occurred between days 7–12, the freshest ones missing in priority. The possible reasons for the disappearance of FP dropped by the does are multiple: they may either crumble in the nest wetted by pups urinating copiously after suckling or alternatively they may be ingested by the pups, a possibility that is examined below.

## 3 Pup Responsiveness to Faecal Pellets

Do the pups' come across FP as a result of apparently random movements in the nest after each suckling episode or do they contact them as a result of directional actions drawn by FP-related odour stimuli? The next experiment aimed to establish whether FP are olfactorily attractive to pups and how consistent this attraction is during postnatal development. It also aimed to evaluate whether young rabbits ingest the FP.

### 3.1 Attraction Inside the Nest

Two groups of pups were followed. Seventy-six pups from 8 litters followed from day 0 to 10 constituted the experimental group, the control group being composed of 39 pups from 4 litters. Pups of both groups were marked with scentless ink for individual recognition. They were tested in their nest box, where they were confronted with a cupule of plastic mesh affixed against a wall (Fig. 3). For the experimental pups, this cupule contained hard FP collected after previous suckling episodes and stored at −80 °C until testing; these pups were exposed to a pool of 5 FP matched to their age (i.e. collected from does on day 5 postpartum), taken from five different females

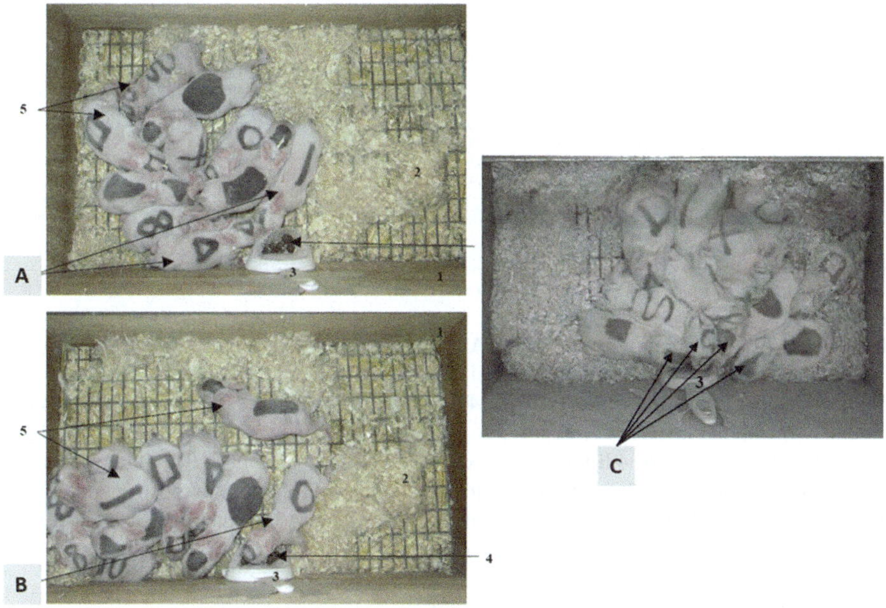


**Fig. 3** The plastic mesh cupule used to present the FP to the pups within their nest box (1, nest box; 2, nest elements (hay, shavings and maternal hair); 3) the cupule; 4) faecal pellets (FP); 5) 5-day-old rabbit pups). (**A**) Pups in contact with the cupule containing FP; (**B**) pup in direct oral contact with the FP within the cupule. (C) Four pups competing to access the FP within the cupule

(excluding own mother). Control pups were exposed only to the empty plastic mesh cupule (Fig. 3). The tests began after the daily suckling, after the removal of the nest material to avoid pups burying into it (so that they remained visible for video-recording). The average temperature was 22 °C ± 2 °C in the breeding building. The 5 hard FP were thawed to 38 °C (± 1 °C) before being introduced into the test cup. Experimenters did not intervene during the 30-min tests unless a FP was dragged out of the cupule or contaminated with pups' urine (it was then replaced). When the test was over, the nest material was returned to the nest box, FP discarded, and the test cup cleaned with alcohol and distilled water, and dried.

The 30-min tests were videotaped for offline coding of target behaviours using the Observer software (Noldus, Wageningen, NL). The onset/offset of each occurrence of rabbit pups' contact with the cupule or FP was recorded (Fig. 3). From these individual data, an average duration of contact per rabbit with the cupule and with the FP within it was calculated. The behaviour of the experimental vs. control groups towards the cupule was compared only on days 1, 4, 7 and 10. Pup responsiveness to FP within the cupule were measured daily between days 1 to 10.

Analyses of the videos indicated that the control pups did not approach the empty cupule, whatever their age (cf. Fig. 4a). In comparison, the experimental pups contacted the FP-containing cupule increasingly from day 1 and significantly longer than the control pups on days 4 and 7 (Fig. 4a). The average duration of contact of experimental pups with the FP within the cupule fluctuated thus over the 10-day follow-up (Fig. 4b),

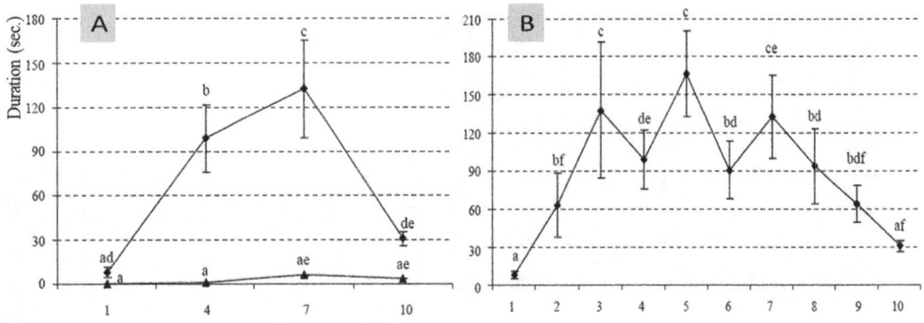


**Fig. 4** Attraction of rabbit pups to the FP. (**A**): Average duration of contact (s) of the rabbit pups of the experimental (◆) and control (▲) groups with the FP cupule during 30-min tests occurring after the suckling bouts of postnatal days 1, 4, 7 and 10. Pups of the experimental group ($n = 76$, 8 litters, 8–10 rabbits/litters) were exposed to 5 FP in the cupule, while those in the control group ($n = 39$, 8–10 rabbits/litters, 4 litters) were exposed to an empty cupule. (**B**) Average duration of contact (s) of the pups ($n = 76$) with the device containing 5 FP for 30 min post-feeding on postnatal days 1 to 10 (Means ± SE denoted by different letters indicate significant differences between days at $p < 0.05$; Fischer's post hoc test)

indicating important variations in neonatal attraction towards age-matched FP as a function of postnatal development. On day 1, these pups contacted the FP on average 0.92 times for a mean duration of 11 s/pup. Day 2 pups touched the FP 1.8 ± 0.6 times for an average duration of 60 ± 12 s. The tests of day 3 to 8 revealed more frequent (on average 3.7 ± 1.2 times) and longer contacts with FP, ranging from 90 ± 22 to 150 ± 34 s. Beyond day 9, frequency and duration of direct contact with the FP declined to reach the level of days 1–2. In sum, the pups approached only the cupule with FP inside, indicating that their movements were driven by attractive odour cues, and more so between days 3 to 8.

A qualitative behavioural analysis was run on ten pups of the experimental group (from eight litters, one to two per litter, 5-day-old) for a duration of 60 s from the moment when they contacted the FP cupule in the nest. Insistent searching-rooting movements were then displayed without latency. Particular movements of forepaw scratching occurred on average after 9.8 ± 7.8 s and finally oral snapping actions were displayed at 27.4 ± 11.2 s after contact. The forepaw scratching was directed to the FP in unilateral or bilateral bursts of wriggling segmental movements, often interspersed with oral snapping towards the FP. The duration of these non-exclusive items was 39.8 ± 11.2 s for searching-rooting towards the FP, 24.6 ± 8.3 s of forepaws' scratching, and 14.8 ± 5.6 s of oral snapping, indicating the urge of attraction to, and direct action of pups on, the FP.

## 3.2 Ingestion of FP

The visual examination of the FP at the end of the above experiment clearly indicated that they were teared by the direct action of pups and that crumbles of them were detached. The appearance of FP before and after the test can be seen in Fig. 5a, b. Direct observation as well as video coding revealed that fractions of the FP were shredded by

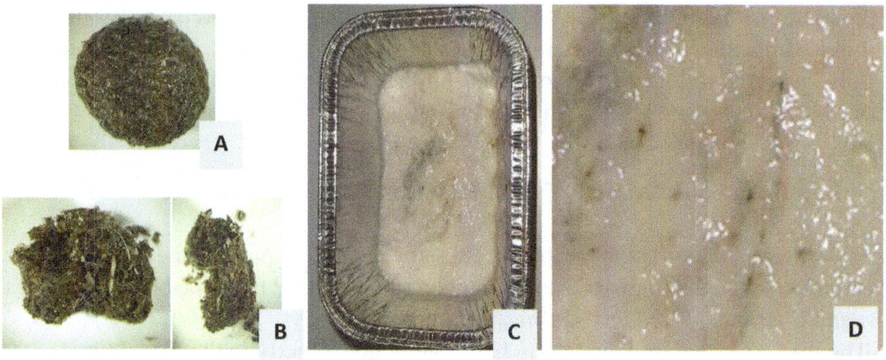


**Fig. 5** (**A**) A fresh faecal pellet (hard type) collected right after maternal emission into the nest; it is clearly coated with a mucous film of unknown composition, probably originating in secretions from the digestive tract and/or accessory glands. (**B**) Faecal pellets after being presented to pups in the test cupule: the secretive film has disappeared, and the pellets are in part disintegrated. (**C**) The gastric content of a rabbit pup after exposure to faecal pellets. (Courtesy of Dr. Laurence Fortun-Lamothe). (**D**) Magnification of **C** showing non-milk particles in the milk coagulate

the pups' oral action on them. They also displayed the characteristic foreleg scratching movements, directed to the cupule when they were not in direct oral contact with the FP. Such foreleg scratching was seen only when the pups were in contact with the FP in the nest in the context of the littermates' huddle. They could not be systematically elicited by presenting a single FP to an individual pup outside of the multisensory background of the nest and huddle (Fig. 6a).

As the crumbles detached from the FP were not found at the bottom of the test cupule, we inferred that they could have been ingested by pups. The pups were clearly seen to display oral snapping movements directed to the FP. Finally, the visual examination of the gastric content of several pups made by Dr. L. Fortun-Lamothe confirmed this point: dark FP particles were conspicuously visible in the milk coagulate obtained from pups 1 h after suckling (Fig. 5c, d) (It may be noted that the nest boxes are lined with clear wood chips and not with darker hay; these are thus not made of hay elements). The same observation of gastric content of pups filled with bits of disaggregated FP was made by Kovacs et al. (2006).

The recency of FP emission is an important factor in adult rabbits' reactions to FP (Schalken, 1976), and it appeared essential as well in pups' attraction towards the FP. This was shown in a test assessing the longevity of FP in the nest as a function of their freshness. Ten nests/litters (8–10 pups/litter) were examined on postnatal day 7 at two points: (i) *before suckling* in order to count all 'old' marked FP remaining from previous days and (ii) *right after suckling* to count and colour-mark freshly emitted FP. Fresh and old FP were then counted every hour over 6 h to track their fading in the nest. Before the day 7 suckling, an average (± SD) of 9.1 ± 4.6 old FP persisted from previous days, and the does dropped 1.2 ± 0.9 fresh FP. It came out that the fresh FP decreased by 67% within the first hour following the 7-day suckling episode, whereas old FP remained untouched. No further reduction in the number of either old or new FP occurred during the next 5 h of the follow-up.

In sum, freshly emitted FP were more attractive and disappeared first from the nest within 1 h after suckling, suggesting further ingestion by the pups. These tests indicated that the attractive odour factor conveyed on the FP is highly labile as its behavioural effect was almost gone within 1 h after emission.

### 3.3 Pup Behaviour Towards Faecal Pellets Outside the Nest

To better assess the olfactory responsiveness of individual pups to the odour of FP, we attempted to decontextualize the encounters of pups and FP from the background of the nest and littermates. Two test settings were considered.

The first consisted in confronting hand-held pups directly to the FP odour (Fig. 6a), capitalizing on their immediate head extension, sniffing and oral searching/grasping behaviour elicited when they were exposed to the odours of milk or a lactating doe's nipple, or to the mammary pheromone (Hudson & Distel, 1983; Keil et al., 1990; Moncomble et al., 2005; Schaal et al., 2003). However, this testing situation did not work with the FP odour as it did with milk odour or the mammary pheromone: In 3- and 7-day-old pups, a 30-s presentation of FP caused merely approach and active sniffing, but surprisingly not the expected typical searching-oral grasping response. In particular, the foreleg scratching and the oral snapping actions that were released when the pups were exposed to fresh FP in the nest context could not be reproduced when the odour stimulus was delivered to handheld pups without contact with the source and outside the nest background.

We thus opted for a double-choice odour test that relied on neonate pups' ability to creep over small distances towards a stimulus of interest in either relative (target stimulus vs. reference stimulus) or absolute (target stimulus vs. control). Earlier validated with newborn rabbit pups (Coureaud et al., 2001), a similar testing device showed that rabbit newborns could display well-differentiated choice behaviour.

The pups were individually introduced into the polypropylene arena (29 × 21 × 20 cm; placed on a heating plate set at 37 °C) divided in two equal compartments, with a wire mesh bottom under which odour stimulations were placed (Fig. 6b). The stimuli consisted of hard FP (pool of 10 FP from 5–10 does; renewed every five tests) from

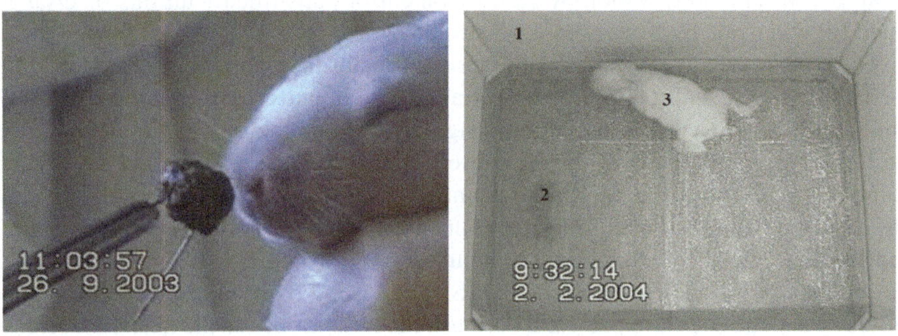


**Fig. 6** The isolated faecal pellet test (left) and the two-odour choice arena (1), with a pup (3) roaming on the wire-mesh bottom over the compartment with faecal pellets (2)

lactating females (excluding those of own mother), sampled after suckling on postpartum days 3 or 7, and kept at −80 °C until testing. The FP odour compartment was contrasted with a control compartment devoid of FP. After thawing and placement of the FP, each individual pup was released from the middle of the arena, and the test began when the pup had passed over each compartment. Each pup was tested once, being free to roam in the arena for 2 min. Every 5 tests, the arena was rotated 180° to control for potential lateral biases of pups. After each test, the arena was cleaned (detergent, alcohol, water). The tests were videotaped for offline analysis of the duration of pup roaming over each compartment of the choice arena (a pup was considered oriented to a side when the base of both ears was on that side). Only those pups who roamed over both sides of the arena during the first 30 s of the test were included in the analyses, and those who were inactive for a continuous 30-s period were excluded. The testing was carried out with 3- and 7-day-old rabbit pups ($n = 40$/age group, from 8 litters, 5/litter), half of each age group being tested 2 h before a feed, the other half being tested within 1 h after a feed.

Regardless of their age group ($n = 20$/age), the pups tested 2 h before being suckled roamed for equivalent durations over both compartments of the arena [3-day pups: FP vs. control: 54.6 vs. 65.9 ± 42.5 s; $t(19) = 1.21$, $p > 0.05$ (Student t test); 7-day pups: 62.4 vs. 57.6 ± 30.8 s; $t(19) = 0.70$, $p > 0.05$; Fig. 7]. In contrast, the pups tested within 60 min after having been suckled stayed longer over the FP compartment. This differential behaviour was noted in 3-day pups ($n = 23$) [80.5 vs. 39.5 ± 16.7 s; $t(16) = 5.06$; $p < 0.001$] as well as in 7-day pups ($n = 19$) [76.9 vs. 43.1 ± 17.7 s; $t(18) = 4.17$; $p < 0.001$; Fig. 7]. In sum, 2 h *before a feed*, pups aged 3 or 7 days were not differentially attracted towards FP presented against the control stimulus, but they did so within 60 min *after a feed*. Thus, regardless of their age, the pups' motivated responsiveness to FP odour appeared to be conditional upon a full belly. In any case, this experiment also validated a two-choice arena to further evaluate the conditions by which adult emitters render their FP attractive to pups *after a feed*.

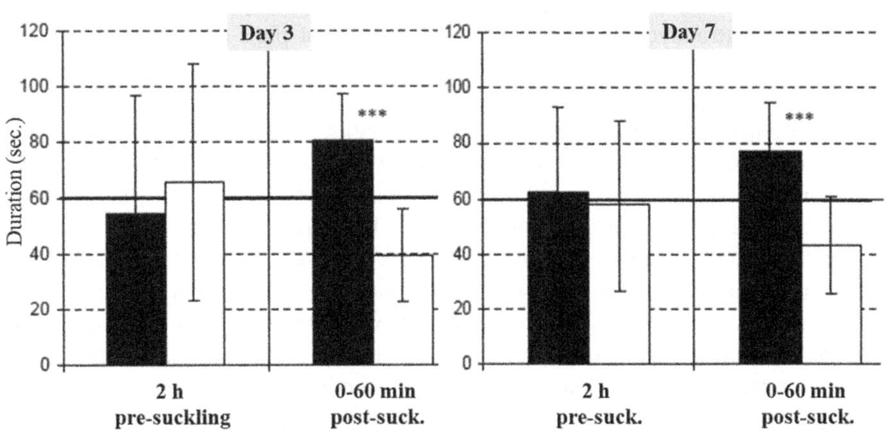


**Fig. 7** Duration (mean ± SD) of 3- and 7-day-old pups' roaming in the choice arena over the side presenting faecal pellets (full bars) against the scentless control side (empty bars), when tested 2 h before or 60 min after suckling (***: $p < 0.001$, Student $t$ test)

# 4 Pup Responsiveness to Faecal Pellets as a Function of Donors' Condition

The two-choice arena implemented above was applied to further understand biological factors susceptible to explain the odour properties of FP perceptible by rabbit pups. Specifically, does the mother or any lactating female release, through the mediation of FP, volatiles that are particularly attractive to newborn pups, as noted in other species (e.g. rat, Leon & Moltz, 1972; Moltz & Lee, 1981; pig, Aviles-Rosa et al., 2019; koala, Osawa et al., 1993). The two-choice test was thus used to weigh the attractive potency of FP from does in early lactation relative to FP from females in other physiological states. The effects of sex or individuality factors of the FPs' donors will also be considered. Additionally, within a same donor, the effect of the source of the FP or their state relative to anal evacuation will be assessed.

## 4.1 General Methods

Capitalizing on above experiments, the present tests were run on pups aged 6–8 days, 25–45 min after a feed and for a duration of 5 min. The choice arena was increased in size (42.5 × 29.5 × 30 cm) and placed on top of heating plates maintained at 37 ± 0.5 °C. The stimuli consisted in a pool of ten hard FP from ten different lactating does (FP collected after the feeds of days 1–7, and deep frozen until testing). Other conditions (lateral position and renewal of stimuli, exclusion criteria of pups, cleaning) were similar to those of above experiments. Additional conditions specific to each test are detailed below.

## 4.2 Females' Physiological State

To determine whether the reproductive stage of females was discriminant of the degree of FP odour's attractiveness to pups, four groups of 24 pups (from 8 litters, 3/litter) were exposed to 4 two-choice tests opposing, on one side, the odour of FP of *early lactating* does (days 1–8 postpartum, non-gestating) with, on the other side, the odours of FP from either: (i) *virgin* females (aged 16 weeks, sexually mature); (ii) *pregnant* females (on gestational days 25–26, non-lactating); (iii) *non-lactating* females (10 days after their pups' weaning, non-gestating), or (iv) *late lactating* females (late lactational stage on days 22–23 postpartum, non-gestating). The hard FP of the does in early lactation (days 1–8) were sampled in the nest boxes after daily suckling episodes, while those of the other categories of female donors were collected under their home cage, avoiding urine contamination. Pups were never exposed to the FP of their own mother.

In the test pairing FP from early lactating does vs. *virgin* females, the 23 pups retained for the analyses roamed equivalently over both compartments of the arena (153.3 vs. 146.7 ± 18.2 s, respectively; $t(22) = 0.86$, $p > 0.05$, Student t-test). In the test opposing FP from early lactating does to FP from *pregnant* (*non-lactating)* females, 22 pups oriented indiscriminately towards either compartments of the arena (160.1 vs.

139.9 ± 34.4 s, respectively; $t(21) = 1.37$, $p > 0.05$). In the next test exposing pups to FP from early lactating does against FP from *non-lactating* (*non-pregnant*) females, 22 pups (retained among 24 tested) did not differentiate both stimuli (142.3 vs. 157.7 ± 33.8 s, respectively; $t(21) = 1.07$, $p > 0.05$). The final test contrasted FP from *early lactating* does and FP from *late lactating* does. The 23 analysable pups did not show differential behaviour towards either compartments of the arena (142.9 vs. 157.1 ± 33.7 s; $t(22) = 1.01$, $p > 0.05$). Overall, FP from females were olfactorily attractive to newborn pups irrespective of the physiological state of female donors.

## 4.3 Female Individuality

We then assessed whether the pups' mother conveys individual-specific faecal cues that are especially attractive to pups. Accordingly, 24 pups (from 8 litters, 3/litter) were exposed to FP from their own mother and to FP from unfamiliar lactating females (both being non-pregnant). Adult rabbit discriminating individuals (i.e. own doe vs. stranger) based on their FP odour (Schalken, 1976), one may expect pups being able to differentiate own mother's vs. unfamiliar does' FP. However, it came out that the pups ($n = 21$) roamed indiscriminately above the maternal FP and the FP from unfamiliar lactating females (134.7 vs. 165.3 ± 40 0.9 s, respectively; $t(20) = 1.72$, $p > 0.05$), suggesting that individual odour cues in FP, if any, are less salient to newborn pups than cues of other social categories (sex, lactation).

## 4.4 Sex

To assess if the attractive faecal odour is exclusively produced by lactating does, we exposed pups to hard FP from females and from males. Twenty-four pups (from 8 litters, 3/litter) were confronted a choice between the odours of FP from lactating non-gestating females ($n = 10$) and of FP from male donors (age > 30 weeks, $n = 7$). The duration of pups' ($n = 21$) roaming over females' FP vs. males' FP compartments of the arena was 145.2 vs. 154.8 ± 52.3 s, respectively ($t(20) = 0.42$, $p > 0.05$). Thus, to our surprise, FP from lactating females were not differentiated from those of males.

## 4.5 Functional Level of FP Production

The present tests evaluated whether the odour of the FP is differentiable to pups as a function of the level of sampling along the digestive tract. A first test assessed pups' differential attraction to hard FP originating in the colon vs. soft FP (caecotrophs) deriving from the caecum (Davies & Davies, 2003). The composition of both types of FP being radically different in terms of nutriments (e.g. Carabano et al., 1988), rabbit pups were expected to differentiate them based on olfaction. These two types of FP were sampled between postpartum days 1 and 7 (immediately deep-frozen until testing). Twenty-four pups (from 8 litters, 3/litter) were simultaneously presented the odour of 5 g of hard FP (from 10 females) against the odour of 5 g of caecotrophs (from 5 females), both stimuli being produced by non-familiar lactating does. To our surprise, it came out that the pups roamed indiscriminately over the hard FP or the soft FP, 154.7 vs. 145.3 ± 38.1 s, respectively ($t(19) = 0.55$, $n = 20$, $p > 0.05$).

In a second test, we evaluated whether the FP odour could be altered during the final stage of their intestinal excretion, when they are drawn through the distal rectum where the ducts of the bilateral anal glands open. These glands are indeed involved in adult rabbit chemocommunication, their secretion coating the hard FP during rectal evacuation (Hesterman & Mykytowycz, 1968). Assuming that the anal glands could produce odour cues that are salient in mother-pup relationships, the present test aimed to compare the attractive power of excreted FP coated with anal glands' secretion and of FP that were not yet coated by anal glands' secretion. Twenty-four pups (from 8 litters, 3/litter) were let to choose between excreted vs. non-excreted FP from the same females. As attractive faecal cues are emitted by any kind of female (see above), hard FP were sampled from reproductively latent does ($n = 3$). Excreted FP were gathered under their cages 1–2 days before slaughtering, while non-excreted FP, which have not passed through the anal gland level, were collected post-mortem in the distal colon. FP were stored at −80 °C until testing. The stimuli presented in the two-choice arena consisted in ten excreted vs. ten non-excreted FP. The 20 analysable pups oriented for significantly longer duration towards the odour of excreted FP rather the odour of non-excreted FP (170.3 vs. 129.7 ± 36.4 s, $t(19) = 2.49$, $p = 0.019$). Thus, the FP which came across the outlets of the anal gland were clearly preferred to FP gathered before their rectal excretion.

## 4.6 Interim Discussion

Taken together, the series of tests in this section revealed that the odour of FP from adult rabbit females are attractive to week-old pups regardless of the females' physiological state or identity. Surprisingly, pups were also equivalently attracted by the odour of FP produced by female and male adult rabbits. Even more, the FP of females in early lactation were as attractive to them than FP from females in late lactation. It appears therefore that being post-parturient and lactating is not a necessary condition for producing FP that are olfactorily appealing to newborn pups. Thus, contrary to expectations, no physiological explanation in terms of FP donor's sexual status or physiological/hormonal state emerged to clarify the source of the attractive odour conveyed on FP. Likewise, the hard vs. soft type of excreted FP, despite their clearly different (bio) chemical and microbiological composition (e.g. Carabano et al., 1988; de Blas & Wiseman, 1990), did not differentiate newborn pups' olfactory responsiveness. The only condition which appeared significant for the FP to become olfactorily attractive to the newborn pups was that they were anally excreted. As the anal glands discharge at the junction of the rectal mucosa and cutaneous epidermis (Schaffer, 1940; Mykytowycz, 1966; El-Ghazali & Ali, 1978 also them as 'rectal glands'), we hypothesize that their secretion may contribute to the reactogenic potency of hard FP to rabbit pups.

## 5 Chemical Correlates of Excreted FP Attractiveness to Rabbit Pups

The behavioural roles of FP in adult, mainly male, rabbits have attracted the merged interest of ethologists and chemists (Hesterman & Mykytowycz, 1968; Goodrich, 1983; Hesterman et al., 1976, 1981). Chemical analyses of the effluvium of excreted FP led to identify some the behaviourally and physiologically impactful volatiles, mainly ketones, esters, phenols, heterocycles and alcohols, including 3-methylbut-2-en-1-ol (Goodrich et al., 1978, 1981). Additional alcohols, aldehydes and ketones were detected in the headspace developing over homogenized preparations of anal glands, such 2- and 3-methylbutan-1-ol, 2- and 3-methylbut-2-en-1-ol, and 3-methylbut-2-enal (Goodrich et al., 1978). The mammary pheromone 2MB2-al was apparently not detected in these two substrates. However, its reduced form 2-methyl-but-2-enol (2MB2-ol) being detected in the anal glands, and its presence in undetected trace amounts cannot be excluded.

We thus engaged gas chromatography-mass spectrometry (GC-MS) analyses of FP sampled from four reproductively latent females, which were equivalently attractive than FP from lactating does (see section "Females' Physiological State"). Excreted hard FP (~15 g/female) were taken under the cages (avoiding urine contamination) 1–2 days before slaughtering, after which non-excreted hard FP were collected from the distal colon (~20 g/female). The anal glands (~3 g/female) were dissected from these same females. These samples were immediately deep-frozen and kept at −80 °C until chemical analyses. The still-frozen samples were then put into a 500-mL glass vial with a doubly sintered-glass bottom allowing dynamic headspace extraction. A high purity nitrogen flow entering the glass vial below the bottom entrained the volatiles for 30 min towards a trap affixed on top of it. This trap consisted in a glass tube containing Tenax (140 mg TA 20/35 mesh; Chrompack, Middleburg, Netherlands). This Tenax trap was then heated up to 250 °C for desorption (TCT thermodesorber, Chrompack, Middleburg, Netherlands) and its effluent condensed into a silica tube at −130 °C. After 15 min of desorption, the silica tube was heated within 20 s from −130° to 250 °C to entrain (carrier gas: helium, HP45; speed: 50 cm/s) the volatiles into the fused silica capillary DB-Wax column (30 m, Ø: 0.32 mm, phase coating: 0.5 μm; J&W Scientific, Folsom, CA) of the GC (Agilent 6890, Palo Alto, California) with a temperature program set from 40 to 220 °C (increment: 5 °C/min). The GC-separated compounds were then ionized in a quadripolar mass spectrometer (Agilent Technologies) that recorded their mass, abundance and spectrum. In a first approach, the resulting TIC (total ionization current) spectra of the analytes of interest were then compared to those of the mass spectra libraries (local bank of CSGA and Wiley bank, New York) for tentative identification. In a second approach, the spectrometer was used in SIM (single ion monitoring) mode to detect the characteristic fragmented ions of 2-methylbut-2-enal (ions of mass 55 and 84 a.m.u., 2-methyl-but-2-enoic acid and 2-methyl-but-2-enol (ions of mass 55 and 100 a.m.u., and 71 and 86 a.m.u., respectively). The final identification of the analytes of interest was made by comparing their ionic fragmentation pattern and retention index with those of corresponding synthetic commercial compounds. Pure 2 MB2-al and 2-methyl-but-2-enoic acid were commercially available (Aldrich, Saint-

Quentin-Fallavier, France), while 2 MB2-ol was graciously synthetized by Pr. Wittko Francke (Institute of Organic Chemistry, University of Hamburg, Germany).

The chromatograms of pure 2MB2-al, 2MB2-ol and 2-methyl-but-2-enoic acid were compared with those obtained following the injection of the extracts from intestinal and excreted FP and from anal glands (Fig. 8). It came out that (1) 2-methyl-but-2-enoic acid was not detected in any of the samples; (2) the two other target compounds, 2MB2-al and 2MB2-ol, could not be traced in the non-excreted FP from the four females; however, (3) 2MB2-al and 2MB2-ol were detected in all samples of anal glands (the corresponding profile of one female is not shown in Fig. 8 due to technical problem at Tenax trap desorption, but both compounds were detected); and (4) in all samples of excreted FP (Fig. 8). In sum, 2MB2-al and 2MB2-ol being undetected in the headspace from the FP sampled in the distal colon, and being detected in the effluvia of anal glands and of excreted FP, we logically deduced that the hard FP are loaded with 2MB2-al and 2MB2-ol during their passage through the region where the anal glands discharge.

When 3- to 5-day-old rabbit pups were presented the odour of pure 2MB2-ol on a glass-stick (concentration: $10^{-6}$ g/mL in water), no other reaction but sniffing was elicited. We therefore concluded that 2MB2-ol is not the reactogenic compound responsible for the attractiveness of FP to pups.

# 6 Summing Up and Raising Questions

## 6.1 A Suckling Chemosignal in Faecal Pellets

The aldehyde 2MB2-al and its alcohol form 2MB2-ol were detected in the effluvia of excreted FP and anal glands from rabbit females. As these compounds could not be traced in non-excreted FP, the mucus coating the FP is assumed to have been laden with them during their transit through the rectum where anal glands discharge. 2MB2-ol being inactive in releasing pups' searching-oral grasping responses, 2MB2-al may explain on its own why pups are attracted to FP. The presence of 2MB2-ol over FP was not a surprise, as Goodrich et al. (1981) detected it in male rabbits, but 2MB2-al had apparently escaped earlier CG-MS detection (although 3-methylbut-2-enal was detected; Goodrich et al., 1978).

Thus, the same chemosignal 2MB2-al occurs in two distinct biological substrates – milk and FP – that are both attractive and appetitive to newborn pups. However, pups' response to 2MB2-al differs as a function of the carrying substrate. While fresh rabbit milk odour (much as diluted pure 2MB2-al) provokes immediate oral seizing responses (Schaal et al., 2003), fresh FP odour effect is deferred of several minutes in releasing approach and oral actions. These temporally distinctive responses to 2MB2-al in both substrates may depend on the matrix in which it is embedded, on its concentration in their respective effluvia, and/or on co-occurring volatile and non-volatile compounds, all point awaiting future analyses. Nevertheless, despite temporally different reactions towards milk and FP, pups generally conclude in approaching and orally seizing, and eventually ingesting them.

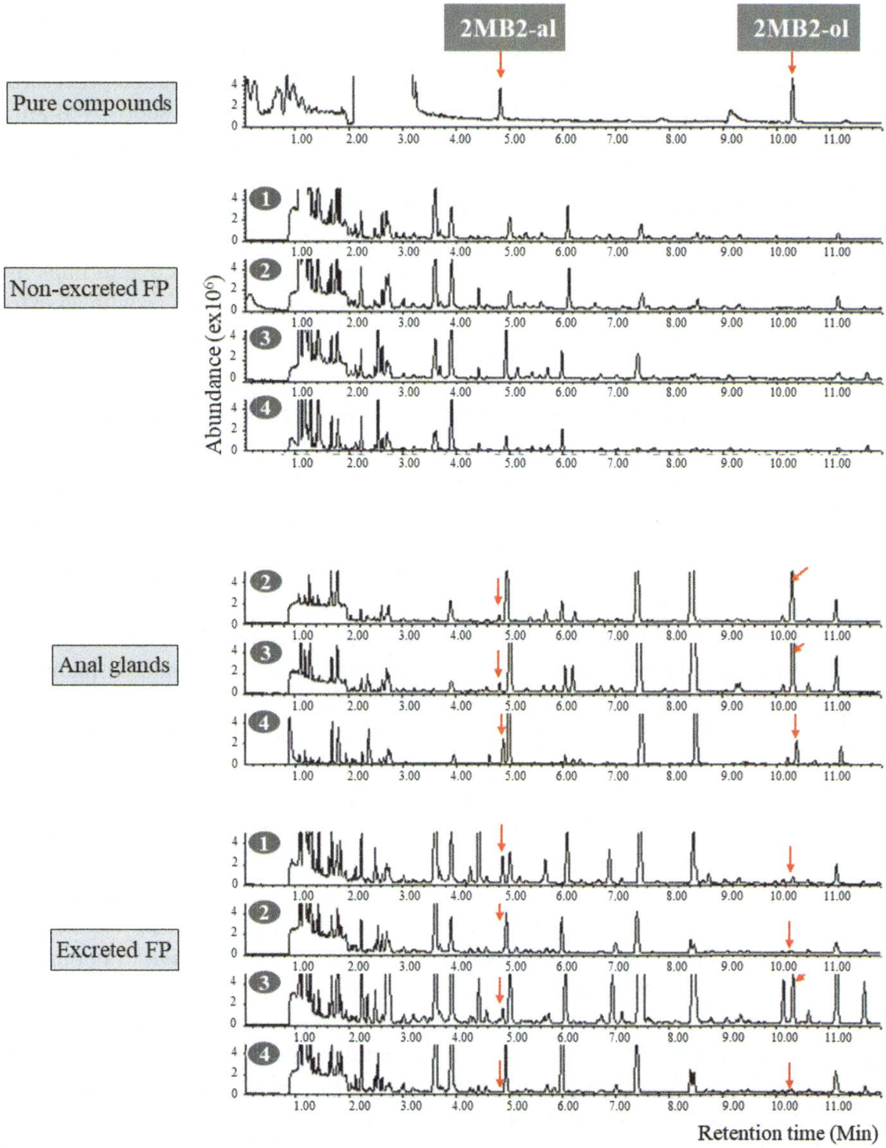


**Fig. 8** Overlaid chromatographic profiles (obtained in SIM mode) showing the detection of target volatiles 2-methylbut-2-enal (2MB2-al) and 2-methylbut-2-enol (2MB2-ol) in faecal pellets (FP) sampled in the distal colon (non-excreted FP), in excised anal glands and in excreted FP from four females (numbers in dark circles; the anal glands' profile of female 1 is lacking, see the text). Red arrows indicate the retention time of 2MB2-al and 2MB2-ol

An additional point comforts 2MB2-al as a correlate of olfactory attractiveness of FP to pups. FP odour loses part of its attractiveness for pups within 1 h after excretion (or decongelation, when FP were sampled right after excretion and immediately deep-frozen), suggesting that the impactful volatile(s) are instable. The same degressive kinetics of pup response occurred when fresh rabbit milk was let to stand at ambient temperature, in correlation with decreasing 2MB2-al concentration in the headspace (Schaal et al., 2003). By analogy, enzymatic and/or oxidative degradation processes of 2MB2-al (Williams & Ringsdorf, 2020) may explain the relatively short-lived attraction effect of freshly excreted FP odour. The evidence of the behavioural effect of 2MB2-al on the FP remains so far correlational. To conclusively establish it as the exclusive or combined cause of pup attraction to fresh FP, 2MB2-al needs to be titrated in the secretions of the anal glands and on the surface of fresh FP, and direct behavioural tests should verify if behaviourally-inactive, 'old' FP become attractive to pups after being imbibed with synthetic 2MB2-al at physiologic concentrations or with an extract of fresh FP containing 2MB2-al.

## 6.2 Adaptive Functions of Faecal Pellets-Associated 2-Methylbut-2-enal?

As a potentially salient chemosensory constituent of both rabbit milk and FP, 2 MB2-al relates with a cascade of events that are critical for the adaptative onset of neonate pups' behaviour and physiology. Beyond guiding pups to the source of milk, 2MB2-al may initiate attraction and ingestion of maternal FP by newborn pups, exposing them at once to multiple assets conveyed by FP.

*First*, neonatal ingestion of FP contributes to the developmental conditioning of gastrointestinal functionality. Maternal FP carry low-digestible fibres and non-pathogenic microbiota which intake impacts the pups' digestive tract (Fortun-Lamothe & Boullier, 2004). A precocious bacterial colonization prepares the offspring's gut to process vegetal ingesta which will massively enter from about 15 days of age. Ingestion of fibres earlier than day 15 accelerates maturation of gut mucosal immunity and of caecal peristalsis leading to differentiated excretion of hard and soft FP and to the development of caecotrophy. In adult rabbits, the deprivation of FP ingestion (specifically of soft FP) is deleterious to general fitness (e.g. Li et al., 2022), and data in young rabbits also indicate that FP ingestion impinges on survival and digestive competence. Kovacs et al. (2006) described indeed faster colonization of the caecum by *Bacterioides* in pups who consumed maternal FP compared to those who did not. Likewise, Combes et al. (2014) showed that neonatal FP ingestion influenced the establishment of caecal microbiome, with delayed microbiological profiles in weanlings deprived of FP (relative to FP-ingesting pups), who also endured higher mortality (but see McNitt & Moody, 1987). Thus, among some other mammals studied so far (e.g. rodents, Leon & Moltz, 1972; Leon, 1974; Moltz & Lee, 1981; Breen & Leshner, 1977; Porter & Doane, 1976; Galef, 1979; ungulates, Crowell-Davis & Houpt, 1985; Nowland et al., 2021; Aviles-Rosa et al., 2020; marsupials, Osawa et al., 1993; reviewed in Bienenstock et al., 2018; Ge et al., 2021; Rosenberg & Rosenberg, 2021), the European rabbit has developed, in supplement to pan-mammalian means of transmission by contact or milk, a chemocommunication system engaging neonates to rapidly self-administer the mother's faecal microbiome.

*Secondly*, FP carry odour cues representative of pregnant or lactating does' diet, which are perceivable to young rabbits (Bilko et al., 1994; Altbäcker & Hudson, 1995). These dietary cues in PF extend the same cues received in the womb and in milk, leading to developmentally redundant exposure under powerful reinforcement conditions

(Hudson et al., 1999). As in milk, such dietary cues in FP are paired with 2MB2-al, which inherent reinforcing properties (Coureaud et al., 2006) may assign them stable referential value canalizing the upcoming development of selective choices of foods.

*Thirdly*, odour learning amplified by FP-associated 2MB2-al may also influence social development beyond the food domain. Anal glands' secretions convey indeed cues related to individual and group identity, sex or dominance status of the emitter (Mykytowycz, 1985). In the same way as for dietary cues, early learning of such social odour from FP may serve later to discriminate mother and littermates from other individuals, as well as to locate oneself in the space around the maternal burrow marked with their FP (Mykytowycz & Ward, 1971; Mykytowycz, 1985). By staying within this olfactorily familiar and safe perimeter during their first nest exits, young rabbits receive less aggression by other adults (Mykytowycz & Dudziński, 1972; Mykytowycz & Hesterman, 1974).

*Fourthly*, the fact that does use FP as a vehicle to emit 2MB2-al emancipates its signal function from their own physical presence. Accordingly, 2MB2-al prolongs its effects after the does' departure, within the limits of its physical and chemical stability, giving time to pups to locate, fragment and ingest freshly emitted FP. During every post-suckling search for freshly emitted FP in the nest, satiated pups are re-exposed to the reinforcing properties of 2MB2-al which may (re)condition them to the current nest chemosphere and so update their memory of an evolving odour background (e.g. Serra et al., 2009). As 2MB2-al on FP vanishes by consumption and/or oxidation, the odour mixture from 'older' accumulating FP, littermates and vegetal elements may assign *familiarity* to the nest, contributing to attenuate the pups' suckling and post-suckling overexcitation (Hudson & Distel, 1982). By such cognitive effects, 2MB2-al participates in a 24-h cyclic conservative strategy that optimizes quiescent states in the litter, and hence pups' metabolic conversion of milk into biomass. Thus, from the very first postnatal days, the on-and-off 2MB2-al signal in milk and on FP is inscribed in a regular sequence of sensory, motor, neural, cognitive and physiological events that may orchestrate the engagement of biological rhythms attuned to the rabbit's once-per-day suckling regimen (Caba & González-Mariscal, 2009; Montúfar-Chaveznava et al., 2013; Trejo-Muñoz et al., 2014).

*Finally*, fresh FP in the nest may participate in another aspect of the conservative strategy of maternal behaviour mentioned above. After the pups' hyperexcitation directed towards the nursing does' abdomen and the post-nursing hyperexcitation leading to huddle disturbance (Hudson & Distel, 1982), fresh FP's odour may tone down the pups' agitation by a transitory maternal stimulation of lower intensity devoid of nutritive and other maternal reinforcers. The sudden maternal depart after suckling is followed by omnidirectional, apparently chaotic searching by pups, actualized in rooting and digging in the nest material with vigorous 'paddling of forepaws' (Hudson & Distel, 1982). We suggest these rooting movements with typical forepaws' paddling (which we monitored in our own tests exposing pups to fresh FP in their nest) are in fact aimed at 'finding pellets in a haystack,' and cornering them before ripping them into ingestible pieces. Several pups assemble then around a FP (see Fig. 3c) and seem to both compete and cooperate in its fragmentation, with the net advantage that the huddle appears to reform around the FP to ingest bits of FP. This gathering effect of fresh FP can only be beneficial in terms of individual and litter thermoregulation (Bautista et al.,

2008; Gilbert et al., 2012) at a time when potential thermal stress may occur due to generalized urination by pups. In sum, the pups' search for fresh, 2MB2-al-emitting FP might also be seen as a buffer sequence to bring the huddle together between the phase of dispersal stress after maternal departure and the phase of quiescence until the next suckling.

Lastly, we would like to revise a bit the interpretation of an earlier study on the prognostic value of pups' reactivity to 2 MB2-al for their long-term viability (Coureaud et al., 2007). This study consisted in probing 1- and 3-day-old pups with highly diluted pure 2MB2-al and following them up to day 21 (when they access to non-milk fodder). Among the 288 pups tested on postnatal days 1 and 3, 8.7 and 7.2%, respectively, did not respond for unknown reasons. Those unresponsive pups were then also those who died most until day 21, especially if they were lightweights. This effect was then attributed to a reduced ability to locate nipples and get milk in these pups. In light of the above results, early mortality may also be credited to the pups' inability to locate the 2MB2-al on FP in the nest and to ingest them.

### 6.3 Some Additional Issues Raised by 2-Methylbut-2-enal on Faecal Pellets

That a same chemosignal, 2MB2-al, appears to function in attracting newborn pups towards milk and towards fresh FP raises unanswered questions and speculations worthy to stir future research on neonatal adaptive behaviour.

*First*, the (bio)chemical processes by which 2MB2-al and other odorous compounds are produced in the anal glands and the physiological processes of their emission and regulation remain unexplored. 2MB2-al might be inherent to native apocrine secretions of the gland or the result of fermentive activity on these secretions by microbiota harboured in the gland. The behavioural roles of skin and glandular microbiomes are increasingly recognized in many vertebrates (e.g. Albone, 1984; Osawa et al., 1993; Archie & Theis, 2011; Ezenwa & Williams, 2014; Whittaker et al., 2016; Carthey et al., 2018; Noonan et al., 2019; Theis et al., 2020; Zhou et al., 2021), involving them in the biosynthesis of species-specific chemosignals by microbe-assisted modification of gland substrates or by microbe-specific signals. Metagenomic studies are awaited to characterize the microbiome of the rabbit skin glands, particularly the specificity of anal glands' microbiome and its degree of covariance with gut microbiota. Measuring the consequences of targeted antibiotics applications on the gland's and excreted FP's volatilomes (and on the development of pups' behaviour, survival and gut microbiome development) would be one way to identify the specific microbes involved. Otherwise, the possible microbial origin of 2MB2-al in the anal gland questions its similar origin in the mammary tract, a point already raised by Moncomble et al. (2005). Such double source of 2MB2-al may help verifying if common symbiotic microbes are involved in its production. With its well-defined maternal-neonatal behavioural system, easy-to-manage mother-offspring interaction and one well-identified chemosignal, the rabbit would be an optimal model to further understand how microbial symbionts manufacture chemosignals and contribute to social behaviour in mammals.

*Secondly*, how does it come that a double source of highly reactogenic 2MB2-al on the small abdominal surface of rabbit females does not confuse pups in their vital quest for nipples? Is it possible that, right at birth, neonate pups sense first the anal source of

2MB2-al and then its mammary source? One can indeed occasionally see neonate rabbit pups searching towards the nursing females' rear abdomen where anal (and inguinal) glands' secretion may leak. Although neonate pups' odour-based anterior-posterior differentiation of the does' abdomen has not been examined per se (comparison of pups' attendance to fore/hind nipples along the nipple lines was made according to milk yield rather than to investigation time; Bautista et al., 2005), the evidence prevails that the mammary area, including nipples, is most attractive to them (Hudson & Distel, 1983; Coureaud et al., 2001; Bautista et al., 2005; Val-Laillet & Nowak, 2008). Either is the mammary odour factor inherently more potent than the anal odour factor (due to quantitative or qualitative specificities) or is its release potentiated by local surface properties (e.g. higher temperature of bare skin areas on anterior-medial abdomen vs. furry areas closed by the tail in posterior abdomen). However, the fact that pups react to female rabbits' abdominal fur outside lactation reveals sources of arousing/attractive odours other than nipples/milk. Hudson and Distel (1983) mentioned indeed that '…not only the skin around nipples, but the entire belly, and indeed the whole body, produces a search releasing odour' (p. 271). Specifically, the anal gland source of 2MB2-al appears to be effective when the mammary source of 2MB2-al is silenced. Rabbit females do indeed actively groom themselves (Gonzales-Mariscal et al., 1994, 2013; Krohn et al., 1999; Hudson & Distel, 1983), distributing odour factors from all potential sources (including anal glands) over their belly. These 'permanent' (i.e. independent of lactation) odour sources are thus probably responsible for the rabbit pups' ubiquitous attraction to the ventrum (Hudson & Distel, 1983; Coureaud & Schaal, 2000; Coureaud et al., 2001; Hudson et al., 2008) or FP (present study) of adults of any sex or physiological state. Thus, one may hypothesize that while anal glands emit a background tonus of attractive 2MB2-al in any adult rabbit, higher amounts of 2MB2-al accompanied with obvious reinforcers (milk) may be emitted from the mammae of lactating females, avoiding directional confusion of pups searching for milk. New chemoanalytical techniques monitoring online emission of volatiles in living animals should shed some light on this soon. Otherwise, rabbit anal glands and FP emit a bunch of other reactive aldehydes that are structurally close to 2MB2-al (Goodrich, 1983), and pups' responsiveness to which would be worth examining separately and in mixtures (e.g. Hanser et al., 2017).

*Thirdly*, the correlate between 2MB2-al on FP and the pups' behaviour raises issues about the evolution of this mother-to-offspring chemosignalling system. Members of the Leporidae family share multiple natural history traits with *Oryctolagus cuniculus*, specifically those relating to ethological-physiological constraints around mother-offspring relationships (Cowan & Bell, 1986). For example, *Sylvilagus floridanus*, *S. cunicularius* or *Brachylagus idahoensis* females do all build grass nests wherein they deliver highly altricial neonates which they visit once per day for a suckling episode lasting only minutes (Verts et al., 1997; Elias et al., 2006; Rodriguez-Martinez et al., 2014; Smith et al., 2018). Their previsual neonates should thus all have coevolved olfaction-based behavioural strategies to swiftly locate nipples, efficiently suck and acquire a non-pathogenic microbiome without delay. Like in the European rabbit, females of these leporid species are endowed with multiple scent glands (Bell, 1985), among which anal glands might have homologous FP-mediated odour-dispersing functions (whether they drop FP during suckling remains to be documented). Thus, how far do Leporidae having similar ways in maternal-neonatal functioning converge in the design of their chemosignalling system? Do they bear similar odourigenic microbes in their glands covarying with similar odour profiles of their secretions contributing to similar chemosignals? Or do they produce similar chemosignals despite dissimilar microbiomes in their anal glands? Does the evolution of semiochemicals regulating leporid mother-offspring interactions reflect selection on compounds produced by the mother emitter or on compounds perceivable to the neonate receiver? Do mothers exploit pre-existing chemosensory biases in neonate receivers (e.g. Endler & Basolo, 1998; Schaal & Al Aïn, 2014)? In this context, it might be of interest to examine whether females of different *O. cuniculus* subspecies (e.g. *O. c. cuniculus* vs. *O. c. algirus*) produce FP conveying 2MB2-al despite they have different ways of social life and diverse gut microbiomes (Funosas et al., 2021), and whether corresponding neonates are equally reactive to 2MB2-al and cross-reactive to FP odour from females of the other subspecies. Otherwise, it is not unusual that distant mammalian taxons meeting similar ecological-physiological requirements evince chemical commonalities embedded in otherwise different mixtures of compounds (Apps et al., 2015). For example, carbon disulphide engages immediate attention and odour learning in pups of both mice and rats (Bean et al., 1988; Galef et al., 1988; Munger et al., 2010); or the urine of adult domestic rabbits is attractive to cottontails (Young & Henke, 1999). One cannot exclude that different leporid females bearing homologous glandular functions could emit chemosignals relevant to neonates from another leporid taxon who have a similar 'receiver psychology' (Guilford & Dawkins, 1993; Schaal & Al Aïn, 2014). For example, how would a 2MB2-al-containing substrate of *Oryctolagus* females affect neonates from *Brachylagus* or *Sylvilagus* neonates? Chemosignalling strategies of decisive survival value may evolutionarily opt for similar chemical solutions among different genus or species within a same clade. The Leporidae family with its 10 genus and 44 species (Cowan & Bell, 1986) may offer a fertile ground for comparative studies in the mechanisms and evolution of mother-offspring chemosignalling.

Finally, another possible property of 2MB2-al of evolutionary significance relates to its potential function as an allelochemical regulating interactions with unattended receivers, in particular parasites and disease vectors. An allelochemical can operate at the advantage of either the parasite or the host (working then as a kairomone or an allomone, respectively; Wyatt, 2014). In the first instance, the behavioural impact of 2MB2-al can indeed turn maladaptive in case of maternal infection or infestation by parasites whose receptors are themselves attuned to 2MB2-al or if emitted in association with milk or faeces to which pups are attracted, leading to snowballing pullulation of parasites and lethality in hosts (e.g. Gruaz et al., 2020). In the second case, 2MB2-al may be rather protective against vectors. Rabbit burrows are characterized by a concentrated $CO_2$ atmosphere (Hayward, 1966), which makes them attractive spots for parasites or disease vectors, such as ticks, fleas, flies or mosquitoes who partly rely on $CO_2$ to trace their hosts (e.g. Cooke, 1990, 2023; Asgari et al., 1998). While the rabbit breeding stop is open during the suckling episodes, such $CO_2$-driven vectors have some minutes to slip into to attain the litter. Recent in vitro evidence suggests, however, that the odorant 2MB2-al might disrupt the host-seeking behaviour of *Anopheles* mosquitoes, major vectors of the myxoma virus. In abnormally prolonging the firing rate of neuroreceptors to $CO_2$, 2MB2-al might interfere with the main directional cue of mosquitoes (Tauxe et al., 2013; Turner et al., 2011). Whether rabbit females deliver suitably and sufficiently protective substances into their nests, through their selection of nest materials, of aromatic foods, or through the release of 2MB2-al-enriched FP, remains to be investigated in vivo. Such protective mechanisms of progeny against parasites and vectors may not be ecologically anecdotic, as "stable populations would be maintained if only 2-3% of rabbits born survived into the following year" (Cooke, 2023).

Lastly, mother-to-infant chemosignalling through 2MB2-al fulfils multiple vital roles in young European rabbits during the first 10 days post-birth: (1) it has strong releasing power on searching for nipples and intake of milk; (2) it constitutes a strong reinforcer of learning during a sensitive period, promoting the fast development of olfactory cognition required for adjusted individual and collective behaviour within the nest and then at the burrow's entrance vicinity; (3) through its regular association with the mother, 2MB2-al potentially operates as a synchronizer setting circadian synchronicity in mother-litter functioning; and (4) the present study suggests that 2MB2-al favours ingestion of FP potentially conveying cues or substrates relating indissociably to organic fitness (incorporation of symbiotic microbiota, initiating gut immunity, and physiological maturation), environmental adaptation (preferences for flavours of mother's diet) and social integration (learning of group or kin odour). Like other well-defined mammalian chemosignals (e.g. Zhang et al., 2008; Martin-Sanchez et al., 2015), the so-called mammary pheromone of the rabbit is thus clearly a *multitask chemosignal* in its wide-ranging effects on the progeny, in addition to its effect on inclusive fitness of emitters. It is additionally a *multi-sourced chemosignal*, opening new questions on what a same chemosignal can mean when it is embedded in different chemical complexities and received in different physiological or behavioural backgrounds.

**Acknowledgements** We are thankful to Pr. Guy Perrier for facilitating years of stimulating and amicable discussions and taking part in intensive research in his rabbit breeding facility at ENESAD, to Jean-Pierre Drouet and Monique Jouanno for their expert everyday management of the experimental rabbits, to Brigitte Quennedey, Bruno Patris and Delphine Montigny for their assistance with animal testing. Finally, we are deeply indebted to Pr. Wittko Francke for long discussions on chemical communication in mammals and for graciously synthetizing 2-methylbut-2-enol in the Hamburg Institute of Organic Chemistry. This chapter is a tribute to him.

## References

Altbäcker, V., Hudson, R., & Bilkó, A. (1995). Rabbit-mothers' diet influences pups' later food choice. *Ethology, 99*(1-2), 107–116.

Albone, E. S. (1984). Mammalian semiochemistry: the investigation of chemical signals between mammals. Wiley.

Apps, P. J., Weldon, P. J., & Kramer, M. (2015). Chemical signals in terrestrial vertebrates: Search for design features. *Natural Product Reports, 32*(7), 1131–1153.

Archie, E. A., & Theis, K. R. (2011). Animal behaviour meets microbial ecology. *Animal Behaviour, 82*, 425–436.

Asgari, S., Hardy, J. R., Sinclair, R. G., & Cooke, B. D. (1998). Field evidence for mechanical transmission of rabbit haemorrhagic disease virus (RHDV) by flies (Diptera: Calliphoridae) among wild rabbits in Australia. *Virus Research, 54*(2), 123–132.

Aviles-Rosa, E. O., Rakhshandeh, A., & McGlone, J. J. (2019). Preliminary study: Depriving piglets of maternal feces for the first seven days post-partum changes piglet physiology and performance before and after weaning. *Animals, 9*(5), 268.

Aviles-Rosa, E. O., Surowiec, K., & McGlone, J. J. (2020). Identification of faecal maternal semiochemicals in swine (*Sus scrofa*) and their effects on weaned piglets. *Scientific Reports, 10*(1), 5349.

Bautista, A., Mendoza-Degante, M., Coureaud, G., Martínez-Gómez, M., & Hudson, R. (2005). Scramble competition in newborn domestic rabbits for an unusually restricted milk supply. *Animal Behaviour, 70*(5), 1011–1021.

Bautista, A., García-Torres, E., Martínez-Gómez, M., & Hudson, R. (2008). Do newborn domestic rabbits *Oryctolagus cuniculus* compete for thermally advantageous positions in the litter huddle? *Behavioral Ecology and Sociobiology, 62*, 331–339.

Bean, N. J., Galef, B. G., Jr., & Mason, J. R. (1988). The effect of carbon disulfide on food consumption by house mice. *Journal of Wildlife Management, 52*, 502–507.

Bell, D. J. (1985). The rabbits and hares: Order lagomorpha. In R. E. Brown & D. W. MacDonald (Eds.), *Social odours in mammals, volume 2* (pp. 507–530). Clarendon Press.

Bienenstock, J., Kunze, W. A., & Forsythe, P. (2018). Disruptive physiology: Olfaction and the microbiome–gut–brain axis. *Biological Reviews, 93*(1), 390–403.

Bilkó, Á., Altbäcker, V., & Hudson, R. (1994). Transmission of food preference in the rabbit: The means of information transfer. *Physiology & Behavior, 56*(5), 907–912.

Bilkó, Á., Petróczi, I., Bárdos, B., Nagy, I., & Altbäcker, V. (2022). Composition of the wild rabbit nest and its implication for domestic rabbit breeding. *Applied Sciences, 12*(4), 1915.

Björnhag, G. (1994). Adaptations in the large intestine allowing small animals to eat fibrous foods. In D. J. Chivers & P. Langer (Eds.), *The digestive system in mammals: Food, form and function* (pp. 287–309). Cambridge University Press.

Björnhag, G., & Snipes, R. L. (1999). Colonic separation mechanism in lagomorph and rodent species—a comparison. *Zoosystematics and Evolution, 75*(2), 275–281.

Breen, M. F., & Leshner, A. I. (1977). Maternal pheromone: A demonstration of its existence in the mouse (*Mus musculus*). *Physiology & Behavior, 18*(3), 527–529.

Caba, M., & González-Mariscal, G. (2009). The rabbit pup, a natural model of nursing-anticipatory activity. *The European Journal of Neuroscience, 30*(9), 1697–1706.

Carabano, R., Fraga, M. J., Santoma, G., et al. (1988). Effect of diet on composition of cecal contents and on excretion and composition of soft and hard feces of rabbits. *Journal of Animal Science, 66*, 901–910.

Carthey, A. J. R., Gillings, M. R., & Blumstein, D. T. (2018). The extended genotype: Microbially mediated olfactory communication. *Trends in Ecology & Evolution, 33*(11), 885–894.

Combes, S., Gidenne, T., Cauquil, L., Bouchez, O., & Fortun-Lamothe, L. (2014). Coprophagous behavior of rabbit pups affects implantation of cecal microbiota and health status. *Journal of Animal Science, 92*(2), 652–665.

Cooke, B. D. (1990). Rabbit burrows as environments for European Rabbit Fleas, *Spilopsyllus-cuniculi* (Dale). *Arid South-Australia Australian Journal of Zoology, 38*(3), 317–325.

Cooke, B. D. (2023). Fifty-year review: European rabbit fleas, *Spilopsyllus cuniculi* (Dale, 1878) (Siphonaptera: *Pulicidae*), enhanced the efficacy of myxomatosis for controlling Australian rabbits. *Wildlife Research, 50*(1), 4–15.

Coureaud, G., & Schaal, B. (2000). Attraction of newborn rabbits to abdominal odors of adult conspecifics differing in sex and physiological state. *Developmental Psychobiology, 36*(4), 271–281.

Coureaud, G., Schaal, B., Langlois, D., & Perrier, G. (2001). Orientation response of newborn rabbits to odours of lactating females: Relative effectiveness of surface and milk cues. *Animal Behaviour, 61*(1), 153–162.

Coureaud, G., Langlois, D., Sicard, G., & Schaal, B. (2004). Newborn rabbit responsiveness to the mammary pheromone is concentration-dependent. *Chemical Senses, 29*(4), 341–350.

Coureaud, G., Moncomble, A. S., Montigny, D., Dewas, M., Perrier, G., & Schaal, B. (2006). A pheromone that rapidly promotes learning in the newborn. *Current Biology, 16*(19), 1956–1961.

Coureaud, G., Fortun-Lamothe, L., Langlois, D., & Schaal, B. (2007). The reactivity of neonatal rabbits to the mammary pheromone as a probe for viability. *Animal, 1*(7), 1026–1032.

Cowan, D. P., & Bell, D. J. (1986). Leporid social behaviour and social organization. *Mammal Review, 16*(3–4), 169–179.

Crowell-Davis, S. L., & Houpt, K. A. (1985). Coprophagy by foals: Effect of age and possible functions. *Equine Veterinary Journal, 17*(1), 17–19.

Davies, R. R., & Davies, J. A. R. (2003). Rabbit gastrointestinal physiology. *Veterinary Clinics: Exotic Animal Practice, 6*(1), 139–153.

de Blas, C., & Wiseman, J. (Eds.). (1990). *Nutrition of the rabbit* (2nd ed.). Cabi International.

El-Ghazali, H. M., & Ali, S. A. (1978). Rectal, anal gland: A corrective morphological study in New Zealand Rabbit (*Oryctolagus cuniculus*). *Benha Veterinary Medical Journal, 34*(2), 221–233.

Elias, B. A., Shipley, L. A., Sayler, R. D., & Lamson, R. S. (2006). Mating and parental care in captive pygmy rabbits. *Journal of Mammalogy, 87*(5), 921–928.

Endler, J. A., & Basolo, A. L. (1998). Sensory ecology, receiver biases and sexual selection. *Trends in Ecology & Evolution, 13*, 415–420.

Ezenwa, V. O., & Williams, A. E. (2014). Microbes and animal olfactory communication: Where do we go from here? *BioEssays, 36*(9), 847–854.

Fortun-Lamothe, L. & Boullier, S. (2004). Interactions between gut microflora and digestive mucosal immunity, and strategies to improve digestive health in young rabbits. In: *Proceedings of 8th World Rabbit Congress* (pp. 7–10)

Funosas, G., Triadó-Margarit, X., Castro, F., Villafuerte, R., Delibes-Mateos, M., Rouco, C., & Casamayor, E. O. (2021). Individual fate and gut microbiome composition in the European wild rabbit (*Oryctolagus cuniculus*). *Scientific Reports, 11*(1), 1–13.

Galef, B. G. (1979). Investigation of the functions of coprophagy in juvenile rats. *Journal of Comparative and Physiological Psychology, 93*(2), 295.

Galef, B. G., Mason, J. R., Preti, G., & Bean, N. J. (1988). Carbon disulfide: A semiochemical mediating socially-induced diet choice in rats. *Physiology & Behavior, 42*(2), 119–124.

Ge, Y., Zhu, W., Chen, L., Li, D., Li, Q., & Jie, H. (2021). The maternal milk microbiome in mammals of different types and its potential role in the neonatal gut microbiota composition. *Animals, 11*(12), 3349.

Gidenne, T., Combes, S., Fidler, C., & Fortun-Lamothe, L. (2013). Comportement d'ingestion de fèces dures maternelles par les lapereaux au nid. 1. Quantification de la production maternelle de fèces et de leur ingestion par les lapereaux au nid. In *Comptes-Rendus des 14èmes Journées de la Recherche Cunicole* (pp. 33–48). ITAVI.

Gilbert, C., McCafferty, D. J., Giroud, S., Ancel, A., & Blanc, S. (2012). Private heat for public warmth: How huddling shapes individual thermogenic responses of rabbit pups. *PLoS One, 7*(3), e33553.

González-Mariscal, G., Díaz-Sánchez, V., Melo, A. I., Beyer, C., & Rosenblatt, J. S. (1994). Maternal behavior in New Zealand white rabbits: Quantification of somatic events, motor patterns, and steroid plasma levels. *Physiology & Behavior, 55*(6), 1081–1089.

González-Mariscal, G., Lemus, A. C., Vega-Gonzalez, A., & Aguilar-Roblero, R. (2013). Litter size determines circadian periodicity of nursing in rabbits. *Chronobiology International, 30*(5), 711–718.

González-Mariscal, G., Hoy, S., & Hoffman, K. L. (2022). Rabbit maternal behavior: A perspective from behavioral neuroendocrinology, animal production, and psychobiology. In G. González-Mariscal (Ed.), *Patterns of parental behavior. Advances in neurobiology* (Vol. 27, pp. 131–176). Springer.

Goodrich, B. S. (1983). Studies of the chemical composition of secretions from skin glands of the rabbit *Oryctolagus cuniculus*. In D. Müller-Schwarze & R. M. Siverstein (Eds.), *Chemical signals in vertebrates, Volume 3* (pp. 275–290). Plenum Press.

Goodrich, B. S., & Mykytowycz, R. (1972). Individual and sex differences in the chemical composition of pheromone-like substances from the skin glands of the rabbit, *Oryctolagus cuniculus*. *Journal of Mammalogy, 53*(3), 540–548.

Goodrich, B. S., Hesterman, E. R., Murray, K. E., Mykytowycz, R., Stanley, G., & Sugowdz, G. (1978). Identification of behaviorally significant volatile compounds in the anal gland of the rabbit, *Oryctolagus cuniculus*. *Journal of Chemical Ecology, 4*, 581–594.

Goodrich, B. S., Hesterman, E. R., Shaw, K. S., & Mykytowycz, R. (1981). Identification of some volatile compounds in the odor of fecal pellets of the rabbit, *Oryctolagus cuniculus*. *Journal of Chemical Ecology, 7*, 817–827.

Gruaz, M., van Praag, A., & van Praag, E. (2020). *When bacteria infect nursing does and/or their kits...* http://www.medirabbit.com/. Consulted on 25/03/2023.

Guilford, T., & Dawkins, M. S. (1993). Receiver psychology and the design of animal signals. *Trends in Neurosciences, 16*(11), 430–436.

Hanser, H. I., Faure, P., Robert-Hazotte, A., Artur, Y., Duchamp-Viret, P., Coureaud, G., & Heydel, J. M. (2017). Odorant-odorant metabolic interaction, a novel actor in olfactory perception and behavioral responsiveness. *Scientific Reports, 7*(1), 1–12.

Hayward, J. S. (1966). Abnormal concentrations of respiratory gases in rabbit burrows. *Journal of Mammalogy, 47*(4), 723–724.

Hesterman, E. R., & Mykytowycz, R. (1968). Some observations on the odours of anal gland secretions from the rabbit, *Oryctolagus cuniculus* (L.). *CSIRO. Wildlife Research, 13*(1), 71–81.

Hesterman, E. R., Goodrich, B. S., & Mykytowycz, R. (1976). Behavioral and cardiac responses of the rabbit, *Oryctolagus cuniculus*, to chemical fractions from anal gland. *Journal of Chemical Ecology, 2*, 25–37.

Hesterman, E. R., Goodrich, B. S., & Mykytowycz, R. (1981). Effect of volatiles collected above fecal pellets on behavior of the rabbit, Oryctolagus cuniculus, tested in an experimental chamber. I. Total volatiles and some chemically prepared fractions. *Journal of Chemical Ecology, 7*, 795–811.

Hirakawa, H. (2001). Coprophagy in leporids and other mammalian herbivores. *Mammal Review, 31*(1), 61–80.

Hudson, R., & Distel, H. (1982). The pattern of behaviour of rabbit pups in the nest. *Behaviour, 79*(2–4), 255–271.

Hudson, R., & Distel, H. (1983). Nipple location by newborn rabbits: Behavioural evidence for pheromonal guidance. *Behaviour, 85*(3–4), 260–274.

Hudson, R., & Distel, H. (1994). On the nature and action of the rabbit nipple-search pheromone: A review. In R. Apfelbach and D. Müller-Schwarze, eds., *Chemical signals in vertebrates, 7*, 223–232. Pergamon Press.

Hudson, R., Bilko, A., & Altbäcker, V. (1996). Nursing, weaning and the development of independent feeding in the rabbit (*Oryctolagus cuniculus*). *Zeitschrift für Säugetierkunde, 61*, 39–48.

Hudson, R., Schaal, B., & Bilko, A. (1999). Transmission of olfactory information from mother to young in the European rabbit. In H. O. Box & K. R. Gibson (Eds.), *Mammalian social learning: Comparative and ecological perspectives* (pp. 141–157). Cambridge University Press.

Hudson, R., Garay-Villar, E., Maldonado, M., & Coureaud, G. (2003). Rabbit pups can orient to the nest by smell from birth. *Chemical Senses, 28*, 55.

Hudson, R., Rojas, C., Arteaga, L., Martinez-Gomez, M., & Distel, H. (2008). Rabbit nipple-search pheromone versus rabbit mammary pheromone, revisited. In J. L. Hurst, R. J. Beynon, S. C. Roberts, & T. Wyatt (Eds.), *Chemical signals in vertebrates, Volume 11* (pp. 315–324). Springer Verlag.

Jouhanneau, M., Schaal, B., & Coureaud, G. (2016). Mammary pheromone-induced odour learning influences sucking behaviour and milk intake in the newborn rabbit. *Animal Behaviour, 111*, 1–11.

Keil, W., von Stralendorff, F., & Hudson, R. (1990). A behavioral bioassay for analysis of rabbit nipple-search pheromone. *Physiology & Behavior, 47*(3), 525–529.

Kovács, M., Szendro, Z., Milisits, G., Bóta, B., Bíró-Németh, E., Radnai, I., et al. (2006). Effect of nursing methods and faeces consumption on the development of the bacteroides, lactobacillus and coliform flora in the caecum of the newborn rabbits. *Reproduction Nutrition Développement, 46*(2), 205–210.

Krohn, T. C., Ritskes-Hoitinga, J., & Svendsen, P. (1999). The effects of feeding and housing on the behaviour of the laboratory rabbit. *Laboratory Animals, 33*(2), 101–107.

Leon, M. (1974). Maternal pheromone. *Physiology & Behavior, 13*(3), 441–453.

Leon, M., & Moltz, H. (1972). The development of the pheromonal bond in the albino rat. *Physiology & Behavior, 8*(4), 683–686.

Li, Z., He, H., Ni, M., Wang, Z., Guo, C., Niu, Y., et al. (2022). Microbiome-metabolome analysis of the immune microenvironment of the cecal contents, soft feces, and hard feces of Hyplus rabbits. *Oxidative Medicine and Cellular Longevity, 2022*, 5725442.

Martín-Sánchez, A., McLean, L., Beynon, R. J., Hurst, J. L., Ayala, G., Lanuza, E., & Martinez-Garcia, F. (2015). From sexual attraction to maternal aggression: When pheromones change their behavioural significance. *Hormones and Behavior, 68*, 65–76.

McNitt, J. L., & Moody, G. L. (1987). Nest box behaviour of the domestic rabbit. *Journal of Applied Rabbit Research, 10*(4), 159–162.

Melo, A. I., & González-Mariscal, G. (2010). Communication by olfactory signals in rabbits: Its role in reproduction. *Vitamins and Hormones, 83*, 351–371.

Merritt, G. C., Goodrich, B. S., Hesterman, E. R., & Mykytowycz, R. (1982). Microflora and volatile fatty acids present in inguinal pouches of the wild rabbit, *Oryctolagus cuniculus*, in Australia. *Journal of Chemical Ecology, 8*, 1217–1225.

Moltz, H., & Lee, T. M. (1981). The maternal pheromone of the rat: Identity and functional significance. *Physiology & Behavior, 26*(2), 301–306.

Moncomble, A. S. (2006). *Contribution à l'étude des mécanismes olfactifs indicateurs de l'ingestion lactée et post-lactée chez le lapereau: Analyses éthologiques, histologiques et chimiques*. Doctoral dissertation,. Université de Bourgogne.

Moncomble, A. S., Quennedey, B., Coureaud, G., Langlois, D., Perrier, P., & Schaal, B. (2004). Newborn rabbit attraction towards maternal faecal pellets. *Developmental Psychobiology, 45*(4), 277.

Moncomble, A. S., Coureaud, G., Quennedey, B., Langlois, D., Perrier, G., & Schaal, B. (2005). The mammary pheromone of the rabbit: From where does it come? *Animal Behaviour, 69*(1), 29–38.

Montúfar-Chaveznava, R., Trejo-Muñoz, L., Hernández-Campos, O., Navarrete, E., & Caldelas, I. (2013). Maternal olfactory cues synchronize the circadian system of artificially raised newborn rabbits. *PLoS One, 8*(9), e74048.

Munger, S. D., Leinders-Zufall, T., McDougall, L. M., Cockerham, R. E., Schmid, A., Wandernoth, P., et al. (2010). An olfactory subsystem that detects carbon disulfide and mediates food-related social learning. *Current Biology, 20*(16), 1438–1444.

Mykytowycz, R. (1966). Observations on odoriferous and other glands in the Australian Wild Rabbit, *Oryctolagus cuniculus* (L.) and the Hare, *Lepus europaeus*. I. The Anal Gland. *CSIRO Wildlife Research, 1*, 11–29.

Mykytowycz, R. (1968). Territorial marking by rabbits. *Scientific American, 218*, 116–126.

Mykytowycz, R. (1970). Role of skin glands in mammalian communication. In J. W. Johnson, D. G. Moulton, & A. Turk (Eds.), *Advances in Chemoreception* (pp. 327–360). Meredith Corp.

Mykytowycz, R. (1985). Odour signals in the life of the wild rabbit. *Australian Science Magazine, 1*, 44–50.

Mykytowycz and Hesterman. (1974). An experimental study of aggressin captive European rabbits. *Behaviour, 52*, 104–123.

Mykytowycz, R., & Dudziński, M. L. (1972). Aggressive and protective behaviour of adult rabbits *Oryctolagus cuniculus* (L.) towards juveniles. *Behaviour, 43*, 97–120.

Mykytowycz, R., & Ward, M. (1971). Some reactions of nestlings of the wild rabbit, *Oryctolagus cuniculus* (L.) when exposed to natural rabbit odours. *Forma et Functio, 4*, 137–148.

Mykytowycz, R., Hesterman, E. R., Gambale, S., & Dudzinski, M. L. (1976). A comparison of the effectiveness of the odors of rabbits, *Oryctolagus cuniculus*, in enhancing territorial confidence. *Journal of Chemical Ecology, 2*(1), 13–24.

Noonan, M. J., Tinnesand, H. V., Müller, C. T., Rosell, F., Macdonald, D. W., & Buesching, C. D. (2019). Knowing me, knowing you: Anal gland secretion of European badgers (Meles meles) codes for individuality, sex and social group membership. *Journal of Chemical Ecology, 45*, 823–837.

Nowland, T. L., Kirkwood, R. N., Plush, K. J., Barton, M. D., & Torok, V. A. (2021). Exposure to maternal feces in lactation influences piglet enteric microbiota, growth, and survival preweaning. *Journal of Animal Science, 99*(7), skab170.

Osawa, R., Blanshard, W. H., & Ocallaghan, P. G. (1993). Microbiological studies of the intestinal microflora of the koala, *Phascolarctos cinereus*. 2. Pap, a special maternal feces consumed by juvenile koalas. *Australian Journal of Zoology, 41*, 611–620.

Patris, B., Perrier, G., Schaal, B., & Coureaud, G. (2008). Early development of filial preferences in the rabbit: Implications of nursing- and pheromone-induced odour learning? *Animal Behaviour, 76*(2), 305–314.

Pickard, D. W., & Stevens, C. E. (1972). Digesta flow through the rabbit large intestine. *The American Journal of Physiology, 222*(5), 1161–1166.

Porter, R. H., & Doane, H. M. (1976). Maternal pheromone in the spiny mouse (*Acomys cahirinus*). *Physiology & Behavior, 16*(1), 75–78.

Rodríguez-Martínez, L., Hudson, R., Martínez-Gómez, M., & Bautista, A. (2014). Description of the nursery burrow of the Mexican cottontail rabbit *Sylvilagus cunicularius* under seminatural conditions. *Acta Theriologica, 59*, 193–201.

Rosenberg, E. (2021). Transmission of hologenomes between generations: Mothers matter most. In E. Rosenberg (Ed.), *Microbiomes: Current knowledge and unanswered questions* (pp. 161–194). Springer.

Schaal, B., & Al Aïn, S. (2014). Chemical signals 'selected for' newborns in mammals. *Animal Behaviour, 97*, 289–299.

Schaal, B., Coureaud, G., Langlois, D., Ginies, C., Sémon, E., & Perrier, G. (2003). Chemical and behavioural characterization of the rabbit mammary pheromone. *Nature, 424*(6944), 68–72.

Schaal, B., Coureaud, G., Moncomble, A. S., Langlois, D., & Perrier, G. (2008). Many common odour cues and (at least) one pheromone shaping the behaviour of young rabbits. In P. C. Alves, N. Ferrand, & K. Hackländer (Eds.), *Lagomorph biology: Evolution, ecology, and conservation* (pp. 189–209). Springer Verlag.

Schaffer, J. (1940). *Die Hautdrüsenorgane der Saügetiere mit besonderer Berücksichtigung ihres histologischen Aufbaues und Bemerkungen über die Proktodialdrusen.* Urban & Schwarzenberg.

Schalken, A. P. M. (1976). Three types of pheromones in the domestic rabbit, *Oryctolagus cuniculus* (L). *Chemical Senses, 2*(2), 139–155.
Schley, P. (1977). Die Ausschaltung des Geruchsvermögens und sein Einfluß auf das Saugverhalten von Jungkaninchen. *Berliner-Münchener Tierärztliche Wochenschrift, 90*, 382–385.
Schley, P. (1985). *Kaninchen*. Eugen Ulmer.
Serra, J., & Nowak, R. (2008). Olfactory preference for own mother and litter in 1-day-old rabbits and its impairment by thermotaxis. *Developmental Psychobiology, 50*(6), 542–553.
Serra, J., Ferreira, G., Mirabito, L., Lévy, F., & Nowak, R. (2009). Post-oral and perioral stimulations during nursing enhance appetitive olfactory memory in neonatal rabbits. *Chemical Senses, 34*(5), 405–413.
Smith, A. T., Johnston, C. H., Alves, P. C., & Hackländer, K. (2018). *Lagomorphs: Pikas, Rabbits and Hares of the World*. Johns Hopkins University Press.
Takagi, S., & Tagawa, M. (1961). Hard and soft faeces of the rabbit and their relation to the anal gland secretion and the contents of the caecum. *Japanese Journal of Zoology, 70*, 248–252.
Tauxe, G. M., MacWilliam, D., Boyle, S. M., Guda, T., & Ray, A. (2013). Targeting a dual detector of skin and CO2 to modify mosquito host seeking. *Cell, 155*, 1365–1379.
Theis, K. R., Whittaker, D. J., & Rojas, C. A. (2020). A Hologenomic approach to animal behavior. In W. Banzhaf et al. (Eds.), *Evolution in action: Past, present and future. Genetic and evolutionary computation* (pp. 247–263). Springer.
Trejo-Muñoz, L., Navarrete, E., Montúfar-Chaveznava, R., & Caldelas, I. (2014). Temporal modulation of the canonical clockwork in the suprachiasmatic nucleus and olfactory bulb by the mammary pheromone 2MB2 in pre-visual rabbits. *Neuroscience, 275*, 170–183.
Turner, S. L., Li, N., Guda, T., Githure, J., Carde, R. T., & Ray, A. (2011). Ultra-prolonged activation of CO2-sensing neurons disorients mosquitoes. *Nature, 474*, 87–91.
Val-Laillet, D., & Nowak, R. (2008). Early discrimination of the mother by rabbit pups. *Applied Animal Behaviour Science, 111*(1–2), 173–182.
Varga, M. (2013). Rabbit basic science. In *Textbook of rabbit medicine* (2nd ed., pp. 3–108). Butterworth-Heinemann. https://doi.org/10.1016/C2011-0-05821-0
Verts, B. J., Carraway, L. N., & Green, R. L. (1997). Sex-bias in prenatal parental investment in the eastern cottontail (*Sylvilagus floridanus*). *Journal of Mammalogy, 78*(4), 1164–1171.
Wales, N. A. M., & Ebling, F. J. (1971). The control of the apocrine glands of the rabbit by steroid hormones. *The Journal of Endocrinology, 51*(4), 763–770.
Whittaker, D. J., Gerlach, N. M., Slowinski, S. P., Corcoran, K. P., Winters, A. D., Soini, H. A., et al. (2016). Social environment has a primary influence on the microbial and odor profiles of a chemically signaling songbird. *Frontiers in Ecology and Evolution, 4*, 90.
Williams, J., & Ringsdorf, A. (2020). Human odour thresholds are tuned to atmospheric chemical lifetimes. *Philosophical Transactions of the Royal Society B, 375*, 20190274.
Wyatt, T. D. (2014). *Pheromones and animal behavior: Chemical signals and signatures*. Cambridge University Press.
Young, J. G., & Henke, S. E. (1999). Effect of domestic rabbit urine on trap response in cottontail rabbits. *Wildlife Society Bulletin*, 306–309.
Zarrow, M. X., Denenberg, V. H., & Anderson, C. O. (1965). Rabbit: Frequency of suckling in the pup. *Science, 150*(3705), 1835–1836.
Zhang, J. X., Liu, Y. J., Zhang, J. H., & Sun, L. (2008). Dual role of preputial gland secretion and its major components in sex recognition of mice. *Physiology & Behavior, 95*(3), 388–394.
Zhou, W., Qi, D., Swaisgood, R. R., Wang, L., Jin, Y., Wu, Q., et al. (2021). Symbiotic bacteria mediate volatile chemical signal synthesis in a large solitary mammal species. *International Society For Microbial Ecology Journal, 15*(7), 2070–2080.

# Tips from the Nose: Odor-Driven Visual Categorization in the Developing Human Brain

Diane Rekow[1] and Arnaud Leleu[1] (✉)

[1] Development of Olfactory Communication and Cognition Laboratory, Centre des Sciences du Goût et de l'Alimentation, CNRS, Université Bourgogne, Inrae-Institut Agro Dijon, Dijon, France
arnaud.leleu@u-bourgogne.fr

**Abstract** While humans are usually considered as overly visual, human vision is poor at the beginning of life. In contrast, olfaction is an early-developing sense that promotes adaptive responses and may support visual development. Accordingly, this chapter reviews recent studies that we conducted showing that the mother's body odor selectively facilitates face categorization in the 4-month-old human brain. We also reveal that as infants grow up, their ability to categorize faces improves until 12 months of age and becomes progressively independent of the influence of maternal odor. We open with findings in adults, demonstrating that odor-driven visual categorization is still observed for ambiguous stimuli, aligning with the inverse effectiveness principle of multisensory integration. Overall, this body of research exposes the powerful impact of odors on visual categorization, especially how social chemical cues emanating from the mother can drive the categorization of conspecifics from visual inputs in the developing human brain. These findings suggest that odors act as disambiguating cues fostering visual categorization as an inverse function of its effectiveness and point toward a critical role of multisensory inputs for the acquisition of categories in humans, relying on the ordered development across the senses.

**Keywords** Human infant · Face categorization · Maternal body odor · Multisensory perception · Inverse effectiveness · Electroencephalography (EEG) · Frequency-tagging

## 1 Introduction

Smell has long been neglected in humans, while vision is generally considered the dominant sense (Broca, 1879; Darwin, 1871; Hutmacher, 2019). However, a recent surge in research has demonstrated that humans possess a keen sense of smell, with performance sometimes akin to that of other species reputed for their olfactory prowess (e.g., McGann, 2017). In fact, olfaction plays a fundamental role in how humans perceive their environment, as recently highlighted by the COVID-19 pandemic and



B. Schaal et al. (eds.), *Chemical Signals in Vertebrates 15*,
https://doi.org/10.1007/978-3-031-35159-4_18

associated anosmia (Parma et al., 2020). A growing body of research indicates that olfaction can even influence perception in the other senses, including visual perception (e.g., Demattè et al., 2006; Seigneuric et al., 2010; W. Zhou et al., 2010).

The key function of olfaction in the human species is particularly clear at the beginning of life and further extends during different sensitive periods throughout development (Schaal et al., 2020). Olfaction is indeed an early-maturing sense reaching operative functionality from the beginning of the third gestational trimester (Browne, 2008; Schaal et al., 2004). Chemically, odor compounds in the amniotic fluid and in the mother's secretions (e.g., colostrum/milk, sweat from the breast area, etc.) partially overlap (Loos et al., 2019; Porter & Winberg, 1999). Neonates show a strong responsiveness to these odor cues (e.g., Klaey-Tassone et al., 2020), enabling stability and familiarity in a postnatal environment full of sensory novelty.

Unlike olfaction, vision is a late-developing sense in infancy. Visual perception mainly starts to develop from birth onward when newborns are exposed for the first time to a rich and complex flow of visual information. The anatomical development of the visual system in human infants takes longer than the other senses to install (Graven & Browne, 2008) and preterm infants have no pupillary response to light, showing a delayed development compared to the other senses, already reacting to stimuli in utero. Human visual function is thus limited at birth (e.g., poor visual acuity and color vision until 2–3 postnatal months) and follows a protracted development over the first postnatal year (Braddick & Atkinson, 2011).

This ordered sequence of sensory development, common to most vertebrate species, may actually be far from an arbitrary developmental timing across independent and parallel sensory systems (Turkewitz & Devenny, 1993). Sensory ontogeny may be adaptive for the offspring and benefit to perceptual learning, early-maturing senses sustaining later-developing senses. The early onset of the olfactory system therefore puts olfaction in an ideal position to mediate perception in the less mature sense of vision.

This chapter aims to outline recent work on the influence of one of the most relevant odors for young infants, their mother's body odor, on their ability to categorize a variety of visual stimuli as faces. We will first describe how maternal chemosignaling actively participates in orienting and regulating infant physiology and behavior early in life. Next, we will portray how one of the most important functions of visual perception—*categorization*—takes time to develop from brain maturation and accumulated visual experience, focusing on the most relevant visual stimuli for young infants: faces. Then, we will survey recent research conducted in 4- to 12-month-old infants to delineate the extent to which the mother's body odor shapes face categorization in the developing brain, as well as experiments conducted in adults to determine whether body odors still influence visual categorization in the mature brain. Finally, we will discuss which mechanisms may underlie the integration between olfaction and vision and what they may tell us about (multi)sensory development. We will conclude with remaining open questions and suggest directions for future research.

### 1.1 Mother-to-Infant Chemical Communication

The chemical senses have an early functional onset during the third gestational trimester (Browne, 2008; Schaal et al., 2004). In particular, the olfactory system is a primary medium of communication from the mother to her infant already before birth. The amniotic fluid is filled with odorous compounds conveying information about the mother's physiological state, including both stable (e.g., genetic) and transient information (e.g., emotional reactions) and about external compounds she consumes (e.g., food, cosmetics) (Schaal et al., 2002). Human fetuses use and retain the odor properties of their prenatal environment such that neonates then differentially respond to the odor of their own amniotic fluid (Schaal et al., 1998) or to flavors they have learned from their mother's diet (Faas et al., 2000; Schaal et al., 2000), some of these early preferences lasting long after birth (Hepper et al., 2013; Wagner et al., 2019). The uterine chemical environment therefore provides an early olfactory experience that benefits the infant after birth. Many odorants experienced in utero are also found in the mother's body secretions, especially from the areolar-nipple and axillary areas (i.e., colostrum, milk, sweat), allowing a transnatal transfer of information (Hartmann et al., 2012; Porter & Winberg, 1999; Spahn et al., 2019).

During the first postnatal months, maternal secretions elicit adaptive behaviors to satisfy physiological needs and promote regulatory functions in her young infant. A few minutes after birth, maternal odor cues from an unwashed vs. washed breast guide the neonate toward the nipple to feed (Varendi et al., 1994). Similarly, the mother's breast, presented entirely or partly covered, stimulates breathing and orofacial reactions, while this behavior disappears when the breast is fully covered by a transparent film, i.e., blocking olfactory but not visual cues (Doucet et al., 2007, 2009). The odor of human colostrum/milk is obviously highly attractive in the first hours of aerial life, eliciting positive emotional facial reactions, head orientation, and appetitive behavior (Klaey-Tassone et al., 2020; Marlier et al., 1998; Soussignan et al., 1997). Human milk is also attractive to bottle-fed infants who never experienced breastfeeding, revealing its potency to provoke adaptive behaviors even in the absence of postnatal exposure (Marlier & Schaal, 2005).

The mother's scent does not only convey food-related cues but has a soothing and distress-relieving effect (Zhang et al., 2018, for review). For instance, physiological (e.g., cortisol level, heart rate, and oxygen saturation) and behavioral (e.g., crying intensity and duration, motor agitation) stress reactions observed during medical procedures endured in neonatology are reduced in neonates exposed to the odors emanating from the mother. These effects are shown with either pre- (e.g., amniotic fluid) or postnatal odors (e.g., breast milk) and suggest an analgesic effect (Akcan & Polat, 2016; Jebreili et al., 2015; Nishitani et al., 2009; Rattaz et al., 2005). Maternal body odors also convey identifying cues about the emitter that can be actively used by infants. Indeed, while many effects previously described have been obtained using secretions from either the own mother or (an)other mother(s), infants are also able to differentiate between them. For example, neonates preferentially orient toward their own rather than another mother's odor, whether the scent comes from her breast (Macfarlane, 1975; Russell, 1976), armpit (Cernoch & Porter, 1985; Marin et al., 2015), or uterus (i.e., amniotic fluid; Schaal et al., 1998). Neonates also exhibit more appetitive behavior when exposed to a hospital gown worn by their mother rather than by another mother (Sullivan & Toubas, 1998). Newborns do not only perform odor-based discrimination of familiar vs. unfamiliar individuals (i.e., their own vs. another mother), they also recognize categories of conspecifics, such as lactating women vs. nonlactating women (Makin & Porter, 1989), and species-specific odors, such as the odor of human vs. nonhuman milk (Marlier & Schaal, 2005; Russell, 1976).

At the brain level, preterm and full-term neonates presented with a cloth worn by their mother show an activation of the orbitofrontal cortex (OFC) as assessed by functional near-infrared spectroscopy (Frie et al., 2019). This neural response has been interpreted as an odor-induced state of arousal, in line with the observation that the mother's breast odor triggers eye opening in 3-day-old newborns and may increase visual attention (Doucet et al., 2007). The OFC is also activated by the odor of colostrum in the youngest neonates and this response then progressively decreases as a function of age (Bartocci et al., 2000), suggesting a relation with their behavioral attraction to colostrum during the first postnatal days (Klaey-Tassone et al., 2020). Similarly, breast milk odor activates the OFC more than formula milk odor irrespective of the infants' experience with formula milk (Aoyama et al., 2010), corroborating behavioral findings (Marlier & Schaal, 2005). Later in the first year, the frontal region of 2- to 9-month-old infants still presents with a larger response to breast milk odor than other odors as measured by scalp electroencephalography (EEG) (Gellrich et al., 2021). Exposure to the mother's odor also decreases the brain response to unfamiliar fearful faces in 7-month-old infants, interpreted as if they were reassured by chemical cues signaling the presence of their mother in an anxious/stressful context (Jessen, 2020). At the same age, a recent study has shown greater brain-to-brain synchrony between infants and unfamiliar women when infants are exposed to their mother's body odor compared to a baseline odor, in association with greater looking time at the unfamiliar woman's face, suggesting that maternal odor promotes infants' interest toward unknown conspecifics (Endevelt-Shapira et al., 2021).

## 1.2 The Early Tuning of Visual Categorization

In contrast to early olfactory learning, the human infant brain only starts to experience a flurry of visual inputs from birth onward. To avoid being overwhelmed by the myriad of inputs coming from the visual surroundings, the infant visual system must progressively go beyond low-level cues (e.g., luminance, contrast, spatial frequencies) and organize visual events into discrete entities according to their meaning. This fundamental function, *visual categorization*, deals with environmental diversity by putting some stimuli into distinct categories (between-category discrimination), while grouping others into the same category (within-category generalization). However, visual categorization is challenging for the infant brain as stimuli belonging to different categories can share some low-level cues, whereas different exemplars from a single category can be widely variable in appearance. Categorizing the visual environment is therefore one of the most fundamental abilities that infants must acquire during the first months of postnatal life.

### 1.2.1 An Early Ability to Categorize Visual Stimuli in Infancy

As early as 3 months of age, behavioral studies have documented human infants' abilities to visually categorize stimuli. For instance, when infants are first exposed to several pictures of cats (i.e., familiarization) and then presented with paired pictures of another cat and a dog, they look more at the dog (i.e., novelty preference), suggesting that they recognize the cat as belonging to the category they have been familiarized to (Quinn et al., 1993). It is therefore suggested that 3- to 4-month-old infants are already able to rapidly extract regularities among unfamiliar visual objects (i.e., during the course of an experiment, namely *online category learning*) (Eimas & Quinn, 1994). Over the last three decades, a large body of research using the familiarization/novelty preference paradigm has argued that young infants have either broad (e.g., furniture vs. animals) or subtler (e.g., cats vs. tigers) categorization abilities, either for abstract/unfamiliar objects or for already-encountered real-life objects (Quinn, 2011, for review).

The neural underpinnings of visual categorization have also been investigated in infancy. In particular, developmental scientists have routinely used scalp EEG as a non-invasive, efficient and relatively inexpensive method to isolate neural markers of visual categorization in human infants (Hoehl, 2016, for review). Electrophysiological studies using event-related potentials (i.e., ERPs) have reported a larger negative ERP component at central scalp locations for novel exemplars when comparing two categories of living beings (Marinović et al., 2014; Quinn et al., 2006) or one living vs. one nonliving object category (Elsner et al., 2013; Grossmann et al., 2009), interpreted as sustained attention to novelty (Reynolds & Richards, 2005). It has also been evidenced that the brain response to different exemplars of the same category evolves throughout an experiment, as a neural signature of online category learning (Quinn et al., 2006).

### 1.2.2 Categorizing Conspecifics from Visual Inputs: The Development of Face Categorization

The ability to categorize visual objects culminates in infants' ability to categorize the face of their conspecifics early on, for which the first year represents a highly sensitive period (Pascalis et al., 2020). Newborns already prefer to track slowly moving stimuli organized with more features in their upper part, such as faces (Johnson et al., 1991; Valenza et al., 1996; see Buiatti et al., 2019, for a neural signature in EEG). However, whether this truly reflects face selectivity, or simply an early preference for stimuli with more elements in the upper visual field, namely top-heavy patterns, has been debated (Macchi Cassia et al., 2004; Turati et al., 2002; Simion et al., 2007, for review). Infants need a few additional months to exhibit preferential looking for faces over other objects (de Heering et al., 2016; Durand et al., 2013). The top-heavy bias becomes selective to faces at 3 months of age (Turati et al., 2005), mirroring the well-known face inversion effect described in adults (Yin, 1969; Rossion, 2008, for review). When faces are embedded in complex visual arrays (Di Giorgio et al., 2012; Gliga et al., 2009; Kwon et al., 2016) or in naturalistic scenes and sequences (Amso et al., 2014; Frank et al., 2014; Kelly et al., 2019), the ability to detect them mainly improves after 6 months of age (see Leppänen, 2016, for review). Head-mounted cameras provide in situ measures showing that faces are not only the most present visual object in the infant everyday scenes, but that they also benefit from a qualitative advantage compared to the exposure to other visual objects through reciprocal face-to-face interactions (Jayaraman et al.,

2015). Thus, maturation and experience are required to tune and refine the visual categorization of conspecifics from their faces (Nelson, 2001; Pascalis et al., 2020; Simion & Di Giorgio, 2015, for reviews).

At the brain level, the emergence of the "face" category has been widely explored by comparing the ERPs elicited by faces from those elicited by other objects. A large body of research has identified distinct neural activity for faces vs. meaningless (Halit et al., 2004; Kouider et al., 2013) or meaningful (Conte et al., 2020; de Haan & Nelson, 1999; Gliga & Dehaene-Lambertz, 2007; Peykarjou & Hoehl, 2013) stimuli after at least 3–4 months. However, it is fair to say that the standard ERP approach is limited to document the development of face categorization in the infant brain.[1] Hence, infant ERP studies have often recorded weak and inconsistent neural responses to faces, which are generally dominated by a medial occipital activity driven by low-level image statistics rather than the face category *per se* (see de Haan et al., 2003; Hoehl, 2016, for reviews).

To circumvent these issues, a recent EEG approach, termed frequency-tagging EEG, has been introduced. It capitalizes on the old observation that a periodic sensory input elicits a same-frequency response over corresponding sensory cortices which can be expressed at the exact stimulation frequency and its harmonics (i.e., integer multiples) in the frequency spectrum (Adrian & Matthews, 1934; Norcia et al., 2015, for review; see Sect. 2.1 for details). This approach, actually used for decades in seminal infant studies of the visual system (e.g., Harris et al., 1976), has been recently adapted to investigate high-level vision, especially markers of face perception abilities in the infant brain (e.g., de Heering & Rossion, 2015; Peykarjou et al., 2017, 2022; Poncet et al., 2022). With this approach, a face-selective response has been evidenced at 4–6 months of age over the right occipito-temporal cortex, individual responses were quantified, and face selectivity was significant in a majority of infants (de Heering & Rossion, 2015). However, at this age, the face-selective response is low and captured in a single harmonic. The response then increases, becomes bilateral and distributed on several harmonics in 5- (Lochy et al., 2019) and 10-year-old children (Vettori et al., 2019), before being observed on many more harmonics in adults, with the reappearance of the right hemisphere advantage (e.g., Retter & Rossion, 2016; Rossion et al., 2015). This nonlinear evolution of face-selective neural activity reveals the complex developmental pattern of the face categorization system over the first years of life.

[1] ERPs suffer from a low signal-to-noise ratio and are constrained by their sensitivity to (e.g., muscular) artifacts in awake infant studies. ERPs also use post-hoc subtraction between absolute brain responses to transient stimuli, such that face-selective neural activity is indirectly obtained and can be masked by large noise in the subtracted responses, or driven by a subset of stimuli. In addition, the time-windows of ERP components are often difficult to define objectively from brain responses that are heterogenous in terms of shape, latency and polarity across individuals and age groups. Finally, ERP studies generally use a limited set of homogenous and edited stimuli (i.e., segmented from their natural background with identical exposure conditions across category exemplars; see footnote 2), and compare faces to only a few nonface categories.

# 2 Odor-Driven Face Categorization in the Developing Human Brain

To sum up previous sections, the mother's body odor, henceforth designated as "maternal odor," refers to a mixture of excretions/secretions from the axillary and areolar-nipple areas that elicits differentiable physiological, behavioral, and neural responses in infants. Maternal odor is particularly involved in nursing and caregiving, having a specific weight in the sensory bubble surrounding the offspring during the first postnatal months. Aside from these functions, maternal odor is a major sensory control of early social interactions and provides information about what constitutes the body odor of conspecifics. Interestingly, maternal odor also boosts arousal and visual attention early on, allowing the newborn to sample visual cues concurrent to the presentation of the chemical cues. This may be particularly frequent in the very beginning of life, when maternal odor could be a powerful reinforcement during exposure to salient visual information. As a systematically available and co-occurring cue during the infant's experience with conspecifics, whether it is the mother or other conspecifics, maternal odor is therefore in an ideal position to ease the categorization of faces, which takes several years to develop.

A few pieces of evidence tend, indeed, to support a tutoring function of maternal odor in the development of face perception. Seven-month-olds display a reduced neural response to fearful faces in the presence of maternal odor (Jessen, 2020) and infants exposed to maternal odor vs. a baseline odor look more at their mother's face at 4 months of age (Durand et al., 2020). Most importantly for our purpose, maternal odor influences the visual exploration of paired pictures of an unfamiliar female face and a car in 4-month-old infants (Durand et al., 2013). In this latter study, infants exposed to maternal odor spent more time looking at the face than infants exposed to a baseline odor whereas equal gazing was found for the car, suggesting improved categorization of the face due to maternal odor. However, although this behavioral effect is important enough to inform about the final output of the cognitive system, it can be due to reorientation of attention following face categorization and it has been described only during pairwise discrimination of a few stimuli. Hence, we aimed at producing direct evidence that maternal odor shapes face categorization (i.e., incorporating both discrimination of faces from many nonface categories and generalization across widely variable individual faces) in the infant brain.

## 2.1 Measuring Maternal Odor's Influence on Face Categorization

To measure face categorization in the infant brain and its sensitivity to maternal odor, we conducted several studies relying on scalp EEG and a frequency-tagging approach, which already successfully isolated a face-selective neural response in 4- to 6-month-old infants (de Heering & Rossion, 2015; see Sect. 1.2.2). We exposed infants to a fast periodic visual stimulation presenting streams of various living and nonliving objects (i.e., 170 images depicting plants, animals, and man-made objects) at a base rate of 6 Hz (6 images/second; Fig. 1a left) and face images (i.e., 66 or 68 images of individual

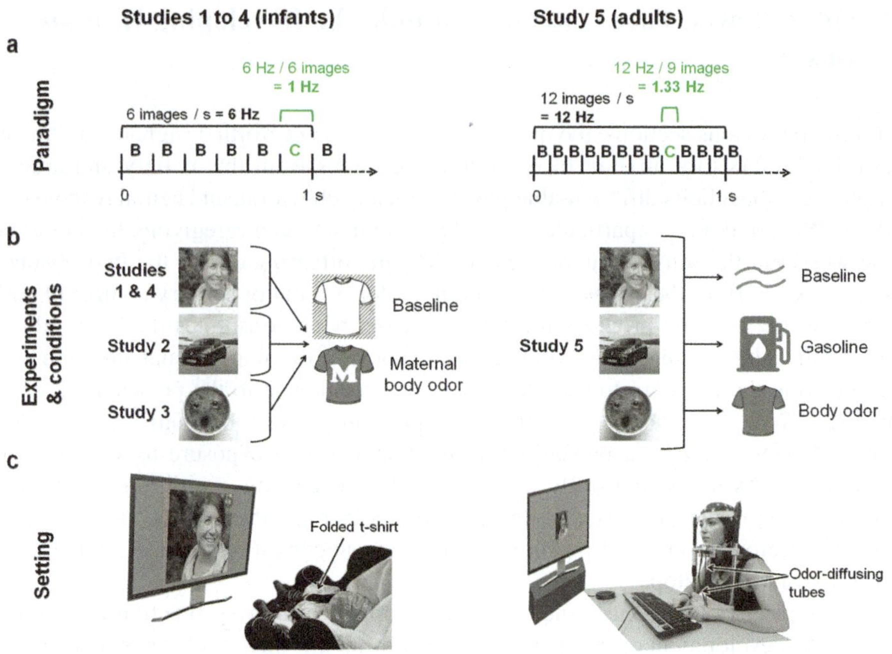


**Fig. 1** Measuring body odor influence on face categorization in the infant (*left*) and adult (*right*) brains using a frequency-tagging approach in scalp electroencephalography (EEG). (**a**) A variety of pictures of visual objects were presented at a rapid base rate (e.g., 6 stimuli/s = 6 Hz; referred to as "B") and various stimuli from a target category were periodically inserted in between (e.g., every sixth stimulus at 6 Hz, i.e., at 6/6 = 1 Hz; referred to as "C") to isolate a category-selective neural response at the same frequency and its harmonics in the EEG spectrum. (**b**) In infants, Studies 1 and 4 used human faces as the target category, Study 2 used cars and Study 3 used face-like objects (i.e., nonface objects that elicit the illusory perception of a face in adults). Infants were exposed to maternal odor collected on a t-shirt for three nights or to a baseline odor (unworn t-shirt). In Study 5 (adults), the three visual categories were used and participants were exposed to three odor contexts (scentless mineral oil (baseline), gasoline or axillary sweat from 8 donors (body odor)). (**c**) During visual stimulation, odors were delivered by disposing the folded t-shirt on the infant's upper chest in Studies 1–4, and using hidden tubes delivering an air-flow under the participants' head in Study 5. (Copyright D. Rekow)

male and female adult faces) were interspersed every sixth image (i.e., at 6 Hz/6 = 1 Hz). Two distinct neural responses elicited during a single stimulation sequence were thus tagged in the EEG frequency spectrum: a general response measured at 6 Hz and harmonics (i.e., integer multiples) and a face-selective response measured at 1 Hz and harmonics. The general visual response is common to both face and nonface stimuli, reflecting the brain synchronization to every cue that changes six times per second. To emerge, this response to the fast train of images simply requires that infants look at the visual stimulation. In contrast, the face-selective response translates in the frequency spectrum only if the infant brain categorizes faces. It reflects a *differential* response to faces (discrimination from various nonface objects) that is reliably elicited by a majority of face stimuli (generalization across face exemplars). Importantly, we used naturalistic stimuli[2] (see examples in Fig. 1), such that the face-selective response is a signature of rich face categorization devoid of low-level confounds, controlling for category-irrelevant properties by variability rather than elimination or homogenization, as imposed in natural viewing conditions. This is an important aspect to ensure that infants are able to categorize a variety of human faces against many other biological (i.e., animals, plants) or manufactured objects, all embedded in their natural backgrounds, which is not a trivial achievement for their brain.

Frequency-tagged responses thereby provide many advantages when testing young infants. First, brain responses are measured at pre-defined frequencies, making the approach highly objective. Second, stimuli are presented at a fast rate, allowing measurable responses in a few minutes of recording. Third, frequency-tagged responses are of high signal-to-noise ratio with relative immunity to artifacts, reducing experimental attrition. This is due to the frequency resolution obtained by analyzing "long" stimulation sequences (e.g., we used 32-second-long sequences in our studies, leading to a resolution of 1/32 = 0.03125 Hz, i.e., one value every 0.03125 Hz). Therefore, noise spreads to broad frequency ranges while the signal is isolated in tiny frequency bins associated with little noise. The approach is thus resistant to high-amplitude artifacts such as those generated by infant movements, which usually lead to the rejection of a large number of trials in standard ERP studies and often result in few reliable data during the short testing duration imposed by infants. Fourth, various measurements, including statistical indices, can be derived from the amplitude at the defined frequencies and surrounding noise, allowing individual analysis of a single response. Finally, with frequency-domain representation of the brain responses, there is no assumption as to their properties (e.g., form, temporal dynamics), increasing the robustness of the approach in case of high variability across individuals and age groups. For all these benefits with a difficult-to-test population such as young infants, we used frequency-tagging EEG as a powerful approach to measure face categorization using the same methodology in infants and adults.

During visual stimulation, odors were concurrently presented as context. Such odor delivery does not need an olfactometer inserting tubes in infants' nostrils, avoiding a considerable challenge to infant compliance, and suits well the 32-second-long stimulation sequences we used. Maternal odor was collected from a t-shirt the mother had worn for three consecutive nights preceding the experiment, to reduce activity-related individual differences. During the days, this t-shirt was stored in a hermetic bag left at ambient temperature. Maternal odor was contrasted to a baseline odor consisting in an unworn t-shirt with the same storage procedure (Fig. 1b left). During testing, the t-shirts were disposed on the infants' upper chest and folded to optimally expose infants to the

[2] Naturalistic stimuli depict the main item of each stimulus embedded in the original background the photography was taken in, without controlling for viewpoints, luminance, contrast, colors, etc. This is opposed to edited (sometimes referred to as standardized) stimuli which are commonly used in visual experiments, where the item is isolated from its background and generally homogenized across the different exemplars to control for inter-stimulus variability.

parts that were in contact with the most odorous areas, i.e., the axillary and breast regions (Fig. 1c left).

To determine whether and how maternal odor selectively shapes face categorization in the infant brain, we conducted a series of four studies (Fig. 1b left). Study 1 investigated the modulation of face-selective neural activity in 4-month-old infants exposed to the maternal vs. baseline odor contexts (Leleu et al., 2020). In Study 2, face images were replaced by pictures of cars to test for the selectivity of the maternal odor effect to face categorization (Rekow, Leleu, et al., 2020b). Study 3 explored whether maternal odor is even able to initiate face categorization for ambiguous nonface stimuli, but resembling faces (i.e., objects that elicit illusory face perception—*face pareidolia*—in adults), which are hard to categorize by 4 months of age (Rekow et al., 2021). In addition, to investigate whether the maternal odor effect progressively changes as face categorization develops, Study 4 evaluated the evolution of the face-selective response from 4 to 12 months and its sensitivity to maternal odor (Rekow et al., in preparation).

Finally, we designed a fifth study to elucidate whether the presence of a body odor still influences face categorization in the *adult* brain (Rekow, Baudouin, Durand, & Leleu, 2022b). We built on previous frequency-tagging EEG studies investigating rapid visual categorization in adults (e.g., Retter & Rossion, 2016) and used a base stimulation rate of 12 Hz rather than 6 Hz, with either faces, cars, or face-like objects inserted every ninth stimulus (at 12 Hz/9 = 1.33 Hz) in dedicated sequences (Fig. 1a right). The sets of images from Studies 1, 2, and 3 were used. Three odor contexts were presented during visual stimulation: baseline (i.e., scentless mineral oil), gasoline (diluted by 1‰ in mineral oil), and body (axillary sweat pooled from 8 donors) odors (Fig. 1b right). Odors were presented by hidden tubes conducting a low-pressured air-flow diffused just under the participants' head rested on a chinrest (Fig. 1c right). Odor intensity was selected following pilot studies that established peri-liminal thresholds and iso-intensity across odors, so that participants remained unaware of the odor diffusion throughout the experiment, as confirmed by a post-testing interview.

## 2.2 Maternal Odor Shapes Face-Selective Activity in the 4-Month-Old Brain

In Study 1 (Leleu et al., 2020), we took advantage of the former frequency-tagging EEG study showing rapid face categorization from natural images in 4- to 6-month-old infants (de Heering & Rossion, 2015) and extended this observation in a younger age group of 4-month-old infants ($N = 18$) with a new set of highly variable face images. A face-selective neural response was recorded at 1 Hz (i.e., rate of face presentation in the stimulation stream) over the occipito-temporal cortex, with a larger amplitude in the right hemisphere. Importantly, this response is focal (i.e., recorded over a restricted location of the scalp) and of low amplitude in the baseline odor context, but it is largely increased by the presence of the mother's body odor (Fig. 2a left). This larger activation is particularly found over a right occipital electrode (O2 according to the 10–10 classification system), where 17 out of 18 infants show a positive odor effect, i.e., an enhanced face-selective response in the maternal vs. baseline odor context. In contrast, the general visual response elicited by the rapid 6-Hz stream of face and nonface images, which results in a medial occipital neural activity, is not affected by

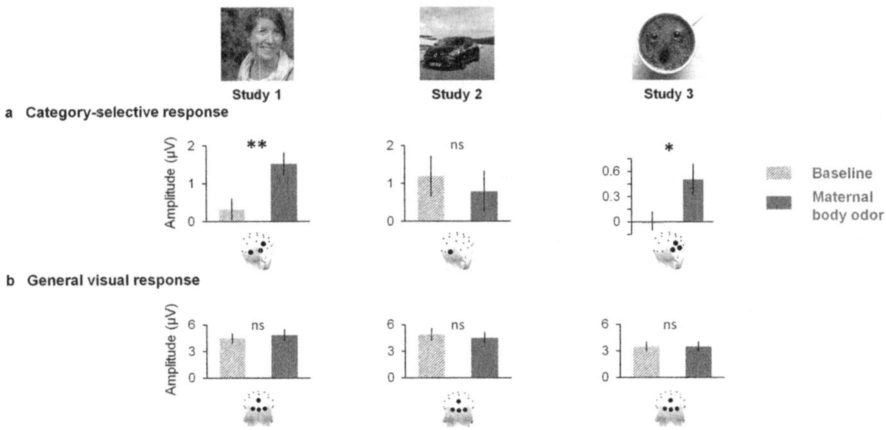


**Fig. 2** Maternal odor effect on visual categorization in the 4-month-old brain. (**a**) Studies 1–3 revealed that exposure to the maternal body odor compared to the baseline odor elicits a larger category-selective neural response to both human faces (*left*) and face-like objects (*right*) over right occipito-temporal brain regions (*$p < 0.05$, **$p < 0.01$), while maternal odor does not influence the categorization of cars (*middle*, ns (nonsignificant): $p > 0.05$). (**b**) The general visual response to the rapid stream of stimuli (i.e., recorded at the base rate of stimulus presentation over the middle occipital region) is immune to the presence of maternal odor in the three studies (ns: $p > 0.05$). Error bars are standard errors of the mean (SEM)

concomitant maternal odor cues (Fig. 2b left), excluding a mere increase in arousal or visual attention that typically leads to larger brain activity in response to periodic visual stimulation. Overall, this first study reveals that face-selective activity in the 4-month-old brain is actively shaped by the presence of a (maternal) body odor, providing strong support for the fact that chemical cues can tune rapid face categorization in young infants.

Following-up on Study 1, an outstanding issue concerns the selectivity of the olfactory–visual interaction. One could indeed argue from these first observations that the temporal stability of the odor stimulation during a 32-second-long rapidly changing visual stimulation would help the infant brain to detect any visual regularity within the fast stream of pictures, so that any periodically presented visual object could be better categorized in the presence of maternal odor. Thus, in Study 2 (Rekow, Leleu, et al., 2020b), we tested this alternative hypothesis.

To that aim, we recorded scalp EEG in another group of 4-month-old infants ($N = 18$), but this time, cars were used as the target category presented at 1 Hz within the rapid 6-Hz stream of natural images. Following the same procedure, we exposed each infant to the two odor contexts while recording their visual brain responses. A relatively weak car-selective response was observed at 1 Hz over the right occipital channel O2 in both odor conditions (Fig. 2a middle), revealing the 4-month-old infant brain's ability to rapidly assign a series of highly variable exemplars of cars to a single category. However, contrary to the face-selective response measured in Study 1, no significant modulation of the neural categorization response to cars was found in the maternal odor context, which was rather slightly diminished. Similarly, the general visual response to the 6-Hz train of visual stimuli was also immune to the presence of maternal odor, reflecting a similar level of visual attention or arousal in both odor contexts (Fig. 2b middle). These findings indicate that the odor effect for faces was not accounted for by the mere temporal stability of the olfactory stimulation and rather suggest a selective tuning of the infant visual system to categorize congruent inputs such as faces.

The two first studies have demonstrated a powerful influence of maternal odor on the brain's ability to rapidly categorize a variety of faces at 4 months of age. Does that mean that maternal chemical cues can even *initiate* neural face categorization within the immature infant visual system? In other words, is maternal odor able to trigger a face-selective neural response to stimuli that are hardly categorized as faces from the sole visual input? To answer this question, we used a special type of stimuli instead of genuine human faces: a variety of nonface objects categorized as faces by adults (i.e., inducing face *pareidolia*), namely face-like objects. To our knowledge, this kind of stimuli resembling faces, but depicting many different nonface categories, had not yet been associated with face-selective neural activity in young infants (see e.g., Kobayashi et al., 2012 for evidence of face-like perception in Arcimboldo paintings at 7–8, but not 5–6, months of age).

In Study 3 (Rekow et al., 2021), we recorded scalp EEG in another group of 4-month-olds ($N = 20$) exposed to the same two odor conditions (maternal vs. baseline). Many variable face-like objects (e.g., a coffee cup where the foam makes out a face as in Fig. 2, fried eggs of which the yolks evoke eyes, etc.) were used as the target category presented at 1 Hz in the fast 6-Hz streams of other nonface objects. Face-like and nonface stimulus sets depicted the same object categories (e.g., a regular version of a coffee cup without any illusory face was used as a nonface stimulus) to control for the natural properties of the object categories (e.g., shape, color). Hence, a face-like-selective neural response would emerge only if exemplars from different categories (e.g., a face-like plug and a face-like coffee cup) elicit similar neural responses according to their face likeness, and, at the same time, if exemplars from the same categories (e.g., face-like and nonface coffee cups) elicit dissimilar neural responses.

In the baseline odor context, we found a weak neural categorization response to face-like stimuli at 1 Hz in the EEG spectrum over bilateral occipito-temporal regions. Critically, this face-like-selective response is amplified and becomes right-lateralized in the maternal odor context (Fig. 2a right). This observation extends our two previous studies, revealing that chemical cues co-occurring with human faces in the infant's experience also shape the categorization of face-like objects as faces in the right hemisphere, which is dominant for this function in humans (Rossion & Lochy, 2021, for review). Importantly, at the individual level, the odor effect is particularly effective in infants with no sign of face-like categorization in the absence of maternal body odor, as they display an odor effect twice as large as the group. In other words, this demonstrates that maternal chemical cues are able to initiate face-selective activity in the infant visual cortex in response to ambiguous stimuli that are hardly categorized as faces based on the sole visual input. In line with our two previous studies, no odor effect was observed on the general response to the fast train of visual stimuli (i.e., the 6-Hz response, Fig. 2b right), ruling out a general effect on visual attention or arousal when exposed to maternal odor.

### 2.3 The Effect of Maternal Odor on Face Categorization Fades during Development

Our first three studies have disclosed the early tuning of face(-like)-selective activity by chemical stimuli in the 4-month-old developing brain, showing a maternal odor effect on face categorization which is broad enough to occur for any visual input interpretable as a face. They have also shown that the maternal odor effect appears particularly effective when categorization is demanding for the sole visual system, as revealed by the individual odor effects in Study 3. Thus, given that these studies were conducted in 4-month-old infants, they also suggest that maternal odor is especially effective in fostering face-selective neural activity at an age when face categorization is not fully developed. This would be in line with our main hypothesis that the developmental heterochrony between the two senses favors olfactory-to-visual modulations (see Sect. 1), the early-maturing olfactory system driving perception in the later-developing visual system. Accordingly, we wondered if maternal odor still exerts such an effect on face categorization later on, as the first year of life is marked by dramatic progress in this function (Pascalis et al., 2020). In other words, is the mother's body odor still useful in channeling the infant visual system to respond to faces as face categorization becomes more efficient by itself?

In Study 4, we addressed this question with a cross-sectional approach (Rekow et al., in preparation). We recorded scalp EEG in 4- to 12-month-old infants ($N = 50$, aged from 121 to 374 days old) tested in the same conditions as in Study 1: infants watched 6-Hz streams of natural images with human faces periodically inserted once per second (at 1 Hz) and were exposed to their maternal odor or to a baseline odor using t-shirts. In a first step, we evaluated the development with age of both the face-selective response recorded at 1 Hz and harmonics and the general visual response recorded at 6 Hz and harmonics. This first set of analyses was conducted in the baseline odor condition to measure the visual response without the effect of the congruent maternal odor. It revealed that the face-selective response increases in amplitude with age (Fig. 3a left) and becomes more complex, as indicated by a higher number of harmonics in the oldest infants compared to the youngest. This qualitative and quantitative change of the response is not found in the case of the general visual response which remains stable in amplitude and complexity over development (Fig. 3b left). In a second step, we analyzed whether maternal odor influences both visual responses (face-selective and general) as a function of age. For the face-selective response, as in Study 1, a positive odor effect was found in the youngest infants, resulting in a stronger amplitude over the right hemisphere in the maternal compared to the baseline odor context (Fig. 3a right). However, this effect linearly declines with development, and the oldest infants do not display any positive influence of maternal odor on face-selective neural activity. In contrast, the general visual response remains immune to the presence of the mother's body odor at any age (Fig. 3b right), as described at 4 months in our previous studies.

Overall, this study demonstrates a direct relation between the strength of the face-selective visual response and the strength of the maternal odor effect. This suggests a developmental trade-off between olfaction and vision, as if the mother's body odor gradually loses its effect over face categorization in infants as faces become more and more readily categorized from the sole visual input. This supports the view that multi-

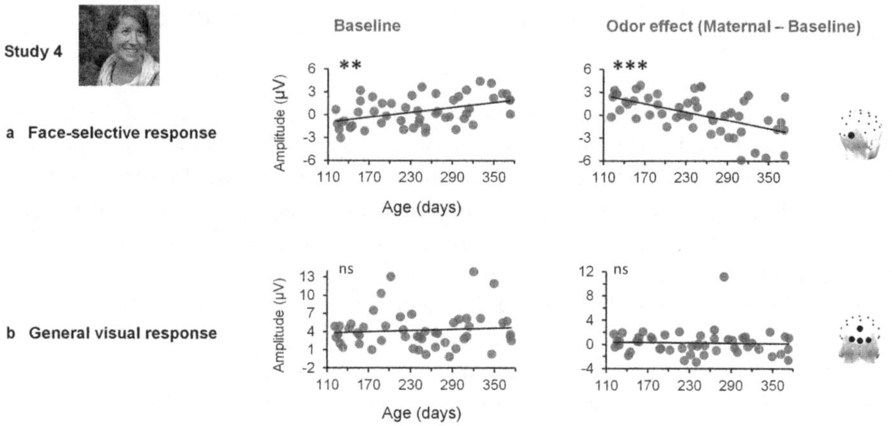


**Fig. 3** Maternal odor effect on face categorization between 4 and 12 months of age. (**a**) The face-selective neural response recorded in Study 4 over the right occipital region in the baseline odor context (without maternal odor) significantly increases (**: $p < 0.01$) between 121 days (≈ 4 months) and 374 days (≈ 12 months) of age (*left*). At the same time, the maternal odor effect (amplitude difference of the face-selective response between the maternal and the baseline odor contexts) significantly decreases (***: $p < 0.001$) over the same region (*right*). (**b**) The general visual response recorded over the middle occipital region does not increase with age and remains immune to the presence of maternal odor at any age (ns: $p > 0.05$). Circles represent individual data points

sensory inputs help perceiving stimuli as coherent early in infancy, before the age when unisensory inputs become sufficient to elicit accurate perception. This result is more generally in line with the *inverse effectiveness* principle of multisensory integration, whereby the strength of the integration between two senses is higher when unisensory responses are scarce on their own (Regenbogen et al., 2016; Stein & Meredith, 1993).

## 2.4 Face Categorization Still Relies on Body Odors Under Certain Circumstances in the Adult Brain

The aforementioned four studies conducted in infants revealed that mother's body odor improves face categorization in the developing brain in the form of an increased face(-like)-selective response. They also indicated that this intersensory facilitation of visual categorization depends on the brain's ability to readily categorize faces from the sole visual input, as we found an inverse relation between the strength of the visual response and the strength of the odor effect, both across individual infants (Study 3) and across ages (Study 4). These latter observations suggest that the inverse effectiveness principle of multisensory integration (Regenbogen et al., 2016; Stein & Meredith, 1993) applies during perceptual development, in line with our main hypothesis assuming that olfaction is particularly prone to support the visual system early in life (see Sect. 1). In this perspective, one could argue that the human typical adult brain does not rely on odors to visually categorize its environment anymore, as visual categorization is then fully developed and extremely efficient.

Indeed, studies investigating visual categorization with frequency-tagging EEG in adults have shown strong category-selective neural responses to different categories despite the tight constraints put on the visual system (i.e., a single glance at variable naturalistic category exemplars contrasted to many other categories; Jacques et al., 2016; Hagen et al., 2020; Rekow, Baudouin, et al., 2020a). For face categorization in particular, a brain response is reliably found, whether stimuli are greyscale (Or et al., 2019), negatively contrasted (Liu-Shuang et al., 2022), low-pass filtered (Quek et al., 2018), or presented for very short durations (Retter et al., 2020). Thus, it was still unknown whether such robust face categorization can be sensitive to the addition of concurrent odor cues. We investigated this question in Study 5 and hypothesized that odors influence face categorization provided the face-selective neural response is not saturated (i.e., at its ceiling level), according to the inverse effectiveness principle. In other words, when the visual input is scarce or ambiguous and therefore difficult to categorize by the adult brain, face categorization would be improved by the addition of congruent body odor cues.

Previous evidence in adults tends to support this hypothesis, as body odors modify our perception and appreciation of conspecifics (e.g., Damon et al., 2021; Spence, 2021, for reviews). For example, in a continuum of faces morphed from neutrality to anger (Mujica-Parodi et al., 2009) or from happiness to fear (W. Zhou & Chen, 2009), a stress odor (i.e., body odors collected during an induced stressful event) reduces the threshold of visual information needed to classify the face as angry (threatening) or fearful (threatened) compared to odors sampled during physical exercise. Likewise, it decreases the proportion of "happy" responses to ambiguous happy faces (i.e., morphed with a neutral face; Zernecke et al., 2011). At the brain level, stress odor also increases an EEG response to ambiguous angry faces and makes it of similar amplitude to the response elicited by full-blown angry expressions (Rubin et al., 2012). Accordingly, Study 5 explored whether body odors act as disambiguating cues for difficult-to-categorize faces in adults.

One way to tackle this issue is to rely on face-like objects that trigger face pareidolia (i.e., the illusory perception of a face), as we did in Study 3 with infants. When various face-like objects are contrasted to similar (i.e., belonging to the same object categories) and equally variable nonface objects in a stream of periodically appearing images, they elicit a lower neural categorization response than human faces (Rekow, Baudouin, Brochard, et al., 2022a). Furthermore, face-like objects elicit individual differences, with some participants spontaneously reporting having seen illusory faces, when others struggle to perceive a face from these ambiguous stimuli. Interestingly, this overt report of face pareidolia is related to the amplitude of the face-like-selective brain response, being larger in "aware" than in "unaware" participants (Rekow, Baudouin, Brochard, et al., 2022a, Experiment 1). We thus chose the face-like category to investigate the body odor effect on face categorization in adults, since face pareidolia is not systematic despite a mature and efficient visual system.

The putative modulation of visual categorization by odors was tested implicitly, i.e., without any instruction and with participants being left unaware of the actual aim of the study. According to earlier studies investigating rapid face categorization in adults (e.g., Retter & Rossion, 2016), images were displayed at 12 Hz and variable exemplars of either human faces, cars, or face-like objects (in dedicated sequences) were interleaved every ninth stimulus to tag category-selective responses at 1.33 Hz and harmonics in the EEG frequency spectrum. We asked 26 participants to detect briefly appearing white crosses that were superimposed on the stimulation stream. Their scalp EEG was recorded while they were blindly exposed to three odor contexts: a composite body odor, a gasoline odor, and a baseline odor (see Sect. 2.1 for details). Since face-like objects were the only ambiguous visual category, we expected to observe that the body odor context enhances the categorization of face-like objects and that this effect depends on the reported awareness of illusory faces at the end of the experiment.

In average, across odor conditions, clear category-selective responses to human faces, cars and face-like objects were measured over the occipito-temporal cortex at 1.33 Hz and harmonics. The largest response was found for human faces, then cars, and the lowest for face-like objects (Fig. 4a). When considering the impact of odor contexts, we observed that odors do not modulate the face- and car-selective responses. In contrast, as compared to the gasoline and baseline odors, the body odor boosts the response to the ambiguous face-like objects over the right occipital brain region (Fig. 4a right). Critically, this increased neural activity in response to face-like objects in the body odor context is systematically found for the 9 out of 26 participants who reported having seen objects resembling faces at the end of the experiment ("aware" participants; Fig. 4b). Finally, the general visual response recorded at 12 Hz and harmonics is immune to the different categories presented at 1.33 Hz, and also the different odor contexts (Fig. 4c), ruling out any general effect of a given visual category, odor, or visual category/odor combination on the attention paid to the rapid visual stimulation. Altogether, these findings disclose a specific influence of congruent odors on the visual categorization of ambiguous stimuli in the adult brain and point toward a relation between this odor effect on visual categorization and visual awareness.

## 3 Discussion

Across the five studies summarized above, we have demonstrated the powerful effect concurrent conspecifics' chemical cues can exert during visual perceptual development, especially the development of face categorization. At 4 months of age, the mother's body odor tunes the visual system of her infant selectively toward face(-like) stimuli, enhancing the neural activity recorded over the right occipito-temporal region, known to be dominant for face categorization. This maternal odor effect then diminishes progressively over the course of the first year and is not found in adulthood when face categorization is fully effective on its own. However, in the case of ambiguous stimuli resembling faces, but not always reported as such by adult observers, exposure to conspecifics' body odor facilitates the perception of a face. Overall, this influence of (maternal) body odor on face categorization reveals that combining socially relevant congruent stimuli from different senses plays a key role in early perceptual development and is still effective later in life in case of ambiguous perception. This indicates the importance of multisensory inputs within an ordered developmental timing, when a late-developing sensory system, such as vision, can find support from an early-maturing system, such as olfaction.

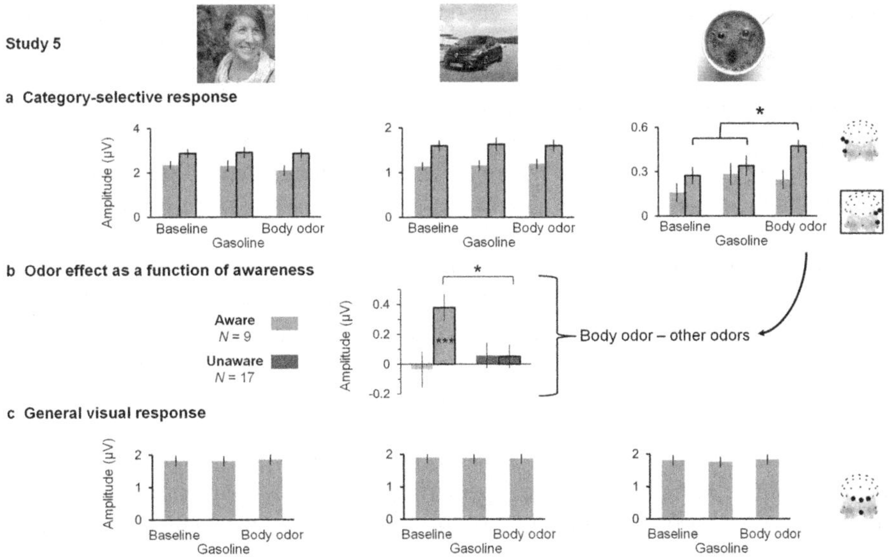


**Fig. 4** Odor effect on visual categorization in the adult brain. (**a**) The amplitude of the category-selective neural response recorded over left and right occipito-temporal regions in Study 5 is large for human faces (*left*), lower for cars (*middle*), and the lowest for face-like objects (*right*). The two former responses are immune to the presence of odor contexts, whereas the face-like-selective response significantly increases over the right hemisphere (*: $p < 0.05$) in the body odor context as compared to the gasoline and baseline odor contexts. (**b**) The body odor effect on the face-like-selective response (amplitude difference between the body and the two other odor contexts) is driven by 9 out of 26 participants who reported having noticed face-like objects after testing ("aware" participants vs. "unaware" participants; *: $p < 0.05$, ***: $p < 0.001$). (**c**) The general visual response recorded over the middle occipital region is insensitive to the different visual categories and the different odor contexts. Error bars are SEM

## 3.1 An Early Perceptual Attunement to Conspecifics' Odor and Face

Our studies have shown that exposure to the mother's body odor selectively shapes the ability to categorize a variety of unknown human faces against many other living and nonliving categories in the 4-month-old infant brain, an effect that is broad enough to occur for other (nonliving) visual inputs interpretable as faces (face-like objects). How is the perception of the mother's body odor able to improve the infant's visual categorization of so diverse and unfamiliar face(-like) stimuli? Perceiving our conspecifics is of vital importance from the very beginning of life. In that endeavor, mammalian infants are first equipped with a sensitive olfactory system tuned to attend specific signals or cues from their mother (e.g., Klaey-Tassone et al., 2020; Schaal et al., 2003; Schaal & Al Aïn, 2014). Human fetuses are already exposed to many odors in the womb (Schaal et al., 2002), some of them being also found in the mother's body secretions after birth (Loos et al., 2014), ensuring transnatal continuity and familiarity (e.g., Porter & Winberg, 1999). Maternal secretions elicit selective physiological, behavioral, and neural responses in neonates (e.g., Aoyama et al., 2010; Marlier et al., 1998; Soussignan et al., 1997) and attenuate their stress reactions (e.g., Rattaz et al., 2005). Maternal odor then benefits from positive reinforcement through nursing and caregiving during the first months, and therefore accompanies the early social interactions between the mother and her infant (Schaal et al., 2020).

In this context, maternal odor stimulates visual arousal at birth (Doucet et al., 2007), thus favoring optimal exposure to the mother's face. Later on, developing infants gradually broaden their social network and are exposed to more and more people, especially to their face during the first year of life (Jayaraman et al., 2015; Rennels & Davis, 2008). Importantly, infants are often held in their mother's arms during social interactions, being simultaneously exposed to her odor and to more or less familiar faces. Exposure to the mother's body odor also reduces the infant's reaction to a fearful face (Jessen, 2020) and appears to promote visual attention toward, and inter-brain synchrony with, unfamiliar conspecifics (Endevelt-Shapira et al., 2021). Therefore, at a perceptual level, sustained co-exposure to both types of stimuli could create, maintain, and reinforce a relevant association between chemical and visual stimuli, through associative learning. Being able to learn regularities in the environment and to combine stimuli according to their repeated co-occurrence is a core mechanism of perceptual development. Such statistical learning (Saffran & Kirkham, 2018, for review) is particularly involved in the development of multisensory perception, as infants are extremely sensitive to the spatiotemporal coincidence of events across the senses (e.g., Lewkowicz, 1996). Interestingly, however, odor perception is relatively stable in space and time compared to other sensory perceptions (Sela & Sobel, 2010). Olfactory–visual interactions could thus operate with less spatiotemporal constraints than other intersensory interactions. Rather, odor cues are ideally suited to sustain the development of visual categorization by facilitating the generalization of variable inputs into a single category thanks to the relative stability of odor percepts.

As a result, by 4 months of age, human infants have already been so intensively co-exposed to the mother's body odor and to a variety of faces that the presence of maternal odor cues favors the categorization of visual inputs as faces, even in common nonface objects configured as faces. Years later, children and adults exhibit great sensitivity to fine-grained social cues conveyed by the body odor of conspecifics (de Groot et al., 2017; Lübke & Pause, 2015; Weisfeld et al., 2003) and associations between olfactory and visual information refine. Their combination facilitates the perception of congruent information in adults, such as recognizing the fearful or angry emotion expressed by a face when exposed to a stress odor, especially when the facial expression is ambiguous (Mujica-Parodi et al., 2009; W. Zhou & Chen, 2009). We have shown that this remains also effective for the generic ability to categorize a visual stimulus as a face, since a human body odor still improves the categorization of ambiguous face-like objects in adults. In that case, body odor cues could signal the potential presence of a person (Cecchetto et al., 2020), and even trigger visual imagery (Parma et al., 2017, for review), orienting more systematically toward the "face" interpretation of the visual input.

### 3.2 Putative Mechanisms Subtending Odor-Driven Priming of the Face Category

A main observation in our studies is the fact that (maternal) body odor does neither merely stimulate visual arousal/attention (no effect on the general response to the rapid visual stimulation stream) nor facilitate the categorization of any stable visual object (e.g., periodically presented cars), but selectively enhances the neural categorical response to face(-like) stimuli. This indicates an intersensory association relying on the congruency between simultaneous olfactory and visual inputs, like a priming effect of body odor on face categorization, as previously described for other congruent olfactory–visual pairings (e.g., Seigneuric et al., 2010; W. Zhou et al., 2010). This is more generally in line with the view that sensory cues channeled by different senses are combined when they complete each other (Bahrick & Lickliter, 2012; Ernst & Bülthoff, 2004). Interestingly, we did not find odor effects for incongruent associations (i.e., body odor effect for cars and gasoline effect for face(-like) stimuli), contrary to previous reports of odor-driven interference on visual categorization (e.g., Hörberg et al., 2020). Since interference is often observed for an *explicit* decision and the neural activity related to this controlled behavior, the automatic visual categorization we measured in our studies without any explicit behavior may be immune to odor interference. Alternatively, since category-selective responses reflect the direct contrast between the neural activities elicited by the target category and other visual categories, and since incongruency with the target category also means incongruency with the other categories (e.g., a body odor presented during a car categorization sequence is incongruent with both cars and other nonface objects), incongruent associations may cancel each other. Accordingly, it is also worth reminding that category-selective responses include both between-category discrimination and within-category generalization. Hence, regarding the face-selective response, exposure to body odor may improve both the ability to discriminate faces from other visual categories because it is congruent only with faces and to generalize variable unfamiliar faces into a single category because they share the congruent association with the odor. Moreover, as incongruency with body odor is also shared across the different nonface categories against which faces are contrasted, it may further improve their discrimination from faces.

At the brain level, the repeated co-occurrence of body odors and faces in the social environment could shape a dedicated connectivity between the olfactory and visual systems, so that a body odor would become able to mediate neural activity in the face-selective network and enhance its responsiveness when a face(-like) stimulus appears. This is in line with the reentrant signaling hypothesis in the theory of neuronal group selection (e.g., Edelman, 1993). Accordingly, systematic co-exposure to different sensory inputs reinforces bidirectional connections between distinct neuronal groups responding to each input. Experience therefore progressively shapes connections within a large-scale and broad connectivity established during early development, some connections disappearing due to scarce or random co-activation, while others strengthen from repeated and reinforced co-activation.

The existence of such an underlying connected network is supported by a convergent line of evidence demonstrating odor-induced neural activity in brain areas typically considered as visual. For instance, activity in the adult right occipital cortex has been consistently reported during the presentation of odors without visual stimuli (Djordjevic et al., 2005; Royet et al., 2001; Zatorre et al., 2000). More specifically, the sole presentation of body odors activates the lateral fusiform gyrus, a well-known face-selective region of the ventral occipito-temporal cortex (Prehn-Kristensen et al., 2009; Zheng et al., 2018; W. Zhou & Chen, 2008), which is also a downstream recipient of the primary olfactory cortex as revealed by functional connectivity analysis (G. Zhou et al., 2019). More generally, this concurs with large-scale connectivity between distant and typically unisensory brain regions dedicated to the same semantic domain (Mahon & Caramazza, 2011). Recent studies in congenitally blind people (Mattioni et al., 2020) and early-deaf individuals (Benetti et al., 2021) illustrate how such circuitry shapes the activity of a deprived sensory system. For example, category-selective regions of the ventral occipito-temporal cortex are organized in the same fashion in blind people stimulated with auditory inputs as in sighted individuals stimulated with visual inputs (Mattioni et al., 2020). Hence, it is tempting to speculate that this cross modal plasticity reflects how perceptual categorization organizes during development, category-selective regions within a given sensory system receiving direct inputs from the other senses that constrain their progressive tuning toward their preferred category.

Another recurrent observation in our studies pertains to hemispheric asymmetry. Both face-selective (in infants) and face-like-selective (in infants and adults) neural responses increase only over the right hemisphere in the presence of (maternal) body odor. This indicates that adding the odor may strengthen the well-known right lateralization of face categorization (Grill-Spector et al., 2017; Hagen et al., 2020; Jonas et al., 2016), even in the infant brain for which the advantage of the right hemisphere is disputed (see Behrmann & Plaut, 2020; Rossion & Lochy, 2021). Right-hemispheric dominance has been related to single-glance perception of the face as a whole, especially for face-like stimuli which are configured as faces without genuine facial features (Caharel et al., 2013; Rossion et al., 2011). Thus, body odor could trigger the perception of a whole face pattern from a single glance at a wide variety of visual stimuli as soon as they are configured as a face. To a larger extent, since the right hemisphere has been suggested to be particularly involved in identifying the odor source, at least in adults (Brand et al., 2001; Royet, 2004), it might be a "hotspot" for integrating social information across the senses with facilitative effects on their categorization. Finally, our findings are also in line with increased brain-to-brain synchrony in the right occipito-temporal region of 7-month-old infants exposed to their mother's body odor while interacting with unfamiliar women (Endevelt-Shapira et al., 2021), suggesting that maternal odors can promote various prosocial functions, including the right-hemispheric ability to categorize conspecifics from olfactory or visual cues.

### 3.3 Implications for the Development of (Multi)Sensory Perception

Collectively, our studies support our initial hypothesis that olfaction is particularly prone to sustain the development of visual categorization because it is an early-maturing sense compared to vision (Lecanuet & Schaal, 1996), in line with the view that the ordered sequential ontogeny across sensory systems actively contributes to perceptual development (Turkewitz & Devenny, 1993). Indeed, from birth onward, neonates respond to a variety of chemical cues and signals emanating from their mother (e.g., Porter & Winberg, 1999; Schaal, 1988), while visual perception is limited and follows a protracted development over the first year (Braddick & Atkinson, 2011). The mere ability to categorize a visual input as a face is already difficult for young infants, and it takes time to develop. For example, human infants need about 6 months to readily detect faces in naturalistic scenes (e.g., Kelly et al., 2019; see Leppänen, 2016, for review), and the face-selective response tagged in EEG in the 4- to 6-month-old brain is largely immature compared to the response measured in children and adults (de Heering & Rossion, 2015; Lochy et al., 2019; Rossion et al., 2015, respectively). In Study 4, we also informed this neural development, revealing a dramatic quantitative and qualitative evolution of the response between 4 and 12 months. As a result, the maternal odor effect we observed only in the youngest infants[3] appears to be particularly effective when face categorization is not fully developed.

Such immature face categorization means that the face category is broadly defined in the 4-month-old infant brain, and face(-like) stimuli are still difficult to disambiguate, leading to a lack of sensitivity (some face(-like) stimuli are missed) and of specificity (some nonface stimuli are erroneously categorized as faces). The same applies to adults exposed to face-like objects within a fast train of other objects belonging to the same categories, who are inclined to miss some face-like objects and erroneously report faces in nonface objects (Rekow, Baudouin, Brochard, et al., 2022a, Experiment 2). Therefore, (maternal) body odor can be seen as a disambiguating cue that reduces sensory noise induced by the physical variability of the environment, a key function of multisensory integration (Ernst & Bülthoff, 2004), which has already been reported on several occasions (e.g., for odor effects on facial expression perception; Leleu et al., 2015; Mujica-Parodi et al., 2009; Poncet et al., 2021; W. Zhou & Chen, 2009). This would explain why no odor effect was found for human faces in adults, faces being already easily categorized without the need of another sensory cue (such as a body odor) and under tight visual constraints (e.g., Retter et al., 2020). Likewise, over the

[3] It should be also noted that with frequency-tagging EEG, infants are tested under the same conditions and using the same stimuli as for adults, making infants' data comparable to that of adults but increasing categorization difficulty at such an early stage of development. The maternal odor effect could thus have been also favored by this demanding paradigm for their immature visual system. More generally, one may question the ecological validity of our paradigm since infants rarely see faces so rapidly in their everyday experience. However, while slower modes of visual display better match the natural timing of exposure to a conspecifics' face, they are rather limited in terms of measurement validity, especially for visual categorization, as they often rely on a limited set of edited visual stimuli depicting only a few visual categories (see Sect. 2.1 for a discussion of this issue; see also Kominsky et al., 2022, for a recent discussion about the difficult balance between ecological and measurement validity in infant research). Moreover, as mentioned above, we aimed to investigate the development of face categorization using adult performance as reference. Neurotypical adults reach a *plateau* for categorizing natural views of faces after only a short exposure of about 80 ms ($\approx$ 12 Hz of fast periodic visual stimulation; see e.g., Retter et al., 2020). Hence, despite its limitations in ecological terms, our paradigm is well designed to measure the development of adult-like rapid face categorization in the infant brain. Admittedly, future studies should investigate odor-driven visual categorization in other experimental and/or natural designs to further improve the generalizability of our findings.

first year of life, the odor effect gradually decreases as face perceptual abilities increase, suggesting that maternal odor becomes less relevant for the visual system.

In sum, the odor effect is not observed when face categorization is sufficiently developed to be readily achieved through the sole visual input. This relates to the inverse effectiveness principle of multisensory integration, first described at the level of single neurons in animals (Stein & Meredith, 1993), and further noted in behavioral data of human adults (Regenbogen et al., 2016). This principle states that as the strength of unisensory responses increases, the strength of multisensory integration decreases, following a reverse function. In other words, cues from the different senses compensate one another, helping disambiguate inputs to reach optimal perception, unless unisensory inputs are already sufficient to elicit a robust percept. In our work, we repeatedly found such an inverse relation between the odor effect and the strength of the sole visual response, either for individual infants (Study 3), as a function of age (Study 4), or depending on the ambiguity of the visual category (Study 5). In a recent study, we have further addressed the inverse effectiveness of maternal odor on human face categorization at 4 months (Kiseleva et al., in preparation). In a first group of 4-month-old infants, we have replicated our initial observation that maternal odor increases a right occipito-temporal neural response to natural face images (Study 1). In a second group, we replaced the variable naturalistic stimuli by homogenous edited stimuli (i.e., stimuli segmented from their background and always depicted centered and from a full-front view; see footnote 2) to increase both similarity across individual faces and dissimilarity between face and nonface stimuli, thus making rapid face categorization less demanding. As expected, we found that the face-selective EEG response to edited stimuli is larger and more complex (with more harmonics) than to naturalistic stimuli, indicating more efficient categorization. Critically, the response is equivalent in both odor contexts, suggesting that the maternal odor effect has been concealed by an effective categorization from the sole visual input.

From a broad developmental perspective, our findings orient toward a multisensory shaping of knowledge, where inputs from the different senses are merged together into unified percepts, providing the building blocks for the development of semantic memory, language, and concepts (e.g., Goldstone & Barsalou, 1998; Murphy, 2002). In this context, we propose that the development of perception, and by extension, memories and concepts, relies on the ability to disambiguate a rich and complex environment by recognizing invariant categories, the combination of information across the senses actively participating in the progressive acquisition of such categories. Later on, as unisensory perception becomes effective on its own, the other senses would be useful when inputs are ambiguous. In other words, perception would function through multisensory integration throughout the life course, but the latter would become evident only when one sense has difficulties to categorize an input by itself. This multisensory account of perception is in line with the observation that many brain regions usually considered as unisensory can respond to stimuli from other sensory modalities (Driver & Noesselt, 2008; Ghazanfar & Schroeder, 2006). It is worth noting that the ability to apprehend and use multisensory inputs along development has long been a matter of debate. On the one hand, the "integration" view suggests that unisensory categories develop first before infants become gradually able to integrate them into multisensory objects (Birch & Lefford, 1963; Piaget, 1952). On the other hand, the "differentiation" view argues that infants build multisensory categories first, before being able to differentiate between unisensory percepts (Gibson, 1969). More recently, intermediate views have been formulated, suggesting that infants are already able to form multisensory percepts early in life and that multisensory integration further continues to develop (e.g., Bahrick & Lickliter, 2012; Lewkowicz, 2010). The present set of studies favors such a developmental lifespan perspective of multisensory perception and suggests to consider how unisensory responses are effective on their own to reveal interactions across the senses.

### 3.4 Conclusions and Perspectives

In this chapter, we have reviewed five of our studies showing that a human body odor is able to foster neural activity in face-selective brain regions in response to any stimulus configured as a face, provided that this stimulus is not effectively categorized by the sole visual system (i.e., due to a lack of maturity or to a scarce or ambiguous stimulus). This suggests that odor acts as a disambiguating cue, which reduces the impact of sensory variance and favors visual discrimination and generalization, following an inverse relation with the strength of the sole visual response (inverse effectiveness). At the neural level, this is probably subserved by reentrant signaling between olfactory and visual areas through a large-scale connectivity that is shaped during development by the repeated co-exposure to (maternal) body odor and faces. This accumulated experience with co-occurring olfactory and visual stimuli would create an associative learning, such that body odor and faces would become mutually congruent. Accordingly, future studies should test whether faces modulate body odor perception to elucidate how the two sensory systems support each other. Given that olfaction matures earlier than vision, we speculate that visual-to-olfactory influence is not functional early in life, but could certainly gradually appear as the visual system develops. Likewise, it would be interesting to explore whether face categorization is already odor-driven in younger infants. Since maternal odor induces eye opening in neonates (Doucet et al., 2007), maternal odor could trigger visual attention indiscriminately, an effect that could be captured by an enhanced general response to any fast train of visual stimuli in frequency-tagging EEG.

Which odor cues carried by maternal body odor contribute to the odor effects should also be disentangled in future studies. Maternal odor is a complex mixture of compounds pertaining to familiarity/identity, maternity, femininity, humanity, etc., which partially overlap across body regions and conspecifics. Human infants are able to differentially respond to these cues. They discriminate their own from another mother's odor, with odors stemming from different body areas across studies (Marin et al., 2015; Russell, 1976). They also differentiate lactating from nonlactating women (Makin & Porter, 1989), human from nonhuman milk (Marlier & Schaal, 2005), or early lactation milk from late lactation milk (Klaey-Tassone et al., 2020). However, there is some evidence that infants' looking behavior toward faces is insensitive to the familiarity of maternal odor or to lactating status (Durand et al., 2020). Thus, we can only speculate that some cues, combined or in isolation, are more potent than others (how effective is the father's odor or the odor of a stranger nulliparous woman?). In adults, we used a composite body odor pooled from males and nulliparous females. Again, a wealth of information related to a variety of states (e.g., emotions) or traits (e.g., identity) emanates from such biological odors and are detected by human adults (de Groot et al., 2017, for review), but we do not know which cues are necessary/sufficient for the odor to be effective on visual perception. At this stage, we can merely speculate that some cues signal that the odor belongs to a human being. Future studies should thus be designed to properly delineate which odor cues sustain the visual system in categorizing faces. More generally, it would be worth determining whether nonbody odor cues repeatedly associated with conspecifics during social interactions are also able to shape face categorization. For instance, in our studies, we were only interested in the effect of body odor and asked the mothers to not use cosmetics during odor sampling, but cosmetics can be part of maternal odor in the infants' everyday experience. Future studies should thus investigate whether the odor of cosmetics learned during mother–infant interactions facilitates the infants' ability to categorize faces. If confirmed, such idiosyncratic olfactory–visual integration would further support our interpretation that the individual experiences of co-occurring sensory inputs drive the development of multisensory perception.

Relatedly, one may also wonder whether our findings can be generalized to any congruent association between an odor and a visual category. According to the title of this chapter, we suggest that odor-driven visual categorization is not restricted to social stimuli but extends to other congruent combinations. Actually, we first used maternal odor and faces in 4-month-old infants because they are arguably the most relevant inputs encountered by infants (Jayaraman et al., 2015; Schaal et al., 2020; Sugden et al., 2014), making them the best candidates to reveal potent odor effects on visual categorization at such early age. Furthermore, there is no other visual category that is familiar enough to infants to be tested with a large and variable stimulus set (see e.g., Werchan et al., 2018 for an intersensory influence in the infant brain with nonsocial simple stimuli). Unlike infants, adults have become familiar with many odors and visual categories, and a large body of research has described nonsocial odor effects on visual perception (e.g., Seigneuric et al., 2010; W. Zhou et al., 2010). Hence, in future studies, it would be worth to evaluate, for instance, whether the gasoline odor affects car categorization by using degraded visual stimuli or a challenging mode of stimulus presentation that both hamper visual categorization in adults. This would also further assess the inverse effectiveness of odor cues on vision.

Another outstanding issue stems from the relation between odor-driven face(-like)-selective responses and participants' perceptual awareness, i.e., their subjective experience of a face. In adults, it is remarkable that the odor effect is stronger for participants who are "aware" of the face-like stimuli (who reported face pareidolia at the end of the experiment). This suggests that body odor, by priming face-selective areas to respond to face(-like) stimuli, could facilitate the face percept "popping out" of the ambiguous stimulus. This is in line with a strong positive relation between the magnitude of frequency-tagged category-selective responses and participants' awareness of the stimulus (Rekow, Baudouin, Brochard, et al., 2022a; Retter et al., 2020). However, given that aware participants were identified post-experimentally, an alternative interpretation could be that they are more sensitive to face pareidolia, in general, and then more prone to associate body odor and face-like perception. Several elements in our data point toward the first interpretation (e.g., the face-like selective response is of similar amplitude in both aware and unaware participants without body odor), but future studies should directly manipulate awareness in a single group of participants to reach a definitive conclusion. If confirmed, such odor-driven visual awareness would broadly participate in rehabilitating the function of the olfactory system in human cognition. In this context, our findings suggest that the long-granted alerting function of olfaction toward environmental hazards (Herrick, 1933) is not simply a primitive mechanism of survival, but an integral part of an evolved system generating knowledge, and even consciousness. More work is still needed to elaborate on this issue, to shed full light on how chemical cues and signals participate in high-level cognition in human beings.

**Acknowledgments** The studies presented in this chapter have received financial support from the Centre National de la Recherche Scientifique (CNRS), the "Conseil Régional Bourgogne Franche-Comté" (PARI grant), the FEDER (European Funding for Regional Economic Development), the French 'Investissements d'Avenir' program, project ISITE-BFC (contract ANR-15-IDEX-03) and the French National Research Agency (contract ANR-19-CE28-0009). The authors wish to thank the co-authors of the original papers of the studies described here (in alphabetical order: J.-Y. Baudouin, F. Damon, K. Durand, F. Poncet, B. Rossion & B. Schaal), along with Sylviane Martin for her help in recruiting participants and Vincent Gigot for his assistance in preprocessing the EEG signals.

## References

Adrian, E. D., & Matthews, B. H. C. (1934). The Berger rhythm: Potential changes from the occipital lobes in man. *Brain, 57*(4), 355–385. https://doi.org/10.1093/brain/57.4.355

Akcan, E., & Polat, S. (2016). Comparative effect of the smells of amniotic fluid, breast milk, and lavender on newborns' pain during heel lance. *Breastfeeding Medicine, 11*(6), 309–314. https://doi.org/10.1089/bfm.2015.0174

Amso, D., Haas, S., & Markant, J. (2014). An eye tracking investigation of developmental change in bottom-up attention orienting to faces in cluttered natural scenes. *PLoS ONE, 9*(1). https://doi.org/10.1371/journal.pone.0085701

Aoyama, S., Toshima, T., Saito, Y., Konishi, N., Motoshige, Ishikawa, N., Nakamura, K., & Kobayashi, M. (2010). Maternal breast milk odour induces frontal lobe activation in neonates: A NIRS study. *Early Human Development, 86*(9), 541–545. https://doi.org/10.1016/j.earlhumdev.2010.07.003

Bahrick, L. E., & Lickliter, R. (2012). The role of intersensory redundancy in early perceptual, cognitive, and social development. In A. Bremner, D. J. Lewkowicz, & C. Spence (Eds.), *Multisensory development* (pp. 183–205). Oxford University Press. https://doi.org/10.1093/acprof:oso/9780199586059.003.0008

Bartocci, M., Winberg, J., Ruggiero, C., Bergqvist, L. L., Serra, G., & Lagercrantz, H. (2000). Activation of olfactory cortex in newborn infants after odor stimulation: A functional near-infrared spectroscopy study. *Pediatric Research, 48*(1), 1. https://doi.org/10.1203/00006450-200007000-00006

Behrmann, M., & Plaut, D. C. (2020). Hemispheric organization for visual object recognition: A theoretical account and empirical evidence. *Perception, 030100661989904*. https://doi.org/10.1177/0301006619899049

Benetti, S., Zonca, J., Ferrari, A., Rezk, M., Rabini, G., & Collignon, O. (2021). Visual motion processing recruits regions selective for auditory motion in early deaf individuals. *NeuroImage, 230*, 117816. https://doi.org/10.1016/j.neuroimage.2021.117816

Birch, H. G., & Lefford, A. (1963). Intersensory development in children. *Monographs of the Society for Research in Child Development, 28*(5), 1–48. https://doi.org/10.2307/1165681

Braddick, O., & Atkinson, J. (2011). Development of human visual function. *Vision Research, 51*(13), 1588–1609. https://doi.org/10.1016/j.visres.2011.02.018

Brand, G., Millot, J.-L., & Henquell, D. (2001). Complexity of olfactory lateralization processes revealed by functional imaging: A review. *Neuroscience and Biobehavioral Reviews, 25*(2), 159–166. https://doi.org/10.1016/S0149-7634(01)00005-7

Broca, M. P. (1879). Recherches sur les centres olfactifs. *Revue d'Anthropologie, 2*(385).

Browne, J. V. (2008). Chemosensory development in the fetus and newborn. *Newborn and Infant Nursing Reviews, 8*(4), 180–186. https://doi.org/10.1053/j.nainr.2008.10.009

Buiatti, M., Di Giorgio, E., Piazza, M., Polloni, C., Menna, G., Taddei, F., Baldo, E., & Vallortigara, G. (2019). Cortical route for facelike pattern processing in human newborns. *Proceedings of the National Academy of Sciences, 116*(10), 4625–4630. https://doi.org/10.1073/pnas.1812419116

Caharel, S., Leleu, A., Bernard, C., Viggiano, M.-P., Lalonde, R., & Rebaï, M. (2013). Early holistic face-like processing of Arcimboldo paintings in the right occipito-temporal cortex: Evidence from the N170 ERP component. *International Journal of Psychophysiology, 90*(2), 157–164. https://doi.org/10.1016/j.ijpsycho.2013.06.024

Cecchetto, C., Fischmeister, F. P. S., Gorkiewicz, S., Schuehly, W., Bagga, D., Parma, V., & Schöpf, V. (2020). Human body odor increases familiarity for faces during encoding-retrieval task. *Human Brain Mapping*. https://doi.org/10.1002/hbm.24920

Cernoch, J. M., & Porter, R. H. (1985). Recognition of maternal axillary odors by infants. *Child Development, 56*(6), 1593–1598. JSTOR. https://doi.org/10.2307/1130478

Conte, S., Richards, J. E., Guy, M. W., Xie, W., & Roberts, J. E. (2020). Face-sensitive brain responses in the first year of life. *NeuroImage, 116602*. https://doi.org/10.1016/j.neuroimage.2020.116602

Damon, F., Mezrai, N., Magnier, L., Leleu, A., Durand, K., & Schaal, B. (2021). Olfaction in the multisensory processing of faces: A narrative review of the influence of human body odors. *Frontiers in Psychology, 12*, 750944. https://doi.org/10.3389/fpsyg.2021.750944

Darwin, C. (1871). *The descent of man and selection in relation to sex*. John Murray.

de Groot, J. H. B., Semin, G. R., & Smeets, M. A. M. (2017). On the communicative function of body odors: A theoretical integration and review. *Perspectives on Psychological Science, 12*(2), 306–324. https://doi.org/10.1177/1745691616676599

de Haan, M., Johnson, M. H., & Halit, H. (2003). Development of face-sensitive event-related potentials during infancy: A review. *International Journal of Psychophysiology, 51*(1), 45–58. https://doi.org/10.1016/S0167-8760(03)00152-1

de Haan, M., & Nelson, C. A. (1999). Brain activity differentiates face and object processing in 6-month-old infants. *Developmental Psychology, 35*(4), 1113–1121. https://doi.org/10.1037/0012-1649.35.4.1113

de Heering, A., Goffaux, V., Dollion, N., Godard, O., Durand, K., & Baudouin, J.-Y. (2016). Three-month-old infants' sensitivity to horizontal information within faces: Infant preference for face horizontal information. *Developmental Psychobiology, 58*(4), 536–542. https://doi.org/10.1002/dev.21396

de Heering, A., & Rossion, B. (2015). Rapid categorization of natural face images in the infant right hemisphere. *eLife, 4*, 14. https://doi.org/10.7554/eLife.06564

Demattè, M. L., Sanabria, D., & Spence, C. (2006). Cross-modal associations between odors and colors. *Chemical Senses, 31*(6), 531–538. https://doi.org/10.1093/chemse/bjj057

Di Giorgio, E., Leo, I., Pascalis, O., & Simion, F. (2012). Is the face-perception system human-specific at birth? *Developmental Psychology, 48*(4), 1083–1090. https://doi.org/10.1037/a0026521

Djordjevic, J., Zatorre, R. J., Petrides, M., Boyle, J. A., & Jones-Gotman, M. (2005). Functional neuroimaging of odor imagery. *NeuroImage, 24*(3), 791–801. https://doi.org/10.1016/j.neuroimage.2004.09.035

Doucet, S., Soussignan, R., Sagot, P., & Schaal, B. (2007). The '"smellscape"' of mother's breast: Effects of odor masking and selective unmasking on neonatal arousal, oral, and visual responses. *Developmental Psychobiology*. Wiley, *49*, 129–138. https://doi.org/10.1002/dev

Doucet, S., Soussignan, R., Sagot, P., & Schaal, B. (2009). The secretion of areolar (Montgomery's) glands from lactating women elicits selective, unconditional responses in neonates. *PLoS ONE, 4*(10), e7579. https://doi.org/10.1371/journal.pone.0007579

Driver, J., & Noesselt, T. (2008). Multisensory interplay reveals crossmodal influences on 'sensory-specific' brain regions, neural responses, and judgments. *Neuron, 57*(1), 11–23. https://doi.org/10.1016/j.neuron.2007.12.013

Durand, K., Baudouin, J.-Y., Lewkowicz, D. J., Goubet, N., & Schaal, B. (2013). Eye-catching odors: Olfaction elicits sustained gazing to faces and eyes in 4-month-old infants. *PLoS ONE, 8*(8), 1–8. https://doi.org/10.1371/journal.pone.0070677

Durand, K., Schaal, B., Goubet, N., Lewkowicz, D. J., & Baudouin, J.-Y. (2020). Does any mother's body odor stimulate interest in mother's face in 4-month-old infants? *Infancy, 25*(1). https://doi.org/10.1111/infa.12322

Edelman, G. M. (1993). Neural darwinism: Selection and reentrant signaling in higher brain function. *Neuron, 10*(2), 115–125. https://doi.org/10.1016/0896-6273(93)90304-A

Eimas, P. D., & Quinn, P. C. (1994). Studies on the formation of perceptually based basic-level categories in young infants. *Child Development, 65*(3), 903–917. https://doi.org/10.1111/j.1467-8624.1994.tb00792.x

Elsner, B., Jeschonek, S., & Pauen, S. (2013). Event-related potentials for 7-month-olds' processing of animals and furniture items. *Developmental Cognitive Neuroscience, 3*, 53–60. https://doi.org/10.1016/j.dcn.2012.09.002

Endevelt-Shapira, Y., Djalovski, A., Dumas, G., & Feldman, R. (2021). Maternal chemosignals enhance infant-adult brain-to-brain synchrony. *Science Advances, 7*(50), eabg6867. https://doi.org/10.1126/sciadv.abg6867

Ernst, M. O., & Bülthoff, H. H. (2004). Merging the senses into a robust percept. *Trends in Cognitive Sciences, 8*(4), 162–169. https://doi.org/10.1016/j.tics.2004.02.002

Faas, A. E., Spontón, E. D., Moya, P. R., & Molina, J. C. (2000). Differential responsiveness to alcohol odor in human neonates: Effects of maternal consumption during gestation. *Alcohol, 22*(1), 7–17. https://doi.org/10.1016/S0741-8329(00)00103-8

Frank, M. C., Amso, D., & Johnson, S. P. (2014). Visual search and attention to faces during early infancy. *Journal of Experimental Child Psychology, 118*, 13–26. https://doi.org/10.1016/j.jecp.2013.08.012

Frie, J., Bartocci, M., & Kuhn, P. (2019). Neonatal cortical perceptions of maternal breast odours: A fNIRS study. *Acta Paediatrica*, 1–8. https://doi.org/10.1111/apa.15114

Gellrich, J., Breuer, A. S., Han, P., Güdücü, C., Hummel, T., & Schriever, V. A. (2021). Central nervous system processing of floral odor and mother's milk odor in infants. *Chemical Senses, 46*, bjab024. https://doi.org/10.1093/chemse/bjab024

Ghazanfar, A., & Schroeder, C. (2006). Is neocortex essentially multisensory? *Trends in Cognitive Sciences, 10*(6), 278–285. https://doi.org/10.1016/j.tics.2006.04.008

Gibson, E. J. (1969). *Principles of perceptual learning and development*. Appleton-Century-Crofts.

Gliga, T., & Dehaene-Lambertz, G. (2007). Development of a view-invariant representation of the human head. *Cognition, 102*(2), 261–288. https://doi.org/10.1016/j.cognition.2006.01.004

Gliga, T., Elsabbagh, M., Andravizou, A., & Johnson, M. H. (2009). Faces attract infants' attention in complex displays. *Infancy, 14*(5), 550–562. https://doi.org/10.1080/15250000903144199

Goldstone, R. L., & Barsalou, L. W. (1998). Reuniting perception and conception. *Cognition, 65*(2), 231–262.

Graven, S. N., & Browne, J. V. (2008). Visual development in the human fetus, infant, and young child. *Newborn and Infant Nursing Reviews, 8*(4), 194–201. https://doi.org/10.1053/j.nainr.2008.10.011

Grill-Spector, K., Weiner, K. S., Kay, K., & Gomez, J. (2017). The functional neuroanatomy of human face perception. *Annual Review of Vision Science, 3*(1), 167–196. https://doi.org/10.1146/annurev-vision-102016-061214

Grossmann, T., Gliga, T., Johnson, M. H., & Mareschal, D. (2009). The neural basis of perceptual category learning in human infants. *Journal of Cognitive Neuroscience, 21*(12), 2276–2286. https://doi.org/10.1162/jocn.2009.21188

Hagen, S., Jacques, C., Maillard, L., Colnat-Coulbois, S., Rossion, B., & Jonas, J. (2020). Spatially dissociated intracerebral maps for face- and house-selective activity in the human ventral occipito-temporal cortex. *Cerebral Cortex*. https://doi.org/10.1093/cercor/bhaa022

Halit, H., Csibra, G., Volein, Á., & Johnson, M. H. (2004). Face-sensitive cortical processing in early infancy. *The Journal of Child Psychology and Psychiatry and Allied Disciplines, 45*(7), 1228–1234. https://doi.org/10.1111/j.1469-7610.2004.00321.x

Harris, L., Atkinson, J., & Braddick, O. (1976). Visual contrast sensitivity of a 6-month-old infant measured by the evoked potential. *Nature, 264*(5586), 570–571. https://doi.org/10.1038/264570a0

Hartmann, C., Doucet, S., Niclass, Y., Dittrich, R., Cupisti, S., Schaal, B., Buettner, A., & Starkenmann, C. (2012). Human sweat odour conjugates in human milk, colostrum and amniotic fluid. *Food Chemistry, 135*(1), 228–233. https://doi.org/10.1016/j.foodchem.2012.04.077

Hepper, P. G., Wells, D. L., Dornan, J. C., & Lynch, C. (2013). Long-term flavor recognition in humans with prenatal garlic experience. *Developmental Psychobiology, 55*(5), 568–574. https://doi.org/10.1002/dev.21059

Herrick, C. J. (1933). Morphogenesis of the brain. *Journal of Morphology, 54*(2), 233–258. https://doi.org/10.1002/jmor.1050540202

Hoehl, S. (2016). The development of category specificity in infancy—what can we learn from electrophysiology? *Neuropsychologia, 83*. https://doi.org/10.1016/j.neuropsychologia.2015.08.021

Hörberg, T., Larsson, M., Ekström, I., Sandöy, C., Lundén, P., & Olofsson, J. K. (2020). Olfactory influences on visual categorization: Behavioral and ERP evidence. *Cerebral Cortex*. https://doi.org/10.1093/cercor/bhaa050

Hutmacher, F. (2019). Why is there so much more research on vision than on any other sensory modality? *Frontiers in Psychology, 10*. https://doi.org/10.3389/fpsyg.2019.02246

Jacques, C., Retter, T. L., & Rossion, B. (2016). A single glance at natural face images generate larger and qualitatively different category-selective spatio-temporal signatures than other ecologically-relevant categories in the human brain. *NeuroImage, 137*, 21–33. https://doi.org/10.1016/j.neuroimage.2016.04.045

Jayaraman, S., Fausey, C. M., & Smith, L. B. (2015). The faces in infant-perspective scenes change over the first year of life. *PLoS ONE, 10*(5), e0123780. https://doi.org/10.1371/journal.pone.0123780

Jebreili, M., Neshat, H., Seyyedrasouli, A., Ghojazade, M., Hosseini, M. B., & Hamishehkar, H. (2015). Comparison of breastmilk odor and vanilla odor on mitigating premature infants' response to pain during and after venipuncture. *Breastfeeding Medicine, 10*(7), 362–365. https://doi.org/10.1089/bfm.2015.0060

Jessen, S. (2020). Maternal odor reduces the neural response to fearful faces in human infants. *Developmental Cognitive Neuroscience, 45*, 100858. https://doi.org/10.1016/j.dcn.2020.100858

Johnson, M. H., Dziurawiec, S., Ellis, H., & Morton, J. (1991). Newborns' preferential tracking of face-like stimuli and its subsequent decline. *Cognition, 40*, 1–19. https://doi.org/10.1016/0010-0277(91)90045-6

Jonas, J., Jacques, C., Liu-Shuang, J., Brissart, H., Colnat-Coulbois, S., Maillard, L., & Rossion, B. (2016). A face-selective ventral occipito-temporal map of the human brain with intracerebral potentials. *Proceedings of the National Academy of Sciences, 113*(28), E4088–E4097. https://doi.org/10.1073/pnas.1522033113

Kelly, D. J., Duarte, S., Meary, D., Bindemann, M., & Pascalis, O. (2019). Infants rapidly detect human faces in complex naturalistic visual scenes. *Developmental Science*, e12829. https://doi.org/10.1111/desc.12829

Klaey-Tassone, M., Durand, K., Damon, F., Heyers, K., Mezrai, N., Patris, B., Sagot, P., Soussignan, R., Schaal, B., & the MILKODOR Consortium. (2020). Human neonates prefer colostrum to mature milk: Evidence for an olfactory bias toward the "initial milk"? *American Journal of Human Biology*. https://doi.org/10.1002/ajhb.23521

Kobayashi, M., Otsuka, Y., Nakato, E., Kanazawa, S., Yamaguchi, M. K., & Kakigi, R. (2012). Do infants recognize the Arcimboldo images as faces? Behavioral and near-infrared spectroscopic study. *Journal of Experimental Child Psychology, 111*(1), 22–36. https://doi.org/10.1016/j.jecp.2011.07.008

Kominsky, J. F., Lucca, K., Thomas, A. J., Frank, M. C., & Hamlin, J. K. (2022). Simplicity and validity in infant research. *Cognitive Development, 63*, 101213. https://doi.org/10.1016/j.cogdev.2022.101213

Kouider, S., Stahlhut, C., Gelskov, S. V., Barbosa, L. S., Dutat, M., de Gardelle, V., Christophe, A., Dehaene, S., & Dehaene-Lambertz, G. (2013). A neural marker of perceptual consciousness in infants. *Science, 340*(April), 376–380. https://doi.org/10.1126/science.1232509

Kwon, M.-K., Setoodehnia, M., Baek, J., Luck, S. J., & Oakes, L. M. (2016). The development of visual search in infancy: Attention to faces versus salience. *Developmental Psychology, 52*(4), 537–555. https://doi.org/10.1037/dev0000080

Lecanuet, J.-P., & Schaal, B. (1996). Fetal sensory competencies. *European Journal of Obstetrics, Gynecology, and Reproductive Biology, 68*, 1–23.

Leleu, A., Demily, C., Franck, N., Durand, K., Schaal, B., & Baudouin, J.-Y. (2015). The odor context facilitates the perception of low-intensity facial expressions of emotion. *PLoS ONE, 10*(9), e0138656. https://doi.org/10.1371/journal.pone.0138656

Leleu, A., Rekow, D., Poncet, F., Schaal, B., Durand, K., Rossion, B., & Baudouin, J.-Y. (2020). Maternal odor shapes rapid face categorization in the infant brain. *Developmental Science, 23*(2), e12877. https://doi.org/10.1111/desc.12877

Leppänen, J. M. (2016). Using eye tracking to understand infants' attentional bias for faces. *Child Development Perspectives, 10*(3), 161–165. https://doi.org/10.1111/cdep.12180

Lewkowicz, D. J. (1996). Perception of auditory–visual temporal synchrony in human infants. *Journal of Experimental Psychology. Human Perception and Performance, 22*, 1094–1106. https://doi.org/10.1037/0096-1523.22.5.1094

Lewkowicz, D. J. (2010). The ontogeny of human multisensory object perception: A constructivist account. In J. Kaiser & M. J. Naumer (Eds.), *Multisensory object perception in the primate brain* (pp. 303–327). Springer.

Liu-Shuang, J., Yang, Y.-F., Rossion, B., & Goffaux, V. (2022). Natural contrast statistics facilitate human face categorization. *ENeuro, 9*(5). https://doi.org/10.1523/ENEURO.0420-21.2022

Lochy, A., Schiltz, C., & Rossion, B. (2019). The right hemispheric dominance for face perception in preschool children depends on the visual discrimination level. *Developmental Science*. https://doi.org/10.1111/desc.12914

Loos, H. M., Doucet, S., Soussignan, R., Hartmann, C., Durand, K., Dittrich, R., Sagot, P., Buettner, A., & Schaal, B. (2014). Responsiveness of human neonates to the odor of 5α-androst-16-en-3-one: A behavioral paradox? *Chemical Senses, 39*(8), 693–703. https://doi.org/10.1093/chemse/bju041

Loos, H. M., Reger, D., & Schaal, B. (2019). The odour of human milk: Its chemical variability and detection by newborns. *Physiology & Behavior, 199*, 88–99. https://doi.org/10.1016/j.physbeh.2018.11.008

Lübke, K. T., & Pause, B. M. (2015). Always follow your nose: The functional significance of social chemosignals in human reproduction and survival. *Hormones and Behavior, 68*, 134–144. https://doi.org/10.1016/j.yhbeh.2014.10.001

Macchi Cassia, V., Turati, C., & Simion, F. (2004). Can a nonspecific bias toward top-heavy patterns explain newborns' face preference? *Psychological Science, 15*(6), 379–383. https://doi.org/10.1111/j.0956-7976.2004.00688.x

Macfarlane, A. (1975). Olfaction in the development of social preferences in the human neonate. *CIBA Foundation Symposium, 33*, 103–117. https://doi.org/10.1002/9780470720158.ch7

Mahon, B. Z., & Caramazza, A. (2011). What drives the organization of object knowledge in the brain? *Trends in Cognitive Sciences, 15*(3), 97–103. https://doi.org/10.1016/j.tics.2011.01.004

Makin, J. W., & Porter, R. H. (1989). Attractiveness of lactating females' breast odors to neonates. *Child Development, 60*(4), 803–810. JSTOR. https://doi.org/10.2307/1131020

Marin, M. M., Rapisardi, G., & Tani, F. (2015). Two-day-old newborn infants recognise their mother by her axillary odour. *Acta Paediatrica, 104*(3), 237–240. https://doi.org/10.1111/apa.12905

Marinović, V., Hoehl, S., & Pauen, S. (2014). Neural correlates of human–animal distinction: An ERP-study on early categorical differentiation with 4- and 7-month-old infants and adults. *Neuropsychologia, 60*, 60–76. https://doi.org/10.1016/j.neuropsychologia.2014.05.013

Marlier, L., & Schaal, B. (2005). Human newborns prefer human milk: Conspecific milk odor is attractive without postnatal exposure. *Child Development, 76*(1), 155–168. https://doi.org/10.1111/j.1467-8624.2005.00836.x

Marlier, L., Schaal, B., & Soussignan, R. (1998). Neonatal responsiveness to the odor of amniotic and lacteal fluids: A test of perinatal chemosensory continuity. *Child Development, 69*(3), 611–623. https://doi.org/10.1111/j.1467-8624.1998.00611.x

Mattioni, S., Rezk, M., Battal, C., Bottini, R., Cuculiza Mendoza, K. E., Oosterhof, N. N., & Collignon, O. (2020). Categorical representation from sound and sight in the ventral occipito-temporal cortex of sighted and blind. *eLife, 9*, e50732. https://doi.org/10.7554/eLife.50732

McGann, J. P. (2017). Poor human olfaction is a 19th-century myth. *Science, 356*(597). https://doi.org/10.1126/science.aam7263

Mujica-Parodi, L. R., Strey, H. H., Frederick, B., Savoy, R., Cox, D., Botanov, Y., Tolkunov, D., Rubin, D., & Weber, J. (2009). Chemosensory cues to conspecific emotional stress activate amygdala in humans. *PLoS ONE, 4*(7), e6415. https://doi.org/10.1371/journal.pone.0006415

Murphy, G. L. (2002). Conceptual combination. In *The big book of concepts*. MIT Press.

Nelson, C. A. (2001). The development and neural bases of face recognition. *Infant and Child Development, 10*(1–2), 3–18. https://doi.org/10.1002/icd.239

Nishitani, S., Miyamura, T., Tagawa, M., Sumi, M., Takase, R., Doi, H., Moriuchi, H., & Shinohara, K. (2009). The calming effect of a maternal breast milk odor on the human newborn infant. *Neuroscience Research, 63*(1), 66–71. https://doi.org/10.1016/j.neures.2008.10.007

Norcia, A. M., Appelbaum, L. G., Ales, J. M., Cottereau, B. R., & Rossion, B. (2015). The steady-state visual evoked potential in vision research: A review. *Journal of Vision, 15*(6), 1–16. https://doi.org/10.1167/15.6.4

Or, C. C.-F., Retter, T. L., & Rossion, B. (2019). The contribution of color information to rapid face categorization in natural scenes. *Journal of Vision, 19*(5), 20–20. https://doi.org/10.1167/19.5.20

Parma, V., Gordon, A. R., Cecchetto, C., Cavazzana, A., Lundström, J. N., & Olsson, M. J. (2017). Processing of human body odors. In A. Buettner (Ed.), *Springer handbook of odor* (pp. 127–128). Springer. https://doi.org/10.1007/978-3-319-26932-0_51

Parma, V., Ohla, K., Veldhuizen, M. G., Niv, M. Y., Kelly, C. E., Bakke, A. J., Cooper, K. W., Bouysset, C., Pirastu, N., Dibattista, M., Kaur, R., Liuzza, M. T., Pepino, M. Y., Schöpf, V., Pereda-Loth, V., Olsson, S. B., Gerkin, R. C., Rohlfs Domínguez, P., Albayay, J., et al. (2020). More than smell—COVID-19 is associated with severe impairment of smell, taste, and chemesthesis. *Chemical Senses, 45*(7), 609–622. https://doi.org/10.1093/chemse/bjaa041

Pascalis, O., Fort, M., & Quinn, P. C. (2020). Development of face processing: Are there critical or sensitive periods? *Current Opinion in Behavioral Sciences, 36*, 7–12. https://doi.org/10.1016/j.cobeha.2020.05.005

Peykarjou, S., & Hoehl, S. (2013). Three-month-olds' brain responses to upright and inverted faces and cars. *Developmental Neuropsychology, 38*(4), 272–280. https://doi.org/10.1080/87565641.2013.786719

Peykarjou, S., Hoehl, S., Pauen, S., & Rossion, B. (2017). Rapid categorization of human and ape faces in 9-month-old infants revealed by fast periodic visual stimulation. *Scientific Reports, 7*(1), 1–12. https://doi.org/10.1038/s41598-017-12760-2

Peykarjou, S., Langeloh, M., Baccolo, E., Rossion, B., & Pauen, S. (2022). Superior neural individuation of mother's than stranger's faces by five months of age. *Cortex, 155*, 264–276. https://doi.org/10.1016/j.cortex.2022.07.011

Piaget, J. (1952). *The origins of intelligence in children*. International University Press.

Poncet, F., Leleu, A., Rekow, D., Damon, F., Durand, K., Schaal, B., & Baudouin, J.-Y. (2021). Odor-evoked hedonic contexts influence the categorization of facial expressions in the human brain. *Biological Psychology, 158*, 108005. https://doi.org/10.1016/j.biopsycho.2020.108005

Poncet, F., Leleu, A., Rekow, D., Damon, F., Dzhelyova, M. P., Schaal, B., Durand, K., Faivre, L., Rossion, B., & Baudouin, J.-Y. (2022). A neural marker of rapid discrimination of facial expression in 3.5- and 7-month-old infants. *Frontiers in Neuroscience, 16*. https://doi.org/10.3389/fnins.2022.901013

Porter, R. H., & Winberg, J. (1999). Unique salience of maternal breast odors for newborn infants. *Neuroscience and Biobehavioral Reviews, 23*(3), 439–449. https://doi.org/10.1016/S0149-7634(98)00044-X

Prehn-Kristensen, A., Wiesner, C., Bergmann, T. O., Wolff, S., Jansen, O., Mehdorn, H. M., Ferstl, R., & Pause, B. M. (2009). Induction of empathy by the smell of anxiety. *PLoS ONE, 4*(6), e5987. https://doi.org/10.1371/journal.pone.0005987

Quek, G. L., Liu-Shuang, J., Goffaux, V., & Rossion, B. (2018). Ultra-coarse, single-glance human face detection in a dynamic visual stream. *NeuroImage, 176*, 465–476. https://doi.org/10.1016/j.neuroimage.2018.04.034

Quinn, P. C. (2011). Born to categorize. In U. Goswami (Ed.), *The Wiley-Blackwell handbook of childhood cognitive development* (Vol. 2, 2nd ed., pp. 129–152). Wiley-Blackwell.

Quinn, P. C., Eimas, P. D., & Rosenkrantz, S. L. (1993). Evidence for representations of perceptually similar natural categories by 3-month-old and 4-month-old infants. *Perception, 22*(4), 463–475. https://doi.org/10.1068/p220463

Quinn, P. C., Westerlund, A., & Nelson, C. A. (2006). Neural markers of categorization in 6-month-old infants. *Psychological Science, 17*(1), 59–66. https://doi.org/10.1111/j.1467-9280.2005.01665.x

Rattaz, C., Goubet, N., & Bullinger, A. (2005). The calming effect of a familiar odor on full-term newborns. *Journal of Developmental and Behavioral Pediatrics, 26*(2), 86–92. https://doi.org/10.1097/00004703-200504000-00003

Regenbogen, C., Johansson, E., Andersson, P., Olsson, M. J., & Lundström, J. N. (2016). Bayesian-based integration of multisensory naturalistic perithreshold stimuli. *Neuropsychologia, 88*, 123–130. https://doi.org/10.1016/j.neuropsychologia.2015.12.017

Rekow, D., Baudouin, J.-Y., Brochard, R., Rossion, B., & Leleu, A. (2022a). Rapid neural categorization of facelike objects predicts the perceptual awareness of a face (face pareidolia). *Cognition, 222*, 105016. https://doi.org/10.1016/j.cognition.2022.105016

Rekow, D., Baudouin, J.-Y., Durand, K., & Leleu, A. (2022b). Smell what you hardly see: Odors assist visual categorization in the human brain. *NeuroImage, 255*, 119181. https://doi.org/10.1016/j.neuroimage.2022.119181

Rekow, D., Baudouin, J.-Y., Poncet, F., Damon, F., Durand, K., Schaal, B., Rossion, B., & Leleu, A. (2021). Odor-driven face-like categorization in the human infant brain. *Proceedings of the National Academy of Sciences, 118*(21). https://doi.org/10.1073/pnas.2014979118

Rekow, D., Baudouin, J.-Y., Rossion, B., & Leleu, A. (2020a). An ecological measure of rapid and automatic face-sex categorization. *Cortex, 127*, 150–161. https://doi.org/10.1016/j.cortex.2020.02.007

Rekow, D., Leleu, A., Poncet, F., Damon, F., Rossion, B., Durand, K., Schaal, B., & Baudouin, J.-Y. (2020b). Categorization of objects and faces in the infant brain and its sensitivity to maternal odor: Further evidence for the role of intersensory congruency in perceptual development. *Cognitive Development, 55*, 100930. https://doi.org/10.1016/j.cogdev.2020.100930

Rennels, J. L., & Davis, R. E. (2008). Facial experience during the first year. *Infant Behavior & Development, 31*(4), 665–678. https://doi.org/10.1016/j.infbeh.2008.04.009

Retter, T. L., Jiang, F., Webster, M. A., & Rossion, B. (2020). All-or-none face categorization in the human brain. *NeuroImage, 213*, 116685. https://doi.org/10.1016/j.neuroimage.2020.116685

Retter, T. L., & Rossion, B. (2016). Uncovering the neural magnitude and spatio-temporal dynamics of natural image categorization in a fast visual stream. *Neuropsychologia, 91*, 9–28. https://doi.org/10.1016/j.neuropsychologia.2016.07.028

Reynolds, G. D., & Richards, J. E. (2005). Familiarization, attention, and recognition memory in infancy: An event-related potential and cortical source localization study. *Developmental Psychology, 41*(4), 598–615. https://doi.org/10.1037/0012-1649.41.4.598

Rossion, B. (2008). Picture-plane inversion leads to qualitative changes of face perception. *Acta Psychologica, 128*(2), 274–289. https://doi.org/10.1016/j.actpsy.2008.02.003

Rossion, B., Dricot, L., Goebel, R., & Busigny, T. (2011). Holistic face categorization in higher order visual areas of the normal and prosopagnosic brain: Toward a non-hierarchical view of face perception. *Frontiers in Human Neuroscience, 4*, 1–30. https://doi.org/10.3389/fnhum.2010.00225

Rossion, B., & Lochy, A. (2021). Is human face recognition lateralized to the right hemisphere due to neural competition with left-lateralized visual word recognition? A critical review. *Brain Structure & Function*. https://doi.org/10.1007/s00429-021-02370-0

Rossion, B., Torfs, K., Jacques, C., & Liu-Shuang, J. (2015). Fast periodic presentation of natural images reveals a robust face-selective electrophysiological response in the human brain. *Journal of Vision, 15*(1), 1–18. https://doi.org/10.1167/15.1.18

Royet, J.-P. (2004). Lateralization of olfactory processes. *Chemical Senses, 29*(8), 731–745. https://doi.org/10.1093/chemse/bjh067

Royet, J.-P., Hudry, J., Zald, D. H., Godinot, D., Grégoire, M. C., Lavenne, F., Costes, N., & Holley, A. (2001). Functional neuroanatomy of different olfactory judgments. *NeuroImage, 13*(3), 506–519. https://doi.org/10.1006/nimg.2000.0704

Rubin, D., Botanov, Y., Hajcak, G., & Mujica-Parodi, L. R. (2012). Second-hand stress: Inhalation of stress sweat enhances neural response to neutral faces. *Social Cognitive and Affective Neuroscience, 7*(2), 208–212. https://doi.org/10.1093/scan/nsq097

Russell, M. J. (1976). Human olfactory communication. *Nature, 260*(5551), 5551. https://doi.org/10.1038/260520a0

Saffran, J. R., & Kirkham, N. Z. (2018). Infant statistical learning. *Annual Review of Psychology, 69*(1), 181–203. https://doi.org/10.1146/annurev-psych-122216-011805

Schaal, B. (1988). Olfaction in infants and children: Developmental and functional perspectives. *Chemical Senses, 13*(2), 145–190.

Schaal, B., & Al Aïn, S. (2014). Chemical signals 'selected for' newborns in mammals. *Animal Behaviour, 97*, 289–299. https://doi.org/10.1016/j.anbehav.2014.08.022

Schaal, B., Coureaud, G., Langlois, D., Giniès, C., Sémon, E., & Perrier, G. (2003). Chemical and behavioural characterization of the rabbit mammary pheromone. *Nature, 424*(6944), 68–72. https://doi.org/10.1038/nature01739

Schaal, B., Hummel, T., & Soussignan, R. (2004). Olfaction in the fetal and premature infant: Functional status and clinical implications. *Clinics in Perinatology, 31*(2), 261–285. https://doi.org/10.1016/j.clp.2004.04.003

Schaal, B., Marlier, L., & Soussignan, R. (1998). Olfaction function in the human fetus: Evidence from selective neonatal responsiveness to the odor of amniotic fluid. *Behavioral Neuroscience, 112*(6), 1438–1449. https://doi.org/10.1037/0735-7044.112.6.1438

Schaal, B., Marlier, L., & Soussignan, R. (2000). Human foetuses learn odours from their pregnant mother's diet. *Chemical Senses, 25*(6), 729–737. https://doi.org/10.1093/chemse/25.6.729

Schaal, B., Saxton, T. K., Loos, H., Soussignan, R., & Durand, K. (2020). Olfaction scaffolds the developing human from neonate to adolescent and beyond. *Philosophical Transactions of the Royal Society B: Biological Sciences, 375*(1800), 20190261. https://doi.org/10.1098/rstb.2019.0261

Schaal, B., Soussignan, R., & Marlier, L. (2002). Olfactory cognition at the start of life: The perinatal shaping of selective odor responsiveness. In C. Rouby, B. Schaal, D. Dubois, R. Gervais, & A. Holley (Eds.), *Olfaction, taste, and cognition* (pp. 421–440). Cambridge University Press.

Seigneuric, A., Durand, K., Jiang, T., Baudouin, J.-Y., & Schaal, B. (2010). The nose tells it to the eyes: Crossmodal associations between olfaction and vision. *Perception, 39*(11), 1541–1554. https://doi.org/10.1068/p6740

Sela, L., & Sobel, N. (2010). Human olfaction: A constant state of change-blindness. *Experimental Brain Research, 205*(1), 13–29. https://doi.org/10.1007/s00221-010-2348-6

Simion, F., & Di Giorgio, E. (2015). Face perception and processing in early infancy: Inborn predispositions and developmental changes. *Frontiers in Psychology, 6*(July), 1–11. https://doi.org/10.3389/fpsyg.2015.00969

Simion, F., Leo, I., Turati, C., Valenza, E., & Dalla Barba, B. (2007). How face specialization emerges in the first months of life. In *Progress in brain research* (Vol. 164, pp. 169–185). Elsevier. https://doi.org/10.1016/S0079-6123(07)64009-6

Soussignan, R., Schaal, B., Marlier, L., & Jiang, T. (1997). Facial and autonomic responses to biological and artificial olfactory stimuli in human neonates: Re-examining early hedonic discrimination of odors. *Physiology & Behavior, 62*(4), 745–758. https://doi.org/10.1016/S0031-9384(97)00187-X

Spahn, J. M., Callahan, E. H., Spill, M. K., Wong, Y. P., Benjamin-Neelon, S. E., Birch, L., Black, M. M., Cook, J. T., Faith, M. S., Mennella, J. A., & Casavale, K. O. (2019). Influence of maternal diet on flavor transfer to amniotic fluid and breast milk and children's responses: A systematic review. *The American Journal of Clinical Nutrition, 109*(Supplement_1), 1003S–1026S. https://doi.org/10.1093/ajcn/nqy240

Spence, C. (2021). The scent of attraction and the smell of success: Crossmodal influences on person perception. *Cognitive Research: Principles and Implications, 6*(1), 46. https://doi.org/10.1186/s41235-021-00311-3

Stein, B. E., & Meredith, M. A. (1993). *The merging of the senses*. The MIT Press.

Sugden, N. A., Mohamed-Ali, M. I., & Moulson, M. C. (2014). I spy with my little eye: Typical, daily exposure to faces documented from a first-person infant perspective. *Developmental Psychobiology, 56*(2), 249–261. https://doi.org/10.1002/dev.21183

Sullivan, R. M., & Toubas, P. (1998). Clinical usefulness of maternal odor in newborns: Soothing and feeding preparatory responses. *Neonatology, 74*(6), 402–408. https://doi.org/10.1159/000014061

Turati, C., Simion, F., Milani, I., & Umiltà, C. (2002). Newborns' preference for faces: What is crucial? *Developmental Psychology, 38*(6), 875–882. https://doi.org/10.1037/0012-1649.38.6.875

Turati, C., Valenza, E., Leo, I., & Simion, F. (2005). Three-month-olds' visual preference for faces and its underlying visual processing mechanisms. *Journal of Experimental Child Psychology, 90*(3), 255–273. https://doi.org/10.1016/j.jecp.2004.11.001

Turkewitz, G., & Devenny, D. A. (1993). *Developmental time and timing*. Lawrence Erlbaum Associates.

Valenza, E., Simion, F., Cassia, V. M., & Umiltà, C. (1996). Face preference at birth. *Journal of Experimental Psychology. Human Perception and Performance, 22*(4), 892–903. https://doi.org/10.1037/0096-1523.22.4.892

Varendi, H., Porter, R. H., & Winberg, J. (1994). Does the newborn baby find the nipple by smell? *Lancet, 344*(8928), 989–990. https://doi.org/10.1016/S0140-6736(94)91645-4

Vettori, S., Dzhelyova, M., Van der Donck, S., Jacques, C., Steyaert, J., Rossion, B., & Boets, B. (2019). Reduced neural sensitivity to rapid individual face discrimination in autism spectrum disorder. *NeuroImage, 21*, 101613. https://doi.org/10.1016/j.nicl.2018.101613

Wagner, S., Issanchou, S., Chabanet, C., Lange, C., Schaal, B., & Monnery-Patris, S. (2019). Weanling infants prefer the odors of green vegetables, cheese, and fish when their mothers consumed these foods during pregnancy and/or lactation. *Chemical Senses, 44*(4), 257–265. https://doi.org/10.1093/chemse/bjz011

Weisfeld, G. E., Czilli, T., Phillips, K. A., Gall, J. A., & Lichtman, C. M. (2003). Possible olfaction-based mechanisms in human kin recognition and inbreeding avoidance. *Journal of Experimental Child Psychology, 85*(3), 279–295. https://doi.org/10.1016/S0022-0965(03)00061-4

Werchan, D. M., Baumgartner, H. A., Lewkowicz, D. J., & Amso, D. (2018). The origins of cortical multisensory dynamics: Evidence from human infants. *Developmental Cognitive Neuroscience, 34*, 75–81. https://doi.org/10.1016/j.dcn.2018.07.002

Yin, R. K. (1969). Looking at upside-down faces. *Journal of Experimental Psychology, 81*(1), 141. https://doi.org/10.1037/h0027474

Zatorre, R. J., Jones-Gotman, M., & Rouby, C. (2000). Neural mechanisms involved in odor pleasantness and intensity judgments. *NeuroReport, 11*(12), 2711–2716. https://doi.org/10.1097/00001756-200008210-00021

Zernecke, R., Haegler, K., Kleemann, A. M., Albrecht, J., Frank, T., Linn, J., Brückmann, H., & Wiesmann, M. (2011). Effects of male anxiety chemosignals on the evaluation of happy facial expressions. *Journal of Psychophysiology, 25*(3), 116–123. https://doi.org/10.1027/0269-8803/a000047

Zhang, S., Su, F., Li, J., & Chen, W. (2018). The analgesic effects of maternal milk odor on newborns: A meta-analysis. *Breastfeeding Medicine, 13*(5), 327–334. https://doi.org/10.1089/bfm.2017.0226

Zheng, Y., You, Y., Farias, A. R., Simon, J., Semin, G. R., Smeets, M. A. M., & Li, W. (2018). Human chemosignals of disgust facilitate food judgment. *Scientific Reports, 8*(1), 1–10. https://doi.org/10.1038/s41598-018-35132-w

Zhou, G., Lane, G., Cooper, S. L., Kahnt, T., & Zelano, C. (2019). Characterizing functional pathways of the human olfactory system. *eLife, 8*, e47177. https://doi.org/10.7554/eLife.47177

Zhou, W., & Chen, D. (2008). Encoding human sexual chemosensory cues in the orbitofrontal and fusiform cortices. *The Journal of Neuroscience, 28*(53), 14416–14421. https://doi.org/10.1523/JNEUROSCI.3148-08.2008

Zhou, W., & Chen, D. (2009). Fear-related chemosignals modulate recognition of fear in ambiguous facial expressions. *Psychological Science, 20*(2), 177–183. https://doi.org/10.1111/j.1467-9280.2009.02263.x

Zhou, W., Jiang, Y., He, S., & Chen, D. (2010). Olfaction modulates visual perception in binocular rivalry. *Current Biology, 20*(15), 1356–1358. https://doi.org/10.1016/j.cub.2010.05.059

# Part V

# Interspecific Communication

# Specific Antipredator Response of Leopard Geckos (*Eublepharis macularius*) to the Smell of Snake Exuvia

Eva Landová[1], Petra Hnidová[1], Aleksandra Chomik[1], Jitka Jančúchová-Lásková[1], Petra Frýdlová[1] (✉), and Daniel Frynta[1]

[1] Ecology and Ethology Research Group, Department of Zoology, Faculty of Science, Charles University, Prague, Czech Republic
petra.frydlova@natur.cuni.cz

**Abstract** This study aimed to establish a new bioassay for the chemosensory discrimination of snake predators by leopard geckos, *Eublepharis macularius*. It quantified tongue flicking and antipredator behaviour of leopard geckos using several tests analysing reactions to: (1) exuviae from six snake predators selected in terms of gecko predation, (2) intact exuviae against exuviae extracted with polar and non-polar solvents, and (3) exuviae against various scentless and scented control stimuli. We found that geckos discriminate the odour of snake exuviae against controls: snake exuviae trigger antipredator responses that take the form of either active (high postures, attacks) or passive (low postures, avoidance, and escape) antipredator reactions. Geckos reacted to the odour of all snake exuviae but with variations in form and intensity between snake species. Geckos reacted to exuviae after odour removal by a polar solvent, but the reaction was abolished after non-polar solvent application. Tongue flicking appeared to be indiscriminate although it may reflect the animal's interest towards unfamiliar stimuli. Finally, the geckos' antipredator reactions to snake exuviae were comparable to those elicited by live snakes. This study validates a new bioassay for predator scents in geckos.

**Keywords** Leopard geckos (*Eublepharis macularius*) · Snake · Predator recognition · Antipredation · Chemoreception · Semiochemicals · Exuvia

## 1 Introduction

Reptiles have developed chemoreception allowing them to explore the environment, detect prey, or even recognize predators as well as perform many social behaviours like aggregation and recognition of a familiar and unfamiliar individual or mate partner (Martín & López, 2006; Gabirot et al., 2010; Labra, 2011; Font et al., 2012; reviewed in Mason & Parker, 2010; Martín & López, 2014). The main system for

 
B. Schaal et al. (eds.), *Chemical Signals in Vertebrates 15*,
https://doi.org/10.1007/978-3-031-35159-4_19

chemoreception includes olfaction when volatile odorants are perceived by the main olfactory epithelium, the signal being then transmitted to the main olfactory bulb and further processed in the brain (Halpern, 1992). Many vertebrate taxa also possess a vomeronasal (Jacobson's) organ (VNO) located above the palate with an opening into the roof of the mouth (Burghardt, 1980). This accessory olfactory system has become very important for the diverse taxa of squamates and tuataras, its perceptual functioning being actualized by tongue-flicking behaviour (Meredith & Burghardt, 1978; Halpern & Martínez-Marcos, 2003). Tongue-flicking behaviour is indeed elicited by any airborne odorants, although the discrimination of biologically significant odorants requires functional vomerolfaction, probably due to motivational components mediated by the amygdala (Zuri & Halpern, 2003). In fact, both main olfaction and vomerolfaction are anatomically and functionally interconnected and cooperate in reptilian chemoreception (Halpern & Martínez-Marcos, 2003).

Numerous studies explored the function of VNO (Meredith & Burghardt, 1978), showing considerable between-taxa diversity in its contribution to chemoreception in squamates. Certain genera, like *Anoles*, have poorly developed olfaction and vomerolfaction, while in Agamidae, Phrynosomatidae, Lacertidae, Scincidae, Cordylidae, and Viperidae both chemosensory systems are highly developed. Contrastedly, in Varanidae, Anguidae, Amphisbaenidae, and most snakes, vomerolfaction is the predominant chemical sense (Halpern, 1992). Recent studies analysing the diversity of genes for olfactory receptors in reptiles (Vandewege et al., 2016) confirmed such variability in chemosensory abilities as they found only one coding gene for the green anole (*Anolis carolinensis*) and 598 for the Burmese python (*Python molurus*).

The chemical detection ability of snake predators has been demonstrated in lizards as well as snakes (reviewed in Landová et al., 2016; McElroy, 2019). Snake predators have a substance in their skin that is recognized by rattlesnakes using vomerolfaction (Miller & Gutzke, 1999). The fully functional VNO is essential for several species of North American pit vipers performing antipredator reactions to a snake-eating predator, the common kingsnake (*Lampropeltis getula*). When the vomeronasal duct of pit vipers was sutured, they did not perform any snake-typical antipredator behaviour like avoiding, fleeing, bridging, or head hiding, although they could still see the predator (Miller & Gutzke, 1999). Similarly, desert iguanas (*Dipsosaurus dorsalis*) are able to detect and distinguish between a saurophagous kingsnake and snakes that mainly eat arthropods (e.g. *Chionactis occipitalis*). After chemoreception, they respond by performing antipredator behaviours (typical body postures, slow motion) and increasing tongue-flicking activity (Bealor & Krekorian, 2002). Multiple studies showed that lizards were able to chemically recognize saurophagous snake predators, especially those living in sympatry (Thoen et al., 1986; Dial et al., 1989; Downes & Shine, 1998; López & Martín, 2001; Amo et al., 2004 but see Balderas-Valdivia & Ramírez-Bautista, 2005; Landová et al., 2016; Ortega et al., 2018; Hammond et al., 2020). Surprisingly, Simon et al. (1981) did not find any difference in the tongue-flicking activity of the mountain spiny lizard (*Sceloporus jarrovi*) that was presented against either a clean paper or paper with semiochemicals from the Arizona mountain kingsnake (*Lampropeltis pyromelana*), a lizard-eating predator. Regarding the ability of predator detection, nocturnal species, in comparison to many diurnals, visually orientated organisms, rely on olfactory perception and chemoreception. The chemical sense facilitates orientation in the environment, intraspecific communication, or prey localization (Dial et al., 1989; Cooper, 1994; Balderas-Valdivia & Ramírez-Bautista, 2005; Barbosa et al., 2006) and allows the animal to differentiate between the current and previous presence of predators (Bramley et al., 2000; Borowski & Owadowska, 2010). While actively foraging predators (e.g. mammals) can be heard or spotted in time, detecting the presence of a lurking snake through other senses than chemoreception can be particularly difficult (Punzo, 2005).

In our previous two-choice experiment (Landová et al., 2016), we studied the antipredator response of wild-caught and laboratory-born leopard geckos (*E. macularius*, Eublepharidae) towards nine species of live snakes from two families (Colubridae, Boidae) varying in foraging ecology and sympatric/allopatric distribution with the geckos. We thoroughly described for the first time the antipredator repertoire (active vs. passive defence) of this gecko towards live snakes. We did not find any effect of predator sympatry/allopatry or food specialization. Predator-naive individuals showed general wariness, explored both chemically (by tongue flicking) and visually, and performed antipredator postures towards a majority of snake predators.

In many studies, a cotton swab that was rolled on the dorsal and abdominal skin of a snake is used as a source of the odour (Cooper Jr, 1990; Mori & Hasegawa, 1999; Cooper Jr & Sherbrooke, 2009; Martín et al., 2015). In other designs, a retreat site selection is used, when the tested animals avoid cages or shelters with predator scent marks (van Damme et al., 1995; Van Damme & Castilla, 1996; Amo et al., 2006; Hammond et al., 2020; Van Moorleghem et al., 2020; Daniell et al., 2021). Both methods are widely used, however, the concentration of scent was reduced considerably and differs in the situational context. Presentation of odour on a cotton swab in front of the animal requires an immediate reaction as the experiments are usually short (e.g. 60 s), whereas retreat site selection experiments last longer and are more affected by the subject's exploration activity, motivation, and general decision-making strategy.

For these reasons, we tried to introduce a new bioassay consisting of pieces of shed snakeskin (exuviae). We assume that these stimuli contain indeed full chemical information so the experiment should be relatively short and elicit a wide variety of antipredator responses. We selected again the leopard gecko as our model species due to our earlier knowledge of its antipredator repertoire (Landová et al., 2016). Moreover, previous research revealed that leopard geckos employ well-developed vomerolfaction in the reproduction context. Males use skin semiochemicals as sex pheromones, and they react aggressively to other males or by courting to females (Mason & Gutzke, 1990; but see Schořálková et al., 2018, for the role of steroid hormones). They also increase the frequency of tongue flicking when discriminating sex (Mason & Gutzke, 1990; Cooper & Steele, 1997) or familiarity with the female (Steele & Cooper Jr, 1997; Ferkin & LaDage, 2006).

We aimed to introduce a new bioassay that can be used for eliciting antipredator behaviour and the chemical discrimination of snake predators. Advantageously, snakes shed their whole skin (exuvia) in one moment in comparison to lizards which do it rather gradually piece by piece. We used a piece of exuvia as a source of skin chemicals, hypothesizing that it contains complete chemical information from the donor. Presentation of exuvia triggers tongue flicking together with antipredator behaviours (same as elicited by live snakes, see Landová et al., 2016). We experimentally checked this new method by including several types of controls. Specifically, we asked three main questions: (1) Can geckos detect and discriminate the smell of snake exuvia from scentless controls? (2) Can we use snake exuviae as stimuli eliciting antipredator behaviour? (3) Can geckos recognize different snake predators based on their exuviae?

According to these questions, we had several specific aims. We would like to verify that the snake exuviae convey chemical cues or signals eliciting antipredator behaviour. To test this, we compared the antipredator reaction and tongue flicking to exuviae against scentless control stimuli, predicting higher frequency of tongue flicking towards the exuviae and typical antipredator behaviour towards snakes. Moreover, we were interested in beginning to assess the nature of exuvial odour molecules, in removing them with polar and non-polar solvents (extracting respectively acids, alcohols, and aldehydes with short carbon chains, and fatty acids, cholesterol, waxes, and long carbon chain molecules); we predicted a lower frequency of antipredator responses towards the properly deodorized exuvia. We also compared the antipredator reaction and tongue flicking to an unfamiliarly scented stimulus (perfume), allowing us to exclude that geckos display antipredator behaviour to any unfamiliar scents. As most of our geckos were naïve in terms of experience with predators, we were further interested if they can differentiate exuviae from various snake predators, as well as between sympatric and allopatric snake species. Last but not least, we compared the antipredator reaction of wild-caught vs. captive-bred geckos (later called effect of generation), assuming that wild-caught animals ($F_0$) would be more specific towards predator cues than their captive descendants ($F_1$ and $F_2$).

## 2 Animals, Materials, and Methods

### 2.1 Animals

The model species, leopard gecko *Eublepharis macularius* (Blyth 1854) is a nocturnal and ground-dwelling lizard from the Eublepharidae family inhabiting arid areas with shrubby plants in eastern and southern Afghanistan, Pakistan, and north-western India (Seufer et al., 2005; Sindaco et al., 2008). During the day, they hide in crevices and holes. At night, they become active foragers (Seufer et al., 2005; Abramjan et al., 2020), searching for insects by vomerolfaction actualized in tongue flicking (Cooper Jr, 1995). In the wild, the most significant predators of the leopard gecko are mammals (foxes) and snakes (Szczerbak & Golubev, 1986; Khan, 2004). We used animals captured in the wild in Pakistan in 2002 (12 individuals), as well as their $F_1$ and $F_2$ descendants born in the laboratory (for details, see Jančúchová-Lásková et al., 2015; Frynta et al., 2018). In total, we tested 263 animals (23 males) in various behavioural tests (see below).

Geckos were kept individually or in couples (male-female or two females) in glass terraria (30 ×30 × 20 cm) separately from snakes (in different breeding room). The experimental animals were adults. Wild individuals ($F_0$ generation) were 8–15 years old. Their $F_1$ and $F_2$ captive-bred descendants were 3–10 years old and naïve (they were not familiar with the odour of snakes). All animals were fed ad libitum with crickets and mealworms dusted with vitamins ($AD_3$ and E) and minerals (Nutri Mix, Trouw Nutrition Biofaktory, Amersfoort, Netherland, respectively). The air temperature in the terrariums was kept around 27–28 °C.

## 2.2 Stimuli and Experimental Procedures

The snake stimuli consisted of exuviae removed from a snake terrarium shortly after shedding (maximum within 24 h). They were conditioned in a glass container to be stored in a freezer (−20 °C) for a maximum of 2 months. To avoid the effect of an individual snake, we aimed to obtain exuviae from several individuals of the same species. We sampled exuviae from saurophagous snakes, including four Colubridae species: *Hemorrhois ravergieri* ($n = 3$ individuals), *H. hippocrepis* ($n = 2$), *Spalerosophis atriceps* ($n = 2$), *S. diadema* ($n = 4$), and two sand boas (Erycidae): *Eryx johnii* ($n = 5$) and *E. colubrinus* ($n = 4$). We also used exuviae of a saurophagous lizard (*Pseudopus apodus*, family: Anguidae), a European anguid legless lizard that visually resembles a snake predator and preys on leopard geckos (for details on snake ecology and distribution, see Landová et al., 2016). We did not investigate the sex of exuvia donors. For certain treatments, exuviae were selectively deodorized by immersion for 24 h into either a polar (technical 96% alcohol $C_2H_5OH$, Severochema, Liberec, Czech Republic) or a non-polar (technical petrol, Severochema, Liberec, Czech Republic) solvent at ambient temperature. After solvent extraction, exuviae were dried in fresh air and stored again in a glass container in a freezer (−20 °C). By this procedure, we gained two new types of stimuli (alcohol and petrol deodorized exuviae) that were assayed separately. Finally, the control stimuli consisted of pieces of unused polythene bags (transparent) that were stored in a glass container in a freezer (−20 °C).

The experiments were conducted out of the intensive breeding season (April to July), during the post-breeding season (August to October) and after short hibernation (December and January). The life cycle of our captive geckos is a bit shifted by laboratory conditions. Tests were conducted from 6 to 11 pm, when the activity of geckos is at its highest level. We realized tests in familiar home cages in presenting alternately a series of stimuli consisting of, depending on the test (see below), pieces of intact or deodorized exuvia, or pieces of scentless or scented plastic controls (all stimuli were 2 × 3 cm in size). The stimuli were presented with tweezers held in front of the geckos' snout without contact (at 1 cm).

## 2.3 Experimental Design

The typical experimental session, also named as 'tests', consisted of three trials each presenting a different stimulus during 90 s. Each experiment was thus composed of a sequence of three trials, where the focal stimulus is preceded and followed by control stimuli: i.e. control 1 (piece of plastic bag to assess novelty effect) – focal stimulus under test – control 2 (piece of plastic bag to assess sensitization effect). An interval of 60–90 min was set between each trial. Whenever the animals were tested repeatedly, they rested for at least 2 weeks between consecutive sessions. Different tests were designed to assess the responsiveness of geckos to the various odour substrates of interest (see Table 1).

The *Snake detection test* aimed to assess the geckos' detection of a potential snake predator presented against control stimuli. We presented in semi-random order exuviae from three sympatric (*H. ravergieri, S. atriceps, E. johnii*) and three allopatric (*H. hippocrepis, S. diadema, E. colubrinus*) saurophagous snake species. 137 individuals were tested in 395 sessions consisting of 1185 trials. Each gecko was used once per snake species, which means no more than six times per test. We compared reactions between trials.

**Table 1** List of tests, number of tested geckos (N individuals), sessions, trials; sequence of stimuli in trials, and statistical comparisons. Exuviae used as focal stimuli were from six snake species (*E. colubrinus*, *E. johnii*, *S. atriceps*, *S. diadema*, *H. ravergieri*, and *H. hippocrepis*) and one saurophagous reptile (*P. apodus*)

| Name of the test | N Individuals | N Sessions | N Trials | Sequence of stimuli in trials: Control 1 | Focal stimulus (treatment) | Control 2 | Statistical comparison of focal stimulus with |
|---|---|---|---|---|---|---|---|
| Absolute control test | 64 | 64 | 192 | Plastic | Plastic (−) | Plastic | Control 1 and 2 |
| Snake detection test | 137 | 395 | 1185 | Plastic | Snake exuvia (−) | Plastic | Control 1 and 2 |
| Odour removal test | 26 | 26 | 78 | Plastic | Intact exuvia of *S. atriceps* (−) | Plastic | – |
| | 26 | 26 | 78 | Plastic | Exuvia of *S. atriceps* (alcohol-extracted) | Plastic | Intact exuvia of *S. atriceps* |
| | 26 | 26 | 78 | Plastic | Exuvia of *S. atriceps* (petrol-extracted) | Plastic | Intact exuvia of *S. atriceps* |
| Perfume test | 26 | 26 | 78 | Plastic | Plastic (perfume) | Plastic | Control 1 and 2 |
| Squamate predator discrimination test[a] | 147 | 421 | 1263 | Plastic | Squamate exuvia | Plastic | Focal stimulus from absolute control test |
| Snake species recognition test[a] | 147 | 421 | 1263 | Plastic | Squamate exuvia | Plastic | Exuvia from *P. apodus* |

[a]The Snake detection test was completed with a non-snake predator (*Pseudopus apodus*). The *Absolute control test* intended to analyse the geckos' response to the experimental setting (tweezers, experimenter's hand) and the control stimuli in three consecutive trials. We tested 64 individuals in 64 sessions consisting of 192 trials, each animal was used once per test. We compared reactions between trials

The *Odour removal test* evaluated the geckos' reaction to a sample of intact exuvia (*S. atriceps*) in comparison with a sample of exuvia (*S. atriceps*) deodorized with (1) alcohol and (2) petrol. This test compared exuviae after three treatments: intact, deodorized with alcohol, and deodorized with petrol, the presentation order of both latter treatments being counterbalanced. Each test consisted of three trials, with control stimuli before and after the focal stimuli. We tested 78 individuals in 78 sessions consisting of 234 trials. Each gecko was used three times per treatment (nine times per test). We compared geckos' reactions to the focal stimuli of the three treatments. Repeated measures were considered in the statistical analyses (see below).

The *Perfume test* intended to assess the geckos' reaction towards an arbitrarily chosen unfamiliar scent (perfume Calvin Klein IN2U) applied on the control substrate. The amount of perfume was approximately 0.05 ml. We tested 26 individual geckos in 26 sessions consisting of 78 trials. We compared reactions between trials.

The *Squamate predator discrimination test* assessed the geckos' reaction to the exuviae of various squamate predators against the control. We used the dataset from the Snake detection test and completed it with the geckos' reaction to one saurophagous sympatric lizard (*P. apodus*). In total, we compared the behaviour of 147 geckos in 421 sessions consisting of 1263 trials. Each gecko was used once per squamate species, which means no more than seven times per test. Reactions to the exuviae of squamate predators were later statistically compared with those seen at the middle trial (i.e. the focal stimulus; plastic) of the Absolute control test. Repeated measures were considered in the statistical analyses (see below).

The *Snake species recognition test* is intended to assess the difference in geckos' reaction towards the exuviae of distinct squamate predators. We used a dataset from the Squamate predator discrimination test. The reaction to the exuviae of snake predators was statistically compared with the reaction towards the exuvia of the European legless lizard. We also evaluated the difference in reactions (active vs. passive defence response) towards sympatric and allopatric snake species, as well as the ability to discriminate different snake genera (*Eryx*/ *Spalerosophis*/ *Hemorrhois*). Moreover, we compared the reactions of wild and captive-bred geckos.

During our experimental procedure, it was inevitable that the experimenter is knowledgeable of the nature and origin of the stimuli. To avoid later biasing of the results, all video-recordings of the tests were randomized and coded by one observer who was blind to the exuvia donor and/or treatment of focal stimulus.

## 2.4 Behavioural Analyses

All experiments were video recorded to be later analysed for occurrence, frequency, latency, and duration of behaviour using the OBS30 software (Noldus Information Technology, 1993). We measured the geckos' behavioural antipredator reactions and the number of tongue flicks directed to the stimulus.

Two behavioural patterns characterized the geckos' antipredator behaviour: (1) *Active defence*: the gecko stands high on its legs ('high posture', Fig. 1a, d) with raised or slowly waving tail ('tail wave', Fig. 1a, c), sometimes with an open mouth followed by an attack directed to the stimulus (Fig. 1e); (2) *Passive defence*: geckos crouch down

**Fig. 1** Antipredator behavioural elements of leopard gecko towards snake exuvia. We observed (**a**) high posture when the gecko is standing on tight legs with raised or slowly waving tail; (**b**) low posture when the gecko is crouched with their legs bent and pressing the abdomen against the ground; in (**c**) there is a detail on slowly waving tail; (**d**) high posture without tail wave; (**e**) high posture with open mouth and tail raised (drawing: Silvie Lišková)

with their legs bent, pressing the abdomen against the ground ('low posture', Fig. 1b), sometimes also with tail waving; this pattern is usually combined with freezing when the gecko remains motionless, the abdomen may be tightly pressed to the ground and then it ends by escaping or avoiding the stimulus. The last two behavioural elements were combined with low posture to the summary variable called '*passive defensive behaviour*' (reporting the presence of at least one of these elements). High posture, mouth open, and attack were further combined into a summary variable (reporting the presence of at least one of these elements), which is referred to as '*active defensive behaviour*'. As trials were rather short and multiple antipredator behaviours were displayed, we checked for the presence/absence of any antipredator behaviour. We performed a detailed behavioural analysis to check for the relationship between the active and passive defensive responses and the tongue flicking behaviour. For most of the analyses reported in this article, we scored the reaction towards the stimulus as present (active or passive defence) or absent (no defence, only tongue flicking).

### 2.5 Statistical Analyses

We measured two main dependent variables: (1) the antipredator response towards the stimulus, which was coded in a binary way (present/absent) and (2) the number of tongue flicks, which was square-root transformed to improve the normality of the distribution. We employed the GLM model with binomial distribution in R project version 4.1.1 (R Core Team, 2021) to test for the effect of stimulus on the geckos' reaction. The GLM model with a quasi-Poisson distribution was employed to analyse the number of tongue flicks towards the stimulus. In the Odour removal test, we employed the geeglm function under the geepack package version 1.3.4 (Højsgaard et al., 2006) in R with a quasi-binomial model to assess the effect of the treatment (alcohol/petrol). We controlled for the identity of the tested gecko, which was included as an animal identity argument in the marginal model. In the Snake species recognition test, we employed the geeglm function as well in a quasi-binomial model to assess the effect of the species of the tested exuviae and the generation of the tested geckos. We controlled for the identity of the individual exuvia donors, which was included as an identity argument in the marginal model. In both models, the resulting correlation parameter alpha was negligible (0.0048, S.E. = 0.0719 and −0.0124, S.E. = 0.0148). Therefore, we removed it from the model and returned to the glm approach.

## 3 Results

### 3.1 Detection of the Odour of Exuviae

Out of 192 trials of the Absolute control test, only one antipredator reaction (high posture) was observed to the plastic bag (in the first trial), leading us to consider it as a reliable control stimulus in the following experiments.

In the Snake detection test, a negligible amount of geckos' antipredator reactions was noted towards the control stimuli in comparison to the exuviae (Control 1: 4 reactions/395 trials; Exuviae: 291/395; Control 2: 1/395; $\chi^2 = 514.07$, $p < 0.0001$). However, tongue flicking was not significantly affected by the exuviae against the controls (Control 1: 11.10 ± 7.39; Exuvia: 7.06 ± 9.70; Control 2: 8.09 ± 5.99). The frequency of tongue flicking was highest towards the first control stimulus (ANOVA: $F(2,1182) = 96.85$XX, $p < 0.0001$).

The Odour removal test led to a highly significant effect of the solvent extraction of the exuviae on the geckos' antipredator responses ($\chi^2 = 23.742$, $p < 0.0001$). In comparison to the intact exuviae (Intercept: 2.485, Standard Error (S.E.) = ±0.736), exuviae extracted with petrol abolished antipredator responses (Estimate = −5.704, S.E. = ±1.214, $p < 0.0001$; Fig. 2a). In contrast, the level of antipredator reactions towards the exuviae extracted with ethanol (Estimate = −1.281, S.E. = ±0.935, $p = 0.171$) remained comparable to the level induced by the intact exuvia. Finally, the geckos did not react to the artificial perfume by any antipredator reaction (Fig. 2a).

The tongue-flicking response was intense towards all stimuli (Fig. 2b). Our results revealed a significant effect of solvent extraction on tongue flicking in the Odour removal test (ANOVA, $F(2,75) = 32.95$, $p < 0.0001$): the geckos flicked their tongue significantly more (Estimate ± S.E.) to the petrol (1.002 ± 0.222, $p < 0.0001$) than to the alcohol-extracted exuviae (0.671 ± 0.233, $p = 0.005$) in comparison to the intact exuvia (Intercept: 0.421 ± 0.189, Fig. 2b). We also observed a high amount of tongue flicks towards the perfume (Fig. 2b).

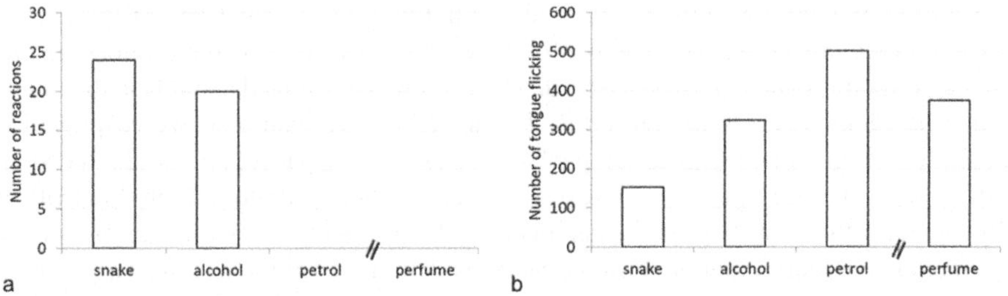


**Fig. 2** Occurrence of (active/passive) antipredator reactions (**a**) and number of tongue flicks (**b**) of leopard geckos (*Eublepharis macularius*) to various stimuli in the Odour removal and Perfume tests [snake (intact exuvia of *Spalerosophis atriceps*), alcohol (exuvia of *S. atriceps* after alcohol extraction), petrol (exuvia of *S. atriceps* after petrol extraction), and perfume (control added with perfume)]

## 3.2 Predator Discrimination and Recognition

The Squamate predator discrimination test yielded a significant effect of the stimulus on the geckos' antipredator reaction ($\chi^2 = 478.03$, $p < 0.0001$). The mean occurrence of geckos' antipredator reactions towards all squamate exuviae were significantly different from those towards the controls (all $p$ values <0.0001). Furthermore, using the Snake species recognition test, we found a significant effect of the snake species on the antipredator reaction of geckos, $F(6,414) = 33.64$, $p < 0.0001$: their reactions were significantly different towards the exuviae of all snake species (except *E. colubrinus*) in comparison with those of the European legless lizards (Fig. 3).

The generation of geckos, i.e. whether they were wild-caught or raised in the laboratory, reached also significance on the rate of their antipredator reaction, $F(1,413) = 36.06$, $p < 0.0001$: the geckos hatched in nature ($F_0$ generation) evinced significantly less antipredator reactions than those bred in captivity ($F_1$ and $F_2$ generations).

Interestingly, the exuviae triggering the highest antipredator reaction in the geckos were not those eliciting most tongue flicking. The main effect of the stimulus on the rate of tongue flicks was indeed significant, $F(6,414) = 186.90$, $p = 0.008$, with all snakes' exuviae (except *E. colubrinus*) eliciting significantly less frequent tongue flicking relative to European legless lizards' exuviae (Fig. 4).

We also found a significant effect of generation, $F(1,413) = 582.32$, $p < 0.0001$, in that the geckos hatched in nature ($F_0$ generation, $n = 12$ individuals) had a significantly higher number of tongue flicks (Mean: $13.57 \pm 12.99$) than those bred in captivity ($F_1$ and $F_2$ generations, $n = 135$ individuals; Mean: $5.78 \pm 8.00$).

The effect of the snake genus of the exuviae was highly significant on the occurrence of antipredator response ($\chi^2 = 39.461$, $p < 0.0001$). Thus, taxonomically related species of snakes induced a similar antipredator response in the geckos. However, we did not find any evidence that geckos can discriminate exuviae of sympatric vs. exuviae from allopatric snake species ($\chi^2 = 2.55$, $p = 0.11$).

Regarding the geckos' passive defensive behaviour, we found a significant effect of snake species, $F(6,414) = 55.40$, $p < 0.0001$. The rate of passive defensive reaction differed significantly only towards the exuviae of fossorial sand boas (*E. colubrinus* and

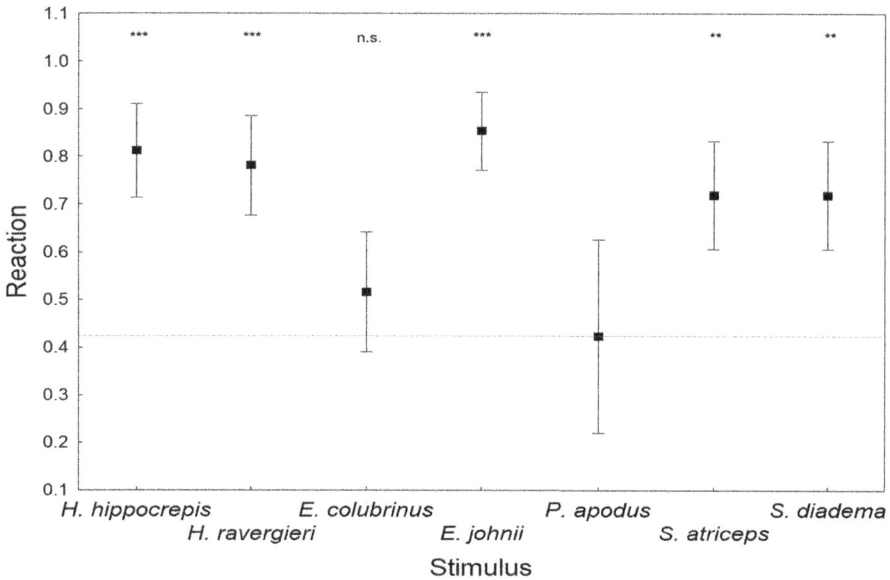


**Fig. 3** Mean occurrence of the gecko's (active/passive) antipredator reaction towards various snake exuviae in comparison to the exuvia from the European legless lizard (*P. apodus*) in the Snake species recognition test (***: $p < 0.0001$, **: $p < 0.001$, as revealed by glm model; mean estimates and standard errors from glm model). The dashed line is a baseline (mean reaction to *P. apodus*) for statistical comparison. Abbreviations: (*H.*) *Hemorrhois*, (*E.*) *Eryx*, (*P.*) *Pseudopus*, and (*S.*) *Spalerosophis*

*E. johnii*) in comparison with those of the European legless lizards (Fig. 5). The geckos adopted 'low posture' and avoided more frequently the chemostimuli associated with sand boas. Likewise, the geckos' active defensive behaviour to the exuviae differed significantly as a function of the snake species, $F(6,414) = 61.30$, $p < 0.0001$. The geckos displayed significantly more frequent high postures and attacks towards the exuviae of all colubroid snakes (but not sand boas) in comparison with the European legless lizards (Fig. 6).

# 4 Discussion

## 4.1 Chemosensory Responses in Geckos: Tongue Flicking vs. Antipredator Behaviour

Our results clearly showed that leopard geckos detect chemicals emitted by squamate exuviae. They reliably expressed antipredator behaviour in response to exuviae, but not to the scentless control stimuli. However, the control stimuli presented before and after the exuviae still elicited tongue flicking. Thus, tongue flicking appears to reflect non-discriminative chemosensory exploration of any object. This notion is supported by the

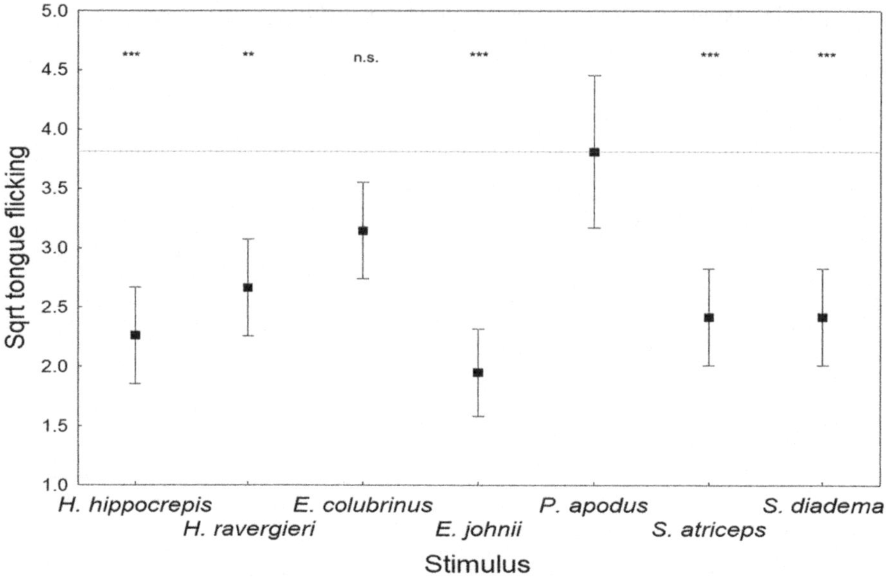


**Fig. 4** Mean number of geckos' tongue flicks towards various exuviae in the Snake species recognition test in comparison to the exuvia from European legless lizard (*P. apodus*). Mean estimates and standard errors from glm model; ***: $p < 0.0001$, **: $p < 0.001$, *: $p < 0.01$. The dashed line is a baseline (mean number of tongue flicks to *P. apodus*) for statistical comparison. Abbreviations: (Sqrt) square root transformation, (*H.*) *Hemorrhois*, (*E.*) *Eryx*, (*P.*) *Pseudopus*, and (*S.*) *Spalerosophis*

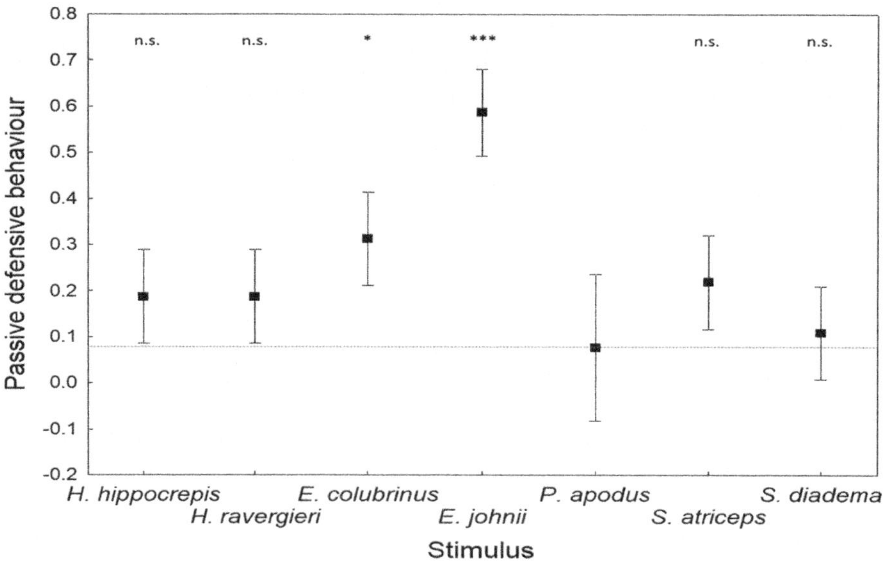


**Fig. 5** Mean gecko's passive defensive behaviour towards various snake exuviae in comparison to the exuvia from the European legless lizard (*P. apodus*) in the Snake species recognition test (as revealed by glm model; mean estimates, and standard errors from glm model; ***: $p < 0.001$, *: $p < 0.05$). The dashed line is a baseline (mean reaction to *P. apodus*) for statistical comparison. Abbreviations: (*H.*) *Hemorrhois*, (*E.*) *Eryx*, (*P.*) *Pseudopus*, and (*S.*) *Spalerosophis*

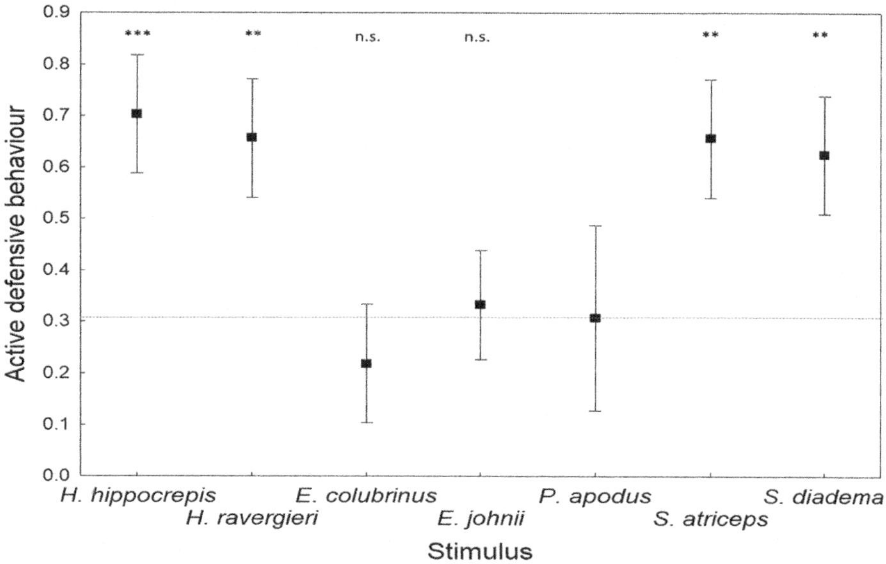


**Fig. 6** Active defensive strategy. Mean gecko's active defensive behaviour (high posture, mouth open, attack) towards various snake exuviae in comparison to the exuvia from the European legless lizard (*P. apodus*) in the Snake species discrimination test (as revealed by glm model; mean estimates and standard errors from glm model; ***: $p < 0.001$, **: $p < 0.01$). The dashed line is a baseline (mean reaction to *P. apodus*) for statistical comparison. Abbreviations: (*H.*) *Hemorrhois*, (*E.*) *Eryx*, (*P.*) *Pseudopus*, and (*S.*) *Spalerosophis*

Perfume test and the Odour removal test: geckos tongue flicked actively to the arbitrarily odorized exuviae as well as to deodorized exuviae. The rate of tongue flicking tended to be inverse to the rate of antipredator behaviour (i.e. the highest to the lowest tongue-flicking rate being elicited by: exuvia deodorized by non-polar solvent > perfume > exuvia deodorized by polar solvent > intact exuvia). In contrast, geckos showed antipredator behaviour only in response to intact exuviae and to a lesser extent to exuviae deodorized by ethanol. This may be interpreted as whenever geckos detect potentially dangerous odours, they engage in antipredator behaviours instead of lengthily exploring the stimulus by tongue flicking. On the contrary, Cooper and Burghard (Cooper & Burghardt, 1990) and Dial and Schwenk (1996) showed that, in some species of squamate reptiles (*Eumeces laticeps* and *Coleonyx brevis*), the change of tongue-flicking activity is accompanied by specific behaviour, such as attacking the prey or performing a defensive tail display in response to the predator odour. These specific behaviours determine the context of chemical detection in squamate reptiles. In this study, the exploration of a chemical stimulus initially increased tongue-flicking activity, especially in situations when the animal appears to perceive uncertainty about it, but it tended to decrease in the presence of exuviae.

In the present experiments, the shed skin scent was clearly altered with petrol, resulting in a loss of geckos' antipredator behaviour, although tongue flicking was still elicited. Petrol may indeed extract the skin lipids and other compounds contained in snake exuviae, which evolved from an initial water retention function to a secondary semiochemical function (Martín & López, 2014; Torri et al., 2014). When skin lipids of different species of squamates were analysed (Roberts & Lillywhite, 1980; Weldon & Bagnall, 1987), a complex pattern was identified constituted of free fatty acids, sterol esters, mainly cholesterol, wax esters, and other chemicals such as alcohols, aldehydes, methyl ketones, di- and tri-acylglycerols, phospholipids including sphingomyelin and glycolipids (reviewed in Weldon et al., 2008). But skin lipid chemistry was also shown to differ amongst snake species (e.g. within the genus *Chilabothrus*; Tolson, 1987; or strait-chain hydrocarbons typical of the indigo snake, *Drymarchon corais;* Ahern & Downing, 1974), leading to differentiated responses.

## 4.2 Antipredator vs. Tongue-Flicking Responses Towards Specific Snake Species

Our results revealed three main types of behavioural responses towards the exuviae. Geckos explored it by vomerolfaction (tongue-flicking) or performing active (high posture, attack) or passive (avoidance, low posture, freezing,) antipredator behaviour. When comparing the geckos' antipredator responses to the exuviae from snakes vs. one non-snake species (*P. apodus*), their rate was significantly higher towards all snake stimuli (except *E. colubrinus*) than to the non-snake stimuli. Thus, geckos can discriminate exuviae as a function of their snake vs. non-snake origin, suggesting that snake's exuviae convey common volatile compounds and potentially shared odour information. More specifically, the fact that the gecko's antipredator responses to the exuviae of snake species were more similar when the source animals belonged to the same genus suggests that related volatile compounds are genus- or species-differentiated. Thus, the gecko's antipredator reaction has a genus-specific component. Geckos demonstrated more passive antipredator strategy (low postures, avoidance) towards fossorial sand boas (genus *Eryx*), while they exhibited more active defensive behaviour (high postures, attacks) towards colubrid snakes (*Hemorrhois*, *Spalerosophis*). From the ecological point of view, it may be adaptive to stay motionless in presence of fossorial sand boas, but it is certainly a successful strategy not to enter into confined spaces with poor visibility of burrows. Similar behaviour was observed in Lesueur's geckos (*Amalosia lesueurii*) which avoid crevices scented with snake chemicals (Webb et al., 2010). On the other hand, active defensive behaviour can be beneficial in interaction with colubroid snakes, which are fast, agile predators that can actively chase their prey. It was demonstrated in another closely related eye-lid gecko of the genus *Coleonyx* that active defence exhibited in direct confrontation with a snake is beneficial (Dial & Schwenk, 1996).

Again, the occurrence of geckos' tongue flicking to the stimuli did not follow the same rules as their antipredator response. Tongue flicking was indeed more frequent to the non-snake exuvia (*P. apodus*) than to the different snakes' exuviae. Tongue flicking may reflect a general chemosensory exploration of an unfamiliar stimulus before deciding on whether to engage in an antipredator response or not. Otherwise, when the geckos responded by antipredator behaviour (either active or passive defence), they did not spend so much time in chemically exploring the stimulus, and the tongue-flicking rate was accordingly lower.

The gecko's cognitive appraisal of the presented odour stimuli may be expressed in three graded levels of behavioural response: (1) Detection of the odour against background noise through an increased frequency of tongue flicking and related activation of vomerolfaction (Conversano et al., 2014; Wyatt, 2014). (2) Discrimination amongst chemostimuli, leading to differentiation of odour categories and adjusted behavioural responses (Lloyd et al., 2009); for example, in this study, tongue-flicking appeared inversely related to antipredator behaviours depending on the source of odour: the more 'threatening' the category, the less it elicited tongue flicking and the more it elicited antipredator responses. (3) Recognition of a threat-related chemosignal and selection of appropriate behavioural responses: leopard geckos olfactorily discriminated predators from controls and between predators and reacted by the distinctive and intensive antipredator response. This recognition performance of snake predators may be innate (Landová et al., 2016) or learned (Epp & Gabor, 2008).

In a previous study on gecko's antipredator behaviour towards live snakes (Landová et al., 2016), we used a wider array of stimulus species including those of the current study and the same non-snake species. At first qualitative sight, it appeared that the geckos expressed a similar repertoire of antipredator responses to live snakes than to exuviae, although they expressed a higher level of antipredator behaviour with live snakes. But geckos displayed also other behaviours towards live snakes, such as binocular visual fixation and freezing. Frequency of tongue flicking appeared also slightly lower towards live snakes in comparison to the live non-snake species (*P. apodus*). This discrepancy between the earlier and present studies may be due to the condition of presentation of the stimuli, in terms of duration (30 min vs. 1.5 min), emitter-receiver distance and visibility (live snakes were separated by a wire-mesh, whereas the exuviae were presented directly), the chemical composition of live animals vs. exuviae, and the perception of multisensory cues associating odours with visual movement. The discrepancy in the intensity of tongue flicking between previous and current experiments may also be interpreted as reflecting different levels of risk perception by the geckos: while more distant stimuli were first explored by tongue flicking, danger assessment from proximally presented exuviae led to an immediate antipredator response (Landová et al., 2016).

The captive-bred geckos reacted to the odour of the exuviae without previous direct experience with any snake predator, which supports the idea that they possess an innate general antipredator reaction to snakes. Also, wild geckos ($F_0$ generation) displayed lower antipredator reaction in comparison with captive-bred ($F_1$ and $F_2$) animals. Wild geckos may indeed have undergone previous experience with real predation episodes, and their reactions may thus have become more specific, less intensive, and adjusted only to real danger. Finally, we did not find any difference in the response towards sympatric and allopatric snake species, which is consistent with previous results (Landová et al., 2016).

### 4.3 Conclusions: A New Bioassay for Chemosensory Discrimination in Geckos

According to Mackintosh (1985) who reviewed the properties of reliable olfactory bioassays in mammals, this study validated a new bioassay for squamates. We met most of the proposed criteria. We found a clearly defined behavioural response to the odour of snake exuvia, which we were able to record and measure properly. We examined an adequate test population, and the situation for testing was close to the natural. In future, we can focus on the last criterion concerning chemical analysis. Geckos' antipredator behaviour was triggered by chemostimuli released by snake exuviae. Within the range of stimuli used, the geckos' antipredator reactions were exclusive to the exuviae of snake/lizard predators relative to several scentless or scented (perfume) controls. These reactions were attributable neither to general arousal or alert nor to sensitization effects. Furthermore, the geckos differentiated olfactorily the exuviae from different predators, with more pronounced reactions towards snake predators relative to a saurophagous lizard predator. Finally, the geckos' antipredator repertoire to snake exuviae is representative of similar responses reported previously with live snakes (Landová et al., 2016). Finally, we would like to emphasize that the tongue-flicking response is not discriminant and should therefore be used with caution in chemosensory experiments with squamate reptiles.

**Acknowledgements** A. C. and P. H. were supported by the SVV project (260571/2022), and P. F. was supported by Charles University Research Centre program No. 204069. We are grateful to Jakub Polák who kindly improved English. For providing the technical facilities and for the financial foundation of this project we thank D. F.

## References

Abramjan, A., Baranová, V., Frýdlová, P., Landová, E., & Frynta, D. (2020). Ultraviolet reflectance and pattern properties in leopard geckos (*Eublepharis macularius*). *Behavioural Processes, 173*, 104060. https://doi.org/10.1016/j.beproc.2020.104060

Ahern, D. G., & Downing, D. T. (1974). Skin lipids of the Florida indigo snake. *Lipids, 9*(1), 8–14. https://doi.org/10.1007/BF02533208

Amo, L., López, P., & Martín, J. (2004). Chemosensory recognition and behavioral responses of wall lizards, *Podarcis muralis*, to scents of snakes that pose different risks of predation. *Copeia, 2004*(3), 691–696. https://doi.org/10.1643/CE-03-220R1

Amo, L., Lopez, P., & Martín, J. (2006). Can wall lizards combine chemical and visual cues to discriminate predatory from non-predatory snakes inside refuges? *Ethology, 112*(5), 478–484. https://doi.org/10.1111/j.1439-0310.2005.01170.x

Balderas-Valdivia, C. J., & Ramírez-Bautista, A. (2005). Aversive behavior of beaded lizard, *Heloderma horridum*, to sympatric and allopatric predator snakes. *The Southwestern Naturalist, 50*(1), 24–31.

Barbosa, D., Font, E., Desfilis, E., & Carretero, M. A. (2006). Chemically mediated species recognition in closely related Podarcis wall lizards. *Journal of Chemical Ecology, 32*(7), 1587–1598. https://doi.org/10.1007/s10886-006-9072-5

Bealor, M. T., & Krekorian, C. N. (2002). Chemosensory identification of lizard-eating snakes in the desert iguana, *Dipsosaurus dorsalis* (Squamata: Iguanidae). *Journal of Herpetology, 36*(1), 9–15. https://doi.org/10.1670/0022-1511(2002)036[0009:CIOLES]2.0.CO;2

Borowski, Z., & Owadowska, E. (2010). Field vole (*Microtus agrestis*) seasonal spacing behavior: The effect of predation risk by mustelids. *Naturwissenschaften, 97*(5), 487–493. https://doi.org/10.1007/s00114-010-0663-1

Bramley, G. N., Waas, J. R., & Henderson, H. V. (2000). Responses of wild Norway rats (*Rattus norvegicus*) to predator odors. *Journal of Chemical Ecology, 26*(3), 705–719. https://doi.org/10.1023/A:1005432322861

Burghardt, G. M. (1980). Behavioral and stimulus correlates of vomeronasal functioning in reptiles: Feeding, grouping, sex, and tongue use. In D. Müller-Schwarze & R. M. Silverstein (Eds.), *Chemical signals in vertebrates and aquatic invertebrates* (pp. 275–301). Springer.

Conversano, J., Tan, E. J., van Wilgenburg, E., & Elgar, M. A. (2014). Background odour may impair detection of chemical signals for social recognition. *Austral Entomology, 53*(4), 432–435. https://doi.org/10.1111/aen.12087

Cooper, W. E. (1994). Chemical discrimination by tongue-flicking in lizards: A review with hypotheses on its origin and its ecological and phylogenetic relationships. *Journal of Chemical Ecology, 20*(2), 439–487. https://doi.org/10.1007/BF02064449

Cooper, W. E., & Burghardt, G. M. (1990). A comparative analysis of scoring methods for chemical discrimination of prey by squamate reptiles. *Journal of Chemical Ecology, 16*(1), 45–65. https://doi.org/10.1007/BF01021267

Cooper, W. E., Jr. (1990). Chemical detection of predators by a lizard, the broad-headed skink (Eumeces laticeps). *The Journal of Experimental Zoology, 256*(2), 162–167. https://doi.org/10.1002/jez.1402560206

Cooper, W. E., Jr. (1995). Prey chemical discrimination and foraging mode in gekkonoid lizards. *Herpetological Monographs, 9*, 120–129. https://doi.org/10.2307/1467000

Cooper, W. E., Jr., & Sherbrooke, W. C. (2009). Prey chemical discrimination by tongue flicking is absent in the Texas horned lizard, *Phrynosoma cornutum*. *Journal of Herpetology, 43*(4), 688–692.

Cooper, W. E., & Steele, L. J. (1997). Pheromonal discrimination of sex by male and female leopard geckos *(Eublepharis macularius)*. *Journal of Chemical Ecology, 23*(12), 2967–2977. https://doi.org/10.1023/A:1022587716051

Daniell, T. L., Hutchinson, M. N., Ainsley, P., & Gardner, M. G. (2021). Recognition of reptile predator scent is innate in an endangered lizard species. *Australian Journal of Zoology, 68*(2), 76–84. https://doi.org/10.1071/ZO20064

Dial, B. E., & Schwenk, K. (1996). Olfaction and predator detection in *Coleonyx brevis* (Squamata: Eublepharidae), with comments on the functional significance of buccal pulsing in geckos. *The Journal of Experimental Zoology, 276*(6), 415–424. https://doi.org/10.1002/(SICI)1097-010X(19961215)276:6<415::AID-JEZ5>3.0.CO;2-Q

Dial, B. E., Weldon, P. J., & Curtis, B. (1989). Chemosensory identification of snake predators (*Phyllorhynchus decurtatus*) by banded geckos (*Coleonyx variegatus*). *Journal of Herpetology, 23*(3), 224–229. https://doi.org/10.2307/1564443

Downes, S., & Shine, R. (1998). Sedentary snakes and gullible geckos: Predator–prey coevolution in nocturnal rock-dwelling reptiles. *Animal Behaviour, 55*(5), 1373–1385. https://doi.org/10.1006/anbe.1997.0704

Epp, K. J., & Gabor, C. R. (2008). Innate and learned predator recognition mediated by chemical signals in *Eurycea nana*. *Ethology, 114*(6), 607–615. https://doi.org/10.1111/j.1439-0310.2008.01494.x

Ferkin, M. H., & LaDage, L. D. (2006). Male leopard geckos (*Eublepharis macularius*) can discriminate between two familiar females. *Behaviour, 143*(8), 1033–1049. https://doi.org/10.1163/156853906778623644

Font, E., Barbosa, D., Sampedro, C., & Carazo, P. (2012). Social behavior, chemical communication, and adult neurogenesis: Studies of scent mark function in Podarcis wall lizards. *General and Comparative Endocrinology, 177*(1), 9–17. https://doi.org/10.1016/j.ygcen.2012.02.015

Frynta, D., Jančúchová-Lásková, J., Frýdlová, P., & Landová, E. (2018). A comparative study of growth: Different body weight trajectories in three species of the genus *Eublepharis* and their hybrids. *Scientific Reports, 8*(1), 1–11. https://doi.org/10.1038/s41598-018-19864-3

Gabirot, M., Castilla, A. M., López, P., & Martín, J. (2010). Chemosensory species recognition may reduce the frequency of hybridization between native and introduced lizards. *Canadian Journal of Zoology, 88*(1), 73–80. https://doi.org/10.1139/Z09-120

Halpern, M. I. (1992). Nasal chemical senses in reptiles: Structure and function. In C. Gans & D. Crews (Eds.), *Hormones, brain and behavior, biology of the reptilia, 18* (pp. 423–523). University of Chicago Press.

Halpern, M., & Martínez-Marcos, A. (2003). Structure and function of the vomeronasal system: An update. *Progress in Neurobiology, 70*(3), 245–318. https://doi.org/10.1016/S0301-0082(03)00103-5

Hammond, J. E., Witkowski, S., Wilson, T., Zouvi, C. A., Goetz, N. L., Eck, N. F., & Clark, R. W. (2020). Know thine enemy: Predator identity influences the response of Western Banded Geckos (*Coleonyx variegatus*) to chemosensory cues. *Journal of Herpetology, 54*(4), 480–484. https://doi.org/10.1670/19-131

Højsgaard, S., Halekoh, U., & Yan, J. (2006). The R package geepack for generalized estimating equations. *Journal of Statistical Software, 15*, 1–11. https://doi.org/10.18637/jss.v015.i02

Jančúchová-Lásková, J., Landová, E., & Frynta, D. (2015). Experimental crossing of two distinct species of leopard geckos, *Eublepharis angramainyu* and *E. macularius*: Viability, fertility and phenotypic variation of the hybrids. *PLoS One, 10*(12), e0143630. https://doi.org/10.1371/journal.pone.0143630

Khan, M. S. (2004). Annotated checklist of amphibians and reptiles of Pakistan. *Asiatic Herpetological Research, 10*(1), 191–201.

Labra, A. (2011). Chemical stimuli and species recognition in *Liolaemus* lizards. *Journal of Zoology, 285*(3), 215–221. https://doi.org/10.1111/j.1469-7998.2011.00832.x

Landová, E., Musilová, V., Polák, J., Sedláčková, K., & Frynta, D. (2016). Antipredatory reaction of the leopard gecko (*Eublepharis macularius*) to snake predators. *Current Zoology, 62*(5), 439–450. https://doi.org/10.1093/cz/zow050

Lloyd, R. A. Y., Alford, R. A., & Schwarzkopf, L. I. N. (2009). Chemical discrimination among predators by lizards: Responses of three skink species to the odours of high-and low-threat varanid predators. *Austral Ecology, 34*(1), 50–54. https://doi.org/10.1111/j.1442-9993.2008.01881.x

López, P., & Martín, J. (2001). Chemosensory predator recognition induces specific defensive behaviours in a fossorial amphisbaenian. *Animal Behaviour, 62*(2), 259–264. https://doi.org/10.1006/anbe.2001.1762

Mackintosh, J. H. (1985). The bioassay of mammalian olfactory signals. *Mammal Review, 15*(2), 57–70. https://doi.org/10.1111/j.1365-2907.1985.tb00388.x

Martín, J., & López, P. (2006). Interpopulational differences in chemical composition and chemosensory recognition of femoral gland secretions of male lizards *Podarcis hispanica*: Implications for sexual isolation in a species complex. *Chemoecology, 16*(1), 31–38. https://doi.org/10.1007/s00049-005-0326-4

Martín, J., & López, P. (2014). Pheromones and chemical communication in lizards. In J. L. Rheubert, D. S. Siegel, & S. E. Trauth (Eds.), *Reproductive biology and phylogeny of lizards and tuatara* (pp. 43–77). CRC Press.

Martín, J., Ortega, J., & López, P. (2015). Experience may allow increasing accuracy of the innate chemosensory recognition of snake predators by Iberian wall lizards. *Behavioral Ecology and Sociobiology, 69*(9), 1565–1572. https://doi.org/10.1007/s00265-015-1968-z

Mason, R. T., & Gutzke, W. H. (1990). Sex recognition in the leopard gecko, *Eublepharis macularius* (Sauria: Gekkonidae) possible mediation by skin-derived semiochemicals. *Journal of Chemical Ecology, 16*(1), 27–36. https://doi.org/10.1007/BF01021265

Mason, R. T., & Parker, M. R. (2010). Social behavior and pheromonal communication in reptiles. *Journal of Comparative Physiology. A, 196*(10), 729–749. https://doi.org/10.1007/s00359-010-0551-3

McElroy, E. J. (2019). Antipredator behavioral mechanisms: Avoidance, deterrence, escape, and encounter. In V. L. Bels & A. P. Russell (Eds.), *Behavior of lizards* (pp. 143–188). CRC Press.

Meredith, M., & Burghardt, G. M. (1978). Electrophysiological studies of the tongue and accessory olfactory bulb in garter snakes. *Physiology & Behavior, 21*(6), 1001–1008. https://doi.org/10.1016/0031-9384(78)90178-6

Miller, L. R., & Gutzke, W. H. (1999). The role of the vomeronasal organ of crotalines (Reptilia: Serpentes: Viperidae) in predator detection. *Animal Behaviour, 58*(1), 53–57. https://doi.org/10.1006/anbe.1999.1126

Mori, A., & Hasegawa, M. (1999). Geographic differences in behavioral responses of hatchling lizards (*Eumeces okadae*) to snake-predator chemicals. *Japanese Journal of Herpetology, 18*(2), 45–56. https://doi.org/10.5358/hsj1972.18.2_45

Noldus Information Technology. (1993). *Observer 3.0. Noldus, Wageningen* [cited 2016 April 28]. Available from: http://www.noldus.com

Ortega, Z., Mencía, A., & Pérez-Mellado, V. (2018). Antipredatory behaviour of a mountain lizard towards the chemical cues of its predatory snakes. *Behaviour, 155*(10–12), 817–840. https://doi.org/10.1163/1568539X-00003504

Punzo, F. (2005). Chemosensory recognition by males of the desert pocket mouse, *Chaetodipus pencillatus* to odors of various species of snakes. *Ethology Ecology and Evolution, 17*(1), 83–89. https://doi.org/10.1080/08927014.2005.9522618

R Core Team. (2021). *R: A language and environment for statistical computing*. R Foundation for Statistical Computing. https://www.R-project.org/

Roberts, J. B., & Lillywhite, H. B. (1980). Lipid barrier to water exchange in reptile epidermis. *Science, 207*(4435), 1077–1079. https://doi.org/10.1126/science.207.4435.1077

Schořálková, T., Kratochvíl, L., & Kubička, L. (2018). Female sexual attractiveness and sex recognition in leopard gecko: Males are indiscriminate courters. *Hormones and Behavior, 99*, 57–61. https://doi.org/10.1016/j.yhbeh.2018.01.007

Seufer, H., Kaverkin, H., & Kirschner, A. (2005). *The eyelash geckos. Care, breeding and natural history*. Kirschner & Seufer Verlag.

Simon, C. A., Gravelle, K., Bissinger, B. E., Eiss, I., & Ruibal, R. (1981). The role of chemoreception in the iguanid lizard *Sceloporus jarrovi*. *Animal Behaviour, 29*(1), 46–54. https://doi.org/10.1016/S0003-3472(81)80150-9

Sindaco, R., Jeremčenko, V. K., Venchi, A., & Grieco, C. (2008). *The reptiles of the Western palearctic: Annotated checklist and distributional atlas of the turtles, crocodiles, amphisbaenians and lizards of Europe, North Africa, Middle East and Central Asia*. Edizioni Belvedere.

Steele, L. J., & Cooper, W. E., Jr. (1997). Investigations of pheromonal discrimination between conspecific individuals by male and female leopard geckos (*Eublepharis macularius*). *Herpetologica, 53*(4), 475–484.

Szczerbak, N. N., & Golubev, M. L. (1986). *Geckos of the USSR fauna and adjoining countries*. UkrSSR: Trudy Zoologicheskogo Instituta Akademii Nauk SSSR. [Proc. Zool. Inst. Acad. Sci.]

Thoen, C., Bauwens, D., & Verheyen, R. F. (1986). Chemoreceptive and behavioural responses of the common lizard Lacerta vivipara to snake chemical deposits. *Animal Behaviour, 34*(6), 1805–1813. https://doi.org/10.1016/S0003-3472(86)80266-4

Tolson, P. J. (1987). Phylogenetics of the boid snake genus *Epicrates* and Caribbean vicariance theory. *Occasional Papers of the Museum of Zoology, 715*, 1–67.

Torri, C., Mangoni, A., Teta, R., Fattorusso, E., Alibardi, L., Fermani, S., Bonacini, I., Gazzano, M., Burghammer, M., Fabbri, D., & Falini, G. (2014). Skin lipid structure controls water permeability in snake molts. *Journal of Structural Biology, 185*(1), 99–106. https://doi.org/10.1016/j.jsb.2013.10.007

Van Damme, R., & Castilla, A. M. (1996). Chemosensory predator recognition in the lizard *Podarcis hispanica*: Effects of predation pressure relaxation. *Journal of Chemical Ecology, 22*(1), 13–22. https://doi.org/10.1007/BF02040196

van Damme, R., Bauwens, D., Thoen, C., Vanderstighelen, D., & Verheyen, R. F. (1995). Responses of naive lizards to predator chemical cues. *Journal of Herpetology, 29*(1), 38–43. https://doi.org/10.2307/1565083

Van Moorleghem, C., Huyghe, K., & Van Damme, R. (2020). Chemosensory deficiency may render island-dwelling lizards more vulnerable to invasive predators. *Biological Journal of the Linnean Society, 129*(1), 128–142. https://doi.org/10.1093/biolinnean/blz142

Vandewege, M. W., Mangum, S. F., Gabaldón, T., Castoe, T. A., Ray, D. A., & Hoffmann, F. G. (2016). Contrasting patterns of evolutionary diversification in the olfactory repertoires of reptile and bird genomes. *Genome Biology and Evolution, 8*(3), 470–480. https://doi.org/10.1093/gbe/evw013

Webb, J. K., Pike, D. A., & Shine, R. (2010). Olfactory recognition of predators by nocturnal lizards: Safety outweighs thermal benefits. *Behavioral Ecology, 21*(1), 72–77. https://doi.org/10.1093/beheco/arp152

Weldon, P. J., & Bagnall, D. (1987). A survey of polar and nonpolar skin lipids from lizards by thin-layer chromatography. *Comparative Biochemistry and Physiology--Part B: Biochemistry and Molecular Biology, 87*(2), 345–349.

Weldon, P. J., Flachsbarth, B., & Schulz, S. (2008). Natural products from the integument of nonavian reptiles. *Natural Product Reports, 25*(4), 738–756. https://doi.org/10.1039/B509854H

Wyatt, T. D. (2014). *Pheromones and animal behavior: Chemical signals and signatures*. Cambridge University Press.

Zuri, I., & Halpern, M. (2003). Differential effects of lesions of the vomeronasal and olfactory nerves on garter snake (*Thamnophis sirtalis*) responses to airborne chemical stimuli. *Behavioral Neuroscience, 117*(1), 169–183. https://doi.org/10.1037/0735-7044.117.1.169

# Reactions of Free-Ranging African Carnivores to Single Components of Carnivore Odours: Evidence that Mammalian Semiochemicals Are Not All Irreducibly Complex

Peter Apps[1] (✉) and J. Weldon McNutt[2]

[1] BioBoundary Project, Botswana Predator Conservation, Maun, Botswana
[2] Botswana Predator Conservation, Maun, Botswana

**Abstract** Despite their potential, very few mammalian olfactory cues or signals have found practical applications, partly because the assumption that mammalian semiochemicals are complex mixtures has deterred or misdirected applied research. Here, we provide evidence that challenges the assumption of complexity; camera trap videos show predators of eight species reacting to single compounds or simple mixtures of components of carnivore odours by rubbing and urine-marking, behaviours that commonly occur in response to natural scent marks. Reactions were selective for specific compounds, and different species reacted to different compounds, but interpretation of the results was complicated by interferences from animals scent-marking the test compound dispensers and possible influences of location, time of year, and wind direction. Our results are consistent with other evidence that mammalian chemical signals are simple mixtures of compounds that could be applied to management and conservation.

**Keywords** Semiochemical · Camera trapping · Human-wildlife conflict · Pheromone · Carnivore · Predator · Repellent · Deterrent

## 1 Introduction

Since the first chemical characterization of a pheromone in 1959 (Butenandt et al., 1961; reviewed in Wyatt, 2009), thousands of pheromones from insects have been identified, but only a few dozen from mammals (Apps, 2013; Apps et al., 2015). There is an even more striking contrast between insect and mammal semiochemicals in how many of them have found practical applications. Apart from synthetic boar pheromones to detect oestrus in sows (Melrose et al., 1971), and synthetic 'appeasing' pheromones for household pets (Pageat & Gaultier, 2003; DePorter et al., 2019), hardly any practical

 
B. Schaal et al. (eds.), *Chemical Signals in Vertebrates 15*,
https://doi.org/10.1007/978-3-031-35159-4_20

applications of synthetic mammalian chemical signals have progressed beyond experimental trials and proofs of concept. We have found no applications of synthetic intraspecific mammalian semiochemicals for pest control or conservation. Even for the Asian elephant (*Elephas maximus*), whose semiochemistry is better understood than any other wild species', the promise of semiochemical 'applications for the conservation of wild elephants and the management of elephants in human care' (Schulte et al., 2007; Schulte & LaDue, 2021) remains elusive.

Practical applications of synthetic interspecific mammalian olfactory cues are similarly rare. There have been major investments into research on the responses of prey to cues from predator odours (Weldon, 1990; Kats & Dill, 1998; Apfelbach et al., 2005; Banks et al., 2014), but promising early results and expectations for commercialization (Boag & Mlotkiewicz, 1994; Lindgren et al., 1995) have not been realized. Synthetic repellents based on predator odours have been ineffective in independent trials (e.g. Schlageter & Haag-Wackernagel, 2012), while effective repellents are not derived from predators' odours (e.g. Mason, 1998; Nolte & Wagner, 2000).

The view that mammalian chemical signals are coded by the relative quantities of the components of complex mixtures (Müller-Schwarze, 2016; Pickett et al., 2014) has been a barrier to the identification of the semiochemically active components of mammal odours (Apps, 2013), and the view that interspecific cues are also complex mixtures (Apfelbach et al., 2015) has had a similar effect on the identification of the active components of interspecific cues. Only five out of 70 tests of predator odours as repellents used synthetic scent (Kats & Dill, 1998), and there have been few attempts to reduce the number of components in formulations.

In both intra- and interspecific semiochemistry, assuming that signals and cues are complex impedes the development of applications for management and conservation. Simulating complex mixtures is technically challenging and prohibitively costly, while using natural scents such as urine or faeces at more than experimental scale requires impractically large quantities of material (Jackson et al., 2012; Wauson & Rogers, 2021) and creates a risk of disease transmission. All else being equal, single compounds are easier to release at controlled rates, and mixtures with fewer components are cheaper to formulate and less likely to contain incompatible components (Kimball et al., 2000). Apps (2013) reviewed the compositions of mammalian semiochemicals, and the most complex of them have about a dozen components. The effects of some mammalian cues and signals with potential as management tools can be replicated by one or a few of their components, simplified versions of fermented lures are as effective as the originals (Mason & Blom, 1998; Kimball et al., 2000), single components of castoreum lure beavers (*Castor canadensis*) out of the water (Müller-Schwarze & Houlihan, 1991), and a single component of a social signal is an effective rodent bait enhancement (Parshad, 2002). The few interspecific cues that have been characterized are simple in composition (Lindgren et al., 1995; Osada et al., 2013; Ferrero et al., 2011). Predators respond equally to whole blood odour and to a single blood odour component (Nilsson et al., 2014). Laboratory mice (*Mus musculus*) avoid single compounds from predator odours, albeit with a striking selectivity according to minor differences in molecular structure (Sievert & Laska, 2016).

We report here on records of predators' scent-marking responses to test substances composed of single components or a two-component mixture of components of predator odours. These records were obtained as camera trap by-catch during pilot experiments on which components of carnivore odours have potential as deterrents to keep predators away from livestock and thereby ameliorate human-predator conflict.

# 2 Methods

## 2.1 Odour Compounds

Components of carnivore odours from urine, faeces, and glandular secretions were identified by gas chromatography – mass spectrometry in the BioBoundary Laboratory (Apps et al., 2012, 2013, 2014 and unpublished), and a selection of pure components that were available commercially were tested for repellent effects on wild, free-ranging African carnivores (Table 1). In the absence of any guidance from published work on which classes of compounds were most likely to generate the desired effect, priority was given to compounds that are species-specific, contain sulphur or nitrogen heteroatoms, are more abundant in predators' than herbivores' odours, and have molecular weights and vapour pressures that generate emission rates of micrograms per hour, either as pure substances or from controlled-release devices.

## 2.2 Controlled Release of Test Substances

The potential predator repellents were dispensed from controlled release devices built from laboratory sample vials with silicone rubber septa (Restek Inc. #21743 Supelco #27021) with the impermeable PTFE face shaved off with a scalpel and apertures of

**Table 1** Test substances, species in which they were identified, and controlled release method

| Test compound(s) | African species where the compound has been identified | Composition of test substance | Controlled release |
|---|---|---|---|
| 2-ethyl-5(6)-methyl pyrazine | African wild dog, lion | Pure | silicone septum |
| dimethyl disulfide | African wild dog | Pure | silicone septum |
| hydrocinnamic acid | African wild dog, lion | Pure | 12 mm aperture |
| Indole | African wild dog, lion, leopard, black-backed jackal, cheetah | Pure | 8 mm aperture |
| methylbutanoic acids | African wild dog, lion, leopard, cheetah | 1:1 mixture of 2- and 3-methylbutanoic acids | 8 mm aperture |
| 2-Piperidinone | African wild dog, black-backed jackal, spotted hyaena, lion, leopard, cheetah, | Pure | 12 mm aperture |
| Skatole | lion | Pure | 8 mm aperture |

different sizes (Fig. 1). Apertures were screened with fine mesh to exclude insects. Emissions were adjusted by septum material and thickness, and also aperture size to provide a detectable odour to the human nose 5–10 cm from the vial, but no detectable odour 1 m away. This odour intensity was subjectively similar to the intensity of odours from fresh scent marks. Emission rates were determined by loss in mass over a measured time, and vials were loaded with sufficient substance to last for 2 months (two cycles of camera services, see below), with an extra allowance for increased emissions when temperatures were high, and to allow for delays in exchanging test substances.

The controlled-release vials were protected by metal housings made from 25 mm plumbing fittings, attached to metal stakes hammered into the ground, with the housing 20–30 cm above the surface (Fig. 1). The stakes were initially 50 or 60 cm lengths of 12 mm rebar, sharpened to a point at one end, with a 25 mm pipe cap welded to one side about 50 mm from the top. The stakes were hammered into the ground for about half their length before the top part of the dispenser housing was attached, with the longer stakes being used on softer, sandy soil and the shorter ones on pans and old termite mounds. The top part of the housing was then attached by screwing it into the cap welded to the stake, with plumber's tape to stop the two parts seizing together with rust. For the last 2 months of the period reported here, the rebar stakes were replaced by 25 × 25 × 1.8 mm angle iron, with the housing bolted on. Dispenser vials were loaded into holders when the camera traps were serviced (see below).

**Fig. 1** *Left*: An example of a dispenser vial with a silicone rubber membrane. *Right*: a dispenser housing in the field. The vial stands upright in the stem of plumbing connector

### 2.3 Field Testing Conditions, Behavioural Monitoring, and Variables

Tests were run on a wildlife ranch in northern Botswana with test stations set up along dirt roads because African predators are known to use roads to travel (Abrahms et al., 2016; Rafiq et al., 2020). Sites were selected according to the following criteria; accessible by vehicle without driving off-road, not on the perimeter of the ranch due to risks of theft or vandalism, approximately evenly spaced, on a circuit that reduced travelling time when cameras and dispensers were serviced, at sites were vegetation at ground level was sparse so that cameras were not triggered excessively by moving vegetation and so that the task of trimming vegetation could be accomplished in a practical time span, where leopards (*Panthera pardus*), the primary target species of the deterrents research, or their spoor had been seen by the land-owner, and not within 500 m of a water source. Ten stations with nearest neighbour distances of 1.56 ± 1.11 km (mean ± SD) each had two dispensers and two camera traps (trail cameras). In a layout developed over several iterations at different study sites since 2015 (Apps et al., 2017, 2019; Apps & McNutt, 2018), a dispenser in its housing was positioned approximately 1 m from the edge of the road, with a camera trap (Reconyx XR6 or Browning Patriot) mounted 30–40 cm off the ground on the opposite side of the road to the dispenser, 7–8 m from the scent dispenser, with the dispenser in the centre of the camera's field of view. Between approximately 5 m and approximately 15 m (depending on vegetation) along the road another dispenser and camera were set up with the dispenser on the opposite side of the road to the first one, so that animals travelling along the road encountered windborne scent irrespective of the direction of the wind. Cameras were set to highest sensitivity, maximum IR light output and the shortest post-capture delay. This arrangement has been validated (Apps & McNutt, 2018) for its ability to capture video of animals approaching and interacting with the dispenser or ignoring it. Video duration was 30 s on the XR6s and 20 s on the Patriots. Cameras were checked approximately monthly, logistics permitting, with secure digital SD memory cards and batteries being replaced, lenses cleaned, and aim adjusted.

Two test odours were in the field simultaneously, one at each of five of the 10 test stations. Each odour was tested for 2 months (with some adjustments to accommodate logistic constraints), with the same odour in both dispensers at one test station and different odours at adjacent stations. Before any of the substances whose results are reported here were released, reactions to empty dispensers were monitored for 2 months at five stations and for 3 months at the other five stations as a check on reactions being elicited not by the test odour but by the dispenser housing being a convenient size and height for rubbing or marking or by the presence of scent marks deposited earlier.

The deterrent screening tests have been running since 2019, with continual refinements to hardware and procedures, and the by-catch results on scent-marking presented here are from December 2021 onwards when the number of stations was increased to 10 and the camera models, settings, and arrangement were standardized.

### 2.4 Data Extraction and Analysis

Videos of predators were examined for the predator's reactions (if any) to the test odour. Rubbing, urinating, defecating, and anal marking are all considered as scent-marking. All else being equal, stations with more activity would be expected to record more reactions, and so the number of reactions was corrected for activity by expressing the number of reactions as a percentage of the total number of video captures for that species and odour.

The data are not suitable for inferential analysis, and even descriptive statistics had to be applied with considerable caution due to the small number of stations, the skewed, truncated, and heterogeneous distribution of reactions (see Results), and also the self-selection bias (Webster & Rutz, 2020) that is intrinsic to camera trap data because they record a selection of behaviours, not a random sample; they select for behaviours that occur in their detection zones, that involve more movement, movement in particular directions (Apps & McNutt, 2018), and that last long enough to be recorded after the camera's trigger delay.

## 3 Results

### 3.1 Camera Trap Captures

The camera traps captured 3720 videos of 15 carnivore species; aardwolf *Proteles cristata,* African wild dog *Lycaon pictus,* black-backed jackal *Lupulella mesomelas,* bat-eared fox *Otocyon megalotis.* African wildcat *Felis silvestris lybica,* caracal *Caracal caracal,* leopard *Panthera pardus,* lion *Panthera leo,* brown hyaena *Parahyaena brunnea,* honey badger *Mellivora capensis,* slender mongoose *Galerella sanguinea,* Cape fox *Vulpes chama*, striped polecat *Ictonyx striatus,* small-spotted genet *Genetta genetta,* and spotted hyaena *Crocuta crocuta.*

The minimum number of video captures per species was three (lions), and the maximum 1801 (black-backed jackals). Six of the species were captured more than once per week on average, and their data were analysed further. The numbers of videos of each species fluctuated markedly with time (Fig. 2).

### 3.2 Reactions to Test Substances

All of the test substances except the mixture of methylbutanoic acids were scent-marked by at least one species, and of the species that were videoed more than once a week, all except spotted hyaenas scent-marked at least one test substance. Scent-marking behaviours were qualitatively the same as those used to mark natural substrates (videos archived at Apps, Peter, 2023, 'African predators scent-marking in response to single compounds', https://doi.org/10.7910/DVN/OGOS1Z, Harvard Dataverse, V1). With the exception of black-backed jackals which reacted unselectively, and spotted hyaenas which did not react at all, predators reacted selectively to different subsets of the test substances, and different test substances elicited reactions from different subsets of species (Table 2).

**Table 2** Percentages of camera trap videos of each species that captured scent-marking (rub, urinate) reactions to each test substance for the species that were videoed more than once per week on average

| | AWC | BBJ | BEF | Caracal | Leopard | SH |
|---|---|---|---|---|---|---|
| 2-ethyl-5(6)-methyl pyrazine | 0.0 | 2.8 | 18.7 | 0.0 | 10.0 | 0.0 |
| dimethyl disulfide | 18.2 | 0.4 | 0.0 | 0.0 | 5.9 | 0.0 |
| hydrocinnamic acid | 4.3 | 2.4 | 0.0 | 0.0 | 0.0 | 0.0 |
| Indole | 0.0 | 5.7 | 0.0 | 0.0 | 30.0 | 0.0 |
| methylbutanoic acids | 0.0 | 0.0 | 0.0 | 0.0 | 0.0 | 0.0 |
| Piperidinone | 6.3 | 2.0 | 16.7 | 10.0 | 0.0 | 0.0 |
| Skatole | 0.0 | 2.4 | 5.1 | 0.0 | 5.9 | 0.0 |
| empty | 2.5 | 2.0 | 0.0 | 5.6 | 4.5 | 0.0 |

*AWC* African wildcat, *BBJ* black-backed jackal, *BEF* bat-eared fox, *SH* spotted hyaena

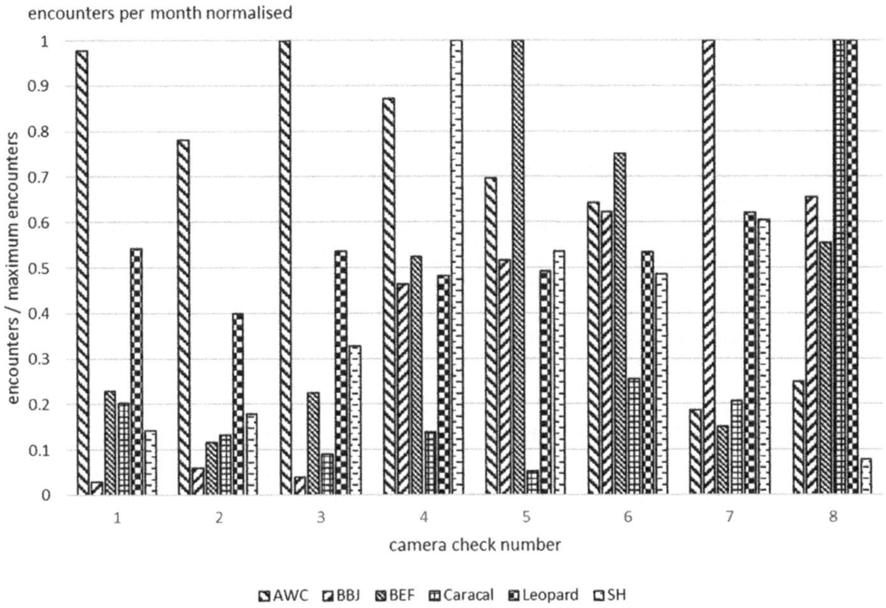


**Fig. 2** Fluctuations in the total number of camera trap videos (sum of those with reactions and without reactions) per month at each camera check for each species that was captured at least once per week on average. To bring all species to a common scale, the raw numbers have been divided by the maximum for that species. AWC African wildcat, BBJ Black-backed jackal, BEF bat-eared fox, SH spotted hyaena

Reactions to test substances were strikingly heterogeneous in both space and time. The number of times that a dispenser housing with a particular compound or the soil at the base of a housing was scent-marked varied from 0 to 19 and the number of scent-marks at a station (sum of the two dispensers) from 1 to 29. Some stations showed similar numbers and temporal distributions of reactions to each dispenser, others showed different reactions (Fig. 3).

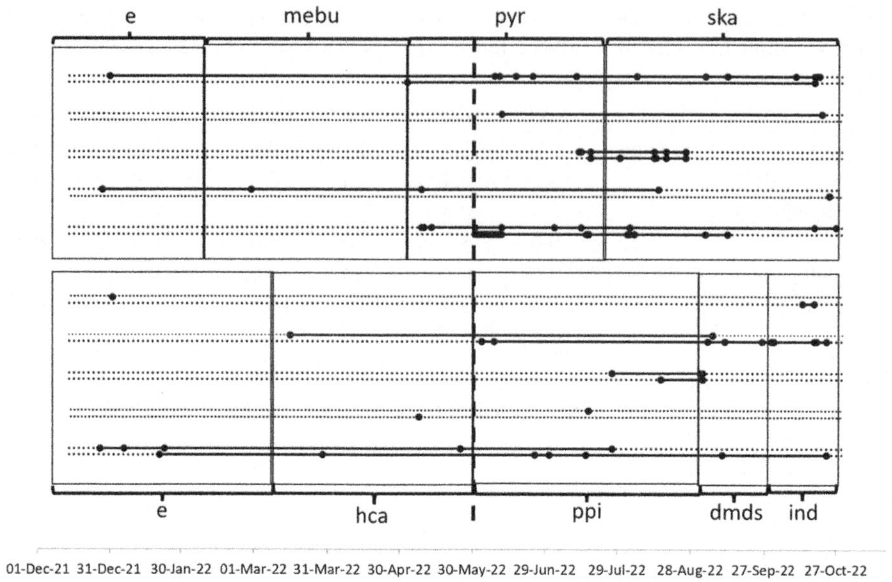


**Fig. 3** Dates on which predators reacted to test substances composed of single components or a two-component mixture of the components of carnivore odours. Each pair of lines is the camera traps at one test station, each videoing one dispenser. A dot marks a reaction, solid lines span the period between the first and last recorded reaction. Each rectangle marks the period during which a test substance was released from that set of dispensers; dmds; dimethyl disulfide, ppi; piperidinone, hca; hydrocinnamic acid, e; empty, ind; indole, ska; statole, pyr; 2-ethyl-5(6)-methyl pyrazine, mebu; 2- and 3-methylbutanoic acids. The dashed vertical line at the end of May shows where all the dispenser housings were exchanged for clean ones

There was a trend for the number of reactions per compound to increase with the number of videos, but the correlation is weak overall, for black-backed jackals, and for bat-eared foxes. The felids show no relationship at all (Fig. 4).

# 4 Discussion

## 4.1 The Nature of Reactions to Test Substances

The rubbing and urine-marking elicited by the test substances were the same as reactions to natural scent marks. There were definite increases in reactions when certain test compounds were released, the reactions by some species were selective to certain compounds, and compounds differed in which species (if any) responded to them (Table 2), as would be expected for different species' reactions to natural scent marks.

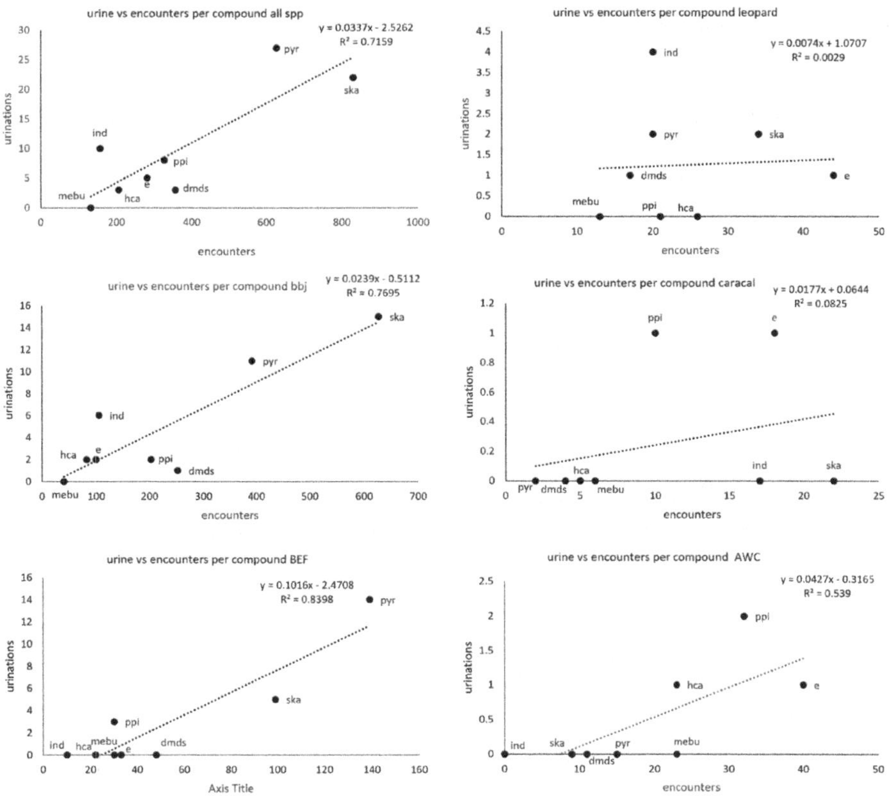


**Fig. 4** Relationship of number of urine-marks against number of camera trap video captures for predators exposed to test substances composed of single components or a two-component mixture of components of carnivore odours. (**a**) all species combined, (**b**) leopard, (**c**) black-backed jackal, (**d**) caracal, (**e**) bat-eared fox, (**f**) African wildcat. Test substance: dmds dimethyl disulfide, ppi piperidinone, hca hydrocinnamic acid, e empty, ind indole, ska statole, pyr 2-ethyl-5(6)-methyl pyrazine, mebu 2- and 3-methylbutanoic acids

Since the advent of modern camera trapping, interspecific scent-marking has been found to be common and widespread (Apps et al., 2017, 2019), and at least some of the scent-marking reactions recorded in this study may have been an individual of one species responding to an essential component of the scent mark of another, evidence that olfactory cues, as well as olfactory signals are carried by single compounds.

Some of the substances that elicited scent-marking, and other substances not reported here, have short-range repellent effects on some species (Apps, unpublished) that are being investigated as the basis for measures to deter predators from killing livestock. Scent-marking's role in predators' spatial organization suggests that it may be worthwhile to investigate synergies between repellents and substances that elicit scent-marking.

### 4.2 Rates of Reaction to Test Substances Compared to Natural Substrates and Natural Scent Marks

The percentages of video captures that involved scent-marking were generally low (Table 2), but comparable results from the literature are scarce; during their mating season, red foxes (*Vulpes vulpes*) marked 65% of artificially scented snow heaps (Whitten et al., 1980). Most results in the literature are from studies that targeted latrines and shared scent-marking sites, where high rates of scent-marking are to be expected. Nonetheless, the rates of scent-marking per visit recorded in this study for substances that elicited scent-marking fall within the range of marking rates for natural scent-marks at marking sites. Mature male pumas (*Puma concolor*) scent-marked on 78.5% of visits to marking sites, and mature females on 13.6% of visits (Allen et al., 2014). Bobcats scent-mark on 25% of visits to scent-marking sites (Allen et al., 2015). Red foxes urine-marked half of translocated lynx (*Lynx lynx*) scats they encountered (Wikenros et al., 2017). Male cheetahs (*Acinonyx jubatus*) scent-marked on 25% of visits to marking sites in their home range cores and on 50% of visits to marking sites in range peripheries (Kusler et al., 2019), and cheetahs scent-marked on 57% of visits to scent-marking sites (Cornhill & Kerley, 2020). Dominant African wild dogs scent-marked every time they visited a multi-pack marking site (Apps et al., 2022). Spotted hyaenas visited their latrines 249 times and scent-marked 50 times (20%), African wild dogs visited hyaena latrines 72 times and scent-marked 15 times (21%), black-backed jackals visited 89 times and marked 14 times (16%), civets (*Civettictis civetta*) visited 35 times and marked four times (11%), and leopards visited 47 times and marked once (2%) (Vitale, 2018). Snow leopards scent-marked on 86 of 145 visits (59%) to a multi-species marking site, wolves (*Canis lupus*) visited the site 42 times and scent-marked five times (12%), domestic dogs visited 30 times and marked three times (10%), brown bear (*Ursus arctos*) marked on 7 out of 12 visits (58%), red foxes marked on 11 of 81 visits (14%), Tibetan foxes (*Vulpes ferrilata*) marked on 3 of 44 visits (7%), and Pallas's cats visited 16 times and marked once (6%) (Li et al., 2013).

In a study similar to this one, that quantified natural scent-marking and reactions to natural scent-marks from camera trap videos of African predators (Apps et al., 2019), only 18% of individual predators scent-marked along 28 m stretches of dirt road and 39% of natural scent-marks were countermarked. These frequencies of responses to natural marks are strikingly similar to the frequency of response of African wildcats to dimethyl disulfide and bat-eared foxes to 2-ethyl-5(6)-methyl pyrazine and piperidinone (Table 2). Leopards in the current study responded more frequently to 2-ethyl-5(6)-methyl pyrazine and indole than they did to natural leopard scent-marks in the earlier work (Vitale, 2018; Apps et al., 2019) (but see discussion below on the effects of dispenser housings being scent-marked).

Some test substances were scent-marked by some species less than empty dispensers were (Table 2), and at lower rates than natural substrates and scent-marks (Li et al., 2013; Allen et al., 2015; Wikenros et al., 2017; Vitale, 2018; Apps et al., 2019, 2022; Kusler et al., 2019; Cornhill & Kerley, 2020). It would be premature to reject the possibility that those compounds inhibited scent-marking by some (or in the case of the mixture of methylbutanoic acids all) of the species, an effect also shown by a synthetic, multi-component 'appeasing pheromone' when presented to domestic cats (Frank et al., 1999).

In the absence of other influences, the number of reactions would correlate with the number of videos. Deviations from this correlation, either positive or negative, when the data are classified by test substance, (Fig. 4) would arise if the test substances (amongst other possible influences) elicited or inhibited scent-marking, as they would if they approximated a chemical signal. Plainly a substantially larger dataset is required for proper analysis, but Fig. 4 is suggestive of the deviations that would be expected if the test substances approximated whole chemical signals or cues.

### 4.3 Alternative Explanations, Interferences, and Refinements to Procedures

Although the videoed reactions, their species specificity and their frequencies in response to different test substances were consistent with single compounds approximating natural chemical signals or cues, interpreting these results as showing that the test substances are semiochemically active components of predators' odours is complicated by potential sources of interference. The heterogeneity of scent-marking reactions to the same test substance at different stations and the differences between the two dispensers at some stations (Fig. 3) show that reactions were influenced by factors other than the substance being tested. Eliminating or reducing these interferences by improvements to hardware and procedures would yield more clear-cut results in tests of the semiochemical activity of artificial scent marks. Interferences and suggestions for how to deal with them are discussed below.

Some species may have found the height and shape of the dispenser housings convenient for scent-marking, independently of any odour they emitted. Nonetheless, reactions to empty dispenser housings were rare, and if the dispenser housings were the focus, reactions would have occurred approximately equally with the different test substances as well as with empty housings (Table 2).

One of the most important of the interferences will have been natural scent marks on the dispenser housings or nearby soil. Both artificial and natural scent marks can trigger cascades of intra- and interspecific scent-marking (Apps et al., 2017, 2019), and attraction to heterospecific odours can be exploited to lure predators to camera traps (Garvey et al., 2017). Reactions to these natural scent marks would have inflated the number of reactions to test compounds to an unknown extent. Empty housings did not trigger marking cascades, but test compounds that triggered cascades at some stations did not do so at others, and at some stations only one of the dispensers triggered a marking cascade.

The current experimental implementation suffers under the paradox that a positive reaction, i.e. scent-marking, interferes with the test by adding the odour of the scent mark to the odour of the test substance, which is certain to affect subsequent reactions to the test substance. As an added complication, mixing the odour of a test substance with the odour of a scent mark may make the natural mark more or less attractive and enhance or inhibit counter-marking.

There are practical steps that could be taken to reduce the interference between test substances and natural scent marks. It may be possible to design dispenser housings that do not induce animals to scent mark them when they are empty. Replacing marked housings with clean ones more frequently on a regular schedule, or after reviewing the videos showed that they had been scent-marked, would interrupt marking cascades; in other studies that showed scent-marking responses to artificial odours in the wild, the scent posts were checked within 3 days of being set (Whitten et al., 1980), or responses were observed directly (Katsir & Crewe, 1980; Müller-Schwarze & Houlihan, 1991), but such frequent checks negate the advantage that camera traps can operate for weeks unattended.

Scent cannot travel upwind, and so a lack of response by an animal upwind of a dispenser may be a false negative. The differences in reactions recorded by the two cameras at some of the locations may have been due to differences in movement of airborne test substance from the two dispensers. The prevailing winds are from the ENE, but movement of vegetation and wind-borne dust in the videos shows substantial variation in the direction and speed of small-scale air movements at ground level. Recording the direction of air movements around the dispensers would detect false negatives, and small wind vanes in the camera traps' fields of view show some promise for this purpose.

Although having dispensers on both sides of a road ensured that animals travelling along the road encountered scent being blown onto the road irrespective of wind direction, it did not ensure that the scent plume crossed the road within the cameras' fields of view. Some videos show animals passing a camera, leaving the field of view, and then reappearing, approaching the dispenser and sniffing it. Presumably they encountered a scent plume that was crossing the road obliquely outside the field of view, and then air-scented upwind to the dispenser. If animals can be attracted to a dispenser from outside a field of view, it is also possible that others are repelled while outside a field of view, and consequently do not trigger the camera trap. If approaching animals were able to detect the test substance before they entered the cameras' detection zones, and either avoided it or were attracted to it, video captures were not independent of the test substance. Putting extra cameras further along the road extends the area over which an animal could encounter an odour plume and be videoed by a camera trap (Apps et al., 2019), but at a higher cost in hardware and time, and additional constraints on where stations can be sited – the road needs to be nearly straight between the furthest cameras, and not have any game trails or other roads intersecting it.

Animals' responses to odours vary with circumstances (Garvey et al., 2020), and this probably accounts at least in part for the heterogeneity of reactions. A station near a den will generate both a high number of encounters, and probably different reactions from one on a home range boundary, which in turn will generate different reactions from one in a range core far from a den. There are additional possible effects; does direction of travel influence scent-marking at a given location? is a central place forager more likely to scent mark on the way out to forage or on the way back to its den? If these spatial effects coincide with differences in test scent they will generate spurious differences in reactions. Spatial effects can be reduced by rotating scent between stations; rather than two odours in each of half the stations, I now recommend a different odour at each station (plus one station with empty dispensers), rotated between stations until each odour has been at each station. With 12 stations and 11 test compounds plus empty dispensers, a cycle with monthly changes would conveniently take a year to complete.

Males and females are certain to respond differently to sex-specific signals and their artificial equivalents, and the nature of the difference will depend on the time of year in seasonal breeders, and the location if males and females live separately. The actual presence of a potential mating partner is also likely to temporarily influence reactions to chemical signals or their approximations; black-backed jackal and bat-eared fox pairs routinely overmark one another's urine, and if one marks a test substance the other will too. Temporal effects could be reduced by changing scents on a cycle markedly shorter or markedly longer than seasonal or biological (e.g. mating, breeding, dispersal) factors that might affect reactions, or by presenting all the scents at the same time, as in the spatial rotation mentioned previously.

The ubiquity and scale of these interferences mean that generating sufficient data to apply currently fashionable big data approaches would be prohibitively costly in time and materials; a more conventional approach of carefully designed small-scale experiments and critical examination of results will prove more productive. In an ideal and properly funded world, reactions to odours would be tested at the same time and in the same place as studies of movements, social behaviour, and reproduction that would provide the background against which to interpret differences in reactions to odours.

## 5 Conclusion

Single compounds eliciting the behavioural reactions expected to whole odours show that single compounds approximate at least some chemical messages emitted by natural scent marks, and contradicts the orthodox view that mammals' chemical signals and cues are too complex to be practically replicated artificially. These results warrant further investment into research on the use of single compounds and simple mixtures of predator odour components for management and conservation.

**Acknowledgements** I would like to thank the Leopardess Foundation, Oppenheimer Generations Research and Conservation, and Stichting Spots for financial support. The work was carried out under research permit number EWT 8/36/4 XXXVIII (15) to Botswana Predator Conservation, for which we thank the Department of Wildlife and National Parks of the Ministry of Environment, Natural Resources, Conservation and Tourism of Botswana. I thank Dr. Marius Viljoen for allowing me to work on his property.

## References

Abrahms, B., Jordan, N. R., Golabek, K. A., McNutt, J. W., Wilson, A. M., & Brashares, J. S. (2016). Lessons from integrating behaviour and resource selection: activity-specific responses of African wild dogs to roads. *Anim Conserv, 19*, 247–255.

Allen, M. L., Wittmer, H. U., & Wilmers, C. C. (2014). Puma communication behaviours: understanding functional use and variation among sex and age classes. *Behaviour, 151*(6), 819–840.

Allen, M. L., Wallace, C. F., & Wilmers, C. C. (2015). Patterns in bobcat (*Lynx rufus*) scent marking and communication behaviors. *J Ethol, 33*(1), 9–14.

Apfelbach, R., Blanchard, C. D., Blanchard, R. J., Hayes, R. A., & McGregor, I. S. (2005). The effects of predator odors in mammalian prey species: a review of field and laboratory studies. *Neurosci Biobehav Rev, 29*(8), 1123–1144.

Apfelbach, R., Parsons, M. H., Soini, H. A., & Novotny, M. V. (2015). Are single odorous components of a predator sufficient to elicit defensive behaviors in prey species? *Front Neurosci, 9*, 263. https://doi.org/10.3389/fnins.2015.00263

Apps, P. J. (2013). Are mammal olfactory signals hiding right under our noses? *Naturwissenschaften, 100*(6), 487–506.

Apps, P. J., & McNutt, J. W. (2018). Are camera traps fit for purpose? A rigorous, reproducible and realistic test of camera trap performance. *Afr J Ecol, 56*, 710–720.

Apps, P., Mmualefe, L., & McNutt, J. W. (2012). Identification of volatiles from the secretions and excretions of African wild dogs (*Lycaon pictus*). *J Chem Ecol, 38*, 1450–1461.

Apps, P., Mmualefe, L., & McNutt, J. W. (2013). A reverse engineering approach to identifying which compounds to bioassay for signalling activity in the scent marks of African wild dogs (*Lycaon pictus*). In M. L. East & M. Denhard (Eds.), *Chemical Signals in Vertebrates 12* (pp. 417–432). Springer.

Apps, P., Mmualefe, L., Jordan, N. R., Golabek, K. A., & McNutt, J. W. (2014). The "tomcat compound" 3-mercapto-3-methylbutanol occurs in the urine of free-ranging leopards but not in African lions or cheetahs. *Biochem Syst Ecol, 53*, 17–19.

Apps, P. J., Weldon, P. J., & Kramer, M. (2015). Chemical signals in terrestrial vertebrates: search for design features. *Nat Prod Rep, 32*(7), 1131–1153.

Apps, P. J., Claase, M., Yexley, B., & McNutt, J. W. (2017). Interspecific responses of wild African carnivores to odour of 3-mercapto-3-methylbutanol, a component of wildcat and leopard urine. *J Ethol, 35*, 153–159.

Apps, P., Rafiq, K., & McNutt, J. W. (2019). Do carnivores have a world wide web of interspecific scent signals? In C. Buesching (Ed.), *Chemical signals in vertebrates 14* (pp. 182–202). Springer.

Apps, P., Claase, M., van Mourik, E., Lostrom, S., Yexley, B., Webster, H., & McNutt, J. W. (2022). A description of a new discovery: African wild dog packs communicate with other packs by posting scent-mark messages at a shared marking site. *Behav Ecol Sociobiol, 76*(3), 1–14.

Banks, P. B., Bytheway, J. P., Carthey, A. J., Hughes, N. K., & Price, C. J. (2014). Olfaction and predator-prey interactions amongst mammals in Australia. In A. Glen & C. Dickman (Eds.), *Carnivores of Australia: past, present and future* (pp. 389–413). CSIRO Publishing.

Boag, B., & Mlotkiewicz, J. A. (1994). Effect of odor derived from lion faeces on behavior of wild rabbits. *J Chem Ecol, 20*(3), 631–637.

Butenandt, A., Beckmann, R., & Hecker, E. (1961). Über den Sexuallockstoff des Seidenspinners, I. Der biologische Test und die Isolierung des reinen Sexuallockstoffes Bombykol. *Hoppe-Seyler's Z Physiol Chem, 324*, 71–83. https://doi.org/10.1515/bchm2.1961.324.1.71

Cornhill, K. L., & Kerley, G. I. (2020). Cheetah behaviour at scent-marking sites indicates differential use by sex and social rank. *Ethology, 126*(10), 976–986.

DePorter, T. L., Bledsoe, D. L., Beck, A., & Ollivier, E. (2019). Evaluation of the efficacy of an appeasing pheromone diffuser product *vs* placebo for management of feline aggression in multi-cat households: a pilot study. *J Feline Med Surg, 21*(4), 293–305.

Ferrero, D. M., Lemon, J. K., Fluegge, D., Pashkovski, S. L., Korzan, W. J., Datta, S. R., Spehr, M., Fendt, M., & Liberles, S. D. (2011). Detection and avoidance of a carnivore odor by prey. *Proc Natl Acad Sci U S A, 108*(27), 11235–11240.

Frank, D. F., Erb, H. N., & Houpt, K. A. (1999). Urine spraying in cats: presence of concurrent disease and effects of a pheromone treatment. *Appl Anim Behav Sci, 61*(3), 263–272.

Garvey, P. M., Glen, A. S., Clout, M. N., Wyse, S. V., Nichols, M., & Pech, R. P. (2017). Exploiting interspecific olfactory communication to monitor predators. *Ecol Appl, 27*(2), 389–402.

Garvey, P. M., Banks, P. B., Suraci, J. P., Bodey, T. W., Glen, A. S., Jones, C. J., McArthur, C., Norbury, G. L., Price, C. J., Russell, J. C., & Sih, A. (2020). Leveraging motivations, personality, and sensory cues for vertebrate pest management. *Trends Ecol Evol, 35*(11), 990–1000.

Jackson, C., McNutt, J. W., & Apps, P. J. (2012). Managing the ranging behaviour of African wild dogs (*Lycaon pictus*) using translocated scent marks. *Wildl Res, 39*, 31–34.

Kats, L. B., & Dill, L. M. (1998). The scent of death: chemosensory assessment of predation risk by prey animals. *Ecoscience, 5*(3), 361–394.

Katsir, Z., & Crewe, R. M. (1980). Chemical communication in *Galago crassicaudatus*: investigation of the chest gland secretion. *S Afr J Zool, 15*(4), 249–254.

Kimball, B. A., Mason, J. R., Blom, F. S., Johnston, J. J., & Zemlicka, D. E. (2000). Development and testing of seven new synthetic coyote attractants. *J Agric Food Chem, 48*(5), 1892–1897.

Kusler, A., Jordan, N. R., McNutt, J. W., & Broekhuis, F. (2019). Cheetah marking trees: distribution, visitation, and behavior. *Afr J Ecol, 57*(3), 419–422.

Li, J., Schaller, G. B., McCarthy, T. M., Wang, D., Jiagong, Z., Cai, P., Basang, L., & Lu, Z. (2013). A communal sign post of snow leopards (*Panthera uncia*) and other species on the Tibetan Plateau, China. *Int J Biodivers, 2013*, 1–8. https://doi.org/10.1155/2013/370905

Lindgren, P. M. F., Sullivan, T. P., & Crump, D. R. (1995). *Review of synthetic predator odor semiochemicals as repellents for wildlife management in the Pacific Northwest*. In: National Wildlife Research Center Repellents Conference 1995. p. 24. https://digitalcommons.unl.edu/nwrcrepellants/24

Mason, J. R. (1998). Mammal repellents: options and considerations for development. In R. O. Baker & A. C. Crabb (Eds.), *Proceedings of the 18th Vertebrate Pest Conference* (Vol. 18). Univ. of Calif.

Mason, J., & Blom, F. (1998). Coyote lure ingredients: what determines success. *Wildl Control Technol, 5*(5), 26–30.

Melrose, D. R., Reed, H. B., & Patterson, R. L. S. (1971). Androgen steroids associated with boar odour as an aid to the detection of oestrus in pig artificial insemination. *Br Vet J, 127*, 497–501.

Müller-Schwarze, D. (2016). Chemical Signals in Vertebrates 13: where we stand and what might be next. In B. A. Schulte, T. E. Goodwin, & M. H. Ferkin (Eds.), *Chemical signals in vertebrates 13* (pp. 11–16). Springer.

Müller-Schwarze, D., & Houlihan, P. W. (1991). Pheromonal activity of single castoreum constituents in beaver, *Castor canadensis*. *J Chem Ecol, 17*(4), 715–734.

Nilsson, S., Sjöberg, J., Amundin, M., Hartmann, C., Buettner, A., & Laska, M. (2014). Behavioral responses to mammalian blood odor and a blood odor component in four species of large carnivores. *PLoS One, 9*(11), e112694.

Nolte, D. L., & Wagner, K. K. (2000). Comparing the efficacy of delivery systems and active ingredients of deer repellents. In F. P. Salmon & A. C. Crabb (Eds.), *Proceedings of the 19th Vertebrate Pest Conference* (Vol. 19). Univ. of Calif.

Osada, K., Kurihara, K., Izumi, H., & Kashiwayanagi, M. (2013). Pyrazine analogues are active components of wolf urine that induce avoidance and freezing behaviours in mice. *PLoS One, 8*(4), e61753. https://doi.org/10.1371/journal.pone.0061753

Pageat, P., & Gaultier, E. (2003). Current research in canine and feline pheromones. *Vet Clin Small Anim Pract, 33*(2), 187–211.

Parshad, V. R. (2002). Carbon disulphide for improving the efficacy of rodenticide baiting and trapping of the house rat, *Rattus rattus* L. *Int Biodeterior Biodegradation, 49*(2–3), 151–155.

Pickett, J. A., Barasa, S., & Birkett, M. A. (2014). Vertebrate pheromones and other semiochemicals: the potential for accommodating complexity in signalling by volatile compounds for vertebrate management. *Biochem Soc Trans, 42*(4), 846–850.

Rafiq, K., Jordan, N. R., Meloro, C., Wilson, A. M., Hayward, M. W., Wich, S. A., & McNutt, J. W. (2020). Scent-marking strategies of a solitary carnivore: boundary and road scent marking in the leopard. *Anim Behav, 161*, 115–126.

Schlageter, A., & Haag–Wackernagel, D. (2012). Evaluation of an odor repellent for protecting crops from wild boar damage. *Journal of Pest Science, 85*, 209–215.

Schulte, B. A., & LaDue, C. A. (2021). The chemical ecology of elephants: 21st century additions to our understanding and future outlooks. *Animals, 11*, 2860. https://doi.org/10.3390/ani11102860

Schulte, B. A., Freeman, E. W., Goodwin, T. E., Hollister-Smith, J., & Rasmussen, L. E. L. (2007). Honest signalling through chemicals by elephants with applications for care and conservation. *Appl Anim Behav Sci, 102*(3–4), 344–363.

Sievert, T., & Laska, M. (2016). Behavioral responses of CD-1 mice to six predator odor components. *Chem Senses, 41*, 399–406.

Vitale, J. (2018). *The olfactory behaviour of spotted hyaenas (Crocuta crocuta) and sympatric mammals in the Okavango Delta, Botswana* (Doctoral dissertation, University of Nottingham)

Wauson, M., & Rogers, W. (2021). A test of the use of gray wolf (*Canis lupus*) urine to reduce coyote (*Canis latrans*) depredation rates on loggerhead sea turtles (*Caretta caretta*) nests. *J Nat Conserv, 63*, 126050.

Webster, M. M., & Rutz, C. (2020). How STRANGE are your study animals? *Nature, 582*, 337–340. https://doi.org/10.1038/d41586-020-01751-5

Weldon, P. J. (1990). Responses by vertebrates to chemicals from predators. In D. W. Macdonald, D. Müller-Schwarze, & S. E. Natynczuk (Eds.), *Chemical signals in vertebrates 5* (pp. 500–521). Oxford University Press.

Whitten, W. K., Wilson, M. C., Wilson, S. R., Jorgenson, J. W., Novotny, M., & Carmack, M. (1980). Induction of marking behavior in wild red foxes (*Vulpes vulpes* L.) by synthetic urinary constituents. *J Chem Ecol, 6*(1), 49–55.

Wikenros, C., Jarnemo, A., Frisén, M., Kuijper, D. P., & Schmidt, K. (2017). Mesopredator behavioral response to olfactory signals of an apex predator. *J Ethol, 35*(2), 161–168.

Wyatt, T. D. (2009). Fifty years of pheromones. *Nature, 457*(7227), 262–263.

# Chemical Pest Defense by the Innate Response to Silver Vine and Catnip Plants in the Domestic Cat

Masao Miyazaki[1] (✉) and Reiko Uenoyama[1]

[1] Department of Bioresources Science, The United Graduate School of Agricultural Sciences, Iwate University, Iwate, Japan
mmasao@iwate-u.ac.jp

**Abstract** Cats show a characteristic response to catnip (*Nepeta cataria*) and silver vine (*Actinidia polygama*), which comprises licking, chewing, rubbing, and rolling. This response is induced by plant iridoids: nepetalactone (catnip) and dihydronepetalactone, isodihydronepetalactone, iridomyrmecin, and isoiridomyrmecin (silver vine). However, its biological significance had remained undetermined. In this chapter, we review the behavioral significance of this feline response. Nepetalactol was isolated as a potent bioactive compound from silver vine. Nepetalactol-induced rubbing and rolling behavior transfers nepetalactol to feline fur. Nepetalactol has also mosquito repellent bioactivity, and as a consequence, its transfer to the feline's fur protects cats from mosquito bites. Licking and chewing the plants damages the leaves, which promotes airborne emission of iridoids from the plants and diversifies the iridoids in silver vine. Although both the amount and composition of iridoids considerably differ between these plants, cats show a comparably prolonged duration of the response to the low level of a complex cocktail of iridoids in damaged silver vine and the high level of nepetalactone in damaged catnip. In conclusion, the silver vine and catnip response contributes to chemical pest defense in cats, which provides an important example of how animals utilize plant metabolites for their pest defense.

**Keywords** Domestic cat · Silver vine · Catnip · Iridoid · Self-anointing · Pest defense · Mosquito · Innate behavior

## 1 Backgrounds

Silver vine (*Actinidia polygama*) and catnip (*Nepeta cataria*), which are a family of kiwi fruits and a European herb, respectively, are completely different plant species (Fig. 1a, b) but have the same bioactivity to domestic cats (*Felis silvestris catus*). When cats find these plants, they show a characteristic response toward the plants, which comprises the combination of licking and chewing of the plants, face and head rubbing



B. Schaal et al. (eds.), *Chemical Signals in Vertebrates 15*,
https://doi.org/10.1007/978-3-031-35159-4_21

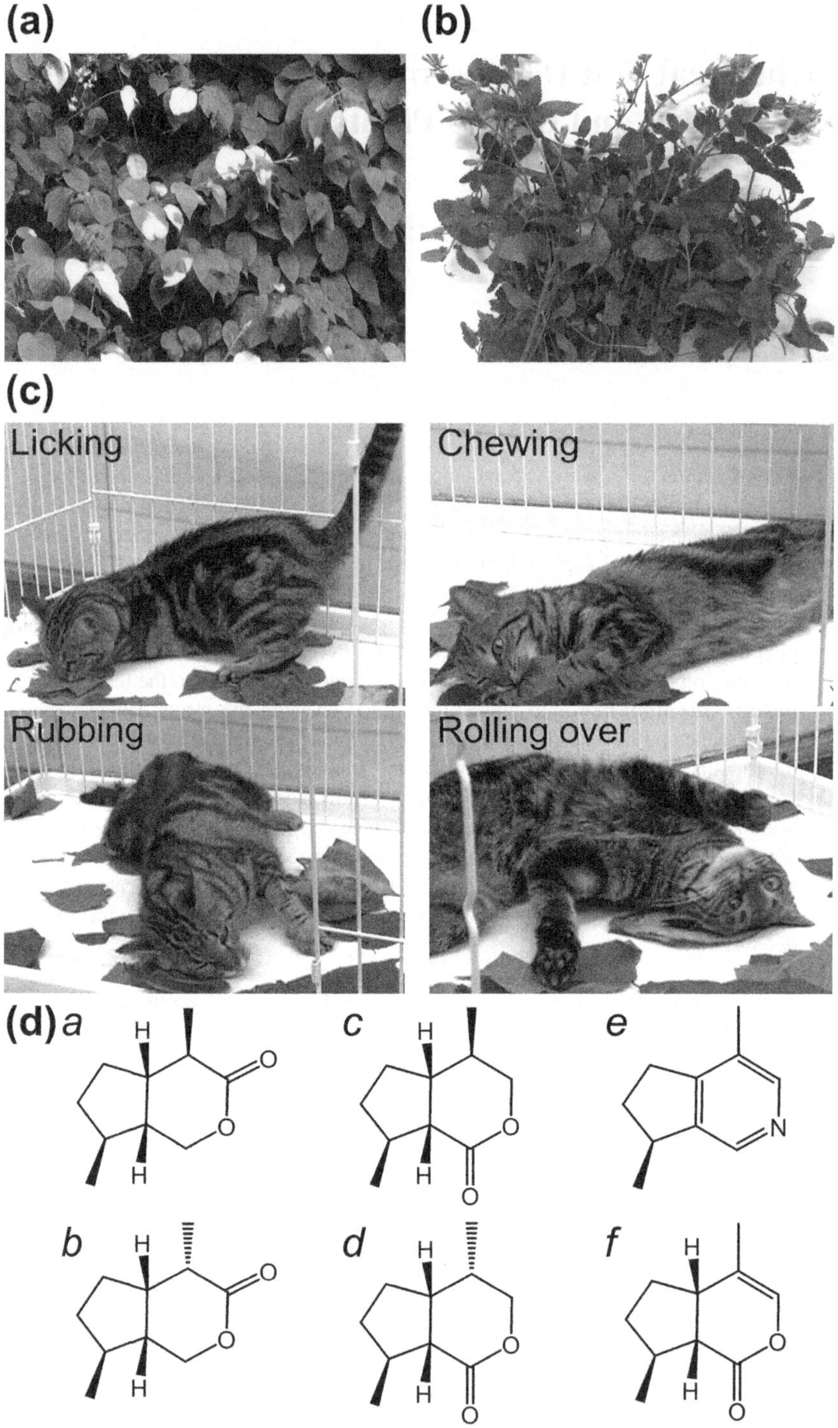


**Fig. 1** The iridoid response in cats. (**a**, **b**) Images of silver vine (**a**) and catnip (**b**). Silver vine has green and white colored leaves. (**c**) Images of cats exhibiting licking and chewing silver vine leaves, face and head rubbing against the leaves, and rolling over on the leaves. The rolling

against the plants, and rolling over on the plants, but does is not fixed to the order of licking, chewing, rubbing, and rolling (Fig. 1c) (Bol et al., 2017; Todd, 1962; Tucker & Tucker, 1988). This feline response usually lasts 5 to 15 minutes, followed by a period of one or more hours when they are non-responsive (Hart & Leedy, 1985). About 30% of cats are negative responders to catnip hence it is thought that this response is inherited as an autosomal-dominant trait in domestic cats (Todd, 1962).

This response is an innate behavior elicited only by second metabolites of catnip and silver vine with no requirement for prior exposure or learning in cats. Bioactive compounds inducing this feline response have been reported as nepetalactone in catnip and isoiridomyrmecin, iridomyrmecin, isodihydronepetalactone, dihydronepetalactone, and actinidine in silver vine (Meinwald, 1954; Sakan et al., 1959a, b, 1969). These compounds have very similar chemical structures with five-membered rings fused to six-membered rings, which are categorized into iridoids (Fig. 1d).

The response to catnip and silver vine (the iridoid response) in cats is well known worldwide. The first reports of the iridoid response date back more than 300 years. Not only domestic cats (*Felis silvestris catus*) but also other felids such as lions (*Panthera leo*) and bobcats (*Lynx rufus*) show the combination of licking, chewing, rubbing, and rolling to these plants (Bol et al., 2017; Hill et al., 1976; Tucker & Tucker, 1988). In contrast, domestic dogs (*Canis lupus familiaris*) are also carnivores but do not respond to plant iridoids (Uenoyama et al., 2021). We hypothesize that only Felidae species have acquired the system for the iridoid response during evolution. Considering that domestic cats diverged from large Felidae species around 10 million years ago (Barnett et al., 2016), we also speculate that common ancestors of Felidae species such as cave lions (*Panthera leo spelaea*) had already acquired the characteristic response. The consistent expression of such a characteristic response to plant iridoids among felids over 10 million years suggests that the response has an important adaptive function for cats, while this response has often been interpreted by pet owners as a playful behavior among cats that appeared to be intoxicated by these specific plant species (Espin-Iturbe et al., 2017; Hatch, 1972). However, the biological significance of the iridoid response had not been understood. Thus, we examined the feline characteristic response to catnip and silver vine using behavioral, neuronal, biochemical, and analytical approaches. In this chapter, we review our recent studies uncovering the behavioral significance of the feline response to catnip and silver vine plants (Uenoyama et al., 2022; Uenoyama et al., 2021).

**Fig. 1** (continued) behavior also enables cats to rub their whole body parts, especially their backs, against the leaves. (**d**) Chemical structures of previously identified bioactive compounds: (*a*) isoiridomyrmecin, (*b*) iridomyrmecin, (*c*) isodihydronepetalactone, (*d*) dihydronepetalactone, and (*e*) actinidine from silver vine, and (*f*) nepetalactone from catnip

## 2 Nepetalactol Is a Potent Bioactive Compound to Cats in Silver Vine

In Japan, silver vine (Matatabi in Japanese) is known to induce the characteristic response to cats, which is called Matatabi dancing in Japanese culture. In fact, silver vine is more famous than catnip to induce the response. Thus, we first used silver vine to examine the feline response. To elucidate the biological significance of the response, it was very important to establish a reliable and reproducible behavioral assay using a synthesized compound that could be controlled, rather than plant materials that emit variable amounts of multiple volatile compounds. Previous studies reported some bioactive iridoids inducing the silver vine response in cats (Sakan et al., 1959a, b, 1969). However, since there was no report which iridoids identified from silver vine is the most potent bioactive compound to cats, our study started from the comparison of the bioactivity (duration of the response) and leaf contents of reported iridoids.

Synthesized by the laboratory of our close collaborator Prof. Toshio Nishikawa (Nagoya University in Japan), bioactivities of compounds were tested by the presentation of one of the reported iridoids-impregnated filter paper with control filter paper to cats simultaneously (Fig. 2a). In behavioral assays, isoiridomyrmecin, iridomyrmecin, isodihydronepetalactone, and dihydronepetalactone, but not actinidine, induced the characteristic response in cats. Quantification of each iridoid by using gas chromatograph-mass spectrometry (GC/MS) showed that a major content was actinidine (14 μg/g of leaf wet weight) without bioactivity and bioactive isoiridomyrmecin (1.42 μg/g), isodihydronepetalactone (<0.18 μg/g). Also dihydronepetalactone (<0.18 μg/g), but not iridomyrmecin, was detectable in silver vine. Since there was a discrepancy between previous reports (Sakan et al., 1959a, b, 1969) and our results, we next purified a potent bioactive compound from silver vine by ourselves.

An organic solvent extract from silver vine leaves was resolved by normal-phase liquid chromatography using a silica gel column. Two fractions induced the characteristic response to cats. One fraction without reported iridoids induced a prolonged response in cats than the other fraction with isoiridomyrmecin, isodihydronepetalactone, and dihydronepetalactone in behavioral assays. This suggested an important contribution of unidentified bioactive compounds to cats in the fraction without reported iridoids. We further purified the unknown bioactive compound by liquid chromatography and finally identified nepetalactol which had been missed in previous reports from silver vine (Sakan et al., 1959a, b, 1969). Nepetalactol is an iridoid which shares a very similar chemical structure with nepetalactone except for lactol and lactone moieties (Fig. 2b).

The synthesized nepetalactol induced the characteristic rubbing and rolling behavior in all of experimental, pet, and free-ranging feral cats. The duration of the feline response to nepetalactol was more prolonged than isoiridomyrmecin, isodihydronepetalactone, and dihydronepetalactone (Fig. 2c). In addition, nepetalactol also induced the characteristic response to non-domesticated captive felids such as a lion, a tiger (*Panthera tigris*), Amur leopards (*Panthera pardus orientalis*), jaguars (*Panthera onca*), and Eurasian lynxes (Fig. 2d), but not to non-Felidae species such as domestic dogs and laboratory mice (*Mus musculus*; C57BL/6 or BALB/cAJcl strains), as far as we

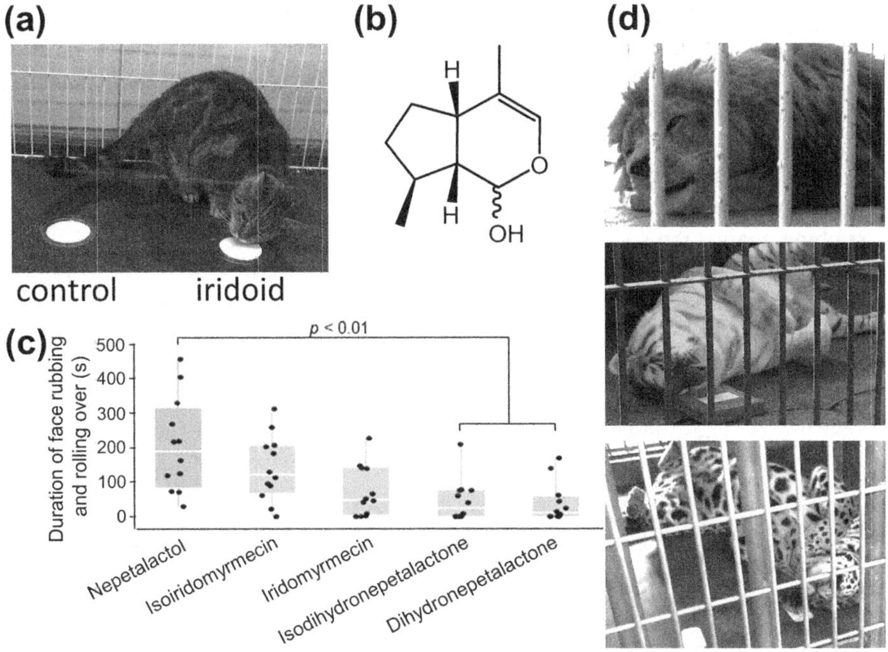


**Fig. 2** Nepetalactol is a potent bioactive compound to felids. (**a**) An image of behavioral assay using cats which were simultaneously presented with an iridoid-impregnated paper and control paper. (**b**) A chemical structure of nepetalactol. (**c**) Duration of face rubbing and rolling over to nepetalactol, isoiridomyrmecin, iridomyrmecin, isodihydronepetalactone, and dihydronepetalactone in cats. $N = 12$. Boxes show median and interquartile range, whiskers are minimum and maximum, dots are individual values. *P* values from Bonferroni–Dunn post-hoc test. (**d**) Images of non-domesticated captive felids exhibiting face rubbing and rolling to nepetalactol papers: a lion (top), a tiger (middle), and a jaguar (bottom). (Original data of **c** and **d** are from our paper (Uenoyama et al., 2021))

examined (Uenoyama et al., 2021). The nepetalactol content of silver vine intact leaves was approximately 20 μg/g of leaf wet weight which was much higher than other bioactive iridoids contained in silver vine. Based on these results, we concluded that nepetalactol is the most potent bioactive compound to cats among iridoids of silver vine, which was a suitable stimulant for behavioral assays to elucidate the biological significance of the feline response.

## 3 Rubbing and Rolling Behavior for the Chemical Defense Against Pest Mosquitoes

Here, we first determined which behavior (among licking, chewing, rubbing, and rolling), especially between rubbing or rolling, is most important in cats by examining the feline response to nepetalactol-impregnated filter papers placed on walls or ceiling of the test cage. The assays were designed by Prof. Jane L Hurst (University of Liverpool in United Kingdom) who is our close collaborator. In this arrangement, cats could rub the papers with their faces, but rolling would not allow rubbing contact with the stimulus. The behavior assay showed that cats rubbed their faces and heads on nepetalactol-paper placed on the cage walls and ceiling (Fig. 3a, b), but no cat rolled on the ground when test papers were on the walls or ceiling. These indicate that the rubbing behavior provides some benefit to cats but not a general euphoric response with no specific adaptive function as proposed in previous studies (Espin-Iturbe et al., 2017; Hatch, 1972).

Rubbing against the source should transfer nepetalactol to the fur, but we failed to detect any nepetalactol in organic solvent extracts from cotton used for wiping the feline face and head fur using GC/MS, probably because it was below the limit of detection. To provide a more sensitive test, we utilized cats as biosensors to detect nepetalactol. Biosensor cats showed the characteristic rubbing response to the paper that had been used for wiping the fur of cats that had rubbed nepetalactol but not the fur of control cats. These indicate that nepetalactol is attached to the fur of cats rubbing

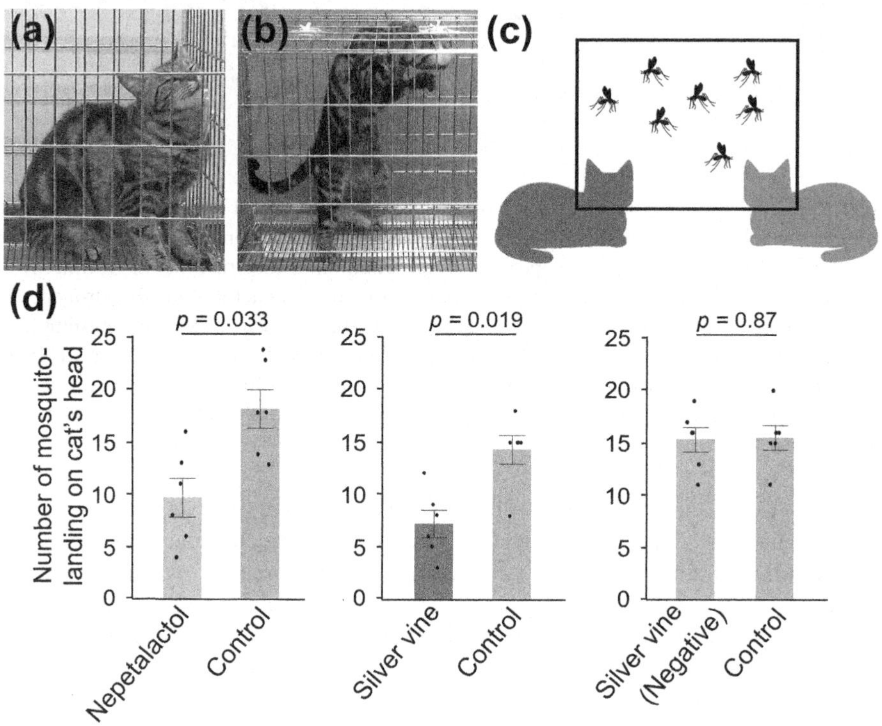


**Fig. 3** Rubbing against silver vine leaves protects cats from mosquitoes. (**a**, **b**) Images of behavioral assays presented nepetalactolpapers on the cage wall (**a**) or ceiling (**b**). Cats rubbed the paper but did not roll over on the floor of the cage. (**c**) A schematic image of mosquito bite assays. To compare mosquito repellency of cats, the heads of two anesthetized cats were put into an acrylic cage with 30 mosquitoes for 10 min. Only one of the cats' heads had been treated with nepetalactol. (**d**) Numbers of mosquito-landing on pairs of cats, when one cat was treated with nepetalactol (left), had rubbed itself on silver vine leaves (center) and was negative responders to the leaves (right). (Original data are from our paper (Uenoyama et al., 2021))

against nepetalactol sources, which biosensor cats can detect. In conclusion, face and head rubbing against the plants is a functional behavior to transfer plant nepetalactol and other iridoids onto the fur in cats.

Considering the function of rubbing behavior in cats, we suggested that plant iridoids have bioactivities other than the induction of the iridoid response to cats. Some of the plant iridoids are known to be repellent to a broad range of insects, including non-herbivores such as *Aedes* and *Culex* mosquitoes and stable flies (*Stomoxys*) (Birkett et al., 2011; Feaster et al., 2009; Gkinis et al., 2014; Melo et al., 2021; Reichert et al., 2019; Zhu et al., 2012). We also found that nepetalactol acts as a repellent against mosquitoes, *Aedes albopictus*, which are the most common mosquito in Japan and bite animals in the early morning and late afternoon when crepuscular cats are mostly at their peak of activity. Thus, we hypothesized that nepetalactol on the fur protects cats from mosquito bites.

This hypothesis was examined by mosquito bite assays in which two anesthetized cats had their heads, with or without nepetalactol treatment, placed into opposite sides of a test cage with about 30 female mosquitoes (*Ae. albopictus*) (Fig. 3c). We tested the feline heads because of the following reasons: (1) it is particularly often observed that cats rub their heads, as compared to other body parts, against the plants and (2) mosquitoes easily bite their mouth, eyelids, ears, and nose, which have relatively little fur. In this assay, the number of mosquito landing on the nepetalactol-treated head was about half of the number landing on the untreated head (Fig. 3d). Cats with rubbing and rolling over on silver vine leaves could avoid mosquito landing as compared to control cats (Fig. 3e), while cats negatively responding to silver vine could not avoid mosquitoes (Fig. 3f). These results showed strong evidence that the characteristic rubbing and rolling response functions to transfer plant chemicals which provide mosquito repellency to cats.

Face and head rubbing against plant sources containing the repellent will help to protect the head area from mosquito bites in cats. In addition, rolling over on the leaves following face rubbing enables cats to pick up repellent iridoids on their whole bodies, which is a functional behavior rather than an indicator of euphoria or extreme pleasure. Since self-anointing with plant iridoids is common in Felidae species, the iridoid response would have first evolved in a common felid ancestor and have been retained. As many felids rely on stealth to stalk and ambush their prey, requiring them to remain cryptic and often unmoving, a repellent that reduces their susceptibility to both the irritation of biting mosquitoes and the diseases that these insect vectors carry is likely to provide a strong selective advantage.

It is reported that nepetalactone has a repellent activity to *Ae. aegypti* which is a common vector of yellow fever, dengue, and Zika viruses (Benelli & Mehlhorn, 2016; Birkett et al., 2011; Gkinis et al., 2014; Zhu & Zeng, 2006). Our studies reveal the repellent effect of nepetalactol on *Ae. albopictus* mosquitos, carrying *Dirofilaria immitis*, which infects the heart and pulmonary arteries of cats (Traversa & Di Cesare, 2014). These indicate that the iridoid response can at least protect cats against mosquito bites. A recent study found that transient receptor potential channel A1 (TRPA1) is the major mediator of repellency of nepetalactone to mosquitoes and other insects, such as *Drosophila melanogaster* and *Formica rufa*. We suggest that plant iridoids transferred to feline fur by rubbing and rolling have broad repellence across a range of insects.

Following these findings, it would be interesting to also examine its protective effect on parasitic insects that eat feline blood such as fleas and ticks.

## 4 Licking and Chewing Silver Vine and Catnip Leaves for Enhancing Bioactivities to Cats and Mosquitoes

Cats exhibit not only rubbing and rolling but also licking and chewing of plant leaves throughout their characteristic response. Although licking and chewing of plants support their intake and digestion in most animals, cats are obligate carnivores and swallow little, if any, of the plant material. We hypothesized that licking and chewing of silver vine and catnip has important adaptive functions rather than consuming either plant material or iridoids. In this section, we show that licking and chewing also relates to insect repellency alongside rubbing and rolling.

Silver vine leaves crumpled and torn by feline licking and chewing has a much stronger aromatic odor compared to intact fresh leaves. GC/MS quantification of the airborne emission and chemical profiles of iridoids from silver vine leaves, damaged by cats or manually (crumpled and torn by the experimenter's hand to simulate feline licking and chewing), showed that leaf damage promoted the immediate emission of total iridoids (feline licking and chewing, median 94.7 ng/g leaf wet weight/4 h; manually crumpled and torn, 550.8 ng/g/4 h), which was 10-fold higher than from intact leaves (9.9 ng/g/4 h). Leaf damage also changed the composition of iridoids in silver vine. Nepetalactol accounted for over 90% of total iridoids in intact leaves, but this dropped to about 45% in damaged leaves as other iridoids such as dihydronepetalactone, isodihydronepetalactone, iridomyrmecin, and isoiridomyrmecin greatly increased (Fig. 4a).

To examine whether these changes in iridoids when silver vine leaves are damaged influence the feline response, synthetic iridoid cocktails corresponding to the ratios found in intact versus damaged leaves were simultaneously presented to cats. The altered iridoid mixture corresponding to damaged leaves promoted a much more prolonged response than those to intact leaves (Fig. 4b). This indicates that the change in iridoid composition when silver vine leaves were damaged by cats greatly increased the feline response. The increased emission of iridoids from damaged leaves and changed chemical composition add together and induce a more extended duration of rubbing and rolling response, allowing cats to transfer more mosquito repellents to their fur as compared to the response to intact leaves. This will help to reduce the health risks and irritation associated with mosquitoes. In conclusion, feline licking and chewing these plant leaves also contributes to chemical pest defense in cats.

Feline licking and chewing of leaves has similar effects on the amount of iridoids in catnip as well as silver vine. Damaged catnip emitted 20-fold more total iridoids (median 11.3 μg/g leaf wet weight/10 min) compared to intact leaves (0.5 μg/g/10 min). However, in contrast to silver vine, leaf damage did not alter the composition of catnip iridoids. Both intact and damaged leaf extracts of catnip consisted almost solely of nepetalactone, which is present at only very low levels in silver vine.

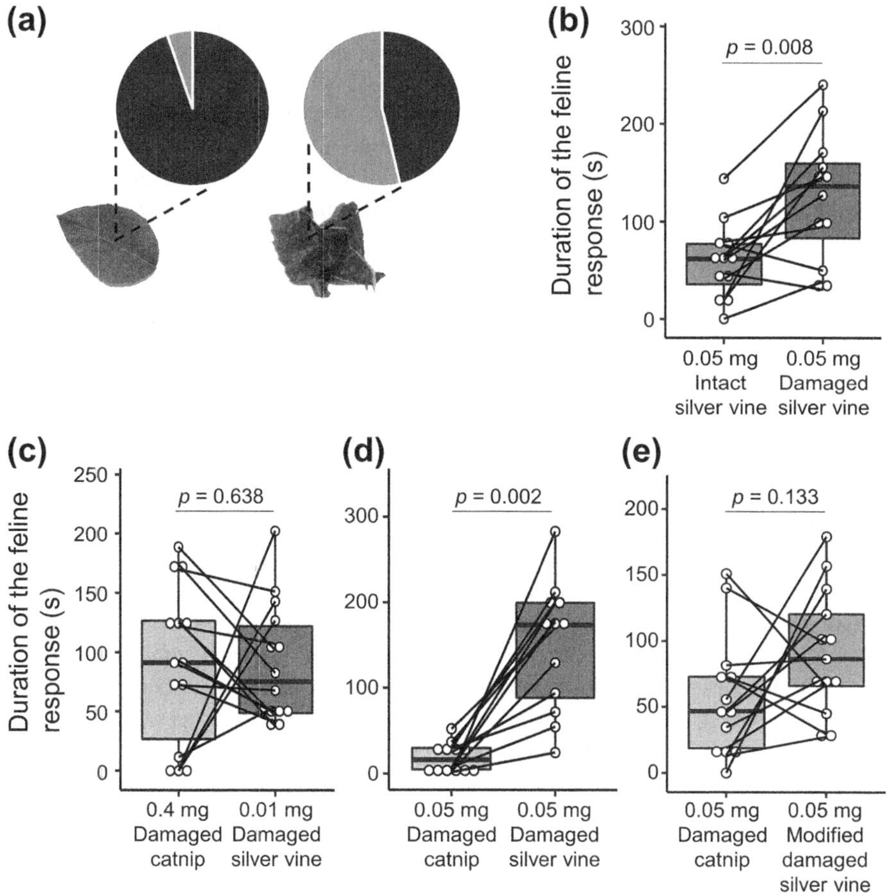


**Fig. 4** Leaf damage changes the amount and composition of iridoids in silver vine and catnip, which enhances bioactivity to cats. (**a**) Composition of iridoids in the extracts of silver vine leaves that were intact (left) and damaged (right). Dark gray and light gray indicate ratios of nepetalactol and other iridoids (isoiridomyrmecin, iridomyrmecin, isodihydronepetalactone, dihydronepetalactone), respectively. (**b–e**) Duration of the feline response to (**b**) 0.05 mg of synthetic iridoid cocktails corresponding to the ratios found in intact and damaged silver vine leaves, (**c**) 0.4 mg damaged catnip-cocktail and 0.01 mg damaged silver vine-cocktail, (**d**) 0.05 mg damaged catnip-cocktail and 0.05 mg damaged silver vine-cocktail, or (**e**) 0.05 mg damaged catnip cocktail and 0.05 mg modified damaged silver vine-cocktail. (Original data are from our paper (Uenoyama et al., 2022))

Total iridoid levels were approximately 40-times higher in damaged catnip leaves (mean ± SE: 343.5 ± 5.3 μg/100 mg leaf wet weight) compared to damaged silver vine leaves (8.9 ± 0.5 μg/100 mg). We examined whether these differences between damaged catnip and silver vine stimulated different levels of the feline response to these two plants. Interestingly, cats showed a comparable duration of response to 0.4 mg of damaged catnip-cocktail and 0.01 mg of damaged silver vine-cocktail, which corresponded to an equivalent amount of damaged catnip and silver vine (100 mg leaf wet weight)

(Fig. 4c). But when the same amount of iridoid cocktails were presented to cats from the two damaged plants, they were much more responsive to the damaged silver vine-cocktail than the damaged catnip-cocktail (Fig. 4d), showing they are sensitive to the silver vine iridoids at a much lower concentration. This was not the case when the damaged silver vine-cocktail was modified by replacing the main silver vine iridoid nepetalactol in the mixture with catnip nepetalactone (Fig. 4e). These indicate that the prolonged feline response to the complex iridoids emitted from damaged silver vine leaves depended on the combination of nepetalactol with other iridoids such as isodihydronepetalactone.

Previous studies have reported mosquito-repelling activity of individual iridoids such as nepetalactone, nepetalactol, and dihydronepetalactone (Birkett et al., 2011; Feaster et al., 2009; Gkinis et al., 2014; Melo et al., 2021; Reichert et al., 2019; Uenoyama et al., 2021; Zhu et al., 2012), but have not examined responses to the iridoid cocktails. Thus, we compared the repellency of the synthetic iridoid cocktails that correspond to extracts from intact and manually crumpled and torn silver vine leaves. Mosquitoes are more sensitive to the complex cocktail of iridoids that is induced by damage to silver vine leaves compared to the nepetalactol-dominated iridoids in intact silver vine. The diversification of iridoids in damaged silver vine leaves provides a stimulus that is more repellent to mosquitoes at low concentration, inducing a faster aversive response than nepetalactol- or nepetalactone-dominated iridoids in plants (Uenoyama et al., 2022). In contrast, a greater amount of damaged catnip-cocktail with nepetalactone-dominated iridoids was required to repel mosquitoes compared to both intact and damaged silver vine-cocktails. These results indicate that all of the iridoid compositions of intact and damaged silver vine leaves and damaged catnip leaves induced mosquito repellency, but the chemical constituent profile of iridoids from the damaged leaves had a faster repellent effect on mosquitoes than the simpler iridoid profile of the intact leaves at low concentration.

Catnip and silver vine have been known as plants that are both unusually attractive to cats but are not closely related taxonomically. These unrelated plants have evolved to produce very similar, but not identical iridoids as secondary plant metabolites. It is suggested that the olfactory and behavioral sensitivity of cats to plant iridoids has been fine-tuned to show a prolonged characteristic response toward either a low level of complex iridoids emitted from damaged silver vine or a much higher level of nepetalactone emitted from damaged catnip. These findings highlight the possibility that plant-damaging behavior, which occurs alongside self-anointing, might play an important role in gaining pest-repellency in other animals as well as cats.

## 5 Feline Neuronal System to Receive Plant Iridoids

Our studies revealed the biological significance of the iridoid response in cats. To understand why only Felidae species have acquired this response contributing to pest defense, it is necessary to fully understand the machinery of chemoreception of iridoids and neuronal circuits responding to its stimulation for the expression of the iridoid response in cats. Cats exhibit the characteristic rubbing and rolling behavior without naso-oral contact with nepetalactol, suggesting that the olfactory system is important for the chemoreception of bioactive iridoids. Considering that no cats exhibit the flehmen response, which is known as a functional behavior to transfer compounds such as pheromones to the vomeronasal organs (Hart & Leedy, 1987), against silver vine and catnip, suggesting that nepetalactol and other bioactive iridoids are detected in the main olfactory system.

Our study demonstrated that chemoreception of nepetalactol in the feline olfactory system stimulates the μ-opioid system, which controls rewarding and euphoric effects in humans (Roth-Deri et al., 2008). Plasma levels of β-endorphin (a peptide hormone and an endogenous opiate) were markedly elevated in cats after the characteristic response induced only by nepetalactol. Inhibition of the μ-opioid system by administration of naloxone, which is an antagonist of μ-opioid receptors, suppressed the rubbing and rolling response in the cats. These results indicate that the μ-opioid system is involved in the induction of the feline behavioral response. These suggest that cats may experience a positive reaction that has often been interpreted as extreme pleasure (Espin-Iturbe et al., 2017; Hatch, 1972), if the μ-opioid system in cats plays the same physiological function as the human system. A previous study reported that the intraperitoneal injection of essential oils from *Nepeta caesarea,* whose major content is nepetalactone, had an analgesic effect involving μ-opioid receptors in rats (Aydin et al., 1998). The other possibility for the activation of the μ-opioid system in cats might further help by providing analgesia to reduce irritation of arthropod bites.

The iridoid response is non-addictive in cats (Abramson et al., 2012)while exogenous opiates develop addiction in mammals (French et al., 1979). This may be because the μ-opioid system is stimulated by an increase in endogenous β-endorphin secretion when olfactory neurons are activated by plant iridoids, which markedly differs from the process induced by exogenous opiates such as morphine, which activates μ-opioid receptors via the bloodstream directly (Oldendorf, Hyman, Braun, & Oldendorf, 1972; Tunblad, Jonsson, & Hammarlund-Udenaes, 2003).

Recently, a nepetalactone receptor which triggers aversive response was identified in insects including mosquitoes (Melo et al., 2021). The catnip extract and nepetalactone activate the transient receptor potential (TRP) ankyrin 1 (TRPA1) ion channel in some insects, including mosquitoes. TRPA1 is commonly known as the "wasabi receptor" (Al-Anzi et al., 2006), which senses noxious and irritating chemicals in the environment. Melo et al. (2021) findings clearly indicate the molecular mechanism underlying the function of plant iridoids as powerful natural insect repellents. In addition, they reported that nepetalactone did not stimulate human TRPA1. Thus, feline TRPA1 is unlikely to be the receptor through which nepetalactone and nepetalactol induce the characteristic response in Felidae species. Further studies are necessary to identify receptors for plant iridoids in cats.

## 6 Conclusions

Despite widespread recognition of the characteristic response to specific plants such as catnip and silver vine by cats, it had long remained a mystery why these plants produce this response only in Felidae species. Our findings provide new insight into this well-known and characteristic plant-induced feline response, for which the biological significance was first questioned in popular science culture more than 300 years ago. We found that nepetalactol identified from silver vine can repel mosquitoes, as a previous report showed (Reichert et al., 2019). Further studies demonstrated that rubbing behavior in the presence of nepetalactol allows cats to transfer nepetalactol from the plant to their fur, thereby reducing the number of mosquitoes that land on the animal and providing protection against bites from these insects, which may carry viruses and parasites. There are few reports about non-human animals exploiting chemicals for protection against insect pests; for example, capuchin monkeys (*Cebus olivaceus*) rub lemon peels containing chemicals that deter ticks and mosquitoes against their back (Weldon et al., 2011). The examples so far uncovered that self-anointing or using prophylactic self-medication (de Roode, Lefèvre, & Hunter, 2013) with secondary plant metabolites to protect against pests and diseases typically occur in individual animal species, even though the same pests and diseases may affect many species. Our unexpected findings against previous knowledge that cats may experience only euphoria from catnip (Espin-Iturbe et al., 2017; Hatch, 1972) provide another important example of how animals use plant metabolites for protection against insect pests via a species-specific behavior expressed without previous exposure or learning.

Our study also provides evidence that the physical damage of silver vine and catnip leaves by feline licking and chewing makes an important contribution to their chemical pest defense in combination with rubbing and rolling when cats are exposed to these plants. Physical damage of silver vine and catnip promotes the immediate emission of plant iridoids. Besides, such damage also changes the composition of plant iridoids in silver vine, though not in catnip. These changes in both the amount of plant iridoid emission (both plant species) and composition (silver vine only) induced significantly extended response to these plants, promoting increased self-anointing behaviors (rubbing and rolling) that transfer plant iridoids to the feline fur. Moreover, the diversification of iridoids in damaged silver vine leaves provides a stimulus that is more repellent to mosquitoes at a low concentration, inducing a faster response than nepetalactol- or nepetalactone-dominated iridoids in plants. Thus, leaf damage by licking and chewing acts in combination with rubbing and rolling as an adaptive response that allows cats to gain effective mosquito repellency from iridoid-producing plants, helping to reduce the health risks and irritation associated with mosquitoes and possibly other arthropod pests that are sensitive to plant iridoids.

The biological significance of the response that we uncovered in cats cannot explain the species-specificity of the response, as other mammals such as dogs also face mosquito-borne diseases, which are not specific to felids. About 30% of cats are insensitive to catnip, and the response is inherited as an autosomal dominant trait in domestic cats (Todd, 1962), strongly suggesting the presence of one or few genes responsible for the iridoid response in felids. By contrast, dogs, mice, and negative responder cats failed to even stop and sniff nepetalactol stimulus. These findings suggest that domestic cats and the non-domestic felids that also respond might have acquired specific olfactory receptor(s) that detect nepetalactol and other iridoids emitted from some plants with high sensitivity. Thus, our next challenge continues to elucidate how and why this characteristic response to silver vine and catnip has evolved specifically in felids; why the characteristic plant-induced response is limited to Felidae species including the domestic cat; and why some cats are insensitive to these plants. A genome-wide association study between positive- and negative-responder cats to identify the olfactory receptor genes and specific neuronal pathways involved in this response could provide invaluable clues.

**Acknowledgments** We thank Prof. Toshio Nishikawa, Prof. Jane L Hurst, Prof. Robert Beynon, Prof. Shuji Kaneko, and Dr. Tamako Miyazaki for our collaboration. This research was funded by JSPS KAKENHI Grant Numbers 18H04602 and 20H04759 (M.M.), and 22 J23343 (R.U.), Suntory Foundation for Life Sciences (M.M.), and the Sasakawa Scientific Research Grant from The Japan Science Society (R.U.). R.U. was supported by a Grant-in-Aid for JSPS Fellows.

## References

Abramson, C. I., Lay, A., Bowser, T. J., & Varnon, C. A. (2012). The use of silver vine (*Actinidia polygama* Maxim, family *Actinidiaceae*) as an enrichment aid for felines: Issues and prospects. *American Journal of Animal and Veterinary Sciences, 7*(1), 21–27.

Al-Anzi, B., Tracey, W. D., & Benzer, S. (2006). Response of Drosophila to Wasabi is mediated by painless, the fly homolog of mammalian TRPA1/ANKTM1. *Current Biology, 16*(10), 1034–1040. https://doi.org/10.1016/j.cub.2006.04.002

Aydin, S., Beis, R., Öztürk, Y., Hüsnü, K., & Baser, C. (1998). Nepetalactone: A new opioid analgesic from *Nepeta caesarea Boiss. Journal of Pharmacy and Pharmacology, 50*(7), 813–817.

Barnett, R., Mendoza, M. L. Z., Soares, A. E. R., Ho, S. Y., Zazula, G., Yamaguchi, N., ... Gilbert, M. T. P. (2016). Mitogenomics of the extinct cave lion, *Panthera spelaea* (Goldfuss, 1810), resolve its position within the *Panthera* cats.

Benelli, G., & Mehlhorn, H. (2016). Declining malaria, rising of dengue and Zika virus: Insights for mosquito vector control. *Parasitology research, 115*(5), 1747–1754.

Birkett, M. A., Hassanali, A., Hoglund, S., Pettersson, J., & Pickett, J. A. (2011). Repellent activity of catmint, *Nepeta cataria*, and iridoid nepetalactone isomers against Afro-tropical mosquitoes, ixodid ticks and red poultry mites. *Phytochemistry, 72*(1), 109–114.

Bol, S., Caspers, J., Buckingham, L., Anderson-Shelton, G. D., Ridgway, C., Buffington, C. A., et al. (2017). Responsiveness of cats (*Felidae*) to silver vine (*Actinidia polygama*), Tatarian honeysuckle (*Lonicera tatarica*), valerian (*Valeriana officinalis*) and catnip (*Nepeta cataria*). *BMC Veterinary Research, 13*(1), 70.

de Roode, J. C., Lefèvre, T., & Hunter, M. D. (2013). Self-medication in animals. *Science, 340*(6129), 150–151.

Espin-Iturbe, L. T., Lopez Yanez, B. A., Carrasco Garcia, A., Canseco-Sedano, R., Vazquez-Hernandez, M., & Coria-Avila, G. A. (2017). Active and passive responses to catnip (*Nepeta cataria*) are affected by age, sex and early gonadectomy in male and female cats. *Behavioural Processes, 142*, 110–115.

Feaster, J. E., Scialdone, M. A., Todd, R. G., Gonzalez, Y. I., Foster, J. P., & Hallahan, D. L. (2009). Dihydronepetalactones deter feeding activity by mosquitoes, stable flies, and deer ticks. *Journal of Medical Entomology, 46*(4), 832–840.

French, E. D., Vasquez, S. A., & George, R. (1979). Behavioral changes produced in the cat by acute and chronic morphine injection and naloxone precipitated withdrawal. *European Journal of Pharmacology, 57*(4), 387–397.

Gkinis, G., Michaelakis, A., Koliopoulos, G., Ioannou, E., Tzakou, O., & Roussis, V. (2014). Evaluation of the repellent effects of *Nepeta parnassica* extract, essential oil, and its major nepetalactone metabolite against mosquitoes. *Parasitology Research, 113*(3), 1127–1134.

Hart, B. L., & Leedy, M. G. (1985). Analysis of the catnip reaction: mediation by olfactory system, not vomeronasal organ. *Behavioral and Neural Biology, 44*(1), 38–46.

Hart, B. L., & Leedy, M. G. (1987). Stimulus and hormonal determinants of flehmen behavior in cats. *Hormones and Behavior, 21*(1), 44–52.

Hatch, R. C. (1972). Effect of drugs on catnip (*Nepeta cataria*)-induced pleasure behavior in cats. *American Journal of Veterinary Research, 33*(1), 143–155.

Hill, J. O., Pavlik, E. J., Smith, G. L., Burghardt, G. M., & Coulson, P. B. (1976). Species-characteristic responses to catnip by undomesticated felids. *Journal of Chemical Ecology, 2*(2), 239–253.

Meinwald, J. (1954). The degradation of nepetalactone. *Journal of the American Chemical Society, 76*(18), 4571–4573.

Melo, N., Capek, M., Arenas, O. M., Afify, A., Yilmaz, A., Potter, C. J., … Stensmyr, M. C. (2021). The irritant receptor TRPA1 mediates the mosquito repellent effect of catnip. *Current Biology, 31*(9), 1988–1994.e1985.

Oldendorf, W. H., Hyman, S., Braun, L., & Oldendorf, S. Z. (1972). Blood-brain barrier: Penetration of morphine, codeine, heroin, and methadone after carotid injection. *Science, 178*(4064), 984–986.

Reichert, W., Ejercito, J., Guda, T., Dong, X., Wu, Q., Ray, A., & Simon, J. E. (2019). Repellency assessment of *Nepeta cataria* essential oils and isolated nepetalactones on *Aedes aegypti*. *Scientific Reports, 9*(1), 1524.

Roth-Deri, I., Green-Sadan, T., & Yadid, G. (2008). Beta-endorphin and drug-induced reward and reinforcement. *Progress in Neurobiology, 86*(1), 1–21.

Sakan, T., Fujino, A., Murai, F., Butsugan, Y., & Suzui, A. (1959a). On the structure of actinidine and matatabilactone, the effective components of *Actinidia polygama*. *Bulletin of the Chemical Society of Japan, 32*(3), 315–316.

Sakan, T., Fujino, A., Murai, F., Suzui, A., & Butsugan, Y. (1959b). The structure of matatabilactone. *Bulletin of the Chemical Society of Japan, 32*(10), 1154–1155.

Sakan, T., Murai, F., Isoe, S., Hyeon, S. B., & Hayashi, Y. (1969). The Biologically Active C9-, C10-, and CCii-Terpenes from *Actinidia polygama* Miq., *Boschniakia rossica* Hult, and *Menyanthes trifoliata* L. *Nippon Kagaku Zasshi, 90*(6), A29–A33. https://doi.org/10.1246/nikkashi1948.90.6_A29

Todd, N. B. (1962). Inheritance of the catnip response in domestic cats. *Journal of Heredity, 53*, 54–56.

Traversa, D., & Di Cesare, A. (2014). Cardio-pulmonary parasitic nematodes affecting cats in Europe: Unraveling the past, depicting the present, and predicting the future. *Frontiers in Veterinary Science, 1*.

Tucker, A. O., & Tucker, S. S. (1988). Catnip and the catnip response. *Economic Botany, 42*(2), 214–231.

Tunblad, K., Jonsson, E. N., & Hammarlund-Udenaes, M. (2003). Morphine blood-brain barrier transport is influenced by probenecid co-administration. *Pharmaceutical research, 20*(4), 618–623.

Uenoyama, R., Miyazaki, T., Hurst, J. L., Beynon, R. J., Adachi, M., Murooka, T., … Miyazaki, M. (2021). The characteristic response of domestic cats to plant iridoids allows them to gain chemical defense against mosquitoes. *Science Advances, 7*(4), eabd9135.

Uenoyama, R., Miyazaki, T., Adachi, M., Nishikawa, T., Hurst, J. L., & Miyazaki, M. (2022). Domestic cat damage to plant leaves containing iridoids enhances chemical repellency to pests. *iScience*, 104455.

Weldon, P. J., Carroll, J. F., Kramer, M., Bedoukian, R. H., Coleman, R. E., & Bernier, U. R. (2011). Anointing chemicals and hematophagous arthropods: Responses by ticks and mosquitoes to Citrus (*Rutaceae*) peel exudates and monoterpene components. *Journal of Chemical Ecology, 37*(4), 348–359.

Zhu, J., & Zeng, X. (2006). Adult repellency and larvicidal activity of five plant essential oils against mosquitoes. *Journal of the American Mosquito Control Association, 22*(3), 515–522. 518.

Zhu, J. J., Berkebile, D. R., Dunlap, C. A., Zhang, A., Boxler, D., Tangtrakulwanich, K., et al. (2012). Nepetalactones from essential oil of *Nepeta cataria* represent a stable fly feeding and oviposition repellent. *Medical and Veterinary Entomology, 26*(2), 131–138.

# Part VI

# Applications

# Canine Detection of Cancer in Humans: Expectations Versus Reality

Tadeusz Jezierski[1] (✉) and Michał Dzięcioł[2]

[1] Institute of Genetics and Animal Biotechnology of Polish Academy of Sciences, Department of Animal Behaviour and Welfare, Jastrzębiec, Poland
t.jezierski@igbzpan.pl
[2] Wrocław University of Environmental and Life Sciences, Department of Reproduction and Clinic of Farm Animals, Wrocław, Poland

**Abstract** Easily accessible screening methods play a crucial role in the early detection of cancer. Over the last few decades, alternative cancer screening methods based on detection of specific volatile organic compounds (VOCs) have been investigated. First reports on canine ability to detect melanoma in humans, based on the odours emitted by the lesions, were published in 1989 and 2001. These reports triggered further attempts for canine detection of different types of cancers using various types of odour samples. Experiments published between 2004 and 2022 revealed promising results that dogs could be routinely used for cancer screening. However, the use of canines for the screening of cancer in a clinical setting still remains in the research phase due to the numerous limitations and methodological concerns that need to be addressed. In order to have a better comprehension of this new experimental method, one must challenge this assessment using a wider range of factors, with a focus on decreasing cancer morbidity and mortality. The following paper describes the various methods currently used and their results. Reasons why canine detection of cancer is still in the research phase, without using it for practical cancer screening on a regular basis, are discussed.

**Keywords** Cancer odour · Dog · *Canis familiaris* · Screening · Training

## 1 Introduction

Cancer is one of the most common human diseases worldwide, accounting for nearly 10 million deaths in 2020. Various physical, chemical, and biological cancerogenic factors, related to environmental pollution and people's unhealthy lifestyle, contribute to an increasing tendency in cancer morbidity and mortality (WHO).[1]

[1] https://www.who.int/news-room/fact-sheets/detail/cancer. Consulted on November 04, 2022.



B. Schaal et al. (eds.), *Chemical Signals in Vertebrates 15*,
https://doi.org/10.1007/978-3-031-35159-4_22

It is commonly acknowledged that cancer mortality can be reduced by early detection and reliable diagnosis (Yu et al., 2012). With this, an early diagnosis makes cancer therapy more likely to be successful, less invasive, less stressful and painful and in a shorter period of time. Unfortunately, early stages of cancer are often accompanied with no symptoms or nonspecific symptoms which are thus often ignored or overlooked. Ideally, the population as a whole should be routinely examined for any cancerous developments. However, systematically examining the whole population may not be feasible, as the process of ultimate identification of the disease from its symptoms takes into account clinical health history, physical exam findings and various blood tests, imaging modalities, and invasive biopsies followed by histopathological assay. The throughput of diagnostic procedures is very limited and can only be applied to cases with serious suspicion for the development of cancer. However, the use of highly sensitive screenings in a larger population may be able to circumvent these challenges at the risk of an increased detection for false positive findings. Different screening methods have been developed for particular types of cancer for example, mammography for breast cancer, colonoscopy for colorectal cancer, prostate-specific antigen (PSA) measurement for prostate cancer, X-ray for lung cancer, blood tests, ultrasonography, or computer tomography (CT) for bladder or ovarian cancer. These screening methods require specific equipment and qualified staffing limiting their accessibility, especially in resource limited countries. Moreover, conventional cancer screening methods give only limited information about the presence, size, and location of real cancerous lesions (Haick et al., 2014) and some inconsistencies exist between the tests (Yu et al., 2012). High resolution imaging screening methods, as an example, may lead to an over-diagnosis due to unsatisfactory specificity resulting in patient's anxiety, exposure to other invasive procedures, and unnecessary financial expenditure (Henschke et al., 2006). Hence, searching for a screening method which would be non-invasive, simple, painless, inexpensive, easily accessible, and sufficiently reliable at early cancer stages remains a desirable goal. Such an ideal screening method does not exist so far.

In the last 20 years, numerous attempts were undertaken to find new, alternative cancer screening methods to those based on organ or tissue imaging. One new approach is based on the detection of emitted volatile organic compounds (VOCs), which may be hypothetically produced by cancer cells (Amann et al., 2014). The association of unique odours with various diseases has been established since ancient times; however, modern medicine has paid little attention to the potential clinical utility of body odours (Shirasu & Touhara, 2011).

Generally, animals are able to distinguish sick conspecifics from healthy ones, presumably by odours emitted by these individuals, but possibly also by changes in their behaviour or body posture. For example, it has been well documented that mice can distinguish between sick and healthy individuals of their own species by odour. This includes parasite infections (Ehman & Scott, 2002; Kavaliers et al., 2003), influenza (Penn et al., 1996), mouse mammary tumour virus (Yamazaki et al., 2002), and inflammatory processes (Arakawa et al., 2010). Odour from sick conspecifics usually evokes an avoidance response in odour-recipient individuals. The ability of animals to recognize odour cues associated with infection and sickness has developed during the course of evolution as it reduces the risk of contagion (Thomas et al., 2005; Arakawa et al., 2009; see Schwambergová et al. in the present volume).

The ability of dogs to detect melanoma in humans was first reported in 1989 in *The Lancet* (Williams & Pembroke, 1989), with similar cases described by Church and Williams (2001) and later by Campbell et al. (2013). In all three reports, described cases of untrained companion dogs, which contributed to detection of melanoma on the leg or ear of their owners due to a spontaneous and peculiar behaviour by their dog(s) like sniffing and licking persistently at an asymptomatic lesion, and in case of the leg, even through the patient's trousers (Williams & Pembroke, 1989).

It is, however, puzzling why the odour of melanoma in humans evokes spontaneous interest and not avoidance reaction in dogs (Williams & Pembroke, 1989; Church & Williams, 2001). Church and Williams (2001) hypothesized that abnormal metabolisms of cancer cells may produce specific VOCs sniffed out by dogs due to their exceptional sense of smell. However, no plausible explanation exists as to why the dogs were interested in the odour emitted by melanoma cells. The cited authors only generally mentioned that dogs often sniff and lick wounds or lesions. It is unclear if canines would be able to detect recurrent melanoma on their owners' body or if they are also able to detect melanoma or other cancers in humans they are unfamiliar with. As to other human diseases, it is worth mentioning that some dogs are able to react or to alert to hypoglycaemia episodes in their owners or family members with type 1 diabetes. It was reported that such reactions of dogs to hypoglycaemia can be both spontaneous (Chen et al., 2000) or trained (Rooney et al., 2013). Epilepsy in humans is another disease to which the dogs react or alert prior to the seizure (Dalziel et al., 2003). While reacting to the seizure may be evoked by visual cues, it is still debated if during alerting prior to the seizure the dogs use visual cues, e.g. subtle changes in patient's behaviour or specific odour cues. Catala et al. (2019), and Maa et al. (2021), however, proved that during a seizure, humans exude a specific odour that can be identified by trained canines. In addition to this, it has been reported that trained canines are able to detect Parkinson's disease in humans when testing odours (Mischley & Farahnik, 2020).

A milestone for canine detection of cancers in humans occurred during a conference on Medical Scent Detection (Saunderton, UK, 4–7 October 2002), resulting in an increase in the number of experimental studies on cancer detection on the basis of different odour samples by specially trained dogs. A number of experimental studies on cancer detection by trained dogs were published beginning from 2004 (see Tables). Subsequent studies, as well as reviews (e.g. Pirrone & Albertini, 2017; Schoon et al., 2022), only confirm previous findings and/or expand upon other kinds of cancers and some methodological concerns. In addition, in the above-mentioned reviews, some explanations were given as to the chemical and biological mechanisms underlying canine detection of various cancers.

The aim of this chapter is to provide an update regarding the state of canine detection of cancer in humans in the context of chemical signalling, from the existing literature as well as our own experimental studies. A thorough search was done through the Web of Science database by using keywords such as: 'dogs', 'cancer' and 'detection' in order to review and survey a wide range of experiments in relation to this subject matter. We analysed 33 peer-reviewed papers published in the time frame of 2004–2022. Our analysis included number and breed of dogs, a pre-selection of dogs for the cancer detection training, training method and duration, transition criteria from training to testing dog, kind of odour samples and storage of said samples, number of positive and negative odour samples used in training and testing, detection sensitivity and specificity, as well as the ratio of positive to negative odour samples, and also any limitations if applicable.

## 2 Methodological Issues in the Reviewed Literature

### 2.1 Canines

In the 33 journals that were examined, a total of 88 dogs were used. However, due to the fact that some pool samples had the same dog (e.g. Urbanova et al., 2015; Pacik et al., 2018; Sonoda et al., 2011; Yamamoto et al., 2020), the accurate total number of dogs was 86. Given the heterogeneity of existing dog breeds, understanding which breed is most suitable for cancer detection is fundamentally important. In Table 1 the number of purebred dogs representing 24 breeds and 11 dogs of different types of crossbreds are given. Labrador Retrievers (19 dogs) and German Shepherds (15 dogs) have been the most commonly utilized breeds in experimental studies for cancer detection (Table 1). The other breeds and types of crossbreds were represented by 1–7 individuals, however, mostly by only one.

As to reports on spontaneous detection of melanoma, different breeds were mentioned, e.g. Border Collie × Doberman cross (Williams & Pembroke, 1989), Labrador Retriever (Church & Williams, 2001), and German Shepherd (Campbell et al., 2013). It seems that using specific breeds for experimental training was connected with availability or individual preference for particular breed or mixes by the trainer. Dog breed may be of importance due to differences in trainability and different olfactory ability. It is known that so-called working breeds, for example retrievers or shepherds, are generally more suitable for operant conditioning training than typical pet breeds such as Chihuahua, Italian Greyhound, or Papillon, which were never selected for their trainability. Polgár et al. (2016) demonstrated that scent breeds selected over generations for scent work (Basset Hound, Beagle, German Pointer, Wire-haired Vizsla, Bracco Italiano, Grand Basset Griffon Vendeen, Transilvanian hound) were superior at detecting odours, compared to both non-scent breeds not selected for scent work (Bichon Havanese, Chinese Crested Powder Puff, English Greyhound, Hungarian Greyhound, Bichon Bolognese, Whippet) and short-nosed breeds (Cavalier King Charles Spaniel, Boston terrier, Boxer, American Bulldog, English Bulldog, Bullmastiff, Pug).

The number of dogs used by particular authors for the training ranged from 1 to 10. Utilizing more adequately and homogeneously trained dogs, rather than a single canine, allows for improved assessment for individual variation in their capability and performance, providing improved estimates in validity and reliability of detection methods. The number of studies describing how intensively the pre-selection of canine candidates for the training was done is scarce (Table 1). Such pre-selection would be advisable since some individual dogs of the same breed may be more suitable for training than others, due to increased drive for sniffing, greater focus, and motivation to earn a reward, which is a prerequisite for effective operant conditioning. Furthermore, persistence and reduced tendency for distractions are desirable for effective training and

**Table 1** Dogs used for experimental training of cancer detection in the 33 different studies

| Authors | Number of dogs *(selected from)* | Breeds | Cancer type for which the dogs were trained |
|---|---|---|---|
| Pickel et al. (2004) | 2 | 1 Standard Schnauzer, 1 Golden Retriever | Melanoma |
| Willis et al. (2004) | 6 | 3 Cocker Spaniels (working line)<br>1 Labrador Retriever, 1 Papillon, 1 Mix breed | Bladder |
| McCulloch et al. (2006) | 5 *(13)* | 3 Labrador Retrievers, 2 Portuguese Water Dogs | Lung<br>Breast |
| Horvath et al. (2008) | 1 | 1 Giant Schnauzer | Ovarian |
| Gordon et al. (2008) | 10 | 1 Aussie Cocker mix, 1 German Shepherd,<br>1 Boxer, 1 Rhodesian Ridgeback, 1 Chihuahua,<br>1 Italian Greyhound, 1 Border Collie, 1 Collie mix<br>1 Pembroke Welsh Corgi, 1 Goldendoodle | Breast<br>Prostate |
| Cornu et al. (2011) | 1 | 1 Belgian Malinois | Prostate |
| Sonoda et al. (2011) | 1 | 1 Labrador Retriever[b] | Colorectal |
| Ehmann et al. (2012) | 4 | 2 German Shepherds, 1 Australian Shepherd,<br>1 Labrador Retriever | Lung |
| Walczak et al. (2012) | 3 *(6)* | 3 Mix breed | Breast<br>Lung<br>Melanoma |
| Horvath et al. (2013) | 2 | 2 Giant Schnauzers | Ovarian |
| Amundsen et al. (2014) | 4 | 1 Belgian Shepherd, 1 Border Collie, 1 Dachshund, 1 Rottweiler | Lung |
| Elliker et al. (2014) | 2 *(10)* | 1 Labrador Retriever, 1 Border Collie | Prostate |
| Taverna et al. (2015) | 2 | 2 German Shepherds | Prostate |
| Urbanova et al. (2015) | 1 | 1 German Shepherd[a] | Prostate |
| Willis et al. (2016) | 1 | 1 Labrador Retriever | Melanoma |
| Guerro-Flores et al. (2017) | 1 | 1 Beagle | Cervical |
| Kitiyakara et al. (2017) | 1 | 1 Golden Retriever | Hepatocellular |

(continued)

**Table 1** (continued)

| Authors | Number of dogs *(selected from)* | Breeds | Cancer type for which the dogs were trained |
|---|---|---|---|
| Fischer-Tenhagen et al. (2018) | 2 *(4)* | 1 Labrador Retriever, 1 Poodle | Lung |
| Pacik et al. (2018) | 1 | 1 German Shepherd[a] | Prostate |
| Seo et al. (2018) | 2 | 1 Cocker Spaniel, 1 English Springer Spaniel | Breast, Prostate |
| Guirao et al. (2019) | 1 | 1 Crossbred Labrador Retriever × Pitbull | Solitary pulmonary nodules (SPN) |
| Junqueira et al. (2019) | 3 *(4)* | 3 Beagles | Lung |
| Murarka et al. (2019) | 4 | 3 German Shepherds, 1 Labrador Retriever | Ovarian |
| Kybert et al. (2020) | 3 *(4)* | 1 English Springer Spaniel, 1 German Shepherd,<br>1 Labrador Retriever | Ovarian |
| Schoon et al. (2020) | 8 | 3 Labrador × Golden Retrievers, 3 English Cocker Spaniels, 1 Belgian Malinois, 1 Nova Scotia Duck Tolling Retriever | Colorectal |
| Yamamoto et al. (2020) | 1 | 1 Labrador Retriever[b] | Cervical |
| Feil et al. (2021) | 1 | 1 Golden Retriever | Lung |
| Kure et al. (2021) | 1 | 1 Labrador Retriever | Breast |
| Crawford et al. (2022b) | 4 *(7)* | 1 Australian Cattle Dog × Border Collie,<br>1 Labrador Retriever, 1 Jack Russel T. × Beagle,<br>1 Border Collie × NZ Huntaway × Kelpie | Lung |
| Jeong et al. (2022) | 2 | 2 Labrador Retrievers | Prostate |
| Kane et al. (2022) | 5 | 3 German Shepherds, 1 Dutch Shepherd,<br>1 Labrador Retriever | Ovarian |
| Ortal et al. (2022) | 1 *(2)* | 1 Belgian Shepherds Malinois | Osteosarcoma |
| Riedlova et al. (2022) | 2 | 1 Labrador Retriever, 1 Australian Cattle Dog | Lung |

[a]The same dog used by Urbanova et al. (2015) and Pacik et al. (2018)
[b]The same dog used by Sonoda et al. (2011) and Yamamoto et al. (2020)

detection. However, given the paucity of existing literature, it cannot currently be ascertained as to which dog breed is preferential for cancer detection. Nonetheless, it is plausible that working breeds may be more suitable for detection performance for the reasons mentioned above. Serpell and Hsu (2005) found significant differences in trainability between breeds as well as between field and show bred lines within the same breeds. The individual variation in cognition, trainability, and detection performance within breeds and lines should be taken into account and have been discussed in more detail by MacLean and Hare (2018).

## 2.2 Cancer Types and Kinds of Odour Samples

The detection of 11 types of neoplastic diseases was studied using trained canines (Table 1). The majority of studies were dedicated to the detection of lung, prostate, breast, and ovarian cancer (9, 8, 5, and 5 studies, respectively), followed by melanoma (3), colorectal (2), and cervical cancers (2 studies). Some studies found that an individual canine was able to appropriately detect two, even three, types of cancer (e.g. McCulloch et al., 2006; Seo et al., 2018; Walczak et al., 2012). As current screening tests for some types of cancer have unsatisfactory sensitivity or specificity and are costly, finding new alternative methods for detection of such types of cancer would be desirable. For example, non-small cell lung cancer has one of the greatest prevalence, yet early detection is difficult because early-stage lung cancer is often asymptomatic and undetectable with a chest X-ray. In contrast, melanoma can be more easily detected by the visible presence and appearance of a skin lesion.

For the detection of cancers, odour samples of exhaled breath were mostly used (11 studies), but some studies on other types of cancer used urine samples (10 studies), tissue scrapes or cancer cell lines, blood plasma or serum, stool samples, saliva, or surgical bandages (Table 2). Ideally, odour samples for canine training should be collected easily, non-invasively and stored without losing their odorous characteristics over the storage period. In the following sections, *positive odours* relate to samples taken from a patient with fully diagnosed cancer while *negative odours* are sampled from healthy controls.

A prerequisite for canine training is to sniff the samples obtained from the patient rather than the patient directly. It is important to note that the samples and tests have been separated over various durations (e.g. several hours up to 3 months) (Guirao et al., 2019). However, this was contraindicated in an early study done on melanoma, where the volunteer was sniffed directly by the canines (Pickel et al., 2004).

The total number of donors of positive odour samples in particular studies ranged from 5 to 225 and the number of donors of negative odour samples ranged from 30 to more than 500 (Table 2).

However, some authors did not clearly indicate the accurate description implied by 'odour sample'. For instance, in some studies a single sample collected related to a single donor, while other studies had several samples from a sole donor.

Although collection of exhaled breath samples and urine is easy and non-invasive, urine samples cannot be stored over a longer time at room temperature due to possible bacterial growth and decomposition of VOCs. Even though breath samples seem to be

**Table 2** Number of positive and negative odour samples and duration of storage before testing reported if available from the 33 different studies (blank cells denote no data reported)

| Authors | Number of positive samples (donors) used | Number of negative samples (donors) used | Sample freezing/storage duration before testing |
|---|---|---|---|
| Pickel et al. (2004) | 1 melanoma tissue sample<br>7 volunteers' body with melanoma | 34 distractor samples (gauze inside adhesive bandages) | Melanoma tissue frozen at −80 °C |
| Willis et al. (2004) | 27 urine samples for training<br>9 urine samples for testing | 54 urine samples for training<br>54 urine samples for testing | Urine frozen at −40 °C stored <5 months |
| McCulloch et al. (2006) | Lung cancer<br>55 breath samples<br>Breast cancer<br>31 breath samples | 83 breath samples | Breath samples, stored 1–60 days |
| Horvath et al. (2008) | Single blind<br>4 tissue samples from each of 5 donors<br>Double blind<br>20 donors of tissue samples | 80 tissue samples | Tumour tissue frozen at −80 °C |
| Gordon et al. (2008) | Breast cancer<br>53 urine samples for training<br>9 urine samples for testing | Breast cancer<br>134 urine samples for training<br>54 urine samples for testing | Urine frozen at −20 °C |
| | Prostate cancer<br>46 urine samples for training<br>11 urine samples for testing | Prostate cancer<br>120 urine samples for training<br>66 urine samples for testing | |
| Cornu et al. (2011) | 26 urine samples for training<br>33 urine samples for testing | 16 urine samples for training<br>33 urine samples for testing | Urine frozen at −4 °C |
| Sonoda et al. (2011)[b] | 33 breath samples<br>37 stool samples | 132 breath samples<br>148 stool samples | Stool frozen at −80 °C until filtration, then at −20 °C until tests |
| Ehmann et al. (2012) | 35 breath samples for training<br>25 breath samples for testing | 60 breath samples for training<br>50 breath samples for testing<br>50 breath samples with COPD for testing | Breath samples stored at room temperature in a light-tight cabinet |

(continued)

**Table 2** (continued)

| Authors | Number of positive samples (donors) used | Number of negative samples (donors) used | Sample freezing/storage duration before testing |
|---|---|---|---|
| Walczak et al. (2012) | Breast<br>57 breath donors<br>Lung<br>118 breath donors<br>Melanoma<br>45 breath donors | 305 breath donors | |
| Horvath et al. (2013) | Series I<br>42 blood plasma donors<br>Series II<br>10 blood plasma donors | 210 blood plasma donors | Plasma frozen at −80 °C |
| Amundsen et al. (2014) | 63 breath donors | 30 breath donors | Breath at room temperature, urine frozen at −20 °C stored <24 months |
| Elliker et al. (2014) | 50 urine samples for training<br>31 urine samples for testing | 67 urine samples for training<br>93 urine samples for testing | Urine frozen at −20 °C stored <6 months |
| Taverna et al. (2015) | 200 urine samples for training<br>162 urine samples for testing | 230 urine samples for training<br>310 urine samples for testing | Urine frozen at −20 °C |
| Urbanova et al. (2015)[a] | 45 urine samples for testing | 25 urine samples for testing | Urine frozen at −20 °C |
| Willis et al. (2016) | 635 skin lesions | 106 healthy skin | Gauze pads frozen at −40 °C |
| Guerro-Flores et al. (2017) | 50 smear samples for training<br>70 medical bandages for training<br>100 medical bandages for testing | 30 smear samples for training<br>170 medical bandages for testing | |
| Kitiyakara et al. (2017) | 127 breath samples for training<br>37 breath samples for testing | 115 breath samples | Face masks in Ziploc® bags, median storage time 8 days, 1–2 test runs performed 3 weeks after the collection |
| Fischer-Tenhagen et al. (2018) | 21 breath samples for training<br>9 breath samples for testing | 20 breath samples for training<br>10 breath samples for testing | Polypropylene fleece stored at 8 °C |
| Pacik et al. (2018)[a] | 60 artificial urine samples with sarcosine | 120 artificial urine samples without sarcosine | |

(continued)

**Table 2** (continued)

| Authors | Number of positive samples (donors) used | Number of negative samples (donors) used | Sample freezing/storage duration before testing |
|---|---|---|---|
| Seo et al. (2018) | Mammary carcinoma cells<br>Colorectal carcinoma cells | 2 samples of mammary and colorectal tissues of noncancerous females<br>1 culture liquid sample<br>1 empty sample | Cells cultivated for 4 days |
| Guirao et al. (2019) | 90 breath samples from 30 patients with SPN | 305 breath samples from 61 healthy donors<br>90 breath samples from 18 donors with COPD | Exhaled breath samples stored <3 months |
| Junqueira et al. (2019) | 30 blood serum samples total[c]<br>10 blood serum samples for testing | 120 serum samples total[c]<br>40 serum samples for testing | Blood serum frozen at −30 °C stored 1 week |
| Murarka et al. (2019) | 1 cell line for training<br>3 cell lines for testing | RPMI 1640 medium supplemented with 10% foetal bovine serum,<br>Fallopian tube cell lines associated media without cells<br>Sterile filter paper<br>Normal plasma samples– plasma samples from patients with benign ovarian growths | Biological samples stored at −80 °C |
| Kybert et al. (2020) | 10 ovarian cancer tissues[c] | 10 tissues healthy controls[c]<br>10 tissues benign ovarian ca. | Tissues samples frozen at −80 °C |
| Schoon et al. (2020) | 10 stool donors[c] | 60 stool donors[c] | Stool samples frozen at −80 °C |
| Yamamoto et al. (2020)[b] | 83 urine donors with cervical cancer[c] | 49 urine donors with benign uterine diseases[c]<br>63 healthy urine donors[c] | Urine frozen at −80 °C stored <6 months |
| Feil et al. (2021) | 36 urine & breath donors for training<br>41 urine & breath donors for testing | 150 urine & breath donors for training<br>205 urine & breath donors for testing | Urine and breath frozen at −80 °C |

(continued)

**Table 2** (continued)

| Authors | Number of positive samples (donors) used | Number of negative samples (donors) used | Sample freezing/storage duration before testing |
|---|---|---|---|
| Kure et al. (2021) | Several hundred breath & urine donors for training | ≈500 urine & breath donors for training<br>5 benign breast lesions for training | Urine frozen at −20 °C |
| | 40 urine donors for testing | 18 urine donors for testing | |
| Crawford, Perrone, et al. (2022b) | 900 individual breath samples collected in 3 years[d] from 225 patients out of which 400 samples for training[d] | | Breath frozen at −80 °C |
| Jeong et al. (2022) | Exometabolome (extracellular medium) of a cancer cell line, PC3 | RPMI 1640 or DMEM medium,<br>Empty cups,<br>Medium of 10% FBS and 1% penicillin/streptomycin,<br>Various concentrations (0.5–10%) of FBS | Supernatant separated from the pellet, immediately frozen at −20 °C |
| Kane et al. (2022) | 10 blood plasma donors[c] | 10 blood plasma donors (healthy controls)[c] | Blood plasma frozen at −80 °C |
| | Samples were pooled to create novel sample mixes as the study was limited by sample availability | 10 blood plasma donors with benign ovarian cancer | |
| Ortal et al. (2022) | 1 primary culture for training<br>6 osteosarcoma cell lines for testing | 1 primary culture – immortalized for training<br>2 primary cultures for testing | Cell lines frozen at −80 °C |
| | 2 saliva samples for testing | 12 saliva samples for testing | |
| Riedlova et al. (2022) | 115 serum or breath samples for training<br>47 serum or breath samples for testing | 101 serum or breath samples for training<br>44 serum or breath samples for testing | Serum frozen at −20 °C breath at room temperature stored <21 days |

[a]The same dog used by Urbanova et al. (2015) and Pacik et al. (2018)
[b]The same dog used by Sonoda et al. (2011) and Yamamoto et al. (2020)
[c]Did not specify how many samples used for training and testing
[d]Did not specify how many positive and negative

unaffected by temperature, however, they require absolutely airtight containers, to prevent sample contamination from external odours. Usually, breath samples were stored at room temperature, while other kinds of samples such as urine, stool, tissues serum, or cultured cells were frozen (Table 2). Some researchers, however, stored frozen breath samples before tests as well (e.g. Feil et al., 2021; Crawford et al., 2022b). A possibility to store odour samples is important for organizational and logistic reasons, since it is often not possible to have the samples tested by dogs directly after collection. Sample storage duration may have an impact on the quantitative changes in odour and thus on indications by dogs; however, we did not find experimental results on this issue in the reviewed literature. Only some authors indicated how long the samples were stored before tests. For example, breath samples used by McCulloch et al. (2006) were stored 1–60 days at room temperature prior to the tests. Guirao et al. (2019) stored breath samples up to 3 months.

Despite results that are comparable to those achieved with exhaled breath or urine samples, blood or tissue samples are limited by reliance on trained medical staff, ethical permission, and additional storage requirements.

The lack of methodological standardization in experiments on canine detection of cancer is an issue that hinders comparison of the validity and reliability across studies. Standardization of sample collection, storage, and handling would be less challenging compared to standardizing other methodological aspects (e.g. canine training or experimental set-up).

### 2.3 Dog Training

Dog training for cancer detection is based on operant conditioning, where a correct altering reaction, usually sitting or lying down in front of the target odour sample (i.e. the positive odour), is immediately rewarded either by a treat or a favourite toy. In many published studies, the description of the training details is scarce or uneven (Table 3). Some studies used canines trained for purposes other than cancer detection, prior to the cancer training (Table 3).

The intensity of the training differed based on the trainer's skills and experience, as well as subjective assessments in regards to the dog (capability, character, and motivation). The decision on when a canine is able to exit training and enter testing must ultimately be made. The criteria on such a decision includes both quantitative (e.g. minimum % of correct alerts) and qualitative assessments (Table 3). Daily training schedules ranged from 1 day per week to 5 days per week, and the total duration of the training ranged from 1 to 16 months (Table 3). In many publications, there is a lack of information in regards to how many trials (runs in the line-up) the dogs made during a test session.

The alert a dog may give for a positive sample may vary. This includes, sitting (Fig. 1, e.g. Cornu et al., 2011; Walczak et al., 2012; Yamamoto et al., 2020), laying down in front of the target sample (Ehmann et al., 2012; Amundsen et al., 2014; Elliker et al., 2014), and standing still while nose pointing at the target odour sample for a minimum of 3 seconds (Fischer-Tenhagen et al., 2018). Another alert used a canine-centred interface which detects a distinct pressure pattern exerted by a dog's nose on the

**Table 3** Dogs training prior to cancer training, training schedule and duration, and criterion for transition from training to testing phase in the 33 different studies. Blank cells denote no data reported

| Authors | Prior training before cancer training | Training schedule | Total duration of training | Criterion for transition from training to testing phase |
|---|---|---|---|---|
| Pickel et al. (2004) | Bomb detection, utility dog-excellent, obedience, master hunter | 20 min training sessions, <6 trials/day | 100 trials, several weeks | |
| Willis et al. (2004) | | | 7 months | |
| McCulloch et al. (2006) | | 4–5 times/week | 3 months | >30 correct consecutive trials |
| Horvath et al. (2008) | | 2 times/week | 12 months | 'The dog detected tumours with a fair degree of certainty' |
| Gordon et al. (2008) | Tracking, obedience, agility, therapy | 2–7 times/week | | 'When the dogs could consistently identify cancer sample among new control samples most of the time' |
| Cornu et al. (2011) | | 5 days/week | 16 months | |
| Sonoda et al. (2011)[b] | Water rescue training | 2–4 trials/day | 1 month | 'Correct distinguishing between cancer sample and 4 controls consecutively in dozens of trials'[c] |
| Ehmann et al. (2012) | Trained by professional trainer | | | |
| Walczak et al. (2012) | | 4–5 days/week | 9 months 812–1060 trials | >40 correct alerts in 100 consecutive trials |
| Horvath et al. (2013) | | 2 times/week | 12 months | |
| Amundsen et al. (2014) | Tracking/trailing | 2–4 times/week | | 99% sensitivity |
| Elliker et al. (2014) | | | 13 months | 9/10 successive runs correct, more frequently than by chance |
| Taverna et al. (2015) | Explosives detection | | 5 months | |
| Urbanova et al. (2015) | Basic obedience and scent work | 4–5 times/week, 3–6 trials/day | 11 months[a] | |

(continued)

**Table 3** (continued)

| Authors | Prior training before cancer training | Training schedule | Total duration of training | Criterion for transition from training to testing phase |
|---|---|---|---|---|
| Willis et al. (2016) | Basic obedience | 4 times of 30 min/week, 3–4 trials/day | 13 months | 'A success rate, after 20 test sets, greater than by chance alone' |
| Guerro-Flores et al. (2017) | Drug detection | 15 min, 5 days/week | 4 months | 'Training routine, endured before the dog was able to fully and unequivocally identify the odour, showing no signs of hesitation when pointing target samples' |
| Kitiyakara et al. (2017) | Search & rescue, mobility assistance for disabled | 3 training phases | | |
| Fischer-Tenhagen et al. (2018) | | 1 time/week 5–20 trials/day 1st dog 73 days 2nd dog 82 days | 16 months | >80% correct indications |
| Pacik et al. (2018) | Basic obedience and scent work | 4–5 times/week 3–6 trials/day | 11 months[a] | |
| Seo et al. (2018) | | 2 × 30 min/day | 6 months | |
| Guirao et al. (2019) | | | Considered well trained by an experienced dog trainer | |
| Junqueira et al. (2019) | | 3 training phases | 8 weeks | |
| Murarka et al. (2019) | Detection of Universal Detector Calibrant (UDC), a synthetic odour, using a wheel | 10 trials in 1–2 sessions per day | | 80% success across 10 trials for 3 training sessions in a row |
| Kybert et al. (2020) | | 7–10 trials/day 3–4 times/week | Tissue 3 months Plasma 45 days | >90% correct trials >85% correct trials |
| Schoon et al. (2020) | Guide dogs (rejected 3 dogs) | 5 training phases 2–3 times/week | | |
| Yamamoto et al. (2020)[b] | Water rescue dog Different types of cancer | 4 steps | 12 months | 'Correct distinguishing between cancer sample and 4 controls consecutively in dozens of trials'[c] |

(continued)

**Table 3** (continued)

| Authors | Prior training before cancer training | Training schedule | Total duration of training | Criterion for transition from training to testing phase |
|---|---|---|---|---|
| Feil et al. (2021) | | 1–2 times/week | 12 months | |
| Kure et al. (2021) | Water rescue, Different types of cancer | 4 steps | 12 months | 12 consecutive correct trials |
| Crawford et al. (2022b) | Detecting amyl acetate using an apparatus with 17 samples | 5.21–5.75 sessions/day | 44–67 days | 'When dogs' accuracy has reached or approached asymptote' |
| Jeong et al. (2022) | | Training composed of 3 courses. | 9 months | Sensitivity >90% |
| Kane et al. (2022) | Discrimination between plasma samples from patients with malignant ovarian cancer, and those with benign growths or no known growths | >270 training sessions of 10–15 trials, 40–60% of the trials did not contain the odour (blank trials). | 17.2 months (on average) | |
| Ortal et al. (2022) | Search for missing persons | 2 training phases | 16 months (training + testing) | Sensitivity & specificity >80% + regular behaviour in training |
| Riedlova et al. (2022) | | 4 phases 1 session/week | | Sensitivity >70% |

The criterion for transition from training to testing phase, if not given quantitatively, is mentioned as a qualitative description from respective publication
[a]The same dog used by Urbanova et al. (2015) and Pacik et al. (2018)
[b]The same dog used by Sonoda et al. (2011) and Yamamoto et al. (2020)
[c]Refers to each training session but not to the entire training period

interface sensor when sniffing the sample. This could be used to distinguish between correct and false alerts (Mancini et al., 2015; Johnston-Wilder et al., 2015). There were further attempts to assess pressure sensor data by using artificial neural networks (Withington et al., 2021). The purpose of the latter experiment was to assess whether artificial neural networks could be used in an automatic system in order to distinguish positive samples from negative samples by a dog-generated pressure sensor data alone. In doing so, this would not require the dog in training to alert the positive sample by sitting or lying down in front of it. However, it should be taken into consideration that the more sophisticated the methods are, the less of a possibility it is to implement it on a larger scale.

**Fig. 1** Alerting by sitting in front of the target odour sample in container #5 in the line-up. (The photo presents the dog 'Gromit' used by Walczak et al., 2012 during the testing phase) (© Tadeusz Jezierski)

The training and testing were conducted in an odour line-up (e.g. McCulloch et al., 2006; Walczak et al., 2012; Guirao et al., 2019) or circle (e.g. Murarka et al., 2019; Schoon et al., 2020; Jeong et al., 2022) where the dogs are taught to sniff all odour samples and to alert the target odour, i.e. a sample taken from a patient with fully diagnosed cancer. In contrast to most studies, Kane et al. (2022) used a single port wheel. Usually, there was one target odour and the number of decoy odours (i.e. samples from healthy controls) ranged from 1 to 16 (Table 4). In studies conducted by Horvath et al. (2013), Amundsen et al. (2014), Taverna et al. (2015), and Fischer-Tenhagen et al. (2018), there was a variable number of target and decoy odours, meaning that in some trials all samples could be the targets or all of them could be decoys (Table 4).

More recently, automatic devices controlled by a computer have been constructed in order to exclude the influence of the handler who might give unconscious cues to the dog. For instance, Edwards (2019) trained 12 dogs to activate a lever to rotate a carrousel with 17 odour samples (amyl acetate as positive sample and demineralized water as negative controls). The number of target odour samples varied between 0 and 9. For a correct alert, the dogs were automatically rewarded with a treat delivered by a feeder (Edwards, 2019). Five out of the 12 dogs were able to successfully complete training through accurate odour detection. In a subsequent experiment by Crawford et al. (2022b), the apparatus described earlier by Edwards (2019) was used. Four out of seven dogs were able to complete the training of detecting lung cancer using breath samples. However, potential limitations with using such sophisticated devices for practical cancer screening are related with the associated financial costs of the technology as well as the necessity to complete the training of two tasks: (1) activation of the lever to get access to the next odour sample and (2) discriminating cancer odours from non-cancer controls in particular odour samples. Obviously, there were individual differences between dogs in training performance. Dogs trained by Edwards (2019) needed 7–19 half-day training sessions for habituation, shaping, and training with prompts to

**Table 4** Cancer type, odour samples, detection sensitivity/specificity, and the ratio positive: negative samples in the 33 different studies

| Authors | Cancer type | Odour samples | Detection sensitivity % | Detection specificity % | Ratio positive: negative samples in line-up or circle |
|---|---|---|---|---|---|
| Pickel et al. (2004) | Melanoma | Tissue | 100 | | 1:0–9 |
| | | Volunteers' body | 1st dog detected 6 out of 7 volunteers<br>2nd dog detected 3 out of 4 volunteers | | |
| Willis et al. (2004) | Bladder | Urine | 41<br>(23–58, c.i. 95%) | | 1:6 |
| McCulloch et al. (2006) | Lung<br>Breast | Breath<br>Breath | 99<br>88 | 99<br>98 | 1:4 |
| Horvath et al. (2008) | Ovarian | Tissue | 100 | 97.5 | 1:4 |
| Gordon et al. (2008) | Breast<br>Prostate | Urine<br>Urine | None of 6 dogs better than chance,<br>2 out of 6 dogs better than chance | | 1:6<br>1:6 |
| Cornu et al. (2011) | Prostate | Urine | 91 | 91 | 1:5 |
| Sonoda et al. (2011)[b] | Colorectal | Breath<br>Faeces | 91<br>97 | 99<br>99 | 1:4<br>1:4 |
| Ehmann et al. (2012) | Lung | Breath | 71 | 93 | 1:4 |
| Walczak et al. (2012) | Breast<br>Lung<br>Melanoma | Breath<br>Breath<br>Breath | 91<br>86<br>72 | 86<br>80<br>82 | 1:4<br>1:4<br>1:4 |
| Horvath et al. (2013) | Ovarian | Blood plasma | 70–97 | 92–99 | Training 0–3: 4–7<br>Testing 1[f]:5 + 1[e] |
| Amundsen et al. (2014) | Lung | Breath<br>Urine | 56.0–64.7<br>64.2–73.6 | 8.3–33.3<br>25.0–29.2 | 0:6–6:0<br>0:6–6:0 |
| Elliker et al. (2014) | Prostate | Urine | 13–25 | 71–75 | 1:3 |
| Taverna et al. (2015) | Prostate | Urine | 98–100 | 97–98 | 0:6–6:0 |

(continued)

**Table 4** (continued)

| Authors | Cancer type | Odour samples | Detection sensitivity % | Detection specificity % | Ratio positive: negative samples in line-up or circle |
|---|---|---|---|---|---|
| Urbanova et al. (2015)[a] | Prostate | Urine | 93.5 | 91.6 | 1:2 |
| Willis et al. (2016) | Melanoma | Lesion VOCs collected prior to excision or biopsy | 45 (23–68, c.i.95%) | | 1:9 |
| Guerro-Flores et al. (2017) | Cervical | Fresh CC scraps Cervical smears | 92.78 (87.1–98.4, c.i.95%) | 99.10 (98.4–99.8, c.i. 95%) | 1:9 |
| | | Surgical bandages | 96.36 (95.5–100, c.i.95%) | 99.55 (98.8–100, c.i.95% | 1:9 |
| Kitiyakara et al. (2017) | Hepatocellular | Breath | 78 (62–90, c.i.95%) | | 1:3 |
| Fischer-Tenhagen et al. (2018) | Lung | Breath | 88–100 | 40–80 | 0:2–4:0 (2–4 cones with 0–4 positives) |
| Pacik et al. (2018)[a] | Prostate | Sarcosine in artificial urine, | 90 | 95 | 1:2 |
| | | Urine | 93.5 | 91.6 | 1:2 |
| Seo et al. (2018) | Breast<br>Prostate | Cancer cells<br>Cancer cells | 94.4<br>92.6 | 98.6<br>98.1 | 1:4<br>1:4 |
| Guirao et al. (2019) | Solitary pulmonary nodules (SPN) | Breath | 97 (96–98, c.i.95%) | 99 (99–99.6, c.i.95%) | 1:4 |
| Junqueira et al. (2019) | Lung | Serum | 96.7 | 97.8 | 1:4 |
| Murarka et al. (2019) | Ovarian | Blood plasma ca. cell line | | | 1:7 |
| Kybert et al. (2020) | Ovarian | Blood plasma | 93.3–98.7 ± 5.1–20.1 | | Wheel 1:11<br>Wall 1:7<br>Discs 1:3 |
| Schoon et al. (2020) | Colorectal | Stool | 70.6–88.2[c] | 79.8–95.1[c] | 0:8, 1:7, 2:6 |

(continued)

**Table 4** (continued)

| Authors | Cancer type | Odour samples | Detection sensitivity % | Detection specificity % | Ratio positive: negative samples in line-up or circle |
|---|---|---|---|---|---|
| Yamamoto et al. (2020)[b] | Cervical | Urine | 100 | 100 | 1:4 |
| Feil et al. (2021) | Lung | Breath | 78.0 (62.4–89.4 c.i.95%) | | 1:5 |
| | | Urine | 87.8 (73.8–95.9 c.i.95%) | | 1:5 |
| Kure et al. (2021) | Breast | Urine | 100 | 100 | 1:4 |
| Crawford et al. (2022b) | Lung | Breath | 0–100 | 15–100 | 1:16–9:8 |
| Jeong et al. (2022) | Prostate | Extracellular medium of ca. cell line | 95.3–97.5 | 97.8–97.4 | 0:10–1:9 |
| Kane et al. (2022) | Ovarian | Plasma derived VOCs in positive blood sample | 85 | 77 | 1:1[d] |
| Ortal et al. (2022) | Osteosarcoma | Osteosarcoma cell lines, | 1st dog 97.6 (94.3–100, c.i.95% 2nd dog 90.9 (71.8–100, c.i.95%) | 1st dog 98.6 (97.3–100, c.i.95%) 2nd dog 84.8 (70.0–99.9, c.i.95%) | 0:4–1:3 |
| | | Saliva | 1st dog 100 (100–100, c.i.95%) | 1st dog 100 (100–100, c.i.95%) | 0:4–1:3 |
| Riedlova et al. (2022) | Lung | Serum | 1st dog 69 2nd dog 62 | 1st dog 67 2nd dog 97 | 1:3 |
| | | Breath | 1st dog 62 2nd dog 75 | 1st dog 71 2nd dog 90 | 1:3 |

Blank cells denote no data reported; sensitivity and specificity data were not reported in a unified manner

[a]The same dog used by Urbanova et al. (2015) and Pacik et al. (2018)

[b]The same dog used by Sonoda et al. (2011) and Yamamoto et al. (2020)

[c]Concerns 'new' samples (patients)

[d]One single sample on the wheel spoke

[e]Reference sample with a high concentration of VOCs

[f]Test sample

**Table 5** Confusion matrix for calculation of detection sensitivity and specificity.

| | Disease exists | Disease does not exist |
|---|---|---|
| Canine test outcome positive | True positive | False positive |
| Canine test outcome negative | False negative | True negative |

meet the criterion and 4–7 half-day (19–37 training sessions) to meet the criterion to work with the device. Walczak et al. (2012) found that it is difficult to predict the performance of fully trained dogs on the basis of their performance during the training since correlations between training results (correct target indications and false alerts) and results achieved in the working phase were negative and non-significant.

## 3 Detection Sensitivity and Specificity

Canine detection sensitivity and specificity were calculated according to the common formula based on signal detection theory (Table 5).

- Sensitivity (True Positive Rate) = True Positive / (True Positive + False Negative),
- Specificity (True Negative Rate) = True Negative / (True Negative + False Positive)

Canine detection sensitivity in some studies approached or even achieved 100% (Horvath et al., 2008). Although specificity varied, the majority of studies were able to attain specificity rates that were relatively high with only two studies suggesting that detection performance was not better than chance (Gordon et al., 2008; Elliker et al., 2014; see Table 4).

However, studies are limited by a lack of information on whether the odour line-up or circle was considered as sequential or simultaneous, which may impact how the detection sensitivity and specificity were calculated. The sequential line-up assumes that the dog is making separate and independent alerting decisions (yes/no) to each odour sample in the line-up or circle. Thus, if, for example, the dog is making a false alert to an odour sample, it is allowed to continue sniffing the remaining samples in the line-up and if it correctly alerts to one of the remaining samples, both false alert and correct alert will be scored in this trial. The simultaneous line-up, on the other hand, assumes that after checking all samples in the line-up, the dog should make one decision on which sample it alerts. Thus, each false alert before or after a correct alert means that the whole trial is scored as a mistake. Our latest experiments (unpublished) showed that the sensitivity and specificity calculated assuming the simultaneous line-up may be up to 20–30% lower than compared to a sequential line-up. Further studies need to implement standardization of experimental designs to correctly identify which setup produces the most valid and reliable detection outcomes.

## 4 Volatile Organic Compounds the Dogs May Alert To

It is suggested that VOC profiles are reliably associated with not only several types of cancer, but with several other human diseases as well, including asthma, cholera, cystic fibrosis, diabetes mellitus, dental diseases, gut diseases, heart allograft rejection, heart diseases, liver diseases, pre-eclampsia, renal disease, tuberculosis, and epilepsia (Corradi et al., 2010; Shirasu & Touhara, 2011; Edwards et al., 2017; Maa et al., 2021). However, odour profiles assessed by the sense of smell are considered only a secondary symptom of the abovementioned diseases and have not been used for regular screening or as a diagnostic method.

There is an abundance of scientific literature on VOCs analysed by advanced chemical analytical methods, such as solid phase microextraction (SPME) or gas chromatography – mass spectrometry (GCMS). Using these advanced extractive or analytical methods, many different VOCs have been identified as candidates for cancer markers, in turn becoming useful in screening methods (e.g. Phillips et al. 1999, b, 2003, 2007; Miekisch et al., 2004; Ligor et al., 2009; Amann et al., 2014).

Some hypotheses have been proposed on how the VOCs related to cancer may be produced. For example, Phillips et al. (1999) proposed an explanation assuming that oxygen free-radical activity increases in cancerous cells. This degrades cell membranes by lipid peroxidation and converts polyunsaturated fatty acids to volatile alkanes, which are considered to be the main odorous cancer marker in exhaled breath.

Amann et al. (2014) focused on breath analysis which they considered a promising field of research with great clinical potential. In the breath research, new analytical techniques have been developed, which allow real-time analysis of exhaled breath with breath-to-breath resolution (PTR-MS) to be used in a complementary way to GC-MS. VOCs that could be related to cancer are secreted not only in exhaled breath but also through urine, skin, saliva, and faeces. This was evident by various odour samples used by authors cited in Table 2. One of the problems of finding a VOC or combination of VOCs that could be a reliable and repeatable candidate for cancer odour signature is the great number and inter-/intra-individual variation of such VOCs identified in cancer disease and absent in healthy controls (Phillips, Herrera, et al., 1999). Approximately 1800 different VOCs may appear in human breath, urine, saliva, sweat, or faeces (Amann et al., 2014). Although GC-MS technology could be helpful for identification of VOCs that could be candidates for cancer markers, its use for cancer screening is not feasible due to the procedure of odour sample preparation for analysis in a stationary apparatus and sophisticated evaluation of the results obtained. Recently, attempts at developing portable GC-MS devices that would work in real-time, delivering simple yes/no indication, have been explored. Several researchers have developed 'electronic noses' (e-noses) (e.g. Di Natale et al., 2003; D'Amico et al., 2010; Peng et al., 2010), based on nanosensors array technology which, in the future, may be able to discriminate between cancerous and noncancerous VOCs from breath samples linked to lung cancer. Additionally, Lippi and Cervellin (2011) suggested that instead of using dogs for cancer detection, the canine sense of smell might help to identify candidate biomarkers found by analytic technologies.

Despite promising results obtained using GC-MS or e-nose technologies, to our knowledge analyses of the VOC profiles in human breath, urine, saliva, or faeces have not been used clinically for cancer screening purposes. Boedeker et al. (2012) stated that three decades of research on e-nose as a promising screening technology did not bring forward a clinically applicable device. On the other hand, they suggested combining e-sensing technologies with canine testing of a defined patient population for being positive and negative to lung cancer. Such an approach, according to Boedeker et al. (2012), would be beneficial to the identification of VOCs specific for lung cancer. In addition, Pomerantz et al. (2015) indicate that 'a bio-electronic nose based on canine olfactory receptors' would be used for cancer detection and provided a list of VOCs found by different researchers depending on cancer type and odour samples.

Canine detection of cancer is often labelled as 'black box technology' or 'junk science', because the VOCs to which dogs alert to are not known. To date, no study of canine detection of cancer has been able to precisely determine which compound(s) are detected and which VOCs may be confusing them. To our knowledge, only two studies directly compared canine indications of breath samples from patients with lung cancer, with results obtained by GC-TOF-MS analyses to identify volatile lung cancer biomarkers. Buszewski et al. (2012) found no single VOC or set of VOCs that could be a reliable lung cancer marker indicated by dogs. However, applying two multivariate methods, factor analysis and principal components analysis (PCA) separately for data evaluation, Buszewski et al. (2012) showed that concentrations of ethyl acetate and 2-pentanone positively correlated with the dogs' indications ($r = 0.85$ and $r = 0.97$, respectively), whereas acetonitrile, propanal, and 1-propanol negatively correlated with the dogs' indications ($r = -0.78$, $r = -0.87$, and $r = -0.98$, respectively). This provides preliminary evidence that trained dogs may utilize some VOCs present in the exhaled breath of lung cancer patients to distinguish breath samples of healthy controls and patients with lung cancer. Although Buszewski et al. (2012) indicated that some biomarker candidates influence a dog's indications, the interpretation of such results remains uncertain. Fischer-Tenhagen et al. (2018) tested synthetic air samples that were enriched with 1-butanol, 2-butanone, 2-pentanone, and hexanal, which are compounds found in the breath of lung cancer patients in an observational study conducted by Schallschmidt et al. (2016): canines were able to identify three out of four synthetic samples as positive. It is worth mentioning, however, that only the first choice of the dog sniffing at the synthetic air samples were summarized in this study (Fischer-Tenhagen et al., 2018). By using the same synthetic odour sample, one might avoid collecting novel odour samples from patients for each training stage, and thus the learning process of identifying the specific odour could be expedited. However, it remains unanswered if the dogs trained on synthetic odour samples would be able to generalize such synthetic odour when testing real odour samples which contain a variety of VOCs not specific for cancer.

With regard to prostate cancer, Pacik et al. (2018) successfully trained a dog to alert to artificial urine samples enriched with sarcosine, which is believed to be a urinary biomarker for prostate cancer (Table 5).

## 5 Issues That May Impede Using Trained Dogs for Practical Cancer Screening on a Large Scale

Odours related to human cancers seem to have neither biological importance nor rewarding value for dogs. Therefore, it cannot be expected that dogs would spontaneously show interest or alter their behaviour to such odours. It is commonly known that dogs can be trained to detect various odours that have no biological importance for them (e.g. odour of narcotics or explosives). Alerting to such odours can be produced using operant conditioning by rewarding a correct alert and not rewarding a false alert. An association between odour and reward may be variable over time due to extinction by not exercising the behaviour on a regular basis or by developing another strategy by the dog, e.g. 'trial-and-error', to earn the reward. The reliability of alerting to the trained target odour depends on the dog's capability to acquire and maintain a reliable association, motivation for the reward, and avoid distraction by external stimuli. Such limitations were further discussed by Hackner and Pleil (2017).

A question may arise if dogs can distinguish canine cancer odour(s) from odour(s) of healthy conspecifics and, if any, commonalities between cancer VOCs in humans and dogs exist.

A scarce research on this topic does not allow to answer this question thoroughly. Henry (2010) in their review article concluded that in animals no biomarkers were found that would be unique to cancer, giving 'a binary yes/no response'. Thus, Henry (2010) considers this method as ineffective. Dorman et al. (2017) trained four dogs to distinguish canine urinary tract transitional cell carcinoma (TCC) from healthy controls. They used a very small number of urine samples (1 positive and 1 negative in the first stage of the training, 6 positive and 4 negative in the next training stage and 2 novel positive and 6 novel negative in the testing phase). The dogs in the study of Dorman et al. (2017) were trained to alert by moving with their muzzle the Petri dish placed in an apparatus, either with a positive or with a negative odour sample. Overall, no dog was able to distinguish positive from negative samples better than chance. As far as another macrosmatic species is concerned, Kokocińska-Kusiak et al. (2020) showed that mice can be trained to discriminate urine odour from conspecifics with and without melanoma.

In our own experiments using scent line-up for canine identification of humans on the basis of individual scent (Jezierski et al., 2008) and cancer detection (Walczak et al., 2012), we used a treat as a reward. In contrast to the scent line-up, a favourite object to retrieve and play with is used in drug detection (Jezierski et al., 2014). In searching experimental scenarios for drugs or explosives, the treat-based reward is not applicable since the dogs should be not distracted by any food odour in the searching area that could be deliberately left by the traffickers or terrorists. An odour line-up, on the other hand, should be conducted in a room or place where lures or distracting stimuli are absent or extremely limited. In this case, the treat-based reward works better as opposed to throwing an object to be retrieved. This is because, when dogs are given an object to retrieve, they normally do so in a much more vigorous manner which may cause damage to equipment or injuries to the dog itself. When using the reward method, one must also consider what prize the dog would benefit/react the most with. We observed in our studies (Jezierski et al., 2008, 2014; Walczak et al., 2012) that there is a very clear difference, even between littermates, in their preference for a given kind of reward: for some, retrieving an object had a higher value, whereas others only valued retrieving when there was a treat involved.

A canine's excellent sense of smell is often underlined as a main factor contributing to canine detection. It is less frequently mentioned that the dog's engagement in the operant conditioning training is also a crucial point for cancer detection dogs, which is at least as important as the acuity of their sense of smell. There is no current standardization in regards to the training of dogs for the purpose of cancer detection. Overall the standardization of dog training methods and schedules in future studies seems to be most problematic, since individual differences in dogs' capability and personality impact their trainability, and should be considered (MacLean & Hare, 2018). Researchers should be able to assess which training schedule, rewarding scheme, and amount of daily work are optimal for a given individual dog to maintain its interest and motivation for work and to avoid boredom or demotivation, including mood and health conditions of the dogs. Any methods that have been applied up to now have been designed by the trainers' own experience and intuition. In particular, the transition from training phase to the real testing phase is usually made when the trainer feels there is stability in the dog's performance and no further progress can be achieved (Crawford et al., 2022b). The latter authors recommended a description of the criteria that should be applied to determine when animals are deemed ready for the transition from training to testing. These authors presented several methods of determining such criteria and considered a quantitative model as helpful for deciding when the transition from training to testing should take place. Walczak et al. (2012) found non-significant and negative correlations between the training phases and working phase in the percentage of correct indications and false alerts. This highlights that it is difficult to predict the performance of fully trained dogs on the basis of their results achieved during the training. Furthermore, Walczak et al. (2012) concluded that reducing the rate of false alerts is difficult since the tendency for making false alerts was increasing rather than decreasing in the course of particular training phases.

As an advantage of canines, Hackner and Pleil (2017) mentioned an incredible ability to detect odours and extract them from a complex environment and recognize them, which is a task that is more difficult for current analytical technologies. However, there are two issues that still remain unanswered. One is whether defined VOCs that are produced by altered cell metabolisms in cancer are identical or similar in all humans, and the second is if dogs would alert reliably to such cancer odours independently from all other confounding odours caused by a variety of factors. Such confounding factors could be related with comorbidities, diet, smoking habits, medications, hygiene habits, etc., which vary across individuals and also across given samples of the same individuals (McCulloch et al., 2006). All such confounding odours should be ignored by the canines. These components of the individual odour profile may change on a daily basis and may be confusing to dogs unless the dog is able to generalize reliably the 'odour signature' of cancer.

During detection training, dogs have to memorize the target odour in a 'long-term olfactory memory' and should be able to recall this odour every time during detection. Dogs have indeed an excellent ability to memorize a large number of odours (Waggoner et al., 2022). Odour memorization may have both favourable and unfavourable consequences for the outcome of tests in a scent line-up. The long-term memorization of a trained characteristic odour signature of cancer (if it exists) enables the dog to recognize this common odour signature in consecutively tested samples. On the other hand, the memorization of previous odours to which the dog was exposed to and rewarded may cause the dogs to correctly alert the memorized odours reused in the test and ignore novel odour samples (Elliker et al., 2014). Some authors (e.g. Willis et al., 2004; Gordon et al., 2008; Guerro-Flores et al., 2017; Fischer-Tenhagen et al., 2018; Junqueira et al., 2019; Feil et al., 2021; Kure et al., 2021) mentioned that they used novel odour samples in the testing phase that were not used during the training, but it was not done systematically across the reviewed studies.

More recently, Crawford et al. (2022a) evaluated the effect of sample re-use on dogs' training performance. These authors presented the same individual breath samples from patients with lung cancer to dogs up to four times over a period of 2 years. They found that re-using samples (stored at −60 or −80 °C) decreased detection sensitivity but increased specificity and the 'dogs' bias shifted in the direction of a negative indication bias' (Crawford et al., 2022a, in press, page 1, Abstract). They also reported that 'there were no significant changes in the measure associated with the detectability of the target' and 'all observed changes in performance across sample re-use were small' (in press, page 1 Abstract). In their conclusions, Crawford et al. (2022a accepted ms page 8) suggest that 'breath samples can be re-used at least three times', but the authors relate this suggestion to the training conditions they had and do not indicate precisely if samples used in the training can be re-used in the testing of dogs' performance.

Chemical identification of a single VOC or of a combination of VOCs, to which the trained dogs might react, cannot fully explain why dogs alert to some positive odour samples correctly, while ignoring other positive samples, even if such VOCs are present or why they falsely alert to odour samples of healthy controls when typical cancer VOCs are absent. The odour signature of cancer may be related not only to the presence of specific VOCs but also to a specific odour impression produced by mixtures of volatile compounds. So far, little is known about odour impressions evoked in dogs by particular VOCs or combination of them. In humans, odour impressions can be described using words like e.g. 'fruity', 'floral', or 'herbaceous' (Debnath & Nakamoto, 2022). In dogs, however, we can only infer from their sniffing behaviour that an odour is either 'interesting' or 'uninteresting' or even 'abominable' if dogs show avoidance behaviour. It is also known that odour impressions of the same VOC may be different depending on the molecule concentration. There are several examples for that, especially concerning skatole in boar meat. Dijksterhuis et al. (2000, page 268) concluded that 'different concentrations of, in particular, skatole can exhibit different odour qualities. This will make it difficult to ascribe a perceived odour to a specific attribute'. According to Stuetz et al. (2005, page 4) 'human responses to an odour is highly subjective; different people find different odours offensive at different concentrations'. Also, with respect to synthetic terpenoids, Stepanyuk and Kirschning (2019, page 3595) claim that 'odour impressions may change when concentrations are altered'.

It is commonly agreed that both the training and testing, except for the very first stages of the training, must be conducted following a double-blind procedure, meaning that the dog trainer or any person who is visible to the dog during runs in the line-up must be blind as to the position of the target odour in the line-up (e.g. McCulloch et al., 2006; Cornu et al., 2011; Walczak et al., 2012; Elliker et al., 2014; Seo et al., 2018; Junqueira et al., 2019; DeChant et al., 2020; Feil et al., 2021; Kure et al., 2021; Riedlova et al., 2022). As dogs have excellent ability to read even unconscious body language of people, the handler may unconsciously give some cues to the dog where the alert reaction is expected. This phenomenon is termed 'clever Hans effect' and is well known in animal training (Hediger, 2010; Samhita & Gross, 2013). The double-blind testing procedure requires collaboration of two people during testing, the dog handler and experimenter, the latter deciding the position of the target odour sample in the line-up, and the dog handler being blind to the position of the target odour sample. The experimenter is also responsible for changing the position of the target sample and decoys for the next trial(s). The experimenter should be invisible to the dog and dog handler. He/she should observe the trials through a video camera or through a one-way glass and give an acoustic or visual signal to the dog handler that the dog's alert is correct and the dog can be rewarded. However, these requirements can sometimes cause a problem in practical implementations.

Once a dog is fully trained, it requires appropriate maintenance training and exercising to have opportunities to earn a reward and thus maintaining the dogs' level of performance. If there is no source of reward for long periods of time, the dog's motivation to work in the line-up and to alert correctly would decrease and may lead to an increase in the rate of false alerts. Therefore, the performance of fully trained dogs should be systematically checked by analysing the rate of correct and false alerts, using cancer-positive odour samples from patients with cancer confirmed by medical diagnosis. Due to the confounding interference of the memorization of individual odours, novel odour samples should be used instead of reusing previously learned samples. Using double-blind procedures, the maintenance training and systematic evaluation of the outcome as well as necessity to collect frequently novel odour samples is logistically and organizationally challenging. During testing of odour samples from donors of unknown health status, which typically occurs during real screening scenarii, dogs cannot be rewarded for alerting such samples since it is not known if the dogs are rewarded for a correct or for a false alert. Therefore, trials with cancer-positive samples and samples of unknown status should be combined to maintain the dog's motivation for work. Out of the studies we reviewed, only Hackner and Pleil (2017) propose a plausible way to combine testing of the cancer-positive samples and samples of the unknown status, in order to deploy trained dogs for real cancer screening.

Another problem is the possible contamination of odour samples with ambient odour for which the dogs could become conditioned to during training. In a study conducted by Walczak et al. (2012), for practical and organizational reasons, the majority of cancer positive breath samples were taken in hospitals from patients diagnosed with cancer. Three fully trained dogs were compared based on their alert rates to breath samples obtained from lung cancer patients inside two oncology hospitals (60 patients) and a general hospital (32 patients). Samples were also obtained from 10 healthy volunteers taken inside oncology hospitals and 301 healthy volunteers taken outside hospitals. Ambient air inside the oncology and general hospital rooms were also collected in this study. Walczak et al. (2012) found that dogs made significantly more false alerts towards healthy donor samples collected inside hospitals compared to samples of healthy donors collected outside a hospital setting. Samples of ambient air taken in hospital rooms were indicated by dogs at a rate of 34.47% and 30.60% in oncology and general hospitals, respectively, were significantly higher than samples from healthy donors collected outside hospitals (for all comparisons, $Chi^2 > 6.64$, df = 1, $P < 0.01$).

Also, Guest et al. (2020) found an effect of the location where odour samples were taken. In their study, dogs trained to discriminate between hospital-sourced urine samples taken from patients with prostate cancer and externally sourced controls processed and stored the same way performed with a relatively high specificity of 87.9% (78.2–91.9). After including hospital-sourced controls, the discrimination specificity decreased to 67.3% (43.2–83.3).

Dogs are trained using odour samples from donors with biopsy-confirmed cancer disease (i.e. positive odour samples); however, the ultimate goal of using dogs for cancer screening is not to confirm already diagnosed disease but to discriminate early, preclinical stages of cancer. Thus, the stage of cancer can be regarded as an important factor. Edwards et al. (2017) stated that additional research is necessary to assess the potential value of disease detection in practice using trained animals. Pirrone and Albertini (2017, page 115) proposed that 'future research on canine detection of cancer should seek to develop optimal canine olfactory testing protocols, based on internationally standardised training methods, canine selection criteria, and the inclusion of a greater number of dogs'. These authors presented a checklist for 21 variables that should be included in future studies, grouped in five categories (dog training, testing, sample collection, participants, and other confounders).

However, the implementation of some of the demands formulated by Pirrone and Albertini (2017) could be problematic. First, it has to be shown in many comparative experiments that the testing protocol is 'optimal'. Second, amongst dog trainers there exist adages like: '*The only thing two dog trainers can agree upon is what the third trainer is doing wrong*', or '*there are as many ways to train a dog, as there are dogs.*' With this, 'internationally standardised training methods' should be utilized for a unified training method and an agreement between international trainers must be made. A unified training programme may display some difficulty as dogs have different personalities.

Schoon et al. (2020, 2022), following their projects concerning colon, lung, and breast cancers, discussed important details in collecting and storing odour samples, but also details in preparation and presentation, as well as in procedures employed during the analysis by the dog. These authors advice the following good practices (Schoon et al., 2022, page 690): 'both patients samples as well as [appropriately matched] controls should be selected based on some kind of standardized protocol'; 'new samples from new patients and new controls should be used for a test'; 'the patient samples and control samples should be collected in one location but transported and stored separately to prevent cross contamination'. All these details may be used by dogs as cues and thus may be confounding factors that have an impact on canine detection. In respect to testing, Schoon et al. (2020) proposed establishing detection sensitivity and specificity for individual dogs in order to uphold many dogs to the same level. This can create a positive diagnosis using a team of dogs.

In view of variable and inconsistent results in published experimental results, some doubts remain on whether the above-mentioned measures would suffice to employ trained dogs as cancer screening tools at a large scale. Some published studies seek to elaborate a model which would combine maintenance training to demonstrate correct alerting to positive odours from confirmed cancer patients, non-alerting to confirmed healthy controls, and testing novel odour samples of unknown health status (Hackner & Pleil, 2017). The authors admitted that the drawback of this screening model is that cancer and non-cancer samples should be systematically exchanged to avoid memorization of odours; thus this method is very time-consuming, logistically problematic, and probably not applicable for a universal screening test.

In a test with a random number of positive samples, ranging from 0 to 5, Hackner et al. (2016) found a lower detection rate than other studies, based on positive and negative predictive values of 30.9% and 84.0%, respectively. The authors stated that canine scent detection may not be as powerful as expected in real screening situations. To explain the possible reason for the poor performance of dogs in their setting, Hackner et al. (2016) considered the elevated stress from the lack of positive responses for dogs and handlers in their experimental setting. In the literature, there continues to be some optimism for using trained canines in combination with modern analytical technology for cancer screening based on the VOCs emitted. For instance, Pleil and Giese (2017) suggested that the semi-volatile fraction of the samples should be given more attention. Ultimately, these authors concluded that 'the interaction between dog-based and instrument-based research will be mutually beneficial and accelerate progress towards early detection of cancer by breath analysis' (page 1, Abstract).

In order to update information on actual implementation of trained canines for practical cancer screening, on September 26th, 2021 we sent out an e-mail survey to 10 researchers worldwide, who were involved in canine training for cancer detection and who published papers in peer-reviewed scientific journals on this topic. The survey asked whether the respondents knew any example of practical use of trained dogs for cancer screening in oncology hospitals, clinics, or other medical facilities on a regular basis. We received six responses: from the United States, New Zealand, Germany, China, Thailand, and the Netherlands. None of the respondents reported on such practical application of trained dogs neither in their countries nor elsewhere. All respondents stated that according to their knowledge, cancer detection by dogs has still not gone beyond the experimental phase.

## 6 Conclusion

Despite the significant interest in canine detection of cancers and multiple experimental models highlighting their ability to differentiate odour samples of varying cancers in humans, numerous limitations and methodological concerns need to be addressed prior to their use in a clinical setting, and recommendations for their implementation clinically cannot be made at this time.

To demonstrate a real usefulness of the canine detection of cancer, large-scale standardized clinical trials are needed to show that using trained canines can actually contribute to a decrease in cancer morbidity and mortality. However, conducting such trials will require substantial logistic, organizational, and financial support.

**Acknowledgements** The authors express thanks to Natalie Sadowski, Nahal Sadeghi, and Michaela Sukenikova for the editing of English.

## References

Amann, A., Miekisch, W., Schubert, J., Buszewski, B., Ligor, T., Jezierski, T., Pleil, J., & Risby, T. (2014). Analysis of exhaled breath for disease detection. *Annual Review of Analytical Chemistry, 7*, 455–482. https://doi.org/10.1146/annurev-anchem-071213-020043

Amundsen, T., Sundstrøm, S., Buvik, T., Gederaas, O. A., & Haarverstad, R. (2014). Can dogs smell lung cancer? First study using exhaled breath and urine screening in unselected patients with suspected lung cancer. *Acta Oncologica, 53*, 307–315. https://doi.org/10.3109/0284186X.2013.819996

Arakawa, H., Arakawa, K., & Deak, T. (2009). Acute illness induces the release of aversive odour cues from adult, but not prepubertal male rats and supresses social investigation by conspecifics. *Behavioral Neuroscience, 123*, 964–978. https://doi.org/10.1037/a0017114

Arakawa, H., Arakawa, K., & Deak, T. (2010). Sickness related odour communication signals as determinants of social behaviour in rat: A role for inflammatory processes. *Hormones and Behavior, 57*, 330–341. https://doi.org/10.1016/j.yhbeh.2010.01.002

Boedeker, E., Friedel, G., & Walles, T. (2012). Sniffer dogs as part of a bimodal research approach to develop a lung cancer screening. *Interactive Cardiovascular and Thoracic Surgery, 14*, 511–515. https://doi.org/10.1093/icvts/ivr070

Buszewski, B., Ligor, T., Jezierski, T., Wenda-Piesik, A., Walczak, M., & Rudnicka, J. (2012). Identification of volatile lung cancer markers by gas chromatography – Mass spectrometry: Comparison with discrimination by canines. *Analytical and Bioanalytical Chemistry, 404*, 141–146. https://doi.org/10.1007/s00216-012-6102-8

Campbell, L. F., Farmery, L., George, S. M. C., & Farrant, P. B. J. (2013). Canine olfactory detection of malignant melanoma. *BML Case Reports, 2013*, bcr2013008566. https://doi.org/10.1136/bcr-2013-008566

Catala, A., Grandgeorge, M., Schaf, J.-L., Cousillas, H., Hausberger, M., & Cattet, J. (2019). Dogs demonstrate the existence of an epileptic seizure odour in humans. *Scientific Reports, 9*, 4103. https://doi.org/10.1038/s41598-019-40721-4

Chen, M., Daly, M., Williams, N., Williams, S., Williams, C., & Williams, G. (2000). Non-invasive detection of hypoglycaemia using a novel, fully biocompatible and patient friendly alarm system. *BMJ, 321*(7276), 1565–1566. https://doi.org/10.1136/bmj.321.7276.1565

Church, J., & Williams, H. (2001). Another sniffer dog for the clinic? *Lancet, 358*, 930.

Cornu, J. N., Cancel-Tassin, G., Ondet, V., & Girarget, C. (2011). Olfactory Detection of prostate cancer by dogs sniffing urine: A step forward in early detection. *European Urology, 59*, 197–201. https://doi.org/10.1016/j.eururo.2010.10.006

Corradi, M., Gergelova, P., & Mutti, A. (2010). Exhaled volatile organic compounds in non-respiratory diseases. In *European respiratory monograph* (Exhaled biomarkers) (Vol. 49, pp. 140–151). European Respiratory Society. https://doi.org/10.1183/1025448x.00018809

Crawford, M. A., Chang, C. L., Hopping, S., Browne, C. M., Edwards, T. L., & T.L. (2022a). Influences of breath sample re-use on the accuracy of lung cancer detection dogs. *Journal of Breath Research, 17*, 016001. https://doi.org/10.1088/1752-7163/ac9b7f

Crawford, M. A., Perrone, J. A., Browne, C. M., Chang, C. L., Hopping, S., & Edwards, T. L. (2022b). Transitioning from training to testing with scent detection animals: Application to lung cancer detection dogs. *Journal of Veterinary Behavior*. https://doi.org/10.1016/j.jveb.2022.07.004

Dalziel, D. J., Uthman, B. M., Mcgorray, S. P., & Reep, R. L. (2003). Seizure-alert dogs: A review and preliminary study. *Seizure, 12*(2), 115–120. https://doi.org/10.1016/s105913110200225x

D'Amico, A., Pennazza, G., Santonico, M., Martinelli, E., Roscioni, C., Galluccio, G., Paolese, R., & Di Natale, C. (2010). An investigation on electronic nose diagnosis of lung cancer. *Lung Cancer, 68*, 170–176. https://doi.org/10.1016/j.lungcan.2009.11.003

Debnath, T., & Nakamoto, T. (2022). Predicting individual perceptual scent impression from imbalanced dataset using mass spectrum of odorant molecules. *Scientific Reports, 12*, 3778. https://doi.org/10.1038/s41598-022-07802-3

DeChant, M. T., Ford, C., & Hall, N. J. (2020). Effect of handler knowledge of the detection task on canine search behavior and performance. *Frontiers in Veterinary Science, 7*, 250. https://doi.org/10.3389/fvets.2020.00250

Dijksterhuis, G. B., Engel, B., Walstra, P., Furnols, M. F., Agerhem, H., Fischer, K., Olever, M. A., Claudi-Magnussen, C., Siret, F., Beague, P., Homer, D. B., & Bonneau, M. (2000). An international study on the importance of androstenone and skatole for boar taint: II. Sensory evaluation by trained panels in seven European countries. *Meat Science, 54*(3), 261–269.

Di Natale, C., Macagnano, A., Martinelli, E., Paolesse, R., D'Arcangelo, G., Roscioni, C., Finazzi-Agro, A., & D'Amico, A. (2003). Lung cancer identification by the analysis of breath by means of an array of non-selective gas sensors. *Biosensors & Bioelectronics, 18*, 1209–1218. https://doi.org/10.1016/S09565663(03)00086-1

Dorman, D. C., Foster, M. L., Fernhoff, K. E., & Hess, P. R. (2017). Canine scent detection of cancer: A feasibility study. *Veterinary Medicine, 8*, 69–76. https://doi.org/10.2147/VMRR.S148594

Edwards, T. L. (2019). Automated canine scent-detection apparatus: Technical description and training outcomes. *Chemical Senses, 44*, 449–455. https://doi.org/10.1093/chemse/bjz039

Edwards, T. L., Browne, C. M., Schoon, A., Cox, C., & Poling, A. (2017). Animal olfactory detection of human diseases: Guidelines and systematic review. *Journal of Veterinary Behavior, 20*, 59–73. https://www.researchgate.net/publication/316806086

Ehmann, R., Boedeker, E., Friedrich, U., Sagert, J., Dippon, J., Friedel, G., & Walles, T. (2012). Canine scent detection in the diagnosis of lung cancer: Revisiting a puzzling phenomenon. *The European Respiratory Journal, 39*, 669–676. https://doi.org/10.1183/09031936.00051711

Ehman, K. D., & Scott, M. E. (2002). Female mice mate preferentially with non-parasitized males. *Parasitology, 125*, 461–466. https://doi.org/10.1017/S003118200200224X

Elliker, K., Sommerville, B. A., Broom, D. M., Neal, D. E., & Armstrong, S. (2014). Key considerations for the experimental training and evaluation of cancer odour detection dogs: Lesson learnt from a double blind, controlled trial of prostate cancer detection. *BMC Urology, 14*, 22–30. https://doi.org/10.1186/1471-2490-14-22

Feil, C., Staib, F., Berger, M. R., Stein, T., Schmidtmann, I., Forster, A., & Schimanski, C. C. (2021). Sniffer dogs can identify lung cancer patients from breath and urine samples. *BMC Cancer, 21*, 917. https://doi.org/10.1186/s12885-021-08651-5

Fischer-Tenhagen, C., Johnen, D., Nehls, I., & Becker, R. (2018). A proof of concept: Are detection dogs a useful tool to verify potential biomarkers for lung cancer? *Frontiers in Veterinary Science, 5*, 52. https://doi.org/10.3389/fvets.2018.00052

Gordon, R. T., Schatz, C. B., Myers, L. J., Kosty, M., Gonczy, C., Kroener, J., Tran, M., Kurtzhals, P., Heath, S., Koziol, J. A., Arthur, N., Gabriel, M., Hemping, J., Hemping, G., Nesbitt, S., Tucker-Clark, L., & Zaayer, J. (2008). The use of canines in the detection of human cancers. *Journal of Alternative and Complementary Medicine, 14*(1), 61–67. https://doi.org/10.1089/acm.2006.6408

Guerro-Flores, H., Aprese-Garcia, T., Garay-Viiar, Ó., Sánchez-Pérez, A., Flores-Villegas, D., Sanchez-Peres, Bandera-Calderon, A., Garcia-Palacios, R., Rojas-Sánchez, T., Romero-Morelos, P., Sánchez-Albor, V., Mata, O., Arana-Conejo, V., Badillo-Romero, J., Taniguchi, K., Marrero-Rodriguez, D., Mendoza-Rodriguez, M., Rodriguez-Esquivel, M., Huerta-Padilla, V., Martinez-Castillo, A., Hernández-Gallardo, I., López-Romero, R., Bandala, C., Rosales-Guevara, J., & Salcedo, M. (2017). A non-invasive tool for detecting cervical cancer odor by trained scent dogs. *BMC Cancer, 17*, 79. https://doi.org/10.1186/s12885-016-2996-4

Guest, C. M., Harris, R., Anjum, I., Concha, A. R., & Rooney, N. J. (2020). A lesson in standardization – Subtle aspects of the processing of samples can greatly affect dogs' learning. *Frontiers in Veterinary Science, 7*, 525. https://doi.org/10.3389/fvets.2020.00525

Guirao, A., Molinsa, L., Ramónd, I., Sunyer, G., Viñolas, N., Marrades, R., Sánchez, D., Fibla, J. J., Boad, M., Hernandez, J., Guzmán, R., Libreros, A., Gómez-Caro, A., & Agusti, A. (2019). Trained dogs can identify malignant solitary pulmonary nodules in exhaled gas. *Lung Cancer, 135*, 230–233. https://doi.org/10.1016/j.lungcan.2019.06.008

Hackner, K., Errhalt, P., Mueller, M. R., Speiser, M., Marzluf, B. A., Schulheim, A., Schenk, P., Bilek, J., & Doll, T. (2016). Canine scent detection for the diagnosis of lung cancer in a screening-like situation. *Journal of Breath Research, 10*, 046003. https://doi.org/10.1088/1752-7155/10/4/046003

Hackner, K., & Pleil, J. (2017). Canine olfaction as an alternative to analytical instruments for disease diagnosis: Understanding 'dog personality' to achieve reproducible results. *Journal of Breath Research, 11*, 012001. https://doi.org/10.1088/1752-7163/aa5524

Haick, H., Broza, Y., Mochalski, P., Ruzsanyi, V., & Amann, A. (2014). Assessment, origin, and implementation of breath volatile cancer markers. *Chemical Society Reviews, 43*, 1423–1449. https://doi.org/10.1039/c3cs60329f

Hediger, H. K. P. (2010). The Clever Hans phenomenon from an animal psychologist's point of view. In D. Favareau (Ed.), *Essential readings in biosemiotics, biosemiotics 3.*, Chapter 7 (pp. 237–255). Springer. https://doi.org/10.1007/978-1-4020-9650-1_7

Henry, C. J. (2010). Biomarkers in veterinary cancer screening: Applications, limitations and expectations. *Veterinary Journal, 185*, 10–14. https://doi.org/10.1016/j.tvjl.2010.04.005

Henschke, C. I., Yankelevitz, D. F., Libby, D. M., Pasmantier, M. W., Smith, J. P., & Miettinen, O. S. (2006). Survival of patients with stage 1 lung cancer detected on CT screening. *The New England Journal of Medicine, 355*, 1763–1771. https://doi.org/10.1056/NEJMoa060476

Horvath, G., Andersson, H., & Nemes, S. (2013). Cancer odour in the blood of ovarian cancer patients: A retrospective study of detection by dogs during treatment, 3 and 6 months afterward. *BMC Cancer, 13*, 396–402. https://doi.org/10.1186/1471-2407-13-396

Horvath, G., Järverud, G. K., Järverud, S., & Horvath, I. (2008). Human ovarian carcinomas detected by specific odor. *Integrative Cancer Therapies, 7*, 76–80. https://doi.org/10.1177/1534735408319058

Jeong, S., Kim, D., Kim, W.-J., & Kim, G. (2022). Detection of volatile organic compounds from human prostate cancer cell using canine olfaction. *Journal of Veterinary Behavior, 49*, 80–84. https://doi.org/10.1016/j.jveb.2021.10.010

Jezierski, T., Walczak, M., & Górecka, A. (2008). Information-seeking behaviour of sniffer dogs during match-to-sample training in the scent lineup. *Polish Psychological Bulletin, 39*(2), 71–80. https://doi.org/10.2478/v10059-008-0010-y

Jezierski, T., Adamkiewicz, E., Walczak, M., Sobczyńska, M., Górecka-Bruzda, A., Ensminger, J., & Papet, G. (2014). Efficacy of drug detection by fully-trained police dogs varies by breed, training level, type of drug and search environment. *Forensic Science International, 237*, 112–118. https://doi.org/10.1016/j.forsciint.2014.01.013

Johnston-Wilder, O., Mancini, C., Aengenheister, B., Mills, J., Harris, R., & Guest, C. (2015). Sensing the shape of canine responses to cancer. In *ACE '15: Proceedings of the 12th International Conference on Advances in Computer Entertainment Technology*. article no. 63.

Junqueira, H., Quinn, T. A., Biringer, R., Hussein, M., Smeriglio, C., Barrueto, L., Jordan Finizio, J., & Huang, X. Y. (2019). Accuracy of canine scent detection of non–small cell lung cancer in blood serum. *The Journal of the American Osteopathic Association, 119*(7), 413–418. https://doi.org/10.7556/jaoa.2019.077

Kane, S. A., Lee, Y. E., Essler, J. L., Mallikarjun, A., Preti, G., Plymouth, V. L., Verta, A., DeAngelo, A., & Otto, C. M. (2022). Canine discrimination of ovarian cancer through volatile organic compounds. *Talanta, 250*, 123729. https://doi.org/10.1016/j.talanta.2022.123729

Kavaliers, M., Colwell, D. D., Braun, W. J., & Choleris, E. (2003). Brief exposure to the odour of a parasitized male alters the subsequent mate odour responses of female mice. *Animal Behaviour, 65*, 59–68. https://doi.org/10.1006/anbe.2002.2043

Kitiyakara, T., Redmond, S., Unwanatham, N., Rattanasiri, S., Thakkinstian, A., Tangtawee, P., Mingphruedhi, S., Sobhonslidsuk, A., Intaraprasong, P., & Kositchaiwat, C. (2017). The detection of hepatocellular carcinoma (HCC) from patients' breath using canine scent detection: A proof-of-concept study. *Journal of Breath Research, 11*(4), 046002. https://doi.org/10.1088/1752-7163/aa7b8e

Kokocińska-Kusiak, A., Matalińska, J., Sacharczuk, M., Sobczyńska, M., Góral-Radziszewska, K., Wileńska, B., Misicka, A., & Jezierski, T. (2020). Can mice be trained to discriminate urine odor of conspecifics with melanoma before clinical symptoms appear? *Journal of Veterinary Behavior, 39*, 64–76. https://doi.org/10.1016/j.jveb.2020.04.004

Kure, S., Iida, S., Yamada, M., Takei, H., Yamashita, N., Sato, Y., & Miyashita, M. (2021). Breast cancer detection from a urine sample by dog sniffing: A preliminary study for the development of a new screening device, and a literature review. *Biology, 10*, 517. https://doi.org/10.3390/biology10060517

Kybert, N., Prokop-Prigge, K., Otto, C. M., Ramirez, L., Joffe, E. R., Tanyi, J., Piltz-Seymour, J., Johnson, A. T. C., & Preti, G. (2020). Exploring ovarian cancer screening using a combined sensor approach: A pilot study. *AIP Advances, 10*, 035213. https://doi.org/10.1063/1.5144532

Ligor, M., Ligor, T., Bajtarevic, A., Ager, C., Pienz, M., Klieber, M., Denz, H., Fiegl, M., Hilbe, W., Weiss, W., Lukas, P., Jamnig, H., Hackl, M., Buszewski, B., Miekisch, W., Schubert, J., & Amann, A. (2009). Determination of volatile organic compounds in exhaled breath of patients with lung cancer using solid phase microextraction and gas chromatography mass spectrometry. *Clinical Chemistry and Laboratory Medicine, 47*, 550–560. https://doi.org/10.1515/CCLM.2009.133

Lippi, G., & Cervellin, G. (2011). Canine olfactory detection of cancer versus laboratory testing: Myth or opportunity? *Clinical Chemistry and Laboratory Medicine, 49*, 10. https://doi.org/10.1515/CCLM.2011.672

Maa, E., Arnold, J., Ninedorf, K., & Olsen, H. (2021). Canine detection of volatile organic compounds unique to human epileptic seizure. *Epilepsy & Behavior, 115*, 107690. https://doi.org/10.1016/j.yebeh.2020.107690

MacLean, E. L., & Hare, B. (2018). Enhanced selection of assistance and explosive detection dogs using cognitive measures. *Frontiers in Veterinary Science, 5*, 236. https://doi.org/10.3389/fvets.2018.00236

Mancini, C., Harris, R., Aengenheister, B., & Guest, C. (2015). Re-centering multispecies practices: A canine interface for cancer detection dogs. In *CHI '15: Proceedings of the 33rd Annual ACM Conference on Human Factors in Computing Systems* (pp. 2673–2682). ACM Press.

McCulloch, M., Jezierski, T., Broffman, M., Hubbard, A., Turner, K., & Janecki, T. (2006). Diagnostic accuracy of canine scent detection in early- and late-stage lung and breast cancers. *Integrative Cancer Therapies, 5*, 30–39. https://doi.org/10.1177/1534735405285096

Miekisch, W., Schubert, J. K., & Noelge-Schomburg, G. F. E. (2004). Diagnostic potential of breath analysis – Focus on volatile organic compounds. *Clinica Chimica Acta, 347*, 25–39. https://doi.org/10.1016/j.cccn.2004.04.023

Mischley, L., & Farahnik, J. (2020). Canine aromatic detection of Parkinson's disease: Can dogs identify prodromal parkinsonism? *Movement Disorders, 35*(suppl 1) https://www.mdsabstracts.org/abstract/canine-aromatic-detection-of-parkinsons-disease-can-dogs-identify-prodromal-parkinsonism/. Accessed 14 Oct 2022

Murarka, M., Zoë, I., Vesley-Gross, Z. I., Essler, J. L., Smith, P. G., Hooda, J., Drapkin, R., & Otto, C. M. (2019). Testing ovarian cancer cell lines to train dogs to detect ovarian cancer from blood plasma: A pilot study. *Journal of Veterinary Behavior, 32*, 42–48. https://doi.org/10.1016/j.jveb.2019.04.010

Ortal, A., Rodríguez, A., Solis-Hernández, M. P., de Prado, M., Rey, V., Tornín, J., Estupiñán, O., Gallego, B., Murillo, D., Huergo, C., García-Llano, J. L., Costilla, S., & Rodríguez, R. (2022). Proof of concept for the use of trained sniffer dogs to detect osteosarcoma. *Scientific Reports, 12*, 6911. https://doi.org/10.1038/s41598-022-11013-1

Pacik, D., Plevova, M., Urbanova, L., Lackova, Z., Strmiska, V., Necas, A., Heger, Z., & Adam, V. (2018). Identification of sarcosine as a target molecule for the canine olfactory detection of prostate carcinoma. *Scientific Reports, 8*, 4958. https://doi.org/10.1038/s41598-018-23072-4

Peng, G., Hakim, M., Broza, Y. Y., Billan, S., Abdah-Bortnyak, R., Kuten, A., Tisch, U., & Haick, H. (2010). Detection of lung, breast, colorectal and prostate cancers from exhaled breath using a single array of nanosensors. *British Journal of Cancer, 103*, 542–551. https://doi.org/10.1038/sj.bjc.6605810

Penn, D., Schneider, G., White, K., Slev, P., & Potts, W. (1996). Influenza infection neutralizes the attractiveness of male odor to female mice (Mus musculus). *Ethology, 104*, 685–694. https://doi.org/10.1111/j.1439-0310.1998.tb00102.x

Phillips, M., Altorki, N., Austin, J. H., Cameron, R. B., Cataneo, R. N., Greenberg, J., Kloss, R., Maxfield, R. A., Munawar, M. I., Pass, H. I., Rashid, A., Rom, W. N., & Schmitt, P. (2007). Prediction of lung cancer using volatile biomarkers in breath. *Cancer Biomarkers, 3*, 95–109. https://doi.org/10.3233/cbm-2007-3204

Phillips, M., Cataneo, R. N., Ditkoff, B. A., Fisher, P., Greenberg, J., Gunawardena, R., Kwon, C. S., Rahbari-Oskoui, F., & Wong, C. (2003). Volatile markers of breath cancer in the breath. *The Breast Journal, 9*, 184–191. https://doi.org/10.1046/j.1524-4741.2003.09309.x

Phillips, M., Gleeson, K., Hughes, J. M. B., Greenberg, J., Cataneo, R. N., Baker, L., & McVay, W. P. (1999a). Volatile organic compounds in breath as markers of lung cancer: A cross-sectional study. *Lancet, 353*, 1930–1933. https://doi.org/10.1016/S0140-6736(98)07552-7

Phillips, M., Herrera, J., Krishnan, S., Zain, M., Greenberg, J., & Cataneo, R. N. (1999b). Variation in volatile organic compounds in the breath of normal humans. *Journal of Chromatography B, 729*, 75–88. https://doi.org/10.1016/s0378-4347(99)00127-9

Pickel, D., Manucy, G. P., Walker, D. B., Hall, S. B., & Walker, J. C. (2004). Evidence for canine olfactory detection of melanoma. *Applied Animal Behaviour Science, 89*, 107–116. https://doi.org/10.1016/j.applanim.2004.04.008

Pirrone, F., & Albertini, M. (2017). Olfactory detection of cancer by trained sniffer dogs: A systematic review of the literature. *Journal of Veterinary Behavior, 19*, 105–117. https://doi.org/10.1016/j.jveb.2017.03.004

Pleil, J., & Giese, R. (2017). Integrating exhaled breath diagnostics by disease sniffing dogs with instrumental laboratory analysis. *Journal of Breath Research, 11*, 03200. https://doi.org/10.1088/1752-7163/aa79bc

Polgár, Z., Kinnunen, M., Újváry, D., Miklósi, A., & Gácsi, M. (2016). A test of canine olfactory capacity: Comparing various dog breeds and wolves in a natural detection task. *PLoS ONE, 11*(5), e0154087. https://doi.org/10.1371/journal.pone.0154087

Pomerantz, A., Blachman-Braun, R., Galnares-Olalde, J. A., Berebichez-Fridman, R., & Capurso-García, M. (2015). The possibility of inventing new technologies in the detection of cancer by applying elements of the canine olfactory apparatus. *Medical Hypotheses, 85*, 160–172. https://doi.org/10.1016/j.mehy.2015.04.024

Riedlova, P., Tavandzis, S., Kana, J., Tobiasova, M., Jasickova, I., & Roubec, J. (2022). Olfactometric diagnosis of lung cancer by canine scent – A double-blinded study. *Complementary Therapies in Medicine, 64*, 102800. https://doi.org/10.1016/j.ctim.2022.102800

Rooney, N. J., Morant, S., & Guest, C. (2013). Investigation into the value of trained glycaemia alert dogs to clients with type I diabetes. *PLoS ONE, 8*(8), e69921. https://doi.org/10.1371/journal.pone.0069921

Samhita, L., & Gross, H. J. (2013). The "Clever Hans Phenomenon" revisited. *Communicative & Integrative Biology, 6*(6), e27122. https://doi.org/10.4161/cib.27122

Schallschmidt, K., Becker, R., Jung, C., Bremser, W., Walles, T., Neudecker, J., Leschber, G., Frese, S., & Nehls, I. (2016). Comparison of volatile organic compounds from lung cancer patients and healthy controls – Challenges and limitations of an observational study. *Journal of Breath Research, 10*, 046007. https://doi.org/10.1088/1752-7155/10/4/046007

Schoon, A., De Jonge, D., & Hilverink, P. (2020). How dogs learn to detect colon cancer – Optimizing the use of training aids. *Journal of Veterinary Behavior, 35*, 38–44. https://doi.org/10.1016/j.jveb.2019.10.006

Schoon, A., Hilverink, P., de Jonge, D., & Fjellanger, R. (2022). Role of dogs in detection of human diseases. In L. E. de DeGreef & C. A. Schulz (Eds.), *Canines: The original biosensors.*, Chapter 19 (pp. 686–704). Copyright © 2022 Jenny Stanford Publishing Pte. Ltd. ISBN 978-981-4968-04-1.

Seo, I.-S., Lee, H.-G., Koo, B., Koh, C. S., Park, H.-Y., Im, C., & Shin, H.-C. (2018). Cross detection for odor of metabolic waste between breast and colorectal cancer using canine olfaction. *PLoS ONE, 13*(2), e0192629. https://doi.org/10.1371/journal.pone.0192629

Serpell, J. A., & Hsu, Y. (2005). Effects of breed, sex, and neuter status on trainability in dogs. *Anthrozoös, 18*(3), 196–207. https://doi.org/10.2752/089279305785594135

Shirasu, M., & Touhara, K. (2011). The scent of disease: Volatile organic compounds of the human body related to disease and disorder. *Journal of Biochemistry, 150*, 257–266. https://doi.org/10.1093/jb/mvr090

Sonoda, H., Kohnoe, S., Yazamato, T., Satoh, Y., Morizono, G., Shokata, K., Morita, M., Watanabe, A., Morita, M., Kakeji, Y., Inoue, F., & Maehara, Y. (2011). Colorectal cancer screening with odour material by canine scent detection. *Gut, 60*(6), 814–819. https://doi.org/10.1136/gut.2010.218305

Stepanyuk, A., & Kirschning, A. (2019). Synthetic terpenoids in the world of fragrances: Iso E Super® is the showcase. *Beilstein Journal of Organic Chemistry, 15*(1), 2590–2602.

Stuetz, R. M., Gostelow, P., & Burgess, J. E. (2005). Odor perception. In R. M. Stuetz & F. B. Frechen (Eds.), *Odours in wastewater treatment – Measurement, modelling and control* (p. 4. ISBN electronic: 9781780402932). IWA Publishing. https://doi.org/10.2166/9781780402932

Taverna, G., Tidu, L., Grizzi, F., Torri, V., Mandressi, A., Sardella, P., La Torre, G., Cocciolone, G., Seveso, M., Giusti, G., Hurle, R., Santoro, A., & Graziotti, P. (2015). Olfactory system of highly trained dogs detects prostate cancer in urine samples. *The Journal of Urology, 193*, 1382–1387. https://doi.org/10.1016/j.juro.2014.09.099

Thomas, F., Adamo, S., & Moore, J. (2005). Parasitic manipulation: When are we and where should we go? *Behavioural Processes, 68*, 185–199. https://doi.org/10.1016/j.beproc.2004.06.010

Urbanova, L., Vyhnánková, V., Krisová, Š., Pacík, D., & Nečas, A. (2015). Intensive training technique utilizing the dog's olfactory abilities to diagnose prostate cancer in men. *Acta Veterinaria, 84*, 77–82. https://doi.org/10.2754/avb201585010077

Waggoner, P., Lazarowski, L., Hutchings, B., Angle, C., & Porritt, F. (2022). Effects of learning an increasing number of odors on olfactory learning, memory and generalization in detection dogs. *Applied Animal Behaviour Science, 247*, 105568. https://doi.org/10.1016/j.applanim.2022.105568

Walczak, M., Jezierski, T., Górecka-Bruzda, A., Sobczyńska, M., & Ensminger, J. (2012). Impact of individual training parameters and manner of taking breath samples on the reliability of canines as cancer screeners. *Journal of Veterinary Behavior: Clinical Applications and Research, 7*, 283–294. https://doi.org/10.1016/j.jveb.2012.01.001

Williams, H., & Pembroke, A. (1989). Sniffer dogs in the melanoma clinic? *Lancet, 1*, 734.

Willis, C. M., Church, S. M., Guest, C. M., Cook, W. A., McCarthy, N., Bransbury, A. J., Church, M. R. T., & Church, J. C. T. (2004). Olfactory detection of human bladder cancer by dogs: Proof of principle study. *British Medical Journal, 329*, 712–715. https://doi.org/10.1136/bmj.329.7468.712

Willis, C. M., Britton, L. E., Swindells, M. A., Jones, E. M., Kemp, A. E., Muirhead, N. L., Gul, A., Matin, R. N., Knutsson, L., & Ali, M. (2016). Invasive melanoma in vivo can be distinguished from basal cell carcinoma, benign naevi and healthy skin by canine olfaction: A proof-of-principle study of differential volatile organic compound emission. *BJD Bristish Journal of Dermatology, 175*, 1020–1029. https://doi.org/10.1111/bjd.14887

Withington, L., Diaz Pardo de Vera, D., Guest, C., Mancini, C., & Piwek, P. (2021). Artificial Neural Networks for classifying the time series sensor data generated by medical detection dogs. *Expert Systems with Applications, 184*, 115564. https://doi.org/10.1016/j.eswa.2021.115564

World Health Organisation. Cancer 3 February 2022. https://www.who.int/news-room/fact-sheets/detail/cancer

Yamamoto, A., Kamoi, S., Kurose, K., Ito, M., Takeshita, T., Kure, S., Sakamoto, S., Sato, Y., & Miyashita, M. (2020). The trained sniffer dog could accurately detect the urine samples from the patients with cervical cancer, and even cervical intraepithelial neoplasia grade 3: A pilot study. *Cancers, 12*(11), 3291. https://doi.org/10.3390/cancers12113291

Yamazaki, K., Boyse, E. A., Bard, J., Curran, M., Kim, D., Ross, S. R., & Beauchamp, G. K. (2002). Presence of mouse mammary tumor virus specifically alters the body odor of mice. *Proceedings of the National Academy of Sciences USA, 99*, 5512–5515. https://doi.org/10.1073/pnas.082093099

Yu, C., Fan, S., Sun, Y., & Pickwell-Macpherson, E. (2012). The potential of teraherts imaging for cancer diagnosis: A review of investigations to date. *Quantitative Imaging in Medicine and Surgery, 2*, 33–45. https://doi.org/10.3978/j.issn.2223-4292.2012.01.04

# Chemical Ecology and Elephant Conservation: Examining Chili Pepper Deterrent Fences

R. Lynn Von Hagen[1], Sophia C. Corde[1], Pauline Norris[2], Simon Kasaine[3], Mwangi Githiru[3], Bernard Amakobe[3], Urbanus N. Mutwiwa[4], and Bruce A. Schulte[1] (✉)

[1] Western Kentucky University, Biology Department, Bowling Green, KY, USA
bruce.schulte@wku.edu
[2] Advanced Materials Institute, Western Kentucky University, Bowling Green, KY, USA
[3] Wildlife Works, Voi, Kenya
[4] Jomo Kenyatta University of Agriculture and Technology, Nairobi, Kenya

**Abstract** As human–elephant coexistence is challenged across Africa and Asia, conservationists seek methods to mitigate elephant crop raiding. Farmers rely on their crops for sustenance, and crop raiding events can destroy harvests. To mitigate these conflicts, deterrent fences continue to be developed. Modern deterrent techniques incorporate improved understanding of elephant biology, including their reliance on chemical signals. The construction of deterrent fences with cloths soaked in chili pepper extracts with high Scoville heat units (SHU) and used engine oil has become widespread but with variable effectiveness. However, the active levels of capsaicinoids in these mixtures and the rate of degradation of these compounds remain largely unexplored. In Kenya, our field experiments found that chili fences performed less effectively than desired possibly due to environmental conditions. In the lab, we evaluated SHU from chili peppers in three preparation recipes (vodka, water, and oil) and found that mixing with oil was crucial for increased solubility. We simulated a chili fence in the lab and detected no significant SHU loss over 7 days, which has positive implications for fence maintenance. Continuous exploration of the applications of chemical ecology to major conservation issues such as crop raiding can benefit humans and wildlife globally.

**Keywords** African elephant · Food security · Human–elephant conflict · Integrated pest management · *Loxodonta africana* · LC-MS · Sustainable agriculture

## 1 Introduction

Negative interactions between humans and wildlife can have adverse effects on both parties; such exchanges continue to increase as people encroach on previously wild habitats (Redpath et al., 2013). Coexistence is challenged when there is competition over resources such as food, water, or occupiable space (Treves et al., 2006;



B. Schaal et al. (eds.), *Chemical Signals in Vertebrates 15*,
https://doi.org/10.1007/978-3-031-35159-4_23

Seoraj-Pillai & Pillay, 2017), and agricultural areas are often hotbeds for these contentious interactions. Many farmers in rural communities face challenges to food security such as drought, climate change, poor soil conditions, and various pests, especially in Africa (Clover, 2000; Shiferaw et al., 2014; FAO, 2022; IPCC, 2022). Large mammals integrating crops into their foraging routines can be especially devastating to small shareholder farmers in economically challenged areas (Crespin & Simonetti, 2019). Farmers may be able to manage losses from smaller pests such as baboons, birds, rodents, or insects on a consistent basis, incorporating ideas from integrated pest management. However, more rare events such as visits from elephants or other large or numerous herbivores can be ruinous for harvests and related livelihoods (Ehler, 2006; Witmer, 2007; Warren, 2009; Hill, 2015; Gross et al., 2018; Garvey et al., 2020).

Human–elephant conflicts (HEC) occur across Africa and Asia, with the most prevalent form being crop raiding (Hoare, 1999; Davies et al., 2011; Desai & Riddle, 2015). Elephants enter farms most often at night to forage on cultivated crops (Nyhus et al., 2000; Sitati et al., 2003; Graham et al., 2010) angering farmers who may lash out at what they view as oversized pests (Dublin & Hoare, 2004; Nsonsi et al., 2017) making HEC a major threat for elephant conservation. Communities that suffer the greatest from crop failures often depend on their harvests for sustenance and/or income, but may over-attribute losses to elephants. In a study with Tanzanian farmers, individuals were good at estimating crop damage but often did not attribute the causes appropriately (Hoffmeier-Karimi & Schulte, 2015). Elephants may also injure or kill people (Shaffer et al., 2019). Thus, crop raiding is an ongoing challenge for coexistence between rural farmers and elephants (Goswami et al., 2014; Nyhus, 2016).

Historically, farmers have relied on techniques now referred to as traditional methods because they emanated at the local scale using resources at hand. Depending on the location, these methods might include chasing by dogs or people on foot, burning fires, making loud noises, fences made from acacia branches, and other rudimentary barriers (Sitati & Walpole, 2006; Enukwa, 2017; van de Water & Matteson, 2018). Practices that put people in proximity to elephants, especially at night, are dangerous. Such direct involvement can also reduce human productivity for work or school and create lost opportunity costs (Barua et al., 2013). Over time, traditional methods may lose their effectiveness as elephants are rarely harmed, or elephants learn how to circumvent the measures, and thus begin to ignore the stimuli (O'Connell-Rodwell et al., 2000; Davies et al., 2011). Some of these efforts also generate a conservation cost. Fires using biomass extracted from the wild and fences built from cutting down trees can lead to ecosystem degradation and potentially exacerbate crop raiding as elephants are pushed to cultivated areas because of loss of wild resources. The widespread use of traditional methods points to their perceived affordability in terms of capital outlay, practicality as they can be created relatively easily and quickly, and initial effectiveness as new stimuli are often approached warily or avoided by elephants.

The need to create longer lasting, more effective, and safer deterrent methods has led to what are termed modern methods (Sitati & Walpole, 2006; Hoare, 2012). In contrast to traditional methods, modern approaches work to remove humans from directly interacting with elephants. For example, using sounds or lights that are variable in some manner and/or are only activated by elephants tripping a guide wire or passing through the beam of a motion sensor (Wijayagunawardane et al., 2016; Adams et al., 2020). Some methods bridge traditional and modern such as using bright spotlights and shining them at elephants from a safe distance (Davies et al., 2011). A variety of fences have been invented, including those incorporating hot chili pepper extracts, beehives, and metal strips (Osborn & Parker, 2003; King, 2013; Karidozo & Osborn, 2015; Von Hagen et al., 2021).

Modern deterrent methods also aim to reduce conservation costs. The variable nature of the stimulus and the potential for a negative association (e.g., bright lights, noises associated with danger, annoying chemicals, or stings from bees) are directed at reducing habituation by elephants. The means of deploying these techniques are meant to increase safety and decrease night-time investment by humans. Using solar power and materials not extracted from the wild also reduce ecological costs to the habitat. Issues remain as to whether these newer solutions are affordable and practical to erect and upkeep and if the materials are readily available. Approaches that require funds from outside entities to establish and maintain may not be widely adopted, especially in resource-poor farming communities. As with any preventative technique, it can be difficult to discern between effectiveness and the absence of the problem. For farmers, they must weigh the use of a deterrent against the likelihood of elephants raiding their crops. This can lead to a lack of maintenance of the deterrent or use of its components (e.g., batteries, wires) for other purposes, leading to failure of the method. A counterbalance exists for methods that provide an additional benefit to farmers, such as honey from beehives or chili peppers grown for consumption and market as well as for use in a chili fence.

Chemically based methods to mitigate negative human–wildlife interactions, such as the reduction of crop raiding, have a variable history (Schulte, 2016). Chemical signals are ubiquitous, often undetectable or unrecognized by humans for their ecological function, and like all stimuli are subject to variations in the environment for their effectiveness. Organisms commonly use chemicals in their defense against predators, so investigating how such species and their chemicals might be used as components of deterrents has a long history (Wyatt, 2014). Chili peppers (*Capsicum* spp.) contain capsaicin (CAP) and dihydrocapsaicin (DICAP), the predominant components (ca. 90%) that give peppers their heat (measured in Scoville heat units (SHU) (Scoville, 1912), and along with their related alkaloids are termed capsaicinoids (AOAC, 1995). In a food choice experiment with Sumatran Asian (*Elephas maximus sumatranus*) elephants, chili peppers were avoided (Berliani et al., 2018), likely because of their chemical makeup and how capsaicinoids can stimulate tearing, mucous production, and potentially pain (Rasmussen, 1994). However, in a study with African forest elephants (*Loxodonta cyclotis*), a chili pepper extract coated onto mangos (*Mangifera indica* L.) did not repel elephants from consuming them (Ngama et al., 2018). When used in a fence, elephants may associate the multimodal cues of a chili fence (visual, sound; the flapping of cloths in the wind, smell, and touch) and the activation of the trigeminal nerve. Though which type (or types) of cues that cause(s) an elephant to be deterred has not been validated, once a negative association has been made with a chili fence elephants may circumvent the crop field when they detect the fence.

Chili peppers have been deployed as buffer crops around fields, in sprays, burned as part of dung briquettes, and as a solvent-based mixture for fences. A capsicum oleoresin was shown to be effective at deterring elephants (Osborn & Rasmussen, 1995; Osborn, 2002) and also when mixed with dung and burned (Osborn & Parker, 2002), as well as when released as a cloud of gas near elephants (Le Bel & Taylor, 2010; Le Bel, 2015). The solvent-based mixtures is applied to cloth and rope fences erected around crop fields to prevent raiding by elephants in a chili fence (Shaffer et al., 2019; Snyder & Rentsch, 2020; Montgomery et al., 2022) (Fig. 1). Several benefits occur when using chili fences, including the peppers can be locally grown, harvested, sold, and used in cooking; once erected the fences do not need to be monitored at night; and resources are not taken from the wild. However, capsaicinoids are most soluble in oils. The least expensive, most readily available oil in many locales is used motor oil. Thus, the preparation of the ground chili – motor oil mixture and the process of applying it on cloth and rope can be messy and unhealthy for the people and the environment (Von Hagen et al., 2021). Thus, use of an alternative and more practical solvent to motor oil is an important direction for future research associated with chili fences.

The chili fence has been effective in repelling elephants from crops compared to controls in some studies but efficacy varies (e.g., Sitati & Walpole, 2006; Hedges & Gunaryadi, 2010; Davies et al., 2011; Govind & Jayson, 2013; Chang'a et al., 2016; Branco et al., 2019; Kiffner et al., 2021; Von Hagen et al., 2021). Environmental conditions such as humidity, temperature, wind, and direct sunlight could affect the fence's performance by impacting the chemical ecology of capsaicinoids in chili and oil mixtures. Thus, effectiveness of chili fences can vary, based on local conditions and may need to be trialed before large-scale implementation in a desired location. Another

**Fig. 1** A section of a chili fence deployed in the farming community of Sasenyi, Kenya

set of potentially confounding factors relate to the means of preparing the fence, including the SHU of the peppers, the type of liquid used in the mixture, the amount of mixture used on the rope and cloths, the frequency of reapplication, and the steps in making the mixture (Sitati et al., 2003; Chang'a et al., 2016; Von Hagen et al., 2021). Preparation variation has been minimally investigated as a laboratory technique had not been previously available to assess the level of capsaicinoids. Therefore, factors such as inconsistencies in environmental conditions, the reapplication process, and preparation factors are important directions for research surrounding the efficacy of chili fences.

We focused our study of chili fences in three main areas with associated hypotheses. The first was part of an ongoing study to assess the effectiveness of several deterrent methods including chili fences with the hypothesis that chili fences would perform significantly better than their control. Second, we examined the preparation process due to our ability to control and quantify capsaicinoids in a laboratory setting (Von Hagen et al., 2020) and because the difficulty of preparation can affect farmer uptake (Hill, 2015). In a related study, we developed a Liquid Chromatography and Mass Spectrometry (LC-MS) methodology for determining capsaicinoid concentrations (SHU levels) in a chili and motor oil mixture (Von Hagen et al., 2020) that we used herein to further evaluate the preparation process. In recipes for preparation, the chili peppers are sometimes boiled in water before mixing with motor oil, yet the active ingredients are hydrophobic. Therefore, we hypothesized that omitting the water step and boiling the chili peppers in the solvent would provide an increased extraction of CAP and DICAP, which could ultimately translate into increased fence efficacy for farmers. In addition, given the health impact of used motor oil, we also used two other solvents (rubbing alcohol and vodka) that are available in many rural farming areas. Third, as a start for investigating the reapplication process (free from the aforementioned variation in the natural environment), we designed a lab-scale chili fence experiment to determine how chili concentration in motor oil changed over the course of a week in a controlled setting, hypothesizing a decrease over time. Results from each of these experiments can help inform best practices for uptake, use, and maintenance of chili fences, and lab findings can provide novel insight into how capsaicinoids may perform in the environment as part of chili and oil mixtures in deterrent fences.

## 2 Methods

### 2.1 Field Study

#### 2.1.1 Field Site

This study is part of the Elephants and Sustainable Agriculture in Kenya (ESAK) project within the Kasigau Wildlife Corridor (KWC) located between Tsavo East and Tsavo West National Parks of southeastern Kenya. The KWC is home to community and privately owned ranches, including approximately 100,000 people and 15,000 elephants in the Greater Tsavo Ecosystem. Our current study took place in the Sasenyi farming community that borders a subset of the KWC known as the Rukinga Wildlife Sanctuary

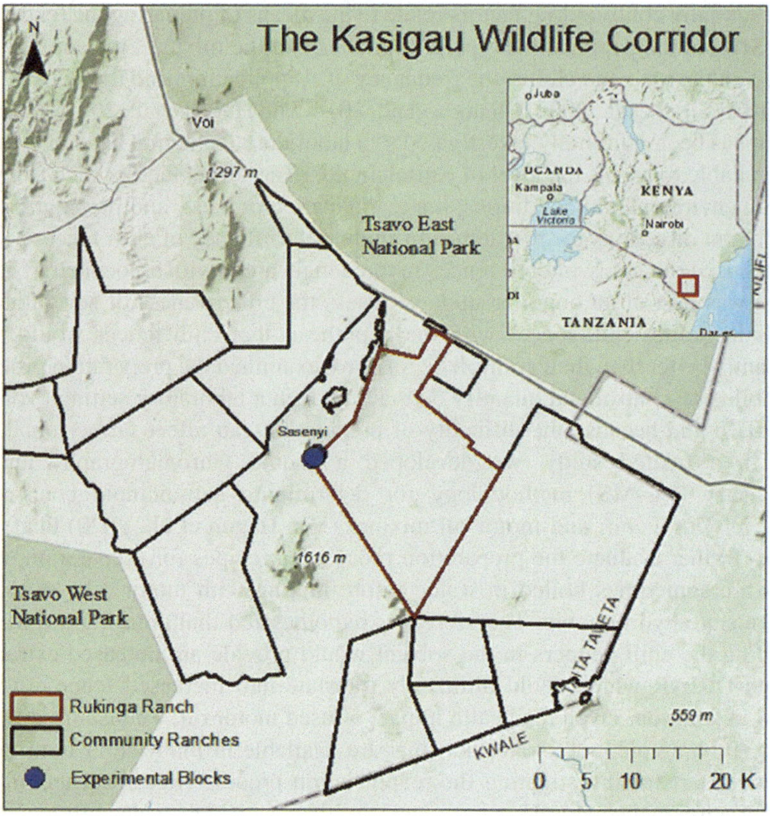


**Fig. 2** Field site in the KWC of southern Kenya (Von Hagen et al., 2021)

(RWS) (Fig. 2) and experiences high levels of crop raiding by elephants (Kagwa, 2011; Von Hagen, 2018; Mukeka et al., 2020; Von Hagen et al., 2021).

### 2.1.2 Deterrent Fence Experimental Design

In 2017, we contracted with farmers from the village of Sasenyi to lease some of their crop fields that bordered RWS for a deterrent fence study. To avoid a location effect, we created four blocks of fields over a 4.4 km distance. Each block contained eight 16 × 32 m fields with 6 m alleys between fields and at each end of a block (16 × 310 m). Deterrent fences were erected around each field with fence posts 8 m apart. We had two control fences per block, one with just fence posts and one using the traditional practice of laying thorny acacia branches on the ground. We tested a variety of fence types, which have been reported in detail (Von Hagen et al., 2021; Corde, 2022), but we only relate findings for the chili fence herein. This fence was used for four growing seasons (trials) in 2017–18 following the design by Chang'a et al. (2016). Briefly, we used black 100% cotton cloths cut to 0.6 m × 0.6 m squares. Sisal rope was used to string the cloths between fence posts at 1.5 m high. Mixtures of crushed chilies and motor oil were applied to the cloths and ropes. A single cloth was attached at the center of the rope between each pair of fence posts (Fig. 1). The control fence omitted the chili from the motor oil (Von Hagen et al., 2021).

#### 2.1.3 Crop-Raiding Events

The effectiveness of a fence was determined by assessing approaches and breaches, where an approach occurred when an elephant was within 12 m of a fence and a breach when a field was entered. We used 12 m as the approach distance because it is the furthest distance where presence of elephants could be reliably defined through camera trap images as associated with a particular fence type. We identified elephants from camera trap images and supplemented with identifying footprints and acquiring first-hand sightings by locals. We mounted between 25–31 Moultrie Spy A-5 Gen2 and A30i series infra-red camera traps on fence posts to include a view of the field and the dirt road that separates the farms from RWS. Camera traps were set to take three consecutive images with 15 second intervals and were motion activated. Thirty-eight elephants approached the chili pepper fences and 22 their matched controls.

#### 2.1.4 Data Analysis

We fit generalized linear mixed models (GLMM) to the data to test for statistical significance in the chili fence's ability to deter elephants from entering crop fields compared to its control using the *lme4* package in R (Bates et al., 2015). Fence type was set as a fixed effect, and trial (the experimental period), block, and interaction between trial and block were set as random effects to account for potential variances. By removing different random effects, models of varying complexities were compared using ANOVA and AIC values (Corde, 2022). To determine if the random effects had a significant impact on the results of the original model, the significance of these effects were tested using a random effect ANOVA (rANOVA). The random effects had no significant effect on the model, and no significant difference was found between the models of differing complexities, so the model containing all random effects tested (trial, block, and interaction between trial and block) was chosen.

### 2.2 Laboratory Study

#### 2.2.1 Chili Pepper Mixture Preparations

The standard recipe used previously in our research area begins with boiling 2 kg of ground chili peppers in 5 L of water for several minutes, then straining out the chili peppers before mixing the infused water with 5 L of used motor oil. Cloths are then soaked in the infused water and oil mixture before being suspended by sisal twine ropes around crop fields. In use as a deterrent fence, cloths are typically 0.6 × 0.6 m (Chang'a et al., 2016; Von Hagen et al., 2021). For our lab study, cloths were 0.18 × 0.18 m with a weight of 3.5 g each. We used bird's eye, piri piri, peppers (*Capsicum frutescens*), similar to what is used locally in Kenya, with SHUs of 130,000 (Frontier Co-Op) and ground them with a commercial bullet blender. In our previous study, we used 40 g of chili and 100 ml of boiled water (2% of standard recipe) (Von Hagen et al., 2020). For this study, we cut this in half (1% standard recipe), using 20 g of ground pepper with

50 ml of boiling (1) water; or (2) 70% isopropyl alcohol; or (3) 40% vodka; or (4) locally obtained used motor oil. Each mixture was heated to boiling at 100 °C on a hot plate for 2 minutes. Once cooled to the touch, the peppers were strained out first with a large mesh strainer, then with a combination of Eppendorf syringe filters and filter papers to ensure that no large particles would remain and potentially damage the LC-MS equipment. Three replicates from each of the samples without motor oil were placed in LC-MS tubes and the remaining 10 ml (isopropyl alcohol) or 22 ml (vodka) or 50 ml (motor oil) were mixed with an additional 50 ml used motor oil.

To separate the infused compound from the viscous oil, we used ca. 40 ml of acetonitrile (ACE) as a solvent for all samples containing motor oil and separated them into three glass tubes. The samples were then moved to an Allegra 6 centrifuge and spun for 15 minutes. This resulted in a top clear layer of ACE containing the infused compound so that 1 ml from each could be placed into LCMS tubes for processing. The same protocols established in Von Hagen et al. (2020) were used for the LC-MS (Agilent Ion Trap 500 LC–MS).

Previously, we published the results on the chili in boiling water, the chili in boiling water plus motor oil, and the chili in oil, without boiling in water (Von Hagen et al., 2020); thus, we do not report these findings here other than to show their values in the results. We had the following five treatments: (i) chili boiled in isopropyl alcohol (C + A) and (ii) mixed with motor oil (C + A + O); (iii) chili boiled in vodka (C + V) and (iv) mixed with motor oil (C + V + O); and (v) chili boiled in motor oil and mixed with additional motor oil (C + O + O).

#### 2.2.2 Chili Fence Simulation Experiment

To quantify how capsaicinoids might persist in the environment, an experimental chili fence was recreated in the laboratory and monitored for 7 days. To create the chili + oil mixture, a beaker containing 500 ml of tap water was mixed with 200.25 g of chili peppers. The same procedures as above were followed for heating and straining, which left 170 ml of chili infused water. This water was mixed with 500 ml of motor oil in a large tub, and the mixture was shaken thoroughly by hand for 30 seconds. Eighteen 100% cotton muslin squares measuring 0.18 × 0.18 m were cut to simulate the cloths used in chili fences. Dry weight was 3.5 grams. The cloths were placed inside the large tub with the chili and oil and stirred with a wooden spoon to evenly coat the cloths. After stirring, the cloths were soaked for an additional 2 minutes. The mixture and cloths were poured into a metal pan in preparation for hanging on the experimental set up.

Dowel rods were placed on wood bases anchored with heavy weights under a laboratory fume hood. The rods had metal attachments for hanging two 1.6 m lengths of jute twine in two layers that did not overlap. The cloths were hung from the jute strings, secured by binder clips, and given an ID number based on the anticipated day of removal and the replicate for that day (Fig. 3). Care was taken not to have cloths overlapping or dripping on the row underneath, as this would artificially increase the weight of the cloths. At this point only, the hood vent was turned on to accelerate the initial dripping of the mixture from the cloths and to pull fumes from the lab. After the cloths were no longer dripping, all cloths were removed sequentially and weighed using a 30 g Pesola

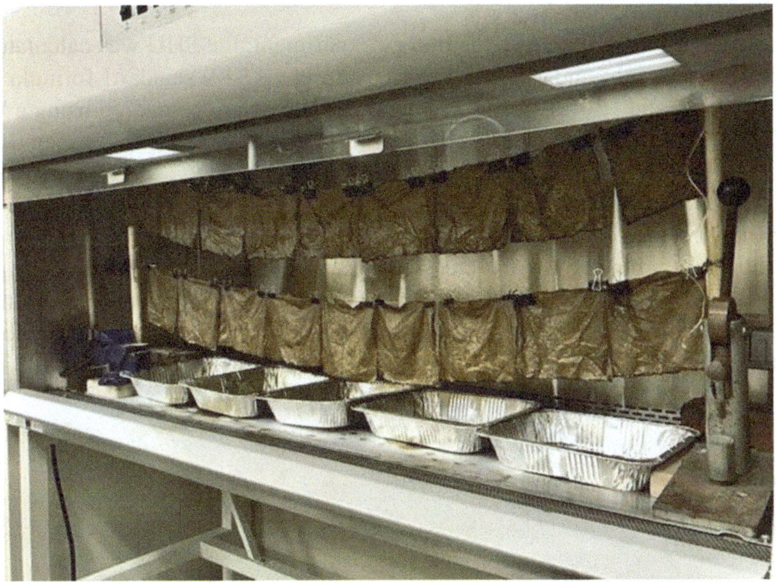

**Fig. 3** The experimental set up for testing rates of capsaicinoid degradation over time. Cloths were suspended from sisal rope attached to wood dowels with drip pans underneath and analyzed over 7 days

scale. The first set of three cloth replicates (Day 0) served as a starting reference point. Each Day 0 replicate was placed in its own clear glass 50 ml capped tube and labeled. The remainder of the saturated cloths had their initial weights recorded and were replaced on the rods. To ensure consistency, the remaining cloths were removed at the same time each day (16:30) and weighed, removing one set each day.

To process the samples, ACE was added to fill and balance each tube. The total weight in grams were taken for each and recorded. Each tube was placed in a 500 ml glass beaker in a water bath in a Leco UC-100 ultrasonic cleaner and sonicated for 30 minutes and centrifuged as described in the mixture preparation section. This resulted in a top clear layer of ACE containing the infused compound with the oil and cloth at the bottom. A 13 mm 0.22 μm PTFE Luer lock inlet syringe filter was used to extract 1 ml for each replicate, which was placed into a LC-MS tube. This process was repeated after each day except for the last day's samples as there was no access to the laboratory over the weekend. Therefore, the last day's (D7) samples sat for 2 days longer before being removed from the twine and processed. The chili + water + oil (C + W + O) samples were processed following the LC-MS protocols established in Von Hagen et al. (2020). One ml of the C + W + O mixture was weighed at 0.74 grams to be used as a conversion factor as well as the mean of the C + W + O reading from the LC-MS of CAP (15.04 μg/ml) and DICAP (2.65 μg/ml).

#### 2.2.3 Data Handling and Analysis

For the chili mixture preparation laboratory experiment, the SHU was calculated from the sample dry weight and LC-MS peaks using the ASTA standard formula (ASTA 1985). The CAP and DICAP SHUs for each treatment were compared using a Welch's ANOVA to determine if there was a statistical difference in the SHU between treatment types.

For the second chili fence simulation laboratory experiment, calculations were performed from the LC-MS readings from means of each of the three replicates from each day to obtain a combined CAP and DICAP measurement of SHU. The CAP and DICAP concentrations in µg/ml for each day were compared to a slope of 0 (null hypothesis of no loss over time) with a regression analysis to determine the relationship between time and concentration.

## 3 Results

### 3.1 Field Study

In our field study, chili fences and their control did not statistically differ in their ability to deter elephants from entering crop fields ($z = -0.434$, $p = 0.664$). When raw numbers were compared to assess percent effectiveness, the fences proved to be 55% (chili fence) and 50% (chili control fence) effective (Fig. 4).

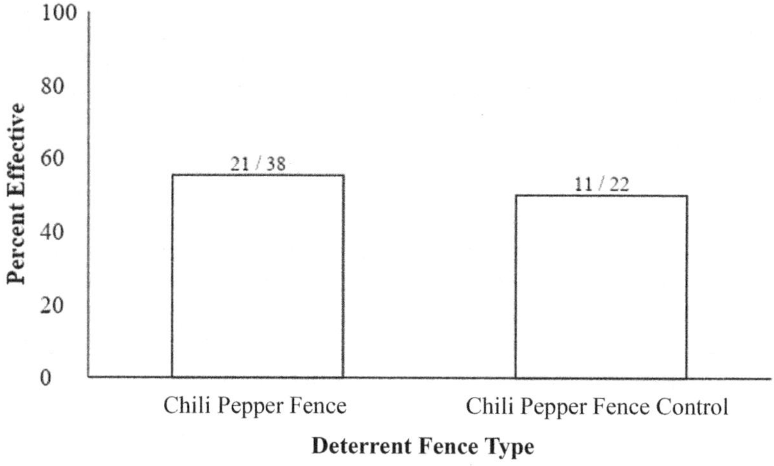


**Fig. 4** Effectiveness of a chili fence (55%) and its control (50%) at experimental crop fields in the study site in southeastern Kenya. Effectiveness = number of elephants deterred/number of elephants approached within 12 m of the fences; sample sizes are shown above each bar

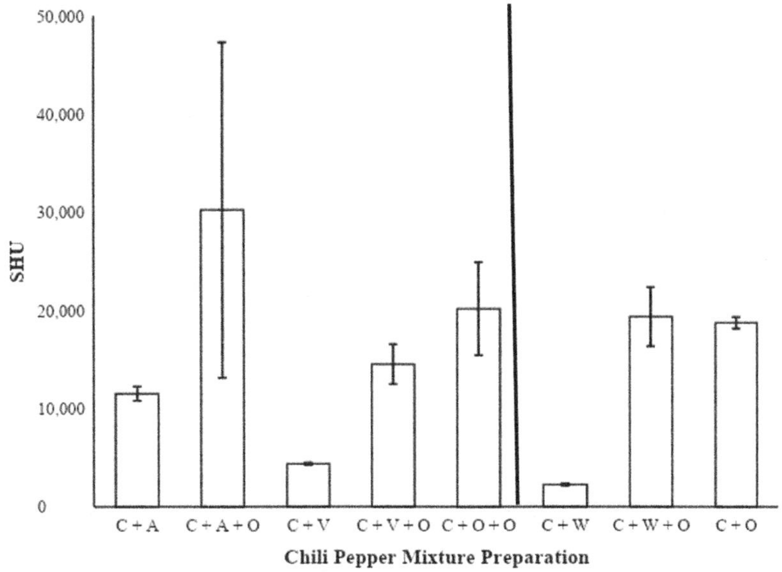


**Fig. 5** The mean (± SD) combined CAP and DICAP SHU of the five different preparation types used in the current study. C = ground chili, A = 70% isopropyl alcohol, V = 40% vodka, O = used motor oil. The three preparations to the right of the vertical line were reported in Von Hagen et al. (2020), W = water

## 3.2 Laboratory Study

### 3.2.1 Chili Pepper Mixture Preparations

We predicted that oil-based solvents would yield greater levels of chili active ingredients (measured by SHU) than those without oil, as capsaicinoids are most soluble in oils (Walpole et al., 1993). As expected, significant differences between treatments occurred (Welch's ANOVA, $F_{4,4.05} = 67.07$, $p < 0.001$; Fig. 5). The two preparations that did not contain oil, C + A (SHU 11,523.92 ± 723.89) and C + V (SHU 4352.80 ± 110.28), had the lowest SHU retentions. The C + V + O (14,513.20 ± 2015.62) and C + O + O (SHU 20,164.93 ± 4735.69) showed intermediate readings with the C + A + O (SHU 30,240.10 ± 17,083) yielding the highest SHU (but also having one outlier value). The C + V + O and C + O + O treatments performed similarly to past experiments of mixtures containing oil (i.e., compared to values to the right of the vertical black line in Fig. 5).

### 3.2.2 Chili Fence Simulation Experiment

Contrary to our hypothesis that concentration would decrease with time, no change was evident in the μg of CAP or DICAP recovered through the LC-MS over the 7-day experimental period (for each regression, $F_{1,16} < 1.5$, $p > 0.2$; Fig. 6). Therefore, we accepted our null hypothesis, in that there was no substantial loss of SHU over the first 7 days of the experiment in a controlled laboratory setting.

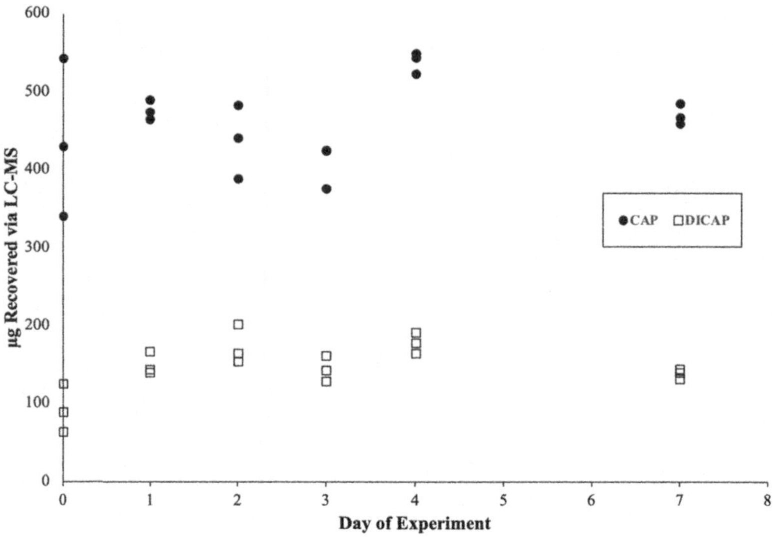


**Fig. 6** The µg of CAP and DICAP measured via LC-MS on each day of the experimental fence study (slopes not different from zero so no lines drawn)

## 4 Discussion

Deterrent methods remain one of the frontline defenses to mitigating negative human–elephant interactions (Branco et al., 2019). In these times of extreme climatic changes, the expansion of successful methods such as deterrent fences are needed to enhance food and financial security for local farmers. Fence designs that provide not only deterrence but the potential for additional income and nutrition are especially valuable. Chemosensory approaches have the potential to contribute to this work, especially considering elephants' superior sense of smell. Using a standard method for preparing chili fences (Chang'a et al., 2016; Von Hagen et al., 2021), we experimentally tested the deterrence ability of chili fences at our study site over four growing seasons. Although there was no significant difference in the performance of the chili fence and its control, the fences were 55% and 50% effective, respectively, at deterring elephants. The similarity in effectiveness between the active and control fences brings into question the driving factor behind the effectiveness of the chili fence in this and other studies. The only difference in our control was the lack of chili peppers in the oil mixture. For example, the flapping of the cloths in our windy study area may be an effective deterrent, in that the sporadic movement could make elephants uncomfortable whether or not chili pepper extract was present. While we used black cloths in the trials reported, we also have used brightly colored cloths as a fence type with no significantly better results (Corde, 2022). These results, in addition to the variability in success of chili fences reported in the literature, prompted our investigation into the preparation methods and SHU retention of the chili fences. Our study followed from using a technique with LC-MS developed by Von Hagen et al. (2020) to show that boiling with water before mixing with the motor oil did not enhance the SHU level. Eliminating this step without a loss in effectiveness of the deterrent is meaningful, as the boiling step is both time and fuel consuming. Given the many hidden stressors of coping with wildlife conflicts (Barua et al., 2013), we sought further steps to benefit farmers and their livelihoods.

This study further examined the practice of boiling chilis in a solvent before mixing the infused water with motor oil. As alternatives to water, we tried 70% isopropyl alcohol, 40% vodka, and used motor oil, each of which is available in our study area. None of the alternatives performed substantially better than water. However, the final step of mixing with motor oil is still highly beneficial, which aligns with our knowledge that capsaicinoids are soluble in oils (Walpole et al., 1993). We had hoped that alternative solvents to water would improve extraction, but the evidence herein did not support our hypothesis. Given the personal and environmental health costs of motor oil, this result is unfortunate.

For our second lab experiment, we hypothesized that the combined CAP and DICAP would decrease over the 7-day period, supporting the advice to farmers that reapplication of mixtures on chili fences is necessary. In contrast to our expectations, we found that the amount of CAP and DICAP in μg that we measured via the LC-MS remained steady. Unfortunately, the logistical limitations of the study did not allow us to conduct analysis longer than 7 days. Future studies could target extending this period to determine at what point these concentrations start to decline. In addition, our lab study did not expose the cloths to UV, wind, or rain, which are likely to reduce SHU over time. Recommendations for reapplication of oil mixtures usually involve shortened time periods with rain. This can be especially problematic for areas of Asia that experience higher rainfall (Chelliah et al., 2010). Therefore, field and lab studies are also needed to quantify how much rain and other environmental variants can affect mixture efficacy as we may have experienced in our field study. Nevertheless, the stability of SHU in the lab is encouraging for farmers as reducing the time and effort defending their crops is valuable to their livelihoods.

The next step toward finding a more practical and sustainable formula for chili fences is to experiment with environmentally friendly oils as seen in the substitution of vegetable oil in lieu of motor oil (Branco et al., 2019). Working with a substance like motor oil can be discouraging and dangerous to farmers and be permanently damaging to the soil if contamination occurs. Variation in solubility also exists for capsaicin in different types of vegetable oils (Amruthraj et al., 2014), which leaves room for further experimentation. Understanding the rate of decrease of the SHU in oil and pepper mixtures under varying conditions and with different preparation techniques can be an important tool for conservationists seeking to maximize the deterrent power of this method and pave the way for a switch to more environmentally friendly solvents.

Overall, our experiments have shown evidence for pathways toward creating more practical and beneficial formulas for chili fences. In addition, we have created new areas of questions surrounding the mechanism behind successful chili fences and local variation in environmental conditions. Adopting such methods can enhance food security for farmers trying to establish more stable livelihoods. Conservationists continue to search for means to address challenges such as crop raiding by elephants and other types of human wildlife conflicts that can impact local stakeholders (Oniba & Robertson, 2019; Smit et al., 2019; Vogel et al., 2020). Through these experiments we have demonstrated how cross-disciplinary evaluative methodologies and technologies can be applied to such seemingly disparate fields as biochemistry, physical chemistry, and conservation. Our methodologies provide a foundation to continue investigations on procedures and practices to help mitigate crop raiding and provide solutions for local farmers including approaches incorporating chemical ecology (Schulte & LaDue, 2021).

**Acknowledgements** This research was partially supported by the Department of Biology, the Center for Biodiversity Studies, the Ogden College of Science and Engineering, the Graduate School, the WKU Sisterhood, and the Advanced Materials Institute all at Western Kentucky University, The Earthwatch Institute, The International Elephant Foundation, the Richard Lounsbery Foundation, and Wildlife Works. We also thank our colleagues at the Elephants and Bees Team from Save the Elephants, the Kenya Wildlife Service, and the farmers of Sasenyi for their participation. All research was conducted in accordance with the following permissions: WKU IACUC Permit #16-02 and #20-05, NACOSTI/P/17/26390/16537 and NACOSTI/P/20/4805 (Kenyan science agency), and KWS (Kenya Wildlife Service) /BRM/5001.

## References

Adams, T., Mwezi, I., & Jordan, N. R. (2020). Panic at the disco: solar-powered strobe light barriers reduce field incursion by African elephants *Loxodonta africana* in Chobe District, Botswana. *Oryx, 55*, 739–746. https://doi.org/10.1017/S0030605319001182

Amruthraj, N. J., Raj, J. P. P., & Lebel, L. L. (2014). Effect of vegetable oil in the solubility of capsaicinoids extracted from capsicum Chinese Bhut Jolokia. *Asian Journal of Pharmaceutical and Clinical Research, 7*, 48–51.

AOAC. (1995). Capsaicinoids in capsicums and their extractives liquid chromatographic method. *AOAC, 995*(03), 18–21.

Barua, M., Bhagwat, S. A., & Jadhav, S. (2013). The hidden dimensions of human–wildlife conflict: Health impacts, opportunity and transaction costs. *Biological Conservation, 157*, 309–316. https://doi.org/10.1016/j.biocon.2012.07.014

Bates, D., Maechler, M., Bolker, B., & Walker, S. (2015). Lme4: Linear mixed-effects models using Eigen and S4. Retrieved from http://cran.r-project.org/package=lme4

Berliani, K., Alikodra, H. S., Masy'ud, B., & Kusrini, M. D. (2018). Food preference of Sumatran elephant (*Elephas maximus sumatranus*) to commodity crops in human-elephant conflict area of Aceh, Indonesia. *Journal of Physics: Conference Series, 116*, 052015.

Branco, P. S., Merkle, J. A., Pringle, R. M., King, L., Tindall, T., Stalman, M., & Long, R. A. (2019). An experimental test of community-based strategies for mitigating human–wildlife conflict around protected areas. *Conservation Letters, 13*, e12679. https://doi.org/10.1111/conl.12679

Chang'a, A., De Souza, N., Muya, J., Keyyu, J., Mwakatobe, A., Malugu, L., Ndossi, H. P., Konuche, J., Omondi, R., Mpinge, A., Hahn, N., Palminteri, S., & Olson, D. (2016). Scaling-up the use of chili fences for reducing human-elephant conflict across landscapes in Tanzania. *Tropical Conservation Science, 9*(2), 921–930.

Chelliah, K., Kannan, G., Kundu, S., Abilash, N., Madhusudan, A., Baskaran, N., & Sukumar, R. (2010). Testing the efficacy of a chilli-tobacco rope fence as a deterrent against crop-raiding elephants. *Current Science, 99*(9), 1239–1243.

Clover, J. (2000). Food security in Sub-Saharan Africa. *African Security Review, 12*(1), 1–15.

Corde, S. C. (2022). *Alleviating human-elephant conflict through deterrent fences and environmental monitoring in southern Kenya*. M.S. Thesis, Western Kentucky University

Crespin, S. J., & Simonetti, J. A. (2019). Reconciling farming and wild nature: Integrating human – wildlife coexistence into the land-sharing and land-sparing framework. *Ambio, 48*, 131–138. https://doi.org/10.1007/s13280-018-1059-2

Davies, T. E., Wilson, S., Hazarika, N., Chakrabarty, J., Das, D., Hodgson, D. J., & Zimmermann, A. (2011). Effectiveness of intervention methods against crop-raiding elephants. *Conservation Letters, 4*, 346–354. https://doi.org/10.1111/j.1755-263X.2011.00182.x

Desai, A.A., Riddle, H.S. (2015) Human-Elephant Conflict in Asia. A report to the US Fish and Wildlife Service.

Dublin, H. T., & Hoare, R. E. (2004). Searching for solutions: The evolution of an integrated approach to understanding and mitigating human–elephant conflict in Africa. *Human Dimensions of Wildlife, 9*, 271–278. https://doi.org/10.1080/10871200490505701

Ehler, L. E. (2006). Integrated pest management (IPM): definition, historical development and implementation, and the other IPM. *Pest Management Science, 62*, 787–789.

Enukwa, E. H. (2017). Human-elephant conflict mitigation methods: A review of effectiveness and sustainability. *Journal of Wildlife and Biodiversity, 1*(2), 69–78. https://doi.org/10.22120/JWB.2017.28260

FAO. (2022). New standards to curb the global spread of plant pests and diseases. https://www.fao.org/news/story/en/item/1187738/icode/. Accessed 19 Jan 2022.

Garvey, P. M., Banks, P. B., Suraci, J. P., Bodey, T. W., Glen, A. S., Jones, C. J., McArthur, C., Norbury, G. L., Price, C. J., Russell, J. C., & Sih, A. (2020). Leveraging motivations, personality, and sensory cues for vertebrate pest management. *Trends in Ecology & Evolution, 35*, 990–1000. https://doi.org/10.1016/j.tree.2020.07.007

Goswami, V. R., Vasudev, D., & Oli, M. K. (2014). The importance of conflict-induced mortality for conservation planning in areas of human-elephant co-occurrence. *Biological Conservation, 176*, 191–198. https://doi.org/10.1016/j.biocon.2014.05.026

Govind, S. K., & Jayson, E. A. (2013). Efficiency of chilli powder (*Capsicum* sp.) to deter wild elephants from the crop fields in a tropical area. *Millennium Zoology, 14*(1), 21–24.

Graham, M. D., Notter, B., Adams, W. M., Lee, P. C., & Ochieng, T. N. (2010). Patterns of crop-raiding elephants, *Loxodonta africana*, in Laikipia, Kenya, and the management of human–elephant conflict. *Systematics and Biodiversity, 8*, 435–445. https://doi.org/10.1080/14772000.2010.533716

Gross, E. M., Lahkar, B. P., Subedi, N., Nyirenda, V. R., Lichtenfeld, L. L., & Jakoby, O. (2018). Seasonality, crop type and crop phenology influence crop damage by wildlife herbivores in Africa and Asia. *Biodiversity and Conservation, 27*, 2029–2050. https://doi.org/10.1007/s10531-018-1523-0

Hedges, S., & Gunaryadi, D. (2010). Reducing human–elephant conflict: Do chillies help deter elephants from entering crop fields? *Oryx, 44*, 139–146. https://doi.org/10.1017/S0030605309990093

Hill, C. M. (2015). Perspectives of "conflict" at the wildlife–agriculture boundary: 10 years on. *Human Dimensions of Wildlife, 20*, 296–301. https://doi.org/10.1080/10871209.2015.1004143

Hoare, R. E. (1999). Determinants of human-elephant conflict in a land-use mosaic. *Journal of Applied Ecology, 36*, 689–700. https://doi.org/10.1046/j.1365-2664.1999.00437.x

Hoare, R. (2012). Lessons from 15 years of human–elephant conflict mitigation: Management considerations involving biological, physical and governance issues in Africa. *Pachyderm*, 51, 60–74.

Hoffmeier-Karimi, R. R., & Schulte, B. A. (2015). Assessing perceived and documented crop damage in a Tanzanian village impacted by human-elephant conflict (HEC). *Pachyderm, 56*, 51–60.

IPCC. (2022). *Climate change 2022-Impacts adaptation and vulnerability*. IPCC.

Kagwa, S. K. (2011). Spatial distribution of human elephant conflict (HEC) and characterization of crop-raiding elephants in Kasigau Region, Kenya. MS thesis, Western Kentucky University.

Karidozo, M., & Osborn, F. (2015). Community based conflict mitigation trials: Results of field tests of chilli as an elephant deterrent. *Journal of Biodiversity and Endangered Species, 3*, 1–6. https://doi.org/10.4172/2332-2543.1000144

King, L. E. (2013). Elephants and bees: Could honeybees be effective deterrents for Asia's crop-raiding elephants? *Asia Sanctuary*, 60–65. www.sanctuaryasia.com

Kiffner, C., Schaal, I., Cass, L., Peirce, K., Sussman, O., Grueser, A., Wachtel, E., Adams, H., Clark, K., König, H. J., & Kioko, J. (2021). Perceptions and realities of elephant crop raiding and mitigation methods *Conservation Science and Practice,* 3, https://doi.org/10.1111/csp2.v3.3 10.1111/csp2.372

Le Bel, S. (2015). Repelling elephants with a chilli pepper gas dispenser: Field tests and practical use in Mozambique, Zambia and Zimbabwe from 2009 to 2013. *Pachyderm, 56*, 87–96.

Le Bel, S., & Taylor, R. (2010). An easy-to-use capsicum delivery system for crop-raiding elephants in Zimbabwe: Preliminary results of a field test in Hwange National Park. *Pachyderm, 47*, 80–89.

Montgomery, R. A., Raupp, J., Mukhwana, M., Greenleaf, A., Mudumba, T., & Muruthi, P. (2022). The efficacy of interventions to protect crops from raiding elephants. *Ambio, 51*, 716–727. https://doi.org/10.1007/s13280-021-01587-x

Mukeka, J. M., Ogutu, J. O., Kanga, E., Roskaft, E. (2020). Spatial and temporal dynamics of human-wildlife conflicts in the Kenya Greater Tsavo Ecosystem. *Human-Wildlife Interactions*, 14, 255–272.

Ngama, S., Korte, L., Johnson, M., Vermeulen, C., & Bindelle, J. (2018). Camera traps to study the forest elephant's (*Loxodonta cyclotis*) response to chilli pepper repellent devices in Gamba, Gabon. *Nature Conservation Research, 3*, 26–35. https://doi.org/10.24189/ncr.2018.027

Nsonsi, F., Heymanns, J.-C., Diamouangana, J., Mavinga, F. B., & Breuer, T. (2017). Perceived human-elephant conflict and its impact for elephant conservation in northern Congo. *African Journal of Ecology, 56*, 208–215. https://doi.org/10.1111/aje.12435

Nyhus, P. J. (2016). Human–wildlife conflict and coexistence. *Annual Review of Environmental Resources, 41*, 143–171. https://doi.org/10.1146/annurev-environ-110615-085634

Nyhus, P. J., Tilson, R., & Sumianto. (2000). Crop raiding elephants and conservation implications at Way Kambas Nat Park, Sumatra, Indonesia. *Oryx, 34*(4), 262–274.

Osborn, F. V. (2002). Seasonal variation of feeding patterns and food selection by crop-raiding elephants in Zimbabwe. *African Journal of Ecology*, 42, 32–327.

Osborn, F. V., & Parker, G. E. (2002). Community-based methods to reduce crop loss to elephants: experiments in the communal lands of Zimbabwe. *Pachyderm*, 33, 22–31.

Osborn, F. V., & Parker, G. E. (2003). Towards an integrated approach for reducing the conflict between elephants and people: a review of current research. *Oryx,* 37, 80–84.

O'Connell-Rodwell, C.E., Rodwell, T., Rice, M., Hart L.A. (2000). Living with the modern conservation paradigm: Can agricultural communities co-exist with elephants? A five-year case study in East Caprivi, Namibia. *Biological Conservation 93*, 381–391.

Oniba, E., & Robertson, M. R. (2019). Trialing a new scent-based repellent to mitigate elephant crop- raiding around Murchison Falls National Park, Uganda. *Pachyderm, 60*, 123–125.

Osborn, F. V., & Rasmussen, L. E. L. (1995). Evidence for the effectiveness of an oleo-resin capsicum aerosol, as a repellent against wild elephants in Zimbabwe. *Pachyderm, 20*, 55–64.

Rasmussen, L. E. L. (1994). Sensory and communication systems. In S.K. Mikota, Sargent, & G.R. Ranglack (Eds.), *Medical management of the elephant* (pp 207–217). Indira Publishing House.

Redpath, S. M., Young, J., Evely, A., Adams, W. M., Sutherland, W. J., Whitehouse, A., Amar, A., Lambert, R. A., Linnell, J. D. C., Watt, A., & Gutiérrez, R. J. (2013). Understanding and managing conservation conflicts. *Trends in Ecology and Evolution, 28*, 100–109. https://doi.org/10.1016/j.tree.2012.08.021

Schulte, B. A. (2016). Learning and applications of chemical signals in vertebrates for human-wildlife conflict mitigation. In B. A. Schulte, T. E. Goodwin, & M. H. Ferkin (Eds.), *Chemical signals in vertebrates 13* (pp. 499–510). Springer.

Schulte, B. A., & LaDue, C. A. (2021). The chemical ecology of elephants: 21st century additions to our understanding and future outlooks. *Animals, 11*, 2860. https://doi.org/10.3390/ani11102860

Scoville, W. L. (1912). Note on *Capsicums*. *Journal of the American Pharmaceutical Association, 1*, 1–2.

Seoraj-Pillai, N., & Pillay, N. (2017). A meta-analysis of human-wildlife conflict: South African and global perspectives. *Sustain, 9*, 1–21. https://doi.org/10.3390/su9010034

Shaffer, L. J., Khadka, K. K., Van Den Hoek, J., & Naithani K. J. (2019). Human-Elephant conflict: A review of current management strategies and future directions. *Frontiers in Ecology and Evolution, 6*. https://doi.org/10.3389/fevo.2018.00235

Shiferaw, B., Tesfaye, K., Kassie, M., Abate, T., Prasanna, B. M., & Menkir, A. (2014). Managing vulnerability to drought and enhancing livelihood resilience in sub-Saharan Africa: Technological, institutional and policy options. *Weather and Climate Extremes, 3*, 67–79. https://doi.org/10.1016/j.wace.2014.04.004

Sitati, N. W., & Walpole, M. J. (2006). Assessing farm-based measures for mitigating human-elephant conflict in Transmara District, Kenya. *Oryx, 40*, 279–286. https://doi.org/10.1017/S0030605306000834

Sitati, N. W., Walpole, M. J., Smith, R. J., & Leader-Williams, N. (2003). Predicting spatial aspects of human and elephant conflict. *Journal of Applied Ecology, 4*, 667–677. https://doi.org/10.1046/j.1365-2664.2003.00828.x

Smit, J., Pozo, R. A., Cusack, J. J., Nowak, K., & Jones, T. (2019). Using camera traps to study the age-sex structure and behaviour of crop-using elephants *Loxodonta africana* in Udzungwa Mountains National Park, Tanzania. *Oryx, 53*, 368–376. https://doi.org/10.1017/S0030605317000345

Snyder, K., & Rentsch, D. (2020). Rethinking assessment of success of mitigation strategies for elephant-induced crop damage. *Conservation Biology, 34*, 829–842. https://doi.org/10.1111/cobi.13433

Treves, A., Wallace, R. B., Naughton-Treves, L., & Morales, A. (2006). Co-managing human–wildlife conflicts: A review. *Human Dimensions of Wildlife, 11*, 383–396. https://doi.org/10.1080/10871200600984265

van de Water, A., & Matteson, K. (2018). Human-elephant conflict in western Thailand: Socio-economic drivers and potential mitigation strategies. *PLoS ONE, 13*, e0194736. https://doi.org/10.1371/journal.pone.0194736

Vogel, S. M., Lambert, B., Songhurst, A. C., McCulloch, G. P., Stronza, A. L., & Coulson, T. (2020). Exploring movement decisions: Can Bayesian movement-state models explain crop consumption behaviour in elephants (*Loxodonta africana*)? *J Anim Ecol, 89*, 1055–1068. https://doi.org/10.1111/1365-2656.13177

Von Hagen, R. L. (2018). *An evaluation of deterrent methods utilized to prevent crop raiding by African Elephants (Loxodonta africana) in the Kasigau Wildlife Corridor, Kenya*. M.S. Thesis, Western Kentucky University

Von Hagen, R. L., Norris, P., & Schulte, B. A. (2020). Quantifying capsaicinoids from chili pepper and motor oil mixtures used in elephant deterrent fences. *Chromatographia, 83*, 1153–1157. https://doi.org/10.1007/s10337-020-03934-8

Von Hagen, R. L., Kasaine, S., Githiru, M., Amakobe, B., Mutwiwa, U. N., & Schulte, B. A. (2021). Metal strip fences for preventing African elephant (*Loxodonta africana*) crop foraging in the Kasigau Wildlife Corridor, Kenya. *African Journal of Ecology, 59*, 293–298. https://doi.org/10.1111/aje.12821

Walpole, C. S., Wrigglesworth, R., Bevan, S., Campbell, E. A., Dray, A., James, I. F., Perkins, M. N., Reid, D. J., & Winter, J. (1993). Analogs of capsaicin with agonist activity as novel analgesic agents; structure-activity studies. 1. The aromatic "A-region". *Journal of Medicinal Chemistry, 36*(16), 2362–2372.

Warren, Y. (2009). Crop-raiding baboons (*Papio anubis*) and defensive farmers: A West African perspective. *West African Journal of Applied Ecology, 14*, 1–11. https://doi.org/10.4314/wajae.v14i1.44705

Wijayagunawardane, M. P. B., Short, R. V., Samarakone, T. S., Nishany, K. B. M., Harrington, H., Perera, B. V. P., Rassool, R., & Bittner, E. P. (2016). The use of audio playback to deter crop-raiding Asian elephants. *Wildlife Society Bulletin, 40*, 375–379. https://doi.org/10.1002/wsb.652

Witmer, G. W. (2007). *The ecology of vertebrate pests and integrated pest management (IPM)*. USDA National Wildlife Research Center - Staff Publications. 730. https://digitalcommons.unl.edu/icwdm_usdanwrc/730

Wyatt, T. D. (2014). *Pheromones and animal behavior: Chemical signals and signatures* (2nd ed.). Cambridge University Press.

# Index

**A**
Aardwolf, 424
Accelerometry, 60
Accessory olfactory bulb, 40
*Acinonyx jubatus*, 428
*Acomys cahirinus*, 284
Acoustic communication, 76
Acoustic cues, 139
Acoustic signals, 76
*Actinidia polygama*, 435
Actinobacteria, 189
Active radio frequency identification (aRFID), 59–60
*Aedes aegypti*, 441
*Aedes albopictus*, 441
African clawed frog, 89
African forest elephants, 491
African wildcats, 424, 425, 427, 428
African wild dogs, 421, 424, 428
Age signature, 114, 125, 130
Ageing, 42, 130
Aggression, 3, 90, 215, 237, 238, 243–245, 248, 349
Alarm cues, 85
Alarm substances, 77, 78, 85
*Allobates femoralis*, 91
Allo-marking, 58, 179
Allometry, 65
Allomone, 352
Alpine Marmots, 63
Alston's brown mouse, 159
Altruistic/helping behaviour, 237
*Amalosia lesueurii*, 412
American bullfrog, 95
American toad, 92
Amniotic fluid, 309, 316, 362, 363
Amphibians, 76, 80, 87, 89, 96
Amur leopards, 438
Amygdala, 40, 221, 236, 239, 240, 243, 400
Anal gland(s), 59, 63, 178, 180, 188, 190, 215, 333, 344–352
Analgesia, 219, 222–224, 239, 445
*Anas platyrhynchos*, 118, 276
*Anaxyrus cognatus*, 90
*Anolis carolinensis*, 400
*Anopheles*, 211, 352
*Anser anser*, 274
Antibiotics, 350
*Antilocapra americana*, 283
Antioxidant, 65
Antipredation, 77, 95, 215, 399–414
Anti-predatory responses, 85
Anura/anurans, 77–84, 86–94
Anxiety, 236, 238–243, 245, 247, 248, 318, 324–326, 454
*Anyxyrus terrestris*, 89
Appeasement behaviour, 58
Applications, 4–5, 13, 59, 87, 160, 169, 243, 350, 419, 420, 480, 492
Areolar glands, 3


B. Schaal et al. (eds.), *Chemical Signals in Vertebrates 15*, https://doi.org/10.1007/978-3-031-35159-4

Arginine vasopressin, 220–222
Arizona mountain kingsnake, 400
*Arvicola terrestris*, 153–160
*Ascaphus truei*, 92
Ash borer, 5
Asian elephant, 420
Australian sea lions, 5
Autonomic nervous system, 237, 244, 248

**B**
*Bacterioides*, 348
*Baiomys taylori ater*, 280
*Balantidium coli*, 215
Banded mongooses, 215
Bank vole, 153, 282
Bat-eared fox (BEF), 424–428, 431
Beavers, 420
Begging behaviour, 91
Behavioural immune system (BIS), 208, 209, 211, 217, 222–224
Behavioural observation research interactive software (BORIS), 44, 157
Bicolored shrew, 282
Bird odour, 125
Black-backed jackal, 421, 424–428, 431
Black-legged kittiwake(s), 118, 122–124, 126, 127, 140, 141, 145
Black-tailed deer, 283, 295
Blood, 61, 143, 154, 185, 197, 210, 214, 215, 219, 332, 420, 442, 454, 459, 461–464, 471
Blue petrels, 116, 118, 119, 122, 123, 126, 127, 140
Bobcats, 428, 437
Bomb calorimetry, 64, 181, 186
Bombykol, 12
Bonding, 237, 238, 241, 246, 249
*Brachylagus idahoensis*, 351
Brazilian guinea pig (*Cavia aperea*), 279
Breath odor, 4
Breeding season signature, 113, 122–123, 126–128
Breeding success, 56
Brilliant-thighed poison frog, 91
Bromadiolone, 155
Brown bear, 428
Brown hyena (*Parahyaena brunnea*), 424
Brown rat, 159
Buck (he-goat), 166, 167, 169–172
*Bufo*, 86
*Bufo americanus*, 92
*Bufo boreas*, 89
Bufonidae, 83, 86
Bullfrogs, 89
Burmese python, 400
Burrows, 349, 352, 353, 412
Butyric acid, 224

**C**
Caecotrophy, 335, 348
*Cairina moschata*, 276
Calcium imaging ($Ca^{2+}$ imaging), 165–172
California see lion, 276
Calorimetry, 64, 181, 186
Camera trapping, 427
Cancer(s), 4, 453–480
*Candida humicola*, 89
Cane toads, 94, 95
*Canis (lupus) familiaris*, 437
*Canis lupus*, 196, 198, 428, 437
Cape fox, 424
*Capra aegagrus hircus*, 276
*Capra hircus*, 167
Capsicum (*Capsicum frutescens*), 491, 495
Capuchin monkeys, 446
Caracal (*Caracal caracal*), 424, 425, 427
Carbograph, 18, 19, 29, 31, 32
Carbon dioxide ($CO_2$), 3, 61, 167, 168, 352
*Castor canadensis*, 420
Castration, 154, 169, 170
Cat (domestic), 435–447
Catnip, 435–447
Cave lions, 437
*Cavia aperea*, 279
*Cavia porcellus*, 276
*Cebus olivaceus*, 446
Cheetahs, 421, 428
Chemical pollution, 87, 93–95
Chemoreception, 5, 237, 245, 314, 315, 399–401, 444, 445
Chemosensory event-related potentials (CSERPs), 244
Chili peppers, 489–502
Chimpanzees, 209, 215
Chinese hamster, 153
*Chionactis occipitalis*, 400
Civets, 428
*Civettictis civetta*, 428

Classical conditioning, 42, 94
Co-evolution, 238
*Coleonyx brevis*, 411
Common sheldrake, 277
Communication, 1, 12, 56, 75, 114, 147, 155, 178, 236, 259, 297, 332, 363, 401
Competition, 56, 90, 257, 489
Connectedness, 235, 245
Contagion, 208, 211, 219, 223, 238, 454
Contaminants, 12–14, 20–24, 28, 33, 34, 36, 215, 223
Coprophagy, 334
Corticotropin releasing hormone, 239
Cottontails, 352
Countermarking, 185, 186, 188
Courtship pheromones, 93, 96
*Cricetulus*, 153
*Cricetus cricetus*, 159
*Crocidura leucodon*, 282
*Crocuta crocuta*, 424
Cross fostering (Cross-fostering), 273, 274, 278–283
Cuban tree frog, 94
Cuban tree frog tadpoles, 94
Cues/signals, 1, 12, 56, 75, 114, 140, 156, 166, 178, 273, 306, 333, 420, 472, 491
Culture (anthropology), 209, 235, 241, 438, 446, 462, 463
Culture (histology), 167
Curlew, 119, 276
Cytokines, 211, 215, 219, 222, 238

**D**

Damage-released alarm substances (DRAS), 77–83, 85–90, 94
Damselfly, 87, 91
Darcin, 2, 3, 210
Dear enemy phenomenon, 59
Deer mouse, 281, 297
Defecation, 59
Degradation, 12, 13, 24, 30, 34–36, 77, 85, 348, 490, 497
Degus, 286, 296
*Dendrobates auratus*, 92
Dengue virus, 3
Desert iguanas, 400
Desirability, 56
Deterrents, 421, 423, 489–502
Development, 5, 39–49, 88, 89, 94, 131, 148, 167, 223, 224, 237, 242, 248, 274, 275, 278, 280, 281, 283, 285–292, 294–296, 298, 299, 306–326, 333, 336, 338, 348–350, 353, 362, 366, 367, 373, 374, 376, 378–385, 420, 454
Diarrhoea, 212, 216
Dicroglossids, 94
Digestion-released substances, 79, 88, 89
Dimethylpyrazine (2,5-dimethylpyrazine), 3
*Dipsosaurus dorsalis*, 400
*Dirofilaria immitis*, 441
Discriminability, 189, 190
Disease-vectors, 352
Disgust, 208, 216–218, 221–224, 236, 245
Dispersion/dispersal, 58, 62, 94, 350, 431
Disturbance pheromones, 86
Disturbance substances, 78, 80, 81, 85–87, 90
Dog (domestic), 189, 202, 428
Dominance, 12, 198, 224, 238, 243, 244, 258, 349, 380
Doubly labelled water (DLW), 61–62
*Drosophila melanogaster*, 441
*Drymarchon corais*, 412

**E**

Echinococcosis, 155
Eggs, 75–96, 372
Eiffinger's tree frog, 91
*Eimeria vermiformis*, 212, 219
Electroencephalography (EEG), 364–373, 375, 376, 381–383, 385
Electronic noses (e-noses), 3, 5, 473
*Elephas maximus*, 420
*Elephas maximus sumatranus*, 491
Embryo(s), 77, 80–82, 88, 94, 221
Emotion(s), 40, 235, 237–242, 248, 378, 384
Emotional contagion, 238, 241, 247
Endorphin (β-endorphin), 445
Energetic cost, 65, 178
Energy budgets, 60, 66
Energy content, 64, 66, 183, 185, 188
Energy expenditures, 55, 57, 61, 66
*Engystomops pustulosus*, 90
Entorhinal cortex, 40
Epileptic seizures, 4
*Eryx colubrinus*, 403, 404, 408, 412
*Eryx johnii*, 403, 404, 409

Estrous/anestrous, 172, 196–203
Ethyloctanal (4-ethyloctanal), 2
*Eublepharis macularius*, 402, 408
*Eumeces laticeps*, 411
*Eumetopias jubatus*, 276, 277
European badger (*Meles meles*), 56, 178–179
European rabbits, 332, 348, 351, 353
European water vole, 153
Evolution, 57, 81, 131, 248, 351, 352, 366, 370, 381, 437, 454
Ewes, 5, 166

**F**

Face categorization, 362, 366–376, 379–384
Facial expressions, 216, 221, 236, 240, 243, 248, 378, 381
Facial glands, 63
Faecal pellet(s) (FPs), 332–353
Faeces (feces), 62, 63, 147, 178, 180, 202, 209, 212, 215–217, 223, 296, 420, 421, 473, 474
Familiarisation, 186
Familiarity, 59, 90, 185, 186, 189, 190, 213, 216, 219, 220, 223, 224, 238, 325, 349, 362, 377, 383, 401
Fear, 4, 40, 42, 218, 236, 238–242, 245, 248, 375
Feather, 115–124, 126–128, 131
Fecal odor, 141, 147
*Felis silvestris catus*, 435, 437
*Felis silvestris lybica*, 424
Female receptivity, 170, 188–189
Fertility, 65, 66
Fetus/foetus, 242
Field study (studies), 4, 5, 493–495, 498, 501
Firmicutes, 189
Fitness advertisement, 56
Flavour(s), 218, 306–317, 319, 320, 323–326, 353
fMRI, 218
Food security, 490, 501
Foraging, 5, 58, 66, 79, 89, 139, 190, 401, 490
*Formica rufa*, 441
Freezing behaviour, 87
Frequency tagging, 366–370, 375, 383
Friendships, 236, 245–249
Froglets, 76, 77, 81, 82, 85, 86, 88–90
Frogs, 75–96

**G**

*Galerella sanguinea*, 424
*Gallus gallus domesticus*, 274
Gas chromatography-mass spectrometry (GC-MS), 19, 20, 35, 63, 64, 116, 125, 181, 184, 185, 189, 345, 473, 474
*Gasterosteus aculeatus*, 147
Gender differences, 241
Genetic distance, 122, 126
Genetic fitness, 56
Genetic signature, 113, 122, 126, 130
*Genetta genetta*, 424
Genital glands, 5
Giant water bugs, 87
Goat(s) (domestic), 2, 165–172, 276, 282, 292, 296, 297
Gonadotropin, 166
Gonococcus, 216
G protein-coupled receptor(s) (GPCRs), 166, 240
Grass frogs, 93
Great barred-frog, 90
Great Plains toad, 90
Green-and-black poison dart frogs, 92
Green anole, 400
Greylag goose, 274, 276
Group cohesion (group-cohesion), 12, 58
Group odour, 58, 179, 186
Grueneberg ganglion (GG), 114, 239, 240
Guinea pigs, 276, 277, 279, 283, 290, 296
Gustatory cues/signals, 208, 211
Gymnophiona, 76

**H**

Habituation, 44–46, 48, 94, 319, 468, 491
*Halobaena caerulea*, 118, 119, 140
Handicaps, 56–57, 66
Happiness, 235, 236, 240, 245, 375
Heckscher's frog, 86
*Heligmosomoides polygyrus*, 213, 216, 219–222
*Hemorrhois hippocrepis*, 403, 404
*Hemorrhois ravergieri*, 403, 404
Hepatitis A/B, 217
Herbicides, 94
Heterozygosity, 122, 126
Home range, 56, 428, 430
*Homo sapiens*, 249

Honest signalling (honesty signals), 56–57, 63, 236
Honey badger, 424
*Hoplobatrachus tigerinus*, 89
Horses, 189, 238
House mouse, 2, 281, 297
House mouse louse, 213
House sparrows, 140
Human (*Homo sapiens*), 3, 58, 114, 155, 170, 186, 202, 208, 235, 258, 274, 309, 361, 420, 445, 453, 489
Human-elephant conflict (HEC), 490
Human neonate, 3
Human-wildlife conflict, 501
Hylidae, 83, 93
*Hymenolepis diminuta*, 212, 213
Hypothalamus, 221, 237, 239
Hypothalamus-pituitary-adrenal (HPA) axis, 239

I
Iberian green frogs, 86
Iberian spadefood, 88
*Ictonyx striatus*, 424
Immune response(s), 65, 208–210, 213, 215, 219, 222, 239
Immunity, 140, 147, 207, 306, 348, 353, 369
Immunosuppress, 210, 219
Imprinting, 273–299
Incubation, 119, 121, 126, 129, 141, 167
Indian bullfrog, 89
Indigo snake, 412
Individual differences, 241, 369, 375, 468, 476
Individual quality, 2, 56
Individual (identity) signature, 122, 128
Infancy, 362, 365, 374
Infection(s), 58, 66, 89, 131, 147, 207–214, 216–219, 221–224, 352, 454
Inflammation, 209, 222, 245
Influenza virus, 210, 214
Infra-red video surveillance, 58
Inguinal gland, 351
Inguinal wax, 5
Innate behaviour, 202, 437
Insecticides, 94
Instrumental conditioning, 40
Integrated pest management (IPM), 490
Interdigital gland, 178
Intraspecific communication, 114, 155, 202, 401
Invasive species, 87, 93–95
Inverse effectiveness, 374, 375, 382–384
Iridoids, 437–439, 441–446

J
Jaguars, 438, 439
Junglefowl (domestic), 276

K
Kairomone(s), 75, 77, 79–83, 87–91, 352
Kingsnake, 400
Kin recognition, 77, 80, 82, 90, 237
Koalas, 63, 342
*Kurixalus eiffingeri*, 91
*Kurixalus odontotarsus*, 86, 92

L
Laboratory mice, 280, 281, 286, 292, 295, 307, 420, 438
Lactation, 307, 308, 313, 315, 316, 322, 342–344, 351, 383
Lambs, 5, 282, 283
*Lampropeltis getula*, 400
*Lampropeltis pyromelana*, 400
Lateral gland, 4
Lateral scent glands (LSG), 154–157, 159
Latrines, 58–60, 62, 63, 66, 179–181, 428
Learning, 2, 39–49, 86, 93–95, 220, 223, 268, 273–299, 306, 307, 332, 349, 352, 353, 362, 365, 378, 383, 437, 446, 474
*Lemur catta*, 147
Lemurs, 5
Leopard(s), 399–414, 423–425, 427, 428
Leopard geckos, 401–403, 406, 408, 409, 413
Leporidae (leporids), 351, 352
*Leptodactylus fallax*, 90
Lesueur's geckos, 412
Life-history trade-offs, 65
Limbic system, 40
Lion(s), 5, 276, 277, 297, 424, 437, 438
Lip-smacking, 197, 199, 202
Liquid chromatography, 63, 438, 493
Liquid Chromatography and Mass Spectrometry (LC-MS), 493, 496–501

*Lithobates*, 86
*Lithobates catesbeianus*, 89, 95
*Lithobates heckscheri*, 86
*Lithobates sylvaticus*, 86, 94
*Loxodonta cyclotis*, 491
*Lupulella mesomelas*, 424
*Lycaon pictus*, 424
Lynx, 428
*Lynx lynx*, 428
*Lynx rufus*, 437

**M**

Macrolides, 93
Magnificent tree frog, 92
Main olfactory bulb, 400
Major histocompatibility complex (MHC), 113, 126, 140–143, 147, 148
Major urinary protein(s) (MUPs), 3, 148, 178, 210
Malaria, 211
Male effect, 166, 167, 172
Mallard duck(s), 118, 126, 127, 276
Mammary gland, 350
Mammary pheromone, 2, 3, 332–353
Mandrill, 63
*Mangifera indica*, 491
Mango, 491
Mantellid, 93
*Mantidactylus multiplicatus*, 93
Mate choice, 12, 140, 147, 148, 209, 212, 213, 215, 224, 237, 287, 307
Maternal behaviour, 297, 333, 335, 349
Maternal body odors, 363, 371, 372, 383
Maternal-offspring interactions, 39
Mate selection, 236
Mating season(s), 58, 126, 179, 185, 292, 428
Meadows, 153, 155, 157
Meadow voles (*Microtus pennsylvanicus*), 160, 214
Melanoma, 4, 455–457, 459, 460, 475
*Meles meles*, 56, 177–191
*Mellivora capensis*, 424
*Melospiza melodia*, 117, 118, 120, 140, 147, 276
Memorability, 189
Memory, 3, 39–49, 207, 217, 275, 293, 306, 349, 382, 423, 476
*Meningococcus*, 217
Menstruation/menstrual cycle, 221
Menthone, 4
Metabolic by-products, 178, 191
2-Methylbut-2-enal (2MB2-al), 2, 345, 347–353
Methyl paraben (MP), 196, 199–202
Microbial signatures, 131
Microbiome(s), 3, 58, 131, 147, 179, 189, 209, 306, 314, 348, 350–353
*Microtus*, 153
*Microtus ochrogaster*, 214
*Microtus pennsylvanicus*, 214
Milk, 2, 5, 306, 308, 309, 314, 316, 324, 332, 339, 340, 346, 348–353, 362–364, 383
*Mixophyes fasciolatus*, 90
Mosquito(es), 3, 88, 211, 352, 440–442, 444–446
Mother-infant interactions, 332
Moths, 12
Motivation, 40, 198, 247, 266, 401, 456, 464, 475, 476, 478
Mountain chicken, 90
Mountain spiny lizard, 400
Mouse/mice, 3, 41, 44, 166, 167, 280–282, 284, 287, 295, 297, 307, 308, 310, 314, 317, 318, 325, 454
Mule deer, 283
Multimodality, 297–298
Multisensory depression, 218
Multisensory enhancement, 217
Multisensory perception, 5, 218, 378, 383, 384
*Mungos mungo*, 215
Muscovy duck, 276
*Mus domesticus*, 282
*Mus musculus*, 276, 280, 420, 438
*Mycobacterium mungi*, 215
*Myodes glareolus*, 282
Myxoma virus, 352

**N**

Navigation, 39, 139
Neanderthals, 249
*Neisseria gonorrhoeae*, 216
Neonate/newborn, 2, 3, 154, 297, 308–316, 332–353, 362–364, 377, 380, 383
*Nepeta cataria*, 435
Nest, 124, 140–142, 145, 147, 153, 275, 276, 282, 287, 292, 332–342, 349–351, 353

Nest sanitation, 139–140
Newborn/neonate, 2, 3, 154, 297, 308–316, 332–353, 362, 364
Newcastle disease virus, 210
Newts, 76
*Nidirana pleuraden*, 92–93
Nipple(s), 287, 288, 291, 316, 332, 340, 350, 351, 353, 363
Non-volatile compounds, 2, 115, 116, 125, 346
Non-volatile organic compounds (non-VOCs), 63
*Numenius arquata*, 276

**O**

*Octodon degus*, 286
*Odocoileus hemionus*, 283
*Odocoileus hemionus columbianus*, 283
Odor/odour, 2–5, 13, 40–44, 48, 58, 80, 86, 90, 114–116, 122, 123, 125, 139–148, 156, 159, 166, 179, 189, 196, 199, 202, 209–224, 236, 239, 243, 246–248, 258–260, 262, 264–268, 278, 280–285, 287, 291–294, 296, 442
Oestrogen/estrogen, 185, 220, 221
Olfactory learning, 40, 273–299, 364
Ontogeny, 209, 224, 275, 306, 362, 380
*Onychomys torridus*, 281
*Oophaga pumilio*, 91
Opioid (μ-opioid), 222, 223
*Oryctolagus cuniculus*, 332, 351
*Osteopilus septentrionalis*, 94
*Otocyon megalotis*, 424
Outbreak cycles, 154
Over-marking, 58, 59, 186, 188, 199
*Ovis aries*, 276
Ovulation, 166, 172
*Owl monkeys* (*Aotus* spp.), 4
Oxidative stress, 65
Oxytocin, 220–222

**P**

Pain, 219, 491
Palatability, 314, 317, 319, 320
Panic attacks, 4
*Panthera leo*, 424, 437
*Panthera leo spelaea*, 437
*Panthera onca*, 438
*Panthera pardus*, 423, 424
*Panthera pardus orientalis*, 438
*Panthera tigris*, 438
*Pan troglodytes troglodytes*, 215
*Parahyaena brunnea*, 424
Parasite load, 129, 147
Parasite(s), 81, 89, 129, 140, 211, 213, 214, 219, 352, 353, 446, 454
Parental care, 12, 80, 91–92, 306
Parental investment, 147
Parent-child bonding, 236
Parturition, 197, 308, 316, 332–334
*Passer dosmesticus*, 140
Pathogens, 57, 89, 207–212, 214–217, 220–224
Pavlovian conditioning, 42
Pectoral glands, 4
*Pelobates cultripes*, 88
*Pelophylax perezi*, 86
Perfume(s), 236, 283, 285, 287, 289, 294, 296, 402, 404, 405, 407, 408, 411, 414
Perinatal learning, 306, 307, 315, 325, 326
*Peromyscus californicus*, 156
*Peromyscus leucopus*, 281
*Peromyscus maniculatus bairdii*, 281
Personality, 245, 306–326, 476, 479
Pest defense, 435–447
Phenotypes, 306, 326
Pheromones, 2, 3, 12, 75, 76, 78, 80, 85, 92, 93, 95, 114, 115, 166, 170, 172, 198, 199, 201, 202, 210, 285, 332, 419, 428, 445
*Phodopus*, 153, 276, 277, 279, 286, 296
*Phodopus campbelli*, 156
*Phodopus sungorus*, 156
Photoperiod, 166
Physiological immune system (PIS), 207, 208, 221–224
Physiological state(s), 333, 342–345, 351, 363
Phytotelmata, 91
Pig, 276, 277, 279, 283, 286, 290, 342
*Pipa pipa*, 92
Piri piri pepper, 495
Piriform cortex, 40, 239
Plant(s), 91, 238, 402, 435–447
*Plasmodium falciparum*, 211
Pleasure, 240, 441, 445
Plethodontid modulating factor (PMF), 93

Poison frogs, 90, 91
Pollution, 87, 93–95, 453
*Polyplax serrata*, 213
Population connectivity, 65
Population control, 155, 160
Population density, 62, 65, 154, 155, 191, 207
Post-traumatic stress disease (PTSD), 4
Prairie voles, 214
Precocial, 276, 278, 279, 285, 286
Precursor-like factor, 93
Predation, 56, 66, 77, 85, 88, 94, 246, 249, 413
Predator avoidance, 39
Predator odours, 77, 220, 411, 420–421, 431
Predator-prey interactions, 85
Predator recognition, 77, 80, 91, 94
Preen gland, 140, 141, 147
Preen oil, 115–131
Preferences (odor/odour/olfactory), 40, 41, 43, 48, 49, 157, 214–216, 224, 278–281, 283–299, 311–312
Preference testing, 40, 48, 160, 212, 284, 288–290, 314
Pregnant/pregnancy, 221, 222, 242, 307, 308, 313, 315, 316, 332, 342, 348
Prenatal, 306–326, 363
Progesterone, 172, 198, 221, 222
Pronghorn, 283, 295
Prosocial behaviour, 220, 237, 380
*Proteles cristata*, 424
Proton transfer reaction mass spectrometry (PTR-MS), 473
*Pseudophryne bibronii*, 92
*Pseudopus apodus*, 403–405, 409–413
Puberty acceleration, 166, 167, 170
*Puma concolor*, 428
Pumas, 428
Pygmy mice/mouse, 280, 281, 292, 295
Pyrazine(s), 156, 185, 240
*Python molurus*, 400

**R**
Rabbit (European), 332, 348, 351, 353
*Rana*, 86
*Rana aurora*, 86
*Rana temporaria*, 93
Ranidae, 83, 86, 93, 95
*Ranitomeya variabilis*, 86, 91
Ranoidea "Litoria" splendida, 92
Rat, 281, 282, 285–292, 295, 325, 342
*Rattus norvegicus*, 281
Receiver psychology, 352
Recognition of predators, 87
Recoverability, 13
Rectal glands, 344
Red foxes, 428
Red-legged frog, 86
Redundancy, 63, 178, 191
Reinforcement, 274, 275, 287–289, 291, 292, 298, 348, 367, 378
Repellent/repellency, 201, 420, 421, 427, 440–442, 444–446
Reproducibility, 13, 18
Reproductive behavior, 165
Reproductive fitness, 56
Reproductive status, 58, 63, 64, 181, 258
Reproductive success, 66, 129, 130, 147, 214–216, 224, 237
Respirometry, 57, 60–61, 65
*Rhinella marina*, 86, 94
*Rhinella ornata*, 87
Ring tailed lemur, 147
*Rissa tridactyla*, 118, 140
Rodenticides, 155
Rolling, 14, 436–442, 444–446
Rubbing, 423, 424, 426, 435–442, 444–446, 493

**S**
Salamanders, 76, 77, 80, 87, 93, 166
Saliva, 62, 222, 332, 459, 463, 473, 474
*Salmonella enterica*, 211, 212
Sample preparation, 13, 473
Sampling methods, 12
*Sceloporus jarrovi*, 400
Scent marking (scent-marking), 5, 56–62, 65, 114, 139, 178–180, 185, 186, 190, 211, 420, 423–425, 427–430
Scent-provisioning, 58, 59, 64
Scent secretions, 63, 191
Schooling behaviour, 90
*Scotinomys teguina*, 159
Screening(s), 236, 423, 454, 459, 468, 473, 474, 478–480
Seasonality, 65, 185, 189
Seasonal signature, 113, 122, 123, 126–130
Self-anointing, 441, 444, 446

Self-medication, 446
Semen, 215
Semiochemicals, 5, 12, 14, 24, 32, 35, 75–96, 129, 154, 333, 352, 400, 401, 412, 419–431
Semi-volatile organic compounds (sVOCs), 19, 34, 35
Sensitive periods, 274–279, 283–286, 289–291, 298, 299, 307, 353, 362, 365
Sensory threshold(s), 239, 325
Septal organ, 114
Serrate-legged small treefrogs, 86
Serrate-legged tree frogs, 92
Sex differences, 45, 46, 48, 118, 123, 124, 127, 279–282, 289, 295, 296, 323, 325, 326
Sex pheromones, 78, 80, 82, 92–93, 196, 199, 200, 202, 210, 401
Sex-ratio, 65, 140, 147
Sex signature, 114, 117, 123, 124, 127, 129
Sexual preference(s), 159, 280–282, 292
Sexual selection, 56
Sheep (domestic), 276, 286
Siberian hamster, 153
Sickness behaviour, 211, 215
Sick/sickness, 209, 211, 212, 215–218, 222, 454
Signal selection, 57
Signalling, 55–66, 332, 351, 352, 455
Signals/cues, 1, 12, 56, 75, 114, 140, 156, 166, 178, 273, 306, 333, 420, 472, 491
Signature mixtures, 115, 128
Silver vine, 435–447
Skin bacteria (microbiota), 3
Skin conductance, 243
Skin microbiota, 3
Slender mongoose, 424
Small-spotted genet, 424
Smell(s), 1, 39, 76, 139, 140, 170, 202, 208, 215, 217, 223, 224, 247, 248, 258, 259, 261, 262, 266, 268, 293, 309, 333, 361, 402, 455, 473, 476, 491, 500
Snake, 88, 147, 399–414
Snake exuvia, 402, 406, 414
Sniffing, 58, 59, 62, 114, 139, 158, 180, 185, 188–190, 201, 202, 214, 258, 285, 290, 318–320, 340, 346, 430, 455, 456, 467, 472, 474, 477
Social capital, 56–57
Social communication, 147, 236, 237, 242
Social networks, 236, 242, 245, 247, 248, 378
Social perception, 242
Social preferences, 273, 280, 292
Social skills, 241, 243, 247, 248
Sodefrin, 93
Solid-phase micro-extraction (SPME), 13, 18–20, 29, 31–33, 35, 116, 473
Solventless extraction, 13, 18, 20, 21, 31–33, 35
Song sparrow(s), 117, 118, 120, 126–128, 140, 276
Southern caracaras, 5
Southern grasshopper mouse, 281
Southern toads, 89
*Spalerosophis atriceps*, 403–405, 408
*Spalerosophis diadema*, 403, 404
Species signatures, 113, 116, 125–126
Spiny mice, 284–286, 290, 291, 297
Splendipherin, 92
Spotted hyenas, 421, 424, 425, 428
Startle reflex, 241, 242
Steller sea lion, 276, 277
Sternal glands, 5, 63
Strawberry poison frog, 91
*Streptopelia risoria*, 276
Stress, 4, 65, 86, 88, 131, 210, 218, 219, 236, 238–246, 248, 318, 350, 363, 375, 378, 480
Striped polecat, 424
Subcaudal, 4, 60, 63, 179, 186
Subcaudal gland, 57, 58, 62–66, 178–180
Suckling, 326, 332–353
Sumatran Asian elephant, 491
Surinam toad, 92
Survival lifetime, 56, 65, 244
Sustainable agriculture, 493
Sustainable/sustainability, 501
Sweat, 4, 209, 240, 241, 243, 244, 362, 363, 368, 370, 473
*Sylvilagus cunicularius*, 351
*Sylvilagus floridanus*, 351

**T**

Tactile signals/cues, 87, 178, 208, 211
*Tadorna tadorna*, 277
Tadpoles, 76, 77, 79–83, 85–92, 94, 95
*Taeniopygia guttata*, 121, 275
Tailed frogs, 92
Tamarins, 4

Tenax, 18, 19, 29, 31–33, 35, 36, 200, 345, 346
Territorial marking, 62, 332
Territories, 63, 65, 178
Testosterone, 90, 129, 154, 169, 185, 210, 214, 220–222, 243, 244
Thermal desorption (TD), 12–14, 17–20, 29, 31–33, 35, 200
Thermoregulation, 61, 349
Three-spined stickleback, 147
Tibetan foxes, 428
Tick-borne encephalitis virus, 214
Tiger(s), 63, 365, 438, 439
Toads, 76, 83, 86, 88, 92
Tongue flicking (tongue-flicking), 400–402, 406–409, 411–413
Trace amine-associated receptors (TAARs), 239, 240
Transnatal, 306, 308–315, 325, 326, 363, 377
Treefogs, 83
*Trichinella spiralis*, 214
Trier-social-stress test (TSST), 240
Trigeminal system, 239
Tungara frog, 90
Turkey vultures, 5
Turtle, 87

U
Urinary volatile compounds, 200
Urination(s), 59, 61, 62, 180, 199, 350
Urine(s), 2–5, 62, 156, 157, 159, 160, 168–172, 178, 180, 181, 184, 185, 188–191, 196–203, 209, 210, 212–221, 258, 279, 281, 285, 287, 295, 296, 298, 332, 337, 342, 345, 352, 420, 421, 431, 459–464, 470, 473–475, 479
*Ursus arctos*, 428

V
Vagina(s), 199, 287
Vaginal secretions, 196, 199, 202
Vandenbergh effect, 166
Variable poison frog, 86
Variety (chemosensory), 94, 155, 235–249, 306–326
*Vibrio cholerae*, 216
Virgin, 342
Vision/visual perception, 361, 362, 366, 373, 376, 380, 383, 384
Visual cues/signals, 80, 91, 218, 363, 367, 380, 455
Vitamin-E, 65
Volatile/non-volatile fraction, 115, 116, 125–128, 131
Volatile organic compound(s) (VOCs), 3, 4, 63, 64, 156, 185, 188, 190, 454, 455, 459, 470, 471, 473–477, 480
Volatilome(s), 129, 131, 350
Vomeronasal organ/system - Vomerolfaction, 40, 114, 166, 199, 400, 445
Vomeronasal type 1 receptors (V1Rs), 166
Vomeronasal type 2 receptors (V2Rs), 166
*Vulpes chama*, 424
*Vulpes ferrilata*, 428
*Vulpes vulpes*, 428

W
Water voles, 4, 153–160
Wax esters, 125–128, 412
Well-being, 235, 248
Western European house mouse, 153
Western toads, 89
White-footed mouse, 281, 296
Wildlife, 4, 141, 423, 489, 491, 501
Wolf, 198
Wood frogs, 86, 88

X
*Xenopus laevis*, 89

Y
Yeast, 89
Yunnan pond frogs, 92

Z
*Zalophus californianus*, 276
Zebra finch, 116, 121, 126, 127, 275
Zika, 3, 441
Zimmermann's poison frog, 91

The manufacturer's authorised representative in the EU is Springer Nature Customer Service Centre GmbH, Europaplatz 3, 69115 Heidelberg, Germany. If you have any concerns regarding our products, please contact ProductSafety@springernature.com

Printed and bound by CPI Group (UK) Ltd, Croydon, CR0 4YY
28/09/2026
02224056-0016